ORGANIC POLLUTANTS TEN YEARS AFTER THE STOCKHOLM CONVENTION – ENVIRONMENTAL AND ANALYTICAL UPDATE

Edited by **Tomasz Puzyn**
and **Aleksandra Mostrag-Szlichtyng**

Organic Pollutants Ten Years After the Stockholm Convention – Environmental and Analytical Update
Edited by Tomasz Puzyn and Aleksandra Mostrag-Szlichtyng

Published by InTech
Janeza Trdine 9, 51000 Rijeka, Croatia

Publishing Process Manager Martina Durovic

Technical Editor Teodora Smiljanic

Cover Designer InTech Design Team

Additional hard copies can be obtained from orders@intechopen.com

Organic Pollutants Ten Years After the Stockholm Convention – Environmental and Analytical Update, Edited by Tomasz Puzyn and Aleksandra Mostrag-Szlichtyng
p. cm.
ISBN 978-953-307-917-2

For EU product safety concerns:
IN TECH d.o.o., Prolaz Marije Krucifikse Kozulić 3, 51000 Rijeka, Croatia,
info@intechopen.com or visit our website at intechopen.com.

Meet the editors

Dr Tomasz Puzyn is the Head of Laboratory of Environmental Chemometrics at the University of Gdansk (Poland). After obtaining his PhD in 2005 at the University of Gdansk (Poland), he spent almost two years working for the Interdisciplinary Centre for Nanotoxicity (Jackson State University, Jackson, MS, USA) and the National Laboratory for Environmental Studies in Tsukuba, Japan (post-doctoral fellowship by the Japan Society for the Promotion of Science). His main area of interest is developing computational methods for assessing the human and environmental risk of novel chemicals. His accomplishments have been widely recognized. Tomasz Puzyn is an author of many contributions devoted to POPs/POPs-like chemicals risk assessment as well as QSAR modeling. In collaboration with Prof. Jerzy Leszczynski from the Interdisciplinary Centre for Nanotoxicity, he has developed the first Nano-QSAR models. The results were recently published in Nature Nanotechnology and Small journals.

M.Sc. Eng Aleksandra Mostrag-Szlichtyng is a Research Associate closely collaborating with Laboratory of Environmental Chemometrics at the University of Gdansk (Poland). Her scientific interests include applying and developing in silico tools for regulatory risk assessment of chemicals (particularly Organic and Persistent Organic Pollutants). She is an author of several contributions devoted to these issues. Aleksandra Mostrag-Szlichtyng has acquired a diverse and in-depth knowledge in the fields of computational chemistry (University of Gdansk and Gdansk University of Technology, Poland) and computational toxicology (European Commission, Ispra, Italy). She cooperates with the Interdisciplinary Center for Nanotoxicity (Jackson State University, Jackson, MS, USA), the National Laboratory for Environmental Studies (Tsukuba, Japan), S-IN Soluzioni Informatiche (Vicenza, Italy) and the governmental Bureau for Chemical Substances (Poland). In January 2012 she starts her Post-Doctoral Researcher Fellowship in Altamira, LLC (Columbus, OH, USA) and Visiting Research Scientist position at Ohio State University (Columbus, OH, USA).

Contents

Preface

More than twenty years ago, Organic Pollutants (OPs), particularly those exhibiting persistence, bioaccumulation and long-range transport potential (so-called Persistent Organic Pollutants, POPs), have been recognized worldwide as the prior environmental problem. International efforts aimed at controlling and eliminating the most hazardous organic pollutants from the natural environment, resulted in a global convention on POPs entering into force. This document, known as the Stockholm Convention, introduced the list of twelve POPs to be limited or banned, due to their adverse impact on humans and the environment. Recently, the list was further extended by the addition of nine new chemicals. This indicates that after so many years organic pollutants are still among the major environmental hazards.

The present book touches three major fields of concern, as far as the environmental impact of Organic Pollutants is considered. The first part focuses on selected pollution sources of various environmental compartments (e.g. soil, water), as well as the considered compounds' possible emission routes to the environment. The pollution sources of increasing meaning, like sewage sludge, dredged sediments application or textile industry, have been widely discussed and characterized. The second part of the book discusses the influence of organic pollutants on living organisms (e.g. OPs bioavailability, exposure assessment), distribution and persistence of OPs in particular environmental compartments (e.g. depositional history, global distillation), as well as novel analytical techniques (e.g. surface-enhanced Raman scattering) useful for identification and monitoring of OPs. In the third part of the book, several methods, including photochemical, chemical, electrochemical, and biological degradation, have been proposed as efficient techniques for decontaminating the environmental compartments from OPs.

We hope that this book will be particularly valuable to environmental scientists and engineers and will contribute to better assessments of the fate of Organic Pollutants in a multimedia environment.

Dr. Tomasz Puzyn and **M.Sc. Eng Aleksandra Mostrag-Szlichtyng**
Laboratory of Environmental Chemometrics
Faculty of Chemistry, University of Gdansk
Poland

Part 1

High Concern Sources of Organic Pollutants

1

The Inputs of POPs into Soils by Sewage Sludge and Dredged Sediments Application

Radim Vácha
Research Institute for Soil and Water Conservation, Prague
Czech Republic

1. Introduction

The use of sewage sludge and dredged sediments in agriculture belong to the most important ways of possible pollutants inputs into agricultural soils in many countries including Czech Republic.

The application of sewage sludge on agricultural soils is connected with following facts:

- increasing amounts of sewage sludge thanks to intensive waste water treatment
- the characteristic of sludge as the material with increased content of organic matter and nutrients

The application of sludge into soil could lead to an increase of the contents of organic matter or macro elements, but the contamination by potentially risky elements and persistent organic pollutants could be relevant also. The problems connecting with increased persistent organic pollutants (POPs) contents in sewage sludge were confirmed by many authors (Markard, 1988; Melcer et al., 1988; Starke, 1992; Oleszczuk, 2007; Clarke et al., 2008; Natal-da-Luz et al., 2009). The load of soil by POPs after sludge application can influence their transfer into food chains (Passuello et al., 2010). Increased contents of polycyclic aromatic hydrocarbons (PAHs) limit not only direct application of sewage sludge on the soil but also the use of sludge in composting processes for example (Rosik-Dulewska et al., 2009). The inputs of POPs into agricultural soils by biosolids use in agriculture plays an important role. This problematic is documented on the example of following study realised in the Czech Republic where the contents of POPs in the soil and plants after sewage sludge and sediments application were observed.

The number of waste water factories increased after implementation of Czech Republic into European Union when the obligation of waste water factory existence in every settlement over 10 000 inhabitants till 2010 year had to be fulfilled. The necessity of legislative regulation existence controlling this process was obvious since the beginning of ninetieth years and the Directive No. 382/2001 was the first version of legislative adaptation. The Directive was modified under the No. 504/2004 Sb. in 2004 year.

The directive of Czech Ministry of Environment No. 504/2004 Sb. regulates the application of the sludge on agricultural soils. The directive determines the conditions of sludge application on agricultural soils, including limit values of potentially risk elements and some persistent organic pollutants (sum of halogenated organically bound substances - AOX, sum of six congeners of polychlorinated biphenyls - PCB6) in sludge. The directive 86/278/EEC regulates the sludge application in EU legislation. Only the contents of 6 potentially risk elements in sludge (Cd, Cu, Hg, Pb, Ni and Zn) are limited in the directive. The proposal of limit values of potentially risk elements and persistent organic pollutants was presented in Working Document on Sludge that was available for professional community, too. This proposal altered existing criteria and installed new criteria for persistent organic pollutants especially. The contents of seven POPs groups were regulated, the sum of halogenated organic compounds (AOX), linear alkylbenzene sulphonates (LAS), di(2-ethylhexyl)phtalate (DEHP), nonylphenol and nonylphenoletoxylates substances with 1 or 2 ethoxy groups (NPE), sum of polycyclic aromatic hydrocarbons (PAHs), the sum of seven congeners of PCB (28+52+101+118+138+153+180) and polychlorinated dibenzo-p-dioxins and dibenzofuranes (PCDD/F). The acceptance of Working Document on Sludge for the legislation was complicated by the lobbies and by economical needs for the determination of the pollutants. The proposal was refused and the directive 86/278/EEC is valid in original form.

The second group of problematic materials including into our research are the sediments dredged from river or pond bottoms. The volumes of dredged river and pond sediments reach huge amounts because of the necessity of periodical maintenance of river channels and water reservoirs. The existence of 97 millions m3 of ponds sediments and 5 millions m3 of river and irrigation channel sediments was reported in Czech Republic (Gergel, 1995). The problem of the liquidation or suitable use of extracted sediments of these amounts is evident. In spit of the traditional use of the sediments as the fertilizers on agricultural soils till to first halve of 20th century is not current approach unified, especially thanks to misgivings of their hygienic standards and environmental merits.

The elaboration of complex methodological approach including the assessment and testing of sediment conditions, the contamination and possible negative effects and the evaluation of positives and negatives of their application is highly needed. This approach must follow current EU politics of soil protection, sewage sludge application and the use of the other wastes (European Parliament, 2003; ISO 15799, 2003; EN 14735, 2006). The complex system should use chemical and biological methods concluded by risk assessment where contact ecotoxicity tests cannot be missing (Domene et al., 2007; Pandard et al., 2006).

The sedimentation of soil particles originated from agricultural soil erosion seems to be the most important way of sediments inputs into water systems. This process is described in Czech Republic also where about 50% of soil fund is endangered by water erosion (Janeček et al., 2005). The accumulation of nutrients and organic matter especially in pond and downstream sediments belongs to the positives of sediment application. The sediment could be valuable substrate useful in soil and landscape reclamation for example (Santin et al., 2009).

The other hand must be accepted that eroded soil particles are under the influence of many factors in water environment resulting to the changeover of their quality especially from the viewpoint of elements and substances sorption. The sediment characteristics are changing by particles sedimentation process in different parts of the stream and this process influences sorption of risky substances (Tripathy & Praharaj, 2006; Fuentes et al., 2008). It could lead to the problems of water eutrofization or sediment contamination. The sediments are known as the "chemical time bomb" thanks to their function of final deposits of pollutants in the river basins (Hilscherová et al., 2007; Holoubek et al., 1998). The sediment load by risky substances is connected with the presence of pollution sources like industrial or urban zones or wastes outputs from mining activities. The negative impact of these sources can be confirmed by chemical methods (Gomez-Alvarez et al., 2007) or by toxicity tests (Riba et al., 2006). The inputs of risky elements into sediments from geochemical anomalous substrates or from the other natural sources respectively can play an important role (Liu et al., 2008). The increased loads of risky substances lead to the complications of sediment use in the same way as the use of sewage sludge and the other organically reach materials (Vácha et al., 2005a).

The potential contamination of the sediments by wide spectrum of hazardous substances could not be eliminated. We accept the fact that fluvisols developed on alluvial sediments in river fluvial zones belong to the most loaded soils in our conditions by risky elements Cd>Hg>Zn>Cu>Pb and Cr (Podlešáková et al., 1994) and by persistent organic pollutants (POPs). Increased contents of polycyclic aromatic hydrocarbons (PAHs), chlorinated pesticides (sum of DDT), petroleum hydrocarbons and polychlorinated biphenyls (PCBs) on some localities were observed (Podlešáková et al., 1994; Vácha et al., 2003). The monitoring of fluvisols load by polychlorinated dibenzo-p-dioxins and dibenzofurans (PCDDs/Fs) resulted into similar trends (Podlešáková et al., 2000; Vácha et al., 2005b). The contamination of sediments from water reservoirs by PCDDs/Fs was confirmed (Urbaniak et al., 2009). At the same time, POPs degradation is strongly influenced by sediment conditions, oxygenation conditions belong to the most important (Devault et al., 2009). The sediment quality was monitored in Labe river basin by germen researchers. They observed increasing water quality in Labe River after collapsing of communist regime in central Europe thanks to increasing number of wastewater factories and the other modern pollution-controlling technologies (Netzband et al., 2002). In spit of this fact the concentrations of several contaminants are still remaining in sediments of Labe River and their use for agriculture is questionable (Heininger et al., 2004). The most problematic are the contents of Cd, Hg, As, Zn, HCB, PCBs and PCDDs/Fs in the sediments (Heise et al., 2005).

The other hand, realised monitoring of pond sediments load by risky elements in the Czech Republic confirmed relatively low contamination (Benešová & Gergel, 2003). The authors did not find the exceeding of risky elements limit values in the Czech Direction for soil protection No. 13/1994 Sb. The database of sediment load by risky elements and some POPs separated into groups following sediment origin (field ponds, village ponds, forest ponds and rivers) is available in the Central Institute for Supervising of Testing of Czech Republic (Čermák et al., 2009). The results of this monitoring show only sporadically increased values of risky elements (Cd and Zn usually) in the sediments but these load can reach extremely increased contents namely in village ponds (1660 mg/kg for Cd or 1630 mg/kg for Zn) in some cases. The contents of risky elements and observed POPs (AOX, PCB_7) were under background values of agricultural soils in the most observed sediment samples.

Long-term prepared legislative regulation (Direction No. 257/2009 Sb.) for sediment application on agricultural soils is valid in the Czech Republic since 2009 year. The Direction regulates selected characteristics and conditions for the application of extracted sediments. The limits of potentially risky elements (As, Be, Cd, Co, Cr, Cu, Hg, Ni, Pb, V and Zn) and persistent organic pollutants (BTEX, sum of PAHs, PCB_7, sum of DDT and C_{10} - C_{40} hydrocarbons) in the sediment and soil of the locality for the application are defined. The limits of risky elements and substances in the soil were derived from the background values of Czech agricultural soils proposed originally (Podlešáková et al., 1996; Němeček et al., 1996). The limits in the Direction use total contents of risky elements only.

The paper shows the results of the research of risky substances contents in the set of sediment samples collected in 2008 year. These contents are compared with sediment characteristics depending on sediment origin and the way of sediment processing. The experiences following from real use of Czech legislative on the field of sediments use in agriculture can contribute to the process of European legislative formulation.

2. Materials and methods

2.1 Sewage sludge analyses

The research focused on the contents of POPs in sewage sludge resulting in the proposal of their recommended maximum contents in the sludge for application on agricultural soils was based on:

The POPs monitoring in 45 wastewater factories in Czech Republic,

the realisation of pot and micro field trial,

the synthesis of the results and their comparison with the proposal of EU directive amendment (EU 2000, Working Document on Sludge), table 1.

The monitoring of POPs in sewage sludge covered the area of the Czech Republic. The waste-water factories were separated into following groups:

- Areas of regional and district towns (including capital city of Prague),
- areas of towns with the presence of industrial activities,
- areas of settlements under 15 000 inhabitants.

The waste-water factories with comparable technologies of wastewater treatment were collected. The contents of polychlorinated dibenzo-p-dioxins and dibenzofurans (PCDDs/Fs) were analysed in the samples from 16 wastewater factories. The example of wastewater characteristics for sludge sampling show table 2.

The list of POPs analyses realised in sludge samples shows table 3.

Two sludge samples form Nord-Moravian region with increased contents of PAHs and PCB6 (table 4) were used in pot and field trials. The application of sludge followed the criteria of Czech directive 382/2001 Sb. and the dose of sludge in trials was derived from the dose of 5 t/ha of dry matter.

2.2 Sewage sludge experiments

Three soil types (typic Chernozem, typic Cambisol and arenic Cambisol) were used in the pot trial (6 kg of soil in Mitscherlich pots). The pot trial was run in three replications.

The field trial was set up on typic Cambisol in the area of Bohemian and Moravian highlands. The field trial was realised in four variants (ploughed and not ploughed, two sludge samples) each in three replications. Ploughed and not ploughed variants were focused on the influence of soil treatment on the decomposition of POPs in the soil (photo degradation, increased input of the air, stimulation of microbial activity). The ploughed variant was treated every two weeks in the layer of humic horizon (cca 20 cm). The characteristics of all used soils are presented in table 5.

The mustard (*Brassica alba*) was used in both (pot and field) trials in first year. The pot trial was sowed by radish (*Raphanus sativus*) and the field trial by parsnip (*Pastinaca sativa*) in the second year. The samples of soil and plants were taken after the harvest, the yield was measured and the contents of POPs in soil and plant samples were analysed. The list of POPs substances and analytical methods for POPs determination in sludge and soil is identical with table 2, except of PCDDs/Fs. The identical analytical methods were used for POPs determination in digested plant samples. The standard elementary statistic methods (file characteristics) were used for the evaluation of the results.

2.3 Sediment sampling

The pond sediment samples from 29 locations were collected in 2008. The samples from pond bottoms and from sediment heaps were used. Field ponds, village ponds and forest ponds were observed. Probe poles with a length of 50 cm for the sampling of bottom sediments and 100 cm for the sampling of heap sediments were used. The individual samples consist of 10 partial samples. The samples were stored in plastic bags and closed jars (for POPs analyse). Closed jars were stored in a deep-freeze condition before chemical analysis. The summary of collected samples is presented (Table 6).

2.4 Sediment analysis

The following characteristics were analysed in sediment samples by the Research Institute for Soil and Water Conservation (RISWC):

- Dry matter content (%)
- Organic matter content (%) - 550°, (CSN EN 12879, 2001)
- pH (H2O), pH (KCl) (CSN ISO 10390, 1996)
- Indicators of the cation exchange capacity CEC (CSN ISO, 13536), BS - the rate of complex saturation adsorption (%)
- Al-exchangeable - titration method (Hraško et al., 1962)

The content and quality of primary organic matter and humus substances were analysed in RISWC using the following approach:

- C_{ox} - organic carbon indicative of the carbon content in primary soil organic matter (SOM). The determination procedure is based on the chromic acid oxidation of organic carbon under the abundance of sulphuric acid and at elevated temperature. Unexpended chromic acid is determined by the iodometric method. This method is a modification of CSN ISO, 14235. The assay of loosely and tightly bound humus materials includes the determination of the humic acid carbon (C-HA), fulvic acid carbon (C-FA), humus matter carbon (C-FA+C-HA) and the assessment of the colour coefficient (Q4/6) indicating the humus quality. The determination procedure is based on the sample extraction method using a mixed solution of sodium diphosphate and sodium hydroxide (Zbíral et al. 2004). Carbon contents (C-FA, C-HA) are determined by titration and the coefficient Q4/6 results from the photometry.
- C_{ws} - water-soluble carbon, indicating the quality of primary SOM (bio available carbon for soil microorganism). Laboratory determination consists of an hour sample extraction using 0.01mol/L CaCl2 solution (1:5 w/V) and the determination of oxidizable carbon in the filtrate evaporation residue by heating the filtrate with chromium sulphuric acid and subsequent titration with Mohr's salt.
- C_{hws} - hot water-soluble carbon, being similar for the assessment purpose to water-soluble carbon. After the soil sample was boild for 1 hour in 0.01mol/L CaCl2 solution (1:5 w/V), the oxidizable carbon in the filtrate evaporation residue through the heating of filtrate with chromium sulphuric acid and subsequent titration with Mohr's salt is determined.

The contents of potentially potentially toxic elements were analysed in sediment samples in RISWC:

- As, Cd, Co, Cr, Cu, Hg, Ni, Pb a Zn in the extract of Aqua regia (ČSN EN, 13346), Hg was analysed by AMA 254 method (Advanced mercury analyser, total content).
- As, Cd, Cu, Pb and Zn in the extract of 1mol/L NH4NO3 (mobile contents). The samples were prepared according to ISO, 11464.

The analysis of the elements in the samples were conducted by the AAS method (AAS Varian), flame and hydride technique.
Persistent organic pollutants were analysed in commercial accredited laboratories Aquatest a.s.:

- BTEX (benzene, toluene, e-benzene and xylene), gas chromatography with mass spectrometry (GS/MS), EPA Method, 8260 B.
- PAHs - polycyclic aromatic hydrocarbons, the contents of 16 substances following EPA, liquid chromatography with fluorescence detector (HPLC), methodology TNV, 75 8055.
- PCB_7 - polychlorinated biphenyls, seven indicator congeners (28, 52, 101, 118, 138, 153, 180), gas chromatography with ECD detector (GC/ECD), EPA Method, 8082.
- DDT sum - sum of DDT, DDE and DDD, gas chromatography with ECD detector (GC/ECD), EPA Method, 8082.
- C_{10} - C_{40} hydrocarbons, gas chromatography with flame-ionisation detector (GC/FID), CSN EN, 14039.

The evaluation of sediment characteristics and the contents of potentially potentially toxic elements and persistent organic pollutants in the sediments separating on the base of their origin and type were done by the use of elementary statistics where median, maximum, minimum, average, standard deviation are presented (Excel). The correlations (Pearson correlation coefficients) between selected sediment properties (pH, CEC, content and quality of soil organic matter) significant at the 0.01 and 0.05 level were processed (SPSS Statistics 17.0).

3. Results and discussion

3.1 Sewage sludge results

The values of POPs (Polycyclic aromatic hydrocarbons - PAHs, monocyclic aromatic hydrocarbons - MAHs, Chlorinated hydrocarbons - ClHs and Petroleum hydrocarbons - PHs) contents are demonstrated in table 7. The sludge samples differentiation follows the type and range of studied area. The overview of POPs contents in sludge in individual years presented table 8.
On the example of tested set of sludge samples it was concluded that fluoranthene reaches the highest average concentrations among PAHs. This finding corresponds with the fact that fluoranthene concentrations in the environment belong to the highest from PAHs group (Holoubek et al., 2003). The phenanthrene concentration with highest maximum values follows fluoranthene. The variability of the values of concentrations of these two substances is the highest among PAHs group. Opposite naphtalene reaches the lowest values of all investigated substances.
The highest average and maximum values from the monocyclic aromatic hydrocarbons (MAHs) were detected in the case of toluene. Contents of toluene in the set of sludge samples were characterised by the highest variability, too. Toluene concentrations influenced predominantly the contents of the sum of MAHs because of very low concentrations of all the other substances.
The contents of chlorinated substances reach relatively low level. The values of PCBs concentrations are characterised by maximum variability. The concentrations of DDE are increased in comparison with DDD and DDT. The persistence of decomposition products of DDT in the environment is still detected (Holoubek et al., 2003; Poláková et al., 2003; Vácha et al., 2003).

Generally the highest contents were found in the case of petroleum hydrocarbons (PHs). The evaluation of the contents is complicated by difficult resolution of substances originated from petroleum contamination and of the substances from the decomposition of organic matter in the sludge.

The comparison between the values of sum of PAHs and values of sum of their toxic equivalent factors (TEF) in 25 samples presents figure 1. Good agreement between these values is evident. It could be concluded increased rate of more nuclei substances respectively substances with higher carcinogenic risk (table 9). This findings confirm the need of PAHs monitoring in sludge used for application on agricultural soils.

The data of the contents of POPs in the set of sewage sludge were processed for the assessment of their "background values". The 90% percentile was used after elimination of outlying values. These background values (table 10) are compared with background values of POPs in agricultural soils (Němeček et al., 1996) in table 11. If we compare obtained "background values" of the content of POPs in the set of sewage sludge with the limit values of POPs in sludge in Czech and European legislative norms we get following results.

The value of the sum of 6 congeners of PCBs is suitable from the viewpoint of Czech (0.6 mg/kg) and European (0.8 mg/kg for 7 congeners) legislation. More problematic seems to be content of the sum of PAHs (9.37 mg/kg) where the overcome of proposed limit of EU directive (6 mg/kg) was observed. No limit value regarding PAHs is included in Czech directive No. 382/2001.

On the base of comparison of background values of POPs in sewage sludge and soil (table 5) emerged following findings. Toluene (MAHs group) shows the maximum difference between the content in the soil and in the sludge from all POPs substances. The concentration in the sludge is cca 243-fold higher than the concentration in the soil. The difference of these contents is significantly lower in the group of PAHs with the maximal difference in the case of benzo(ghi)perylene where sludge content represented 13.7-fold higher value as compared to soil. The contents of PCBs in the sludge are cca 10-fold higher in sludge compared to soil while the contents of DDT (including DDD, DDE) are comparable with the contents in the soil.

The values of I-TEQ PCDD/F fluctuated in the range from 9.2 to 280.2 ng/kg. The value of 280.2 ng/kg was eliminated as outlying by statistic procedure. Resulting average I-TEQ PCDD/F is than 22.5 ng/kg in the set of sludge samples. For 90% percentile I-TEQ PCDD/F reaches the value 37.7 ng/kg. The values of I-TEQ PCDD/F fulfil safely the proposed limit of EU order (100 ng/kg I-TEQ PCDD/F).

The assessment of sludge load on the base of congener analysis of PCDD/F indicates regional differences (with the dominance of octo-chlorinated dibenzodioxins in sludge), which are depending on the wastewater load from the different sources very probably (the rate of communal and industrial wastewater of different type). The data are according with the finding that octo-chlorinated (OCDD) and hepta-chlorinated (HpCDDs) congeners are dominant in the sewage sludge (Holoubek et al., 2002). In spite of this fact the definition of typical general congener pattern of the load of set of sludge samples seems to be complicated considering to regional differences. Congener patterns of individual sludge samples could be used for the localisation of sources of wastewater contamination by PCDD/F (Holoubek et al., 2002).

The proposal of recommended limit values of elected POPs in sludge for the application on agricultural soils (table 12) was derived from the following:

- The background values of selected POPs in set of sludge samples from the wastewater factories of the areas of regional, district and industrial towns and smaller settlements were determined.
- Vegetation experiments did not confirm that sludge application in the dose of 5t/ha of dry matter on the soil influenced POPs contents in the soil and tested plants. Together with these findings we respect the results of the other authors following from long-term experiments about the accumulation of some POPs substances in the soil.
- The proposed limit values in "Working Document on Sludge" were observed.
- The substances from POPs group included in Czech Directive of Soil Protection No. 13/1994 Sb. were selected for the observation.
- Theoretical and simplified balance sheet of the input of POPs into soil by sludge application resulted that the background values of most selected POPs in the soil will be multiplied two times after period of 300 years by sludge application. This balance was not used for PCDD/F.
- Increased limit value was proposed for PAHs in comparison with primary proposal in "Working Document on Sludge". EU primary proposal seems to be not relevant in view of load by Czech sludge by PAHs and from the viewpoint of the strictness of PAHs limit against the other limits of the substances (PCDD/F, PCB_7) in EU primary proposal. The presence of PAHs in the environment in Czech conditions does not correspond with primary EU proposal of PAHs in the sludge and majority of sludge production will be excluded respecting the limit 6 mg/kg. We could not find the explanation for the respecting of this limit by the comparison of limit values of PAHs and PCDDs/Fs in the sludge and their background values in the soil for example. The content of PAHs in sludge is 6 times higher as in the soil but the content of PCDDs/Fs is 100 times higher as in the soil regarding the primary EU proposal.
- The extent of selected POPs substances was adapted for Czech legislative for soil protection (Directive No.13/1994 Sb.). The use of results of the research for the Czech legislation is depending on the confrontation of soil protection and sludge application needs respecting economical site of the problem. The difficulty of this process was documented by the refusal of "Working Document on Sludge" for EU legislation.

The results were derived from the set of sludge samples collected in the territory of the Czech Republic. The international validity could be assumed for European countries thanks to connected markets resulting to similar load of municipal waste waters by potentially toxic substances.

3.2 Dredged sediments results

The limit values of POPs in soil for sediment use in Czech legislation (No. 257/2009 Sb.) shows table 13 where only two POPs groups are limited.

The limit values of POPs in sediments in Czech legislation (No. 257/2009 Sb.) shows table 14 where six POPs groups are limited. The existence of national limits of pollutants in sediments for agricultural use in European countries is recommended

The basic physio-chemical properties of dredged sediments are presented in table 15. The content of dry matter, organic matter, sediment reaction, exchangeable H+ content and adsorption characteristics are defined for the set of sediment samples. The wide range of values of observed parameters is clearly visible in table 15. The differences between individual sediment groups can be observed when the separation of sediments with respect to their origin (the sediments of field, forest and village pounds) is carried out. The differences between sediment acidity were detected primarily. Forest sediments are characterised by higher acidity than the others. The lower values of the saturation of adsorption complex by basic ions (S value) and the values of the rate of adsorption complex saturation (V value) consecutively display an increase in sediment acidity.

The sediments were separated based on the sediment storage method (bottom, heap) due to the tendency to increase acidity during storage, and the comparison of the acidity of separated sediments and adsorption characteristics were observed. The prevailing separate sources (field, village and forest) were accepted also but village sediments were not calculated using this procedure due to missing data (only 1 sample of heap sediment was from a village pond). The results are presented in table 16. The storage of sediments on the heaps before application on agricultural soils is generally used methods in many countries.

The results confirm the trend of sediment acidification during sediment storage in the category of both sediment groups (field, forest). The forest sediments show sharper differences between the reaction of bottom and heap sediments. It was surprising to see, however, that the bottom forest sediments reached the highest pH value. The results demonstrate that decreasing pH value influences the values of adsorption characteristics markedly (S and V values).

The values of content and quality of sediment organic matter are presented in table 17.

The wide range of organic matter content in the set of sediment samples is evident; the sediment application with minimal C_{ox} content seems to not provide economical benefit from the viewpoint of organic matter inputs into agricultural soils. Conversely, the application of sediments with maximal C_{ox} content in a set of sediment samples will lead to increased organic matter input into soils. The lower values of organic matter contents are displayed in village pond sediments. Some countries (Slovakia for example) use minimal limit values of organic matter for sediment use in agriculture.

The quality of primary organic matter (the carbon ability for microbial utilization) when compared by water-soluble and hot-water soluble carbon contents (C_{ws} and C_{hws} values that characterise easily available carbon) reached the highest values in forest pond sediments following by field pond sediments. The lowest values in these parameters were observed in village pond sediments again. The same order can be observed by the evaluation of the content of humus substances where the rise of carbon content of total humus substances in forest pond sediments is distinctly increased. The quality of humus substances compared with the ratio of the carbon of humic and fulvic acid is higher in the field pond sediments compared with forest pond sediments. The lowest values of humus substances quality were observed in village pond sediments. From the comparison of carbon contents of primary organic matter and humus substances it follows that the highest humification degree in organic matter is observed in forest pond sediments. This parameter is comparable in field and village pond sediments. It could be generally resulted that forest sediments are very suitable for application on agricultural soils from the viewpoint of their organic matter quality.

The medians and maximums of POPs contents in field, village and forest sediments are presented in table 18 where the comparison with the Direction No. 257/2009 Sb. is available also.

The median values of PAHs indicate an increased load of village pond sediments and a similar trend can be found in the case of DDT. The contents of the others POPs are comparable between individual sediment types. The maximum limits of PAHs were exceeded in all three sediment types. Very probably, PAHs will be the most problematic of the observed POPs group in the sediments. This trend could be expected generally and the proposed limits for sludge in European proposal (Working Document of Sludge) confirm this fact. From the comparison of sediment load by PAHs with the proposal of PAHs limit values in Czech agricultural soils (Němeček et al., 1996) it was concluded that increased persistence of more nuclei compounds in the sediments was found. The tendency of the substances to accumulate in the sediments was observed in the order benzo(ghi)perylene>benzo(b)fluoranthene, benzo(k)fluoranthene, pyrene>benzo(a)pyrene, benzo(a)anthracene, fluoranthene and chrysene. The order was assessed on the basis of the rate between the individual PAHs substances content in the sediment and proposed soil limit value, and the comparison of the sum of PAHs in the sediments and soil limit value.

Despite the findings of DDT it remains that the increased contents in agricultural soils (Vácha et al., 2001; Čupr et al., 2009) did not exceed limits in sediment samples. The existence of the limit for BTEX in the sediments in Direction No. 257/2009 Sb. must be supported with more data collected, especially from river sediments. The limit for C_{10} - C_{40} hydrocarbons will eliminate their increased contents in sediments for agricultural use from local leaks of petroleum hydrocarbons.

The correlation between the contents of observed POPs groups (except of C_{10} - C_{40} hydrocarbons where a dominant number of values were under detection limit) and content and quality of organic matter was assessed. The data in table 19 confirm only sporadic correlation surprisingly.

The trend of PCB and BTEX accumulation in the dependency on content and quality of humus substances is presented. The PAHs groups did not show any trend of accumulation regarding their properties and affinity to organic carbon. Some authors (Cave et al., 2010) measured bioaccessible PAHs fraction in the soil (varied from 10 - 60%) and the multiple regression showed that the PAHs bioaccessible fraction could be explained using the PAHs compound, the soil type and the total PAHs to soil organic carbon content.

It could be assumed that the sources of the contamination by POPs determined in most POPs groups, except for BTEX, influenced the sediments load stronger than the selected sediment properties in an observed set of sediment samples.

The inputs of potentially toxic substances by sludge and sediment application can play important role in soil hygiene. The easy balance of POPs inputs into soil by sludge and sediments application in accordance with Czech legislative is presented in table 20. It must be accepted that the application of sewage sludge and dredged sediments runs under different conditions. The sludge can be applied once in 3 years in maximal dose of 5 tons of dry matter per hectare. The sediments can be applied once in 10 years in maximal dose of 750 tons of dry matter per hectare. The table presented the dose of sludge and sediments in 10 years. This balance could differ between individual countries following national legislative standards.

The maximum possible increase of POPs content in the soil after sludge and sediment application was derived from their possible maximum inputs (table 21). The values are only tentative because no process of POPs decomposition and migration in the soil was reflected.

4. Conclusion

It is evident that legislative regulation of sewage sludge and dredged sediment application on agricultural soils limits the inputs of risky substances into soils and the other parts of the environment. The uncontrolled application of these materials as well as the other biosolids could lead to serious damage of the soils and their functions. The problem with the limiting of POPs in sewage sludge is still continuing not only in the Czech Republic where only PCB_7 and AOX are limited but in European context especially. The refusal of Working Document on Sludge extended the validity of EU directive 86/278 with the absence for limit values of any POPs substances. At the same time it is known that sludge application significantly increased inputs of PAHs and chlorinated substances (PCBs, PCDDs/Fs) into agricultural soils.

The comparison of POPs inputs by sediment and sludge application demonstrated that the application of dredged sediments loads the agricultural soils more by POPs inputs thanks to use of high possible sediment doses. The European legislative is not available on the field of sediment use in agriculture in present time and the existence of national legislative regulations for sediment application can be highly recommended. The experiences of the practical use of limits application in individual countries can be utilized in the process of European legislative assessment.

5. Annex

5.1 Tables

Organic substances	The value (mg/kg dm)
AOX	500
LAS	2600
DEHP	100
NPE	50
PAHs	6
PCB_7	0,8
Dioxins	The value (ng TE/kg dm)
PCDDs/Fs	100

AOX - Sum of halogenated organic compounds
LAS - Linear alkylbenzene sulphonates
DEHP - Di(2-ethylhexyl)phthalate
NPE - Nonylphenol and nonylphenolethoxylates
PAHs - Sum of polycyclic aromatic hydrocarbons
PCB_7 - Sum of seven indication PCB congeners (28, 52, 101, 118, 138, 153, 180)
PCDDs/Fs - Polychlorinated dibenzodioxins/dibenzofurans

Table 1. The proposed limit values of EU directive 86/278.

Anaerobic and aerobic stabilisation (microbial activity stimulation), sludge dehydration and pressing		
No.	Potential use in agriculture	Characterisation
1	yes	agglomeration, different wastewaters, high technological level of wastewater factory - WF
2	yes	Small area, municipal wastewater, lower technological level of WF
3	yes	Small area, municipal wastewater, lower technological level of WF
4	yes (in use)	Regional town up to 35 000 inhabitants., municipal wastewater predominantly, good technological level of WF
8	yes	Regional town up to 55 000 inhabitants, municipal and industrial wastewater (glass, ceramic), high technological level of WF
9	-	Regional town up to 100 000 inhabitants, municipal and industrial wastewater (food production, chemistry – pre-treatment of wastewater), high technological level of WF
10	yes	Regional town up to 100 000 inhabitants, municipal and industrial wastewater (food and paper production), high technological level of WF
11	yes	settlement up to 7 000 inhabitants, municipal wastewater, good technological level of WF
12	yes	Regional town up to 40 000 inhabitants, municipal and industrial wastewater (food production), high technological level of WF
13	yes	Town up to 15 000 inhabitants, municipal wastewater, lower technological level of WF
14	yes	Regional town up to 170 000 inhabitants, municipal and industrial wastewater (food production), high technological level of WF
15	yes	Regional town up to 20 000 inhabitants, municipal wastewater predominantly, high technological level of WF
16	yes (in use)	Regional town up to 50 000 inhabitants, municipal and industrial wastewater (car production), high technological level of WF
17	yes	Regional town up to 50 000 inhabitants, municipal and industrial wastewater (car production), high technological level of WF
18	yes	Industrial town up to 20 000 inhabitants, municipal and industrial wastewater 50/50 (chemistry), high technological level of WF
19	yes	Regional town up to 80 000 inhabitants, municipal wastewater only, high technological level of WF
20	yes	Town up to 20 000 inhabitants, municipal wastewater, high technological level of WF

21	yes	Town up to 20 000 inhabitants, municipal and industrial wastewater, good technological level of WF
22	no	Regional town up to 100 000 inhabitants, industrial WF, high technological level
23	yes	Regional town up to 100 000 inhabitants, municipal and industrial (lower rate) wastewater, high technological level of WF
24	yes	Settlement up to 5 000 inhabitants, municipal wastewater, lower technological level of WF
25	no	Industrial town up to 20 000 inhabitants, increased rate of industrial wastewater (chemistry), high technological level of WF
Mechanical filtration, cold sludge maturation		
5	yes	Spa town up to 15 000 inhabitants, municipal wastewater
6	yes	Settlement up to 5 000 inhabitants, municipal wastewater
7	yes	Central WF for few small settlements, municipal wastewater

Table 2. The characteristics of selected wastewater factories.

Analyse	Samples
pH, Cox,Ca,Mg, P, K	45 samples
As, Be, Cd, Co, Cr, Cu, Hg, Mn, Ni, Pb, V, Zn (extract of aqua regia)	45 samples
Monocyclic aromatic hydrocarbons benzene, toluene, xylene, ethylbenzene Polycyclic aromatic hydrocarbons naphtalene, anthracene, pyrene, phluoranthene, phenanthrene, chrysen, benzo(b)phluoranthene, benzo(k)phluoranthene, benzo(a)anthracene, benzo(a)pyrene, indeno(c,d)pyrene, benzo(ghi)perylene chlorinated hydrocarbons PCB, HCB, α-HCH, β-HCH, γ-HCH Pesticides DDT, DDD, DDE styrene, petroleum hydrocarbons	45 samples
PCDF 2,3,7,8 TeCDF, 1,2,3,7,8 PeCDF, 2,3,4,7,8 PeCDF, 1,2,3,4,7,8 HxCDF, 1,2,3,6,7,8 HxCDF, 1,2,3,7,8,9 HxCDF, 2,3,4,6,7,8 HxCDF, 1,2,3,4,6,7,8 HpCDF, 1,2,3,4,7,8,9 HpCDF, OCDF PCB 189, PCB 170, PCB 180 PCDD 2,3,7,8 TeCDD, 1,2,3,7,8 PeCDD, 1,2,3,4,7,8 HxCDD, 1,2,3,6,7,8 HxCDD, 1,2,3,7,8,9 HxCDD, 1,2,3,4,6,7,8 HpCDD,OCDD PCB PCB 77, PCB 126, PCB 169, PCB 105, PCB 114, PCB 118+123, PCB 156, PCB 157, PCB 167	16 samples

Table 3. The analyses in sludge samples.

	PAHs												
	A	N	P	Ch	Ph	F	B(a)P	B(b)F	B(k)F	B(a)A	B(ghi)P	I(cd)P	PAHs
S 1	1440	3400	2520	1420	8340	7990	3630	4190	1820	1890	1930	1740	40310
S 2	851	50	2950	2590	7220	9520	6640	7490	3360	3150	3690	2830	50341
	MAHs					ChHs							
	B	T	X	Eb	MAHs	PCB6	αHCH	βHCH	γHCH	HCB	DDT	DDD	DDE
S 1	120	830	7	2300	1043	1090	1.00	1.00	1.00	1.00	1.00	1.00	1.00
S 2	14	90	3	3800	111	57	1.00	1.00	1.00	1.25	1.26	2.08	21.5

S 1 – sludge 1, S 2 – sludge 2
A – anthracene, N – naphthalene, P – pyrene, Ch – chrysene, Ph – phenanthrene, F – fluoranthene, B(a)P – benzo(a)pyrene, B(b)F – benzo(k)fluoranthene, B(a)A – benzo(a)anthracene, B(ghi)P – benzo(ghi)pyrene, I(cd)P – indeno(c,d)pyrene, PAHs – polycyclic aromatic hydrocarbons, B – benzene, T – toluene, X – xylene, EB – ethylbenzene, MAHs – monocyclic aromatic hydrocarbons, PCB6 – sum of 6 polychlorinated biphenyls congeners, HCH – hexachlorcyclohexane, HCB – hexachlorbenzene, DDT – dichlordiphenyltrichloethane, DDD – dichlordiphenyldichlorethane, DDE – dichlordiphenylethane, ChHs – chlorinated hydrocarbons

Table 4. POPs contents in sewage sludge used in pot trial (μg/kg).

Soil type	District of origin	pH (KCl)	Cox (%)	Trial
Arenic Cambisol	Melnik	7.05	1.02	pot
Modal Cambisol	Benesov	6.15	1.29	pot
Modal Chernozem	Nymburk	6.93	2.18	pot
Modal Cambisol	Jihlava	5.85	0.8	field

Table 5. The characteristics of soils used in the experiments.

	Field ponds	Forest ponds	Village ponds	Total
Bottom	6	4	3	13
Heap	7	7	2	16
Total	13	11	5	29

Table 6. The numbers and types of sediment samples.

		PAHs												
		A	N	P	Ch	Ph	Fl	B(a)P	B(b)F	B(k)F	B(a)A	B(ghi)P	I(cd)P	PAHs
Industrial towns	AM	299	279	1176	659	1748	941	316	290	168	481	202	158	6718
	GM	245	110	976	526	1484	783	238	222	129	375	154	124	5580
	std.	144	383	506	301	760	451	165	149	80	223	139	76	2851
	max.	477	1185	1870	984	2570	1500	539	463	245	754	506	254	9714
	min.	49	15	169	75	337	168	28	29	15	45	29	17	976
	med.	343	147	1100	791	2100	851	291	289	191	486	154	189	6771
Regional towns	AM	501	77	1706	1026	2280	1821	606	650	316	802	413	367	10564
	GM	282	26	1338	844	1599	1296	454	486	238	648	324	264	7970
	std.	551	67	1241	676	2074	1588	494	528	254	547	302	306	8454
	max.	1710	203	3850	2190	6700	4880	1410	1500	724	1670	940	918	26528
	min.	96	1	507	371	502	464	216	233	114	310	132	113	3071
	med.	165	64	1130	676	1260	1010	334	342	176	532	260	233	6165
Settlements	AM	215	20	1768	826	1399	1187	415	393	201	685	346	207	7662
	GM	181	5	1394	761	1159	1025	360	353	179	624	268	191	6624
	std.	134	25	1616	338	938	658	218	174	89	317	314	91	4705
	max.	548	69	6490	1570	3810	2810	900	732	368	1460	1240	434	20431
	min.	88	1	638	352	453	345	120	133	59	329	110	113	2741
	med.	188	1	1395	736	1100	1085	338	352	183	631	212	181	6328

PAHs – polycyclic aromatic hydrocarbon

Table 7a. The POPs contents in individual groups of sludge samples - PAHs (μg/kg).

		MAHs (μg/kg)					ChlH (μg/kg)								Sty-rene	PHs	Te
		B	T	X	Eb	MAU	PCB	α-HCH	β-HCH	γ-HCH	HCB	DDT	DDD	DDE			
Industrial towns	AM	39.6	494.9	154.5	856.4	1545.3	336	2.5	62.5	1.0	30.6	25.6	27.6	26.5	12.6	14571	2.6
	GM	10.8	138.0	93.0	79.6	481.9	240	1.8	5.7	1.0	16.0	21.9	5.4	24.9	2.1	11043	1.9
	std.	62.4	603.2	125.7	1977.6	2173.4	254	2.6	98.4	0.0	44.2	13.3	57.8	9.4	12.7	9322	1.6
	max.	191.0	1860.0	370.0	5700.0	6548.1	738	8.7	254.0	1.0	138.0	43.4	169.0	43.9	32.5	29000	6.1
	min.	0.1	5.1	19.9	16.1	61.1	98	1.0	1.0	1.0	4.3	10.6	1.0	14.2	0.1	2200	0.2
	med.	17.9	392.0	164.0	70.8	644.7	183	1.0	1.0	1.0	16.6	25.4	4.6	24.9	8.7	11000	2.3
Regional towns	AM	17.9	1543.8	38.2	25.3	1625.1	144	2.0	1.0	1.2	4.1	10.3	2.7	15.2	0.1	8943	2.3
	GM	7.7	452.6	2.6	6.1	630.5	119	1.7	1.0	1.1	2.5	9.5	2.2	10.6	0.1	8616	2.2
	std.	14.5	2804.5	57.3	22.5	2794.1	99	1.3	0.0	0.6	3.9	4.8	2.0	10.0	0.0	2496	0.7
	max.	44.1	8380.0	154.0	70.7	8437.4	358	4.5	1.0	2.7	11.6	21.6	7.1	35.3	0.1	13000	3.2
	min.	0.1	48.4	0.1	0.1	168.3	64	1.0	1.0	1.0	1.0	6.3	1.0	1.0	0.1	6200	1.1
	med.	13.6	314.0	5.7	20.9	369.9	99	1.5	1.0	1.0	1.7	9.2	2.1	14.6	0.1	7700	2.4
Settle-ments	AM	14.3	2784.4	13.4	20.5	2832.6	1566	15.1	1.6	1.0	11.4	21.2	11.4	35.9	0.1	11810	3.4
	GM	2.3	878.1	0.7	2.9	1141.6	170	2.4	1.2	1.0	7.1	15.2	6.5	25.1	0.1	10517	2.2
	std.	15.2	3463.8	30.7	22.6	3452.0	4345	40.0	1.9	0.0	7.5	16.9	12.3	34.8	0.0	5700	2.5
	max.	38.8	9330.0	104.0	59.8	9369.0	14600	135.0	7.2	1.0	20.1	58.6	43.0	134.0	0.1	21000	8.4
	min.	0.1	55.8	0.1	0.1	151.1	31	1.0	1.0	1.0	1.0	3.4	1.0	4.7	0.1	6000	0.2
	med.	8.3	800.0	0.1	14.2	830.0	123	1.6	1.0	1.0	13.9	16.5	7.3	25.3	0.1	9050	3.2

MAHs – mococyclic aromatic hydrocarbons; ClHs – chlorinated hydrocarbons
PHs – petroleum hydrocarbons Te – tenzides

Table 7b. The POPs contents in individual groups of sludge samples - MAHs, ClHs (μg/kg), Te and PHs (mg/kg)

		PAHs												
		A	N	P	Ch	Ph	Fl	B(a)P	B(b)F	B(k)F	B(a)A	B(ghi)P	I(cd)P	PAHs
2002 (20 sampl.)	AM	164	47	1442	616	1131	2499	709	906	409	595	508	398	9930
	GM	122	26	1219	527	940	2131	634	810	368	517	453	355	8488
	std.	120	43	814	363	704	1511	333	435	187	368	255	190	5964
	max	484	140	3330	1490	2880	6860	1440	1930	806	1900	1210	762	26434
	min.	16	2	363	189	240	596	206	334	141	193	161	156	2631
	med	160	30	1250	552	947	2115	635	798	370	560	413	333	8516
2001 (25 sampl.)	AM	252	45	1096	692	1479	904	316	329	173	504	216	165	5912
	GM	173	11	852	535	1067	709	241	250	133	392	163	123	4646
	std.	188	55	474	333	977	427	156	166	85	223	118	73	2571
	max	791	203	1870	1570	3910	1710	596	732	368	927	506	305	9714
	min.	3	1	14	11	17	18	7	6	4	10	3	1	95
	med	199	23	1115	676	1230	889	319	322	176	497	201	172	6170
Sum (45 sampl.)	AM	201	50	1159	685	1330	1405	540	617	294	508	352	283	6566
	GM	143	17	949	547	1010	1088	400	438	217	425	264	205	5497
	std.	144	55	536	383	887	815	364	454	200	205	229	192	2632
	max	548	207	2550	1750	3910	3630	1440	1930	806	927	940	762	11218
	min.	3	1	14	11	17	18	7	6	4	10	3	1	95
	med	164	30	1170	632	1070	1220	467	455	240	518	311	213	6525

PAHs – polycyclic aromatic hydrocarbons

Table 8a. Elementary statistic of the POPs in sludge samples – PAHs (μg/kg)

		MAHs					ChlHs					PHs
		B	T	X	Eb	MAU	PCB	HCB	DDT	DDD	DDE	(mg/kg)
2002 (20 sampl.)	AM	37.9	3815.3	39.8	12.5	3950	110	6.96	3.72	6.11	17.15	5845
	GM	27.6	1431.3	13.5	7.3	1680	98	5.54	3.13	4.94	13.75	5303
	std.	33.8	4315.9	47.8	10.6	4366	49	4.00	1.97	3.38	9.24	2310
	max.	120.0	16400.0	170.0	40.0	16820	201	15.40	7.85	11.50	36.30	8800
	min.	12.0	70.0	0.1	0.1	111	33	1.00	1.00	1.00	1.00	2300
	med.	20.5	2900.0	26.0	8.0	2974	103	6.70	3.51	6.18	15.50	6450
2001 (25 sampl.)	AM	14.8	498.8	36.5	28.2	608	122	9.13	16.91	4.24	21.54	11309
	GM	3.8	163.1	2.6	6.0	288	98	5.44	12.43	2.98	15.87	7848
	std.	13.8	553.0	59.0	27.3	614	75	7.13	12.56	3.74	12.38	6987
	max.	44.1	2040.0	192.0	82.1	2324	358	20.10	45.00	16.60	45.80	29000
	min.	0.1	0.1	0.1	0.1	0	7	1.00	1.00	1.00	1.00	20
	med.	11.1	350.0	5.7	20.2	360	104	8.23	11.40	3.24	20.75	9100
together (45 sampl.)	AM	22.4	2209.1	38.0	16.5	2293	110	8.17	8.00	5.13	20.33	6827
	GM	8.3	522.4	5.3	5.4	739	95	5.49	5.62	3.79	15.32	6219
	std.	22.8	2998.3	54.4	16.1	2998	52	6.05	6.54	3.69	12.17	2631
	max.	120.0	10200.0	192.0	59.8	10355	234	20.10	25.40	16.60	51.80	13000
	min.	0.1	0.1	0.1	0.1	0	7	1.00	1.00	1.00	1.00	2200
	med.	17.0	617.0	13.0	9.5	675	103	7.53	6.24	4.13	19.80	7000

MAHs – mococyclic aromatic hydrocarbons; ClHs – chlorinated hydrocarbons
PHs – Petroleum hydrocarbons

Table 8b. Elementary statistic of the POPs in sludge samples – MAHs, ClHs (μg/kg) and PHs (mg/kg)

Compound	The toxic equivalent value	Compound	The toxic equivalent value
Benzo(a)pyrene	1	Benzo(k)fluoranthene	0.01
Benzo(a)anthracene	0.1	Dibenzo(a,h)anthracene	1
Benzo(b)fluoranthene	0.1	Indeno(1,2,3-cd)pyrene	0.1

Table 9. The overview regarding the toxic equivalent value for individual PAHs compounds

	PAHs (μg/kg)												
	Fl	P	Ph	B(b)F	B(a)A	A	B(a)P	I(cd)P	B(k)F	B(ghi)P	Ch	N	ΣPAU
90 percentil	2412	1626	2407	1316	759	433	949	535	572	686	1148	132	9371

	MAHs (μg.kg-1)					ChlHs (μg/kg)					PHs
	B	T	X	Eb	ΣMAU	PCB	HCB	DDT	DDE	DDD	mg/kg
90 percentil	50	7300	150	37	7342	183	17.8	19.6	36.1	9.8	9440

Table 10. Background values of POPs in sludge collection, 90 percentil = backgroun value

	PAHs (μg/kg)											
	Fl	P	Ph	B(b)F	B(a)A	A	B(a)P	I(cd)P	B(k)F	B(ghi)P	Ch	N
Background - soil	300	200	150	100	100	50	100	100	50	50	100	50
Background - sludge	2412	1626	2407	1316	759	433	949	535	572	686	1148	132
difference in %	804	813	1605	1316	759	866	949	535	1144	1372	1148	264

	MAHs (μg/kg)				ChlHs (μg/kg)					PHs(mg/kg)
	B	T	X	Eb	PCB	HCB	DDT	DDE	DDD	
Background - soil	30	30	30	40	20	20	15	10	10	100
Background - sludge	50	7300	150	37	183	18	20	36	10	9440
difference in %	167	24333	500	92	917	89	130	361	98	9440

Table 11. The comparison of background values of POPs in sludge and soils.

	Content (µg/kg)							
Parameter	Sum MAHs	Sum PAHs	PCB_7	HCB	DDT	DDE	DDD	I-TEQ* PCDDs/Fs
Recommended limit	10 000	10 000	600	60	60	60	30	80
EU proposal	6000	-	800	-	-	-	-	100
Soil reference value (Czech)	1000	130	20	20	30	25	20	1

PAHs- polyaromatic hydrocarbons,
MAHs-monoaromatic hydrocarbons,
PCB7-sum of 7 congeners of polychlorinated biphenyls,
HCB-hexachlorbenzene,
DDT-dichlordiphenyltrichlorethane,
DDD-dichlordiphenyldichlorethane,
DDE-dichlordifenyldichlorethen,
I-TEQ PCDD/F-toxic equivalent of polychlorinated dibenzo-p-dioxins and dibenzofurans
I-TEQ PCDDs/Fs (ng/kg)

Table. 12. Recommended limit values of elected POPs in sludge, primary EU proposal and reference values in soils of the Czech Republic.

	Content (mg/kg)	
Limited substance	Middle and heavy texture soils	Light texture soils
PAHs	1.0	1.0
PCB_7	0.02	0.02

PAHs – polycyclic aromatic hydrocarbons
PCB7 – seven indication congeners of polychlorinated biphenyls

Table 13. Directive No. 257/2009, sediment use on agricultural soils, POPs limit values in soil.

Limited substance	Content (mg/kg)
PAHs	6
PCB_7	0.2
BTEX	0.4
DDT	1
C_{10}-C_{40}	300

PAHs – polycyclic aromatic hydrocarbons
PCB7 – seven indication congeners of polychlorinated biphenyls (28, 52, 101, 118, 138, 153, 180)
BTEX – sum of benzene, toluene, ethylbenzene and xylene
DDT – sum of DDT, DDD and DDE;
C10-C40 – sum of hydrocarbons - indication of petroleum hydrocarbons

Table 14. Directive No. 257/2009, sediment use on agricultural soils, POPs limit values in sediments (mg/kg).

	Dry matter %	Organic matter %	pH H_2O	pH KCl	Exchangeable H+ (mmol/100g)	CEC (mmol/100g)	BS %
Field (medians, 13 samples)	85.73	8.40	5.22	5.04	8.0	20.01	52.5
Village (medians, 5 samples)	98.06	7.07	5.42	5.23	6.5	18.91	63.5
Forest (medians, 11 samples)	80.08	8.44	4.09	3.82	16	20.88	48
Together (medians, 29 samples)	81.24	8.68	5.13	4.91	11.0	20.9	57.0
Together (A. means)	79.27	9.89	5.17	4.91	13.12	21.33	56.12
Together (St. deviation)	17.1	4.71	1.01	1.06	9.66	7.05	16.41
Maximum	99.06	22.5	7.1	6.95	42.0	41.19	100
Minimum	46.05	2.73	2.86	2.84	<0.5	8.81	28

CEC – cation exchange capacity
pH H2O – sediment pH measured in the extract of H2O
BS – the rate of complex saturation adsorption
pH KCl - sediment pH measured in the extract of 1M KCl

Table 15. Sediment characteristics in the set of 29 samples

	pH H_2O	pH KCl	Exchangeable H+ (mmol/100g)	CEC (mmol/100g)	BS %
Field-bottom	5.26	5.12	5.5	17.99	55.5
Field-heap	5.21	4.94	9.5	22.19	52.5
Forest-bottom	6.28	6.13	10	17.95	71
Forest-heap	3.81	3.63	26.5	25.24	31

CEC – cation exchange capacity
pH H2O – sediment pH measured in the extract of H2O
BS – the rate of complex saturation adsorption
pH KCl - sediment pH measured in the extract of 1M KCl

Table 16. The medians of sediment characteristics separated into sediment groups based on sediment type and storage method.

	C_{ox} %	C_{ws} mg.kg-1	C_{hws} mg.kg-1	HA %	FA %	HS %	Q4/6	HA:FA	HS:C_{ox}
Field	2.54	167.5	404.5	0.41	0.25	0.61	5.50	1.44	0.25
Village	1.72	108.5	324	0.21	0.23	0.47	6.1	0.92	0.27
Forest	2.53	199	558	0.49	0.61	0.97	5.4	1.18	0.39
Together	2.67	175	458.5	0.41	0.29	0.77	5.35	1.1	0.29
Maximum	8.29	515.0	1738.0	2.33	1.67	3.27	24.8	4.04	0.49
Minimum	0.52	72.0	76.0	0.07	0.04	0.11	3.4	0.23	0.16

Cox- organic carbon HA- carbon of humic acids
Cws- water-soluble carbon FA- carbon of fulvic acids
Chws- hot-water-soluble carbon HS- carbon of humus substances Q4/6- colour quotient

Table 17. The medians, maximum and minimum of organic matter content and quality in the sediments

		PAHs 2n	PAHs 3-4n	PAHs 5-6n	PAHs sum	PCB_7	DDT sum	BTEX	C_{10}-C_{40}*
Field	median	48	494	147	694	15.1	9.19	31.2	100
	max	210	4762	1290	6143	40.8	14.5	71.5	580
Village	median	77	2396	780	3386	14.2	15	30.35	105
	max	210	6842	2133	9052	36.9	32.3	43.2	110
Forest	median	41	326	54	517	15.4	8.83	66.3	100
	max	228	10347	2961	13536	1010	16.7	96.2	200
Limit 257/2009		-	-	-	6000	200	100	400	300
Limits exceeded. Field/Village/Forest					1/2/1	0/0/1	0/0/0	0/0/0	1/0/0

* C10-C40 – sum of hydrocarbons, content in mg/kg
PAHs 3-4n - the sum of PAHs with 3 and 4 rings
PAHs 2n - the sum of PAHs with 2 rings
PAHs 5-6n - the sum of PAHs with 5 and 6 rings
PCB7 – sum of seven indication congeners
BTEX – sum of benzene, toluene, ethylbenzene and xylene
DDT sum – sum of DDT, DDD and DDE

Table 18. The medians and maximums of POPs contents in the sediments (μg/kg) and values exceeding limits (Direction No. 257/2009 Sb.)

	PAHs 2n	PAHs 3-4n	PAHs 5-6n	PAHs sum	PCB_7	DDT	BTEX	C_{10}-C_{40}
C_{ox}	0.209	-0.194	-0.196	-0.187	0.347	0.085	**0.749***	0.287
C_{hws}	0.196	0.071	0.069	0.075	0.246	0.183	**0.468***	-0.015
C_{ws}	**0.418**	0.225	0.235	0.237	0.06	0.148	0.268	**0.419**
HS	0.170	-0.151	-0.165	-0.148	**0.462**	0.067	**0.727***	0.180
FA	0.351	-0.041	-0.51	-0.034	0.286	0.064	**0.723***	0.163
HA	0.051	-0.202	-0.217	-0.202	**0.524***	0.063	**0.632***	0.196
HA/FA	-0.299	-0.199	-0.203	-0.202	0.186	0.045	0.298	-0.034
C_{ox}/HS	0.121	0.092	0.083	0.091	**0.414**	0.094	0.268	-0.088

Cox - organic carbon HA - carbon of humic acids PCB7 – sum of seven indication congeners
Cws - water-soluble carbon FA - carbon of fulvic acids DDT – sum of DDT, DDD and DDD
Chws - hot-water-soluble carbon HS - carbon of humus substances PAHs 2n - the sum of PAHs with 2 rings PAHs 3-4n - the sum of PAHs with 3 and 4 rings PAHs 5-6n - the sum of PAHs with 5 and 6 rings C10-C40 – sum of hydrocarbons, content in mg/kg

Table 19. Pearson correlation coefficients between the contents of individual POPs groups and content and quality of organic matter, correlation significant at the 0,01 level (bold*) and 0,05 level (bold).

	POPs inputs (g/ha)	
Limited substance	Sewage sludge, application of 15t d.m. once in 10 years	sediments, application of 750t d.m. once in 10 years
PCB_7	9	150
PAHs	non limited	4500
BTEX	non limited	300
DDT	non limited	75
C_{10}-C_{40}	non limited	225000

(g/ha) PAHs – polycyclic aromatic hydrocarbons PCB7 – seven indication congeners of polychlorinated biphenyls BTEX – sum of benzene, toluene, e-benzene and xylene DDT – sum of DDT, DDD and DDE C10-C40 – sum of hydrocarbons - indication of petroleum hydrocarbons

Table 20. The comparison of POPs inputs by sewage sludge and dredged sediments application into agricultural soils.

	Sewage sludge		Dredged sediments	
Limited substance	Concentration increase (mg/kg)	% of soil background values CR	Concentration increase (mg/kg)	% of soil background values CR
PCB_7	0.002	10	0.03	150
PAHs	-	-	1	100
BTEX	-	-	0.07	54
DDT	-	-	0.02	27
C_{10}-C_{40}	-	-	50	50

PAHs – polycyclic aromatic hydrocarbons PCB_7 – seven indication congeners of polychlorinated biphenyls BTEX – sum of benzene, toluene, e-benzene and xylene DDT – sum of DDT, DDD and DDE C_{10}-C_{40} – sum of hydrocarbons - indication of petroleum hydrocarbons

Table 21. Maximum possible increase of POPs in soil after sewage sludge and dredged sediments application.

5.2 Figures

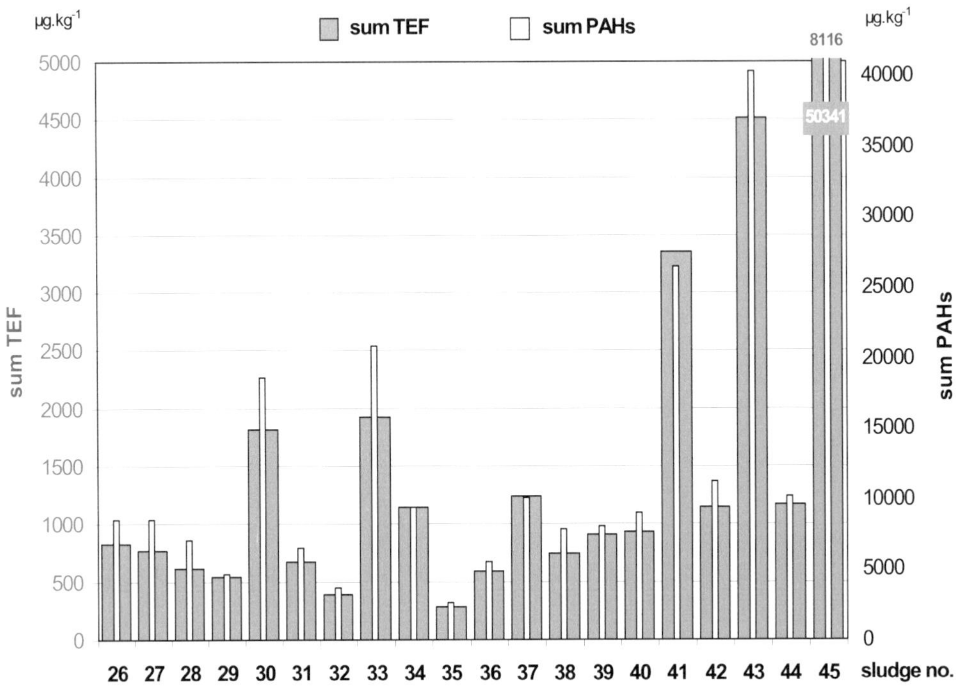

Fig. 1. The comparison of sum TEF PAHs and sum PAHs in sludge (μg/kg).

6. Acknowledgment

The chapter was prepared by the support of the Project of Ministry of Agriculture MZE0002704902 and Project No. QH 82083.

7. References

Benešová, L. & Gergel, J. (2003). Circumstances and relations of sediments agricultural use. *Odpadové fórum*, Vol.9, (September 2003), pp. 14-16, ISSN 1212-7779

Cave M.R.; Wragg, J.; Harrison, I.; Vane, C.H. ; Van de Wiele, T.; De Groeve, E.; Nathanail, C.P.; Ashmore, M.; Thomas, R.; Robinson, J. & Daly, P. (2010). Comparison of Batch Mode and Dynamic Physiologically Based Bioaccessibility Tests for PAHs in Soil Samples. *Environmental Science & Technology*, Vol.44, No.7, (April 201), pp. 2654-2660, ISSN 0013-936X

Clarke, B.; Porter, N.; Symons, R.; Marriott, P.; Ades, P.; Stevenson, G. & Blackbeard, J. (2008). Polybrominated diphenyl ethers and polybrominated biphenyls in Australian sewage sludge. *Chemosphere*, Vol.73, No.6, (October 2008), pp. 980-989, ISSN 0045-6535

Čermák, P.; Budňáková, M. & Kunzová, E. (2009). The utilization of sediments on agricultural farm land in the Czech Republic. *Book of abstracts of 18th CIEC International Symposium*, November 8 - 12th, Roma, Italy.

Default, D.A.; Gerino, M.; Laplanche, C.; Julien, F.; Winterton, P.; Merlina, G.; Delmas, F.; Lim, P.; Sanchez-Perez, J.M. & Pinelli E. (2009). Herbicide accumulation and evaluation in reservoir sediments. *Science of the Total Environment*, Vol.407, No.8, (April 2009), pp. 2659-2665, ISSN 0048-9697

Domene, X.; Alcaniz, J.M. & Andres, P. (2007). Ecotoxicological assessment of organic wastes using the soil collembolan Folsomia candida. *Applied Soil Ecology*, Vol.35, No.3, (March 2007), pp. 461-472, ISSN 0929-1393

EN 14735 (2005). Characterization of waste - Preparation of waste samples for ecotoxicity tests. Available from http://shop.bsigroup.com/ProductDetail/?pid=000000000030152823

European Parlament (2003). Thematic Strategy for Soil Protection. European Parliament resolution on the Commission communication 'Towards a Thematic Strategy for Soil Protection' (COM(2002) 179 - C5-0328/2002 - 2002/2172(COS)). 19/11/2003.

European Union (2000). Working Document on Sludge: An EU-initiative to improve the present situation for sludge management, Brussels, ENV.E.3/LM, p. 19.

Fuentes, A.; Llorens, M.; Saez, J.; Aguilar, M.I.; Ortuno, J.F. & Meseguer, V.F. (2008). Comparative study od six different sludges by sequential speciation of heavy metals. *Bioresources Technology*, Vol.99, No.3, (February 2008), pp. 517-525, ISSN 0960-8524

Gergel, J. (1995). The extraction and the use of sediments from small water reservoirs. *Methodological approach*, 18/1995, Research Institute for Soil and Water Conservation, Prague, 45 p. (in Czech)

Gomez-Alvarez, A.; Valenzuela-Garcia, J.L.; Aquayo-Salinas, S.; Meza-Figueroa, D.; Ramirez-Hernandez, J. & Ochoa-Ortega, G. (2007). Chemical partitioning of sediment contamination by heavy metals in the Pedro River, Sonora, Mexico. *Chemical Speciation and Bioavailability*, Vol.19, No.1, (January 2007), pp. 25-35, ISSN 0954-2299

Heininger, P.; Pelzer, J.; Claus, E. & Pfitzner, S. (2004). Results of long-term sediment quality studies on the river Elbe. *Acta Hydrochimia et Hydrobiologica*, Vol.31, No.4-5, (February 2004), pp. 356-367, ISSN 0323-4320

Heise, S.; Claus E.; Heininger, P.; Krämmer, Th.; Krüger, F.; Schwarz, R. & Förstner, U. (2005). Studie zur Schadstoffbelastung der Sedimente im Elbeeinzugsgebiet - Ursachen und Trends. Im Auftrag von Hamburg Port Authority. Abschlussbericht (Dezember 2005), 169 p.

Hilschnerová, K.; Dušek, L.; Kubík, V.; Klánová, J. & Holoubek, I. (2007). Distribution of organic pollutants in sediments and alluvial soils after major floods. *Journal of Soils & Sediments*, Vol.7, No.3, (June 2007), pp.167-177, ISSN 1439-0108

Holoubek, I.; Čupr, P.; Škarek, M.; Černá, M. & Sáňka M. (2002). The conclusion of analyses of contamination of vicinity of Spolana Neratovice factory by PCDD/F and biphenyl after floods in 2002. TOCOEN, s.r.o. Brno. *TOCOEN REPORT No. 236*, 62 p. (In Czech), Available from http://www.tocoen.cz/zpravy.htm

Holoubek, I.; Adamec, V.; Bartoš, M.; Černá, M.; Čupr, P.; Bláha, K.; Bláha, L.; Demnerová, K.; Drápal, J.; Hajšlová, J.; Holoubková, I.; Jech, L.; Klánová, J.; Kohoutek, J.; Kužílek, V.; Machálek, P.; Matějů, V.; Matoušek, J.; Matoušek, M.; Mejstřík, V.; Novák, J.; Ocelka, T.; Pekárek, V.; Petira, O.; Punčochář, M.; Rieder, M.; Ruprich, J.; Sáňka, M.; Vácha, R. & Zbíral, J. (2003). National stocktaking of Persistent organic pollutants in the Czech Republic. Project GF/CEH/01/003 Enabling activities to facilitate early action on the implementation of the Stockholm Convention on Persistent organic pollutants (POPs) in the Czech Republic. *TOCOEN REPORT No. 249*. Available from http://www.genasis.cz/stockholm-stockholmska_umluva-inventura_pops_2007/

Hraško, J.; Červenka, L.; Facek, Z.; Komár, J.; Němeček, J.; Pospíšil, F. & Sirový, V. (1962). *Soil analysis*. SVPL Bratislava, 342 p. (in Czech)

ISO 15799 (2003). Soil quality - Guidance on the ecotoxicological characterization of soils and soil materials. International Organization for Standardization. Geneve, Switzerland. Available from http://www.iso.org/iso/iso_catalogue/catalogue_tc/catalogue_detail.htm?csnumber=29085

Janeček, M.; Bohuslávek, J.; Dumbrovský, M.; Gergel, J.; Hrádek, F.; Kovář, P.; Kubátová, E.; Pasák, V.; Pivcová, J.; Tippl, M.; Toman, F.; Tomanová, O. & Váška, J. (2005). *The protection of agricultural soil against erosion*. ISV Press Praha, 2nd edition, 195 p. ISBN 80-86642-38-0

Liu, X.; Sun L.; Yin, X. & Wang, Y. (2008). Heavy Metal Distributions and Source Tracing in the Lacustrine Sediments of Dongdao Island, South China Sea. *Acta Geologica Sinica-English Edition*, Vol.82, No.5, Special Issue 4, (2008), pp. 1002-1014, ISSN 1000-9515

Ministry of agriculture and Ministry of Environment of Czech Republic (2009). Direction No. 257/2009 Sb. for sediment use on agricultural soils.

Markard, C. (1988). Organic contaminants in sexage sludge - do they constitute a danger for the food chain. *Korrespondenz Abwasser*, Vol.35, No.5, (May 1988), pp. 449-452, ISSN 0341-1540

Melcer, H.; Monteith, H. & Nutt, S. G. (1988). Variability of toxic trace contaminants in municipal sewage treatments plants. *Water Science and Technology*, Vol.20, No.4-5, (1988), pp. 275-284, ISSN 0273-1223

Ministry of Environment of Czech Republic (1994). The notice of the Ministry of Environment for the management of the soil protection, No. 13/1994 Sb. (in Czech)

Ministry of Environment of Czech Republic (2001). The notice of Ministry of Environment for the sludge application on agricultural soil, No. 382/2001 Sb. (in Czech)

Natal-da-Luz, T.; Tidona, S.; Jesus, B.; Morais, P.V. & Sousa, J. P. (2009). The use of sewage sludge as soil amendment. The need for an ecotoxicological evaluation. *Journal of Soils and Sediments*, Vol.9, No.3, (June 2009), pp. 246-260, ISSN 1439-0108

Němeček, J.; Podlešáková, E. & Pastuszková, M. (1996). Proposal of soil contamination limits for persistent organic xenobiotic substances in the Czech Republic. *Rostlinna Vyroba*, Vol.42, No.2, (February 1996), pp. 49-53, ISSN 0370-663X

Netzband, A.; Reincke, H. & Bergemann, M. (2002). The river Elbe - a case study for the ecological and economical chain of sediments. *Journal of Soils and Sediments*, Vol.2, No.3, (June 2002), pp. 112-116, ISSN 1439-0108

Oleszczuk, P. (2007). Organic pollutants in sewage sludge-amended soil part I. General remarks. *Ecological Chemistry and Engineering-Chemia I Inzynieria Ekologiczna*, Vol.14, No.S1, (2007), pp. 65-76, ISSN 1231-7098

Pandard, P.; Devillers, J.; Charissou, A.M.; Poulen, V.; Jourdain, M.J.; Férard, J.F.; Grand, C. & Bispo, A. (2006). Selecting a battery of bioassays for ecotoxicological characterization of wastes. *Science of the Total Environment*, Vol.363, No.1-3, (June 2006), pp. 114-125, ISSN 0048-9697

Passuello, A.; Mari, M.; Nadal, M.; Schuhmacher, M. & Domingo, J. L. (2010). POP accumulation in the food chain: Integrated risk model for sewage sludge application on agricultural soils. *Environment International.*, Vol.36, No.6, (August 2010), pp. 577-583, ISSN 0160-4120

Podlešáková, E.; Němeček, J. & Hálová, G. (1994). Contamination of fluvisols on Labe floodplains by hazardous elements. *Rostlinna Vyroba*, Vol.40, No.1, (January 1994), pp. 69-80, ISSN 0370-663X

Podlešáková, E.; Němeček, J. & Hálová, G. (1996). Proposal of soil contamination limits for potentially hazardous trace elements in the Czech Republic. *Rostlinna Vyroba*, Vol.42, No.3, (March 1996), pp. 119-125, ISSN 0370-663X

Podlešáková, E.; Němeček, J. & Vácha, R. (2000). Contamination of agricultural soils with polychlorinated dibenzo-p-dioxines and dibenzofurans. *Rostlinna Vyroba*, Vol.46, No.8, (August 2000), pp. 349-354, ISSN 0370-663X

Poláková, Š.; Tieffová, P. & Provazník, K. (2003): DDT and the residues in agricultural soils of the Czech Republic. *Bulletin of Central Institute for Supervising and Testing in Agriculturale.* Brno, Vol.11, No.3 (June 2003), pp. 73-75, ISSN 1212-5458

Riba, I.; DelValls, TA.; Reynoldson, T.B. & Milani, D. (2006). Sediment quality in Rio Guadiamar (SW, Spain) after a tailing dam collapse: Contamination, toxicity and bioavailability. *Environment International*, Vol.32, No.7, (September 2006), pp. 891-900, ISSN 0160-4120

Rosik-Dulewska, C.; Ciesielczuk, T. & Karwaczynska, U. (2009). Polycyclic Aromatic Hydrocarbons (PAHs) Degradation During Compost Maturation Process. *Rocznik Ochrona Srodowiska*, Vol.11, Part 1, (2009), pp. 133-142, ISSN 1506-218X

Santin, C.; de la Rosa, J.M.; Knicker, H.; Otero, X.L.; Alvarez, M.A. & Gonzales-Vila, F.J. (2009). Effects of reclamation and regeneration processes on organic matter from estuarine soils and sediments. *Organic Geochemistry*, Vol.40, No.9, (September 2009), pp. 931-941, ISSN 0146-6380

Starke, U.; Herbert, M. & Einsele, G. (1991). Polyzyklische aromatische Kohlenwasserstoffe (PAK) in Boden und Grundwasser, Teil I Grundlage zur Beurteilung von Schadenfällen. 1680 BOS 9 Lfg., 10: pp. 1-38.

Tripathy, S. & Praharaj, T. (2006). Delineation of water and sediment contamination in river near a coal ash pond in Orissa, India, In: Sajwan, K.S.; Twardowska, I.; Punshon, T.; Ashok, K. & Alva, A.K. (Eds.), *Coal Combustion Byproducts and Environmental Issues*, Springer, New York, pp. 41-49, ISBN 978-0-387-25865-2

Urbaniak, M.; Zielinski, M.; Weselowski, W. & Zalewski, M. (2009). Polychlorinated dibenzo-p-dioxins (PCDDs) and polychlorinated dibenzofurans (PCDFs) compounds in sediments of two shallow reservoirs in central Poland. *Archives of Environmental Protection*, Vol.35, No.2 (2009), pp. 125-132, ISSN 0324-8461

Vácha, R.; Poláček, O. & Horváthová, V. (2003). State of contamination of agricultural soils after floods in August 2002. *Plant, Soil and Environment*, Vol.49, No.7, (July 2003), pp. 307-313, ISSN 1214-1178

Vácha, R.; Horváthová, V. & Vysloužilová, M. (2005a). The application of sludge on agriculturally used soils and the problem of persistent organic pollutants. *Plant, Soil and Environment*, Vol.51, No.1, (January 2005), pp. 12-18. ISSN 1214-1178

Vácha, R.; Vysloužilová, M. & Horváthová, V. (2005b): Polychlorinated dibenzo-p-dioxines and dibenzofurans in agricultural soils of Czech Republic. *Plant, Soil and Environment*, Vol.51, No.10 (October 2005), pp. 464-468, ISSN 1214-1178

2

Textile Finishing Industry as an Important Source of Organic Pollutants

Alenka Majcen Le Marechal[1], Boštjan Križanec[2], Simona Vajnhandl[1] and Julija Volmajer Valh[1]
[1]*University of Maribor, Faculty of Mechanical Engineering, Maribor,*
[2]*Environmental Protection Institute, Maribor,*
Slovenia

1. Introduction

The textile finishing industry is, among all industries in Europe, the greatest consumer of high quality fresh water per kg of treated material and with the natures of their production processes significantly contributing to pollution. Wastewater from the textile industry is also a significant environmental pollution source of persistent organic pollutants.

Not only textile wastewater but also textile products often contain chemicals such as formaldehyde, azo-dyes, dioxins, pesticides and heavy metals, that might pose a risk to humans and the environment. Some of these chemicals found in finished products are there as residues from the production of dyes and auxiliary chemicals (the synthesis of dyes involves a large variety of chemicals with complex synthesis paths, during which toxic, carcinogenic and persistent organic compounds can be formed, such as dioxins, and traces can be found in commercial dyes), others are added to give certain characteristics to the products (colour, flame retardancy, anti wrinkling properties *etc.*) (Križanec & Majcen Le Marechal, 2006), or are already present in the raw textile material. The mentioned compounds have been found in wastewater after home washing, in organic solvent after dry-cleaning and also in the atmosphere after incineration. Possible sources of organic pollutants are also wastewater treatment methods and the incineration of textile materials.

The formation of dioxins can occur via dyeing and textile finishing processes with conditions favourable for their generation (high temperature, alkaline conditions, ultraviolet (UV) radiation, and other radical initiators). Textile dyes are designed to be resistant to microbial, chemical, thermal and photolytic degradation. After the dyeing process, a lot of non-bonded dyes are released into the wastewater, which can also be treated by Advanced Oxidation Processes (AOPs) in order to destroy the dye molecule and to decolourise the wastewater and reduce organic pollution. It is well-known that under the experimental conditions of such methods, which can be very useful because of the short-time of treatment, hazardous compounds can be formed due to very powerful oxidizing agents such as hydroxyl radicals ($OH^{\bullet}$).

In line with the improvement of people's living standard and the growing awareness and need to preserve the environment several regulations were introduced also in the textile industry in order to control the use of chemicals in textile processes. Under REACH regulation (REACH regulation controlled the quality of fabric, apparels, and shoes

and other textile materials, so as to protect human health and the environment) the following main groups of compounds in textiles are under control: Azo Dyes, Phthalates, Formaldehyde, Flame-retardants, Pentachlorophenol, Carcinogenic dyes, Sensitizing disperse dyes, Hexavalent chromium, Polychlorinated biphenyls, Heavy metals, Nickel release, Total lead content, Organic tin compounds, Total cadmium content, Organic chlorine carrier, Nonylphenol, Octylphenol, and Nonyl phenol ethoxylate, and several established directives (e.g. Azo dyes - Directive 2002/61/EC, Pentachlorophenol (PCP)- Directive 94/783/EC)) regulated/banned the use of these substances throughout the textile production chain (www.cirs-reach.com/textile/).
Market pressure, on the other hand, leads to the introduction of an extensive range of new products, especially dyes, and for many of them environmental and health impact data are still lacking. The quantification of chemicals within the environment is leading to the development of sensitive analytical methods in order to effectively detect and control pollution by organic pollutants.

2. Textile raw material

2.1 Fibres

Two general categories of fibres are used in the textile industry: natural and chemical (man-made) fibres. Man-made fibres encompass both purely synthetic materials of petrochemical origin, and regenerated cellulose material from wood fibres. A detailed classification of fibres is presented in Table 1.

Natural fibre		
Animal origin	*Vegetable origin*	*Mineral*
Raw wool Silk fibres Other hair fibres Alpaca Camel Cashmere Horse Llama Mohair Rabbit Vicuna	Cotton Flax Jute Linen Ramie Hemp	Asbetos Glass Metalic Copper Steel
Chemical (man-made) fibres		
Natural polymers fibres	*Synthetic polymer fibres*	
Viscose, Cupro, Lyocell Cellulose Acetate Triacetate	Inorganic polymer Glass for fibre glass Metal for metal fibre Organic polymers Polyester (PES) Polyamide (PA) Polyacrylonitrile (PAC) Polypropylene (PP) Elastane (EL)	

Table 1. Classification of fibres

2.2 Organic pollutants in textile raw materials

Different kinds and amounts of organic pollutants (contaminants) can already be present in fibres before they arrive at the textile mill. The potential contaminants may be released from the raw materials into the water or air during processing, when the fabric is heated or scoured. Due to the large amount of the fibre used during textile manufacturing, even trace contaminants can produce large amounts of pollutants. Many textile operations lack an incoming quality-control system for fibres. Testing of the natural or chemical fibres for organic pollutants is very rarely done in textile mills.

Textile raw materials contain: natural impurities from cotton, wool, silk, etc., fibre solvents (when chemical fibres are produced by dry-spinning or solvent-spinning processes), monomers (caprolactam ex polyamide 6), catalysts (antimony trioxide in polyesters' fibres), sizing agents (woven textiles esp. cotton and cotton blends), and preparation agents (esp. woven and knitted textiles made of man-made fibres). (Lacasse & Baumann, 2004)

2.3 Natural fibres

Natural fibres are acquired from animal, mineral, and plant sources. Several types of organic pollutants are found in natural fibre and all have the potential to create significant pollution problems. For example, waxes, oils, fats, and grease from animal fibre can contribute to biochemical oxygen demand (BOD), and chemical oxygen demand (COD). Pesticide residues from plant fibre can contribute to aquatic toxicity. Metals can accumulate in sludge or during the treatment system itself, thus causing potential long-term problems.

Cotton is the most significant natural fibre. Cotton fibre contains 88-96 % of cellulose and the residue is pectin substances, wax, proteins, ash, and other organic components (less than 1%). Chemicals such as pesticides, herbicides, and defoliants can be used during the production of cotton. These chemicals may remain as a residue on raw cotton fibres that reach the textile mill. Tests of cotton samples from growing regions worldwide, performed from 1991 to 1993, reported levels of pesticides below the threshold limit value for foodstuffs (US EPA 1995).

Although the content of pesticides in raw cotton fibres was negligible, almost half of the insecticides used in agriculture were used in cotton production. For this reason the reduction of insecticides for economic, environmental, and human health reasons was necessary. Insect-resistant (Bt) cotton (GM IR) started being used in 1996. So, the production of Bt cotton during the first ten years (1996-2005) reduced the total volume of those active insecticide ingredients by 94.5 million kilograms, representing a 19 % reduction in insecticides (Naranjo, 2009).

In 2001 it was reported that cotton was contaminated by pentachlorophenol (PCP) when being used, not only as a defoliant, but also as a fungicide during transportation and storage. (UK, 2001).

Wool is another significant commercial natural fibre. It is an animal hair from the bodies of sheep, sheared once or sometimes twice a year, and its quality and quantity vary widely, depending on the breed of sheep and their environment. Raw wool contains natural impurities (wool grease, suint, dirt) and residues of pesticides. Pesticides are applied onto sheep in order to control external parasites such as lice, blowflies, mites etc.

The used pesticides generally fall into four main groups (BAT, 2003):

- **organochlorine insecticides (OCs):** γ-hexachlorocyclohexane (lindane), pentachlorophenol (PCP), dieldrin, DDT,
- **organophosphorous insecticides (OPs):** diazinon, propetamphos, chlorfenvinphos, chloryriphos, dichlorfenthion,

- **synthetic pyrethroids insecticides (SPs):** cypermethrin, deltamethrin, fenvalerate, flumethrin, cyhalothrin and
- **insect-growth regulators (IGRs):** cyromazine, dicyclanil, diflubenzuron, triflumuron.

Biocides, organohalogen and organophosphorus compounds are among the priority substances listed for emission-control in the IPPC Directive. Diazinon (OP), propetamphos (OP), cypermethrin (SP) and cyromazine (IGR) are the most commonly-used ectoparasiticides for treating sheep. Insect-growth regulators such as dicyclanil, diflubenzuron and triflumuron are registered only in Australia and New Zealand. Lindane, which is the most toxic of the hexachlorocyclohexane isomers and also the most active as a pesticide, is still found in wool coming from the former Soviet Union, the Middle East and some South American countries (BAT, 2003). Wools from South America exhibited the highest levels of organochlorine insecticides, whilst wools from Australia and New Zealand exhibited the lowest levels of OCs (Shaw, 1989).

Pesticides such as OCs, OPs, and SPs have a lipophilic nature, so they associate strongly with the natural oils within the wool and as removed using these oils during wool scouring operations. The chemical stability of organochlorine pesticides is reflected in their resistance to microbial degradation. The lipophilic nature, hydrophobicity, and low chemical and biological degradation rates of organochlorine pesticides have led to their accumulation in biological tissues, and subsequent magnification of concentrations in organisms progressing up the food-chain.

The fate of the Ectoparasiticides (antiparasitic drug) in the wool scouring process is following. 96 % of the pesticides are removed from the wool (4 % is retained on the fibre after scouring). Of this 96 % around 30 % or less is retained on-site in recovered grease and the remaining fraction is discharged into the effluent and submitted to wastewater treatment.

Levels of organochlorine insecticides pentachlorophenol (PCP) in textile products are usually too low to be quantified accurately by traditional methods. PCP has been found at levels as high as 100 ppm in consumer products such as wool carpets (Wimbush, 1989). Wimbush tested 140 wool carpets and found that 88 % of them had a PCP content below 5 ppm. Only 3 carpets contained more than 50 ppm of PCP (US EPA, 1996).

Silk accounts for only 0.2 % of total fibre production. It is a protein fibre like wool and derived from the silkworm. The silk fibre is composed of fibroin filaments wrapped with sericine (silk gum), which has to be removed during pretreatment. (BAT, 2003)

2.4 Chemical fibres

Synthetic fibres may contain several types of impurities existing in the fibres before they reach the textile mill. These kinds of impurities are imparted onto the fibres during fibre manufacturing and fall into the categories of finishes, polymer synthetic by-products, and additives. Impurities associated with synthetic fibres are:

- Finishes: antistatic, lubricant,
- Polymer synthesis by-products: non-reactive monomers, low-molecular-weight oligomers, residual catalyst, and
- Additives to facilitate processing: antistatic agents, lubricants, humectants, and others.

Table 2 presents those compounds typical of synthetic fibre extracts. The compounds in Table 2 were not only found in the process wastewater from synthetic fibre dyeing and finishing operations but were also detected in synthetic fibre extracts.

Fibre	Compounds
Polyester	*tetra* hydro-2,5-dimethyl *cis* furan, ketones (methyl isobutyl ketone, 3-methyl cycle pentanone, hexanone, diethyl ketone), dodecanol, alcohols (C_{14} and C_{18}), esters of carboxylic acids (C_{14}-C_{24}), hydrocarbons (C_{14}-C_{18}), carboxylic acids (C_{16}-C_{24}), phthalate esters
Acrylic	hydrocarbons (C_{15}-C_{18}), esters of carboxylic acids (C_{17}-C_{22}), alcohols, phthalate esters, *N*, *N*-dimethyl acetamide
Nylon 6	diphenyl ether, hydrocarbons (C_{16}-C_{20}), carboxylic acids (C_{14}-C_{18}) and dicarboxylic acids, esters of carboxylic acids (C_{10}-C_{18}), alcohols (C_{20}-C_{22})

Table 2. Compounds typical of synthetic fibre extracts

Some of these impurities are produced during the polymerization of synthetic fibres and some are added to control the surface and electrical properties of the fibre. These impurities can create pollution problems.

The above review of possible organic pollutants in textile raw materials indicates the necessity for quality control regarding incoming fibre. Some textile companies have endorsed standard testing methods for fibres and have actively worked to ensure that the fibre user and the producer exchange information about quality control. It is important to set up good incoming fibre quality-control based on performance-testing, statistical sampling, and the analysis of extractable materials, in order to identify potential pollution problems before they arise.

3. Finishing processes within the textile industry

3.1 Basic process within the textile industry

The textile chain begins with the production or harvesting of raw fibres. These basic processes are schematically presented in Fig. 1. Treatments that are broadly referred to as 'finishing processes' are pretreatment, printing, dyeing, finishing, and coating, including washing and drying.

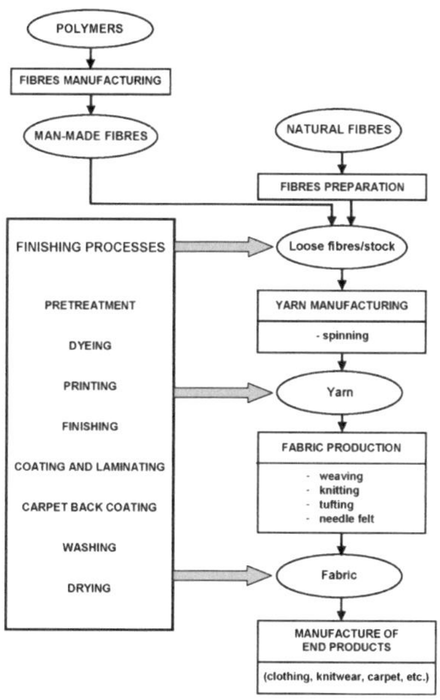

Fig. 1. Schematic presentation of textile processes

3.2 Possible organic pollutants in textile finishing processes

Most textile finishing operational units use chemical specialties. The major specialty consumption operations are pretreatment processes (desizing, scouring, bleaching), dyeing, printing, and finishing. (Mattioli, et al, 2002)
So, this chapter focuses on those textile finishing processes that might produce organic pollutants.

3.2.1 Pretreatment processes

Pretreatment processes depend on the kind and form of the treated fibre, as well as the amount of material to be treated. These processes should ensure the removal of foreign materials from the fibres in order to improve their uniformity, hydrophilic characteristics, and affinity for dyestuffs and finishing treatments. The pretreatment processes are desizing, scouring, bleaching, and mercerizing.

3.2.1.1 Desizing

Desizing is the process for removing size-chemicals from textiles. The possible organic pollutants in effluents after desizing, are presented in Table 3.

Fibres	Organic substances
Cotton	carboxymethyl cellulose, enzymes, fats, hemicelluloses, modified starches, non-ionic surfactants, oils, starch, waxes
Linen	
Viscose	
Silk	carboxymethyl cellulose, enzymes, fats, gelatine, oils, polymeric sizes, polyvinyl alcohol, starch, waxes
Acetates	
Synthetics	

Table 3. Possible organic pollutants in effluents after the desizing process

The washing water from desizing may contain up to 70 % of the total COD in the final effluent. Synthetic esters' oils are less problematic because they are emulsified or soluble in water and easily biodegraded. More problems are caused by compounds such as silicon oils, because they are difficult to emulsify, and poorly biodegradable. Silicon oils are found in elastane blends with cotton or polyamide.

3.2.1.2 Scouring

Scouring is the cleaning process for removing impurities from both natural and synthetic materials. In natural fibres, impurities can be present such as oils, fats, waxes, minerals, and plant-matter. Synthetic fibres can contain spinning, finishing, and knitting oils.
Scouring is performed in an alkali medium together with auxiliaries that include:

- non-ionic surfactants (alcohol ethoxylates, alkyl phenol ethoxylates) and anionic surfactants (alkyl sulphonates, phosphates, carboxylates),
- compounds for removing metal ions (nitrilotriacetic acid (NTA), ethylenediaminetetraacetate (EDTA), diethylene triamine pentaacetate (DTPA), gluconic acid, phosphonic acids as complexing agents),
- polyacrylates and phosphonates as special surfactant-free dispersing agents and
- sulphite and hydrosulphite as reducing agents (to avoid the risk of the formatting of oxycellulose when bleaching with hydrogen peroxide).

Table 4 presents those possible organic pollutants derived from scouring.

Fibers	Organic substances
Cotton	anionic surfactants, cotton waxes, fats, glycerol, hemicelluloses, non-ionic surfactants, peptic matter, sizes, soaps, starch
Viscose Acetates	anionic detergents, fats, non-ionic detergents, oils, sizes, soaps, waxes
Synthetics	anionic surfactants, anti static agents, fats, non-ionic surfactants, oils, petroleum spirit, sizes, soaps, waxes
Wool (yarn and fabric)	anionic detergents, glycol, mineral oils, non-ionic detergents, soaps
Wool (loose fiber)	acetate, anionic surfactants, formate, nitrogenous matter, soaps, suint, wool grease, wool wax

Table 4. Possible organic pollutants in waste-effluents after the scouring process

3.2.1.3 Bleaching

Bleaching is a chemical process for removing unwanted coloured matter from materials. Several different types of chemicals are used as bleaching agents and selection depends on the type of fibre. The most frequently used oxidizing agents for cellulose fibres are hydrogen peroxide (H_2O_2), sodium chlorate(I) (NaClO), sodium chlorate(III) ($NaClO_2$). Peracetic acid and optical brightening agents are also applicable.

Bleaching with H_2O_2 is connecting with the use of H_2O_2 stabilisers. During the bleaching process, hydroxyl radicals attack the cellulose fibre starting with oxidation of the hydroxyl groups, and eventually ending with cleavage of the cellulose molecules, thus decreasing the degree of polymerisation. Reaction is catalysed by heavy metals such as iron, copper, and cobalt. H_2O_2 stabilisers (EDTA, NTA, DTPA, gluconates, phosphonates and polyacrylates) inhibit these reactions. NTA, EDTA and DTPA form very stable metal complexes. EDTA and DTPA are also poorly eliminated compounds, and they could pass non-degraded through the common wastewater treatment system. Their ability to form a very stable complex with metal makes the problem even more serious because they can mobilise those heavy metals present in the effluent, and release them into the receiving water. Other auxiliaries used in hydrogen peroxide bleaching are surfactants with emulsifying, dispersing, and wetting properties.

Sodium chlorate(I) was, for a long time, one of the more widely-used bleaching agents throughout the textile finishing industry. Nowadays bleaching with sodium chlorate(I) is limited in Europe for ecological reason. NaClO leads to secondary reactions that form a number of organic halogen compounds, such as carcinogenic trichloromethane.

3.2.1.4 Mercerizing

Mercerizing is a chemical process that improves the strengths, lustres and affinities of dyes for cotton fabrics. Several organic pollutants can be used. Possible organic pollutants in mercerizing effluents are alcohol sulphates, anionic surfactants, and cyclohexanol.

3.2.2 Dyeing

Dyeing is used to add colour to textile materials. Textiles may be dyed at various stages of production.

Dyeing can be carried out in a batch or in a continuous mode. The choice between the two processes depends on the type of make-up, the chosen class of dye, the equipment available, and the cost involved. Both continuous and discontinuous dyeing involve preparation of the dye, dyeing, fixation, washing, and drying. Quite a large amount of the non-fixed dye leaves the dyeing units.

Besides dyes, other auxiliary substances can be added to the dyeing process, and may give rise to water pollution. Possible pollutants are: fatty amine ethoxylates (levelling agent), alkylphenol ethoxylates (levelling agent), quaternary ammonium compounds (retarders for cationic dyes), cyanamide-ammonia salt condensation products (auxiliaries for fastness improvement), acrylic acid-maleic acid copolymers (dispersing agent), ethylenediamine tetraacetate (EDTA), diethylenetriaminepentaacetate (DTPA), ethylenediaminetetra-(methylenephosphonic acid) (EDTMP), diethylenetriaminepenta(methylenphosphonic acid) (DTPMP).

All these are water soluble and non-biodegradable compounds, which can pass non-transformed or partially-degraded through a wastewater treatment system. Some of them are toxic (quaternary amines) or can give rise to metabolites that may affect the reproduction-chain within an aquatic environment (BAT, 2003).

Organic dyes are presented in Chapter 4.

3.2.3 Printing

Printing, like dyeing, is a process for applying colour to a substrate. The printing techniques are: rotary screen, direct, discharge, resist, flat screen, and roller printing. Pigments cover about 75-85 % of all printing operations, do not require washing steps and generate a small amount of waste. Compared to dyes, pigments are typically insoluble and have a high affinity for fibres. Printing paste residues, wastewater from wash-off and cleaning operations, and volatile organic compounds from drying and fixing, are typical emission sources from printing processes.

Printing process wastewater is small in volume, but the concentration of pollutants is higher than that in wastewater from dyeing. Wastewater after printing contains the following organic pollutants: urea, dyes or pigments, and organic solvents.

These pollutants are likely to be encountered in wastewater after printing, and are presented in Table 5 (BAT, 2003).

Additionally, organic pollutants after the finishing processes, such as aliphatic hydrocarbons (C_{10}-C_{20}) from binders, monomers (acrylates, vinylacetates, styrene, acrylonitrile, acrylamide, butadiene), methanol from fixation agents, other alcohols, esters, polyglycols from emulsifiers, formaldehyde from fixation agents, ammonia, *N*-methylpyrrolidone from emulsifiers, phosphoric acid esters, phenylcyclohexene from thickeners and binders, might also be distributed in the exhaust air.

3.2.4 Finishing

The term 'finishing' covers all those treatments that improve certain properties or the serviceability of the fibre. Finishing may involve those mechanical/physical and chemical treatments performed on fibre, yarn, or fabric, in order to improve appearance, texture, or performance.

Organic pollutants found during finishing processes are:

- cross-linking agents in easy-care finishing,
- flame-retardant agents,
- softening agents,
- antistatic agents,
- hydrophobic/oleophobic agents and
- biocides.

Pollutant	**Source**
Organic dyestuff	Non-fixed dye
Urea	Hydrotropic agent
Ammonia	In pigment printing pastes
Sulphates and sulphites	Reducing agents by-products
Polysaccharides	Thickeners
CMC derivates	Thickeners
Polyacrylates	Thickeners Binder in pigment printing
Glycerin and polyols	Anti-freeze additives in dye formulation Solubilising agents in printing pastes
m-nitrobenzene sulphonate and its corresponding amino derivative	In discharge printing of vat dyes as oxidising agent Direct printing with reactive dyes inhibits the chemical reduction of the dyes
Polyvinyl alcohol	Blanket adhesive
Multiple substituted aromatic amines	Reductive cleavage of azo dyestuff in discharge printing
Mineral oils/aliphatic hydrocarbons	Printing-paste thickeners (half-emulsion pigment printing pastes occasionally)

Table 5. Pollutants in wastewater after the printing process.

Cross-linking agents in easy-care finishing are mainly based on formaldehyde. They can be formaldehyde-rich, formaldehyde-poor or formaldehyde-very poor. Formaldehyde-rich cross-linking agents are: "self-crosslinking "agents such as hydroxymethyl urea, bis(methoxymethyl)urea, hydroxymethyl melamine, and bis(methoxymethyl)melamine. 'Reactant cross-linking' agents are formaldehyde-poor or have very poor cross-linking based on the derivatives of the bis(hydroxymethyl)-dihydroxyethylene urea. All these products may potentially produce emissions of free formaldehyde and methanol. Formaldehyde's presence in these finishing agents represents a potential risk, not only for wastewater and exhausted air, but also to workers in the textile mills, and the final users of the textile as well. Formaldehyde is also suspected of carcinogenity. These are the reasons for much effort being put into the production of free formaldehyde cross-linking agents. One cross-linking agent that is formaldehyde-free is also available on the market and is based on dimethyl urea and glyoxal [BASF, 2000]. Formaldehyde-free cross-linkers are considerably more expensive than formaldehyde cross-linkers, so this is the reason that formaldehyde-free cross-linkers have never been widely used in the textile industry.

The flame-retardant agents most commonly-used within the textile sector, and belong to the organic flame-retardant agents, are both halogenated organic and organo-phosphorus compounds.

As halogenated organic compounds only, brominated and chlorinated flame-retardant agents are used in practice. Brominated compounds are the most effective. Bromine can be bound aliphatically or aromatically, the aromatic derivatives are widely-used because of their high thermal stability. Chlorinated flame-retardant agents include chlorinated aliphatic and cycloaliphatic compounds. They are less expensive, less stable, and more corrosive to the equipment when compared to the brominated compounds. Polybrominated flame-retardants include the following compounds:

- polybrominated diphenyl ethers (PBDE); pentabromodiphenyl ether (penta-BDE), octabromodiphenyl ether (octa-BDE), decabromodiphenyl ether (deca-BDE),
- polybromo biphenyls (PBB); decabromobiphenyl and
- tetrabromobisphenol A (TBBA).

A PDBE, which is of major use as a flame-retardant agent, is a deca-BDE. Deca-BDE and octa–BDE could break down into penta–BDE and tetra–BDE after release into the environment. Penta-BDE is a persistent substance liable to biocumulate.

Organo-phosphorus flame-retardant is represented by molecule phosphonic acid, (2-((hydroxylmethyl)carbamyl)ethyl)-dimethyl ester. Phosphorus containing flame-retardant agents is non readily biodegradable and water-soluble. According to one source, this product is not toxic or harmful to aquatic organisms and shows no potential to bioaccumulate, whilst another source concludes that knowledge about the toxicology of this compound is insufficient (BAT, 2003).

Softening agents are water-based emulsions or dispersions of water-insoluble active materials. There are four groups of softeners:

- non-ionic surfactants (fatty acids, fatty esters, and fatty amides),
- cationic surfactants (quaternary ammonium compounds, amido amines, imidazolines),
- paraffin and polyethylene waxes and
- organo-modified silicones.

Softening agents are molecules with high molecular weight but their volatility is low. The performance of each type of softener varies and each has advantages and disadvantages. Non-ionic surfactants are biodegradable, whilst paraffin and polyethylene waxes are non-biodegradable. Cationic surfactants have high aquatic toxicity. Paraffin waxes are still used although these types of softeners emit smoke when heated and producing air emissions from dryers.

Antistatic agents are applied as functional finishes to selected textile materials for use within static-sensitive environments. From the chemical point of view, antistatic agents are based on quaternary ammonium compounds and phosphoric acid ester derivatives.

Hydrophobic/Oleophobic agents fall under the following categories:

- wax-based repellents (paraffin-metal salt formulations),
- resin-based repellents (fatty modified melamine resins),
- silocone repellents and
- fluorochemical repellents (copolymers of fluoroalkyl acrylates and methacrylates).

Organic pollutants arise from silicone repellents, fluorochemical repellents and from resin-based repellents.

Silicone repellents contain polysiloxane-active substances (dimethylpolysiloxane and modified derivates), emulsifiers, hydrotropic agents (glycols), and water.

Fluorochemical repellents are copolymers of fluoroalkyl acrylates and methacrylates. Market formulations contain active agents together with emulsifiers (ethoxylated fatty alcohols and acids, but also fatty amines and alkylphenols) and other by-products that are often solvents, such as:

- acetic acid esters (butyl/ethyl acetate),
- ketones (methylethyl keton and methylisobutyl ketone),
- diols (ethandiol, propandiol), and
- glycolethers.

Resin-based repellents are produced by condensing fatty compounds (acids, alcohols or amines) with methylolated melamines.

Biocides are used for antimicrobial finishes regarding hospital textile material or as odour suppressants for socks. For this purpose, the following active substancesare used:

- zinc organic compounds,
- tin organic compounds,
- dichlorophenyl(ester) compounds,
- benzimidazol derivatives,
- triclosane, and
- isothiazolinones (the most commonly used today).

Biocides are also applied in the carpet sector to impart wool fibre lifetime protection against a range of textile pests. These auxiliaries are usually known as mothproofing agents. As mothproofing agents are used, such as permethrin (synthetic pyrethoid), gyfluthrin (synthetic pyrethroid), and sulcofuron (halogenated diphenylurea derivative).

All biocides give rise to environmental concerns when they are discharged in wastewater, because of their toxicity to aquatic life.

4. Organic dyes

Dyes make the world more beautiful through coloured products, but cause a lot of problems. Organic dyes contain a majority of substituted aromatic and heteroaromatic groups. The colour results from the conjugated chains or rings can absorb different wavelengths of the visible spectrum. Organic dyes contain chromophoric and auxochromic groups. Chromophoric groups are responsible for colouring properties, and the salt-forming auxochromes for the dying properties. The chromophores are usually composed of double-bonds (carbon-carbon, nitrogen-nitrogen, carbon-nitrogen), aromatic and heteroaromatic rings (containing oxygen, nitrogen or sulphur).

4.1 Classification

Textiles are dyed using many different colorants, that may be classified in several ways. With regard to the methods and domains during usage, the dyes are classified into acid, reactive, direct, basic, disperse, metal complex, vat, mordant, and sulphur dyes. Reactive and direct dyes are commonly in use for cotton and viscose-rayon dyeing, whilst disperse dyes are used for dyeing polyester. Dyes can be classified with regard to: their chemical structures, the methods and domains of usage, and chromogen (a dye or pigment precursor, containing chromophores), as presented in Table 6.

Azo dyes contain one or more azo bonds, and can be used for dyeing natural fibres (cotton, silk, and wool) as well as synthetic fibres (polyesters, polyacrylic, rayon, etc.). Azo dyes are the most-widely used synthetic dyes and are present in 60-70 % of all textile dyestuffs produced (ETAD, 2003). They are mainly used for yellow, orange, and red colours. The biodegradation of more than 100 azo dyes were tested and only a few of them were degraded aerobically. The degree of stability of azo dyes under aerobic conditions depends on the structure of the molecule. C. I. Acid Orange 7 is one of the rare aerobically biodegradable azo dyes (Vandervivere, 1998). Under anaerobic conditions, azo dyes are cleaved into aromatic amines, which are not further metabolized under anaerobic conditions but readily biodegraded within an aerobic environment (Fig. 2).

	Classification	
	Subclass	Characteristic
With regard to chemical structure (C.I.)	E.g. azo, anthraquinone, triphenylmethane, indigo,…	The classification of a dye by chemical structure into a specific group is determined by the chromophore.
With regard to method and domain of usage (C.I.)	E. g. direct, acid, basic, reactive, reductive, sulphuric, chromic, metal complex, disperse, pigments,…	Dyes used in the same technological process of dyeing and with similar fastness are classified into the same group.
With regard to chromogen $n \rightarrow \pi^*$	E. g. absorptive, fluorescent and dyes with energy transfer,…	This classification is based on the type of excitation of electrons, which takes place during light adsorption.
With regard to the nature of donor - acceptor couple	E. g. 1-aminoanthraquinone, *p*-nitroaniline,…	These chromogenes contain a donor of electrons (non-bound electron couple), which directly bonds to the system of conjugated π electrons.
With regard to the nature of polyenes:		
a) Acyclic and cyclic	E. g. polyolefins, annulenes, carotenoids, rhodopsin,…	Polyene chromogen contains sp^2 (or sp) hybridised atoms. The molecules enclose single and double-bonds that form open chains, circles, or a combination of both.
b) Cyanine	E. g. cyanines, amino substituted di- and tri-arylmethane, oxonols, hydroxyarylmethanes,…	Cyanine chromogens have a system of conjugated π electrons, in which the number of electrons matches the number of p orbitals.

Table 6. Classification of dyes

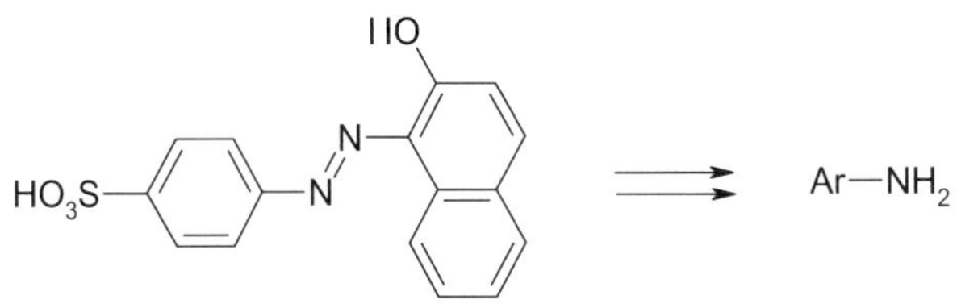

Fig. 2. Degradation of the azo dye C. I. Acid Orange 7 under anaerobic conditions

Carcinogenic amines that can be formed by the cleavage of certain azo dyes are:
4-aminodiphenyl, benzidine, 4-chloro-*o*-toluidine, 2-naphthylamine, *o*-aminoazotoluene, 2-amino-4-nitrotoluene, *p*-chloroaniline, 2,4-diaminoanisol, 4,4′-diaminodiphenylmethane, 3,3′-dichlorobenzidine, 3,3′-dimethoxybenzidine, 3,3′-dimethylbenzidine, 3,3′-dimethyl-4-4′-diaminodiphenylmethane, *p*-cresidine, 4,4′-methylene-*bis*-(2-chloraniline), 4,4′-oxydianiline, 4,4′-thiodianiline, *o*-toluidine, 2,4-diaminotoluene, 2,4,5-trimethylaniline, 4-aminobenzene, and *o*-anisidine (BAT, 2003).

More than 100 azo dyes with the potential to form carcinogenic amines are still available on the market (Euratex, 2000). The usage of these azo dyes that may cleave into one of the potentially carcinogenic aromatic amines from Table 7, is banned according to the 19th amendment of Directive 76/769/EWG on dangerous substances.

Fifty years ago, large amounts and numbers of azo colorants based on benzidine, 3,3-dichlorobenzidine, 3,3-dimethylbenzidine (*o*-toluidine), and 3,3-dimethoxybenzidine (o-dianisidine), have been synthesized, especially within the German chemical industry. 447 of the azo colorants from a list of 2000 in The Colour Index (1987), were based on 2-naphthylamine, benzidine or benzidine derivatives. The manufacturing of these kinds of azo dyes was stopped in 1971, with the exception of the dye Direct Black 4, and the manufacturing of this dye was continued until 1973 (Golka, 2004). The problem of carcinogenicity regarding azo dyes was first officially addressed by *the German Commission for Investigating Health Hazards of Chemical Compounds within the Work Area* ("MAK-Commission") (DFG, 1988). According to current EU regulations, azo dyes based on benzidine, 3,3-dimethoxybenzidine, and 3,3-dimethylbenzidine, are classified as carcinogens of category 2 as 'substances which should be regarded as if they are carcinogenic to man'. This is not the case for 3,3′-dichlorobenzidine-based azo pigments (BIA-Report, 2/2003).

The second more important class of textile dyes is anthraquinone dyes. They have a wide-range of colours over almost the whole visible spectrum, but they are more-commonly used for violet, blue, and green colours. Anthraquinone dyes are more resistant to biodegradation due to their fused aromatic structures, and thus remain coloured for a long time in wastewater.

Acid dyes are water-soluble compounds applied to wool, nylon, silk, and some modified acrylic textiles. Acid dyes have one or more sulphonic or carboxylic acid groups in their molecular structure. The dye-fibre affinity is the result of ionic bonds between the sulphonic acid part of the dye and the basic amino groups in wool, silk, and nylon fibres.

Reactive dyes are water soluble and are mainly used for dyeing cellulosic fibres such as cotton and rayon, and also for wool, silk, nylon, and leather. Reactive dyes form covalent chemical bonds with the fibre and become part of it. They are used extensively within textile industries in regard to their favourable characteristics of bright colour, water-fastness, and simple application techniques with low-energy consumption.

Metal-complex dyes show great affinity towards protein (wool) and polyamide fibres. Generally they are chromium or cobalt complexes (Zollinger, 2003). Chromium complex dyes are formed through chemical reactions between Cr_2O_3 and a variety of azo organic compounds (Zhao et al., 2005). Chromium occurs primarily in the trivalent state (III) and in the hexavalent state (VI). Hexavalent chromium is known to be toxic, can irritate the nose, throat, and lungs, provoke permanent eye damage, cause dermatitis and skin ulcers, and exhibit carcinogenic effects (Baral & Engelken, 2002). When reduced to chromium(III), it may be significantly less-harmful. As chromium compounds were used in dyes and paints and the tanning of leather, these compounds are often found in soil and groundwater at abandoned industrial sites, now needing environmental clean-ups and remediation regarding the treatment of brownfill land.

Disperse dyes have a very low water solubility, so they are applied as a dispersion of fine grounded powders in the dye-bath (Banat et al., 1996). Disperse dyes are used for oleophyllic fibres (polyester and other synthetics) that reject water-soluble dyes. Disperse dye-inks are used in the ink-jet textile printing of polyester fabrics. They have good fastness to light, perspiration, laundering, and dry cleaning. Some disperse dyes also have a tendency to bioaccumulate. Skin sensitisation risks are likely to be within acceptable limits.

Various intermediates used in the manufacture of disperse dyes are given below (Science tech entrepreneur, 2003): *p*-amino acetanilide, 1-amino-4-bromo-2-anthraquinone-sulphonic acid, *p*-amino phenol, 4-amino xanthopurpurin, aniline, anilino methane sulphonic acid, 1-benzamido-4-chloro-anthraquinone, 3-bromo benzanthrone, 1-bromo-4-methyl amino anthraquinone, 1-chloro-2, 4-dinitrobenzene, 2-chloro-4-nitroaniline, 4-chloro-3-nitrobenzene sulfonyl chloride, 2, 2-(*m*-chlorophenylamino)-diethanol, cresidune, *o*-cresol, *p*-cresol, 1,4-diamino, anthraquinone, 1,5-diamino anthraquinone, 2,6-dichloro-4-nitroanioline, 1,5-dihydroxy-4, 8-dinitro anthraquinone, *N*, *N*-dimethyl aniline, 1,5-dinitro anthraquinone, 1,8-dinitro anthraquinone, diphenylamine, ethanolamine, 2-(*N*-ethykamino)ethanol, 1-hydroxy anthraquinone, 1-hydroxy-4-nitro anthraquinone, leuco-1, 4, 5, 8-tetrahydroxy anthraquinone, methylamine, 1-methoxy anthraquinone, 3-methyl-1-phenyl-5-pyrazolone, 1-naphthylamine, *p*-nitroaniline, phenol, *p*-phenyl azoaniline, quinizarin, *p*-toluene sulfoanamide, 2, 2-(*m*-tolyamino) diethanol.

Basic dyes have high brilliance and intensity of colour, and are highly visible even in a very low concentration (Chu & Chen, 2002). Basic dyes are the most acutely toxic dyes for fish, especially those with tri-aryl-methane structures. Occasionally, some people develop a contact allergy to dyes in clothing. This is ample evidence that there is reason to be cautious in the use of some basic dyes. The absence of a known hazard does not prove that the substance is non-carcinogenic or mutagenic.

5. Possible sources of persistent and hazardous organic pollutants in textile chain

Persistent organic pollutants are a wide group of compounds with specific properties: persistency, high bioaccumulative coefficient, ability for transmission to large distances. POPs emission inventory has a great significance as a starting-point for pollutant flows' modelling, impact and risk assessment. The most toxic among POPs are polychlorinated dibenzo-*p*-dioxins and furans (PCDD/Fs), often simply termed as dioxins. There are 75 isomers of dibenzo-*p*-dioxins (PCDDs) and 135 isomers of polychlorinated dibenzofurans differentiated from each other by the number and location of the chlorine atom addition. Chemical and biological properties (including the toxicological) depend on the positions of the chlorine atoms. In this group, isomers with chlorine atoms at positions 2,3,7,8 are especially toxic. The so-called 'dirty group' comprises 17 isomers of PCDD/Fs, among which 2,3,7,8-tetrachlorodibenzodioxin (2,3,7,8-TCDD) is the most toxic (Fig. 3). There are also brominated dioxins (PBDD/Fs), fluoro dioxins and mixed dioxins PXDD/Fs (X = Cl, Br, F). PBDD/Fs are contaminants with properties similar to PCDD/Fs, together with their persistence and toxicity. Fluorinated congeners of dioxins are taken to be less dangerous to humans and the environment, due to their short life and low toxicity.

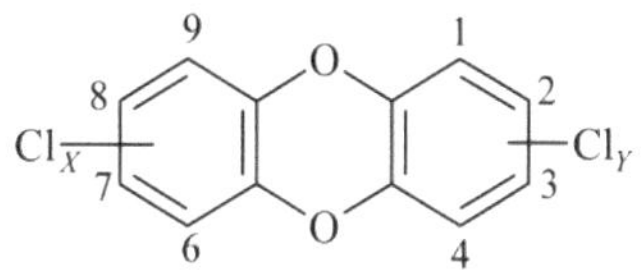

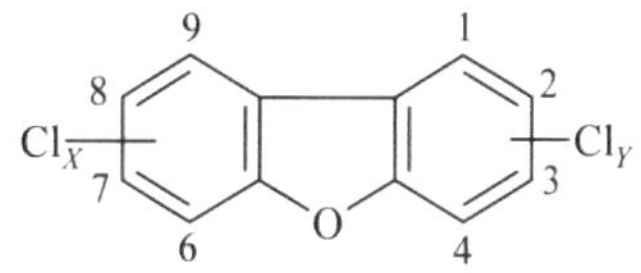

Fig. 3. Molecular structures of polychlorinated dibenzo-*p*-dioxins and dibenzofurans

Possible sources of dioxins among the textile chains will be presented in the following chapters.

5.1 Natural fibre

As already-mentioned in chapter 2.2.1, pentachlorophenol and other organochlorine pesticides are used during the production and transportation of natural textile fibres. The production and use of PCP is a significant source of PCDD/Fs. During the synthesis of pentachlorophenol from chlorinated phenols, PCDD/Fs are formed as by-products. Furthermore, PCDD/Fs are formed from chlorinated phenols during natural UV radiation conditions.

5.2 Textile dyes

The main source of dioxins in the textile industry are dioxazine and antraquinone dyes and pigments. These classes of dyes are produced from chloranil. PCDD/Fs are formed during the synthesis of chloranil from chlorinated phenols. Dioxins are also found in other classes of dyes. Considerable levels of PCDD/Fs have also been determined in some phtalocyanine dyes, and in printing inks (Križanec, 2006). Recently some dispersive aromatic azo dyes were found as sources of dioxins. In dual-black disperse dyes (mixtures of antraquinone and azo substances); concentrations of 50 and 170 ng PCDD/F TEQ/kg have been determined. Octachlorodibenzodioxin (OCDD) was the dominant compound (Križanec, 2005). Disperse azo dyes produced from chlorinated anilines or chlorinated nitro anilines contain dioxins as by-products during synthesis. Dioxins are formed after the dediazonation of aromatic diazonium salts via chlorinated phenols and chlorinated nitrophenols. The hydrolysis of diazonium salts leads to phenols (Križanec, 2007). Chlorinated phenols and chlorinated nitrophenols are present in dyes' formulations as impurities, and dioxins may form further during dyeing processes (Fig.4) (Križanec, 2007).

5.3 Halogenated organic compounds

Halogenated organic compounds, especially aromatic halogenated compounds, are also precursor compounds for the synthesis of dioxins. Halogenated organic compounds and metals such as copper are involved in the thermal formation processes of dioxins. During the combustion processes of waste textile materials, the formation of dioxins is possible via the 'The Precursor Concept' or 'De novo Mechanism' (Križanec, 2006).

All the halogenated flame-retardant agents presented in chapter 3.2.4 are involved in the formation of dioxins and furans, when submitted to high temperature treatments. Dioxins and furans can be formed in small amounts during the syntheses of these compounds, and as a side-reaction when they are subject to combustion/burning for disposal.

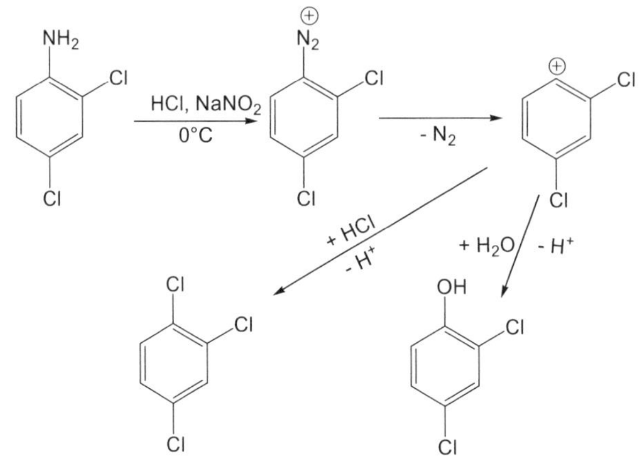

Fig. 4. Heterolytic pathway during the dediazonation of chlorinated aromatic diazonium salts, and the formation of chlorinated benzenes and chlorinated phenols

5.4 Formation of dioxins during finishing processes

The generation of persistent organic pollutants in the textile industry during finishing processes is caused by (Križanec, 2006):

- high temperatures (>150 °C);
- alkaline conditions;
- UV radiations or other radical starters;
- the presence of chlorinated organic compounds;
- the presence of metal catalysts.

The distribution of dioxins and their fate during textile dyeing processes was investigated by Križanec (Križanec, 2005). Two dyeing experiments were conducted at laboratory scale using disperse dye contaminated with PCDD/Fs. After the dyeing and finishing process, the PCDD/Fs, and especially OCDD, increased compared to the input. The authors concluded that PCDD/Fs are formed during the textile process, most probably from precursors present in the dyestuffs (Križanec, 2005).

5.5 Incineration of textile materials

Incineration processes are a known source of dioxins. Several textile products are also potential sources for the formation of dioxins during waste incineration. As described in previous sections, dioxins, and especially dioxin precursor compounds, may be present in textile products. According to the dioxin mass-balance present in a Spanish municipal waste incinerator, the incineration of textiles results in the highest dioxin emission levels (Abad et al., 2000).

6. Textile wastewater

6.1 Characteristics of wastewater after finishing processes

The diversity of textile finishing processes results in variable compositions of textile wastewater. In general, the wastewater is loaded with complex mixtures of organic and inorganic chemicals (Volmajer Valh et al., 2011). Moreover, textile finishing processes are, despite the gradual reduction of fresh water consumption by introducing, for example, jet dyeing machines and short liquor-dyeing systems, still very water intensive, leading to large volumes of produced wastewater that have to be treated before being discharged. For illustration purposes, 70 L of freshly-softened water is used per kg of cotton material during dark-colour reactive dyeing processes (data from Slovene textile company). With further optimization, it is possible to reach values between 25 to 40 L of fresh water per kilogram of cotton for a dark reactive dyeing, in contrast to the 100 to 150 L water consumption for this type of dyeing, as used in the mid-nineteen seventies.

When considering both the volume generated and the effluent composition, textile industry wastewater is considered to be the most polluted of all the industrial sectors.

The pollutants of major concern are recalcitrant or hazardous organics, such as dyes or some surfactants, metals, and salts. During the textile dyeing process, dyes are always used in combination with other chemicals such as acids, alkali, salts, fixing agents, carriers, dispersing agents, and surfactants that are partly or almost completely discharged in the wastewater. Dye-fixation rates vary considerably among the different classes of dyes and may be especially low for reactive dyes (in the case of cotton 20-50 % residual dyestuff) and for sulphur dyes (30-40 % residual). Moreover, large variations are round even within a given class of colorants. This is particularly significant in the case of reactive dyes.

As already-mentioned, from among all the dye classes, azo dyes are the group of colorants most used. Whilst most azo dyes themselves are non-toxic, a significantly larger portion of their metabolites are (Isik & Sponza, 2007; Van der Zee & Villaverde 2005). Brown and DeVito (Brown & DeVito, 1993) postulated that azo dyes may be toxic only after the reduction and cleavage of the azo linkage, producing aromatic amines. Azo dyes with structures containing free aromatic amine groups that can be metabolically oxidized without azo reduction, may cause toxicity. Azo dye toxic activation may occur following direct oxidation of the azo linkage, producing highly reactive electrophilic diazonium salts.

Most dyes shown to be carcinogenic are no longer used; however, a complete investigation of all dyestuffs available on the market is impossible.

Substituted benzene and naphthalene rings are common constituents of azo dyes, and have been identified as potentially carcinogenic agents (IARC, 1982).

Other concerns are the impurities within commercial dye products and the additives used during the dyeing process. Understanding the dye structures and how they are degraded is crucial to understanding how toxic by-products are created. The colour of wastewater is one of the major problems facing industries involved in dyeing processes. Wastewaters from dye-houses often carry high concentrations of excess, unfixed dye.

A study conducted on 45 combined effluents from textile finishing plants showed that 27 percent of the wastewater samples were mutagenic during the Ames test (McCarthy, 1997). The potential for toxic effects to humans, resulting from exposure to dyes and dye metabolites, is not a new concern.

6.2 Wastewater treatment processes

Textile wastewater is the main source of organic contamination regarding pollution within the textile industry. Several cleaning processes may be used to remove organic pollutants from textile wastewater. In general, we distinguish between physical methods (adsorption, filtration methods, coagulation and flocculation processes), chemical methods (oxidation, advanced oxidation, Fenton´s reagent) and, more recently, more and more attractive biological treatment (anaerobic, aerobic) as an effective option for relatively inexpensive effluent decolouration. Non-destructive physical techniques just transfer the pollutants to other mediums (sludge, concentrate in filtration techniques), and cause secondary pollution. From this point of view, the chemical destruction of pollutants is more desirable, but could have some drawbacks such as the formation of aromatic amines, when the degradation/mineralisation is incomplete. An overview of the treatment methods for textile wastewater treatment, as well as their advantages and disadvantages, are gathered in 'Decolouration of textile wastewaters' (Volmajer Valh & Majcen Le Marechal, 2009). The following paragraph focuses on chemical methods, especially on advanced oxidation processes (AOPs).

According to their definition, AOPs combine ozone (O_3), ultraviolet (UV) irradiation, hydrogen peroxide H_2O_2 and/or a catalyst in order to offer a powerful water treatment solution for the reduction (removal) of residual organic compounds, as measured by COD, BOD or TOC, without producing additional hazardous by-products or sludge, which requires further handling (Arslan-Alaton, 2004). All AOPs are designed to produce hydroxyl radicals that act with high-efficiency to destroy organic compounds. The most widely-applied advanced oxidation processes (AOP) are: H_2O_2/UV, O_3/UV, H_2O_2/O_3, H_2O_2/O_3/UV (Kurbus et al. 2003), and ultrasound (US) (Vajnhandl & Majcen Le Marechal, 2007) have several advantages such as: rapid reaction rates, small foot-print, reduction of toxicity, and complete mineralization of treated organics, no concentration of waste for further treatment (as membranes), no production of materials that require further treatment, such as 'spent carbon' from activated carbon absorption, no creation of sludge as with physical-chemical processes or biological processes (wasted biological sludge), and a non-selective pathway allows for the treatment of multiple organics, at once. On the contrary, these processes are capital-intensive and, in the case of complex chemistry, must be tailored to specific applications (Slokar & Majcen Le Marechal, 1998).

AOPs are marked as treatment methods for the effective removal of organic pollutants in terms of total organic carbon (TOC), COD, and BOD reduction, but less information is available regarding the chemical structures of formed degradation products and their toxicity aspect. Moreover, with modern advanced oxidation processes, caution is necessary when dioxins and other halogenated persistent organic-pollutants could be present. During UV or US irradiation, for example, high chlorinated dioxins and other halogenated pollutants can be de-chlorinated. Lower chlorinated dioxins are known as more toxic compounds compared to high chlorinated pollutants. The very efficient processes for removing persistent organic pollutants from wastewater are those coagulation and adsorption processes used in municipal-waste plants (Križanec, 2007).

6.3 Toxicity of organic pollutants within textile wastewater

The toxicity of textile wastewater varies depending on the different processes applied within the textile industry. The wastewaters of some processes have high aquatic toxicity, whilst others show little or no toxicity. An identification of all toxic compounds used within the textile industry is impossible due to the huge variety of used chemicals and a lack of data about their toxicities. Usually overall toxicity is determined by toxicity-testing the whole effluent stream of aquatic organisms, which is a cost-effective method.

The sources of organic pollutants that can cause aquatic toxicity can be dyes, surfactants, toxic organic chemicals, or biocides. Examples of compounds in each of these classes and their sources, are shown in Table 7.

Textile wastewaters contain different polar and non-polar compounds, but the polar ones are predominant. Polar organic pollutants are non-biodegradable and their elimination is often incomplete.

Agent	Chemical example	Source
Surfactants	Ethoxylated phenols	Multiple processes
Organics	Chlorinated solvents	Scour, machine cleaning
Biocides	Pentachlorophenol	Wool fibres contaminant

Table 7. Typical causes of aquatic toxicity

Non-polar organic pollutants such as dioxins, may be present in textile wastewater in trace amounts. Due to their high toxicity and persistent organic pollutant properties, the restricted limits on wastewater are relatively low. According to EU regulations there are no restrictive limits for dioxins in textile wastewater. For wastewater from incineration plants the restrictive limit is 0.3 ng TEQ-ITF/L. According to our study the concentrations of dioxins in textile wastewater with disperse dyes are considerable (Križanec, 2007). The concentration of PCDD/Fs found in the wastewater sample polluted with disperse dyes was 0.44 ng TEQ-ITF/L. The dominant PCDD/F congener in the disperse dyes' wastewater sample was 2,3,4,6,7,8-HxCDF, the contribution of which to the TEQ was more than 85 % (Fig. 5). The concentration of PCDD/F in this selected wastewater polluted with disperse dyes exceeded the limit of 0.3 ng TEQ-ITF/L, as determined by the European regulation for wastewaters from incineration plants.

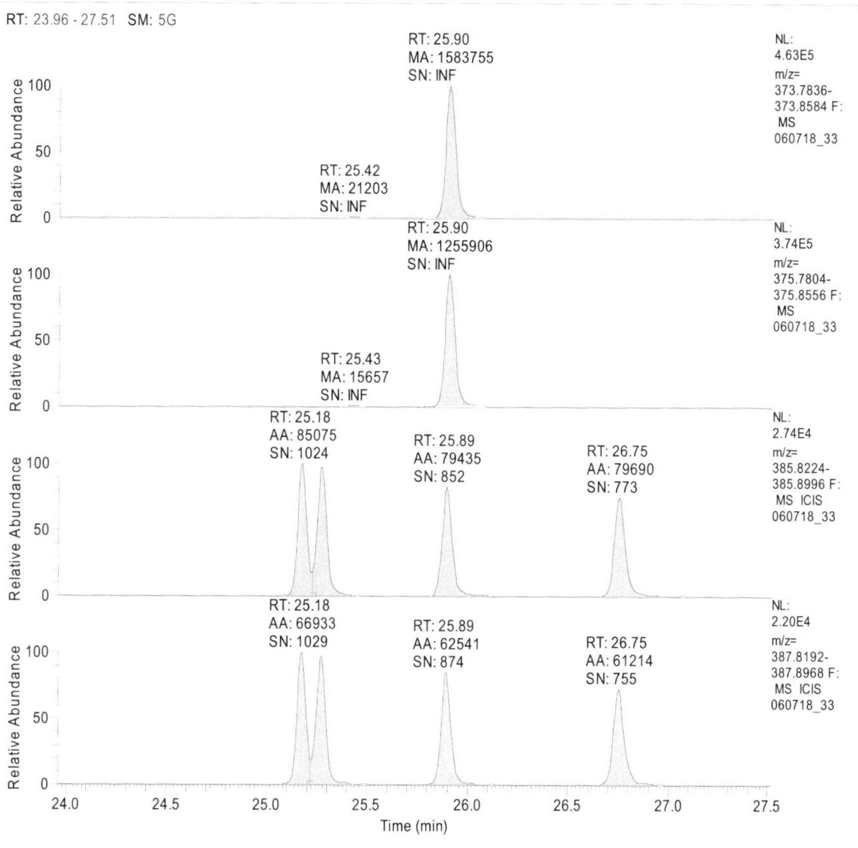

Fig. 5. Dioxin homologue profile found in textile wastewater. There is evident that 2,3,4,6,7,8-HxCDF congener is dominant

7. Tracking persistent organic pollutants in textile finishing and AOP processes

Information about the emissions of persistent organic pollutants from the textile industry is limited due to a lack of complete and accurate emission data. Indirect measurements are necessary regarding the reduction of persistent organic pollution from the textile industry. The presence and concentrations of pollutants should be controlled throughout the textile industry regarding incoming materials and also for textile products and wastewater.

Several instrumental and analytical approaches are in use for determining persistent organic pollutants. For the determination of semi-volatile organic pollutants, the most appropriate are gas chromatographic methods connected with appropriate detectors. For tracking low-level halogenated organic pollutants, the use of gas chromatography with an electron capture detector (ECD), is a useful and relatively cost-effective method.

In order to determine the trace-levels of organic pollutants (dioxins), gas chromatography with mass spectrometry or high-resolution mass spectrometry is required (UEPA method 1613, 1994). Liquid chromatography with mass spectrometry is used to determine water solubility and highly-volatile organic pollutants(Dolman & Pelzing, 2011).

In our pilot studies, two samples of textile wastewater (a mix of disperse dyes' wastewater and mix of metal-complex dyes' wastewater) were analyzed for persistent organo-halogen compounds (POPs). Polychlorinated dibenzo-*p*-dioxins (PCDDs), polychlorinated dibenzofurans PCDFs), dioxin-like PCBs, and polybrominated diphenyl ethers (PBDEs), were determined using the HRGC/HRMS technique. Further decolouration experiments of these wastewaters were performed using the advanced oxidation process (AOP). The POPs' analyses were performed on the samples after the AOP process, and the results were then compared with the initial values.

Wastewater samples were collected from a local textile-dyeing facility, the main activity of which is the dyeing (wet processes) of polyester yarns with disperse dyes. Samples of wastewater were collected at the out-flow of the facility, and stored in pre-cleaned 2.5 L borosilicate amber bottles at 5 °C until analysis or experiment. The disperse dyes' and metal complex dyes' wastewaters were a mix of at least four different dispersals, and at least four different metal complex dyes, respectively. The exact composition of the wastewater was unknown. Wastewater samples (0.5 L) were transferred in separate funnels and spiked with an internal standard mixture containing ^{13}C-labelled isomers of analytes. Hexane (100 mL) was used as the liquid-liquid extraction solvent. The sample extract clean-up was performed according to the USEPA method 1613 (USEPA method 1613, 1994). By using adsorption chromatography on a graphitized carbon column (Carbopack C), the PCBs and PBDEs (60 mL dichloromethane-direct flow),were separated from the PCDD/Fs (60 mL of toluene-opposite flow). The obtained fractions were concentrated to a final volume of 20 µL, and analyzed with GC/HRMS.

Extracts were analyzed on a HP 6890 gas chromatograph GC (Hewlett-Packard, Palo Alto, CA, USA) coupled to a Finnigan MAT 95 XP high resolution mass spectrometer. Aliquot (1.5 µL) of PCDD/F toluene sample extracts were injected into the GC system, equipped with a Restek Rtx Dioxin2 capillary column (60 m x 0.25 mm i.d., film thickness 0.25 µm) in splitless mode. The mass spectrometer operated within the electron impact ionization mode using selected ion-monitoring (SIM), at a minimum resolution of 10,000. The samples were analyzed for PCDD/Fs concentrations, using the isotope dilution method based on USEPA 1613 protocol (USEPA method 1613, 1994). A similar determination/quantification approach was also used for the determining of PCBs and PBDEs in the dichloromethane extracts. In each set of experiments the blank samples were analyzed, and in cases of some PCBs and PBDEs congeners, the blank values were subtracted from the sample values.

H_2O_2/UV experiments were carried out in a batch photo-reactor using a low-pressure mercury UV lamp emitting at 254 nm. The high-frequency plate type URS 1000 L-3 communications system ELAC Nautik was used for ultrasonic irradiation. A system consisting of an AG 1006 LF generator/amplifier (200 W maximum output) and USW 51 ultrasonic transducer (817 kHz) with a working volume of 500 mL. The experimental parameters for the power of the UV lamp was 1600 W, and the reaction time was 60 minutes. Ultrasonic irradiation was performed at 820 kHz.

The concentrations of PCDD/Fs in the wastewater samples at different stages of the experiment, are presented in Table 8. The concentrations of PCDD/Fs in the metal complex dyes wastewater samples were low, and close to both the quantification limits and the concentrations of the background levels. As expected, the concentrations of PCDD/Fs in the samples of disperse dyes' wastewater, were relatively high (Fig. 5).

After AOP treatment the concentrations of PCDD/Fs were similar to those of the original sample, for both types of wastewater. This indicates that PCDD/Fs are stable enough in these types of wastewater to resist those conditions used during our AOP experiments. Also, no new PCDD/F congener was observed.

The concentrations of dioxin-like PCBs in the wastewater samples at different stages of the experiment, are presented in Table 9. The concentrations of dioxine-like PCBs and also PBDEs after AOPs experiments are incomparable with the PCDD/F concentrations of the original wastewater samples. This is because PCBs and PBDEs are not as resistant as PCDD/Fs when some transformations were observed

The concentrations in the samples of H_2O_2/UV-treated wastewater were, for most of the dioxine-like PCBs and the selected PBDEs, higher in comparison with the original samples. We suggest that this is due to the dechlorination or debromination of the higher halogenated congeners and/or the cleavages of the polychlorinated terphenyls. After the sonification AOP method, the concentrations of dioxine-like PCBs and selected PBDEs were relative low. These results suggest that local rigorous conditions (high-pressure, high- temperature) cause the destruction of PCBs and PBDEs via a thermal and/or radical mechanism.

The HGC/HRMS analytical method was used for tracking POP during this pilot-study. This analytical method is expensive and time-consuming but gives us very important information regarding POP's congener distribution at trace levels. In addition, other analytical methods should be tested on samples from the textile industry. When tracking POPs in the textile industry, appropriate screening methods (for example GC/ECD) should be tested and used during routine controls, in order to prevent POPs' pollution from this sector.

With the aim of reducing organic pollution within the textile industry, all the important steps towards sustainable thinking and the introduction of green chemistry are necessary at all stages of the textile-chain, in-line with the recommendations of industrial platforms (WSSTP, Textile Platform). The development of new wet-processing equipment is recommended in order to minimize the amount of fresh water and, consequently, to reduce the volume of generated wastewater and production costs. New wastewater treatment approaches should be implemented through pollution prevention in order to prevent environmental problems, such as waste minimization and reuse of treated water during the production processes. Finally, regulation of pollution by developing strategies for characterization and monitoring (IPPC) the most dangerous pollutants, should certainly also be expanded to cover the textile sector.

	Concentration (ng/L)					
Congener/Group	1	2	3	4	5	6
TCDD	< 0.01	0.04	< 0.01	< 0.01	0.01	< 0.01
PeCDD	0.02	0.03	0.03	< 0.01	< 0.01	< 0.01
HxCDD	0.08	0.19	0.10	0.01	0.03	0.01
HpCDD	0.20	< 0.01	0.23	0.01	0.05	0.02
OCDD	0.25	0.51	0.23	0.06	0.05	0.02
TCDF	0.15	1.02	0.50	< 0.01	0.06	< 0.01
PeCDF	0.22	1.54	0.24	< 0.01	0.03	< 0.01
HxCDF	3.77	3.05	6.44	< 0.01	0.08	0.30
HpCDF	0.12	0.10	0.09	< 0.01	0.01	< 0.01
OCDF	0.05	0.03	0.02	< 0.01	0.01	< 0.01
2.3.7.8-TCDD	< 0.01	0.01	< 0.01	< 0.01	< 0.01	< 0.01
1.2.3.7.8-PeCDD	< 0.01	0.01	0.01	< 0.01	< 0.01	< 0.01
1.2.3.4.7.8-HxCDD	< 0.01	0.12	< 0.01	< 0.01	< 0.01	< 0.01
1.2.3.6.7.8-HxCDD	0.01	0.06	0.01	< 0.01	< 0.01	< 0.01
1.2.3.7.8.9-HxCDD	0.01	0.03	0.01	< 0.01	< 0.01	< 0.01
1.2.3.4.6.7.8-HpCDD	0.14	0.58	0.17	0.01	0.02	0.01
1.2.3.4.6.7.8.9-OCDD	0.25	0.51	0.23	0.06	0.05	0.02
2.3.7.8-TCDF	< 0.01	0.07	< 0.01	< 0.01	0.01	< 0.01
1.2.3.7.8-PeCDF	0.01	0.04	< 0.01	< 0.01	0.01	< 0.01
2.3.4.7.8-PeCDF	0.01	0.04	0.01	< 0.01	0.01	< 0.01
1.2.3.4.7.8-HxCDF	0.01	0.02	< 0.01	< 0.01	0.01	< 0.01
1.2.3.6.7.8-HxCDF	0.04	0.01	< 0.01	< 0.01	0.01	< 0.01
2.3.4.6.7.8-HxCDF	4.23	4.00	6.73	0.01	0.07	0.30
1.2.3.7.8.9-HxCDF	< 0.01	< 0.01	< 0.01	< 0.01	< 0.01	< 0.01
1.2.3.4.6.7.8.-HpCDF	0.09	0.08	0.07	< 0.01	0.01	< 0.01
1.2.3.4.7.8.9-HpCDF	< 0.01	0.01	0.01	< 0.01	< 0.01	< 0.01
1.2.3.4.6.7.8.9-OCDF	0.05	0.03	0.02	< 0.01	0.01	< 0.01
Sum TEQ-ITF	**0.44 ± 0.09**	**0.47 ± 0.09**	**0.69 ± 0.14**	**< 0.01**	**0.02 ± 0.01**	**0.03 ± 0.01**

Legend: 1- Wastewater with disperse dyes. 2- Wastewater with disperse dyes treated with H_2O_2/UV. 3- Wastewater with disperse dyes treated with US. 4- Wastewater with metal complex dyes. 5- Wastewater with metal complex dyes treated with H_2O_2/UV. 6- Wastewater with metal complex dyes treated with US

Table 8. Mass-balance of PCDD/Fs in wastewater polluted with disperse and metal complex dyes treated with H_2O_2/UV and US processes

	Concentration (ng/L)					
PCB congener	1	2	3	4	5	6
PCB 81	< 0.05	< 0.05	< 0.05	< 0.05	< 0.05	< 0.05
PCB 77	0.35	< 0.05	0.37	0.15	0.41	< 0.05
PCB 126	< 0.05	< 0.05	< 0.05	< 0.05	< 0.05	< 0.05
PCB 169	< 0.05	< 0.05	< 0.05	< 0.05	< 0.05	< 0.05
PCB 105	0.36	< 0.05	2.81	0.38	< 0.05	< 0.05
PCB 114	0.05	< 0.05	0.17	< 0.05	< 0.05	< 0.05
PCB 118	0.88	< 0.05	5.52	1.10	< 0.05	< 0.05
PCB 123	0.11	< 0.05	0.45	0.08	< 0.05	< 0.05
PCB 156	0.08	< 0.05	1.20	0.16	< 0.05	< 0.05
PCB 157	< 0.05	< 0.05	0.29	< 0.05	0.28	< 0.05
PCB 167	< 0.05	< 0.05	0.46	0.07	< 0.05	< 0.05
PCB 189	< 0.05	< 0.05	< 0.05	< 0.05	< 0.05	< 0.05

Legend: 1- Wastewater with disperse dyes. 2- Wastewater with disperse dyes treated with H2O2/UV. 3- Wastewater with disperse dyes treated with US. 4- Wastewater with metal complex dyes. 5- Wastewater with metal complex dyes treated with H2O2/UV. 6- Wastewater with metal complex dyes treated with US

Table 9. Mass-balance of PCB congener in wastewater polluted with disperse and metal complex dyes treated with H_2O_2/UV and US processes.

8. References

Abad. E.; Adrados M. A.; Caixach J.; Fabrellas B. & Rivera J. (2000). Dioxin mass balance in a municipal waste incinerator. *Chemosphere*. Vol. 40. pp. 1143-1147. ISSN 0045-6535

Arslan-Alaton; I. (2004); Advanced oxidation of textile industry dyes. In: Advanced Oxidation Processes for Water and Wastewater Treatment. Simon Pearsons. 302-323. IWA Publishing. ISBN 1-84339-017-5. Tunbridge Wells.

Banat. I. M.; Nigam. P.; Singh. D. & Marchant. R. (1996). Microbial decolourization of textile-dye containing effluents: a review. *Bioresour Technol*. Vol. 58. pp. 217-227. ISSN 0960-8524

Baral. A. & Engelken. R.D. (2002). Chromium-based regulations and greening in metal finishing industries in the USA. *Environmental Science & Policy*. Vol. 5. pp. 121-133. ISSN 1462-9011

BASF. (2000) "Technical Information about BASF Products for Resin Finishing". TI/T 344.

BAT (2003) Integrated Pollution Prevention and Control (IPPC) Reference Document on Best Available Techniques for the Textiles Industry

BIA-Report (2003) 2/2003. Grenzwerteliste 2003. HVBG. Hauptverband der gewerblichen Berufsgenossenschaften. Sankt Augustin.

Brookes. G. & Barfoot P. (2008). Global impact of biotec crops: socio-economic and environmental effects 1996-2007. *AgBioForum*. Vol.11. pp. 21-38. ISSN 1522-936X

Brown. M.A.; DeVito. S.C. (1993). Predicting azo dye toxicity. *Critical Reviews in Environmental Science and Technology*. Vol. 23. No. 3. pp. 249-324. ISSN 1064-3389.

Chemical Inspection & Regulation Service. (www.cirs-reach.com/textile/)

Chu. H. C. & Chen. K. M. (2002). Reuse of activated sludge biomass: I. Removal of basic dyes from wastewater by biomass. *Process Biochem*. Vol. 37. pp.595-600. ISSN 0032-9592

DFG (Deutsche Forschungsgemeinschaft). (1988) List of MAK and BAT Values. VCH Publishers. Weinheim.

Dolman. S. & Pelzing. M. (2011) An optimized method for the determination of perfluorooctanoic acid perfluorooctane sulfonate and other perfluoro-chemicals in different matrices using liquid chromatography/ion-trap mass spectrometry. *Journal of Chromatography B*. Vol. 879. pp. 2043-2050. ISSN 0378-4347.

ETAD (2003). ETAD information on the 19th amendment of the restriction on the marking and use of certain azocolourants. ETAD-Ecology and Toxicology Association of Dyes and Organic Pigments Manufacturers.

EURATEX. E.-D. (2000) "Textile Industry BREF document (Chapter 2-3-4-5-6)"

Golka. K.; Kopps. S. & Myslak. Z. W. (2004). Carcinogenicity of azo colorants: influence of solubility and bioavailability. *Toxicology Letters*. Vol. 151. pp.203-210. ISSN 0378-4274

Gupta. G. S.; Prasad G. & Singh V. H. (1990). Removal of chrome dye from aqueous solutions by mixed adsorbents: fly ash and coal. *Water Research*. Vol. 24. pp. 45-50. ISSN 0043-1354

Humans. "Some Industrial Chemicals and Dyestuffs." Vol. 29. Lyon. France.

IARC (1982). World Health Organization International Agency for Research on Cancer. Monographs on the Evaluation of the Carcinogenic Risk of Chemicals to

Isik. M.; Sponza. T.S. (2007). Fate and toxicity of azo dye metabolites under batch long-term anaerobic incubations. *Enzyme and Microbial Technology*. 40. pp. 934-939. ISSN 0141-0229.

Križanec. B. (2007) *Obstojne organske halogenirane spojine v tekstilni industriji*. PhD Thesis.

Križanec. B.; Majcen Le Marechal. A. (2006). Dioxins and Dioxin-like Persistent Organic Pollutants in Textiles and Chemicals in the Textile Sector. *Croatica Chemica Acta.* Vol. 79. No. 2. pp. 177-186. ISSN 0011-1643

Križanec. B.; Majcen Le Marechal. A.; Vončina. E. & Brodnjak Vončina. D. (2005). Presence of dioxins in textile dyes and their fate during the dyeing processes. *Acta Chimica Slovenica.* Vol.52. (January 2005). pp. 111-118. ISSN 1318-0207

Kurbus. T.; Majcen Le Marechal. A. & Brodnjak Vončina. D. (2003). Comparison of H_2O_2/UV. H_2O_2/O_3 and H_2O_2/Fe^{2+} processes for the decolorisation of vinylsulphone reactive dyes. *Dyes and Pigments.* Vol. 58. pp. 245-252. ISSN 0143-7208.

Lacasse, K. & Baumann, W. (2004). *Textile Chemicals Environmental Data and Facts*, Springer, ISBN 3-540-40815-0, Berlin.

Mattioli. D.; Malpei, F.; Borone, G.; Rozzi, A. (2002). Water minimisation and reuse in the textile industry. In: Water Recycling and Resource Recovery in Industry. Piet Lens, Look Hulshoff Pol, Peter Wilderer and Takashi Asano. IWA Publishing. ISBN 1-84339-005-1. Cornwall.

McCarthy. B.J. (1997). Biotechnology and Coloration. *Coloration Technology.* Vol. 27. No.1. pp. 26-31. ISSN 1472-3581.

Naranjo. E. S. (2009). Impact of Bt crops on non-target invertebrates and insecticide use patterns. CAB Reviews: *Perspectives in Agriculture. Veterinary Science. Nutrition and Natural Resources.* Vol.4. No.11. (January 2009). pp. 1-23. ISSN 1749-8848

Science tech entrepreneur (May 2003) http://www.techno-preneur.net/information-desk/sciencetech-magazine/2006/may06/Disperse_dyes.pdf

Shaw. T. (1989). Environmental issues in wool processing. International Wool Secretariat (IWS) Development center monograph. IWS. West Yorkshire. England

Slokar. Y. M. & Majcen Le Marechal. A. (1998). Methods of decoloration of textile wastewater. *Dyes and Pigments.* Vol. 37. No. 4. pp.335-356. ISSN 0143-7208.

UK. (2001) "Comments made by UK to the First Draft of the BREF Textiles"

US EPA. (1996) "Manual - Best Management Practices for Pollution Prevention in the Textile Industry"

USEPA method 1613 (1994). Tetra-trough Octa-Chlorinated Dioxins and Furans by isotopic Dilution HRGC-HRMS. USEPA. Washington.

Vajnhandl. S. & Majcen Le Marechal. A. (2007). Case study of the sonochemical decolouration of thextile azo dye Reactive Black 5. *Journal of hazardous materials.* Vol. 141. pp. 329-335. ISSN 0304-3894.

Van der Zee. F.; Villaverde. S. (2005). Combined anaerobic-aerobic treatment of azo dyes-Ashort review of bioreactor studies. *Water Research.* 39. pp.1425-1440. ISSN 0043-1354.

Vandervivere. P. C.; Bianchi. R. & Verstraete. W. (1998). Treatment and reuse of wastewater from the textile wet-processing industry: Review of emerging technologies. *J. Chem. Technol. Biotechnol.* Vol. 72. pp.289-302. ISSN 0268-2575

Volmajer Valh. J.; Majcen Le Marechal A. (2009). Decolouration of Textile Wastewaters. In: *Dyes and Pigments New Research.* Arnold R. Lang. 175-199. Nova Science Publishers. Inc.. ISBN 978-1-60692-027-5. New York.

Volmajer Valh. J.; Majcen Le Marechal. A.; Vajnhandl. S.; Jerič. T.; Šimon E. (2011) Water in the Textile Industry. In: Peter Wilderer (ed.) *Treatise on Water Science.* vol. 1. pp. 685-706 Oxford: Academic Press.

Zollinger. H. (2003). *Color chemistry syntheses, properties, and applications of organic dyes and pigments* (3th edition). Wiley-VCH. ISBN 3-906390-23-3. Zürich

3

Textile Organic Dyes – Characteristics, Polluting Effects and Separation/Elimination Procedures from Industrial Effluents – A Critical Overview

Zaharia Carmen and Suteu Daniela
'Gheorghe Asachi' Technical University of Iasi,
Faculty of Chemical Engineering and Environmental Protection,
Romania

1. Introduction

The residual dyes from different sources (e.g., textile industries, paper and pulp industries, dye and dye intermediates industries, pharmaceutical industries, tannery, and Kraft bleaching industries, etc.) are considered a wide variety of organic pollutants introduced into the natural water resources or wastewater treatment systems.

One of the main sources with severe pollution problems worldwide is the textile industry and its dye-containing wastewaters (i.e. 10,000 different textile dyes with an estimated annual production of $7 \cdot 10^5$ metric tonnes are commercially available worldwide; 30% of these dyes are used in excess of 1,000 tonnes per annum, and 90% of the textile products are used at the level of 100 tonnes per annum or less) (Baban et al., 2010; Robinson et al., 2001; Soloman et al., 2009). 10-25% of textile dyes are lost during the dyeing process, and 2-20% are directly discharged as aqueous effluents in different environmental components.

In particular, the discharge of dye-containing effluents into the water environment is undesirable, not only because of their colour, but also because many of dyes released and their breakdown products are toxic, carcinogenic or mutagenic to life forms mainly because of carcinogens, such as benzidine, naphthalene and other aromatic compounds (Suteu et al., 2009; Zaharia et al., 2009). Without adequate treatment these dyes can remain in the environment for a long period of time. For instance, the half-life of hydrolysed Reactive Blue 19 is about 46 years at pH 7 and 25°C (Hao et al., 2000).

In addition to the aforementioned problems, the textile industry consumes large amounts of potable and industrial water (Tables 1, 2 and Fig. 1) as processing water (90-94%) and a relatively low percentage as cooling water (6-10%) (in comparison with the chemical industry where only 20% is used as process water and the rest for cooling). The recycling of treated wastewater has been recommended due to the high levels of contamination in dyeing and finishing processes (i.e. dyes and their breakdown products, pigments, dye intermediates, auxiliary chemicals and heavy metals, etc.) (Tables 3, 4 and 5) (adapted from Bertea A. and Bertea A.P., 2008; Bisschops and Spanjers, 2003; Correia et al., 1994; Orhon et al., 2001).

Type of finishing process	Water consumption, 10^{-3} m^3/kg textile product		
	Minimum	Medium	Maximum
Raw wool washing	4.2	11.7	77.6
Wool finishing	110.9	283.6	657.2
Fabric finishing			
• Short process	12.5	78.4	275.2
• Complex processing	10.8	86.7	276.9
Cloth finishing			
• Simplified processing	8.3	135.9	392.8
• Complex process	20	83.4	377.8
• Panty processing	5.8	69.2	289.4
Carpet finishing	8.3	46.7	162.6
Fibre finishing	3.3	100.1	557.1
Non-fabrics finishing	2.5	40	82.6
Yarn finishing	33.4	212.7	930.7

Table 1. Specific water consumption in textile finishing processes (adapted from Bertea A. & Bertea A.P., 2008)

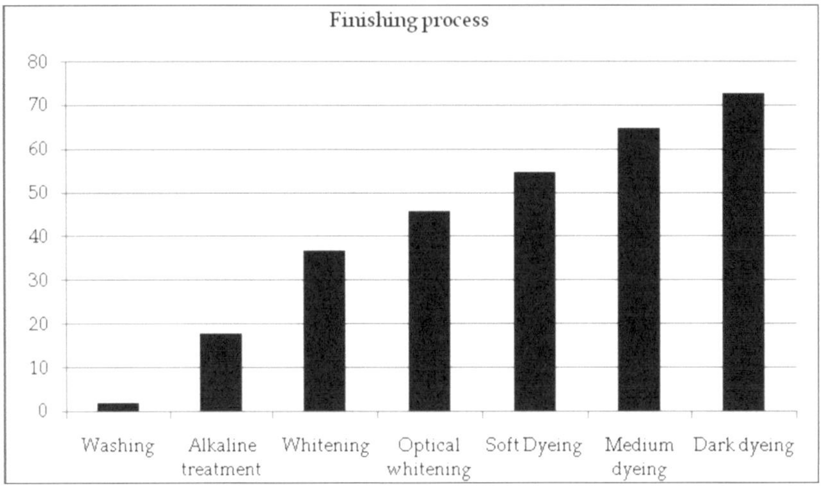

Fig. 1. Specific water consumption in different operations of textile finishing (EPA, 1997)

Operation/Process	**Water consumption** (% from total consumption of the textile plant)			**Organic load** (% from total organic load of the textile plant)		
	Minimum	Medium	Maximum	Minimum	Medium	Maximum
General facilities	6	14	33	0.1	2	8
Preparation	16	36	54	45	61	77
Dyeing	4	29	53	4	23	47
Printing	42	55	38	42	59	75
Wetting	0.3	0.4	0.6	0	0.1	0.1
Fabrics washing	3	28	52	1	13	25
Finishing	0.3	2	4	0.1	3	7

Table 2. Water consumption and organic load in different textile finishing steps (EWA, 2005)

The most common textile processing technology consists of desizing, scouring, bleaching, mercerizing and dyeing processes (EPA, 1997):

- *Sizing* is the first preparation step, in which sizing agents such as starch, polyvinyl alcohol (PVA) and carboxymethyl cellulose are added to provide strength to the fibres and minimize breakage.
- *Desizing* is used to remove sizing materials prior to weaving.
- *Scouring* removes impurities from the fibres by using alkali solution (commonly sodium hydroxide) to breakdown natural oils, fats, waxes and surfactants, as well as to emulsify and suspend impurities in the scouring bath.
- *Bleaching* is the step used to remove unwanted colour from the fibers by using chemicals such as sodium hypochlorite and hydrogen peroxide.
- *Mercerising* is a continuous chemical process used to increase dye-ability, lustre and fibre appearance. In this step a concentrated alkaline solution is applied and an acid solution washes the fibres before the dyeing step.
- *Dyeing* is the process of adding colour to the fibres, which normally requires large volumes of water not only in the dye bath, but also during the *rinsing step*. Depending on the dyeing process, many chemicals like metals, salts, surfactants, organic processing aids, sulphide and formaldehyde, may be added to improve dye adsorption onto the fibres.

In general, the textile industry uses a large quantity of chemicals such as:

- *Detergents and caustic*, which are used to remove dirt, grit, oils, and waxes. Bleaching is used to improve whiteness and brightness.
- *Sizing agents*, which are added to improve weaving.
- *Oils*, which are added to improve spinning and knitting.
- *Latex and glues*, which are used as binders.
- *Dyes, fixing agents, and many in-organics*, which are used to provide the brilliant array of colours the market demands.
- *A wide variety of special chemicals*, which are used such as softeners, stain release agents, and wetting agents.

Many of these chemicals become part of the final product whereas the rest are removed from the fabric, and are purged in the textile effluent.

Type of finished textile product	**Dyes, g/kg textile product**	**Auxiliaries, g/kg textile product**	**Basic chemical compounds, g/kg textile product**
Polyester fibres	18	129	126
Fabrics from synthetic fibres	52	113	280
Fabrics from cotton	18	100	570
Dyed fabrics from cellulose fibres	11	183	200
Printed fabrics from cellulose fibres	88	180	807

Table 3. Principal pollutants of textile wastewaters (EWA, 2005)

The annual estimated load with pollutants of a textile wastewater is of: 200,000-250,000 t salts; 50,000-100,000 t impurities of natural fibres (including biocids) and associated materials (lignin, sericine, etc.); 80,000-100,000 t blinding agents (especially starch and its derivatives, but also polyacrylates, polyvinyl alcohol, carboxymethyl cellulose); 25,000-

30,000 t preparation agents (in principal, mineral oils); 20,000-25,000 t tensides (dispersing agents, emulsifiers, detergents and wetting agents); 15,000-20,000 tonnes carboxylic acids (especially acetic acid); 10,000-15,000 t binders; 5,000-10,000 t urea; 5,000-10,000 t ligands, and < 5,000 t auxiliaries (EWA, 2005). The environmental authorities have begun to target the textile industry to clean up the wastewater that is discharged. The principal quality indicators that regulators are looking for polluting effect or toxicity are the high salt content, high Total Solids (TS), high Total Dissolved Solids (TDS), high Total Suspended Solids (TSS), Biological Oxygen Demand (BOD) and Chemical Oxygen Demand (COD), heavy metals, colour of the textile effluent (ADMI color value - American Dye Manufacturer Institute color value), and other potential hazardous or dangerous organic compounds included into each textile processing technological steps (Tables 4 and 5).

Process	**Textile effluent**
Singering, Desizing	High BOD, high TS, neutral pH
Scouring	High BOD, high TS, high alkalinity, high temperature
Bleaching, Mercerizing	High BOD, high TS, alkaline wastewater
Heat-setting	Low BOD, low solids, alkaline wastewater
Dyeing, Printing & Finishing	Wasted dyes, high BOD, COD, solids, neutral to alkaline wastewater

Table 4. Wet processes producing textile wastewater (adapted from Naveed S. et al., 2006)

Process	**COD, g O_2/L**	**BOD, g O_2/L**	**TS, g/L**	**TDS, g/L**	**pH**	**Colour (ADMI)**	**Water usage, L/kg product**
Desizing	4.6-5.9	1.7-5.2	16.0-32.0	-	-	-	3-9
Scouring	8.0	0.1-2.9	7.6-17.4	-	10-13	694	26-43
Bleaching	6.7-13.5	0.1-1.7	2.3-14.4	4.8-19.5	8.5-9.6	153	3-124
Mercerising	1.6	0.05-0.1	0.6-1.9	4.3-4.6	5.5-9.5	-	232-308
Dyeing	1.1-4.6	0.01-1.8	0.5-14.1	0.05	5-10	1450-4750	8-300

Table 5. Principal characteristics of a cotton wet processing wastewater (Cooper, 1995)

The wastewater composition is depending on the different organic-based compounds, inorganic chemicals and dyes used in the industrial dry and wet-processing steps. Textile effluents from the dyeing and rinsing steps represent the most coloured fraction of textile wastewaters, and are characterized by extreme fluctuations in many quality indicators such as COD, BOD, pH, colour, salinity and temperature.

The colour of textile wastewater is mainly due to the presence of textile dyes, pigments and other coloured compounds. A single dyeing operation can use a number of dyes from different chemical classes resulting in a complex wastewater (Correia et al., 1994). Moreover, the textile dyes have complex structures, synthetic origin and recalcitrant nature, which makes them obligatory to remove from industrial effluents before being disposed into hydrological systems (Anjaneyulu et al, 2005).

The dye removal from textile effluent is always connected with the decolourization treatment applied for textile wastewater in terms of respectation the local environmental quality requirements and standards (Table 6) (i.e. removal values of COD, BOD, TS, TSS, TDS, colour, total nitrogen, and total phosphorus from textile wastewater higher than 70-

85% or concentration values of the specific quality indicators under the imposed or standard limits) (Zaharia, 2008).

The decolorization treatments applied for different textile effluents include current and also advanced non-biological (i.e. specific mechano-physical, chemical, electrochemical processes, etc.) and also biological processes (Suteu et al., 2009; Zaharia, 2006; Zaharia et al, 2011).

Quality indicator	**M.A.C.***, mg/L		
	Discharge directly in water bodies	*Discharge in urban WW sewerage network*	*Water bodies quality, class I -natural non-polluted state*
pH	6.5-8.5	6.5-8.5	6.5-8.5
BOD_5	25	300	3
COD	125	500	10
TSS	35	350	-
TDS	2000	250	< 500
Total N	10	15	1.5
Total P	1	5	0.1
Sulphates, SO_4^{2-}	600	600	80
Chlorides, Cl^-	500	500	<100
Sulphides (S^{2-}) + H_2S	0.5	1.0	<0.5
Synthetic detergents	0.5	25	<0.5
Others (Oil & grease)	20	30	<0.1
* M.A.C. - Maximum Admissible Concentration			

Table 6. Romanian national wastewater and water quality standard adapted to European and international standards (adapted from Zaharia, 2008)

Studies on the behavior of textile organic dyes in water and wastewater treatment processes refer predominantly to laboratory tests or investigations of semi-technical plants, sometimes under conditions related to waterworks practice. In addition, textile operators, water supply companies, local environmental authorities have collected a lot of data on the behavior of textile dyes during textile wastewater treatment, but have seldom published their results.

However, information on the behavior of textile organic dyes is needed, because the limited number of reports available that are based on realistic operating conditions or which reproduce practical conditions that are already several years old.

2. Textile organic dyes – Classification and characteristics

The dyes are natural and synthetic compounds that make the world more beautiful through coloured products. The textile dyes represent a category of organic compounds, generally considered as pollutants, presented into wastewaters resulting mainly from processes of chemical textile finishing (Suteu et al., 2011a; Zaharia et al., 2009).

The textile coloration industry is characterised by a very large number of dispersed dyehouses of small and medium size that use a very wide range of textile dyes.

2.1 Textile organic dye classification

The nature and origin are firstly considered as criteria for the general classification in natural and synthetic textile dyes.

The natural textile dyes were mainly used in textile processing until 1856, beginning in 2600 BC when was mentioned the use of dyestuff in China, based on dyes extracted from vegetable and animal resources. It is also known that Phoenicians were used Tyrian purple produced from certain species of crushed sea snails in the 15th century BC, and indigo dye produced from the well-known indigo plant since 3000 BC. The dyes from madder plants were used for wrapping and dyeing of Egyptian mummies clothes and also of Incas fine textures in South America.

The synthetic dyes were firstly discovered in 1856, beginning with ‚mauve' dye (aniline), a brilliant fuchsia colour synthesed by W.H. Perkin (UK), and some azo dyes synthesed by diazotisation reaction discovered in 1958 by P. Gries (Germany) (Welham, 2000). These dyes are aromatic compounds produced by chemical synthesis, and having into their structure aromatic rings that contain delocated electrons and also different functional groups. Their color is due to the chromogene-chromophore structure (acceptor of electrons), and the dyeing capacity is due to auxochrome groups (donor of electrons). The chromogene is constituted from an aromatic structure normally based on rings of benzene, naphthaline or antracene, from which are binding chromofores that contain double conjugated links with delocated electrons. The chromofore configurations are represented by the azo group (-N=N-), ethylene group (=C=C=), methine group (-CH=), carbonyl group (=C=O), carbon-nitrogen (=C=NH; -CH=N-), carbon-sulphur (=C=S; ≡C-S-S-C≡), nitro ($-NO_2$; -NO-OH), nitrozo (-N=O; =N-OH) or chinoid groups. The auxochrome groups are ionizable groups, that confer to the dyes the binding capacity onto the textile material. The usual auxochrome groups are: $-NH_2$ (amino), -COOH (carboxyl), $-SO_3H$ (sulphonate) and -OH (hydroxyl) (Suteu et al, 2011; Welham, 2000). Five examples of textile dyes are presented in Fig. 2.

The textile dyes are mainly classified in two different ways: (1) based on its application characteristics (i.e. CI Generic Name such as acid, basic, direct, disperse, mordant, reactive, sulphur dye, pigment, vat, azo insoluble), and (2) based on its chemical structure respectively (i.e. CI Constitution Number such as nitro, azo, carotenoid, diphenylmethane, xanthene, acridine, quinoline, indamine, sulphur, amino- and hydroxy ketone, anthraquinone, indigoid, phthalocyanine, inorganic pigment, etc.) (Tables 7 and 8).

Excepting the colorant precursors such as azoic component, oxidation bases and sulphur dyes, almost two-third of all organic dyes are azo dyes (R_1-N=N-R_2) used in a number of different industrial processes such as textile dyeing and printing, colour photography, finishing processing of leather, pharmaceutical, cosmetics, etc. The starting material or intermediates for dye production are aniline, chloroanilines, naphthylamines, methylanilines, benzidines, phenylenediamines, and others.

Considering only the general structure, the textile dyes are also classified in anionic, nonionic and cationic dyes. The major anionic dyes are the direct, acid and reactive dyes (Robinson et al., 2001), and the most problematic ones are the brightly coloured, water soluble reactive and acid dyes (they can not be removed through conventional treatment systems).

The major nonionic dyes are disperse dyes that does not ionised in the aqueous environment, and the major cationic dyes are the azo basic, anthraquinone disperse and reactive dyes, etc. The most problematic dyes are those which are made from known carcinogens such as benzidine and other aromatic compounds (i.e. anthroquinone-based dyes are resistant to degradation due to their fused aromatic ring structure). Some disperse dyes have good ability to bioaccumulation, and the azo and nitro compounds are reduced in sediments, other dye-accumulating substrates to toxic amines (e.g. $R_1\text{-}N{=}N\text{-}R_2 + 4H^+ + 4e^- \rightarrow R_1\text{-}NH_2 + R_2\text{-}NH_2$).

The organic dyes used in the textile dyeing process must have a high chemical and photolytic stability, and the conventional textile effluent treatment in aerobic conditions does not degrade these textile dyes, and are presented in high quantities into the natural water resources in absence of some tertiary treatments.

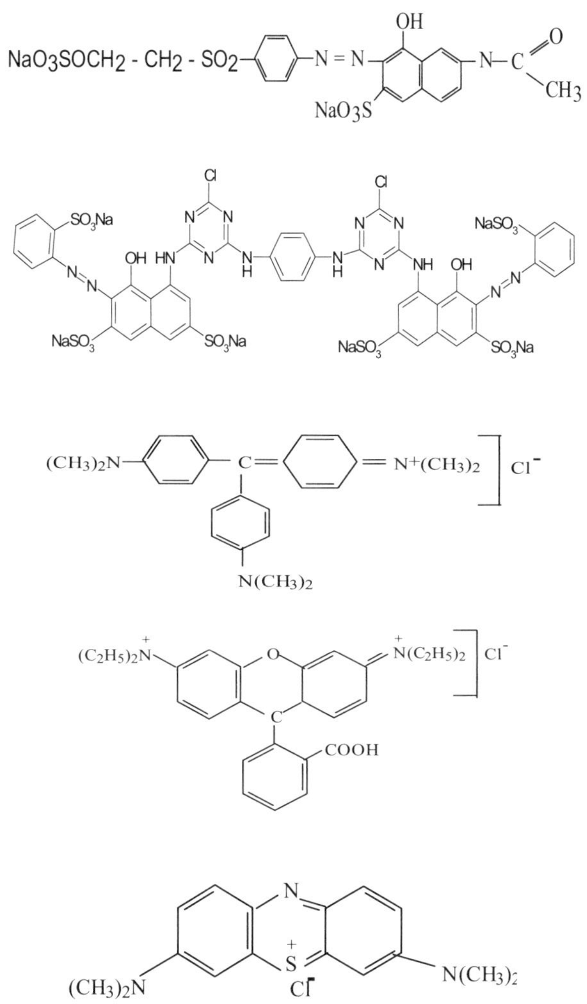

Reactive Orange 16
C.I. 18097
Anionic monoazo reactive dye;
MW = 617.54 g/mol;
λ_{max} = 495 nm

Brilliant Red HE-3B
(Reactive Red 120)
C.I. 25810
Anionic, bifunctional azo reactive dye;
MW = 1463 g/mol;
λ_{max} = 530 nm

Crystal Violet
(Basic Violet 3)
C.I. 42555
Cationic triphenylmethane dye;
MW = 407.99 g/mol;
λ_{max} = 590 nm

Rhodamine B
(Basic Violet 10)
C.I. 45170
Cationic, Xantenic dye;
MW =479.2 g/mol;
λ_{max} = 550 nm

Methylene Blue
(Basic Blue 9)
C.I. 52015
Cationic, phenothiazine dye;
MW =319.85 g/mol;
λ_{max}= 660 nm

Fig. 2. Chemical structure and principal characteristics of different textile dyes (Suteu et al., 2011a)

Chemical class	C.I. Constitution numbers	Chemical class	C.I. Constitution numbers
Nitroso	10000-10299	Indamine	49400-49699
Nitro	10300-10099	Indophenol	49700-49999
Monoazo	11000-19999	Azine	50000-50999
Disazo	20000-29999	Oxazine	51000-51999
Triazo	30000-34999	Thiazine	52000-52999
Polyazo	35000-36999	Sulphur	53000-54999
Azoic	37000-39999	Lactone	55000-56999
Stilbene	40000-40799	Aminoketone	56000-56999
Carotenoid	40800-40999	Hydroxyketone	57000-57999
Diphethylmethane	41000-41999	Anthraquinone	58000-72999
Triarylmethane	42000-44999	Indigoid	73000-73999
Xanthene	45000-45999	Phthalocyanine	74000-74999
Acridine	46000-46999	Natural	75000-75999
Quinoline	47000-47999	Oxidation base	76000-76999
Methine	48000-48999	Inorganic pigment	77000-77999
Thiazole	49000-49399		

Table 7. Colour index classification of dye chemical constituents (Cooper, 1995)

Chemical class	Distribution between application ranges, %								
	Acid	Basic	Direct	Disperse	Mordant	Pigment	Reactive	Solvent	Vat
Unmetallised azo	20	5	30	12	12	6	10	5	-
Metal complex	65	-	10	-	-	-	12	13	-
Thiazole	-	5	95	-	-	-	-	-	-
Stilbene	-	2	98	-	-	-	-	-	-
Anthraquinone	15	2	-	25	3	4	6	9	36
Indigoid	2	-	-	-	-	17	-	-	81
Quinophthalene	30	20	-	40	-	-	10	-	-
Aminoketone	11	-	-	40	8	-	3	8	20
Phtalocyanine	14	4	8	-	4	9	43	15	3
Formazan	70	-	-	-	-	-	30	-	-
Methine	-	71	-	23	-	1	-	5	-
Nitro, nitroso	31	2	-	48	2	5	-	12	-
Triarylmethane	35	22	1	1	24	5	-	12	-
Xanthene	33	16	-	-	9	2	2	38	-
Acridine	-	92	-	4	-	-	-	4	-
Azine	39	39	-	-	-	3	-	19	-
Oxazine	-	22	17	2	40	9	10	-	-
Thiazine	-	55	-	-	10	-	-	10	25

Table 8. Distribution of each chemical class between major application ranges (adapted from Cooper, 1995)

The major textile dyes can be included in the two high classes: azo or anthraquinone (65-75% from total textile dyes). The azo dyes are characterised by reactive groups that form covalent bonds with HO-, HN-, or HS- groups in fibres (cotton, wool, silk, nylon). Azo dyes are mostly used for yellow, orange and red colours. Anthraquinone dyes constitute the second most important class of textile dyes, after azo dyes, and have a wide range of colours in almost the whole visible spectrum, but they are most commonly used for violet, blue and green colours (Fontenot et al., 2003). Considering the nature of textile fibres that are dyeing, the textile dyes can be classified as into Table 9.

Class	Subclass	PES	CA	PAN	PA	Silk	Wool	Cotton
Disperse		+++	+++	++	++	-	-	-
Cationic		-	~	+++	++	-	-	-
Acid	Standard	-	-	~	+++	+++	+++	-
	1 : 1	-	-	-	P	+	+++	-
	1 : 2	-	-	-	++	+	+++	-
Reactive		-	-	-	~	++	++	+++
Direct		-	-	-	++	++	P	+++
Stuff		~	-	~	~	~	–	+++
Indigoid		-	-	–	~	~	~	P
Sulphur		-	+	–	–	–	–	+++
Insoluble azo		+++	-	~	~	~	–	+++
Legend: +++ very frequent; ++ frequent; + sometimes; ~ possible; P especially printing								

Table 9. Dye classes and dyeing textile substrates (adapted from Bertea A. & Bertea A.P., 2008)

The textile dyeing process is due to physico-chemical interactions developed at contacting of textile material with dye solution or dispersion, which contains a large variety of chemicals (salts, acids) and dyeing auxiliaries (tensides, dispersing agents, etc.).

2.2 Textile dye characterisation

The identification of individual unknown dyes in a coloured effluent or watercourse is difficult to be done and implies advanced analytical methods (i.e. individual and/or coupled spectrophotometry, G/L chromatography and mass spectrometry procedures), and also the colour determination and appreciation in different operating situations.

The characterisation and identification data of the textile dyes as main chemicals in dyeing process must consist of:

- dye identity data (i.e. name, C.I. or CAS number, molecular and structural formula; composition, degree of purity, spectral data; methods of detection and determination) (e.g., some examples illustrated in Fig. 2),
- dye production information (i.e. production process, proposed uses, form, concentration in commercially available preparations, estimated production, recommended methods and precautions concerning handling, storage, transport, fire and other dangers, emergency measures, etc.) (e.g. some indications in Table 3),
- dye physico-chemical properties (i.e. boiling point (b.p.), relative density, water solubility, partition coefficient, vapour pressure, self-ignition, oxidising properties, granulometry, particle size distribution, etc.) (e.g., for the first synthetic discovered dye: Aniline - 184 (b.p.), k_H= 2.05E-01, C_{sat}^{w}= 3.6E+04, pK_a= 4.6, log K_{ow}= 0.90; for 4,4'-

Methylenedianiline - 398 (b.p.), k_H= 5.67E-06, C_{sat}^w= 1000, pK_a= n.s., log K_{ow}= 1.59; for 4-Aminodiphenylamine - 354 (b.p.), k_H= 3.76E-01, C_{sat}^w= 1450, pK_a= 5.2, log K_{ow}= 1.82; where, k_H - the Henry's constant at 1013 hPa and 25°C ($Pa \cdot m^3/mol$), C_{sat}^w - the water solubility (mg/L) at 25°C, pK_a – dissociation constant of the protonated azo dye at 25°C, K_{ow} – n-octanol/water partition coefficient, n.s. – not specified)

- toxicological studies (i.e. acute toxicity-oral, inhalation, dermal, skin or eyes irritation, skin sensitisation, repeated dose toxicity-28 days, mutagenicity, toxicity to reproduction, toxicokinetic behaviour),
- ecotoxicological studies (i.e. acute toxicity to fish, daphnia, growth inhibition on algae, bacteriological inhibition, degradation: ready biodegradability, abiotic degradation-hydrolysis as a function of pH, BOD, COD, BOD/COD ratio).

These characteristics and indentification data are given obligatory by the dye producers or distributors of textile products on the free market of homologated colorants, and also exist in the library data of some operating programs of advanced analysis apparatus.

The textile azo dyes are characterized by relatively high polarity (log K_{ow} up to 3) and high recalcitrance. Recalcitrance is difficult to evaluate because of the dependence of degradation on highly variable boundary conditions (e.g., redox milieu or pH). For example, aniline (the first synthetic discovered dye) is known to be easily degradable, but under specific anoxic conditions it has been proven to be easily stable (Börnick and Schmidt, 2006). Furthermore, the azo dyes are relevant in terms of eco- and human toxicity, industrially produced in high quantities, and known to occur in hydrosphere.

The azo dyes can accept protons because of the free electron pair of the nitrogen, and the free electron pair of nitrogen interacts with the delocalized п-orbital system.

Acceptor substituients at the aromatic ring such as –Cl or $-NO_2$ cause an additional decrease in the basic character of aminic groups. Donor groups such as $-CH_3$ or –OR (in metha and para position) lead to an increase in the basicity of aromatic aminic groups. However, donor substituients in the ortho position can sterically impede the protonation and consequently decrease the basicity of aminic groups. The azo dyes are characterized by amphoteric properties when molecules contain additional acidic groups such as hydroxyl, carboxyl or sulfoxyl substituents.

Depending on pH value, the azo dyes can be anionic (deprotonation at the acidic group), cationic (protonated at the amino group) or non-ionic. Accordingly, knowledge of the acidity constants is indispensable for the characterization of the behavior of azo dyes. Environmental partitioning is influenced by substituents as well as the number of carbon atoms and aromatic structure of the carbon skeleton. The presence of an amino group causes a higher boiling point, a higher water solubility, a lower Henry's law constant, and a higher mobility in comparison with hydrocarbons (the amino group can also reduce the mobility by specific interactions with solids via covalent bonding to carbonyl moieties or cation exchange) (Börnick and Schmidt, 2006). The volatility of azo dyes in aqueous solution is in most cases very low. Some colour characteristics from different studies, tests and literature data are presented in Table 10, especially for reactive dyes.

The chromophore distribution in reactive dyes indicated that the great majority of unmetallised azo dyes are yellow, orange and red. Contrarily, the blue, green, black and brown contain a much more proportion of metal-complex azo, anthraquinone, triphenyldioxazine or copper phthalocyanine chromophores (Table 10).

Chemical class	Yellow	Orange	Red	Violet	Blue	Green	Brown	Black	Proportion of all reactive dyes
Unmetallised azo	97	90	90	63	20	16	57	42	66
Metal complex azo	2	10	9	32	17	5	43	55	15
Anthraquinone	-	-	-	5	34	37	-	3	10
Phthalocyanine	-	-	-	-	27	42	-	-	8
Miscellanous	1	-	1	-	2	-	-	-	1

Table 10. Distribution of chemical classes in reactive dye range (adapted from Cooper, 1995)

In general, colour in wastewater is classified in terms of true/real colour (i.e. colour of turbidity-free water sample), or apparent colour (i.e. colour of non-treated water sample). The most common methods to measure the colour of dye solution or dispersion, and/or wastewater are visual comparison and spectrophotometry, although there is still a lack of an universal method to classify coloured wastewater discharges. By *visual comparison*, colour is quantified by comparing the colour of sample with either known concentrations of coloured standards (normally a platinum-cobalt solution), or properly calibrated colour disks, and is less applicable for highly coloured industrial wastewaters. In the *spectrophotometric method*, colour-measuring protocols differ between the methodologies, of which the most commonly used are Tristimulus Filter Method, American Dye Manufacturer Institute (ADMI) Tristimulus Filter Method, and Spectra record (Table 11).

Spectrophotometric method	Description
Tristimulus	Three tristimulus light filters combined with a specific light source (i.e. tungsten lamp) and a photoelectric cell inside a filter photometer. The output transmittance is converted to trichromatic coefficient and colour characteristic value.
ADMI Tristimulus	The ADMI colour value provides a true watercolour measure, which can be differentiated in 3 (WL) ADMI (i.e. the transmittance is recorded at 590, 540 and 438 nm) or 31 (WL) ADMI (i.e. the transmittance is determined each 10 nm in the range of 400-700 nm).
Spectra record	The complete spectrum is recorded, and the entire spectrum, or a part of it, is used for comparison. A modified method has been suggested in which areas beneath an extinction curve represent the colour intensity, being expressed as space units.

Table 11. Spectrophotometric methods for colour determination in dye solution or dispersion, water and wastewater (adapted from Dos Santos et al., 2004).

2.3 Dye fixation on textile fibres

In general, textile fibres can catched dyes in their structures as a result of van der Waals forces, hydrogen bonds and hydrophobic interactions (physical adsorption). The uptake of the dye in fibres depends on the dye nature and its chemical constituents. But the strongest dye-fibre attachment is a result of a covalent bond with an additional electrostatic interaction where the dye ion and fibre have opposite charges (chemisorption).

In alkaline conditions (i.e. pH 9-12), at high temperatures (30-70°C), and salt concentration from 40-100 g/L, reactive dyes form a reactive vinyl sulfone ($-SO_3-CH=CH_2$) group, which creates a bond with the fibres. However, the vinyl sulfone group undergoes hydrolysis (i.e. a spontaneous reaction that occurs in the presence of water), and because the products do not have any affinity with the fibres, they do not form a covalent bond (Dos Santos et al., 2004). Therefore, a high amount of dye constituents are discharged in the wastewater.
The fixation efficiency varies with the class of azo dye used, which is around 98% for basic dyes and 50% for reactive dyes (Table 12) (Bertea A. & Bertea A.P., 2008; O'Neill et al., 1999). Large amounts of salts such as sodium nitrate, sodium sulphate and sodium chloride are used in the dyebath, as well as sodium hydroxide that is widely applied to increase the pH to the alkaline range. It is estimated that during the mercerising process the weight of these salts can make up 20% of the fibre weight (EPA, 1997).

Dye class	Fibre type	Fixation degree, %	Loss in effluent, %
Acid	Polyamide	80-95	5-20
Basic	Acrilic	95-100	0-5
Direct	Cellulose	70-95	5-30
Disperse	Polyester	90-100	0-10
Metal complex	Wool	90-98	2-10
Reactive	Cellulose	50-90	10-50
Sulphur	Cellulose	60-90	10-40
Dye-stuff	Cellulose	80-95	5-20

Table 12. Fixation degree of different dye classes on textile support (EWA, 2005).

The problem of high coloured effluent or dye-containing effluent has become identified particularly with the dyeing of cellulose fibres (cotton – 50% of the total consumed fibres in the textile industry worldwide), and in particular with the use of reactive dyes (10-50% loss in effluent), direct dyes (5-30% loss in effluent), vat dyes (5-20% loss in effluent), and sulphur dyes (10-40% loss in effluent).
The research for dynamic response and improved dyeing productivity has served to focus the attention of the textile coloration industry on right-first-time production techniques that minimise wastes, make important contribution to reduce colour loads in the effluent by optimisation of processes, minimising of dye wastage, and control automatically the dyeing and printing operation.
After the textile dyeing and finishing processes, a *predicted environmental concentration* of a dye in the receiving water can be estimated based on the following factors: (*i*) daily dye usage; (*ii*) dye fixation degree on the substrate (i.e. textile fibres or fabrics); (*iii*) dye removal degree into the effluent treatment process, and (*iv*) dilution factor in the receiving water.
Some scenario analyses were mentioned the values of dye concentration in some receiving rivers of 5-10 mg/L (average value, 50 days each year) or 1300-1555 mg/L (the worst case, 2 days each year) for batchwise dyeing of cotton with reactive dyes, and of 1.2-3 mg/L (average value, 25 days each year) or 300-364 mg/L (the worst case, 2 days each year) for batchwise dyeing of wool yarn with acid dyes (adapted from Cooper Ed., 1995).
Limits on dye-containing organic loads will become more restrictive in the future, which makes cleaning exhausts an environmental necessity.

3. Textile organic dyes – Environmental problems and polluting effects

The environmental issues associated with residual dye content or residual colour in treated textile effluents are always a concern for each textile operator that directly discharges, both sewage treatment works and commercial textile operations, in terms of respecting the colour and residual dye requirements placed on treated effluent discharge (Zaharia et al., 2011).
Dye concentrations in watercourses higher of 1 mg/L caused by the direct discharges of textile effluents, treated or not, can give rise to public compliant. High concentrations of textile dyes in water bodies stop the reoxygenation capacity of the receiving water and cut-off sunlight, thereby upsetting biological activity in aquatic life and also the photosynthesis process of aquatic plants or algae (Zaharia et al., 2009).
The colour in watercourses is accepted as an aesthetic problem rather than an eco-toxic hazard. Therefore, the public seems to accept blue, green or brown colour of rivers but the 'non-natural' colour as red and purple usually cause most concern.
The polluting effects of dyes against aquatic environment can be also the result of toxic effects due to their long time presence in environment (i.e. half-life time of several years), accumulation in sediments but especially in fishes or other aquatic life forms, decomposition of pollutants in carcinogenic or mutagenic compounds but also low aerobic biodegradability. Due to their synthetic nature and structure mainly aromatic, the most of dyes are non-biodegradable, having carcinogenic action or causing allergies, dermatitis, skin irritation or different tissular changes. Moreover, various azo dyes, mainly aromatic compounds, show both acute and chronic toxicity. High potential health risk is caused by adsorption of azo dyes and their breakdown products (toxic amines) through the gastrointestinal tract, skin, lungs, and also formation of hemoglobin adducts and disturbance of blood formation. LD_{50} values reported for aromatic azo dyes range between 100 and 2000 mg/kg body weight (Börnick & Schmidt, 2006).
Several azo dyes cause damage of DNA that can lead to the genesis of malignant tumors. Electron-donating substituents in ortho and para position can increase the carcinogenic potential. The toxicity diminished essentially with the protonation of aminic groups. Some of the best known azo dyes (e.g. Direct Black 38 azo dye, precursor for benzidine; azodisalicylate, precursor for 4-phenylenediamine) and their breakdown derivatives inducing cancer in humans and animals are benzidine and its derivatives, and also a large number of anilines (e.g. 2-nitroaniline, 4-chloroaniline, 4,4'-dimethylendianiline, 4-phenylenediamine, etc.), nitrosamines, dimethylamines, etc.
The main pollution characteristics and category (pollution risk) of the principal products used in processing of textile materials are summarized in Table 13 (EWA, 2005).
In different toxicological studies are indicated that 98% of dyes has a lethal concentration value (LC_{50}) for fishes higher than 1 mg/L, and 59% have an LC_{50} value higher than 100 mg/L (i.e. 31% of 100-500 mg/L and 28% higher than 500 mg/L).
Other ecotoxicological studies indicated that over 18% of 200 dyes tested in England showed significant inhibition of the respiration rate of the biomass (i.e. wastewater bacteria) from sewage, and these were all basic dyes (adapted from Cooper, 1995).
The bioaccumulation potential of dyes in fish was also an important measure estimating the bioconcentration factor (dye concentration in fish/dye concentration in water). No bioaccumulation is expected for dyes with solubility in water higher than 2000 mg/L.
Dyes are not biodegradable in aerobic wastewater treatment processes and some of them may be intactly adsorbed by the sludge at wastewater biological treatment (i.e. bioelimination by adsorptive removal of dyes).

Products used in textile industry	Pollution characteristics	Pollution category
Alkali, mineral acids, salts, oxidants	Inorganic pollutants, relatively inofensive	1
Sizing agents based on starch, natural oils, fats, waxes, biodegradable surfactants, organic acids, reducing agents	Easy biodegradables; with a moderate - high BOD_5	2
Colorants and optic whitening agents, fibres and impurities of polymeric nature, synthetic polymeric resins, silicones	Difficult to be biodegraded	3
Polyvinyl alcohols, mineral oils, tensides resistant to biodegradation, anionic or non-ionic emolients	Difficult to be biodegraded; moderate BOD_5	4
Formaldehyde or N-methylolic reagents, coloured compounds or accelerators, retarders and cationic emolients, complexants, salts of heavy metals	Can not be removed by conventional biological treatment, low BOD_5	5

Table 13. Characterization of products used in textile industry vs. their polluting effect

Some investigations of adsorption degree onto sludge of some azo dyes indicated typically high levels of adsorption for basic or direct dyes, and high to medium range for disperse dyes, all the others having very low adsorption, which appears to depend on the sulphonation degree or ease of hydrolysis. These tested azo dyes are presented in Table 14.

Group 1		Group 2	Group 3
Dyes unaffected by biological treatment		*Dyes eliminated by adsorption on sludge*	*Dyes with high biodegradability*
CI Acid Yellow 17	CI Acid Red	CI Acid Red 151	CI Acid Orange 7
CI Acid Yellow 23	CI Acid Red 14	CI Acid Blue 113	CI Acid Orange 8
CI Acid Yellow 49	CI Acid Red 18	CI Direct Yellow 28	CI Acid Red 88
CI Acid Yellow 151	CI Acid Red 337	CI Direct Violet 9	
CI Acid Orange 10	CI Acid Black 1		
CI Direct Yellow 4			

Table 14. Fate of water-soluble azo dyes in the activated sludge treatment (Cooper, 1995)

The high degree of sulphonation of azo dyes in Group 1 enhanced their water solubility and limited their ability to be adsorbed on the biomass. The dyes in Group 2 were also highly sulphonated but permitted a relatively good adsorption performance on sludge.

Other information in bioelimination of different reactive dyes mentioned that monoazo dyes are particularly poorly adsorbed, and disazos, anthraquinones, triphendioxazines and phthalocyanines are generally much better adsorbed than monoazos.

It is important to underline that toxic compounds (e.g. toxic aromatic amines, benzidine and its derivatives) can be formed in the environment via transformation of textile dye-precursors (e.g., reduction or hydrolysis of textile azo dyes). The textile dye-precursors are

introduced in water environment due to industrial production of dyes and industrial production of textile fibres, fabrics and clothes via wastewater, sludge, or solid deposits.
The quality problem of dye content and/or colour in the dyehouse effluent discharged in watercourses can be solved by using of a range of advanced decolourisation technologies investigated by the major dye suppliers, textile operators and customers who are under pressure to reduce colour and residual dye levels in their effluents.

4. Textile organic dyes – Separation and elimination procedures from water environment (especially industrial wastewater)

The textile organic dyes must be separated and eliminated (if necessary) from water but especially from industrial wastewaters by effective and viable treatments at sewage treatment works or on site following two different treatment concepts as: (*1*) separation of organic pollutants from water environment, or (2) the partial or complete mineralization or decomposition of organic pollutants. Separation processes are based on fluid mechanics (sedimentation, centrifugation, filtration and flotation) or on synthetic membranes (micro-, ultra- and nanofiltration, reverse osmosis). Additionally, physico-chemical processes (i.e. adsorption, chemical precipitation, coagulation-flocculation, and ionic exchange) can be used to separate dissolved, emulsified and solid-separating compounds from water environment (Anjaneyulu et al., 2005; Babu et al., 2007; Robinson et al., 2001; Suteu et al., 2009a; Suteu et al., 2011a; Zaharia, 2006; Zaharia et al., 2009; Zaharia et al., 2011).
The partial and complete mineralization or decomposition of pollutants can be achieved by biological and chemical processes (biological processes in connection with the activated sludge processes and membrane bioreactors, advanced oxidation with ozone, H_2O_2, UV) (Dos Santos et al., 2004 ; Oztekin et al., 2010 ; Wiesmann et al., 2007 ; Zaharia et al., 2009).
A textile operator will decide on options available to plan forward strategy that will ensure compliance with the environmental regulators' requirements on a progressive basis focused on some options and applied solutions of different separation processes (sedimentation, filtration, membrane separation), and some physico-chemical treatment steps (i.e. adsorption; coagulation-flocculation with inorganic coagulants and organic polymers; chemical oxidation; ozonation; electrochemical process, etc.) integrated into a specific order in the technological process of wastewater treatment for decolourization or large-scale colour and dye removal processes of textile effluents.
To introduce a logical order in the description of treatment methods for textile dye and colour removal, the relationship between pollutant and respective typical treatment technology is taken as reference. The first treatment step for textile wastewater and also rainwater is the separation of suspended solids and immiscible liquids from the main textile effluents by gravity separation (e.g., grit separation, sedimentation including coagulation/flocculation), filtration, membrane filtration (MF, UF), air flotation, and/or other oil/water separation operations.
The following treatment steps are applied to soluble pollutants, when these are transferred into solids (e.g., chemical precipitation, coagulation/flocculation, etc.) or gaseous and soluble compounds with low or high dangerous/toxic effect (e.g., chemical oxidation, ozonation, wet air oxidation, adsorption, ion exchange, stripping, nanofiltration/reverse osmosis). Solid-free wastewater can either be segregated into a biodegradable and a non-biodegradable part, or the contaminants responsible for the non-biodegradable wastewater part that can be decomposed based on physical and/or chemical processes. After an

adequate treatment, the treated wastewater (WW) can either be discharged into a receiving water body, into a subsequent central biological wastewater treatment plant (BWWTP) or a municipal wastewater treatment plant (MWWTP).
Some selected treatment processes for dyes and colour removal of industrial wastewater applied over the time into different textile units are summarized in Table 15. Some of these methods will be further detail and some of authors' results summarized.

Treatment methodology	**Treatment stage**	**Advantages**	**Limitations**
		Physico-chemical treatments	
Precipitation, coagulation-flocculation	Pre/main treatment	Short detention time and low capital costs. Relatively good removal efficiencies.	Agglomerates separation and treatment. Selected operating condition.
Electrokinetic coagulation	Pre/main treatment	Economically feasible	High sludge production
Fenton process	Pre/main treatment	Effective for both soluble and insoluble coloured contaminants. No alternation in volume.	Sludge generation; problem with sludge disposal. Prohibitively expensive.
Ozonation	Main treatment	Effective for azo dye removal. Applied in gaseous state: no alteration of volume	Not suitable for dispersed dyes. Releases aromatic dyes. Short half-life of ozone (20 min)
Oxidation with NaOCl	Post treatment	Low temperature requirement. Initiates and accelerates azo-bond cleavage	Cost intensive process. Release of aromatic amines
Adsorption with solid adsorbents such as:			
Activated carbon	Pre/post treatment	Economically attractive. Good removal efficiency of wide variety of dyes.	Very expensive; cost intensive regeneration process
Peat	Pre treatment	Effective adsorbent due to cellular structure. No activation required.	Surface area is lower than activated carbon
Coal ashes	Pre treatment	Economically attractive. Good removal efficiency.	Larger contact times and huge quantities are required. Specific surface area for adsorption are lower than activated carbon
Wood chips/ Wood sawdust	Pre treatment	Effective adsorbent due to cellular structure. Economically attractive. Good adsorption capacity for acid dyes	Long retention times and huge quantities are required.

Silica gels	Pre treatment	Effective for basic dyes	Side reactions prevent commercial application
Irradiation	Post treatment	Effective oxidation at lab scale	Requires a lot of dissolved oxygen (O_2)
Photochemical process	Post treatment	No sludge production	Formation of by-products
Electrochemical oxidation	Pre treatment	No additional chemicals required and the end products are non-dangerous/hazardous.	Cost intensive process; mainly high cost of electricity
Ion exchange	Main treatment	Regeneration with low loss of adsorbents	Specific application; not effective for all dyes
Biological treatments			
Aerobic process	Post treatment	Partial or complete decolourization for all classes of dyes	Expensive treatment
Anaerobic process	Main treatment	Resistant to wide variety of complex coloured compounds. Bio gas produced is used for stream generation.	Longer acclimatization phase
Single cell (Fungal, Algal & Bacterial)	Post treatment	Good removal efficiency for low volumes and concentrations. Very effective for specific colour removal.	Culture maintenance is cost intensive. Cannot cope up with large volumes of WW.
Emerging treatments			
Other advanced oxidation process	Main treatment	Complete mineralization ensured. Growing number of commercial applications. Effective pre-treatment methodology in integrated systems and enhances biodegradability.	Cost intensive process
Membrane filtration	Main treatment	Removes all dye types; recovery and reuse of chemicals and water.	High running cost. Concentrated sludge production. Dissolved solids are not separated in this process
Photocatalysis	Post treatment	Process carried out at ambient conditions. Inputs are no toxic and inexpensive. Complete mineralization with shorter detention times.	Effective for small amount of coloured compounds. Expensive process.
Sonication	Pre treatment	Simplicity in use. Very effective in integrated systems.	Relatively new method and awaiting full scale application.
Enzymatic treatment	Post treatment	Effective for specifically selected compounds.	Enzyme isolation and purification is tedious.

		Unaffected by shock loadings and shorter contact times required.	Efficiency curtailed due to the presence of interferences.
Redox mediators	Pre/ supportive treatment	Easily available and enhances the process by increasing electron transfer efficiency	Concentration of redox mediator may give antagonistic effect. Also depends on biological activity of the system.
Engineered wetland systems	Pre/post treatment	Cost effective technology and can be operated with huge volumes of wastewater	High initial installation cost. Requires expertise and managing during monsoon becomes difficult

Table 15. Various current and emerging dye separation and elimination treatments applied for textile effluents with their principal advantages and limitations (adapted from Anjaneyulu et al., 2005; Babu et al., 2007; Robinson et al., 2001)

4.1 Physical treatments

4.1.1 Adsorption

One of the most effective and proven treatment with potential application in textile wastewater treatment is adsorption. This process consists in the transfer of soluble organic dyes (solutes) from wastewater to the surface of solid, highly porous, particles (the adsorbent). The adsorbent has a finite capacity for each compound to be removed, and when is 'spent' must be replaced by fresh material (the 'spent' adsorbent must be either regenerated or incinerated).

Adsorption is an economically feasible process for dyes removal and/or decolourization of textile effluents being the result of two mechanisms: adsorption and ion exchange. The principal influencing factors in dye adsorption are: dye/adsorbent interaction, adsorbent surface area, particle size, temperature, pH, and contact time. Adsorbents which contain amino nitrogen tend to have a significantly larger adsorption capacity in acid dyes.

The most used adsorbent is activated carbon, and also other commercial inorganic adsorbents. Some 'low cost' adsorbents of industrial or agricultural wastes (i.e. peat, coal ashes, refused derived coal fuel, clay, bentonite and modified bentonite, red soil, bauxite, ebark, rice husk, tree barks, neem leaf powder, wood chips, ground nut shell powder, rice hulls, bagasse pith, wood sawdust, grounded sunflower seed shells, other ligno-cellulosic wastes, etc.) are also used for removal of dye and organic coloured matter from textile effluents (i.e. a removal of 40-90% basic dyes and 40% direct dyes, with maximum adsorption capacities for basic dyes of 338 mg/g) (Anjaneyulu et al., 2005; Bhattacharyya & Sarma, 2003; Gupta et al., 1992; Nigam et al., 2000; Ozcan et al., 2004; Robinson et al., 2001; Suteu & Zaharia, 2008; Suteu et al., 2009b; Suteu et al., 2011a,b; Zaharia et al., 2011). The use of these materials is advantageous mainly due to their widespread availability and cheapness. Sometimes the regeneration is not necessary and the 'spent' material is conventionally burnt although there is potential for solid state fermentation (SSF) for protein enrichment. The use of 'low cost' adsorbents for textile dye removal is profitable but requires huge quantity of adsorbents, being lower efficient than activated carbon. Some authors' results for dye adsorption are summarized in Table 16.

Adsorptive material	**Operating conditions / Optimal doses** (Adsorption + Sedimentation/Filtration)	**Adsorption efficiency for some tested textile dyes, %**
Peat	pH= 2 (BRed); 5.7 (MB, RhB); C_{dye}= (20-300) mg/L BRed; (19-134) mg/L MB; (28-155) mg/L RhB; $C_{adsorbant}$= 12 g/L	(67.50 - 85.60) BRed (65.70- 89.30) MB (98.30- 99.00 RhB
Wood sawdust	pH= 2 (BRed); 5,7 (MB; CV; RhB); 1 (RO); C_{dye}= (20-150) mg/L BRed; (6-40) mg/L MB; (8-50) mg/L CV; (9-58) mg/L RhB; (24-160) mg/L (RO); $C_{adsorbant}$= 20 g/L (BRed); 4 g/L (MB, CV, RhB); 8 g/L (RO)	(63.90- 83.00) BRed (69.40 - 91.80) MB (66.00 - 80.00) CV (52.20 - 72.00) RhB (38.30 - 50.80) RO
Sunflower seed shell	pH= 1 (RO); 6 (MB); C_{dye}= (25-160) mg/L RO; (25-280) mg/L MB; $C_{adsorbant}$= 8 g/L (RO), 4 g/L (MB)	(80.60 - 82.60) RO (84.40 - 93.00) MB
Corn cobs	pH= 1 (RO) or 6 (MB); C_{dye}= (24-160) mg/L RO; (25-280) mg/L MB; $C_{adsorbant}$= 8 g/L (RO), 4 g/L (MB)	(83.00 - 86.20) RO (87.20 - 98.10) MB
Lignine	pH= 1.5 (BRed); 1 (RO); 6 (MB); C_{dye}= (50-300) mg/L BRed; (25.60-280) mg/L MB; (30-150) mg/L RO; $C_{adsorbant}$= 14 g/L (BRed); 4 g/L (MB); 12 g/L (RO)	(49.50-59.50) BRed (43.50-60.70) RO (55.30-64.30) MB
Cellolignine	pH=6; C_{dye}= (25.6-281.6) mg/L MB; $C_{adsorbant}$= 4 g/L	(96.60 - 98.40) MB
Abbreviations: BRed - Brilliant Red HE-3B (Reactive Red 120)/CI 25810; RO - Reactive Orange 16/CI 17757; MB - Methylene Blue (Basic Blue 9)/CI 52015; RhB - Rhodamine B (Basic Violet 10)/CI 45170; CV - Crystal Violet (Basic Violet 3)/CI 42555		

Table 16. Dye adsorption performance onto some natural adsorptive materials (adapted from Suteu et al., 2009b; Suteu et al., 2011a,b).

Adsorption with activated carbon. Activated carbon have been engineered for optimal adsorption of the contaminants found in dyehouse effluents: large, negatively charged or polar molecules of dyes. Powdered or granular activated carbon (specific surface area of 500-1500 m^2/g; pore volume of 0.3-1 cm^3/g; bulk density of 300-550 g/L) has a reasonably good colour removal capacity when is introduced in a separate filtration step. The activated carbon is used as granulate (GAC) in columns or as powder (PAC) in batchwise treatment into a specific treatment tank or basin. High removal rates are obtained for cationic mordant and acid dyes (Anjaneyulu et al., 2005), and a slightly lesser extent (moderate) for dispersed, direct, vat, pigment and reactive dyes (Cooper, 1995; Nigam et al., 2000) with consumable doses of 0.5-1.0 kg adsorbent/m^3 wastewater (i.e. dye removal of 60-90%).

Most recent studies mentioned that an effective irreversible adsorption of dye molecules onto the adsorbent particles takes place via a combination of physical adsorption of dye onto adsorbent surfaces within the microporous structure of the particles, enhanced by an ion-exchange process wherein the interlayer anions of the adsorbent are displaced by the dye molecules, and also inter-particle diffusion processes. Removal of pollutants can take place at any pH between 2 and 11, and at any temperature between 0 and 100°C (effluent temperature as received, generally 30-40°C).

The adsorption on activated carbon without pretreatment is imposible because the suspended solids rapidly clog the filter, and may be feasible in combination with flocculation-sedimentation treatment or a biological treatment (Masui et al., 2005; Ramesh Babu et al., 2007). The main important disadvantage of this process is attributed to the high cost of activated carbon. Performance is dependent on the type of activated carbon used and wastewater characteristics, and can be well suited for one particular wastewater system and ineffective in another. The activated carbon has to be reactivated otherwise disposal of the concentrates has to be considered (reactivation results in 10-15% loss of the adsorbent).

4.1.2 Irradiation

The irradiation treatment is a simple and efficient procedure for eliminating a wide variety of organic contaminants, and as well disinfecting harmful microorganism using gamma rays or electron beams (e.g., source for irradiation can be a monochromatic UV lamps working under 253.7 nm). A high quantity of dissolved oxygen is required for an organic dye to be effectively broken down by irradiation. The dissolved oxygen is consumed very rapidly and so a constant and adequate supply is required. Irradiation treatment of a secondary effluent from sewage treatment plant reduced COD, TOC and colour up to 64%, 34% and 88% respectively, at a dose of 15 K Gy gamma-rays (Borrely et al., 1998). The efficiency of irradiation treatment increases when is used catalyst as titanium dioxide (Krapfenbauer et al., 1999). A lot of data are reported with the practical results obtained at the simple exposure of different dye solutions or dispersions and dye-containing textile wastewaters to sunlight for a period of a half, one or two months (direct photolysis with natural sunlight into open basins). All these reports indicated high removals of colour (>84%), dye destruction by photooxidation following first order kinetics at treatment of some vat dye effluents. But the direct photolysis of textile organic dye in the natural aquatic environment has proven difficult due to strong dependence of the decay rates on dye reactivity and photosensitivity. Most of all commercial dyes are usually designed to be light resistant. Therefore, the recent researches have been directed towards investigation of organic dye photodegradation by sensitizers or catalysts in aqueous/dispersion systems by UV irradiation. Moreover, there are reported high removal of indigo-colour when is initiated a laser fading process for indigo coloured denim textile mainly based on basic interaction of laser beam with indigo-coloured textile (Dascalu et al., 2000).

4.1.3 Membrane processes

The increasing of water cost and necessity of reduction of water consumption implies treatment process which is integrated with in-plant water circuits rather than subsequent treatment (Baban et al., 2010; Machenbach, 1998). From this point of view, membrane filtration offers potential application in combination with other textile effluent treatments.

Membrane processes for wastewater treatment are pressure-driven processes, capable to clarify, concentrate, and most important, separate dye discontinuously from effluent (Xu & Lebran, 1991). These are new technologies, which can restrict organic contaminants and microorganisms presented in wastewater (i.e. color removal, BOD reduction, salt reduction, Polyvinyl Acetate (PVA) recovery, and latex recovery). The common membrane filtration types are: Micro-Filtration (MF), Ultra-Filtration (UF), Nano-Filtration (NF), and Reverse Osmosis (RO). The choice of the membrane process must be guided by the required quality of the final effluent.

Micro-filtration is mainly used for treatment of dye baths containing pigment dyes as well as for subsequent rinsing baths (Ramesh Babu et al., 2007). Chemicals that can not be filtrated by microfiltration will remain in the dye bath. Microfiltration can be used as a pretreatment for nanofiltration or reverse osmosis (Ghayeni et al., 1998), and also to separate suspended solids, colloids from effluents or macromolecules with pores of 0.1 to 1 micron.
MF performance is typically of >90% for turbidity or silt density index. Microfiltration membranes are made of specific polymers such as Poly (Ether Sulfone), Poly (Vinylidiene Fluoride), Poly (Sulfone), Poly (Vinylidene Difluoride), Polycarbonate, Polypropylene, Poly Tetrafluoroethylene (PTFE), etc. Ceramic, glass, carbon, zirconia coated carbon, alumina and sintered metal membranes have been employed where extraordinary chemical resistance or where high temperature operation is necessary. MF and UF operate at 20 to 100 psi transmembrane pressures (Ptm) (low pressure membrane procees) and velocities of 20 to 100 cm/s (Naveed et al., 2006).
Ultra-filtration is used to separate macromolecules and particles, but the elimination of polluting substances such as dyes is never complete (only 31-76% dye removal). The quality of treated wastewater does not permit its reuse for sensitive processes, such as textile dyeing (Ramesh Babu et al., 2007) but permit recycling of 40% treated wastewater in stages in which salinity is not a problem, such as rinsing, washing, etc. Ultrafiltration can only be used as a pretreatment for reverse osmosis (Ciardelli & Ranieri, 2001) or in combination with a biological reactor (Mignani et al., 1999) or to remove metal hydroxides (reducing the heavy metal content to 1 ppm or less) (Naveed et al., 2006). UF membranes are made of polymeric materials (i.e. polysulfone, polypropylene, nylon-6, polytetrafluoroethylene (PTFE), polyvinyl chlorides (PVC), acrylic copolymer etc.
Nano-Filtration was used for the treatment of coloured effluents from the textile industry, mainly in a combination of adsorption (for decreasing of concentration polarization during the filtration process) and nanofiltration (NF modules are extremely sensitive to fouling by colloidal material and macromolecules). NF membranes are generally made of cellulose acetate and aromatic polyamides, and retain low-molecular weight organic compounds, divalent ions, large monovalent ions, hydrolized reactive dyes, and dyeing auxiliaries. Inorganic materials, such as ceramics, carbon based membranes, zirconia, are also used in manufacturing NF and RO membranes. Typical NF flux rates are 5 to 30 GFD (Gross Flow per Day) (Naveed et al., 2006). A performance of above 70% colour removal for a NF plant was reported working at 8 bar/18°C, with four polyethersulphonate membranes with molecular weight cut offs of 40, 10, 5 and 3 kda for three different effluents coming from dyeing cycle of textile industry (Alves & Pinho, 2000). Values of colour removal higher than 90% were reported for single NF process, and also combination MF and NF, in the case of different effluents from textile fabrics processing. Harmful effects of high concentrations of dye and salts in the dye house effluents were frequently reported (i.e. concentration of dye > 1.5 g/L, and of mineral salts >20 g/L) (Tang & Chen, 2002). An important problem is the acculumation of dissolved solids, which makes discharge of treated effluents in watercourses almost impossible. NF treatment can be an alternative fairly satisfactory for textile effluent decolourization.
Reverse Osmosis is used to remove in a single step most types of ionic compounds, hydrolized reactive dyes, chemical auxiliaries, and produce a high quality of permeate (Ramesh Babu et al., 2007). Like NF, RO is very sensitive to fouling and the influent must be carefully pretreated. RO membranes are generally made of cellulose acetate and aromatic polyamides but also of inorganic materials. The Ptm in RO is typically 500 to 1000 psi, with

cross flows of 20 to 100 cm/s. The range of typical RO fluxes is 5 to 15 GFD (Naveed et al., 2006). In combination with physio-chemical treatment, the membrane processes has advantages over the other conventional treatments, such as the ability to recover materials with valuable recyclable water, reducting fresh water consumption and wastewater treatment costs, small disposal volumes which minimizes waste disposal costs, reduction of regulatory pressure and fine improved heat recovery systems. Membrane processes have many cost-effective applications in textile industry.

4.2 Chemical treatment

4.2.1 Oxidative processes

Chemical oxidation represents the conversion or transformation of pollutants by chemical oxidation agents other than oxygen/air or bacteria to similar but less harmful or hazardous compounds and/or to short-chained and easily biodegradable organic components (aromatic rings cleavage of dye molecules).

The modern textile dyes are resistant to mild oxidation conditions such those existing in biological treatment systems. Therefore, efficient dye and colour removal must be accomplished by more powerful oxidising agents such as chlorines, ozone, Fenton reagents, UV/peroxide, UV/ozone, or other oxidising procedures or combinations.

Oxidative processes with hydrogen peroxide. The oxidation processes with hydrogen peroxide (H_2O_2) (oxidation potential, E°= 1.80 V at pH 0, and E°= 0.87 V at pH 14) can be explored as wastewater treatment alternatives in two systems: (1) homogenous systems based on the use of visible or ultraviolet light, soluble catalysts (Fenton reagents) and other chemical activators (e.g. ozone, peroxidase etc.) and (2) heterogenous systems based on the use of semiconductors, zeolites, clays with or without ultraviolet light, such as TiO_2, stable modified zeolites with iron and aluminium (i.e. FeY_5, $FeY_{11.5}$ etc.) (difficulty encountered in the separation of the solid photocatalysts at the end of the process) (Neamtu et al., 2004; Zaharia et al., 2009).

Fenton reagent is usually hydrogen peroxide (H_2O_2) that is activated by some iron salts (i.e. Fe^{2+} salts) (without UV irradiation) to form hydroxyl radicals (HO·) which are strong oxidants (oxidation potential, E°= 3.06 V) than H_2O_2 and ozone. The Fenton oxidation reactions are detailed in other chapters of this book, and the treatment efficiency depends mainly of effluent characteristics, and operating parameters (e.g., colour removal of 31.10 or 56.20%, at a pH of 4.00, for Fenton oxidation of textile Remazol Arancio 3R, Remazol Rose RB dye-containing effluents working with 0.18-0.35 M H_2O_2 and 1.45 mM Fe^{2+}, after 30 or 120 min) (Zaharia et al., 2011).

Heterogenous catalytic oxidation with 20 mM H_2O_2 and $FeY_{11.5}$ (1 g/L) of Procion Marine H-EXL dye-containing effluents lead to colour removal of 53-83%, COD removal of 68-76% and TOC removal of 32-37% at pH=3-5, after 10 min of oxidation (Neamtu et al., 2004). Working with FeY_5 (1 g/L) and 20 mM H_2O_2, at pH=3 and 5, for the same textile effluent the treatment *efficiency*, was of 95 and 35% for colour and COD removal after 10 min of oxidation, and 97% for colour after 60 min of oxidation; COD removal (60 min) of 64.20% (Zaharia, 2006).

When small quantities of wastewater are involved or when there is no biotreatment available at the textile site, chemical oxidation might be recommendable treatment option instead of installing a central biological WWT plant. Advantages of this oxidative treatment include reduction of effluent COD, colour and toxicity, and also the possibility to be used to remove both soluble and insoluble dyes (i.e. disperse dyes). Complete decolourization was obtained after the complete Fenton reagent stage (generally 24 hours).

Ozonation process. Ozone is a powerful oxidising agent (oxidation potential, E°= 2.07 V) capable of cleavage the aromatic rings of some textile dyes and descomposition of other organic pollutants from industrial effluents. The ozone decomposes the organic dyes with conjugated double bonds forming smaller molecules with increased carcinogenic or toxic properties, and so ozonation may be used alongside a physical method to prevent this (i.e. irradiation, membrane separation, adsorption, etc). Ozone can react directly or indirectly with dye molecules. In the direct pathway, the ozone molecule is itself the electron acceptor, and hydroxide ions (i.e. pH > 7-8) catalyze the auto decomposition of ozone to hydroxyl radicals (·OH) in aqueous effluents (very strong and non-selective oxidants) which react with organic and inorganic chemicals. At low pH ozone efficiently reacts with unsaturated chromophoric bonds of a dye molecule via direct reactions (Adams & Gorg, 2002).
The main advantage is that ozone can be applied in its gaseous state and therefore does not increase the volume of wastewater and sludge. A disadvantage of ozonation is its short half-life, tipically being 20 min, the destabilisation by the presence of salts, pH, and temperature, and the additional costs for the installation of ozonation plant. The improvement of ozonation preformance is obtained in combination with irradiation (Surpateanu & Zaharia, 2004a; Zaharia et al., 2009) or with a membrane filtration technique (Lopez et al., 1999). Treatment of dye-containing wastewater with ozone followed by chemical coagulation using $Ca(OH)_2$ indicated 62% colour removal after ozonation (Sarasa et al., 1998).
Oxidation process with sodium hypochlorite. This treatment implies the attack at the amino group of the dye molecule by Cl^+, initiating and accelerating azo-bond cleavage. The increasing of chlorine concentration favors the dye removal and decolourization process, and also the decreasing of pH. The dye containing amino or substituted amino groups on the naphthalene ring (i.e. dyes derived from amino-naphtol- and naphtylamino-sulphonic acids) are most susceptible for chlorine decolourization (Omura, 1994). This treatment is unsuitable for disperse dyes, and is becoming less frequent due to the negative effects at releasing into watercourses of aromatic amines or otherwise toxic molecules. Moreover, althought about 40% of the pigments used worldwise contain chlorine this corresponds to only less than 0.02% of the total chlorine production (Slokar & Le Marechal, 1997).
Photochemical oxidation process. The UV treatment in the presence of H_2O_2 can descomposed dye molecules to low weight organic molecules, or even to CO_2, H_2O, other inorganic oxides, hydrides, etc. There can be also produced additional by-products such as halides, metals, inorganic acids, organic aldehydes and organic acids depending on initial materials and the extent of decolourisation treatment (Yang et al., 1998). The dye decomposition is initiated by the generated hydroxyl radicals ($H_2O_2 + h\upsilon \rightarrow 2HO\cdot$) and hydroperoxide radicals ($H_2O_2 + HO\cdot \rightarrow HO_2\cdot + H_2O$).
The treatment may be set-up in a batch or continuous column unit, and is influenced by the intensity of the UV radiation, pH, dye structure and the dye bath composition (Slokar & Le Marechal, 1997). The performance of photooxidation treatment in the presence of hydrogen peroxide are high (i.e. >60-90% for colour removal, working with 400-500 mg/L H_2O_2 at pH 3-7, for Red M5B, H-acid and Blue MR dye-containing effluents) (Anjaneyulu et al., 2005) or 81-94% dye removal after 60 min, working with 88 mM H_2O_2 at pH of 4-6, for Acid Red G dye-containing effluent (Surpateanu & Zaharia, 2004b; Zaharia et al., 2009).
Electrochemical oxidation process. As an advanced process, the electrochemical treatment of dye-containing effluents is a potentially powerful method of pollution control, offering high removal efficiencies (Anjaneyulu et al., 2005) especially for acid dyes as well as

disperse and metal complex dyes. The main advantages of this treatment are considered the requirement of simple equipment and operation, low temperature in comparison with other non-electrochemical treatments, no requirement of any additional chemicals, easy control but crucial for pH, the electrochemical reactors (with electrolytic cells) are compact, and prevent the production of unwated by-products. The principal oxidising agent in electrochemical process is hypochlorite ion or hypochlorous acid produced from naturally occuring chloride ions. Hydroxyl radical and other reactive species also participate in electrochemical oxidation of organics (Kim et al., 2002) that can be achieved directly or indirectly at the anode. The breakdown compounds are generally not hazardous being discharged into watercourses without important environmental and health risks.

The electrochemical oxidation is considered an efficient and economic treatment of recycling textile wastewater for the dyeing stage. The environmental advantage mainly achieved is the minimization of all emissions: emission of gases, solid waste, and liquid effluent. Other important advantage is its capacity of adaptation to different volumes and pollution loads. The main disadvantage is the generation of metallic hydroxide sludge (from the metallic electrodes in the cell), that limits its use (Ramesh Babu et al., 2007).

Some studies reported colour removals of about 100% for dyeing wastewater within only 6 min of electrolysis (Vlyssides et al., 2000) (e.g., complete decolourization of textile effluent containing blue-26 anthraquinone dye by electrochemical oxidation with lead dioxide coated anode - Titanium Substract Insoluble Anode (TSIA), at neutral pH, in the presence of sodium chloride, current density of 4.5 A/dm^2, electrolysis time of 220 min, or maximum 95.2% colour and 72.5% COD removal of textile azo dye-containing effluent in a flow reactor working at rate of 5 mL/min and current density of 29.9 mA/cm^2) (Anjaneyulu et al., 2005).

4.2.2 Coagulation-flocculation and precipitation

It is clearly known that the coloured colloid particles from textile effluents cannot be separated by simple gravitational means, and some chemicals (e.g., ferrous sulphate, ferric sulphate, ferric chloride, lime, polyaluminium chloride, polyaluminium sulphate, cationic organic polymers, etc.) are added to cause the solids to settle. These chemicals cause destabilisation of colloidal and small suspended particles (e.g. dyes, clay, heavy metals, organic solids, oil in wastewater) and emulsions entrapping solids (coagulation) and/or the agglomeration of these particles to flocs large enough to settle (flocculation) or highly improve further filtration (Zaharia et al., 2006; Zaharia et al., 2007). In the case of flocculation, anionic and non-ionic polymers are also used.

The mechanism by which synthetic organic polymer removes dissolved residual dyes from effluents is best described in terms of the electrostatic attraction between the oppositely charged soluble dye and polymer molecules. Many of the most problematic dye types, such as reactive dyes, carry a residual negative charge in their hydrolysed dissolved form, and so positively charged groups on the polymers provide the neccesary counter for the interaction and subsequent precipitation to occur. The immediate result of this coprecipitation is the almost instantaneous production of very small coloured particles, having little strength and breaking down at any significant disturbances. The agglomeration of the coloured precipitates by using appropriate high polyelectrolyte flocculants produces stable flocs (Zaharia et al., 2007, 2011). The main disadvantages of this treatment are the process control that is a little difficult, the potential affection of precipitation rate and floc size by impurities such as non-ionic detergents remaining in the effluent, and the sludge production which has to be settled, dewatered and pressed into a cake for subsequent landfilling tipping.

There are reported very effective chemical coagulation-flocculation (C-F) and precipitation of textile wastewater which reduced the load on the biological treatment, working with polyaluminium chloride along with an organic polymer (Lin & Chen, 1997) or ferrous/ferric chloride and a commercial organic coagulant aid at pH of 6.7-8.3 (colour removal > 80%) (Venkat Mohan et al., 1999) or alum at pH=8.2 (54-81% colour removal) with addition of bentonite (3 g/L) for Remazol Violet dye-containing effluent (Sanghi et al., 2001). Other efficient textile treatments mentioned by different textile operators consist in coagulation-flocculation followed by membrane technology (especially for recycling textile effluents).
Some of authors' results in different (C-F) treatments are summarized in Table 17.

Process	**Coagulant/ Flocculant**	**Wastewater characteristics or dye type / Results**
Coagulation-flocculation/ sedimentation/ filtration	Ferrous sulphate (5 mg/L) + Ponilit GT-2 anionic polyelectrolyte (15 mg/L) + bentonite (3 g/L)	*Wastewater characteristics*: pH=6.5-7; TSS=250-1000 mg/L; colour=650 UH; COD_{Cr}=152.7-272 mg O_2/L *Process efficiency*: turbidity removal=70.31-91.34%; colour removal=70.20-90.50% and COD_{Cr} removal=34.90-45.20%
Precipitation/ flocculation/ flotation with dissolved air	NaOH + Na_2CO_3 + $Ca(OH)_2$, Ponilit GT-2 anionic polyelectrolyte	*Wastewater characteristics*: pH=7.0-9.5; total metallic ions= 48.30 mg/L; extractable substances in organic solvents= 980 mg/L *Process efficiency*: Metallic ions removal=78.67-92.33%
Coagulation-flocculation/ sedimentation/ filtration	Ferric sulfate (2-5 mg/L) +Prodefloc CRC 301 (0.25-1.5 mg/L) cationic polyelectrolyte	*Wastewater characteristics*: pH=6.98; T=20°C, Turbidity=556 FTU; COD_{Cr}=152.60 mg O_2/L; colour=1320 UH *Process efficiency*: maximal turbidity removal=95.87% and colour removal= 93.90-98.10%

Table 17. Some applications of coagulation- flocculation in wastewater treatment (adapted from Zaharia, 2006; Zaharia et al., 2006, 2007).

4.2.3 Electrocoagulation

An advanced electrochemical treatment for dye and colour removal is electrocoagulation (EC) that has as main goal to form flocs of metal hydroxides within the effluent to be cleaned by electro-dissolution of soluble anodes. EC involves important processes as electrolytic reactions at electrodes, formation of coagulants in aqueous effluent and adsorption of soluble or colloidal pollutants on coagulants, and removal by sedimentation and flotation. This treatment is efficient even at high pH for colour and COD removals being strongly influenced by the current density and duration of reaction. The EC treatment was applied with high efficiency for textile Orange II and Acid red 14 dye-containing effluents (i.e. > 98% colour removal) (Daneshvar et al., 2003) or Yellow 86 dye-containing wastewater (i.e. turbidity, COD, extractible substances, and dye removal of 87.20%, 49.89%, 94.67%, and 74.20%, after 30 min of operation, current intensity of 1 A, with monopolar electrodes) (Zaharia et al, 2005) when iron is used as sacrificial anode. In general, decolorization performance in EC treatment is between 90-95%, and COD removal between 30-36% under optimal conditions (Ramesh Babu et al., 2007).

4.2.4 Ionic exchange

The ion exchange process has not be widely used for treatment of dye-containing effluents, mainly because of the general opinion that ion exchangers cannot accomodate a wide range of dyes (Slokar & Le Marechal, 1997).

The ionic exchange occurs mainly based on the interaction of ionic species from wastewater with an adsorptive solid material, being distinguished from the conventional adsorption by nature and morphology of adsorptive material or the inorganic structure containing functional groups capable of ionic exchange (Macoveanu et al., 2002). The mechanisms of ionic exchange process are well known, and two principal aspects must be mentioned: (1) ionic exchange can be modeling as well as adsorption onto activated coal; (2) the ion exchangers can be regenerated without modifying the equilibrium condition (e.g., by passing of a salt solution containing original active groups under ion exchanger layer). In the case of wastewater treatment, the effluent is passed over the ion exchanger resin until the available exchange sites are saturated (both cationic and anionic dyes are removed).

The ionic exchange is a reversible process, and the regenerated ion exchanger can be reused. The essential characteristic of ionic exchange that makes distinction of adsorption is the fact that the replace of ions takes place in stoechiometric proportion (Macoveanu et al., 2002).

The effluent treatment by ionic exchange process contributes to the diminishing of energetic consumption and recovery of valuable components under diverse forms, simultaneously with the wastewater treatment. In practice, the ion exchangers are used in combination with other wastewater treatments. The main advantages of ion exchange are removal of soluble dyes, no loss of adsorbent at regeneration, and reclamation of solvent after use. The important disadvantages of this process is the cost, organic solvents are expensive, and ion exchange treatment is not efficient for disperse dyes (Robinson et al., 2001). Our results in batchwise treatment of Brilliant Red HE-3B dye-containing effluents (0.05 - 0.3 mg/mL) using anionic Purolite A-400 and Purolite A-500 indicated dye removal of 48-89% working in the optimal conditions, or dye removal of 56-78% for Crystal Violet (Basic Violet 3) using Purolite C-100 or dye removal between 78-89% for Reactive Blue M-EB using ion exchange celluloses (Suteu et al., 2002).

4.3 Biological treatments

Biological treatments are considered reproduction, artificially or otherwise, of self-purification phenomena existing in natural environment. There are different biological treatments, performed in aerobic or anaerobic or combined anaerobic/aerobic conditions. The processing, quality, adaptability of microorganisms, and the reactor type are decisive parameters for removal efficiency (Börnick & Schmidt, 2006).

Biological treatment process for decolorization of industrial effluents is ambiguous, different and divergent (Anjaneyulu et al., 2005). Previous subchapters indicate that dyes themselves are not biologically degradable since microorganisms do not use the coloured constituents as a source of food. The most currently used biodegradation involve aerobic micro-organisms, which utilize molecular oxygen as reducing equivalent acceptor during the respiration process. But biodegradation in anaerobic environment conditions (anoxic and hypoxic environments) also occurs, and survival of microorganisms is possible by using sulphates, nitrates and carbon dioxide as electron acceptors (Birch et al., 1989).

Research data indicates that certain dyes are susceptible to anoxic/anaerobic decolourization, and also that an anaerobic step followed by an aerobic step may represent a

significant advancement in biological decolourization treatment in future (Ong et al., 2005). The treatment plant that receives dye-containing effluents has high potential to form toxic biodegradation products such as toxic amines, benzidine and its derivates, etc. To avoid that risk, anaerobic/aerobic sequential reactor systems seem to be an efficient procedure (i.e. efficient colour removal takes place during the anaerobic treatment, and high reduction of aromatic amines and other organic compounds occurs during the subsequent aerobic treatment). Under aerobic conditions, most of the azo dye metabolites are quickly degraded by oxidation of the substituents or of the side branches. However, some of them are still rather recalcitrant. Successful removal of poorly degradable amines was often achieved by adaptation of microorganisms (i.e. acclimatization of biological sludge to nitroaniline-containing wastewaters; after gradual adaptation, the microorganisms are able to eliminate 3- and 4-nitroaniline simultaneously) (Börnick & Schmidt, 2006). Difficult to be biodegraded are also the aromatic amines containing sulfo substituents in metha position (e.g., 3-aminobenzenesulfonate) that can be treated with good efficiencies with flow-through bioreactors within 28 days. Contrarily, some experimental results found anaerobic mineralization to be efficient for aromatic amines.
The main advantage of biological treatment in comparison with certain physico-chemical treatments is that over 70% of organic matter expressed by COD_{Cr} may be converted to biosolids (Anjaneyulu et al., 2005).

4.3.1 Aerobic biological treatment

Biological treatment with 'activated sludge' was the most used in large scale textile effluent treatment, and the trickling filter or biological aerated filter (BAF) is an alternative, permitting a 34-44% dye-colour removal for different high dyeing loads of industrial effluents. The main microorganisms contributing to biodegradation of organic compounds are bacteria (e.g. *Bacillus subtilis, Aeromonas Hydrophilia, Bacillus cetreus, Klebsiella pneunomoniae, Acetobacter liquefaciens, Pseudomonas* species, *Pagmentiphaga kullae, Sphingomonas*, etc.), fungi (e.g., white-rot fungi: *Phanerochaete chrysosporium, Hirschioporus larincinus, Inonotus hispidus, Phlebia tremellosa, Coriolus versicolor*, etc.), algae (e.g. *Chlorella* and *oscillotoria* species) etc. Moreover, some bacteria, white-rot fungi, mixed microbial cultures from a wide variety of habitats are found to be able to degrade dyes using enzimes, such as lignin peroxidases (LiP), manganese dependent peroxidases (MnP), H_2O_2-producing enzime such as glucose-1-oxidase and glucose-2-oxidase, along with laccase, and a phenoloxidase. Biological aerated filters involve the growth of an organism on inert media that are held stationary during normal operation and exposed to aeration. In aerobic conditions, the mono- and dioxygenase enzymes catalyse the incorporation of dissolved oxygen into the aromatic ring of organic compounds prior to ring fission. Although azo dyes are aromatic compounds, their nitro and sulfonic groups are quite recalcitrant to aerobic bacterial degradation (Dos Santos et al., 2004). However, in the presence of specific oxygen-catalysed enzymes called azo reductases, some aerobic bacteria are able to reduce azo compounds and produce aromatic amines (Stolz, 2001). The batch experiments with aerobic activated sludge confirmed the biodegradability of sulphonated azo dyes. Only aerobic degradation of the azo dyes is possible by azo reduction (i.e. high colour removal (>90%) of Red RBN azo dye-containing effluents (3000 mg/L) working with *Aeromonas hydrophilla* in the specific optimal conditions of pH (5.5-10), temperature (20°-38°C), and time (8 days)), and mineralization does not occur. The subsequential anaerobic and aerobic bioreactor was able to completely remove the sulphonated azo dye (i.e. MY10) at a maximum loading rate of 210 mg/L per day (Tan et al., 2000). The degradation of azo dyes (i.e. Acid Red 151; Basic Blue 41; Basic Red 46, 16; Basic Yellow 28, 19) in an aerobic biofilm system indicated 80% colour removal (Anjaneyulu et al., 2005). The improvement of dye biodegradation performance (i.e. >90% colour removal) are made by adding activated carbon (PAC) or bentonite in aeration tank.

4.3.2 Anaerobic biological treatment

Anaerobic biodegradation of azo and other water-soluble dyes is mainly reported as an oxidation-reduction reaction with hydrogen, and formation of methane, hydrogen sulphide, carbon dioxide, other gaseous compounds, and releasing electrons. The electrons react with the dye reducing the azo bonds, causing the effluent decolourization. Azo dye is considered an oxidising agent for the reduced flavin nucleotides of the microbial electron chain, and is reduced and decolourized concurrently with reoxidation of the reduced flavin nucleotides (Robinson et al., 2001). An additional carbon organic source is necessary, such as glucose which is a limiting factor in scale set-up technology application. The azo and nitro-components are reduced in anoxic sediments and in the intestinal environment, with regeneration of toxic amines (Banat et al., 1996). A major advantage of anaerobic system along with effluent decolourization is the production of biogas, reusable for heat and power generation that will reduce energy costs. Since textile industry wastewaters are generally discharged at high temperatures (40–70°C), thermophilic anaerobic treatment could serve as an interesting option, especially when closing process water cycles is considered. Anaerobic decolourization of textile effluents (e.g., colour removal of >99% for a Orange II, Black 2HN under anaerobic condition, more than 72 h) is not yet well established although successful pilot-scale and full-scale plants are very well operating (Tan et al., 2000). Among the different studied reactors, anaerobic filter and UASB thermophilic anaerobic reactor gave good colour removals, using or not redox mediator (e.g., antraquinone-2,6, disulphonic acid) as catalyst capable of acceleration the colour removal of azo dye-containing wastewaters.

5. Conclusions

The dyes are natural and synthetic compounds that make the world more beautiful through coloured products but are also considered as pollutants of some water resources.

The textile sector will continue to be vitally important in the area of water conservation due to its high consumption of water resources, and its individual or combined effluents' treatments for no environmental pollution generation (i.e. polluting colourants).

The satisfaction of both discharge criteria for sewerage systems, watercourses and textile reuse standards within economically viable limits implies critical analyses of industrial effluents (total wastewater and raw reusable stream characterisation) and removal of all pollutants from final effluents. The special category of organic pollutants - textile organic dyes - must respect the strict limits in final effluents discharged or not in natural water resources. This fact imposes the colour and/or dye removal from final effluents (especially industrial effluents).

Dye removal from textile effluents in controlled conditions and strict reproductibility is an environmental issue achievable by application of adequate mechano-physico-chemical and also biological treatment procedures.

6. References

Adams, C.D. & Gorg, S. (2002). Effect of pH and gas-phase ozone concentration on the decolourization of common textile dyes. *J.Environ.Eng.*, Vol.128, No.3, pp. 293-298

Alves, B.M.A. & Pinho, D.N.M. (2000). Ultrafiltration for colour removal of tannery dyeing wastewaters., *Desalination*, Vol.1303, pp. 147-154

Anjaneyulu, Y.; Sreedhara Chary, N. & Suman Raj, D.S. (2005). Decolourization of industrial effluents - available methods and emerging technologies - a review. *Reviews in Environmental Science and Bio/Technology*, Vol.4, pp. 245-273, DOI 10.1007/s11157-005-1246-z

Baban, A.; Yediler, A. & Ciliz, N.K. (2010). Integrated water management and CP implementation for wool and textile blend processes. *Clean*, Vol.38, No.1, pp. 84-90

Balchioglu, I.A.; Aslan, I. & Sacan, M.T. (2001). Homogeneous and heterogeneous advanced oxidation of two commercial reactive dyes. *Environ.Technol.*, Vol.22, pp.813-822

Banat, M.E.; Nigam, P.; Singh, D. & Marchant, R. (1996) Microbial decolourization of textile dye containing effluents, a review. *Biores. Technol.*, Vol.58, pp. 217-227

Bhattacharyya, K.G. & Sarma, A. (2003). Adsorption characteristics of the dye, Brilliant Green, on Neem leaf powder. *Dyes Pigments*, Vol.57, pp. 211-222

Birch, R.R.; Biver, C. Campagna, R.; Gledhill, W.E.; Pagga, U.; Steber, J.; Reust, H. & Bontinck, W.J. (1989). Screening of chemicals for anaerobic biodegradability. *Chemosphere*, Vol.19, No.(10-11), pp. 1527-1550

Bisschops, I.A.E. & Spanjers, H. (2003). Literature review on textile wastewater characterisation. *Environmental Technology*, Vol.24, pp. 1399-1411

Bertea, A. & Bertea, A.P. (2008). *Decolorisation and recycling of textile wastewater* (in Romanian), Performantica Ed, ISBN 978-973-730-465-0, Iasi, Romania

Börnick, H. & Schmidt, T.C. (2006). Amines, In: *Organic pollutants in the water cycle. Properties, occurrence, analysis and environmental relevance of polar compounds*, T. Reemtsma & M. Jekel, (Eds.), pp. 181-208, Wiley-VCH Verlag GmbH & Co. KGaA, ISBN 978-3-527-31297-9, Weinheim, Germany

Borrely, S.I.; Cruz, A.C.; Del Mastro, N.L.; Sampa, M.H.O. & Somssari, E.S. (1998). Radiation process of sewage and sludge - A review. *Prog.Nucl.Energy*, Vol.33, No.1/2, pp. 3-21

Carliell, C.M.; Barclay, S.J.; Naidoo, N.; Buckley, C.A.; Mulholland, D.A. & and Senior, E. (1994). Anaerobic decolorisation of reactive dyes in conventional sewage treatment processes. *Water* SA, Vol.20, pp. 341-344

Ciardelli, G. & & Ranieri, N. (2001). The treatment and reuse of wastewater in the textile industry by means of ozonation and electroflocculation. *Water Res.*, Vol.35, pp. 567-572

Cooper P. (1995). *Color in dyehouse effluent*, Society of Dyers and Colourists, ISBN 0 901956 694, West Yorkshire BDI 2JB, England

Correia, V.M.; Stephenson, T. & Judd, S.J. (1994). Characterization of textile wastewaters - a review. *Environmental Technology*, Vol.15, pp.917-929

Daneshvar, N.; Sorkhabi, H.A. & Tizpar, A. (2003). Decolorization of orange II by electrocoagulation method. Sep.Purifi.Technol., Vol.31, pp.153-162

Dascalu T.; Acosta-Ortiz, E.S.; Morales, O.M. & Compean, I. (2000). Removal of the indigo colour by laser beam-denim interanction. *Opt. Lasers Eng.*, Vol.34, pp.179-189

Dos Santos, A.B.; Cervantes, F.J. & Van Lier, J.B. (2004). Azo dye reduction by thermophilic anaerobic granular sludge, and the impact of the redox mediator AQDS on the reductive biochemical transformation. *Applied Microbiology and Biotechnology*, Vol.64, pp.62-69

EPA. (1997). *Profile of the textile industry*. Environmental Protection Agency, Washington, USA

EWA. (2005). *Efficient use of water in the textile finishing industry*, Official Publication of the European Water Association (EWA), Brussels, Belgium

Fontenot, E.J.; Lee, Y.H.; Matthews, R.D.; Zhu, G. & Pavlostathis, S.G. (2003). Reductive decolorization of a textile reactive dyebath under methanogenic conditions. *Applied Biochemistry and Biotechnology*, Vol.109, pp. 207-225

Ghayeni, S.B.; Beatson, P.J.; Schneider, R.P. & Fane, A.G. (1998). Water reclamation from municipal wastewater using combined microfiltration-reverse osmosis (ME-RO): Preliminary performance data and microbiological aspects of system operation. *Desalination*, Vol.116, pp. 65-80

Gupta, G.S.; Singh, A.K.; Tayagi, B.S.; Prasad, G. & Singh, V.N. (1992). Treatment of carpet and metallic effluent by China clay. *J.Chem.Technol.Biotechnol.*, Vol.55, pp. 227-283

Hao, O.J.; Kim, H. & Chang, P.C. (2000). Decolorization of wastewater. *Critical Reviews in Environmental Science and Technology*, Vol.30, pp. 449-505

Kim, T.H.; Park, C.; Lee, J.; Shin, E.B. & Kim, S. (2002). Pilot scale treatment of textile wastewater by combined processes (fluidized biofilm process- chemical coagulation - electrochemical oxidation). *J.Hazar.Mat. B*, Vol.112, pp.95-103

Krapfenbauer, K.F.; Robinson, M.R. & Getoff, N. (1999). Development of and testing of TiO_2-Catalysts for EDTA-radiolysis using γ-rays (1st part). *J.Adv.Oxid.Technol.*, Vol.4, No.2, pp.213-217

Lin, S.H. & Chen, L.M. (1997). Treatment of textile waste waters by chemical methods for reuse. *Water Res.*, Vol.31, No.4, pp.868-876

Lopez, A.; Ricco, G.; Ciannarella, R.; Rozzi, A., Di Pinto, A.C. & Possino, R. (1999). Textile wastewater reuse: ozonation of membrane concentrated secondary effluent. *Water Sci. Technol.*, Vol.40, pp. 99-105

Machenbach, I. (1998). Membrane technology for dyehouse effluent treatment. *Membrane Technol.*, Vol.58, pp.7-10

Macoveanu, M.; Bilba, D.; Bilba, N.; Gavrilescu, M. & Soreanu, G. (2002). *Ionic exchange processes in environmental protection* (in Romanian), MatrixRom Ed., Bucuresti, Romania

Matsui, Y.; Murase, R.; Sanogawa, T.; Aoki, N.; Mima, S.; Inoue, T. & Matsushita, T. (2005). Rapid adsorption pretreatment with submicrometre powdered activated carbon particles before microfiltration. *Water Science and Technology*, Vol.51, pp.249-256

Mignani, M.G.; Nosenzo, G. & Gualdi, A. (1999). Innovative ultrafiltration for wastewater reuse. *Desalination*, Vol.124, pp.287-292

Naveed, S.; Bhatti, I. & Ali, K. (2006). Membrane technology and its suitability for treatment of textile waste water in Pakistan. *Journal of Research (Science)*, Bahauddin Zakariya University, Multan, Pakistan, Vol. 17, No. 3, pp.155-164

Neamtu, M.; Zaharia, C.; Catrinescu, C.; Yediler, A.; Kettrup, A. & Macoveanu, M. (2004). Fe-exchanged Y zeolite as catalyst for wet peroxide oxidation of reactive azo dye Procion Marine H-EXL. *Applied Catalysis B: Environmental*, Vol.78, No.2, pp.287 294

Nigam, P.; Armou, G.; Banat, I.M.; Singh, D. & Marchant, R. (2000). Physical removal of textile dyes and solid-state fermentation of dye-adsorbed agricultural residues. *Biores. Technol.*, Vol.72, pp.219-226

Omura, T. (1994). Design of chlorine - fast reactive dyes - part 4; degradation of amino containing azo dyes by sodium hydrochlorite. *Dyes Pigments*, Vol.26, pp.33-38

Ong, S.A.; Toorisaka, E.; Hirata, M. & Hano, T. (2005). Treatment of azo dye Orange II in aerobic and anaerobic-SBR systems. *Proc.Biochem.*, Vol.40, pp.2907-2914

Orhon, D.; Babuna, F.G. & Insel, G. (2001). Characterization and modelling of denim-processing wastewaters for activated sludge. *Journal of Chemical Technology and Biotechnology*, Vol.76, pp.919-931

Ozcan, A.S.; Erdem, B. & Ozcan, A. (2004). Adsorption of Acid Blue 193 from aqueous solutions onto Na-bentonite and DTMA-bentonite. *J.Coll.Interf.Sci.*, Vol.280, No.1, pp.44-54

Oztekin, Y; Yazicigil, Z.; Ata, N. & Karadayl, N. (2010). The comparison of two different electro-membrane processes' performance for industrial application. *Clean-Soil, Air, Water*, Vol.38, No.5-6, pp.478-484

Ramesh Babu, B.; Parande, A.K.; Raghu, S. & Prem Kumar, T. (2007). Textile technology. Cotton Textile Processing: Waste Generation and Effluent Treatment. *The Journal of Cotton Science*, Vol.11, pp.141-153

Robinson, T.; McMullan, G.; Marchant, R. & Nigam, P. (2001). Remediation of dyes in textile effluent: a critical review on current treatment technologies with a proposed alternative. *Bioresource Technology*, vol.77, pp.247-255

Slokar, Y.M. & Le Marechal, A.M. (1997). Methods of decoloration of textile wastewaters. *Dyes Pigments*, Vol.37, pp.335-356

Sarasa, J.; Roche, M.P.; Ormad, M.P.; Gimeno, E.; Puig, A. & Ovelleiro, J.L. (1998). Treatment of wastewater resulting from dye manufacturing with ozone and chemical coagulation. *Water Res.*, Vol.32, No.9, pp.2721-2727

Soloman, P.A.; Basha, C.A.; Ramamurthi, V.; Koteeswaran, K. & Balasubramanian, N. (2009). Electrochemical degradation of Remazol Black B dye effluent. *Clean, Vol.*37, No.11, pp.889-900

Stolz, A. (2001). Basic and applied aspects in the microbial degradation of azo dyes. *Appl.Microbiol.Biotechnol.*, Vol.56, pp.69-80

Surpățeanu, M. & Zaharia, C. (2004a). Advanced oxidation processes. Decolorization of some organic dyes with hydrogen peroxide. *Environmental Engineering and Management Journal*, Vol.3, No.4, pp.629-640

Surpateanu, M. & Zaharia C. (2004b). Advanced oxidation processes for decolorization of aqueous solution containing Acid Red G azo dye. *Central European Journal of Chemistry*, Vol.2, No.4, pp.573-588

Suteu, D.; Bîlbă, D. & Zaharia, C. (2002). Kinetic study of Blue M-EB dye sorption on ion exchange resins. *Hungarian Journal of Industrial Chemistry*, Vol.30, pp.7-11

Suteu, D.; Zaharia, C.; Bilba, D.; Muresan, A.; Muresan, R. & Popescu, A. (2009a). Decolorization wastewaters from the textile industry - physical methods, chemical methods. *Industria Textila*, Vol.60, No.5, pp.254-263

Suteu, D.; Zaharia, C.; Muresan, A.; Muresan, R. & Popescu, A. (2009b). Using of industrial waste materials for textile wastewater treatment. *Environmental Engineering and Management Journal*, Vol.8, No.5, pp.1097-1102

Suteu, D. & Zaharia, C. (2008). Removal of textile reactive dye Brilliant Red HE-3B onto materials based on lime and coal ash, *ITC&DC, Book of Proceedings of 4th International Textile, Clothing & Design Conference - Magic World of Textiles*, pp.1118-1123, ISBN 978-953-7105-26-6, Dubrovnik, Croatia, October 5-8, 2008

Suteu, D.; Zaharia, C. & Malutan, T. (2011a). Biosorbents Based On Lignin Used In Biosorption Processes From Wastewater Treatment (chapter 7). In: *Lignin: Properties and Applications in Biotechnology and Bioenergy*, Ryan J. Paterson (Ed.), Nova Science Publishers, 27 pp., ISBN 978-1-61122-907-3, New York, U.S.A.

Suteu, D.; Zaharia, C. & Malutan, T. (2011b). Removal of Orange 16 reactive dye from aqueous solution by wasted sunflower seed shells. *Journal of the Serbian Chemical Society*, Vol.178, No.3, pp.907-924

Tan, N.C.G.; Borger, A.; Slenders, P., Svitelskaya, A.; Lettinnga, G. & Field, J.A. (2000). Degradation of azo dye mordent yellow 10 in a subsequential anaerobic and bioaugmented aerobic bioreactor. *Water Sci.Technol.*, Vol.42, No.(5-6), pp.337-344

Tang, C. & Chen, V. (2002). Nanofiltration of textile wastewater for water reuse. *Desalination*, Vol.143, pp.11-20

Vandevivere, p.C.; Bianchi, R. & Verstraete, W. (1998). Treatment and reuse of wastewater from the textile wet-processing industry review of emerging technologies. *J.Chem.Technol.Biotechnol.*, Vol.72, No.4, pp.289-302

Venkat Mohan, s.; Srimurli, M.; Sailaja, P. & Karthikeyan, J. (1999). A study of acid dye colour removal using adsorption and coagulation. *Environ.Eng.Poly.*, vol.1, pp.149-154

Vlyssides, A.G.; Papaioannou, D.; Loizidoy, M.; Karlis, P.K. & Zorpas, A.A. (2000). Testing an electrochemical method fortreatment of textile dye wastewater. *Waste Manag.*, Vol.20, pp.569-574

Xu, Y. & Lebrun, R.E. (1991). Treatment of textile dye plant effluent by nanofiltration membrane. *Separ.Sci.Technol.*, Vol.34, pp.2501-2519

Welham, A. (2000). The theory of dyeing (and the secret of life). *Journal of the Society of Dyers and Colourists*, Vol.116, pp.140-143

Wiesmann, U.; Choi, I.S. & Dombrowski, E.M. (2007). *Fundamentals of Biological Wastewater Treatment*. Wiley-VCH Verlag GmbH&Co. KgaA, Weinheim, Germany

Yang, Y.; Wyatt II, D.T & Bahorsky, M. (1998). Decolorisation of dyes using UV/H_2O_2 photochemical oxidation. *Text.Chem.Color.*, Vol.30, pp.27-35

Zaharia, C. (2006). *Chemical wastewater treatment* (in Romanian), Performantica Ed., ISBN 978-973-730-222-9, Iasi, Romania

Zaharia, C. (2008). *Legislation for environmental protection* (in Romanian), Politehnium Ed., ISBN 978-973-621-219-2, Iasi, Romania

Zaharia, C.; Diaconescu, R. & Surpăţeanu, M. (2006). Optimization study of a wastewater chemical treatment with PONILIT GT-2 anionic polyelectrolyte. *Environmental Engineering and Management Journal*, Vol.5, No.5, pp.1141-1152

Zaharia, C.; Diaconescu, R. & Surpăţeanu, M. (2007). Study of flocculation with Ponilit GT-2 anionic polyelectrolyte applied into a chemical wastewater treatment. *Central European Journal of Chemistry*, Vol.5, No.1, pp.239-256

Zaharia, C.; Surpăţeanu, M.; Creţescu, I.; Macoveanu, M. & Braunstein, H. (2005). Electrocoagulation/electroflotation - methods applied for wastewater treatment. *Environmental Engineering and Management Journal*, Vol.4, No.4, pp.463-472

Zaharia, C.; Suteu, C. & Muresan, A. (2011). Options and solutions of textile effluent decolourization using some specific physico-chemical treatment steps. *Proceedings of 6th International Conference on Environmental Engineering and Management ICEEM'06*, pp. 121-122, Balaton Lake, Hungary, September 1-4, 2011

Zaharia, C.; Suteu, D.; Muresan, A.; Muresan, R. & Popescu, A. (2009). Textile wastewater treatment by homogenous oxidation with hydrogen peroxide. *Environmental Engineering and Management Journal*, Vol.8, No.6, pp.1359-1369

Part 2

Environmental Fate, Effects and Analysis of Organic Pollutants

4

Bioavailability of Polycyclic Aromatic Hydrocarbons Studied Through Single-Species Ecotoxicity Tests and Laboratory Microcosm Assays

Bernard Clément
LEHNA-IPE, Université de Lyon-ENTPE
France

1. Introduction

The bioavailability of a pollutant is defined as its capacity to transfer from the surrounding environment (water and sediments) to organisms and is one of the key factors governing its bioaccumulation and toxicity. Knowledge of the bioavailable fraction of a compound is therefore vital in order to evaluate environmental risks and in particular to study its effects on living organisms. We began works on this topic at the laboratory Transfert et Effets des Polluants sur l'Environnement (TEPE) of Université de Savoie by studying the toxicity of a mixture of three polycyclic aromatic hydrocarbons (PAHs) (phenanthrene, fluoranthene and benzo(k)fluoranthene) in the framework of a PNETOX (programme national d'écotoxicologie) 1998-2001 program (Verrhiest, 2001; Verrhiest *et al.* 2000, 2001, 2002a, 2002b). The aim of this program, which brought together three research laboratories (TEPE of the Université de Savoie, Centre d'Etudes du Machinisme Agricole, du Génie Rural, des Eaux et Forêts (CEMAGREF) Lyon, INERIS, Institut Français pour la Recherche et les Etudes en Mer (IFREMER)), was to contribute to the development of a method of evaluating the risks of freshwater sediments contaminated by PAHs *via* in-depth changes to methods and better knowledge of the fate and effects of PAHs in aquatic ecosystems. The choice of PAHs was motivated by the fact that they are:

- ubiquitous organic contaminants (**table 1**), stemming from incomplete combustion (pyrolitic origin) or from the slow maturation of organic matter (petroleum hydrocarbons), and mostly emitted by anthropic activities (industry, transport),
- hydrophobic compounds which, after being introduced in a water column, rapidly become linked to colloidal and suspended matter, thus they enter the sediment where their effects depend on their bioavailability and fate, in relation with the physicochemical and microbiological properties of the sediment and interstitial water,
- toxic (**tables 2 and 3**) or/and mutagenic and carcinogenic compounds.

2-L microcosm assays (28 days) were performed on different types of sediment (artificial and natural) spiked with a mixture of PAHs (fluoranthene, phenanthrene, benzo(k)fluoranthene, **table 4**). The effects were evaluated on higher organisms and on the bacterial compartment of the sediment. Details on the protocols can be found in Verrhiest *et* *al.* (2000, 2001, 2002a, 2002b). The microcosms consisted in glass beakers filled with artificial sediment (300 g sand + kaolinite + alpha-cellulose + TetraMin®) or natural sediment, synthetic water (2000 mL), and inoculated with daphnids (*Daphnia magna*), micro-algae (*Pseudokirchneriella subcapitata*), duckweeds (*Lemna minor*), chironomids (*Chironomus riparius*) and amphipods (*Hyalella azteca*). The organisms were exposed 28 days to PAH-spiked sediment, measurements were performed on PAH contents, organism development (survival, growth, reproduction), bacterial exo-enzymatic activities involved in the nitrogen (leucine-aminopeptidase) and carbon (β-glucosidase) cycles, and bacterial density (direct counting of bacteria by epi-fluorescence or Colony Forming Units after spreading on agar). The fate and the effects of PAHs were analysed as a function of the sediments' characteristics, especially organic carbon content and sediment type (artificial sediment conditioned or not, natural sediment).Studying the bioavailability of PAHs has been extended through a program on the impact of pyrene (**table 4**) on the organisms of an aquatic ecosystem (Clément *et al.*, 2005b). This study, performed in the framework of the Centre National de la Recherche Scientifique (CNRS) Life and Societies Environment Program (Ecosystems and Environment), was performed in 2001-2002 (Jouanneau *et al.*, 2003). It brought together four laboratories: the Laboratory of Biochemistry and Biophysics of Integrated Systems, the Department of Molecular and Structural Biology, Unité Mixte de Recherche (UMR) 5092, Commissariat à l'Energie Atomique (CEA) Grenoble (Yves Jouanneau), the TEPE Laboratory, Université de Savoie (Pr Gérard Blake), the Laboratoire Chimie Moléculaire et Environnement (LCME) Laboratory, Université de Savoie (Emmanuel Naffrechoux), and the Laboratoire des Sciences de l'Environnement (L.S.E.) (Bernard Clément). During the first phase (2001), the toxicity of pyrene was evaluated at the L.S.E. by single-species assays (daphnia, algae, duckweed, chironomidae, amphipods), and a bioaccumulation measurement protocol for *D. magna* and *H. azteca* was formulated. During the second year, 2-litre microcosm tests were performed in the presence of two lacustrine sediments. As in the case of Verrhiest's (2001) study of PAH mixtures, we completed monitoring of the higher organisms of the microcosm by measuring enzymatic activity and bacterial density in the bacterial compartment of the sediments

	Place	Area studied	PHE (mg/kg)	FLU (mg/kg)	BKFLU (mg/kg)	Pyrene (mg/kg)	Sum of 16 PAHs (mg/kg)	References
Marine sediments	West of the sea Méditerranée	N+C	0.06 to 1.86	0.001 to 3.18	–	dna	1.18 to 20.44	Baumard *et al.*, 1998
	Bassin d'Arcachon	C	–	–	–	dna	5 to 10	Geffard *et al.*, 1999 Raymond *et al.*, 1999
	West of the Beaufort Sea (Alaska)	N	–	–	–	dna	0.16 to 1.1	Valette-Silver *et al.*, 1999
	Baltic Sea	N	0.03 to 0.08	0.05 to 0.27	0.05 to 0.16	dna	0.72 to 1.9	Witt, 1995

	Ain	N *	0.05 to 0.12	0.25 (mean)	0.07 to 0.3	dna	–	Verrhiest, 2001
	Neyrieux	N *	<0.05	<0.05	0.03	dna	–	Verrhiest, 2001
	Confluence of Seine-Marne	N	0.5 (mean)	0.82 (mean)	–	dna	1.5 to 7.4	Garban and Ollivon (1995)
	Rhone, Saone and Ain	N+C	<0.05 to 0.55	< 0.04 to 0.88	< 0.01 to 0.67	dna	0.52 to 14.64	Bonnet, 2000
	Moselle and Meurthe	N+C	<0.53 to 3.44	0.75 to 6.3	0.38 to 1.5	dna	6.68 to 32	Bonnet, 2000
	Seine	N C	– –	– –	– –	dna	2 to 4 60	Ollivon *et al.*, 1995
	Lake Annecy	C	–	–	–	dna	1.43	Naffrechoux *et al.*, 1999
	Niagara river (EU)	N C	0.01 0.4	0.04 0.9	–	dna	0.4 (mean) 3.3 to 5.4	Eisler, 2000
	Black river (EU)	C	52 (mean)	33 (mean)	1.5 (mean)	dna	–	Eisler, 2000
Fresh water sediments	CEBS code B2	C*	0.51	0.87	0.26	0.95	5.86	Bray *et al.*, 2001
	CEBS code B13	C*	1.3	1.6	0.48	1.5	9.58	Bray *et al.*, 2001
	CEBS code B22	C*	0.84	1.45	0.46	1.33	9.59	Bray *et al.*, 2001
	North 13990	C*	0.53	0.68	0.89	1.84	10.09	Bray *et al.*, 2003
	North 17000	C*	3.96	2.53	0.15	7.00	19.11	Bray *et al.*, 2003
	North 12570	C*	1.23	1.64	2.04	3.31	17.04	Bray *et al.*, 2003
	North 12730	C*	2.29	7.36	4.43	6.34	37.64	Bray *et al.*, 2003
	North 12800	C*	2.52	4.67	3.71	5.59	62.54	Bray *et al.*, 2003

Table 1. PAH concentrations of several sediments (according to Verrhiest 2001 and completed by the author) (PHE: phenanthrene; FLU: fluoranthene; BKFLU: benzo(k)fluoranthene; N: normal; C: contaminated; CEBS: Canal de l'Est Branche Sud; dna: data non available; * sediments used in our studies).

PAH	Organism	Type	EC50/LC50 (µg/l overlaying water or mg/kg dry sediment)	Duration of exposure/ Endpoint
Phenanthrene	Hydra *Hydra sp.*	Water	96 µg/l	_/mortality
	Daphnia *Daphnia magna* *Daphnia pulex*	Water	700 µg/l 734 µg/l	_/mortality _/mortality
	Amphipod *Gammarus pseudolimnaeus*	Water & sediment (epibenthic)	126 µg/l	_/mortality
	Insect chironomid *Chironomus tentans*	Sediment (benthic)	490 µg/l	_/mortality
	Annelid *Lumbriculus variegatus*	Sediment (benthic)	> 419 µg/l	_/mortality
Fluoranthene	Hydra *Hydra americana*	Water	70.06 µg/l	_/mortality
	Daphnia *Ceriodaphnia dubia* *Daphnia magna*	Water	45 µg/l 102.8 µg/l 43 to 92 µg/l 4.2 to 15 mg/kg	_/mortality _/mortality 10 days/mortality 10 days /mortality
	Amphipod *Hyalella azteca*	Water & sediment (epibenthic)	97 to 114 µg/l 32 to 54 µg/l 2.3 to 7.4 mg/kg	10 days/mortality 10 days/mortality 10 days/mortality
	Insect *Chironomus tentans* *Chironomus riparius*	Sediment (benthic)	30 to 61 µg/l 3 to 8.7 mg/kg 29 to 41 µg/l 170 mg/kg	10 days/mortality 10 days/mortality 11 days/mortality 28 days /emergence
	Annelid *Lumbriculus variegatus*	Sediment (benthic)	> 178.5 µg/l	mortality

Table 2. Acute toxicity data for phenanthrene and fluoranthene towards water invertebrates (Verrhiest, 2001) (EC/LC50: concentration which produces 50% effect on a given end point or kills half of the organisms initially present).

2. Results and main findings

2.1 Difficulties specific to experiments on PAHs

PAHs are difficult to handle due to their properties. Indeed, their strong affinity for particular phases and low hydrosolubility complicate assays when spiking sediments and studying the fate of these substances in experimental systems. The sediment has to be spiked so that the PAHs introduced become homogenously distributed within the matrix. However, their poor solubility does not allow spiking with an aqueous solution. This makes it necessary to use an apolar solvent which makes it possible to concentrate PAHs, although this may prove toxic for organisms. We used the "wall-coating" method (Ditsworth *et al.*, 1990) for all the assays. This method consists in distributing the PAHs on the wall of a glass flask after adding an acetone solution, evaporating the solvent by rotating the flask horizontally, and then adding the wet sediment that comes into contact with the PAHs inside the flask as it is being rotated. High spiking yields were always obtained during the assays performed on fluoranthene, phenanthrene, benzo(k)fluoranthene and their mixtures (>80%) (Verrhiest *et al.*, 2000, 2001, 2002a). On the contrary, the assays on pyrene provided generally variable and much lower yields (**table 5**). This may have been due to the different characteristics of the sediments used in these assays, as the efficiency of the spiking depends on the way in which sediments come into contact with the flask wall. Chevron (2004) proposed another more classical method that consists in dissolving pyrene in DMSO (dimethylsulfoxide) at a given concentration, then mixing it with an organic matrix. The low toxicity of this solvent to microorganisms means that it can be used, as she did, in biodegradation assays. It remains to be determined, however, whether DMSO has no effect on higher organisms in contact with the sediment.

	Daphnia		Algae		Ceriodaphnia	
µg/L	LC50-48 h survival	IC95%	EC10 growth	IC95%	EC10 reprod 7 d	IC95%
Benzo(k)fluoranthene	> 1.1		> 1.5		> 1.5	
Benzo(a)anthracene	> 9.1		4.1	3.8-4.6	> 13	
Benzo(b)fluoranthene	> 1.1		> 1.5		> 1.5	
Benzo(ghi)perylene	> 0.2		> 0.26		0.124	0-0.17
Benzo(a)pyrene	> 2.7		1.54	1.52-1.57	0.77	0.03-2.2
Dibenzo(a,h)anthracene	> 0.35		0.73	0.52-1.31	> 0.046	
Acenaphtene	958	916-994	308	266-371	64	50-100
Acenaphtylene	1800	1731-1956	595	505-703	95	47-231
Anthracene	> 25		23.3	18.6-27.8	> 4.9	
Fluorene	408	368-449	485	411-540	33	28-45
Chrysene	> 1.3		> 3.9		> 0.13	
Fluoranthene	> 112		33.1	27.8-37.5	1.2	0.2-4.9
Indeno(1,2,3,cd)pyrene	> 357		5.7	3.6-9.8	0.38	0-9.49
Naphtalene	1664	1441-1902	> 8024		999	761-1331
Phenanthrene	> 400		123.5	80.1-170.4	15	4.7-19
Pyrene	24.6	21.6-28.4	12.4	6.7-17.9	2.1	1.3-3.1

Table 3. Toxicity data of PAHs in aqueous phase for three pelagic organisms (Vindimian *et al.*, 2000) (EC10: concentration which produces 10% effect on a given end point; LC50 concentration which kills half of the organisms initially present; IC95: confidence interval at 95%).

The affinity of PAHs for solid surfaces can also bias results when studying their effects in the aqueous phase. PAHs dissolved in water tend to adsorb onto the walls of beakers, even glass ones, as already observed (Clément *et al.*, 2005b), thus confirming the results of other experiments (McCarthy, 1983; Gauthier *et al.*, 1986; Pelletier *et al.*, 1997; Miller, 1999). Therefore a solution of 10 µg/L pyrene contains only 4 µg/L pyrene after 30 days in the absence of light to avoid any photodegradation. Finally, the same problem exists when conserving aqueous samples before analysis. Since we were unable to carry out the first assays on pyrene ourselves, we had to subcontract this task to another laboratory which, due to the lead times given, provided results that greatly underestimated the real values that we were able to check against solutions spiked at known concentration. We strongly recommend strict control over the analytical part (extraction and analyses) when working on PAHs, otherwise there is a risk of coming up against the same problems.

PAH	Structure	Hydrosolubility (µg/L) to 25°C	log Kow	Photosensitivity
Phenanthrene $C_{14}H_{10}$	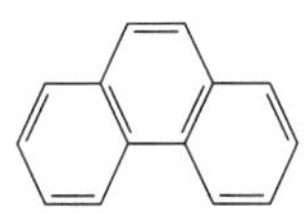	1000 (May, 1980)	4.29 (May, 1980)	Non absorbent (Newsted and Giesy, 1987) UV_C (Huovinen *et al.*, 2001)
Fluoranthene $C_{16}H_{10}$	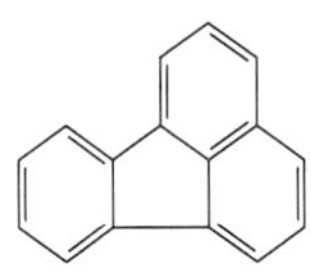	206 (May, 1980)	5 (May, 1980)	UV_A, UV_B (Newsted and Giesy, 1987)
Benzo(k)fluoranthene $C_{20}H_{12}$	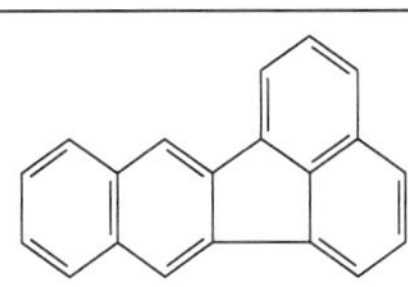	1.5 (Swartz *et al.*, 1995)	6.72 (Pelletier *et al.*, 1997)	UV_A, UV_B and visible (Pelletier *et al.*, 1997)
Pyrene $C_{16}H_{10}$		160	5.18	UV_A and above all UV_B (Huovinen *et al.*, 2001)

Table 4. Physicochemical properties of phenanthrene, fluoranthene, benzo(k)fluoranthene, and pyrene (according to Verrhiest *et al.* (2001) completed by data on pyrene) (Kow: water/octanol partition coefficient; UV: ultra-violet).

		Carbonate sediment		Peat sediment	
μg/g nominal f.w.	Test	μg/g measured f.w.	Spiking rate, %	μg/g measured f.w.	Spiking rate, %
2	ss	4.9 [a]	243	1.5 [a]	73.5
20	ss	15.1 [a]	75.7	11.5 [a]	57.7
50	sp	31.6 [b] ± 7.9	63.2 ± 15.8	23.6 [b] + 7.6	47.3 ± 15.2
50	m1	24.1 [a]	48.2	1.4 [a]	2.7
50	m2	33.6 [a]	67.2	3.9 [a]	7.8
50	m3	29.1 [b]	58.2	7.2 [b]	14.4
200	ss	175.3 [a]	87.7	194.6 [a]	97.3

Table 5. Nominal and measured concentrations of sediments from lake Aiguebelette spiked with pyrene, with spiking rates (p.f.: fresh weight; ss: amphipod test; sp: spiking assay; m1: 1st microcosm assay; m2: 2nd microcosm assay; m3: microcosm assay without organisms; carbonate sediment; peat sediment; [a] concentrations measured on D0, after 7 days underwater ; [b] concentrations measured on 3 samples immediately after spiking the sediment) (source: Clément *et al.*, 2005b).

2.2 Toxicity in the aqueous phase towards microcosm organisms exposed during single-specific bioassays and modulating factors

The results of single-species bioassays in the aqueous phase on *Daphnia magna* and *Hyalella azteca* (**table 6**) confirm the acute toxicity of the PAH studies on these crustaceans, except for benzo(k)fluoranthene, which has very weak hydrosolubility (1.5 μg/L) (Verrhiest *et al.*, 2001; Clément *et al.*, 2000; Clément *et al.*, 2005b).

The absence of acute toxicity of benzo(k)fluoranthene was confirmed by the results of Vindimian *et al.* (2000), who also showed the absence of chronic toxicity (**table 3**). When organisms are exposed in obscurity, toxicity at 24 and 48 h is observed at contents close to solubility. Conversely, in the presence of the fluorescent light usually used in the laboratory, particularly in all microcosm assays, toxicity is higher, though the induction of toxicity is more significant for fluoranthene than for phenanthrene, generally recognised as non phototoxic (Newsted and Giesy, 1987; Swartz *et al.*, 1997; Boese *et al.*, 1998), and pyrene. This type of light (*cool-white fluorescent*) contains a fraction of UV (Clément *et al.*, 2000) which probably explains the increase in toxicity. Phototoxicity mechanisms are recalled in **figure 1**. For the most part, phototoxicity can be attributed to photosensitisation (Arfsten *et al.*, 1996). Furthermore, the phototoxicity of phenanthrene and fluoranthene has been confirmed by the post-exposure of organisms for 2 hours to pure UV radiation (A or C) (according to the method of Wernersson and Dave, 1997). This post-exposure could not be performed with pyrene. Our results strengthen the hypothesis that, although many PAHs are not acutely toxic to aquatic organisms at concentrations corresponding to their range of solubility (National Research Council of Canada (NRCC), 1983), the presence of natural light containing UV rays represents a factor that considerably increases the risk of toxicity (Lyons *et al.*, 2002). This is especially the case for pelagic invertebrates such as *Daphnia magna*, which are positively heliotropic and are thus more exposed in the presence of light and PAHs in the water column (Wernersson *et al.*, 1999).

	24 h lux	48 h lux	48 h lux + UV	24 h dark	48 h dark	48 h dark + UV
			Daphnia magna			
PHE	2500 lux: 678	2500 lux: 604	2500 lux: 273 [a]	854	731	725 [a]
FLU	1500 lux: 56 2500 lux: 63	1500 lux: 36 2500 lux: 34	1500 lux: 29 [b] 2500 lux: <18 [a]	>200 >180	201 >180	80 [b] 20 [a]
BKFLU			non toxic			
PYR	1500 lux: 139 2500 lux: 105 6000 lux: 161	1500 lux: 74 2500 lux: 48 6000 lux: 45	/	167	68	/
			Hyalella azteca			
PYR	1500 lux: 80 2500 lux: 134 6000 lux: 77	1500 lux: 41 2500 lux: 41 6000 lux: 26	/	82	63	/

Table 6. Toxicity in the aqueous phase of PAHs to *Daphnia magna* and *Hyalella azteca* (according to Verrhiest *et al.* (2001), Clément *et al.* (2000), and Clément *et al.* (2005b); lux: exposure to 1500 and 2500 lux 16h/day; dark: obscurity; lux+UV [a]: post-exposure 2 h to UV-A (365 nm, 247 µW/cm²); lux+UV [b]: post-exposure 2 h to UV-C (365 nm, 247 µW/cm²).

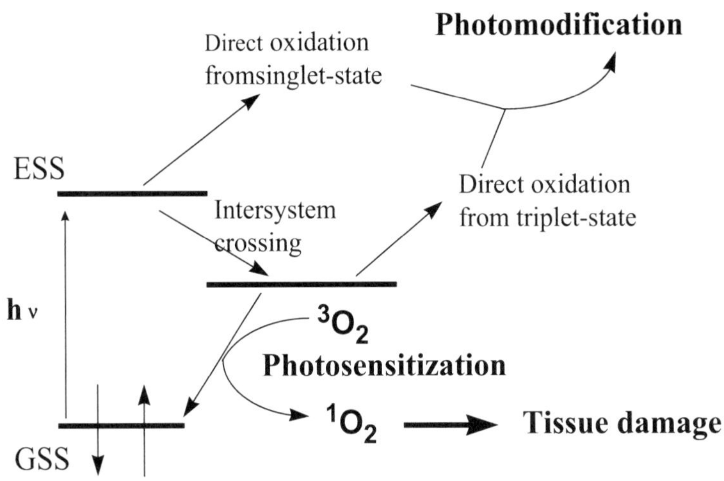

Fig. 1. Phototoxicity mechanisms (according to Krylov *et al.*, 1997).

The absence of phototoxicity of benzo(k)fluoranthène to *D. magna* (Verrhiest *et al.*, 2001) is probably due to the fact that the weak hydrosolubility of this PAH (1.5 µg/L) does not allow sufficient bioaccumulation to cause effects, as suggested by Boese *et al.* (1998) for benzo(b)fluoranthene (hydrosolubility: 6 µg/L).

Although fluoranthene has been shown to be stable in the aqueous phase for 48 h in light or obscurity (Clément *et al.*, 2000), under the same conditions, pyrene is much more sensitive to light (30 to 44% degradation over 24 h).

When algae were added in the beakers in which the daphnia were exposed, significant adsorption of fluoranthene occurred on the algae over 48 h, and there was a considerable reduction of toxicity in the presence of light. This phenomenon of reduced toxicity in the presence of algae is not attributed to adsorption, which remains limited; rather we presume that a possible protective effect may be due to algal pigments of type β-carotene (Bennett *et al.*, 1986), or more simply, that the algae absorb part of the light, though this absorption could not be quantified.

Algae are highly sensitive to pyrene (Clément *et al.*, 2005b), with an EC-72 h < 10 µg/L. According to Vindimian *et al.* (2000), they are also more sensitive to fluoranthene and phenanthrene than *Daphnia magna*. Conditions of exposure to very intense light (6000 lux) to ensure sufficient algal growth can induce significant phototoxicity, thereby explaining this higher sensitivity. Nonetheless, Warshawsky *et al.* (1995) obtained only weak inhibition of the growth of *Selenastrum capricornutum*, another species used by us, exposed to 400 µg pyrene/L for 4 days under fluorescent light similar to ours. According to Lei *et al.* (2001), *Selenastrum capricornutum* is resistant to pyrene. Furthermore, pyrene stimulates its glutathione-S-transferase (GST) activity during exposures of several days to concentrations ranging from 0.1 to 1 mg/L (Lei *et al.*, 2003). This enzymatic activity involves an enzyme, GST, that catalyses the binding reaction between an endogenous bio-molecule and the toxic substance, thereby detoxifying the contaminants. The authors associate the stimulation of GST activity with the degradation of pyrene observed, to suggest that the GST activity is responsible for the metabolization of pyrene. It is possible that the differences in culture and assay conditions between the experiments performed by Lei *et al.* (2001, 2003) and us resulted in very different expressions of GST activity. The same authors observed that pyrene inhibited the growth of another species of Chlorophyceae, *Scenedesmus quadricauda* (Lei *et al.*, 2003). The type of light, containing more or less UV, could have an impact on the cytotoxicity of PAHs to algae, due to its importance in biotransformation, highlighted with benzo(a)pyrene and *Selenastrum capricornutum* (Warshawsky *et al.*, 1995).

Contrary to algae, duckweed is not or only slightly sensitive to PAHs. We observed a value close to the solubility limit for pyrene, which proved to be non toxic in a single-species bioassay for an exposure of 6 days at 180 µg/L (3500 lux 16 h/day). An absence of sensitivity was also observed in another species of duckweed, *Lemna gibba*, by different authors (Huang *et al.*, 1995; Ren *et al.*, 1994). This resistance may be linked to the strong metabolization related to GST activity.

2.3 Toxicity to pelagic organisms in microcosms

In microcosms containing sediments contaminated by PAHs, organisms of the water column were exposed to PAHs present in the latter in dissolved, colloidal and particulate form, and some were also exposed to PAHs of the sediment with which they can be in contact for long periods (sedimented algae) or occasionally (daphnia consuming particles on the surface of the sediment). The results obtained on the daphnia exposed in beakers containing the sediment spiked with fluoranthene or phenanthrene (Verrhiest *et al.*, 2001) showed increased toxicity in comparison to bioassays in the aqueous phase, by taking into account in both cases the contents measured in the supernatant water. This increase can be attributed to different exposure routes, as the daphnia above the sediment can be brought into contact with the sediment or ingest fine particles returned to suspension. Other results take the same direction. During microcosm bioassays on a mixture of fluoranthene, phenanthrene and benzo(k)fluoranthene (Verrhiest, 2001), we observed strong toxicity after a week's exposure to *Daphnia magna*, for a natural sediment (Neyrieux) spiked at 300 mg/kg, whereas the contents measured in the supernatant water after filtration at 0.8 µm did not exceed 1 µg/L. Conversely, in microcosm bioassays on pyrene (carbonate and peat sediments from lake Aiguebelette spiked at 50 mg/kg), the response of *Daphnia magna* was likely related to contents measured in the supernatant water: a significant effect on the survival above the carbonate sediment in water containing 30 ± 9 µg pyrene/L; absence of effect above the peat in water containing 5 ± 1 µg pyrene/L (Clément *et al.*, 2005b). The same correlations were observed in single-species bioassays in small beakers (Jouanneau *et al.*, 2003), where the survival of daphnia above the contaminated sediment was reduced by about 50% to contents close to EC50.

We have never been able to evaluate the effects of PAHs on algae in microcosms for various reasons : the difficulty of taking all the algae into account (in suspension or sedimented), the influence of browsing and/or insufficient growth, and in particular high variability.

As mentioned previously, duckweed is not sensitive to PAHs and, in general, we have not observed effects in microcosms, in the presence of artificial or natural sediments spiked with contents ranging from 4 to 300 mg/kg. This can be explained easily by the fact that PAH contents in the supernatant waters could not exceed the solubility limits of these contaminants. No effect on *Lemna minor* was observed at these values in our works (single-species bioassays) or in the literature. However, there was one noteworthy exception. In a microcosm bioassay on a natural sediment (Neyrieux) spiked with a mixture of PAHs (phenanthrene, fluoranthene, and benzo(k)fluoranthene) at 300 mg/kg, significant inhibition was demonstrated from 19 days onwards (Verrhiest, 2001). However, as the sum of the PAH contents of supernatant waters did not exceed 1 µg/L during the test, it is possible that the inhibition of growth observed was due to a PAH degradation metabolite or the occurrence of an indirect effect. Verrhiest *et al.* (2002a) highlighted the effects on the bacterial compartment of a natural sediment spiked at 300 mg/kg. This disturbance perhaps concerned the sediment of Verrhiest (2001), possibly resulting in modifications of contents and flows of nutritive substances vital for *Lemna minor*. This hypothesis could not be supported by precise measurements.

2.4 Toxicity to benthic organisms in single-species and microcosm bioassays

We were able to measure the sensitivity of the species *Hyalella azteca* (crustacean-amphipod) and *Chironomus riparius* (dipterous insect) to several PAHs (fluoranthene, phenanthrene, benzo(k)fluoranthene, pyrene) alone and in mixture (for the first three), under different experimental conditions : single-species bioassays, microcosm bioassays, artificial sediments and natural sediments. Taking all the results into account, the first observation is that for exposures not exceeding one month, sediments contaminated by PAHs presented relatively weak acute toxicity as the first effects on benthic organisms (mortality and inhibited growth) were observed for contents of several mg/kg (or µg/g), higher than 30 mg/kg for the mixture "phenanthrene + fluoranthene + benzo(k)fluoranthene", and from 20 mg/kg for pyrene. Such contents are those of strongly contaminated sediments (**table 1**) and these results correspond to the threshold values provided by McDonald *et al.* (2000) and Kalf *et al.* (1997), at least if we consider the PEC (probable effect concentration) and MPC (maximum permissible concentration), thresholds from which an effect is very much probable (**table 7**). What is more, it should be also be mentioned that these thresholds result from matching contamination data with data from bioassays on benthic organisms, the effects generally being imputable to all the contaminants present, especially total PAHs, whose effects are generally additive (Munoz and

Tarazona, 1993; Swartz *et al.*, 1995). However, it can be seen that effects are probable from 23 mg/kg in total PAHs, a value comparable to those for which we were able to show the effects in certain cases. The results obtained from microcosm bioassays were not different from those obtained from single-species tests, despite exposure being nearly twice as long. Other more sensitive sublethal effect criteria should be taken into account with these substances, some of which are known to be carcinogenic or mutagenic.

Substance	TEC (µg/g dw)	PEC (µg/g dw)	MPC (µg/g dw)
anthracene	0.0572	0.845	0.12
fluorene	0.0774	0.536	/
naphtalene	0.176	0.561	0.14
phenanthrene	0.204	1.170	0.51
benzo(a)anthracene	0.108	1.050	0.36
benzo(a)pyrene	0.150	1.450	2.70
chrysene	0.166	1.290	10.7
dibenzo(a,h)anthracene	0.033	/	/
fluoranthene	0.423	2.230	2.60
benzo(k)fluoranthene	/	/	2.40
benzo(ghi)perylene	/	/	7.50
indeno(1,2,3-cd)pyrene	/	/	5.90
pyrene	0.195	1.520	/
Total PAHs	1.610	22.800	/

Table 7. Toxicity thresholds in sediment for PAHs (TEC (threshold effect concentration) and PEC (probable effect concentration) taken from MacDonald *et al.*, 2000; MPC (maximum permissible concentration) taken from Kalf *et al.*, 1997).

The toxicity of mixtures of PAHs has rarely been studied, although natural sediments are contaminated by several substances. We approached this section with the study of the toxicity of the mixture of three PAHs (phenanthrene + fluoranthene + benzo(k)fluoranthene, Verrhiest *et al.*, 2001), and were able to highlight synergetic effects for this specific case, which contrasts with the hypothesis of additivity generally accepted.

We also showed (Verrhiest, 2001; Verrhiest *et al.*, 2001) that the effects are more significant in artificial sediments than in natural sediments, a result that we impute to a partition between the particular and aqueous phases (interstitial water) favouring the aqueous phase, the main path of exposure for the organisms studied (Di Toro *et al.*, 1991), more in artificial sediment. These differences of partition are not only due to the quantity of organic matter in the sediments (content of total organic carbon for which PAHs have great affinity), but also seem to affect the quality of organic matter. (Grathwohl, 1990; DePaolis and Kukkonen, 1997; Haitzer *et al.*, 1999).

More generally, the type of sediment (grain size, proportion of clay, sand, silt, etc.) influences the toxicity of PAHs (Landrum and Faust, 1994; Borglin *et al.*, 1996; Landrum *et al.*, 1997; Haitzer *et al.*, 1998). In the tests on the two natural sediments taken from Lake Aiguebelette ("carbonate sediment" and "peat sediment"), that differ in terms of grain size, composition, and above all the quantity of organic matter, we showed that the toxicity of the "peat" sediment was lower but that the partition coefficients between the dissolved fraction adsorbed on the organic carbon (Koc) were close, leading us to conclude that the quantity of organic matter was the main explanatory factor (Clément *et al.*, 2005b). The response of the

amphipod *Hyallela azteca* to pyrene in the two types of sediment can be explained for the most part by the contents measured in the interstitial water.
Toxicity to benthic organisms is also influenced by the fate of PAHs in the sediment. A balance in the partition of PAHs between solid and liquid phases occurs in the sediment after the latter has been spiked. A study performed on this point showed that the equilibration of artificial sediments for 8 hours led to lower PAH toxicity (Verrhiest, 2001). Beyond this period, other phenomena can contribute towards modifying PAH bioavailability. In natural sediments spiked with a mixture of PAHs (Verrhiest *et al.*, 2002a) over a period of 30 days, we highlighted that PAHs degraded with the exception of benzo(k)fluoranthene, a heavier and probably more recalcitrant PAH. We attributed this degradation to endogenous bacteria of the sediments, whose β-glucosidase activity, measured in parallel, was stimulated. In the studies on pyrene, we also observed a considerable reduction of content over 30 days. This reduction was presumed to be due to ageing, although biodegradation could have been partially responsible (Clément *et al.*, 2005b). This ageing, or reduced extractability (Guthrie *et al.*, 1999; Leppänen et Kukkonen, 2000; Alexander, 2000; Conrad *et al.*, 2002), could be due to the migration of PAHs inside particle pores, making their extraction more difficult and, in parallel, reducing their bioavailability to microorganisms (biodegradation) and higher organisms (toxicity). Therefore the tests on spiked sediments raise the question of sediment conditioning to take into account both the arrival at partition balancing between the different phases, by hypothesising that such a balance exists, and the influence of biodegradation and ageing phenomena on the bioavailability of PAHs in such a way as to reproduce the conditions prevailing for contaminated sediments.

2.5 Bioaccumulation of PAHs during tests

In parallel with monitoring effects on *Daphnia magna*, we were able to measure this organism's bioaccumulation of fluoranthene (Clément *et al.*, 2000; **figure 2**) and pyrene (Jouanneau *et al.*, 2003), in single-species tests (fluoranthene and pyrene) and in microcosms (pyrene). In the studies on pyrene, measurements were also performed on other benthic organisms. The method employed consisted in an extraction procedure using acetone, followed by spectrofluorimetric analysis to identify the PAH spectrum and quantify the dose accumulated. The fact of working each time on a single PAH allowed us to use this simple method, without having to separate the compounds by chromatography.

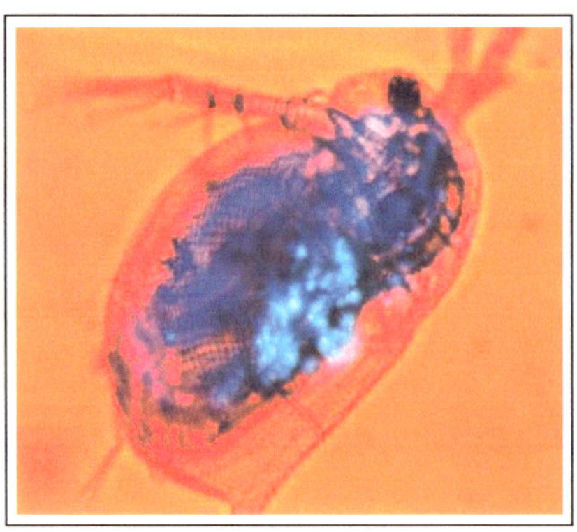

Fig. 2. Bioaccumulation of fluoranthene by *Daphnia magna*, visualised by epifluorescence microscopy (*photo B. Clément*).

In the single-species tests we showed significant bioaccumulation of fluoranthene and pyrene, directly correlated with the PAH content in the water. We found bioconcentration factors of 1000 L/kg (fresh weight) for fluoranthene and 1986 ± 445 L/kg (fresh weight) for pyrene, the latter value being close to those found in the literature (1900 to 2000 L/kg for Nikkilä and Kukkonen (2001) in *D. magna*, 1700 to 3500 L/kg for Akkanen *et al.* (2001) in *D. magna* in river water whose dissolved organic carbon content varied from 0 to 18 mg C/L and spiked at 1 µg pyrene/L, 2702 L/kg for Southworth *et al.* (1978) in *Daphnia pulex*, values expressed in each case on the basis of fresh weight).
The comparison of bioaccumulation and effect data shows good correlation between them (**figure 3**). This is in line with the hypothesis of narcotic effects occurring after a given accumulation in tissues, with narcosis resulting from physical modifications and transformations of the phospholipidic membrane by adsorption of a hydrophobic compound. Disturbances of membrane functions occur when the quantity of compound adsorbed is sufficient (Driscoll *et al.*, 1997). Acute narcosis therefore occurs as a function of the quantity of PAHs bioaccumulated by the organism. According to Landrum *et al.* (1994) and Driscoll *et al.* (1997), EC50 is obtained for different organisms (e.g. the amphipod *Diporeia*) exposed to sediments spiked with PAH for internal doses close to 6 µmol/g (fresh weight of organism). In the tests on fluoranthene and pyrène, we obtained an EC50 of about 0.66 - 0.7 µmol/g (fresh weight), which is consistent with this theory, despite this value being only a tenth of that found by these authors. On the contrary, during the tests with "carbonate sediment" (Aiguebelette) spiked with pyrene on the amphipod *Hyalella azteca* (Jouanneau *et al.*, 2003), we observed that doses leading to effects were of about the same magnitude as the internal dose necessary to achieve 50% mortality in *Diporeia* spp., i.e. 6 µmol/g fresh weight of organism. In the "peat" sediment from Aiguebelette, we also observed an effect on the growth of *Hyalella azteca* for an internal dose of about 13 µmol/g.

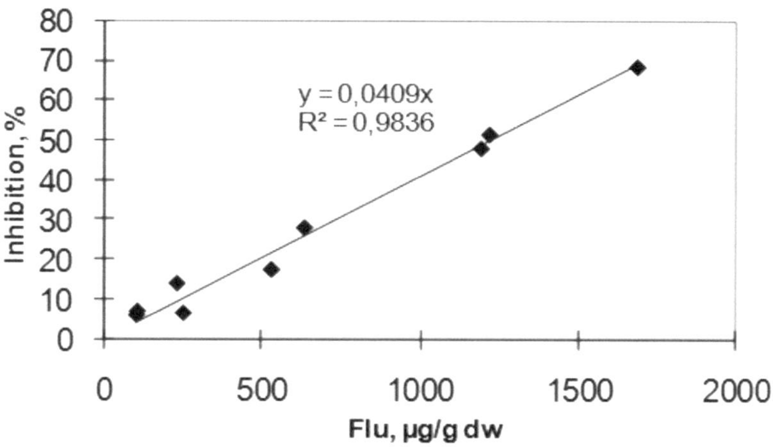

Fig. 3. Relation between the mean fluoranthene dose of *Daphnia magna* and the mean inhibition of mobility following 48 h exposure in darkness (Clément *et al.*, 2000).

During the microcosm bioassays on sediments spiked with pyrene, we were able to monitor the pyrene accumulated by daphnia introduced at the beginning of the assay and by their offspring. The mother daphnia recovered after 12 and 21 days exposure in microcosms probably containing a compound derived from pyrene, as suggested by the differences observed between the fluorescence emission spectra (**figure 4**). It is noteworthy that the measurements performed on the young born in the microcosms and exposed for a maximum of 3 days in the systems, displayed the same type of spectra, whereas there was no modification in the accumulation of pyrene in the young bred by us and introduced in the microcosms for exposure for a maximum of 3 days. These observations suggest that the modified pyrene accumulated by the mothers was transmitted to the embryos and found in the young released during hatching. As with the daphnia in the microcosms, measurements of bioaccumulation in amphipods and chironomidae larvae showed that the modifications of pyrene spectra were similar to those observed for daphnia. Therefore it was not possible to quantify the doses of pyrene accumulated for any of these organisms. The modifications of the pyrene spectra suggest biotransformation processes that could occur in all the organisms of the microcosm, namely daphnia, chironomidae and amphipods. *C. riparius* is known to develop strong pyrene biotransformation activity (Guerrero *et al.*, 2002), and the results of Gourlay *et al.* (2002) showed that although *Daphnia magna* has little effect on fluoranthene, it is capable of biotransforming pyrene and benzo(a)pyrene. As in this study, these authors obtained a different spectrum of bioaccumulated pyrene (with a shift of peaks and an increase of ratio between peaks) that did not result from a matrix effect. As in the case of benzo(a)pyrene, they also highlighted strong fluorescence of the phase recovered in water and less fluorescence of the dichloromethane phase, tending to prove that the product derived from pyrene is more polar and thus clearly the result of biological transformation. This biotransformation of pyrene by *D. magna* was demonstrated by Akkanen and Kukkonen (2003), who showed the involvement of Cytochrome P450 monooxygenases in this elimination route.

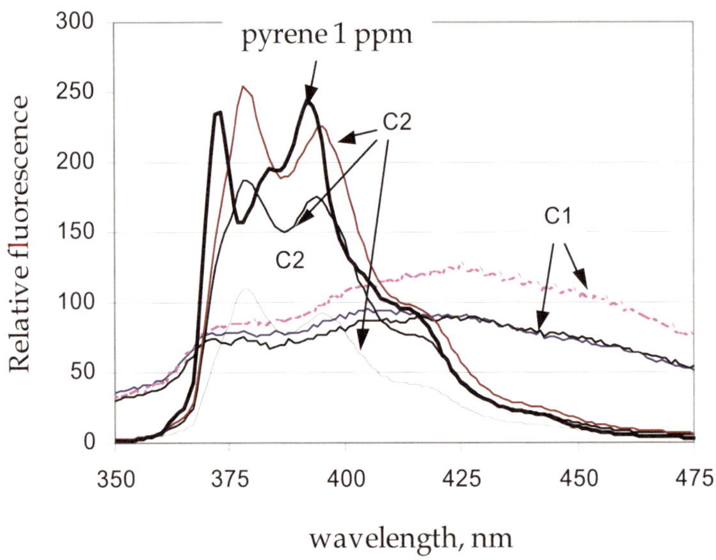

Fig. 4. Fluorescence emission spectra of methanol extracts of mother daphnia exposed for 12 days during microcosm bioassay no. 2 (carbonate sediment contaminated at 50 mg/kg) *versus* the spectrum of pure pyrene at 1 ppm in methanol.

2.6 Responses of the bacterial compartment to PAHs

Microorganisms play a vital role in environmental dynamics as they are involved in biogeochemical cycles acting as a medium through which matter and energy flow. As in any ecosystem, disturbing the microbial communities of a microcosm can impact the entire trophic chain and the balance of the environment. We also considered that it was important to take into account the microbial compartment of sediments in the microcosm bioassays. By using ecologically pertinent parameters (bacterial density and exoenzymatic activity), part of our work consisted in evaluating the responses of indigenous microorganisms in sediments to the contamination of the latter by PAHs.

Furthermore, the capacity of bacteria to biodegrade organic material as well as certain organic compounds, such as PAHs, influences the bioavailability of hydrophobic contaminants. This is why we chose, in addition to monitoring bacterial density, to focus on certain enzymatic activities involved in organic matter transformation processes, namely β-glucosidase (carbon cycle) and leucine-aminopeptidase (nitrogen cycle). In preliminary works on an artificial sediment spiked with fluoranthene (Verrhiest *et al.*, 2000), we also used the activity of INT-reductase by following the protocol of Merlin *et al.* (1995). Fluoranthene had no effect, even at 1000 mg/kg, on any of the parameters monitored (bacterial density of the sediment and supernatant water, INT-reductase and β-glucosidase activity of the sediment and the supernatant water). The study on the mixture of the three PAHs (phenanthrene + fluoranthene + benzo(k)fluoranthene) showed effects at high contents (300 mg/kg) on the bacterial populations of a natural sediment (Ain): reduction of the bacterial density of the sediments, partial inhibition of leucine-aminopeptidase activity, but stimulation of β-glucosidase activity, which it is tempting to parallel with the considerable degradation of fluoranthene and phenanthrene observed during the same bioassay (Verrhiest *et al.*, 2002a). Assays were also performed with pyrene (Jouanneau *et al.*, 2003), first under simple conditions (spiked sediment + water; bacterial density, β-glucosidase and leucine-aminopeptidase activities), then in microcosms (bacterial density and β-glucosidase). Over the range 1, 10, 50, 100 and 200 mg/kg, no significant effect was obtained in the single-species tests or in the microcosms.

In conclusion, PAHs do not appear to lead to effects on the bacterial compartment, at least as seen through the few parameters monitored in these works, except for very high concentrations rarely encountered in the environment. On the other hand, capacities to degrade phenanthrene, fluoranthene and pyrene (Jouanneau *et al.*, 2003) have been demonstrated.

3. General discussion and conclusion on the microcosm study of PAH toxicity

The purpose of the microcosm bioassays performed on sediments contaminated by PAHs was to evaluate the risks for lentic ecosystems related to the presence of these ubiquitous organic contaminants in this sediment compartment. We were able to obtain reasonably realistic results since it was possible to take into account different compartments of these ecosystems, and the relations between the different populations represented and between these populations and their environment.

We first confirmed the possibility of spiking initially non or only slightly contaminated artificial and natural sediments with PAHs, drawing attention to the need to ensure good spike rates and good distribution of PAHs in the sediment, as the physicochemical properties of the latter can influence both rates and distribution. Once the sediment has been spiked, it is vital to observe the equilibration period, evaluated by us at being at one week (Verrhiest, 2001), which generally results in lower toxicity due to the higher adsorption of PAHs in the solid phase and which corresponds better to the situations most usually encountered in the environment (deposited, undisturbed sediment). We also looked into the possibility of extending this equilibration period to take into account ageing (migration of PAHs within particles, reducing their bioavailability) and biodegradation phenomena that we showed could be significant for certain PAHs over a few weeks (Verrhiest *et al.*, 2002a; Clément *et al.*, 2005b). In this framework we think that the use of natural sediments, with microflora more suitable for degrading PAHs and physicochemical and mineralogical characteristics more likely to simulate ageing, is probably preferable to using artificial sediments. The latter have other disadvantages related in particular to the impossibility of pertinently simulating the physicochemical properties impacting on the partition of PAHs between the dissolved and particular (mineral and organic) phases. On several occasions the comparative study of the fate and effects of PAHs in artificial and natural sediments showed large divergences between these two types of sediment, generally expressing an overestimation of risks of exposure and effects in artificial sediment (Verrhiest, 2001; Verrhiest *et al.*, 2001), not only for benthic microorganisms but also for pelagic organisms in contact directly with the sediment or/and *via* the water column. This overestimation is related to the difficulty of representing all the potential adsorption sites, in particular linked to a generally complex, natural organic material. Although we recommend giving up the use of artificial sediment in this type of research, the question of what natural sediment model should be used remains unanswered, given the great diversity of sediments. Here again, the *a priori* wide distribution of physicochemical properties of sediments can lead to behaviours that vary considerably as a function of the sediment chosen. The sediment should therefore be selected according to a site or a specific property under study.

Other procedures of the microcosm test protocol have a significant influence on the fate and effects of PAHs. We showed that the lighting chosen in our tests (classical laboratory fluorescent light) favoured the phototoxicity of certain PAHs (Clément *et al.*, 2000, 2005b). The adsorption of UV by PAHs governs photosensitisation and phototransformation phenomena, varying according to the type of radiation (UVA and UVB). Thus the choice of lighting more or less representative of the solar light spectrum influences the results observed (Wilcoxen *et al.*, 2003). Certain authors have also shown in the laboratory that the penetration of UV down to the sediment could generate phototoxicity for benthic organisms (Ankley *et al.*, 1994, 1995).

However, McDonald and Chapman (2002) questioned the ecological pertinence of phototoxicity, widely studied in the laboratory but, according to them, rarely expressed *in situ*. Indeed, many parameters have to be taken into account to estimate the probability of an organism bioaccumulating PAHs being subject to active solar radiation, as phototoxicity is essentially explained by photosensitisation. McDonald and Chapman (2002) showed that a large number of processes *in situ* allow organisms to avoid this exposure (reduction of bioavailability by organic and particular matter, protection mechanisms against UV in certain organisms, deep burrowing of benthic organisms, shade provided by aquatic plants, etc.). However, certain experimental parameters lead to overestimating exposure (glass flasks facilitate the passage of light in several directions, a shallow water column facilitates the passage of UVs down to the sediment, thin layers of sediment, radiation used at constant intensity non representative of the variations observed during the day, water saturated with oxygen favouring photosensitisation, etc.). In tests performed in the presence of daphnia and algae, we were able to demonstrate their role in reducing the phototoxicity of PAHs to daphnia.

As mentioned earlier, the use of glass recipients does not prevent the adsorption of PAHs on their walls, a phenomenon that, for the generally low contents of PAHs found in the water column, can lead to underestimating the contents expected (Clément *et al.*, 2005b). Conversely, the use of synthetic environments generally free of dissolved organic materials leads to overestimating exposure. Regarding this, the incorporation of a sediment permits reducing this bias by enriching the water column with dissolved organic matter. Sediment also contributes particular and colloidal materials, the latter leading to an increase in the exposure of organisms in the water column to the PAHs adsorbed in it (Baumard, 1997). Similarly, the presence of micro-algae also modifies the exposure of daphnia and other consuming organisms (amphipods), though we did not have the opportunity to specify in which direction since although the algae capture some of the PAHs and thus reduce the dissolved fraction, they introduce an additional route of exposure for the organisms that consume them.

Although we studied monocontamination in most cases, we were also able to evaluate the effects of mixtures of PAHs, a situation closer to reality. Although the hypothesis of effect additivity is generally accepted, we showed that synergetic effects are possible (Verrhiest *et al.*, 2001). It is however necessary to go further by working on a mixture of a greater number of PAHs (for example the 16 priority PAHs of the USEPA), and by incorporating other types of pollutants, such as metals.

The microcosm bioassays made it possible to diversify the exposure routes of organisms, for example, daphnia present in the water column but which can also be in contact with sediment particles. The results of two different studies failed to converge: whereas the effects on daphnia were increased by sediment contaminated by PAHs in one (Verrhiest *et al.*, 2001), the other showed effects were essentially linked to pyrene contents in the water column. This absence of convergence can be explained by the different natures of the sediments used in these two studies and probably other parameters. Specific tests should be performed to study this point.

The toxicity criteria studied do not show that PAHs are very toxic to benthic organisms, even for exposures lasting a month. This is generally due to the high adsorption of PAHs on the particular and dissolved organic matter of the sediments which significantly reduces the bioavailability of these substances, a reduction that continues through time (ageing). It is reassuring in this initial approach to observe that chironomidae are capable of developing and emerging in nonetheless heavily contaminated sediments, a fact corroborated by results obtained on certain natural sediments also heavily contaminated by PAHs and heavy metals. We do not have data on the effects of PAHs on amphipod reproduction, due to the short time in which the tests were performed. It would be interesting to take this biological criterion into account by longer exposure or by the exposure of older individuals at the beginning of the test. Likewise, a study on several generations of chironomidae exposed to PAHs would make it possible to evaluate long term effects, by taking into account the number of hatched larvae and their capacity to pass through their life cycle. The few bioaccumulation measurements that we performed in the test on pyrene and the results in the literature encourage us to persevere along these lines, since the lower toxicity of PAHs could be explained by the capacity of pelagic (daphnia, algae) and benthic (chironomidae and amphipods) organisms to biotransform PAHs. Given the measurements performed on the microbial compartment, it appears that the presence of high PAH contents does not significantly disturb this compartment. This result is important as the functioning of the ecosystem depends in part on the biological activities of sediment which contributes to recycling organic matter and renewing the mineral elements required for primary producers.

Although we were able to perform a global study of the fate of PAHs in microcosms, we were unable to identify the role played by organisms in this fate and in the effects stemming from them, thus this could form the basis for an additional path of research. The bioturbation activity of benthic organisms can contribute towards modifying the distribution of contaminants in sediment and stimulating their biodegradation through better oxygenation, for example, of superficial layers. When studying the fate of fluoranthene in the presence of the marine worm *Capitella*, Madsen *et al.* (1997) observed that bioturbation activity contributed towards burying fluoranthene, but the total loss of fluoranthene in the sediment was higher in the presence of the worms, whose activity increased the transfer of fluoranthene to the supernatant water or/and stimulated the biodegradation of this PAH. The role played on this level by the chironomidae and the amphipods used in our tests remains to be determined. Do they contribute to greater exposure of pelagic organisms or, on the contrary, do they reduce the risks to which they are exposed ? These questions are obviously far-reaching, as answering them requires a large number of microcosm bioassays in which certain populations are present simultaneously in order to highlight the interactions described above.

4. References

Akkanen, J. & Kukkonen, J.V.K. (2003). Biotransformation and bioconcentration of pyrene in *Daphnia magna*, *Aquatic Toxicol.* 64: 53-61.

Akkanen, J.; Penttinen, S., Haitzer, M. & Kukkonen, J.V.K. (2001). Bioavailability of atrazine, pyrene and benzo(a)pyrene in European river waters, *Chemosphere* 45: 453-462.

Alexander, M. (2000). Aging, bioavailability, and overestimation of risk from environmental pollutants, *Environ. Sci. Technol.* 34: 4259-4264.

Ankley, G.T.; Collyard, S.A., Monson, P.D. & Kosian, P.A. (1994). Influence of ultraviolet light on the toxicity of sediments contaminated with polycyclic aromatic hydrocarbons, *Environ. Toxicol. Chem.* 13: 1791-1796.

Ankley, G.T.; Erickson, R.J., Phipps, G.L., Mattson, V.R., Kosian, P.A., Sheedy, B.R. & Cox, J.S. (1995). Effects of light intensity on the phototoxicity of fluoranthene to a benthic macroinvertebrate, *Environ. Sci. Technol.*, 29: 2828-2833.

Arfsten, D.P.; Schaeffer, D.J. & Mulveny, D.C. (1996). The effects of near ultraviolet light radiation on the toxic effects of polycyclic aromatic hydrocarbons in animals and plants: a review, *Ecotoxicol. Environ. Safety* 33: 1-24.

Baumard, P. (1997). Biogéochimie des composés aromatiques dans l'environnement marin, Thèse de Doctorat en Chimie Analytique et Environnement de l'Université de Bordeaux I, 145 pages.

Baumard, P., Budzinski, H. & Garrigues, P. (1998). Polycyclic aromatic hydrocarbons in sediments and mussels of the western Mediterranean Sea. *Environ. Toxicol. Chem.* 17: 765-776.

Bennett, W.E., Maas, J.L., Sweeney, S.A. & Kagan, J. (1986). Phototoxicity in aquatic organisms: the protecting effect of beta-carotene, *Chemosphere* 15: 781-786.

Boese, B.L., Lamberson, J.O., Swartz, R.C., Ozretich, R. & Cole, F. (1998). Photoinduced toxicity of PAHs and alkylated PAHs to a marine infaunal amphipod (*Rhepoxinius abronius*), *Arch. Environ. Contam. Toxicol.* 34: 235-240.

Borglin, S., Wilke, A., Jepsen, R. & Lick, W. (1996). Parameters affecting the desorption of hydrophobic organic chemicals from suspended sediments, *Environ. Toxicol. Chem* 15: 2254-2262.

Bray, M., Babut, M., Vollat, B., Montuelle, B., Devaux, A., Bedell, J.P., Delolme, C., Durrieu, C., Clément, B., Perrodin, Y. & Triffault-Bouchet, G. (2003). Evaluation écotoxicologique de matériaux de dragage: Application to 5 sédiments du Nord-Pas de Calais, rapport Cetmef/Drast/VNF, septembre 2003, 142 pages.

Bray, M., Babut, M., Montuelle, B., Vollat, B., Devaux, A., Delolme, C., Durrieu, C., Bedell, J.P. & Clément, B. (2001). Evaluation écotoxicologique de sédiments contaminés ou de matériaux de dragage. (III). Application au Canal de l'Est Branche Sud, rapport Cetmef/Drast et VNF, avril 2001, 70 pages.

Chevron, N. (2004). Mise en évidence de la biodégradation du pyrène dans une matrice organique, rapport de DEA Sciences et Stratégies Analytiques de Lyon 1, 37 p.

Clément, B., Cauzzi, N., Godde, M., Crozet, K. & Chevron, N. (2005b). Pyrene toxicity to aquatic pelagic and benthic organisms in single-species and microcosm tests, *Polycyclic Aromatic Compounds* 25: 271-298.

Clément, B., Muller, C. & Verrhiest, G. (2000). Influence of exposure conditions on the bioavailability of fluoranthene to *Daphnia magna* (Cladocera), *Polycyclic Aromatic Compounds* 20: 259-274.

Conrad, A.U., Comber, S.D. & Simkiss, K. (2002). Pyrene bioavailability; effect of sediment-chemical contact time on routes of uptake in an oligochaete worm, *Chemosphere* 49: 447-454.

DePaolis, F. & Kukkonen, J. (1997). Binding of organic pollutants to humic and fulvic acids: influence of pH and the structure of humic material, *Chemosphere* 34: 1693-1704.

Di Toro; D.M.; Zarba, C.S., Hansen, D.J., Berry, W.J., Swartz, R.C., Cowan, C.E., Pavlou, S.P., Allen, H.E., Thomas, N.A. & Paquin, P.R. (1991). Technical basis for establishing sediment quality criteria for nonionic organic chemicals using equilibrium partitioning, *Environ. Toxicol. Chem.* 10: 1541-1583.

Ditsworth, G.R.; Schults, D.W. & Jones, J.K.P. (1990). Preparation of benthic substrates for sediment toxicity testing, *Environ. Toxicol. Chem.* 9: 1523-1529.

Driscoll, S.K., Landrum, P.F. & Tigu, E. (1997). Accumulation and toxicokinetics of fluoranthene in water-only exposures with freshwater amphipods, *Environ. Toxicol. Chem.* 16: 754-761.

Eisler, R. (2000). Polycyclic Aromatic Hydrocarbons. In: *Handbook of chemical risk assessment. Health hazards to humans, plants and animals. Organics.* Ed. R. Eisler. CRC Press. Lewis Publishers, pp. 1343-1411.

Garban, B. & Ollivon, D. (1995). Transport et devenir de polluants et micropolluants en Seine. Rôle des matières en suspension et des sédiments dans les processus de transfert, Thèse de Doctorat de l'Université de Paris VI, pp 46-49.

Gauthier, T.D.; Shane, E.C., Guerin, W.F., Seltz, W.R. & Grant, C.L. (1986). Fluorescence quenching method for determining equilibrium constants for polycyclic aromatic hydrocarbons binding to dissolved organic materials, *Environ. Sci. Technol.* 20: 1162-1166.

Geffard, O.; Budzinski, H. & His, E. (1999). The toxicity of phenanthrene (PHE), 2 methylphenanthrene (2MP) and benzo(a)pyrene (BAP) on embryogenesis and larval development oysters (*Crassostrea gigas*) and sea urchins (*Paracentrotus lividus*). Abstract. 17 th International Symposium on Polycyclic Aromatic Compounds, 25-29 October, Bordeaux, France, p 195.

Gourlay, C; Miège, C, Tusseau-Vuillemin, MH, Mouchel, JM & Garric, J. (2002). The use of spectrofluorimetry for the determination of polycyclic aromatic hydrocarbons bioaccumulation and biotransformation in *Daphnia magna. Polycyclic Aromatic Compounds* 22: 3-4, pp 501-516.

Grathwohl, P. (1990). Influence of organic matter from soils and sediments from various origins on the sorption of some chlorinated aliphatic hydrocarbons, implications on Koc correlations, *Environ. Sci. Technol.* 24: 1687-1693.

Guerrero, N.R. V.; Taylor, M.G., Davies, N.A., Lawrence, M.A.M., Edwards, P.A., Simkiss, K. & Wider, E.A. (2002). Evidence of differences in the biotransformation of organic contaminants in three species of freshwater invertebrates, *Environ. Pollut.* 117: 523-530.

Guthrie, E.A.; Bortiatynski, J.M., Van Heemst, J.D.H., Richman, J.E., Hardy, K.S., Kovach, E.M. & Hatcher, P.G. (1999). Determination of [C-13]pyrene sequestration in sediment microcosms using flash pyrolysis GC-MS and C-13 NMR, *Environ. Sci. technol.* 33: 119-125.

Haitzer, M.; Höss, S., Transpurger, W. & Steinberg, C. (1998). Effects of dissolved organic matter (DOM) on the bioconcentration of organic chemicals in aquatic organisms – a review, *Chemosphere* 37: 1335-1362.

Haitzer, M.; Höss, S., Traunspaurger, W. & Steinberg, C. (1999). Relationship between concentration of dissolved organic matter (DOM) and the effect of DOM on the bioconcentration of benzo(a)pyrene, *Aquat. Toxicol.* 45: 147-158.

Huang, X.D.; Dixon, D.G. & Greenberg, G.B.M. (1995). Increased polycyclic aromatic hydrocarbon toxicity following their photomodification in natural sunlight: impacts on the duckweed *Lemna gibba* L.G-3, *Ecotoxicol. Environ. Safety* 32: 194-200.

Huovinen, P.S.; Soimasuo, M.R. & Oikari, A.O.J. (2001). Photoinduced toxicity of retene to *Daphnia magna* under enhanced UV-B radiation, *Chemosphere* 45: 683-691.

Jouanneau, Y.; Blake, G., Clément, B., David, B. & Naffrechoux, E. (2003). Devenir des hydrocarbures aromatiques polycycliques (HAP) dans un écosystème aquatique et impact sur les organismes vivants: exemple du pyrène. Rapport CNRS-PEVS, Ecosystèmes et Environnement, appel d'offres "Dynamique des contaminants", 126 pages.

Kalf, D.F.; Crommentuijn, T. & Van De Plassche, E.J. (1997). Environmental Quality Objectives for 10 Polycyclic Aromatic Hydrocarbons (PAHs), *Ecotoxicol. Environ. Safety* 36: 89–97.

Krylov, S.N.; Huang, X.D., Zeiler, L.F., Dixon, D.G. & Greenberg, B.M. (1997). Mechanistic quantitative structure-activity relationship model for the photoinduced toxicity of polycyclic aromatic hydrocarbons: I. Physical model based on chemical kinetics in a two-compartment system, *Environ. Toxicol. Chem.* 16: 2283-2295.

Landrum, P.F.; Dupuis, W.S. & Kukkonen, J. (1994). Toxicokinetics and toxicity of sediment-associated pyrene and phenanthrene in *Diporeia* spp.: examination of equilibrium-partitioning theory and residue-based effects for assessing hazard, *Environ. Toxicol. Chem.* 13: 1769-1780.

Landrum, P.F. & Faust, W.R. (1994). The role of sediment composition on the bioavailability of laboratory-dosed sediment-associated organic contaminants to the amphipod, *Diporeia* (spp.), *Chem. Speciation Bioavail.* 6: 85-92.

Landrum, P.F.; Gossiaux, D.C. & Kukkonen, J. (1997). Sediment characteristics influencing the bioavailability of nonpolar organic contaminants to *Diporeia* spp., *Chemical Speciation and Bioavailability* 9: 43-55.

Lei, A.P.; Wong, Y.S. & Tam, N.F.Y (2003). Pyrene-induced changes of glutathione-S-transferase activities in different microalgal species, *Chemosphere* 50: 293–301.

Lei, A.P.; Wong, Y.S. & Tam, N.F.Y. (2001). Removal of pyrene by different microalgal species. In: Proceedings of Asian Waterqual 2001, 1st Asia-Pacific Regional Conference, Fukuoka, Japan. pp. 969–974.

Leppänen, M.T. & Kukkonen, J.V.K. (2000). Effect of sediment-chemical contact time on availability of sediment-associated pyrene and benzo(a)pyrene to oligochaete worms and semi-permeable membrane devices, *Aquat. Toxicol.* 49: 227-241.

Lyons, B.P.; Pascoe, C.K. & McFadzen, I.R.B. (2002). Phototoxicity of pyrene and benzo[a]pyrene to embryo-larval stages of the pacific oyster *Crassostrea gigas*, *Marine Environ. Research* 54: 627–631.

MacDonald, D.D.; Ingersoll, C.G. & Berger, T.A. (2000). Development and evaluation of consensus-based sediment quality guidelines for freshwater ecosystems, *Arch. Environ. Contam. Toxicol.* 39: 20-31.

Madsen, S.D.; Forbes, T.L. & Forbes, V.E. (1997). Particle mixing by the polychaete *Capitella* species 1: coupling fate and effect of a particle-bound organic contaminant (fluoranthene) in a marine sediment, *Marine Ecology Progress Series* 147: 129-142.

May, W.E. (1980). The solubility behaviour of polycyclic aromatic hydrocarbons in aqueous systems, *Adv. Chem. Series.* 285 : 143-192.

McCarthy, J.F. (1983). Role of particulate organic matter in decreasing accumulation of polynuclear aromatic hydrocarbons by *Daphnia magna, Arch. Environ. Contam. Toxicol.* 12: 559-568.

McDonald, B.G. & Chapman, P.M. (2002). PAH phototoxicity - an ecologically irrelevant phenomenon ?, *Marine Pollut. Bull.* 44: 1321–1326

Merlin, G.; Lissolo, T. & Morel, V. (1995). Precautions for Routine Use of INT Reductase Activity for Measuring Biological Activities in Soil and Sediments, *Environ. Toxicol. Water Qual.* 10: 185-192.

Miller, J.S. (1999). Determination of polycyclic aromatic hydrocarbons by spectrofluorimetry, *Anal. Chim. Acta* 38 : 27-44.

Munoz, M.J. & Tarazona, J.V. (1993). Synergistic effect to two- and four-component combinations of the polycyclic aromatic hydrocarbons: Phenanthrene, Anthracene, Naphthalene and Acenaphthene on *Daphnia magna, Bull. Environ. Contam. Toxicol.* 50: 363-368.

Naffrechoux, E.; Combet, E., Fanget, B., Paturel, L. & Saber, A. (1999). Occurence and fate of PAHs from motorway runoff in the north drainage basin of Annecy lake, France. Abstract. 17 th International Symposium On Polycyclic Aromatic Compounds, 25-29 October, Bordeaux, France, p 182.

Newsted, J.L. & Giesy, J.P. (1987) Predictive models for photoinduced acute toxicity of polycyclic aromatic hydrocarbons to *Daphnia magna*, Strauss (Cladocera, crustacea), *Environ. Toxicol. Chem.* 6: 445-461.

Nikkilä, A. & Kukkonen, J.V.K. (2001). Effects of dissolved organic material on binding and toxicokinetics of pyrene in the waterflea *Daphnia magna, Arch. Environ. Contam. Toxicol.* 40: 333-338.

NRCC (1983). Polycyclic aromatic hydrocarbons in the aquatic environment: formation, sources, fate and effects on aquatic biota. NRC Associate Committee on the Scientific Criteria for Environmental Quality, National Research Council of Canada, NRCC Publication No. 18981, Ottawa, 209 pp.

Ollivon, D.; Garban, B. & Chesterikoff, A. (1995). Analysis of the distribution of some polycyclic aromatic hydrocarbons in sediments and suspended matter in the river Seine (France), *Water Air Soil Pollut.* 81: 135-152.

Pelletier, M.C.; Burgess, R.M., Ho, K.T., Kuhn, A., McKinney, R.A. & Ryba, S.A. (1997). Phototoxicity of individual polycyclic aromatic hydrocarbons and petroleum to marine invertebrate larvae and juveniles, *Environ. Toxicol. Chem.* 16: 2190-2199.

Raymond, N.; Geoffroy, L., Bourasseau, L., Budzinski, H., Nadalig, T. & Gilewicz, M. (1999). Bacterial aerobic degradation of polycyclic aromatic hydrocarbons by pure strains: characterisation of catabolic abilities using respiration studies. Abstract. 17 th International Symposium on Polycyclic Aromatic Compounds, 25-29 October, Bordeaux, France, p. 213.

Ren, L.; Huang, X.D., Mcconkey, B.J., Dixon, D.G. & Greenberg, B.M. (1994). Photoinduced toxicity of three polycyclic aromatic hydrocarbons (Fluoranthene, Pyrene, and Naphthalene) to the Duckweed *Lemna gibba* L. G-3, *Ecotoxicol. Environ. Safety* 28: 160-171.

Southworth, G.R.; Beauchamp, J.J. & Schmieders, P.K. (1978). Bioaccumulation potential of polycyclic aromatic hydrocarbons in *Daphnia pulex, Water Research*, 12: 973-977.

Swartz, R.C.; Ferraro, S.P., Lamberson, J.O., Cole, F.A., Ozretich, R.J., Boese, B.L., Schults, D.W., Behrenfeld, M. & Ankley, G.T. (1997). Photoactivation and toxicity of hydrocarbon compounds in marine sediment, *Environ. Toxicol. Chem.* 10: 2151-2157.

Swartz, R.C.; Schults, D.W., Ozretich, R.J., Lamberson, J.O., Cole, F.A., Dewitt, T.H., Redmond, M.S. & Ferraro, S.P. (1995). PAHs: a model to predict the toxicity of polynuclear aromatic hydrocarbon mixtures in field-collected sediments, *Environ. Toxicol. Chem.* 14:1977-1987.

Valette-Silver N., Hammed M.J., Efurd D.W., Robertson A. (1999). Status of the contamination in sediments and biota form the western Beaufort Sea (Alaska), *Mar. Poll. Bull.* 38: 702-722.

Verrhiest, G. (2001). Toxicité de sédiments d'eau douce contaminés par des HAPs. Influence de la nature des sédiments sur la biodisponibilité des HAPs, Thèse de Doctorat de l'Université de Savoie (spécialité: Biologie et Biochimie Appliquées).

Verrhiest, G.; Clément, B. & Blake, G. (2001). Single and combined effects of sediment-associated PAHs on three species of freshwater macroinvertebrates, *Ecotoxicology* 10: 363-372.

Verrhiest, G.; Clément, B. & Merlin,G. (2000). Influence of sediment organic matter and fluoranthene-spiked sediments on some bacterial parameters in laboratory freshwater / formulated sediment microcosms, *Aquat. Ecosystem Health Manag.* 3: 359-368.

Verrhiest, G.; Clément, B., Volat, B., Montuelle, B. & Perrodin, Y. (2002a). Interactions between a polycyclic aromatic hydrocarbons mixture and the microbial communities in a natural freshwater sediment. *Chemosphere* 46: 187-196.

Verrhiest, G.; Cortes, S., Clément, B. & Montuelle, B. (2002b). Chemical and bacterial changes during laboratory conditioning of formulated and natural sediments, *Chemosphere* 46: 961-974.

Vindimian, E.; Bisson, M., Dujardin, R., Flammarion, P., Garric, J., Babut, M., Lamy, M.-H., Porcher, J.-M. & Thybaud, E. (2000). Complément au SEQ-Eau: méthode de détermination des seuils de qualité pour les substances génotoxiques. Rapport final", 2000, 151 p, INERIS, Agence de l'eau Rhin-Meuse - Verneuil-en-Halatte - avril 2000.

Warshawsky, D.; Cody, T., Radike, M., Reilman, R, Schumann, B., LaDow, K. & Schneider, J. (1995). Biotransformation of benzo[a]pyrene and other polycyclic aromatic hydrocarbons and heterocyclic analogs by several green algae and other algal species under gold and white light, *Chemico-Biol. Interact.* 97: 131-148.

Wernersson, A.S.; Dave, G.R. & Nilsson, E. (1999). Combining sediment quality criteria and sediment bioassays with photoactivation for assessing sediment quality along the Swedish West Coast, *Aquat. Ecosystem Health Manag.* 2: 379-389.

Wernersson, A-S. & Dave, G.R. (1997). Phototoxicity identification by solid phase extraction and photoinduced toxicity to *Daphnia magna, Arch. Environ. Contam. Toxicol.* 32: 268-273.

Wilcoxen, S.E.; Meier, P.G. & Landrum, P.F. (2003). The toxicity of fluoranthene to *Hyalella azteca* in sediment and water-only exposures under varying light spectra, *Ecotoxicol. Environ. Safety* 54: 105 -117.

Witt, G. (1995). Polycyclic aromatic hydrocarbons in water and sediment of Baltic Sea, *Mar. Poll. Bull.* 31: 237-248.

5

Exposure Assessment to Persistent Organic Pollutants in Wildlife: The Case Study of Coatzacoalcos, Veracruz, Mexico

Guillermo Espinosa-Reyes, Donaji J. González-Mille, César A. Ilizaliturri-Hernández, Fernando Díaz-Barríga Martínez and Jesús Mejía-Saavedra
Universidad Autónoma de San Luis Potosí, Facultad de Medicina-Departamento de Toxicologia Ambiental, Mexico

1. Introduction

Until the early 70s, it was thought that pollution was a phenomenon circumscribed to zones where pollutants were generated. Because of that, in each country concern was limited to regions where pollutant concentration was higher or its danger was greater. However, it has gradually become aware that pollution is a problem that affects everybody and, because of that, everybody is responsible to control it, regardless of the sites distance where pollutants are produced. Therefore, the problem of pollution has become a global phenomenon. Mankind has always depended on natural resources located in the region where they dwell. Nevertheless, the fast population growth coupled with a fast agricultural and industrial development as well as life style changes have increased emissions of pollutants in different ecosystems.

Persistent organic pollutants (POPs) is a group of compounds chemically very stable, able to travel considerable distances and it is resistant to natural degradation processes, most of them were produced to be used as pesticides and certain chemicals to be used as industrial processes, and others are generated as by-products unintentionally from human activities, such as combustion processes or power generation (PNUMA, 2005). Most of these compounds are highly toxic; they bioaccumulate in human and animal tissue, mainly in the fatty tissues, and can damage different organs and systemic targets such as the liver, kidney, hormonal system, nervous system, etc., of both humans and wildlife. According to the Stockholm Convention held in 2001, there are twelve compounds known as POPs: pesticides (DDT, aldrin, chlordane, dieldrin, endrin, mirex, toxaphene and heptachlor), industrial chemicals (hexachlorobenzene and polychlorinated biphenyls -PCB-) and unintentional compounds (dioxins, furans, PCDD-and PCDF-) [Albert, 2004]. In May of 2009 nine new Chemicals were added to the POPs list: alpha hexachlorocyclohexane, beta hexachlorocyclohexane; hexabromodiphenyl heptabromodiphenyl ether and ether tetrabromodiphenyl pentabromodiphenyl ether and ether chlordecone, hexabromobiphenyl, lindane, pentachlorobenzene, perfluorooctane sulfonic acid, its salts and perfluorooctane sulfonyl fluoride.

POPs main route of entry into the organism is food. However, we cannot ignore environmental exposure (inhalation) and dermal exposure (accidents). Because of their properties, POPs are classed as persistent, bioaccumulative and toxic. Therefore, POPs are to be considered as one of the most harmful groups of Toxic environmental pollutants to humans and wildlife. In countries where these compounds have been used are frequently found residuals in food. They are a problem because of their persistence in the environment and characteristics of bioaccumulation and biomagnification along the food chain and because these compounds generate toxic effects in both human population and in biota.

POPs, mainly organochlorine compounds make up a big part of hazardous waste. Mexico annually generates approximately 8 million tons of hazardous waste. Of this amount, only 12 percent is handled properly. The question is: Where does the remaining 88% go? To make matters worse, Mexico's infrastructure for hazardous waste is by far insufficient.

Coatzacoalcos, Veracruz is one of the most commercial and industrialized ports in Mexico. Presently, the Coatzacoalcos River and the areas surrounding it are regarded by many as some of the most heavily polluted sites in Mexico. Several of Mexico's chief petrochemical complexes, such as Cangrejera, Morelos and Pajaritos are based in the region. Furthermore, there have been various toxic substances present in the area which have been stored inside environmental and biological compartments, including persistent organic pollutants (Espinosa-Reyes et al., 2010; Gonzalez-Mille et al., 2010; Stringer et al., 2001), polycyclic aromatic hydrocarbons, volatile organic compounds (Riojas-Rodriguez et al., 2008), polybrominated compounds (Blake 2005), dioxins and metals (Petrlink & DiGangi 2005; Rosales 2005; Vázquez-Botello 2004).

A number of POPs have been registered in Coatzacoalcos. One of these, Hexachlorocyclohexane (HCH), is a manufactured chemical from which there are, theoretically, eight chemical forms or isomers. The three most common isomers are α-HCH, β-HCH and γ-HCH (commonly called lindane). Lindane is used as a pesticide on fruit and vegetable crops as well as on forest plantations; it is also found in medications to treat diseases such as scabies and pediculosis. There are no records indicating that lindane has ever been manufactured in Mexico; however, approximately 20 tons of these compounds are imported and subsequently used in Mexico each year. At present, lindane is authorized for use in Mexico for ectoparasite control in livestock for ticks, fleas, and common fly larvae. It is also registered for use as a seed treatment for oats, barley, beans, corn, sorghum and wheat. Pharmaceutical uses of lindane in Mexico include the formulation of creams and shampoos for scabies and lice treatment (CEC, 2006; ATSDR, 2005). In 1994, the Canadian Environmental Protection Act (CEPA) proclaimed hexachlorobenzene (HCB) as highly toxic. HCB is a manufactured chemical (which was) used as a wood preservative, as a fungicide for treating seeds and as intermediary in organic syntheses. Additionally, hexachlorobenzene can be formed as an unwanted byproduct in preparation processes like in the synthesis of organochlorines from high-temperature sources (Sala et al., 1999; Newhook & Meek 1994). Dichlorodiphenyldichloroethane (DDT) is a synthetic organochloride which is relatively stable with slow degradation rates through sunlight or oxidation, and possesses good absorption capacity and resistance to biodegradation in sediments and soils. It is also insoluble in water (CEC, 2001). In 1945, DDT was used for the first time in Mexico for the control of Malaria, and was widely used in agriculture between the 50s and 70s (CEC, 1997). The use of DDT in the Malaria Control Program was abandoned in the year 2000, when it was replaced by pyrethroids. Polychlorinated biphenyls (PCBs) are allowed in "totally enclosed uses" such as coolants and lubricants, in transformers and capacitors. Dioxins are produced during the combustion of organic materials containing chlorine as well as during the manufacture of various chlorine-containing chemicals, such as ethylene dichloride. Existing involuntary sources of intake include electric arc furnaces, shredders, sinter plants, cement plants, cremation facilities, and coal-based power plants (Lutharddt et al., 2002).

Biomonitoring wildlife can be used to detect chemical pollution and to evaluate the ecosystem's health, using test species as systematic models in the evaluation of risks associated to paths of real exposure. Wildlife species residing in polluted sites are exposed to complex mixtures of pollutants through multiple pathways which could hardly be evaluated in lab studies. The main purpose of this research was to pinpoint exposure levels to POPs in wildlife from different sets of ecosystems throughout the industrial area of Coatzacoalcos, Veracruz, Mexico to obtain a baseline of the ecological condition of this region.

2. Materials and methods

2.1 Test area, sampling sites and species selection

The Coatzacoalcos region is located in the South eastern State of Veracruz, Mexico, in the municipality under the same name, at 18° 8' 56 " N and 94° 24' 41" W. The average altitude is 14 m.a.s.l. The predominant climate is tropical rain [Am (i') gw"], the average annual temperature is 24.5°C and average annual rainfall is 2780.1 mm (García, 2004). The main inland body of water is the Coatzacoalcos River, which has an area of 322 km. It originates above 2000 m in elevation in the State of Oaxaca and over its course is fed by countless other rivers (Jaltepec, Coachapa, Uxpanapa, Calzada) and streams (Teapa, Tepeyac, San Francisco) which inflows contribute to the discharge of pollutants (Páez-Osuna et al., 1986.; Rosales-Hoz & Carranza-Edwards, 1998). The region is comprised of urban, industrial, livestock, riparian and wetland areas. However, its main activity is chemical, namely petrochemical (Ruelas-Inzunza et al., 2007).

In October 2006, six sampling stations for biological sampling were set up at the lower basin of the Coatzacoalcos River (Fig. 1). The selection of sampling sites was based on wind direction, location of industrial zones and urban areas, the presence of organisms and the influence of riparian systems as well as on previous investigations within the area (Páez-Osuna et al., 1986; Rosales-Hoz & Carranza-Edwards 1998; Stringer et al., 2001; Bahena-Manjarrez et al., 2002.).

In this research we selected earthworms, crabs, fish, toads, turtles, iguanas, and crocodiles to measure levels of POPs in muscle or blood. These groups are critical species because they have an important role in the ecosystems dynamics and/or an importance (economic, cultural and scientific) for man. Species were selected according to the following criteria: The kind of pollutant located in the study area. Based on literature, a revision on the pollutant environmental behavior was conducted, considering their physiochemical characteristics as well as environmental parameters (humidity, temperature, pH, type of soil, etc.) that can influence in the environment pollutant levels. Once pollutant groups to evaluate were determined, potential pathways and exposure routes were established. An important criterion that was also considered when selecting animal groups was their biology, as it should be well documented; finally groups that are relatively easy to capture and handle were selected.

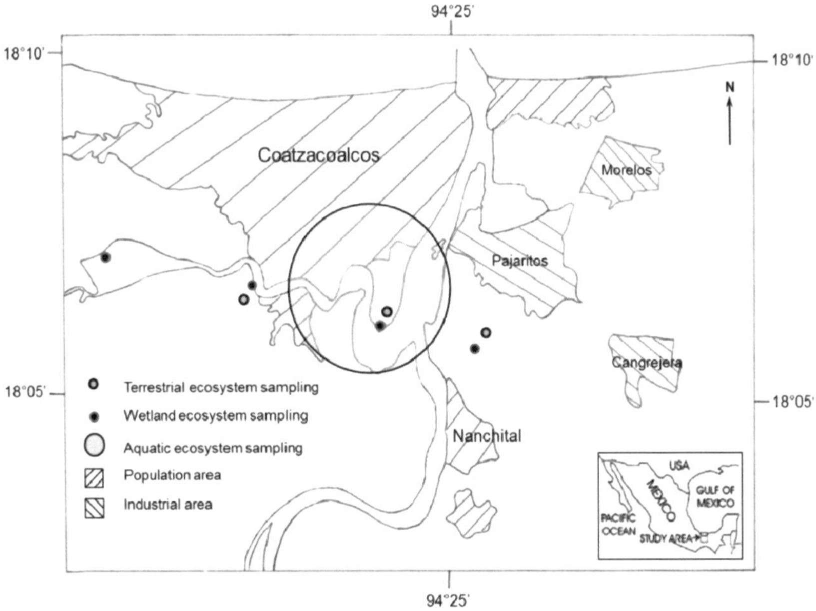

Fig. 1. Study area and location of sampling sites in the region of Coatzacoalcos, Ver.

2.2 Ecological importance of selected species

2.2.1 Earthworms

Earthworms' ecological importance is that they are decomposer organisms (important in biogeochemical cycles) and because of that, they have an important role when adding nutrients to soil (they favor the availability of nitrogen, phosphorus and sulfur) which can be used by vegetable species (Reines et al., 1998; Legall, 2006); they are also an important link in the trophic chain, mainly for some bird species and it has been proven that earthworms can accumulate important metal concentrations (Sánchez-Hernández, 2006). Earthworms can be found in many types of soil and are vulnerable to impacts occurred in soil, their small size represents an advantage to be handled; their distribution is ubiquitous in edaphic horizons with detritus, they are easy to capture, have a close contact with soil, and have a short life-cycle favoring the study of several generations (Ogunseitan, 2002). Persisting organic pollutants have the capacity to bioaccumulate and biomagnify along the trophic chain, as well as animals that belong to decomposers or detritivores levels are very important for the ecosystem functioning, therefore, if animals that are part of soil are affected by pollutants that may show in the ecosystems' health.

2.2.2 Crabs

Crabs are ubiquitous in all temperate and tropical regions in the world. In the wetland ecosystems they are an ecologically important kind, because they play a primary role in the decomposition of organic material and the addition of nutriment to soil. Due to the fact that they build up their galleries by the rivers' basin, lakes or seas, they spend part of their life on land surface and when tides rise they shelter in their burrows (≈ 30 cm. deep) under water. They mainly feed themselves with detritus, so they are excellent filtering organisms, capable to accumulate great pollutant quantities.

2.2.3 Iguanas

There are different kinds of iguanas, however the most common ones in the Coatzacoalcos region are the green iguana (*Iguana iguana*) and the black iguana (*Ctenosaura* spp.); Distribution of both goes from the south of Mexico to South America. According to Lara-Lopez and Gonzalez-Romero, (2002), iguanas are herbivores, the diet of the green iguana is composed mainly as follows: leaves (57.36%), flowers (24.15%) and fruits (3.43%), most of their life is spent on top of the trees, because of the previously said, this specie may be used as a POPs bio-monitor in air. In many coastal communities in Mexico, it is usual to eat iguana as an important source of protein, and it is believed that their blood and eggs contain a lot of energy and help to prevent certain diseases such as anemia. Furthermore people use the skin of this reptile and dissected to sell as ornaments (Alvarez del Toro, 1982). In southern Mexico (Leon & Montiel, 2008), in Central America (FAO, 1997) and in parts of South America, the iguana is one of the most consumed wildlife species. Iguanas have been an important source of protein for humans for over 7000 years FAO (1997). Many of the rural inhabitants of Central America still rely on iguana as a protein source; however, consumption of these species are not uniform during the year because in most cases eating frequency depends on the availability in certain times or seasons that are usually 3 to 4 months per year (Pers. Obs.).

2.2.4 Fish and invertebrates

Fish and aquatic invertebrates are commonly used to monitor pollutants because they bioaccumulate toxic substances and are wile spread, coupled with the diversity and importance of these environments. It has been observed that they are highly sensible to changes in the aquatic environment as well as to low concentrations of environmental pollutants, (Russo et al., 2004; Klobučar et al., 2010). On the other hand, fish have a crucial position in the toxicological field, due to the fact that they have been widely used in studies related to human and ecological health. Fish study includes a wide variety of approaches to detect aquatic pollution impacts from direct measures of mortality, to the analysis of demographical dynamics and the community structure, and to the detection of measures of sub cellular changes (Di Giulio & Hinton, 2008).
At the same time aquatic invertebrates and specifically crustaceans are organisms that have a wide distribution (example: marine, terrestrial and freshwater environment), they are organisms that are in close contact with pollutants in sediment, so they have been used in countless eco-toxicological studies. They have proven to be useful to evaluate effects on different pollutants. They have also served as aquatic pollution indicators. In addition, they may be a source or exposure for local consumers (Nacci et al., 1996; Rinderhagen et al., 2000; Rigonato et al., 2005; Külköylüolu 2004; Regoli et al., 2006). Otherwise, spatial distribution of pollutants on sediments and biota in aquatic ecosystems have been related to a great variety of biological answers in populations and fish/invertebrates communities, with the purpose to determine a possible relation between pollutants in the environment and health in the organisms (Adams et al., 1999).

2.2.5 Giant toads

The giant toad (*Rhinella marina*, after *Bufo marinus*) is a native and geographically widespread species in Mexico and Central America (Zug & Zug 1979). It is an omnivorous and opportunistic species (Zug & Zug 1979), which indicates that toads would integrate different exposure paths due to the ingestion of a wide variety of food items and amphibious living habits. The giant toad is one of the largest amphibians in Mexico (adult body length ranges from 10 to 17 cm), with a life expectancy from 10 to 15 years in the wild. The high lipid-somatic index (2 to 10% compared to less than 0.1% in most anuran species after the spawning period) and the elevated hepatosomatic index (Feder & Burggren 1992) along with its breeding biology make this species prone to bioaccumulation of organic and inorganic pollutants and their toxicological effects (Sparling et al., 2010; Linder et al., 2003). Recently, the giant toad has been used as an aquatic ecosystem biomonitor in the evaluation of air pollution (Dohm et al., 2008), infectious diseases (Zupanovic et al., 1998), organochlorine pesticides (Linzey et al., 2003) and endocrine disruptors (McCoy et al., 2008).

2.2.6 Turtles

Slider turtle (*Trachemys scripta*) is geographically widespread across Mexico and Central America (Burger and Gibbons, 1998). They are eligible species because they have several characteristics associated with their metabolism, life history and ecology (Overmann & Krajicek, 1995). These turtles are generally omnivorous and have temperature-dependent of sex determination and studies have investigated contaminant effects on this process (Selcer 2006), this makes them ideal for studies of chronic exposure of local pollutants. This species has been employed for exposure assessment to metals, radiation, organochloride pesticides and polybrominated biphenyls (Bergeron, et al., 1994; Bickham, et al., 1998; Burger and Gibbons, 1998; Lovelette & Wrigth, 1996; Meyers-Schöne & Walton 1994; Willingham et al., 1999; Willingham et al., 2000). In Mexico, slider turtles are considered an endangered species and are protected by Mexican laws.

2.2.7 Crocodiles

Swamp crocodiles (*Crocodylus moreletii*) are aquatic reptiles living in Mexico's tropical regions. They have a life strategy based on late maturation; they are extremely long-lived animals, show parental care and determine sex depending on temperature (Selcer, 2006). Crocodiles reach the highest levels of food chains, so they are useful for the evaluation of persistent and biomagnifying pollutants. Around the world, their populations are endangered (including Mexico), and because of that the concern on effects (mainly reproductive ones) of pollutants in their populations has increased (Guillette et al., 1999); it must be mentioned that the highest DDE levels registered in wild reptile were found in these organisms (De Solla, 2010). Different Crocodylia species have been used to evaluate heavy metals as well as persistent organic compounds.

2.3 Biological sampling techniques

Wild Earthworms (*Eisenia* sp) were collected by excavation. Crabs were harvested using pitfall traps (*Uca* sp) and traditional fishing gear (*Callinectes* sp). fish (*Aplodinotus* sp, *Ariopsis felis, Centropomus parallelus, Eucinostomus* sp, *Eugerres axillaris, Gobiomorus* sp, *Menticirrhus* sp, *Mugil cephalus* and *Oreochromis* sp) were caught using traditional fishing gear (i.e. cast net) with the help of fishermen. Giant toads (*Rhinella marina*) were collected from each site using nets in nocturnal transects within an area of 10,000 m^2. Crocodiles (*Cocodrylus moreletti*) and Iguanas (*Iguana iguana*) were caught using a noose trap. Turtles (*Trachemys scripta*) were captured with a baited piper trap placed near fallen trees and along the edge of the river during the afternoon and checked early the following morning.
Immediately after capture, organisms were measured, weighed and sorted by type of species; type of ecosystem (terrestrial, aquatic and wetland) and by feeding behaviours (carnivores, omnivores, detritivores and herbivores). Blood samples drawn were obtained using

heparinized syringes on endangered animals (turtles and crocodiles). All organisms were subsequently released. Samples were stored at 4°C for transport and subsequent laboratory analysis. Dissection was performed on each of the specimens from the rest of different species to extract the muscle tissue. The tissue was placed in amber glass containers and frozen at -20°C until analysis. All organisms were collected with a Scientific Collector's Permit (Wild Fauna and Flora Scientific Collector) issued by México's SEMARNAT (Ministry of Environment) or Secretaría de Medioambiente y Recursos Naturales-No. FAUT-0133.

2.4 Analysis of blood and tissue residues

Concentrations of the following compounds were tested for on biological samples: α-, β-, γ-hexachlorocyclohexane (HCH), hexachlorobenzene (HCB), aldrin, dieldrin, mirex, α-, γ-chlordane, oxychlordane, trans-, cis-nonachlor, heptachlor epoxide, p, p'-DDT, p, p'-DDE, polychlorinated biphenyls (PCBs, IUPAC No 28, 52, 99, 101, 105, 118, 128, 138, 153, 156, 187, 180, 183, and 170) and polybrominated diphenyl ethers (PBDE, only on some species of fish). The method of extraction, separation and cleaning of muscle tissue was carried out according to the method established by Jensen et al., (2003) with slight modifications (Gonzalez-Mille et al., 2010) and Dallaire et al., (2006) for blood samples. The endrin-C13 and PCB 14-C13 were used as internal standards and were added to all samples. The chromatographic method (Gas chromatography–mass spectrometry GC-MS) was carried out according to that reported by Trejo-Acevedo et al., (2009). The detection limit for POPs was approximately 0.3 mg/L.

3. Results and discussion

3.1 Terrestrial ecosystem

In Table 1, it can be observed that earthworms have the highest concentrations of polychlorinated compounds biphenyls (PCBs) y persistent organic pollutants (POPs), followed by iguanas and finally by crabs. PCBs congeners that were analyzed were (PCBs 105, 128, 138, 153, 156, 170, 180 y 183).

ECOSYSTEM	SPECIES	α-HCH	β-HCH	γ-HCH	DDT	DDE	Mirex	ΣPCBs	ΣPOPs
	Eisp	12.8		106.3	2.5	13.2		13.4	146.2
	(n=6)*	(5.8 - 31.4)	N.D.	(39.4 - 196.0)	(2.5 - 2.5)	(0.3 - 57.2)	N.D.	(2.3 - 39.2)	(50.3 - 323.8)
	Igig	4.76	0.59	0.55		0.04	0.36	0.11	6.42
TERRESTRIAL	(n=3)	(4.33 – 5.41)	(0.48 – 0.71)	(0.45 – 0.66)	N.D.	(N.D. - 0.06)	(0.27 – 0.49)	(N.D. - 0.22)	(5.97 – 6.82)
	Ucsp	0.24	0.37	0.06	0.02	0.04	0.73	0.04	1.52
	(n=2)*	(0.17 – 0.31)	(N.D. – 0.74)	(0.05 – 0.08)	(0.020- 0.026)	(0.04 – 0.05)	(0.72 – 0.74)	(N.D. – 0.09)	(1.25 – 1.78)

Values represent the mean and range. Eisp: *Eisenia* sp., Igig: *Iguana iguana*, Ucsp: *Uca* sp., * Pool samples, N.D.: non detected

Table 1. Concentrations of persistent organic pollutants (ng/g tissue) from terrestrial wildlife collected in Coatzacoalcos, Veracruz.

Registered results of iguanas (*Iguana iguana*) are relevant because until now there are no studies showing POPs exposure background. In addition, because they are a basic part of rural communities' diet, meat intake may be a potential route of exposure to organic pollutants persistent for humans. Furthermore, because it is a tree species, it may be used to indirectly monitor air quality of some volatile organic compounds (VOCs) and semi-volatile ones as DDT and their metabolites, as well as some congeners of PCBs.

With crabs (*Uca* sp.) there are a few exposure studies to POPs. De Sousa et al., (2008) that registered POPs concentrations in crab eggs (*Chasmagnathus granulata*) in different Brasil stereos. Concentrations of DDE, DDT, γ-HCH, PCBs y total POPs are higher (35.95; 0.38; 3.52; 286.27 and 339.68 ng/g) than those reported in the present work. The previously said may be due to the fact that the matrix analyzed by de Sousa et al., contains a greater quantity of lipids, and crabs evaluated in this study (*Uca* sp.) were not in reproductive stage. Bayen et al., (2005) a study was conducted in a Singapure's mangrove swamp where a thropic web was established and the thunder crab (*Myomenippe hardwicki*) was one of the species presenting high concentrations of POPs. Falandysz et al., (2001) also used crabs (*Carcinus means*) to evaluate the exposure to organochlorine pesticides. It is complicated to make a comparison between crab species because they have different etiology and habitants, however, because most crab species are detritivores, we consider them as a good option to do studies related to POPs exposure.

There are several studies on earthworms with ecotoxicological background mainly focused on effects at population levels (lethality), at both laboratory controlled conditions (Heimbach, 1984; Ma & Bodt, 1993; Kula, 1995, Morrison et al., 2000) as in field ones (Thompson, 1970; Tomlin, 1981; Edwards & Brown, 1982; Haque & Ebing, 1983; Potter et al., 1994; Espinosa-Reyes et al., 2010), however, there are a few studies where exposure to POPs are evaluated, Jones and Hart (1998) revised works related to exposure of different earthworms species to various pesticides–Benomyl, Carbaryl, Carbendazim, Carbofuran, Chlordane, Methiocarb, Parathion, Pentachlorophenol, Phorate, Propoxur, Thiophanateme-. These same authors mentioned that earthworms of the *Eisenia* type are the most resistent to the pesticides previously mentioned. Morrison et al., (2000) evaluated the bioavailibility of DDT, DDE, DDD y Dieldrin in soil samples with different age of use of the mentioned pesticides. They exposed *Eisenia* earthworms. The concentrations registered by Morrison et al., (2000) are high when compared with the registered in this study (DDT 28.3±8.4 & DDE 3.77±0.48 mg/Kg tissue). Based on these backgrounds and results registered in this study, it is possible to postulate the Eisenia earthworms as POPs biomonitors in terrestrial ecosystems because earthworms play a major role in facilitating pivotal interactions within ecosystems through the mixing and translocation of soil constituents, or serving as a conduit for contaminants to predators at higher trophic levels (Harris et al., 2000; Langdon et al., 2003). Earthworms have been used extensively in ecotoxicology like biomonitors (Fitzpatrick et al., 1992; Goven et al., 1993; Reinecke & Reinecke, 1998; Espinosa-Reyes et al., 2010) to assess the effects of diffuse contaminants present in soils.

Finally, when doing a POPs biomonitoring in terrestrial ecosystems, it is important to take into account the following factors: a) species susceptibility difference; b) soil type; c) species behavior; d) exposure time; e) pollutant toxicity; f) when using pesticides consider the applying method.

3.2 Aquatic ecosystem

Because fish are organisms with a wide movement range, the sample was taken within an approximate area of 8 Km^2. Thirty-one fish from five species were caught: *Centropomus parallelus, Mugil cephalus, Eugerres axillaris, Oreochromis* sp, *Ariopsis felis* and thirty organisms of one crustacean species *Callinectes* sp., for quantification of POPs.

From de 29 quantified compounds in the simple, only were detected concentrations of HCB, α-, β-, γ-HCH, DDT, DDE, mirex and 6 congeners of PCBs (52, 101, 105, 118, 138, 153) (Table 2). Most concentrations registered were of β-HCH, α-HCH and mirex. HCB was only detected in 26% of the samples and just on species *Mugil cephalus y Callinectes* sp. El γ-HCH and DDT were registered in a 32% and 10% of the samples respectively. Pollutants α-HCH, β-HCH, DDE, mirex and PCBs were detected in a 100% of the samples. Species where a higher number of compounds were found are *Callinectes* sp, *Ariopsis felis* y *Eugerres axillaris* (8, 7 y 6 respectively), however, most concentrations were found in *Callinectes* sp and *Eugerres axillaris.* According to these results, POPs concentrations in muscular tissue by species decrease in this order β-HCH > α-HCH > mirex > DDE > Total PCBs > γ-HCH > DDT > HCB.

ECOSYSTEM	SPECIES	HCB	α-HCH	β-HCH	γ-HCH	DDT	DDE	Mirex	ΣPCBs	ΣPOPs
AQUATIC	Cepa	N.D.	1.2	0.1	N.D.	N.D.	0.1	0.3	0.2	1,8
	(n=9)		(0.05-5.3)	(0.05-0.5)			(0.05-0.5)	(0.05-0.6)	(0.05-0.4)	(0.5-5,9)
	Muce	0.1	0.2	0.3	N.D.	N.D.	0.2	0.3	0.1	1.0
	(n=7)	(0.05-0.2)	(0.05-0.3)	(0.05-0.7)			(0.05-0.5)	(0.2-0.4)	(0.05-0.2)	(0.6-1,5)
	Euax	N.D.	0.4	1.7	0.1	N.D.	0.1	0.6	0.1	2.9
	(n=7)		(0.03-1.9)	(0.05-3.8)	(0.05-0.3)		(0.05-0.1)	(0.1-0.9)	(0.1-0.2)	(0.9-4.7)
	Orsp	N.D.	0.3	N.D.	N.D.	N.D.	0.2	0.4	N.D.	0.8
	(n=5)		(0.2 -0.4)				(0.1-0.2)	(0.3-0.5)		(0.7-1.0)
	Arfe	N.D.	0.3	N.D.	0.1	0.1	0.1	0.5	0.3	1.3
	(n=3)		(0.2 -0.5)		(0.05-0.1)	(0.05-0.1)	(0.05-0.2)	(0.4-0.7)	(0.1-0.6)	(1.1-1.5)
	All fishes	0.1	0.6	0.9	0.1	0.06	0.2	0.4	0.2	1.7
	(n=31)	(0.05-0.2)	(0.03-5.3)	(0.05-3.8)	(0.05-0.3)	(0.05-0.1)	(0.05-0.5)	(0.05-0.9)	(0.05-0.6)	(0.5-5.9)
	Casp	0.2	1.2	1.6	0.2	0.1	0.4	1.3	0.3	14.8
	(n=4)*	(0.05-0.5)	(0.3 -2.4)	(0.05-5.2)	(0.05-0.2)	(0.05-0.1)	(0.1-0.9)	(0.9-1.7)	(0.1-0.7)	(13.3-17.6)

Values represent the mean and range. Cepa: *Centropomus parallelus,* Muce: *Mugil cephalus,* Euax: *Eugerres axillaris,* Orsp: *Oreochromis* sp, Arfe: *Ariopsis felis,* Casp: *Callinectes* sp. N.D.: non detected

Table 2. Concentrations of persistent organic pollutants in muscle tissue (ng/g wet weight) from aquatic wildlife collected in Coatzacoalcos, Veracruz.

Out of the *Ariopsis felis* fish species the highest concentrations of PCBs were registered and it is the only one that presented DDT, *Mugil cephalus* is the only species in which HCB was detected and it presented the highest concentrations of DDE, *Eugerres axillaris* showed the highest concentrations of lindane and α, β-HCH's, finally *Oreochromis* sp registered the lowest load of POPs in comparison to the other species.

Concentrations of β- HCH and α-HCH were the most abundant among the evaluated species, which matches with what was registered by Lee et al., (1997) and Yim et al., (2005) in other fish species. This can be explained because β-HCH and α-HCH have a greater bio-concentration factor (log BCF 2.8 and 2.5 respectively) in aquatic animals and they are more persistent than γ-HCH (log BCF 1.2) (Willett et al., 1998). Even though technical grade HCH has a greater constitution of α-HCH (60-70%), concentrations of β-HCH and α-HCH were very similar, this can be due to the fact that α-HCH has a high volatility (Henry' law constant 6.68 x 10-6 atm-3/mol) and environment degradation, which can cause low persistency (Yim et al., 2005). Another possible explanation is that HCH metabolized and excreted faster than β-HCH (Takazawa et al., 2005).

DDT was only detected in *Ariopsis Felis* while DDE was detected in all species. Most concentrations were registered in detritivores fish. As it was mentioned, DDT was widely used in Mexico for malaria control and because of its toxic effects, its use was eliminated. Chemical and biological processes transformed DDT into DDD and DDE, particularly DDE has been the most registered in biota (Takazawa et al., 2005). Registered concentrations in fish from the zone indicate that they are mainly exposed to residual DDT and to its degradation products. However, as it was seen, DDT/DDE relation in sediment suggests current use, reason why it is expected to have greater concentrations in the tissues, which suggests that fish from the zone have the capacity to eliminate DDT fastly.

Furthermore, quantification of polybrominated diphenyl ethers (PBDE) was done in 30 fish that belong to different species (*Aplodinotus* sp, *Centropomus parallelus, Mugil cephalus, Eucinostomus* sp, *Gobiomorus* sp, *Menticirrhus* sp and Torombolo -common name-), in these organisms, there were concentrations of 6 congeners registered (PBDE 47, PBDE 100, PBDE 99, PBDE 154, PBDE 153, PBDE 209) (Table 3). Considering all species, the order of concentration of congeners found was 47>41>154>209>153>100, being *Eucinostomus* sp species where the highest concentrations were found and in Torombolo species the lowest ones.

Within this context, aquatic ecosystems are highly vulnerable because of their tendency to accumulate concentrations relatively greater of pollutants coming from terrestrial ecosystems surrounding them, as well as those from direct entries (downloads), in that, regardless of its source of entry into the environment, aquatic systems are frequently deposits for a great variety of chemicals. Pollution in these environments can have negative effects over aquatic life (example: alteration of reproduction and decreased of species) as well as directly or indirectly affect human health and threat food safety. (Jha, 2008).

3.3 Wetland ecosystem

In Table 4 are presented the levels of persistent organic compounds detected in the different animal species from the Coatzacoalcos wetland. When monitoring them, 6 organochlorine pesticides out of the 14 ones analyzed were detected and 13 polychlorinated compounds biphenyls (PCBs) out of 21. All captured organisms from the different species presented detectable levels of at least 3 persistent organic compounds: DDE, Lindane y PCBs (data not shown, found in giant toads' adipose and hepatic tissue). Generally, the pattern of pollutants presence in giant toads' muscular tissue was ∑DDT> ∑HCH> Mirex > HCB; and for the turtles and crocodiles' case, the exposure levels of PCBs and DDE in serum were similar. It must be said that lindane and DDT concentrations in crocodiles were higher.

SPECIES	PBDE 47	PBDE 100	PBDE 99	PBDE 154	PBDE 153	PBDE 209	Σ PBDEs
Mesp (n=7)	3.2	1.9	4.5	0.3	0.7	0.7	11.3
	(0.5-6.5)	(0.5-4.3)	(1.0-8.0)	(0.04-0.5)	(0.3-1.2)	(0.3-1.2)	(3.8-19.7)
*Eu*sp (n=2)	12.03	3.7	2.6	18.0	10.2	13.2	59.7
	(7,7-16,4)	(0,8-6,5)	(1,7-3,5)	(9,5-26,5)	(1,5-9,2)	(8,6-17,7)	(43,1-76,2)
Muce (n=7)	5.5	1.6	5.4	1.3	1.1	0.9	15.8
	(0,2-24,8)	(0,1-2,9)	(0,5-10,6)	(0,03-4,2)	(0,1-2,0)	(0,1-2,2)	(1,0-45,1)
Gosp (n=1)	2.9	2.3	8.8	0.8	1.2	1.4	17.3
Apsp (n=3)	1.9	0.7	2.4	0.2	0.3	0.3	5.7
	(1,1-2,7)	(0,5-1,0)	(1,21-3,7)	(0,1-0,2)	(0,1-0,4)	(0,1-0,6)	(3,2-8,7)
Torombolo (n=2)	2.0	0.4	1.3	0.1	0.2	0.2	4.3
	(1,3-2,8)	(0,2-0,5)	(0,8-1,9)	(0,1-0,2)	(0,1-0,3)	(0,1-0,3)	(2,6-6,0)
Cepa (n=8)	2.2	1.0	3.7	0.2	0.5	0.5	8.1
	(1,0-4,0)	(0,5-1,4)	(0,8-5,5)	(0,1-0,5)	(0,2-0,6)	(0,1-0,8)	(3,1-12,6)
All fish (n=30)	4.2	1.7	4.1	3.0	2.0	2.5	17.5
	(0.2-24.8)	(0.1-6.5)	(0.5-10.6)	(0.03-26.5)	(0.1-9.2)	(0.1-17.7)	(1.0-76.2)

Values represent the mean and range. Mesp: *Menticirrhus sp.*, Eusp: *Eucinostomus sp*, Muce: *Mugil cephalus*, Gosp: *Gobiomorus s*, Apsp: *Aplodinotus sp*, Torombolo: -common name-, Cepa: *Centropomus parallelus*.

Table 3. Concentrations of polybrominated diphenyl ethers (ng/g lipid) in fish muscle tissue from species collected in Coatzacoalcos, Veracruz.

ECOSYSTEM	SPECIES	HCB	α-HCH	β-HCH	γ-HCH	DDT	DDE	Mirex	ΣPCBs	ΣPOPs
WETLAND	Rhma (n=12)	0.08 (N.D.-0.15)	0.75 (0.41-2.05	0.31 (N.D.-1.24)	0.26 (N.D.-0.53)	ND	1.51 (0.05-7.8)	0.39 (N.D. 0.67)	ND	3.1 (0.84-9.56)
	Crmo‡ (n=2)	N.D.	N.D.	N.D.	64.5 (N.D. - 64.5)	59.2 (N.D. - 59.2)	11.5 (N.D. - 11.5)	N.D.	2.7 (2.5 - 2.7)	76.0 (72.7 - 79.2)
	Trsc‡ (n=4)	N.D.	N.D.	N.D.	3.5 (N.D. - 4.58)	N.D.	8.8 (N.D. - 8.78)	N.D.	3.0 (N.D. - 5.83)	7.0 N.D. - 14.61

Values represent the mean and range. Rhma: *Rhinella marina*, Crmo: *Crocodylus moreletii*, Trsc: *Trachemys scripta*. * Pool samples, ‡Blood samples, N.D.: non detected

Table 4. Concentrations of persistent organic pollutants in muscle tissue (ng/g wet weight) and blood serum (ng/ml) from wildlife collected in Coatzacoalcos, Veracruz.

The DDT/DDE relation obtained in toads' tissue was lower than 1.0 which suggests the organism's ability to metabolize the parental compound to DDE. Studies on amphibians, especially the toads' family (Bufonidae), has proven these organisms capacity to accumulate high concentrations of DDT and its metabolites. It should be said that these studies have also shown that proportions of DDE and DDT were higher than those of the parental compound. In our case, DDD concentrations in tissues were not quantified, so exposure of toads from the Coatzacoalcos region to DDT and its metabolites may be underestimated. Registered DDE concentrations in this research are comparable with those found in adult anuran coming from other polluted sites (Table 4 and 5). There are records of ∑DDT concentrations and their metabolites (mainly DDE) up to 3480 µg/g of fat in samples of *R. clamitans* coming from the wetlands from the south of Ontario (Harris et al 1998); in our study, we found concentrations that go up to 3094.5 µg/g of fat in samples of giant toad's liver (data not shown). POPs exposure works in fresh water turtles' populations are scarce; studies done in Terrapin (*Malaclemys terrapin*), Snapping turtle (*Macrochelys temminckii*) and Common musk (*Sernotherus odoratus*) have determined levels of DDE of 0.73-21.7 ng/g wet mass (Basile, 2010, Moss et al., 2009, De Solla et al,. 1998). Several studies on marine turtles have proven levels of DDE of 0.06- 0.73 ng/g wet mass (Stewart et al 2011). Our study shows comparable levels between species of terrestrial and fresh water turtles. Crocodiles present the highest levels of DDT and DDE in plasma; be noted that higher organochlorine pesticides concentrations in the world (mainly DDE) have been found in the Crocodilia species due to their position in the trophic chain. Guillette et al., (1999) Reported DDE levels of 0.9- 17.9 ng/ml and DDT levels of 0.45 to 0.70 in plasma of alligator's residents at Apopka Lake in florida; our study shows comparable levels; even greater to the ones found in Apopka Lake.

The distribution general pattern of Hexaclorociclohexanos (HCH´s) observed in toads' tissues was α-HCH>β-HCH>γ-HCH and for the crocodiles and turtles' case, levels of γ-HCH were detected. In accordance with literature, general distribution patterns of HCH's in mammals, birds and fish are β-HCH>α-HCH>γ-HCH; this distribution pattern is mainly determined by the compound persistency, the exposure path, the species metabolism, and the trophic position (Willet et al., 1998). Our data contrast with the distribution pattern observed in other studies of environmental and biological matrix. Some possible explanations in order to interpret the presence of a higher proportion of γ-HCH is the chronic exposure to isomer of different routes, which can be noted with the found values in crocodile and turtle's plasma where isomer γ-HCH is the only detectable one (Table 4). El γ-HCH is more volatile in comparison with the other isomers, which implies an important transport by air, it is also the isomer with most solubility in water (Walker et al 1999). Coupled to the previously said, it is possible that other important exposure routes exist towards HCH's which have not been explored yet and which may significantly contribute to corporal load and its distribution to these animals. Even though differences in isomers proportions may indicate different sources, routes and times of exposure, for many species it is not clearly understood the influence of processes such as intake, distribution, metabolism and storage in the differences of isomer distribution in tissues (Willet et al., 1998). HCH's concentrations detected in this study are lower than those observed in other studies done on wild amphibians living in agricultural sites (Table 4 and 5). In the case of turtles and crocodiles, there was no useful information found to compare data with.

SPECIE	COMPOUND	CONCENTRATION	TISSUE	REFERENCE
Pseudacris crucifer	DDE	1001	Whole	Russell et al 1995
	DDT	160.6		
	α- HCH	0.37		
	β- HCH	1.37		
	γ- HCH	<DL		
Rana clamitans	DDE	0.58-45.0	Whole	Russell et al 1997
	HCB	0.08-0.49		
	∑DDT	1.24		Guilliland 2001
	∑HCH	0.12		
Rana perezi	DDE	<DL-190	Muscle	Rico et al 1987
	∑DDT	50-550		
	γ- HCH	<DL-10		
	∑DDT	35.4	Whole	Pastor et al 2004
	HCB	2.7		
	α, γ- HCH	0.5		
Rana pretiosa	DDE	91-173	Whole	Kirk 1988
	DDT	563-1750		
Rana mucosa	DDE	17-100	Whole	Fellers et al 2004
	α- HCH	<DL-4.9		
	γ- HCH	<DL-0.7		
Necurus maculosus	DDE	0.3-90.0	Whole	Bonin et al 1995
	DDT	<DL-8.3		
	∑HCH	<DL-10.1		
Necturus lewisi	DDE	60	Whole	Hall et al 1985
Chaunus arenarum	DDE	ND-4.5	Whole	Jofre et al 2008
	α- HCH	ND-5.6		
	β- HCH	ND-2.3		
	γ- HCH	ND-2.7		
Hypsiboas cordobae	DDE	1.1-1.7	Whole	
	α- HCH	3.9-5.4		
	β- HCH	3.0-7.3		
	γ- HCH	4.9-7.2		
Leptodactylus mystacinus	DDE	ND-6.0	Whole	
	α- HCH	3.5-6.9		
	β- HCH	0-8.9		
	γ- HCH	ND-5.7		
Melanophryniscus stelzneri	DDE	10.6	Whole	
	α- HCH	6.99		
	β- HCH	ND		
	γ- HCH	ND		
Odontophrynus occidentalis	DDE	0.8-1.8	Whole	
	α- HCH	1.3-4.1		
	β- HCH	1.0-2.7		
	γ- HCH	1.5-3		
Pleurodema tucumanum	α- HCH	4.7	Whole	
	β- HCH	4.1		
	γ- HCH	4.4		

DL-Detection limit, ND-Not detected

Table 5. Persistent Organic Pollutants concentrations (mg/Kg wet weight) measured in various amphibian species from different studies.

Mirex and HCB were pollutants found in less proportion in giant toads' tissues. There are a few studies related to Mirex and HCB exposure on adult amphibians. Russell y collaborators (2002) reported average concentrations of 0.26-1 ng/g tissue of HCB in cricket frogs (*Acris crepitans*) coming from 5 agricultural locations in Ohio USA; in table (Table 5) other studies are presented with levels of HCB which are higher when compared to the ones obtained in this study (0.14-0.67 ng/g of tissue). Detectable levels of Mirex and HCB in turtles and crocodiles were not found.

PCBs congeners detected in blood and tissue sample were 52,101, 105, 118, 138, 153, 156, 170 and 180, which corresponds with the reported in other studies as the most common for human and biological samples (Table 4). The PCBs congeners presence pattern observed in the giant toad's tissue is consistent with other studies on amphibians from other regions around the world (Loveridge et al., 2007, Russell et al., 1997); as well as in the found concentrations (Table 6). Studies conducted in diverse species of fresh water and terrestrial turtles have reported levels of PCBs totals of 5-414.8 ng/g wet mass (Basile, 2010, Moss et al., 2009, De Solla et al,. 1998) and for alligator populations there have been found levels of 1.54 ± 0.12 ng/ml (Guillette el al 1999); these levels are comparable with the ones obtained in this study. Detected congeners (except 52) correspond in greater proportion (>30%) to aroclor 1254, one of the most sold commercial mixtures in the world, which suggests that the origin of these compounds probably is related to the use of these oils in the region's industrial areas. Detected congeners are characterized for being some of the most persistent ones in the environment and for being absorbed in greater proportion in the organisms.

SPECIE	TISSUE	CONCENTRATION	REFERENCE
Rana clamitans	Carcass	2.8	Loveridge et al 2007
Various anurans	Carcass	151-4470	DeGarady y Halbrook 2003
Rana clamitans	-	7.51	Rusell et al 1997
Necturus maculosus	Carcass	113-1082	Bonin et al 1995
Rana pipiens and *Rana clamitans*	Carcass	50-112	Phaneuf et al 1995
Rana perezi	Muscle	50-1080	Rico et al 1987

Table 6. Total PCB concentrations (ng/g wet weight) measured in various amphibian species from different studies.

Amphibian and reptiles populations are declining at an alarming way in the world (Alford, 2020; Todd et al., 2010); some of the determining concomitant causes in this phenomenon is the exposure to toxic agents; let us note that reptile and amphibian toxicological information is growing, however, it is yet limited in comparison with other vertebrate groups (Sparling et al., 2010). Amphibians and reptiles may be exposed to a wide spectrum of toxic substances; pollutant accumulation in these organisms may be influenced by many factors (physiological, trophic, behavioral, etc.) at the same time, exposure may occur by different routes and in different environments during their life time. It is known that these organisms can accumulate significant pollutant loads in their tissues, mainly of heavy metals and organic compounds. The caused effects by exposure to POPs of greater concern in reptile and amphibian populations are the endocrine disruption, DNA damage, and development abnormalities; some of the studies of greater impact over these effects have been found in these organisms. Ecological importance to maintain viable reptile and amphibian populations is determinant because these organisms are the link between terrestrial and aquatic ecosystems; at the same time they are mainly placed between intermediate links of trophic chains; they have also diversified and occupied a wide spectrum of ecological niches in different types of ecosystems; the presence of highly persistent, bioaccumulative and biomagnifying pollutants is a potentially dangerous situation worth to be evaluated in the Coatzacoalcos Veracruz ecosystems.

3.4 Trophic levels and bioaccumulation of POPs

In Figure 2, the integration of exposure to POPs in tissue (ng/g lip) of the evaluated species in each one of the ecosystems present in Coatzacoalcos; a trophic hypothetical web was established in the region. Results were classified according to the trophic level that species belong to. It can be observed that herbivores are the ones presenting the lowest POPs concentrations followed by carnivorous, omnivorous, and finally the ones with the highest concentrations are the omnivorous. This is explained because detritivores organisms are found in grater contact with contaminated matrix (as soil and sediment), while omnivorous organisms include a greater number of exposure routes (environment and food) and are, therefore, the ones presenting greater POP's concentrations in their tissues.

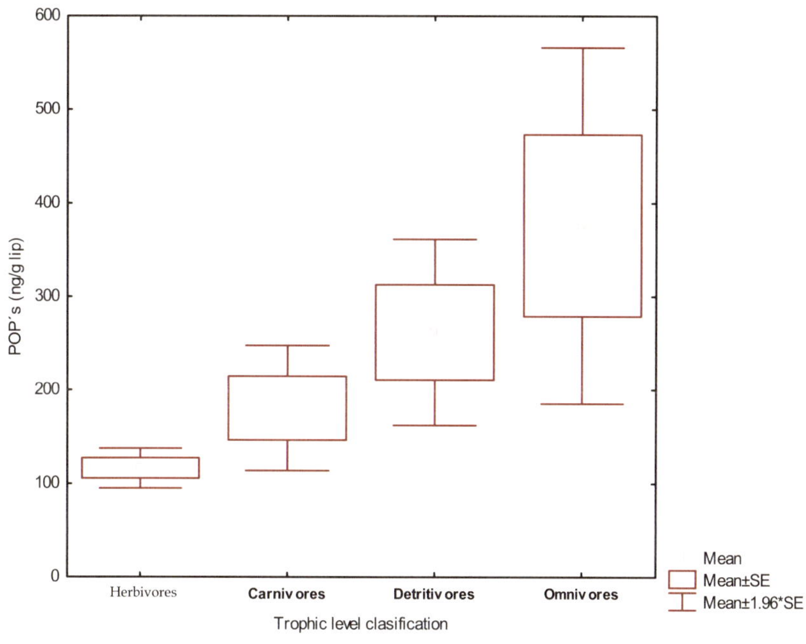

Fig. 2. Concentration Levels of POPs* in tissue taking into account the feeding habits of some species collected at Coatzacoalcos, Ver. *Corresponds to the sum of chloride compounds. Herbivores (iguana); Carnivores (Fish–Cep-); Detritivores (Fish -Arfe, Muce-Crab -Ucsp- and–Casp-); Omnivores (Fish -Tilapia-, Amphibian–Rhma-)

4. Conclusion

With the noted exception of HCB which was not found in the terrestrial ecosystem, traces of all other POPs were identified in species from all three ecosystems. Performing an analysis of feeding habits, it was learned that concentrations of POPs rise as follows: herbivores < carnivores < detritivores < omnivores. With these results it may be established that the region's biota is in fact exposed to diverse Persistent Organic Pollutants. In similar studies it has been found that exposure to POPs may cause several effects (Espinosa-Reyes et al., 2010; González-Mille et al., 2010). In addition, one must consider that Coatzacoalcos organisms are exposed to other types of pollutants in addition to POPs whose collective action could well increase the magnitude of the effects. On the other hand, as far as brominated compounds go, this is the first ever study on fish in the region. The described scenario leads us to ask ourselves what exactly is the risk to humans and wildlife in the region. The Outlook is not promising if you consider that these organisms are of paramount importance in the food chain both to humans and other species of wildlife.

Available data on concentrations of chemical substances in the environment and human beings as well as over effects of exposure to chemicals complex combinations is still scarce. Chemicals generally pose a risk for the environment; they are a rapidly growing pollution load including chemical compounds increasingly complex from which potential effects on public health and environment are probably known. It is estimated that between 70,000 and 100,000 chemicals are available in the market and this number is growing fast. Around 5,000 of these substances are being produced in high volumes, over one million tons a year. The biggest chemical producers are countries members of the OECD, but countries such as India, China, Brazil, South Africa and Indonesia are rapidly increasing their productions. One of the main reasons for the development and adoption of the REACH Regulation is that a great number of substances have been produced and marketed in Europe during many years, at times in very large quantities yet there is not enough information over the hazards for human health and the environment. The European Commission has estimated that in order to fill in information gaps related to toxic effects of that large number of substances it may be required the use of 9 million of lab animals with an approximate cost of 1.3 billion € to perform the necessary tests. A later estimate suggested that required tests would imply 54 million vertebrate animals and costs would grow up to 9.5 billion €. Within this context, wildlife bio-monitoring may be used to detect chemical pollution and evaluate the health of ecosystems, with species as systematic test models during the evaluation of risks associated to actual exposure routes. Wildlife species residing in polluted zones are exposed to pollutant complex mixtures through multiple paths that could hardly be evaluated in laboratory tests.

5. Acknowledgments

This work was supported by a grant from the National Institute of Ecology, SEMARNAT (DGICUR-INE) [No. de convenio INE/A1-047/2007]. We also thank the University of Veracruz, campus Coatzacoalcos. Special thanks to Prof. Jesús Guerrero Cabrera for English language editing of the manuscript.

6. References

Adams, S.M., Bevelhimer, M.S., Greeley, M.S., Levine, D.A., Teh, S.J. 1999. Ecological risk assessment in a large river reservoir: 6. Bioindicators of fish population health. *Environ. Toxicol. Chem.* 18:628–640.

Alford, R.A, 2010. Declines and the global status of amphibians, D.W. Ecotoxicology of Organic Contaminants in amphibians. In Sparling, D.W. Linder, G., Bishop, C.A., Krest; S.K, editors. Ecotoxicology of amphibians and reptiles 2nd ed. Pensacola (FL): SETAC Press p 13-46.

Álvarez del Toro, M. 1982. *Los reptiles de Chiapas*. Talleres Gráficos del Estado. Tuxtla Gutiérrez, Chiapas.

ATSDR. 2005. *Toxicological profile for alpha-, beta-, gamma- and delta-hexachlorocyclohexane.* Department of Health and Human Services. Agency for Toxic Substances and Diseases Registry. 377.

Basile, E.R., Avery, H.W., Bien, W.F., Keller, J.M. 2011. Diamondback terrapins as indicator species of persistent organic pollutants: Using Barnegat Bay, New Jersey as a case study. *Chemosphere* 82:137-44.

Bayen, S.; Wurl, O.; Karuppiah, S.; Sivasothi, N.; Kee Lee, H.; Philip Obbard, J. 2005. Persistent organic pollutants in mangrove food webs in Singapore. *Chemosphere* 61: 303 -313.

Bergeron, J.M., Crews, D., Mc Lachlan, J.A. 1994. PCBs as environmental estrogens: turtle sex determination as a biomarker of environmental contamination. *Environ. Health Perspect.* 102: 780–786.

Bickham, J.W., Hanks, B.G., Smolen, M.J., Lamb, T., Gibbons, J.W. 1988 Flow cytometric analysis of the effects of low-level radiation exposure on natural populations of slider turtles (Pseudemys scripta). *Arch. Environ. Contam. Toxicol.* 17:837–841.

Blake, A. 2005. *The next generation of POP´s: PBDE´s and lindane.* International POP´s Elimination Network (IPEN). Washington D.C. USA. 15.

Bonin, J., DesGranges, J.L., Bishop, C. A. Rodrigue, J., A. Gendron J. Elliott, E. 1995. Comparative study of contaminants in the mudpuppy (amphibia) and the common snapping turtle (reptilia), St. Lawrence River, Canada. *Arch. Environ. Contam. Toxicol.* 28:184-194.

Burger, J., Gibbons, J.W. 1998 Trace Elements in Egg Contents and Egg Shells of Slider Turtles (*Trachemys scripta*) from the Savannah River Site. *Arch. Environ. Contam. Toxicol.* 34:382–386.

CEC. 1997. *Historia del DDT en Norteamérica*. Commission for Environmental Cooperation. Available in:
http://www.cec.org/files/PDF/POLLUTANTS/historiaDDTs_ES.PDF (accessed Jul 2011)

CEC. 2001. *Diagnostico situacional del uso de DDT y el control de la malaria*. Commission for Environmental Cooperation. Available in:
http://www.cec.org/files/PDF/POLLUTANTS/InfRegDDTb_ES_EN.pdf
(accessed Jul 2011)

CEC. 2006. *The North American regional action plan (NARAP) on lindane and other hexachlorocyclohexane (HCH) isomers.* Commission for Environmental Cooperation. 51p.

De Solla, S. 2010.Organic Contaminants in reptiles. In Sparling, D.W. Linder, G., Bishop, C.A., Krest; S.K, editors. *Ecotoxicology of amphibians and reptiles.* 2nd ed. Pensacola (FL): SETAC Press p 289-324.

De Solla, S.R., Bishop, C.A., Van der Kraak, G., Brooks, R.J., 1998. Impact of organochlorine contamination on levels of sex hormones and external morphology of common snapping turtles (Chelydra serpentina serpentina) in Ontario, Canada. *Environ. Health. Perspect.* 106:253-260.

De Souza, A., Machado, J.P., Ornellas, R., Curcio, R., Souto, M., Silveira, C. 2008. Organochlorine pesticides (OCs) and polychlorinated biphenyls (PCBs) in sediments and crabs (*Chasmagnathus granulata*, Dana, 1851) from mangroves of Guanabara Bay, Río de Janeiro state, Brazil. *Chemosphere* 73(1): 186 - 192.

DeGarady, C. J. and Halbrook, R. S. 2003. Impacts from PCB Accumulation on Amphibians Inhabiting Streams Flowing from the Paducah Gaseous Diffusion Plant. *Arch. Environ. Contam. Toxicol.* 45(4): 525-532.

Di Giulio, R.T. and Hinton, D.E. 2008. The toxicology of fishes. Ed. CRC Press, Florida, 1071pp.

Dohm, M.R., Mautz, W.J., Doratt, R.E., Stevens, J.R. 2008 Ozone exposure affects feeding and locomotor behavior of adult *Bufo marinus*. *Environ. Toxicol. Chem.* 27:1209-1216.

Edwards P.J., Brown S.M. 1982. Use of grassland plots to study the effect pesticides on earthworms. *Pedobiol.* 24: 145-150.

Espinosa-Reyes, G.; Ilizaliturri, C.; González-Mille, D.; Costilla, R.; Díaz-Barriga, F.; Cuevas, M.C.; Martínez, M.A.; Mejía-Saavedra, J. DNA Damage in earthworms (*Eisenia* spp.) as indicator of environmental Stress in the industrial zone Coatzacoalcos, Veracruz, Mexico. *J Environ. Science Health A* 45: 49-55.

Falandysz, J.; Strandberg, L.; Zpuzyn, T.; Gucia, M. 2001. Chlorinated cyclodiene pesticide residues in Blue mussel, crab, and fish in the Gulf of Gdansk, Baltic Sea. *Environ. Sci. Technol.* 35: 4163 - 4169.

FAO. 1997. Lista mundial de vigilancia para la diversidad de los animales domésticos. 2a Ed. Organización de las Naciones Unidas para la Agricultura y la Alimentación. Roma, Italia. http://www.fao.org/docrep/V8300S/v8300s00.HTM (última visita diciembre de 2008)

Feder, M.E., Burggren, W.W. 1992. Environmental physiology of the amphibians. University of Chicago Press, Chicago, p 646.

Fellers, G.M., McConnell, L.L., Pratt, D., Datta, S. 2004. Pesticides in mountain yellow-legged frogs (*Rana muscosa*) from the Sierra Nevada Mountains of California, USA. *Environ. Toxicol. Chem.* 23: 2170–2177.

Fitzpatrick, L.C.; Sassani, R.; Venables, B.J.; Goven, A.J. 1992. Comparative toxicity of polychlorinated biphenyls to earthworms *Eisenia fetida* and *Lumbricus terrestris*. *Environ. Pollut.* 77: 65–69.

Gibbons, J.W. (ed). 1990. The slider turtle. In: *Life history and ecology of the slider turtle.* Smithsonian Institution Press, Washington, DC, pp 3–18

Gillilland, C.D., Summer, C.L., Gillilland, M.C., Kannan, K., Villeneuve, D.L., Coady, K., Muzzall, P., Mehne, C., Giesy, J.P. 2001 Organochlorine insecticides, polychlorinated biphenyls and metals in water, sediment and green frogs from southwestern Michigan. *Chemosphere* 44:327–339.

González-Mille, D.J.; Ilizaliturri-Hernández, C.A.; Espinosa-Reyes, G.; Costilla-Salazar, R.; Díaz-Barriga, F.; Ize-Lema, I. and Mejía-Saavedra, J. 2010. Exposure to persistent organic pollutants (POPs) and DNA damage as an indicator of environmental stress in fish of different feeding habits of Coatzacoalcos, Veracruz, Mexico. *Ecotoxicology* 19:1238-1248.

Goven, A.J.; Eyambe, G.S.; Fitzpatrick, L.C.; Venables, B.J.; Cooper, E.L. 1993. Cellular biomarkers for measuring toxicity of xenobiotics: effects of polychlorinated biphenyls on earthworm *Lumbricus terrestris* coelomocytes. *Environ. Toxicol. Chem.* 12: 863–870.

Guillette, L.J., Brock, J.W., Rooney, A.A., Woodward, A.R. 1999. Serum concentrations of various environmental contaminants and their relationship to sex steroid concentrations and phallus size in juvenile American alligators *Arch. Environ. Contam. Toxicol.* 36:447-455.

Hall, R.J, Driscoll, C.T., Likens, G.E., Pratt, J.M.1985. Physical, chemical and biological consequences of episodic aluminium addition to a stream. *Limnol. Oceanogr.* 30, 212-220.

Haque, A. Ebing, W. 1983. Toxicity determinationof pesticides to earthworms in the soil substrate. *J. Plant Dis. Prot.* 90: 395-408.

Harris, M.L., Bishop, C.A., Struger, J., Van Den Heuvel, M.R., Van Den Kraak, M.R., Dixon, G.J., Ripley, B., Bogart, J.P. 1998. The functional integrity of northern leopard frog (*Rana pipiens*) populations in orchard wetlands. I. Genetics, physiology, and biochemistry of breeding adults and young-of-the-year. *Environ. Toxicol. Chem.* 17:1338–1350.

Harris, M.L.; Wilson, L.K.; Elliott, J.E.; Bishop, C.A.; Tomlin, A.D.; Henning, K.V. 2000. Transfer of DDT and Metabolites from Fruit Orchard Soils to American Robins (*Turdus migratorius*) Twenty Years after Agricultural Use of DDT in Canada. *Arch. Environ. Contam. Toxicol.* 39, 205-220.

Heimbach, F. 1984. Correlations between three methods for determining the toxicityof chemicals to earthworms. *Pest. Sci.* 15: 605-611

Jha, A.N. 2008. Ecotoxicological applications and significance of the comet assay. *Mutagenesis* 23:207-221.

Jofré, M.B., Antón, I.R. and Caviedes-Vidal, E. 2008. Organochlorine Contamination in Anuran Amphibians of an Artificial Lake in the Semiarid Midwest of Argentina. *Arch. Environ. Contam. Toxicol.* 55 (3): 471-480.

Jones, A. and Hart, D.M. 1998. Comparison of laboratory toxicity tests for pesticides with field effects on earthworm population: a review. Pp: 247-267. In: Sheppard, S.C; Bembridge J.D.; Holmstrup, M.; Posthuma, L. 1998. *Advances in earthworm ecotoxicology.* Proceedings from the second international Workshop on earthworm ecotoxicology. 2 - 5 april 1997. Amsterdam, The Netherlands. Pensacola FL: Society of environmental Toxicology and chemistry (SETAC) 472 p.

Kirk, J.J. 1988. Western spotted frog (*Rana pretiosa*) mortality following forest spraying of DDT. *Herp. Review* 19:51-53.

Klobučar, G. I. V., Štambuk, A., Pavlica, M., Sertić, P.M. Kutuzović, H., Hylland. 2010. Genotoxicity monitoring of freshwater environments using caged carp (*Cyprinus carpio*). *Ecotoxicology* 19:77–84.

Kula, H. 1995. Comparision of laboratory and field testing for the assessment of pesticide side effects on earthworms. *Acta Zool. Fennica,* 196: 338-341.

Külköylüolu, O. 2004. On the usage of ostracods (Crustacea) as bioindicator species in different aquatic habitats in the Bolu region, Turkey. *Ecological Indicators* 4: 139-147

Langdon, C.J., Piearce, T.G., Meharg, A.A., Semple, K.T. 2003. Interactions between earthworms and arsenic in the soil environment: a review. *Environ. Pollut.* 124, 361-373.

Lara-López, M.S., González-Romero, A. 2002. Alimentación de la iguana verde *Iguana iguana* (Squamata: Iguanidae) en la Mancha, Veracruz, México. *Acta Zool Mex* 85: 139-152.

Lee R.F., Steinert S. 2003. Use of the single cell gel electrophoresis/comet assay for detecting DNA damage in aquatic (marine and freshwater) animals. *Mutation Res.* 544:43-64.

Legall, J.R.; L.E., Dicovskiy, Valenzuela, Z.I. 2006. *Manual Básico de lombricultura para condiciones tropicales.* Escuela de Agricultura y Ganadería de Estelí. Estela, Nicaragua. 16 p.

León, P., Montiel, S. 2008. Wild Meat Use and Traditional Hunting Practices in a Rural Mayan Community of the Yucatan Peninsula, Mexico. *Hum. Ecol.* 36: 249–257.

Linzey, D., Burroughs, J., Hudson, L., Marini, M., Robertson, J., Bacon, J., Nagarkatti, M., Nagarkatti, P. 2003. Role of environmental pollutants on immune functions, parasitic infections and limb malformations in marine toads and whistling frogs from Bermuda. *Int. J. Environ. Health Res.* 13:125-148.

Lutharddt, P.; Mayer, J.; Fuchs, J. 2002. Total TEQ emissions (PCDD/F and PCB) from industrial sources. *Chemosphere*. 46, 1303-1308.

Ma, W., Bodt, J. 1993. Differences in toxicity of the insecticide chlorpyrifos to six species of earthworms (oligochaeta, lumbricidae) in standardized soils tests. *Bull. Environ. Cont. Toxicol*. 50: 864-870.

McCoy, K.A., Bortnick, L.J., Campbell, C.M., Hamlin, H.J., Guillette, L.J., St Mary, C.M. 2008. Agriculture alters gonadal form and function in the toad *Bufo marinus*. *Environ. Health Perspect.* 116:1526-1532.

Morrison, D.E.; Robertson, B.K.; Alexander, M. 2000. Bioavailability to earthworms of aged DDT, DDE, DDD, and Dieldrin in soil. *Environ. Sci. Technol.* 34: 709 - 713.

Moss, S., Keller, J.M., Richards, S., Wilson, T.P. 2009. Concentrations of persistant organic pollutants in plasma from two species of turtle from the Tennessee River Gorge. Chemosphere 76:194-204.

Nacci, D.E., Cayulab S., Jackim E. 1996. Detection of DNA damage in individual cells from marine organisms using the single cell gel assay. *Aquat. Toxicol.* 35:197-210.

Newhook, R., Meek, M.E. 1994. Hexachlorobenzene: evaluation of risks to health from environmental exposure in Canada. *Environ. Carcin. Ecotox. Rev.* 12(2), 345-360.

Ogunseitan, O.A. 2002. *Microbial proteins as biomarkers of ecosystem health.* 217-232 pp. In: Integrated Assessment of Ecosystem Health. Editado por: Scow, K. M.; Fogg, G.E.; Hinton, D.E.; Jonson, M.L. Lewis Publishers. Boca Raton, Florida, U.S.A. 340 p.

Overman, S.R., Krajicek, J.J. 1995. Snapping turtles (Chelydra serpentina) as biomonitors of lead contamination of the Big River in Missouri old lead belt. *Environ. Toxicol. Chem.* 14:689–695.

Pastor, D., Sanpera, C., González-Solís, J., Ruiz, X., Albaigés, J. 2004. Factors affecting the organochlorine pollutant load in biota of a rice field ecosystem (Ebro Delta, NE Spain). *Chemosphere* 55:567–576.

Petrlink, J., J. DiGangi. 2005. *The egg report.* International POPs Elimination Network (IPEN). Washington D.C. USA. 52.

Phaneuf, D., DesGranges, J.L., Plante, N., Rodrigue, J. 1995 Contamination of local wildlife following a fire at a polychlorinated biphenyls warehouse in St. Basile le Grand, Quebec, Canada. Arch Environ Contam Toxicol 28:145–153.

Portter, D.A., Spicer, P.G., Redmond, C.T., Powel,l A.L. 1994. Toxicity of pesticides to earthwormsin Kentucky bluegrass turf. *Bull. Environ. Cont. Toxicol.* 52: 176-181.

Regoli, F., Gorbi. S., Fattorini D., Tedesco S., Notti A., et al.,. 2006 Use of the land snail *Helix aspersa* as sentinel organism for monitoring ecotoxicologic effects of urban pollution: an integrated approach. *Environ. Health Perspect.* 114:63-69.

Reinecke, A.J. and Reinecke S.A. 1998. *The use of earthworms in ecotoxicological evaluation and risk assessment: new approaches* In: Earthworm ecology. Edwards, C.A. (Ed.) St. Lucie Press. Boca Raton. U.S. 273-293

Reines, M., Rodríguez, C.,. Sierra, A., Vázquez, M. 1998. *Lombrices de tierra con valor comercial: Biología y técnicas de cultivo*. Universidad de Quintana Roo. Chetumal, Quintana Roo, México. 60 p.

Rico, M.C., Hernandez, L.M., Gonzalez, M.J., Fernandez, M.A., Montero, M.C. 1987. Organochlorine and metal pollution in aquatic organisms sampled in the Doñana National Park during the period 1983–1986. *Bull. Environ. Contam. Toxicol.* 39:1076–1083.

Rigonato, J., Mantovani, M.S., Quinzani, J.B. 2005. Comet assay comparasion if different *Corbicula fluminea* (Mollusca) tissues for detection of genotoxicity. *Genet. Mol. Biol.* 28: 464-468.

Rinderhagen, M., Ritterhoff, J. & Zauke, G.P. 2000. Crustaceans as bioindicators. In *Biomonitoring of polluted water: reviews on actual topics* (A. Gerhardt, ed.). Trans Tech Publications; Environmental Research Forum, Uetikon, p. 161-194.

Riojas-Rodriguez, H., Baltazar-Reyes, M.C., Meneses, F. 2008. Volatile organic compound presence in environmental samples near a petrochemical complex in Mexico. *Abstracts Epidemiology* 19 (1), S219.

Rosales, L. Carranza, E. Estudio geoquímico de metales en el estuario del río Coatzacoalcos. *In: Golfo de México contaminación e impacto ambiental: diagnóstico y tendencias.* Vázquez-Botello, A.; Rendón-Von Osten, J.; Gold-Bouchot, G., Agraz-Hernández, C., Eds.; Universidad Autónoma de Campeche, Universidad Autónoma de México, Instituto de Ecología. 2005, 389-406.

Russell, R., Lipps, G., Hecnar, S., Haffner, D. 2002. Persistent Organic Pollutants in Blanchard's Cricket Frogs (Acris crepitans blanchardi) from Ohio. *Ohio J Sci.* 102 (5):119-122.

Russell, R.W., Gillan, K.A., Haffner, G.D. 1997 Polychlorinated biphenyls and chlorinated pesticides in southern Ontario, Canada, green frogs. *Environ. Toxicol. Chem.* 1:2258–2263.

Russell, R.W., Hecnar, S.J., Haffner, G.D. 1995. Organochlorine pesticide residues in southern Ontario spring peppers. *Environ. Toxicol. Chem.* 14:815–817.

Russo, C., Rocco, L., Morescalchi, M.A., Stingo, V. 2004. Assessment of environmental stress by the micronucleus test and the comet assay on the genome of teleost populations from two natural environments. *Ecotoxicol. Environm. Saf.* 57:168-174.

Sala, M.; Sunyer, O.; Otero, R.; Santiago-Silva, M.; Ozalla, D.; Herrero, C.; To-Figueras, J.; Kogevinas, M.; Anto, J.; Camps, C.; Grimalt, J. 1999. Health effects of chronic high exposure to hexachlorobenzene in a general population sample. *Arch. Environ. Health.* 54(2), 102-109.

Sánchez-Hernández, J.C. 2006. Earthworm biomarkers in ecological risk assessment. *Rev. Environ. Contam. Toxicol.* 188: 85-126

Selcer, K.W. 2006.Reptile ecotoxicology: studying the effects of contaminants on populations . In Gardner S.C, Oberdösrter, E., editors. *Toxicology of reptiles*. Boca raton (FL): Taylor and Francis Press p 267-297.

Sparling, D.W. Linder, G., Bishop, C.A., Krest; S.K,. 2010. Recent advancements in amphibian and reptile ecotoxicology. In Sparling, D.W. Linder, G., Bishop, C.A., Krest; S.K, editors. *Ecotoxicology of amphibians and reptiles* 2nd ed. Pensacola (FL): SETAC Press p 1-12.

Stringer, R.; Labunska, I.; Bridgen, K. 2001. *Organochlorine and heavy metals contaminants in the environmental around the Complejo Petroquimicos Paharitos, Coatzacoalcos, México.* Technique note Greenpeace. University of Exeter. U.K. 60.

Takazawa, Y., Kitamura, K., Yoshikane, M., Shibata, Y., Morita, M., Tanaka, A. 2005. Distribution patterns of hexachlorocyclohexanes and other organochlorine compounds in muscles of fish from a japanese remote lake during 2002–2003. *Bull. Environ. Contam. Toxicol.* 74:652–659. Thompson AR. 1970. Effects of nine insecticides on numbers and biomass of earthorms in pasture. *Bull. Environ. Cont. Toxicol.* 5: 577-585.

Vázquez-Botello, A.; Villanueva-Fragoso, S.; Rosales-Hoz, L. 2004. *Distribución y contaminación por metales en el Golfo de México.* In: Diagnostico ambiental del Golfo de México. Caso, M. Pisanty, I.; Ezcurra, E., Eds.; SEMARNAT-INE. 682-712.

Venne, L., Anderson, T., Zhang, B., Smith, L., McMurry, S. 2008 Organochlorine pesticide concentrations in sediment and amphibian tissue in playa wetlands in the southern high plains, USA. *Bull. Environ. Toxicol.* 80: 497-501.

Willett K.L., Ulrich E.M., Hites R.A. 1998. Differential toxicity and environmental fates of hexachlorocyclohexane isomers. *Environ. Sci. Technol.* 32:2197-2207.

Willett, K.L., Ulrich, E.M., Hites, R.A. 1998 Differential toxicity and environmental fates of hexachlorocyclohexane isomers. *Environ. Sci. Technol.* 32:2197-2207.

Willingham, E.J., Crews, D. 1999. Organismal effects of environmentally relevant pesticide concentrations on the red-eared slider turtle. *Gen. Comp. Endocrinol.* 113:429–435.

Willingham, E.J., Crews, D. 2000. The red-eared slider turtle: an animal model for the study of low doses and mixtures. American Zool. 40:421–428.

Yim, U.H., Hong, S.H., Shim, W.J., Oh, J.R. 2005 Levels of persistent organochlorine contaminants in fish from Korea and their potential health risk. *Arch. Environ. Contam. Toxicol.* 48:358–366.

Zug, G.R., Zug, P.B. 1979. The marine toad, *Bufo marinus:* a natural history resumé of native populations. Smithsonian Institution Press, Washington, D.C. 284 pp. Available in: http://hdl.handle.net/10088/5188

Zupanovic, Z., Musso, C., Lopez, G., Louriero, C.L., Hyatt, A.D., Hengstberger, S., Robinson, A.J. 1998. Isolation and characterization of iridoviruses from the giant toad *Bufo marinus* in Venezuela. *Dis. Aquat. Org.* 33:1-9.

6

Depositional History of Polycyclic Aromatic Hydrocarbons: Reconstruction of Petroleum Pollution Record in Peninsular Malaysia

Mahyar Sakari
Water Research Unit & School of Science and Technology,
Universiti Malaysia Sabah,
Malaysia

1. Introduction

In the last century, the world has experienced huge and various types of environmental threats. An important group of them is generated from the wide use of fossil fuel such as petroleum as the source of energy in industries, urban development and transportation. Hydrocarbons are the main constituents of fossil fuels thus petroleum hydrocarbons are possible and important source of pollution worldwide. Petroleum hydrocarbons enter the environment from accidental oil spill, natural leaks, industrial releases, vehicles or as by-products from commercial or domestic uses (Ou *et al.*, 2004). Hydrocarbons in petroleum include several types and categories of normal alkanes (saturated, n-alkane), unsaturated hydrocarbons, non-symmetric cyclic hydrocarbons (terpanes) and polycyclic aromatic hydrocarbons (PAHs). Predominance of these compounds in the environmental compartments or samples may indicate petroleum pollution. Petroleum hydrocarbon may disperse in the environment via atmospheric transportation and/or lateral transport. Petroleum contaminants are subject to several processes and changes after production or release such as degradation, photooxidation and decay. The trend over hydrocarbon changes in the environment depends on their chemical characteristics and depositional locations. Locations such as depository sediments under the sea bed surface usually keep hydrocarbon contents unchanged.

2. Polycyclic Aromatic Hydrocarbon (PAHs)

An important class of petroleum hydrocarbons is polycyclic aromatic hydrocarbons (PAHs). PAHs and their derivatives are ubiquitous in the environment such as air, water, soil, sediments and living organisms. PAHs are group of chemicals with more than 10,000 compounds that consist of two or more fused benzene rings (Fig. 1) in different arrangements (Blumer, 1976).

Among PAHs compounds, some have potential for being carcinogen, mutagen and disturbing human endocrine systems (Neff, 1979). Therefore they are categorized as environmental high priority contaminants. PAHs are lipophilic compounds consist of 2 to 7 benzene rings; the 2-4 rings are classified as Lower Molecular Weight (LMW) since 5-7 as Higher Molecular Weight (HMW). The LMW PAHs are more soluble in water and are acutely toxic to human and living organisms whereas HMW are highly soluble in lipid and more carcinogenic, mutagenic with more time period effects (Neff, 1979). The hydrophobic and lipophilic properties of some HMW PAHs make them relatively insoluble in water and tend to accumulate on surfaces or in non-polar matrices.

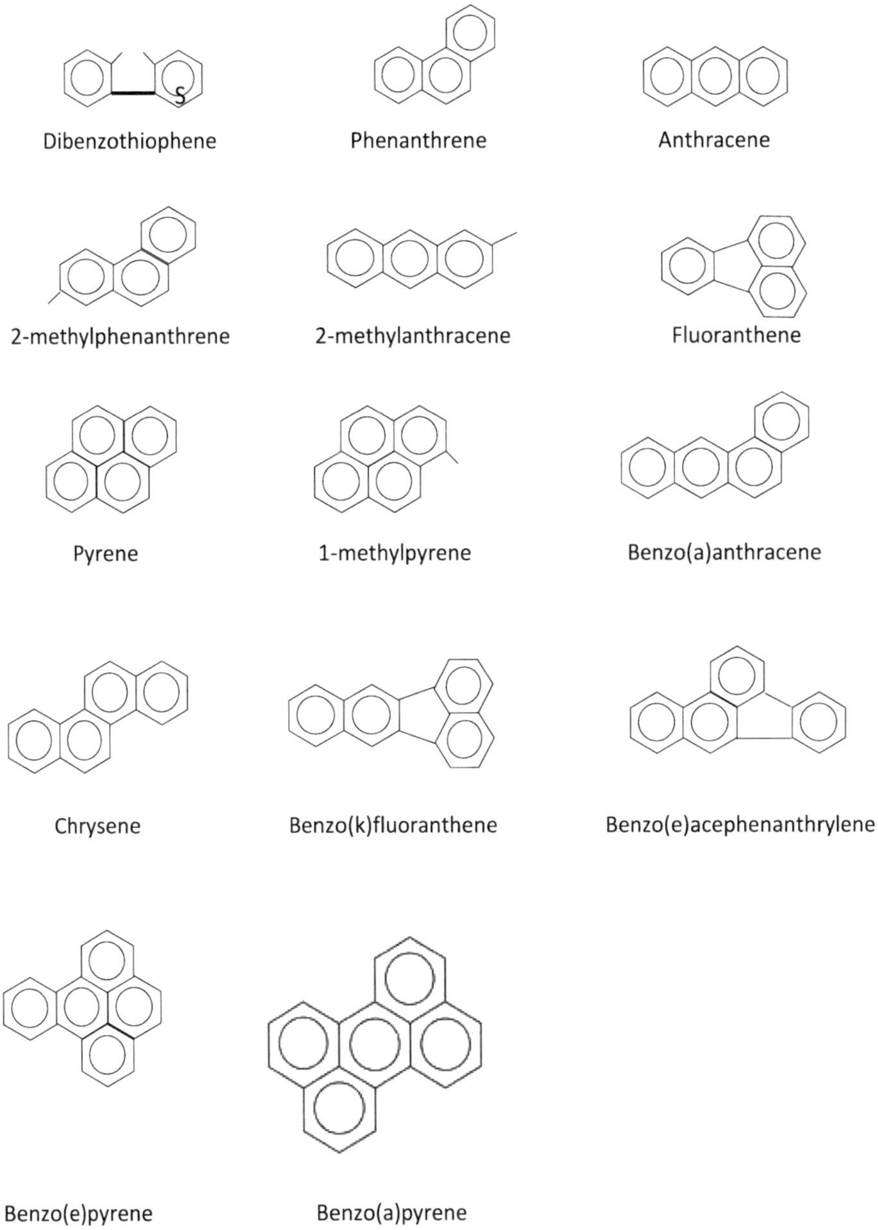

Fig. 1. Some of Polycyclic Aromatic Hydrocarbons Compounds.

The sorptive properties of PAHs are largely controlled by the organic particulate fraction of suspended and deposited sediments (Baker *et al.*, 1986). Particle bound PAHs have a short residence time in the water column before they are settled to the bottom sediments where they may be re-suspended, degraded or subjected to long-term retention. The extent of any release back to the water column depends on the degree of bioturbation, physical re-suspension and the physico-chemical properties of the compound (Wong *et al.*, 1995).
Concerns over PAHs compounds in the environment arise since they are persistent in the environment for a long period of time. PAHs are generated from anthropogenic source as well as nature. In the environment, natural products of PAHs are limited to few types such as Perylene, Phenantherene and Retene where there is no health effect on human and the environment (Neff, 1979).
Anthropogenic PAHs are widespread in the environment as pollutants produced from incomplete combustion of fossil fuel and biomass burning. Anthropogenic PAHs enter marine environment from two primary sources of Pyrogenic and Petrogenic. Pyrogenic source PAHs come from pyrolytic processes such as combustion of fossil fuel, urban and industrial activities, natural fire and biomass burning that produce high molecular weight and less or non-alkylated PAHs. Combusted PAHs after production attach into soot particles, move in far distances and get deposited on soil, terrestrial plants or surface layers of sediments at sea bottom. Some of pyrogenic products of PAHs such as fine particles from charcoals also are washed out from the place of production via sewage plants or precipitations to the marine environment.
Petrogenic sources of PAHs are mainly derived from the release of crude oil and petroleum products such as lubricating oil, diesel fuel, gasoline, asphalt and kerosene. This class of PAHs enters the environment via oil spill, tanker accident, routine tanker operation such as ballast water discharge and discharge from vehicle workshops (NAS, 2002).

3. Pentacyclic triterpanes

Major class of pentacyclic triterpanes, hopanes, is derived from precursor in bacterial membrane (microbial origins) of bacteriohopentetrol. Hopanes are the constituents of crude oil and some petroleum products. Hopanes are believed to be synthesized in the nature by cyclization of squalene precursor during the diagenesis (Fig. 2).

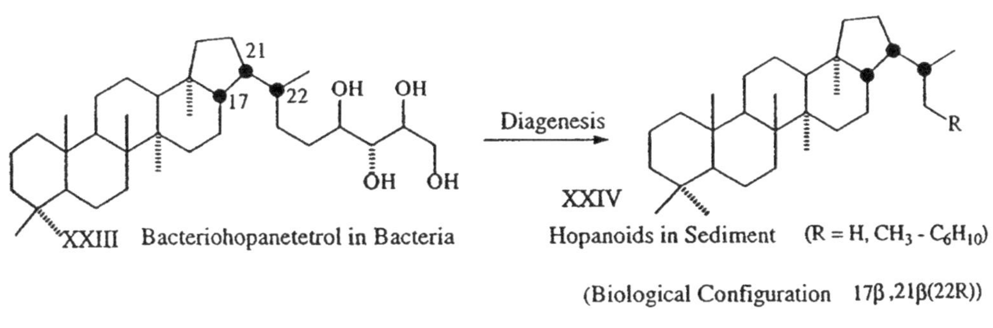

Fig. 2. Diagenesis process that converts bacteriohopentetrol in bacteria to ββ (22R) stereochemistry of hopane. This unstable configuration changes to more stable αβ and βα hopane during the same process.

Classification of hopanes is based on the degree of oil maturation from specific source rock. Hopane itself is not categorized as pollution however they are distributed in the environment with petroleum hydrocarbon pollution and consequently are ubiquitous. They resist degradation processes thus are persistent component in crude oil and petroleum products.Due to this, they are widely used as the sources identifier of oil pollution in the environment. They are relatively existed in small amounts (usually <1% by weight) among other hydrocarbons. Hopanes are commonly found in C_{29}-C_{35}, together with two C_{27} species called regular steranes.

Homohopanes are the name of hopane series which the number of carbon arises by 30. The hopanes are composed of three sterioisomeric series, namely 17α(H),21β(H)- Hopanes, 17β(H),21β(H)- Hopanes and 17β(H),21α(H)- Hopanes. Hopanes with αβ configuration range from C_{27} to C_{35} are characteristics of petroleum because of their greater thermodynamic stability compare to other epimeric (ββ and βα) series. In geologically mature samples, αβ epimeric isomers are greater predominant over βα isomer (moretane). However ββ-isomers are commonly found in living organisms. Ts (18α(H)-22,29,30-trisnorneohopane) and Tm (17α(H)-22,29,30-trisnorhopane) can be a sensitive indicator of thermal maturity when capering oil or sediment samples from the same source. In addition, Hopanes with predominance of 17α(H), 21β(H)- stereochemistry indicates a substantial contribution from petroleum pollution.

Hopane distributions are usually recorded using the m/z 191 in mass chromatograms. An unusually high proportion of the C_{29} hopane is often associated with oil derived from carbonate source rock oil which includes most of those from the Middle East. Those oils also show a slightly enhanced abundance of the C_{35} extended hopane compare with the C_{34} homohopane. In C_{31}-C_{35} hopanes the biologically conferred 22R configuration is preserved during the initial stages of diagenesis. Subsequent isomerization results in a final equilibrium mixture containing approximately equal amount of 22R and 22S isomers. Oleanane is another triterpanes commonly associated with oil derived predominantly from higher plant sources. In conclusion, pentacyclic triterpanes are useful biomarker to identify plant and petroleum input of PAHs into the aquatic environment as well as sediments (Wakeham *et al.*, 1980; Tan and Heit, 1981; Bouloubassi and Saliot, 1993; Yunker and McDonald, 1995; Chandru et al, 2008).

4. Source, distribution and fate of PAHs in aquatic environment

PAHs are released into the environment via natural and anthropogenic sources. Natural source includes oil seeps, volcanoes, grass fires, chlorophyllous and nonchlorophyllous (bacteria and fungi) plants. Anthropogenic sources of PAHs include discharge from routine oil transportation, oil spill, power plants based on fossil fuel consumption, biomass burning, pyrolysis of wood and internal combustion in industrial and vehicle engines.

PAHs enter into the marine environment usually by anthropogenic sources while natural sources have less contribution in this process. Among many possible sources for PAHs contamination in the marine environments; municipal and industrial wastes, city runoff, riverine discharge and atmospheric input have higher proportion.

Petrogenic and pyrogenic based PAHs usually show similar behaviors and fates after entering the environment. Pyrogenic PAHs that are produced via combustion processes have a high and strong affinity to airborne organic particles that may move in greater distances by wind and other atmospheric phenomena. PAHs associated with airborne particles reach to the top layer of the water column in the marine environment, moving to the water column and the bottom of the sea where settles in the sediment. Petroleum and petroleum products which are originated from concentrated hydrocarbon sources enter the marine environment and subjects to dispersion, evaporations, settlement in the bottom on the sediments, weathering, chemical changes, sunlight effects (photooxidation) and microbial degradation (bacteria, yeast and fungi) in short and long term period (Neff, 1979). Petrogenic sources of PAHs on the sediment stick into the particles and consequently is subjected to different chemical and biological changes. Heavier and more complex compounds of crude oil and its products are more resistant to microbial degradation. Regardless of the origin of PAHs, in the marine environment, they adhere to the particles (clay, silt, organisms, detritus and microbes) and settle on the sediments, where a variety of microbes metabolize it into some simple and light compound structure. Accumulation and bioaccumulation of PAHs in the marine environment and organisms are inversely correlated to the potential and ability of hydrocarbons to metabolize them either chemically or biologically.

Finding the source of hydrocarbon pollution is a great concern for many scientists all around the world. Although the first track in this line had been started in 1970s, many researchers are currently try to identify the source of hydrocarbon pollution in the marine environment. In Southeast Asia, the pioneering studies on specific compound analysis have been started by intensive survey in the Straits of Malacca (Zakaria *et al.*, 1999, 2000, 2001, 2002, 2006) and followed laterally in a study in Gulf of Thailand (Boonyatumanoond *et al.*, 2006 and 2007).

In order to identify the sources of hydrocarbon pollution in the environment, there are many techniques such as use of isomer pair ratios (Yunker, 2002), individual compound ratios (Hase and Hites, 1976; Laflamme and Hites, 1978; Baumard *et al.*, 1998; Zakaria *et al.*, 2000) and biomarkers (Volkman *et al.*, 1997; Zakaria *et al.*, 2002 and Wang and Fingas, 2005).

Some molecular ratios of specific hydrocarbons were developed to distinguish differences between PAHs originating from various origins and sources. Among those, the ratio of Phenantherene/anthracene (Ph/An) and flouranthene/pyrene (Fl/Py) were widely used by scientists (Steinhauer and Boehm, 1992; Budzinski *et al.*, 1997; Baumard *et al.*, 1998, 1999). The ratio of F1/Py (fluoranthene/pyrene) has been used to identify fuel sources, showing values < 1.4 for coal combustion (Lee *et al.*, 1977) and < 1.0 for wood (Lee *et al.*, 1977; Knight *et al.*, 1983). In sediments, value for this ratio was 1.3-1.7 at remote sites and < 1.0 near to urban centers (Gschwend and Hites, 1981; Helfrich and Armstrong, 1986). The phenanthrene/anthracene ratio also applies as an indicator for measuring the remoteness (>15) or vicinity (<10) of PAHs sources to urban areas (Zhang *et al.*, 1993).

High-temperature processes such as combustion of organic matters generates PAHs characterized by low ratio of Ph/An (<10), whereas slow maturation during catagenesis, reach to higher values (Ph/An <15). Same trend observed in ratio of flouranthene/pyrene (Fl/Py), where values greater than 1 come from pyrogenic sources and less than unity is indicative of petroleum input. Another ratio which is summarized by Yunker *et al.*, (2002) is Flu over Flu plus Pyr (Flu/(Flu+Pyr)) that is generally greater than 0.5 in grass, wood or coal combustion, and the petroleum boundary ratio appears closer to 0.40 than 0.50, whereas the Flu/(Flu+Pyr) ratios between 0.40 and 0.50 are more characteristic of liquid fossil fuel combustion. Above values were shown to be relatively less reliable in different geographical locations due to the various combustion material sources (Budzinski *et al.*, 1997).

PAHs in the environment have definite behaviors which are controlled by several processes. Processes which can control the transport and degradation of PAHs in sediment include: 1) partition of the compounds between aqueous (pore-water) and particulate phase, 2) microbial degradation, 3) uptake, metabolisms and depuration of PAHs by the benthoses 4) photo-oxidation (surface sediment), chemical oxidation and 5) biosynthesis. Moreover the compounds specific selections for above mentioned processes are absolutely selective.

It is now well established that microbial degradation of PAH occurs primarily in the aerobic zone (Bauer and Capone, 1985) with highest rates occurring with low molecular weight homologues (Lee *et al.*, 1977; Gardner *et al.*, 1979; Readman *et al.*, 1982).

Consequently, any degradation should result in selective losses of, anthracene relative to benzo[a]pyrene, and so affect the ratio of residual PAH. Readman *et al.*, (1987) calculated that up to 80% of anthracene and 40% of benzo[a]pyrene could theoretically be degraded during the approximately 2 year particle/PAH passage through the aerobic layer at the laboratory condition (Table 1).

PAHs	Degradation Rate (10^{-3})	Half Life (years)	PAH Percentage Aerobically Degraded (2 years)
Anthracene	2.18	0.87	80
Fluoranthene	1.67	1.14	71
Benzo(a) Antheracene	1.17	1.63	57
Benzo(a)Pyrene	0.67	2.73	40

Table 1. Theorical impact of degradation on sedimentary PAHs (after Readman *et al.*, 1987).

At the same time, the anthracene/benzo[a]pyrene ratio would be expected to decrease from surface sediment around one third in the anaerobic sediments. Jones *et al.*, (1986), show that oil-derived aromatic hydrocarbons can be rapidly biodegraded in sediments, but combustion-derived aromatic hydrocarbons in the same sediments are relatively resistant to degradation. Similar anomalous behavior of PAH has also been reported by Farrington *et al.*, (1983) where it was suggested that petroleum-derived PAH are more available for uptake by mussels than are pyrogenic PAH. Another process that controls the characteristics and concentration of PAHs is the phenomenon of photo oxidation. Photooxidation is a process which starts from the beginning stage of PAHs production in surface layers of soil, sediment, water or during the transportation in the air. There are selective photooxidation for specific PAHs such as Benz[a]anthracene which is more labile to photooxidation than chrysene + triphenylene (Kamens *et al.*, 1986, 1988), therefore the benz[a]anthracene/(benz[a]anthracene + chrysene + triphenylene) ratio are supposed to be lower in summer than winter samples (Fernandez *et al.*, 2002). Benzo[a]pyrene is photochemically less stable than benzo[e]pyrene where light exposure transforms unstable BaP to more stable BeP (Nielsen, 1988). The indeno[1,2,3-cd]pyrene/(indeno[1,2,3- d]pyrene + benzo[ghi] perylene) ratio is a prior and more stable to photooxidation than the ratios discussed above. Therefore there is expectation for Southeast Asian countries environment to show less concentrations of the low stable PAHs due to heavy and continued sunshine. Interesting to know that PAHs are penetrating in sediment layers after the deposition, where in upper layers of surface sediments PAHs are rich in 5-6 rings PAHs rather than 2-3 rings which migrate downward from upper to deeper layers due to integrity with fine particles and their physical migration with fine particles (Curtosi *et al.*, 2007).

Malaysia, which is located in Southeast Asia, has a unique tropical environment and climate. It is surrounded by the Straits of Malacca in the west and the South China Sea in the west of Peninsular Malaysia. The western part of peninsular Malaysia has been experiencing rapid development during the last half century. On the other hand, the strategic location of this country has made Malaysia as one of the busiest shipping route in the world due to huge petroleum demand from the Middle East to Japan and China (Fig. 3).

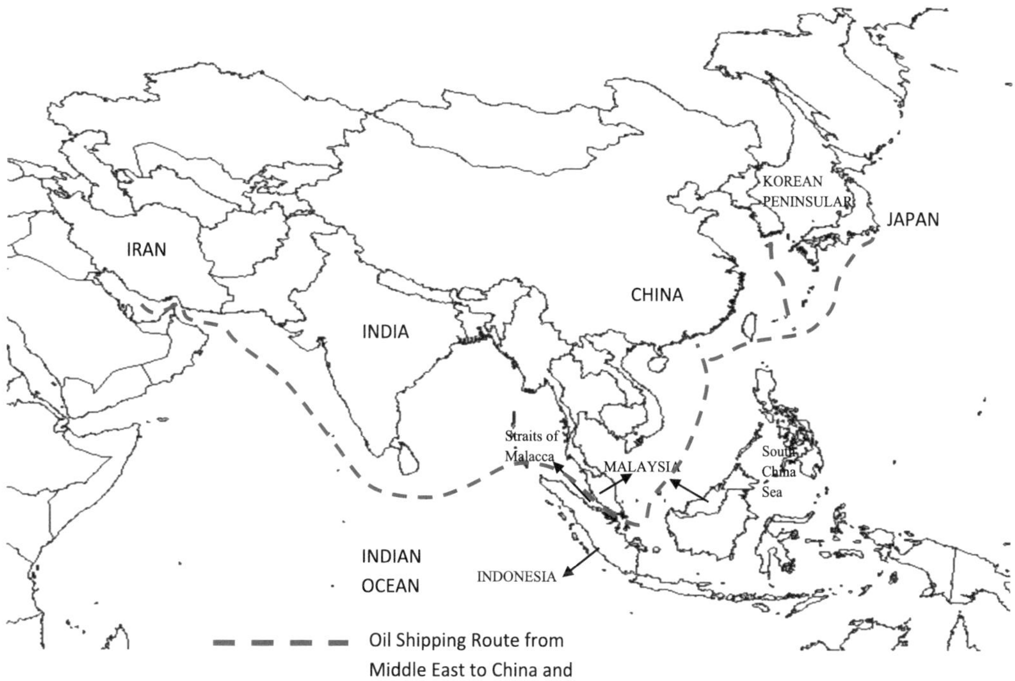

Fig. 3. Oil shipping route from the Middle East to the Far East via Straits of Malacca.

While Malaysia is experiencing extraordinary economic and population growth, it is also developing fast in industrialization, urbanization and motorization in last few decades. As a result of this development, the environment of this country is receiving more threats and hazards especially from the main source of energy which is petroleum. In Malaysia, the concentration and sources of hydrocarbon pollution vary according to locations. For instance, in western P. Malaysia, existence of rapid urban development and the establishment of several industrial areas, the hydrocarbon pollution is introduced throughout non-point and pointed sources. In the eastern P. Malaysia the pollution mostly comes from the urban area and less from industries, due to less industrial developments.

Besides that, Malaysian Marine Department reported 127 oil spill incidents since 1976 due to heavy oil tankers traffic in Straits of Malacca (Malaysian Marine Department, 2003). Zakaria and Takada (2003) believe that the Malaysian environment is under increasing threat of petroleum pollution; although this is not well-documented and recorded. To understand the petroleum hydrocarbon pollution levels in the environment, scientists usually study different environmental samples such as water, sediment, bio-monitoring agents, particles and aerosols. In Malaysia, few researchers have studied petroleum hydrocarbon pollution and were used one or more types of above mentioned samples to demonstrate the status of hydrocarbon pollution in the country.

5. A brief on global historical records of PAHs

Studies on hydrocarbon pollution in historical trend were started alongside studies on recent and modern sediment hydrocarbon pollution. Among the first records, Hites *et al.*, (1977) studied PAHs concentration in Buzzards Bay, Massachusetts for the period from 1900 to 1970. Later on, Wakeham *et al.*, (1979) reported PAHs from sedimentary records of several lakes in Switzerland and Washington, indicative of high levels of PAHs pollution in modern input. In 1984, Prahl and Carpenter published data on PAHs and aliphatic hydrocarbons from Washington coastal sediments indicated that naturally derived aliphatic hydrocarbons are very frequent in ancient sediments while anthropogenic PAHs show high concentrations in recent deposited sediments. In Lake Michigan, the PAHs from sedimentary record reported by Christensen and Zhang (1993) showing a constant trend with fossil fuel consumption. Zhang *et al.*, (1993) reported PAHs maxima (maximum concentration) for early 1950s and 1985 for cores from Green Bay and Lake Michigan. Taylor and Lester (1995) reported significant decrease in PAHs concentrations since 1966 because of less coal combustion. This study is confirmed by Gevao *et al.*, (1997) where they reported PAHs sub-surface maxima in late 1960s and early 1970s from a small rural lake of Cambria, UK. This sub-surface maxima reported by Pereira *et al.*, (1999) from San Francisco Bay in California during 1950s.

There are many more studies on investigation of hydrocarbon pollution in sedimentary record. For instance, Hostettler *et al.*, (1998) studied the trace of biomarker profiles in San Francisco Bay dated sediments, showing anthropogenic input of hydrocarbons in recent decades depositions. Okuda *et al.*, (2002) reported high PAHs concentrations from Chidorigafuchi Moat, Japan, in 1960s. Furthermore, Yim *et al.*, (2005) reported high flux of PAHs in 1950s and 1980 from Masan Bay in Korea. Liu *et al.*, (2005) reported the PAHs fluxes from Pearl River Estuary, South China shows first sharp peak of PAHs levels in 1950s and consequently 1990s. Hartmann *et al.*, (2005) were reported depositional history of organic contaminants from Narragansett Bay, Rhode Island in the United States of America, showing highest PAHs fluxes in modern sediments from Apponaug Cove and sub-surface maxima in Seekonk River core sdiment.

There is an unpublished report of historical record of aliphatic and aromatic hydrocarbons in the straits of Malacca during 1980's by Law in a curtsey communication. Unfortunately, no more details are available since the scientist is deceased. In South East Asia the first available and published study pioneered by Boonyatumanond *et al.*, (2007) from Gulf of Thailand. In this report PAHs fluxes show high levels in 1950s and 1970s because of rapid increase in numbers of vehicles and their usage in Thailand. Also the molecular marker of Hopane shows high contribution of Petroleum and anthropogenic hydrocarbons in dated sediments from the studied area.

Coastal sediment containing a mixture of natural and anthropogenic PAHs presents two important problems for assessing the fate and effect of PAHs in the environment (Wakeham *et al.* 1980). At first, anthropogenic PAHs should be evaluated by site-specific background of the PAHs derived naturally to the studied area. Secondly, parent PAHs data, solely, does not deliver reliable information due to overlap of source composition. This paper investigates the reconstruction of PAHs history and hopanes in one of the highly developing country in tropical Asia that experiences the rapid industrialization, motorization and urbanization.

PAHs derive to the environment via natural and anthropogenic processes. Natural processes of PAHs production occur during diagenesis and microbial activities as well as natural seeps of hydrocarbons and forest fires. Anthropogenic PAHs productions were consistent entering to the environment since mankind used fire for any purposes. The most recent PAHs derive to the environment so called modern input, have been increased since industrial revolution, when man used fossil fuel in industrial wheels. Beginning of the 20th century was in conjunction with rapid increase in PAHs flow to the environment when oil production contributed in fast development of the globe. PAHs come from oil origin enter the environment via petroleum and petroleum products such as gasoline and lubricating oils and their combustion. Combustion derived PAHs are dominated by the un-substituted moieties, whereas PAHs in petroleum are dominated by the alkylated homolog (LaFlamme and Hites, 1978). Lower formation temperatures such as in the formation of petroleum during the diagenesis preserve a higher degree of alkylated compounds (Youngblood and Blumer, 1975). Alkylated and non-alkylated PAHs are the basic knowledge of source identification of hydrocarbon and petroleum pollution in the environment. The degree of alkylation and alkyl homologs existed in environmental samples provides information on the sources of pollution. Youngblood and Blumer (1975) have proposed natural combustion such as natural fires in the forest as primary source of PAHs in deep layers of long core sediments. This idea is criticized by Wakeham *et al.*, (1979) where forest fire might have constant input to the marine environment. In addition to the recent arguments, PAHs from natural fire does not occur often, since forest fire in the environment is not a predominant event. The concentrations of PAHs in the core sediments do not always correspond the unity of sources and input. This is due to various sources and concentrations that possibly interfere in a single layer along sedimentary intervals. Complex mixtures of different sources usually demonstrate irregular ratios that interfere with the results of source identification. Due to the complexity of different PAHs sources, PAHs compound specific ratios are still the most valuable tool for determination of pollution origin (Yunker *et al.*, 1999).

Applications of different ratios are based on molecular structure of specific PAHs compounds in the environment. Lower Molecular Weight (LMW) PAHs is categorized by 2-3 benzene rings while 4-7 rings are known as Higher Molecular Weight (HMW). Pyrogenic PAHs (combusted) are characterized by high abundance of HMW compounds (4-6 benzene rings) and un-substituted (parent) compounds, whereas petrogenic PAHs are dominated by alkyl substituted and abundance of LMW (2-3 benzene rings) PAHs (Garrigues *et al.*, 1995; Budzenski *et al.*, 1997). The ratio of LMW over HMW PAHs is often applied for source discrimination of PAHs in environmental studies. The ratio of LMW/HMW PAHs for values lower than unity indicates pyrogenic source while 2 to 6 is an indicative for petrogenic input into the marine environment. The relative ratio of Methylphenanthrenes over Phenanthrene (MP/P) is also frequently applied for source identification of PAHs. The MP/P is another valuable ratio, as discussed earlier, based on un-substituted moieties and alkylated homolog frequencies in the sample. The MP/P fluctuates among values such as

0.5-1 for combustion derived PAHs in the sediments and 2-6 in sediment dominate by fossil fuel direct release (Prahl and Carpenter, 1983; Garrigues *et al.*, 1995; Budzenski *et al.*, 1997). The MP/P ratio around 4.0 is reported to be derived from crankcase oil (Pruel and Quinn, 1988), close to 1.0 for street and urban dust samples (Takada *et al.*, 1990, 1991) and around 0.5 for atmospheric fallout (Takada *et al.*, 1991). This ratio is higher for coal combustion sources than petroleum (Lee *et al.*, 1977; Takada *et al.*, 1990, 1991).

Among other ratios, scientists use other permanent ratios such as the relevant concentrations of sum of methylfluoranthenes and methylpyrenes over fluoranthene (Youngblood and Blumer, 1975; Laflamme and Hites, 1978; Gustafson *et al.*, 1997). In this ratio, values above the unity (>1) indicate the petroleum sources of pollutions. In addition, the results of these two recent ratios (MP/P and (MFl+MPy)/Fl) are not necessarily same but in some studies deliver similar trends (Pereira *et al.*, 1999).

Some specific compounds are well known in their characteristics. Among those compounds, Benzo(*a*)pyrene are proven carcinogens material to living organisms (Neff, 1979). Characteristics of specific compound PAHs are derived in the environment are strongly associated with the origins. One of these ratios is benzo(ghi)perylene to indeno(1,2,3-cd)pyrene (B*ghi*P/IPy) where the high values come from the automotive exhaust particles (Marr *et al.*, 1999; Nielsen *et al.*, 1996 and Tuominen *et al.*, 1987).

Okuda *et al.*, (2002) reported specific compound PAHs from the core collected from Chidorigafuchi Moat in Japan. They showed that in surface sediments (0-8 cm) there are significant and high values of ratio of benzo(ghi)perylene to indeno(1,2,3-cd)pyrene than lower sections of the core, whereas the ratio showed a constant value around unity for depth below 20 cm and increasing for higher levels up to the sediments from the surface. This strongly suggests that since 1990 the PAHs are more influenced by automotive exhausts. This ratio shows a relatively high value in automotive exhaust particles as it is constant with the socio-economic condition of the study area.

Yim *et al.*, (2005) reported specific hydrocarbon compounds for source identification of the PAHs in a study conducted in Masan Bay, Korea. The ratio that used including Phenantherene/anthracene (Ph/An) and fluorenthene/pyrene (Fl/Py) indicates pyrolitic origins. In the ratio of Fl/Py, scientists reported lower values than those reported from same source of American and European coal. Although Budzinski *et al.*, (1997) showed this low value from coal sources of Australia (Fl/Py: 0.3-0.7).

The amounts of HMW and LMW PAHs in environmental samples are possibly indicative of pollution sources. For instance, coal usually produces high amounts of PAHs than other fuel materials such as petroleum and natural gas. The highest concentration of PAHs (maxima) in core samples collected from Masan Bay, Korea, indicates this issue, where the sharpest peak appeared from layers corresponds 1950s-1980s. During the above era the country was widely used coal for various energy purposes (Yim *et al.*, 2005). Although previous studies found PAHs maximum concentration (maxima) in the sedimentary environment during 1940-1950, as an indication of fuel type changes from coal to oil and gas. The fuel type changes usually reveal increase or decrease in the total concentration of PAHs in the environment (Gschwend and Hites, 1981; Bates *et al.*, 1984; Barrick and Prahl, 1987). Later on, in Masan Bay, Korea this pollution input trends have been decreasing due to Pollution Prevention Act established in 1963 and strong environmental control and monitoring conducted by local and national authorities (Yim, 2005).

Beside the PAHs analysis to determine the concentration and sources of pollution in the environment, there are other tools such as measurements of magnetic susceptibility. This is a fast and cost effective method based on the presence of magnetic-rich spherules that forms during the combustion processes by oxidation of pyrite to magnetic. This method successfully applied by Morris *et al.*, 1994 in Hamilton Harbour, Western Lake Ontario, Canada to compare the method efficiency with PAHs analysis, approve the pyrolytic sources of PAHs pollution in the studied area.

6. PAHs in depositional record, Malaysia

The depositional record of PAHs in peninsular Malaysia is studied during a 4 years scientific investigation. Eight sedimentary core samples were collected from 4 identical coastal areas. Each area represented a historical background of development and socio-economic events of peninsular Malaysia. They are consisted of Klang (Port and Offshore), Malacca (Near and Offshore), Johor (Near and Offshore) and Tebrau (No. 1 and 2). Thus there were 2 core samples taken from each location mainly from near shore and offshore locations to evaluate the distance factor effect of distribution and concentration of PAHs compounds in the environment (Fig. 4).

Fig. 4. Map of sampling locations.

The study elucidated the "Distribution, Sources and Depositional History of PAHs and Hopanes in Selected Locations in Peninsular Malaysia" using chemical molecular markers such as PAHs and Pentacyclic Triterpanes (Hopane) in deposited sediments. The cores lengths were ranged from 21 to 56 cm. The ^{210}Pb was used to reconstruct the pollution history of collected cores, revealed a time period of 60 to 280 years in different cores. Table 2 shows the concentration of PAHs in the sedimentary core of the study areas.

The core that was collected from the Klang City station showed that since 1945, there was an increasing trend in total deposited PAHs (Sakari *et al.*, 2010a). The highest concentration of total PAHs was observed during the era of 1990 to 1998 (2442 ng/g d. w.) as a sub-surface maxima which is interestingly followed by minimum PAHs level of 33 (ng/g d. w.) for the era of 1999 to 2007. Although in lower layers, the total PAHs of 161 (ng/g d. w.) was reported from the period of 1954-1962. In all sedimentary layers and intervals, specific compounds such as BkF, BeP and BaP were the leading PAH among others. This trend of PAHs increase is highly correlated to population increase of surrounded area, increase in registered cars and economic data of the study area. The rapid and sudden drop of total PAHs was interpreted as a joint function of physical phenomena, as well as weather condition, improvements of vehicle engine performance and enforcement of law and legislations.

In Offshore Klang station, the results showed very much depleted concentrations compared to Klang City core. Except for the recent decades deposited sediments, that showed lower concentration usually the core revealed homogenized concentration of PAHs fluctuating from 20 to 32 (ng/g d. w.). It is reported that the highest level of 32 ng/g d.w. happened at the beginning of the 20th century. The PAHs input in this core was not correlated to any of above mentioned socio-economic data, indicative of constant input via atmospheric fallout, where the BkF and BeP were the leading compounds throughout the core intervals. In Offshore Klang core, again the signature of pyrogenic input of PAHs was observed as those MP/P and L/H ratios. For both cores of Klang area, it is found that pyrogenic input from vehicle's emission and asphalt are the main contributor of PAHs into the marine environment of this area, although none of hopane signatures showed definite sign of any specific oil sources mainly due to combustion effect of pyrogenic sources on molecular structure of hopane.

In Malacca, the first core was collected from the near shore location showed the highest concentration of total PAHs in the entire study areas. The highest concentration of total PAHs (4195 ng/g d. w.) was reported from 1977 to 1983 while the lowest were observed at the beginning of the 20th century (Sakari *et al.*, 2011). Interesting to see that very severe depletion of HMW exists in this core and in all layers. The predominant of compound in this core was shown to be Phenanthrene and most of the sediment intervals revealed pyrogenic sources with MP/P ratio below the unity.

Offshore Malacca core was shown lower concentration than near shore but still Phenanthrene and its derivatives are the main PAHs contributor to the total PAHs. The highest and lowest concentrations of PAHs were revealed during 1963 to 1969 and 1914 to 1920, respectively. The signature of PAHs likewise near shore station was shown pyrogenic. This is reconfirmed by MP/P values. Identification of the PAHs origin using hopane marker showed street and urban dusts of Malacca City as the main contributor as observed in near shore Malacca.

In Johor, the first core collected from a location near the city where the connecting bridge (causeway) commutes Malaysia to Singapore. In this location, along the core, the PAHs concentration has ranged from the minimum of 44 (ng/g d. w.) to the maximum concentration of 1129 (ng/g d. w.). The highest concentration of PAHs was observed during the era from 1922 to 1969. Moreover, the lowest concentration observed in recent deposited sediments (Sakari *et al.*, 2011b). The PAHs signature showed a mixture of pyrogenic and petrogenic input where most modern input showed more combusted materials than old sediments with pyrogenic signature. This statement is evidenced by MP/P and LMW/HMW ratios. The leading PAHs compounds along core intervals were BeP, BkF and Phe and its alkyl substitutes.

Klang City Core	Concentration (ng/g d.w.)						
Sediment age (year)*	1945-1953	1954-1962	1963-1971	1972-1980	1981-1989	1990-1998	1999-2007
[a]Total PAHs (ng/g d.w.)	488.99	161.35	238.11	1032.52	1572.97	2422.93	33.85
[b]L/H PAHs	0.55	0.60	0.56	0.75	0.35	0.45	0.92
[c]MP/P	0.63	1.11	0.92	0.66	0.83	0.83	1.11
[d]TOC mg/g	61.23	51.40	51.37	55.93	59.45	71.04	64.96

Offshore Klang Core	Concentration (ng/g d.w.)													
Sediment age (year)	1910-1916	1917-1923	1924-1930	1931-1937	1938-1944	1945-1951	1952-1958	1959-1965	1966-1972	1973-1979	1980-1986	1987-1993	1994-2000	2001-2007
[a]Total PAHs (ng/g d.w.)	29.92	32.05	10.34	24.66	23.29	24.05	20.04	11.21	12.67	7.31	20.10	9.54	15.42	11.91
[b]L/H PAHs	0.55	0.82	0.49	0.49	0.36	0.96	0.41	0.34	0.38	0.34	1.26	0.34	0.99	0.40
[c]MP/P	0.23	0.24	0.22	0.22	0.34	0.46	0.32	0.59	0.37	0.52	0.51	0.54	0.64	0.83
[d]TOC mg/g	69.09	85.67	85.15	65.21	77.23	72.13	45.60	61.23	51.40	51.37	55.98	59.45	71.04	64.96

Near Shore Malacca	Concentration (ng/g d.w.)																	
Sediment age (year)	1879-1885	1886-1892	1893-1899	1900-1906	1907-1913	1914-1920	1921-1927	1928-1934	1935-1941	1942-1948	1949-1955	1956-1962	1963-1969	1970-1976	1977-1983	1984-1990	1991-1997	1998-2005
[a]Total PAHs (ng/g d.w.)	378.02	358.08	402.13	275.10	508.48	431.91	603.70	266.29	967.16	910.25	176.61	3111.47	4447.14	999.16	4195.07	2660.95	335.36	451.87
[b]L/H PAHs	96.84	23.31	126.59	39.66	22.75	12.93	22.87	8.98	29.36	26.70	174.55	144.23	195.50	204.02	122.70	505.00	24.16	20.22
[c]MP/P	0.78	1.26	3.17	0.77	0.84	0.76	0.88	1.09	2.39	1.03	0.72	0.80	0.71	1.31	0.93	0.86	3.19	8.58
[d]TOC mg/g	124.20	117.00	126.90	117.30	113.40	135.30	129.80	149.70	143.50	145.00	137.00	143.90	148.90	143.40	134.40	147.60	133.70	137.20

Offshore Malacca Core	Concentration (ng/g d.w.)																
Sediment age (year)	1886-1892	1893-1899	1900-1906	1907-1913	1914-1920	1921-1927	1928-1934	1935-1941	1942-1948	1949-1955	1959-1962	1963-1969	1970-1976	1977-1983	1984-1990	1991-1997	1998-2005
[a]Total PAHs (ng/g d.w.)	48.00	2.63	61.03	2.92	1.71	325.84	60.42	73.60	10.19	107.11	71.70	714.37	15.26	70.64	19.28	126.74	98.37
[b]L/H PAHs	3.96	NA	3.90	0.32	0.32	0.61	2.13	3.64	0.78	3.79	7.28	0.25	0.62	5.79	0.83	8.60	7.06
[c]MP/P	0.77	0.92	0.79	1.29	2.06	0.90	22.84	0.91	4.95	0.85	0.74	0.80	5.11	0.90	9.92	0.83	0.98
[d]TOC mg/g	143.30	144.60	139.80	137.90	154.20	149.90	141.00	66.90	138.50	136.10	105.60	141.30	129.20	159.60	172.60	156.10	184.10

Johor City Core	Concentration (ng/g d.w.)										
Sediment age (year)	1874-1885	1886-1897	1898-1909	1910-1921	1922-1933	1934-1945	1946-1957	1958-1969	1970-1981	1982-1993	1994-2005
[a]Total PAHs (ng/g d.w.)	580.51	1005.40	648.67	478.73	1129.52	935.25	725.76	920.98	89.11	44.64	171.95
[b]L/H PAHs	0.95	0.15	1.93	0.60	4.27	0.60	0.44	0.54	2.03	1.02	3.05
[c]MP/P	2.45	2.39	2.41	1.93	1.87	1.66	1.61	1.04	0.72	1.59	0.74
[d]TOC mg/g	227.30	218.60	221.20	202.60	207.60	221.20	108.20	256.00	178.40	167.20	206.10

Offshore Johor Core	Concentration (ng/g d.w.)											
Sediment age (year)	1886-1895	1896-1905	1906-1915	1916-1925	1926-1935	1936-1945	1946-1955	1956-1965	1966-1975	1976-1985	1986-1995	1996-2005
[a]Total PAHs (ng/g d.w.)	304.64	321.69	92.06	68.58	411.00	363.70	521.15	215.28	286.63	221.57	126.65	99.60
[b]L/H PAHs	107.31	108.39	73.57	47.71	291.92	33.05	9.01	8.60	3.36	9.62	12.20	10.24
[c]MP/P	1.38	0.96	1.87	1.55	0.90	1.00	1.00	2.43	3.08	0.90	1.41	1.70
[d]TOC mg/g	77.90	82.10	153.50	53.40	50.30	55.40	44.80	51.70	45.40	52.70	52.20	51.20

Tebrau Core I	Concentration (ng/g d.w.)													
Sediment age (year)	1728-1347	1748-1767	1768-1787	1788-1807	1808-1827	1828-1847	1848-1867	1868-1887	1888-1907	1908-1927	1928-1947	1948-1967	1968-1987	1988-2007
[a]Total PAHs (ng/g d.w.)	6.86	11.83	25.42	24.87	23.99	3.51	11.50	21.82	18.02	20.41	31.02	67.23	48.13	310.92
[b]L/H PAHs	NA	NA	NA	NA	NA	NA	NA	NA	NA	NA	8.26	0.67	1.78	0.40
[c]MP/P	0.29	2.00	0.95	0.95	0.99	0.00	0.72	0.90	0.76	0.87	0.93	0.85	0.74	0.97
[d]TOC mg/g	125.30	126.70	156.70	162.80	109.80	160.50	183.00	103.00	213.80	140.00	177.90	146.70	126.80	126.50

Tebrau Core II	Concentration (ng/g d.w.)											
Sediment age (year)	1862-1874	1875-1884	1886-1897	1898-1909	1910-1921	1922-1933	1934-1945	1946-1957	1958-1969	1970-1981	1982-1994	1995-2006
[a]Total PAHs (ng/g d.w.)	10.50	9.65	7.68	12.36	9.53	4.63	13.56	14.59	8.03	10.25	12.64	38.72
[b]L/H PAHs	6.16	23.23	NA	8.05	NA	NA	17.88	2.22	NA	34.33	4.05	8.10
[c]MP/P	0.86	0.92	1.00	0.80	0.51	0.93	0.89	0.59	0.85	0.86	0.92	0.13
[d]TOC mg/g	95.80	96.10	104.80	95.20	95.30	92.20	112.60	110.70	102.40	98.80	103.60	75.40

aTotal PAHs: sum of 18 PAHs ranging from Dibenzothiophene to Dibenzo (a,h) antharacene; bL/H PAHs: ratio of LMW over HMW PAHs; cMP/P: ratio of sum of 3-Methylphenanthrene, 2-Methylphenanthrene, 9-Methylphenanthrene and 1-Methylphenanthrene to Phenanthrene; dTOC: Total Organic Carbon.

Table 2. The Concentrations of Polycyclic Aromatic Hydrocarbon (PAHs) and TOC and ratios of L/H and MP/P in cores collected from the study area.

The second core was collected from Johor strait. This core showed generally lower concentration than those observed in Johor City core. The highest concentration of PAHs (521 ng/g d. w.) observed soon after the WWII and during the independency. The main source of PAHs in this core showed petrogenic signature using MP/P and hopane ratio.

Tebrau Strait is the main gateway connecting Singapore and Malaysia to the waters of South China Sea. The cores from Tebrau Strait were collected from eastern part of Johor-Singapore waterway. The first core revealed the highest concentration during the modern era (1988-2005; 311 ng/g d. w.) and the lowest concentration during the ancient time (1827-1847; 3.51 ng/g d. w.). Since the study was not revealed significant HMW PAHs, the ratio of L/H was not mathematically available in this core however MP/P ratio showed pyrogenic input to the marine environment of the study area (Sakari, 2009).

The second core in Tebrau Strait likewise showed same increasing trend where the highest concentration observed in the recent deposited sediments. In general, the concentration in this core is lower than the first Tebrau core. The sources of PAHs again indicate that there is pyrogenic input received in this location where MP/P was shown values below the unity. The hopane ratio showed that mostly Southeast Asia Crude Oil is the main contributor of PAHs in these cores. This statement is confirmed by ratios such as C_{29}/C_{30} from the hopane compounds.

In conclusion, the concentration of PAHs and hopane in all cores showed that the increase in populations, number of cars, socio-economic indicators such as GDP and GNP, industries, urbanizations, oil production and transportation accelerate the pollution trend. The overall view of PAHs concentration showed that near shore locations demonstrate higher PAHs contribution than offshore stations.

The total concentration of PAHs in this study ranged from 1.7 to 4447 (ng/g d. w.) with a mean value of 381 (n=105). The results of all source identification tools have been shown that a range of highly pyrogenic to extremely petrogenic PAHs are existed in the study area where a zero value of other PAHs is observed in conjunction with a minimum Methyl Phenantherene concentration that possibly indicates negligible nature derived compounds. Total organic carbon (TOC) in this study were fluctuated from 44 to 256 (mg/g) with an average of 117 mg/g (n=105) that statistically showed to be in a very low to negative correlations with total PAHs. The source identification parameters that has applied in this research were ratios such as Ph/(Ph+An), Ph/An, Fluo/Pyr, Fluo/(Fluo+Pyr), BeP/(BeP+BaP), Phe/(Phe+An) and BaA/(BaA+Chry). The application of these ratios revealed vicinity of sources such as adjacent cities, vehicles and industries to the study areas. This study has concluded that these sources emit gasses and particle based materials that transfers via lateral movements by daily rain wash and flushing into the marine environment thru canals, rivers, and drainage and finally settle down to the estuaries and straits. It is also emphasized that shipping and oil transportation play an important role in releasing PAHs into the study areas where the daily heavy ocean going vessels transport goods and oil.

7. Sources and origins of PAHs in deposited environment

Several studies around the world were conducted to understand and determine the sources of hydrocarbon pollution in sedimentary cores. One of the most pioneering studies is conducted by Hites *et al.,* (1977), where three stages of hydrocarbon deposition were reported from 1850 till 1970 from the Buzzards Bay, Massachusetts. This report indicates the sources that were almost constant from 1850 till 1900. The constant source of PAH pollution has been determined as combustion processes, regardless to its origin from natural or anthropogenic sides.

In the UK, Readman *et al.*, (1987) reported PAHs from Tamar Estuary, showing predominant of parent compound rather than alkyl homologues, a clear indication of pyrogenic input correlate with increased motor vehicle activity and road runoff. This is remarkable that compositional uniformity of PAHs throughout the polluted sedimentary core characterize biogeochemical transformation and exchange processes (sorption/leaching; microbial breakdown; photodegradation; etc). Thus it has been concluded that the majority of un-substitute PAHs comes from combusted fossil fuel and/or street dust. Rapid reduction in PAHs concentrations since 1940s may come due to fuel consumption changes from coal to petroleum (Gschwend and Hites, 1981; Bates *et al.*, 1984; Barrick and Prahl, 1987).

Industries are one of the most important contributors of PAHs input into the environment. Appearances of pollutions such as PAHs depends on the time and location of production and deposition. Martel *et al.*, (1987) reported considerable increase of PAHs concentration since 1930 from Saguenay Fjord, Quebec in Canada where two aluminum reduction plants increased the PAHs concentrations in the studied area. The above statement was approved after a couple of years by Cranwell and Koul, (1989) where anthropogenic PAH input that peaked 1900-1920 in Windermere North Basin is tentatively attributed to local industrial input. The decline in post-1975 flux values may result from replacement of coal as the source of energy by oil or gas however flux values remain ten times higher than in the pre-industrial age.

The sources of pollution are always not unique or with a same pattern. It can be a contribution of different sources such as natural and anthropogenic. Christensen and Zhang, (1993) identified a combination of sources including coal, petroleum and wood from four sediment cores collected in Lake Michigan for Source identification. In this study, the sedimentary record of PAHs high flux is reported with petroleum origins (oil and gas during 1985) but the high PAHs flux for 1950s was clarified when coal was used. As the background data, the concentration of PAHs was zero during 1900 for petroleum derived PAHs.

In another study Su *et al.*, (1998) analyzed 6 cores from Green Bay, Wisconsin in order to identify the PAHs concentration and sources. This study showed the same trend in source combinations for PAHs in the studied area. The total concentrations were reported from 0.46 to 8.04 ppm with combination of combustion sources from coal, wood and petroleum hydrocarbon.

Based on the regulations and the availability of different sources, in the energy markets, some of those are decreased in consumption or fully stopped. For instance, Taylor and Lester, (1995) showed that since 1966 that coal combustion had been banned; the coal derived PAHs has decreased and shifted to the oil and gas sources. Although the usage of coal is limited in many countries all around the world, there are still footprints of its application in many countries. For example, Liu *et al.*, 2005 reported 30% of coal combusted PAHs from air particles collected from Guangzhou, China atmospheric environment, due to wide use as energy source.

The historical profile of PAHs from the sedimentary cores collected from Lake Michigan, USA, showed the Wisconsin coal profile exhibit similar trends with peaks for 1946-1951 and 1968-1973, indicative of coal combustion source material (>36%) and petroleum sources (>76%) in various samples using Factor Analysis (FA) model (Rachdawong *et al.*, 1998).

Although the atmospheric environment distributes the PAHs in a homogenized concentrations, lateral transportation such as movements via rivers demonstrate irregular and high amounts of PAHs concentration in the environment. Witt and Trost, (1999) indicate significant contribution of river discharge of the petrogenic hydrocarbon to the sediment with predominant of higher molecular weight PAHs due to its stability in German coastal waters. The highest concentration of the PAHs occurred in recent sediments presented from 1 to 8 cm of surface, indicative of modern input. Petrogenic PAHs pollution which are mainly enter into the marine environment via lateral transport contribute to the pollution history of the world since past centuries. This petrogenic PAHs are abundant in riverine systems due to wash out phenomenon from the city run off. Liu *et al.*, (2000) reported the sources of PAHs from core collected in Yangtze Estuary, China; that mainly was petrogenic origin. However, PAHs concentration in sedimentary records may be affected due to physico-chemical conditions during sedimentation, the nature of inputs, biodegradation, and bioturbation (surface sediments).

As petrogenic PAHs affect the marine environment in short distances, the pyrogenic PAHs are subject to long range transportation via atmospheric movements (Prahl and Carpenter, 1983). This model of transport is able to influence remote and pristine areas. Rose and Rippey (2002) were reported low concentration in recent PAHs deposition via atmospheric movement into a remote lake in the north-west England. Specific compounds analysis for the definition of ratios tested for this study (phenantherene:anthracene and fluorenthene:pyrene) do not identify and clarify any specific reason while shows less urban discharges to the lake comparing to the era of pre-1830.

Natural disasters affect the concentration of the PAHs deposited in the marine environment. Flood as a natural disaster contributes in PAHs irregular concentrations where it washes out city surface to water bodies such as rivers and streams. Ikenaka *et al.*, (2005) reported the highest PAHs of core layers with multiple pyrogenic sources from Lake Suwa, Japan when the heavy rain and consequently flood had been historically consistent. Discharges from natural disasters and local input are characterized by irregular distributions of PAHs. Since PAHs enter locally in mass amounts distribute according to the physical and chemical properties of the destination points. In a study (Moriwaki et al, 2005) on historical trend of PAHs in reservoir sediment core of Osaka, Japan; however the sources of PAHs in the sedimentary record is found a combination of grass, wood and coal for pre-industrial era, for early 20th century and petroleum and its combusted derived materials in recent and mostly modern input.

As discussed earlier in previous sections, there are possibilities for natural inputs of PAHs entering into the enviroment. PAHs naturally derive from higher plant detritus and degradations products (Simoneit and Mazurek, 1982; Yunker and McDonald, 1995). Four and 5 benzene rings PAHs can be produced from microbial breakdown of plant wax and woody tissue. Prahl and Carpenter (1983) were reported natural sources of PAHs in Washington coastal sediments with a constant input of clay samples represents the era of Pleistocene. Quiroz *et al.*, (2005) reported 50 years of PAHs depositions into the Laja Lake from south central Chile, were showed relatively low PAHs concentration (226 to 620 ng/g d. w.) with mostly natural origins. One of the most important and mostly natural PAHs is Perylene. Perylene is 5 benzene rings PAHs which there are doubts in its origin. High temperature combustion of the internal engines produces perylene however other sources originate. This is strongly believed among scientist that perylene can produce naturally in the deep sediments via diagenesis. Interestingly, there are several reports that indicate the

high perylene concentration from surface sediments (Zakaria *et al.*, 2002; Tolosa *et al.*, 2004) and throughout the core (Barra *et al.*, 2006).
PAHs studies in core samples are subject to scientific debates. Core samples have different properties than surface sediments samples thus the fate of contaminants are various from surface. In surficial sediment, there are chemical, physical and biological properties which affect the targeted compounds in analytical analysis and interpretation of their data. In sediments collected from a core, there is no expectation of aerobic condition whereas the anoxic characteristics are notable for any possible chemical and biological changes. The profile of individual PAH concentrations with depth in sediment often reflect changes in source input over time rather than significant *in situ* biological degradation (Hites et el, 1977; Prahl and Carpenter, 1979). Although some organisms were capable to biosynthesize naturally the hydrocarbons (Graef and Diehl, 1966; Hancock *et al.*, 1970), other studies reveal the bioaccumulation effects of organisms in the core sediments (Hase and Hites, 1976). Moreover, some PAHs are generated by post-deposition transformations of biogenic precursors over a short period of time. This subject was confirmed for Perylene in research conducted by scientists (Aizenshtat, 1973; LaFlamme and Hites, 1978).
While aliphatic hydrocarbons are subjected to dissolution and microbial degradation, PAHs remain less or unchanged (Yunker *et al.*, 1999). Although Wakeham *et al.*, (1979) believe that lower molecular weight PAHs contribute in lower concentration in the total PAHs comparing high molecular weight in the core sediments. This interprets as a faster degradation of the PAHs that enter into the marine sediments. PAHs are always associated with organic carbon in sedimentary environment and are integrated with those values, but sometimes greater values of PAHs are not associated with TOC, indicative of soot particle existence in the sediments. For example Richardson Bay studies in the United States showed high values of PAHs with low values of TOC. This is an indicative of soot particle associated in the sediments come from the atmospheric transportations of combusted fossil fuel and organic contents that are less available in partitioned PAHs in the organic carbons (Pereira *et al.*, 1999). PAHs associated with soot particles are less biologically available for uptake than the PAHs derived from the petroleum and oil spill (Farrington *et al.*, 1983; Gustafson *et al.*, 1997). PAHs studies through the core samples are usually consistent with gradual and little changes either in concentration or in ratios in a normal condition but environmental disasters such as oil spills show significant changes. Therefore, core sections always do not show sudden changes in PAHs ratio values.

8. Application of biomarkers in petroleum pollution studies

The forensic chemistry techniques fingerprint pollutants in environmental samples. This technique is based on existence and abundance of the biomarkers. Terpanes are a group of biomarkers that are ubiquitous in the environment together with hydrocarbons. Pentacyclic triterpanes (hopanes) are the group of biomarkers that existed in crude oil and some petroleum products. As discussed earlier, hopane as a fingerprinting biomarker delivers from precursor of bacterial membrane (microbial origins) of bacteriohopentetrol (Fig. 2). The production of hopane in the nature is due to cyclization of bacteriohopanetetrol during the diagenesis. Basically, the stereochemistry of hopane makes them thermodynamically unstable (Peters and Moldowan, 1993). Hopanes in their biological origins present 17β(H), 21β(H) compounds that is known as biological stereochemistry.

Instability against temperature increase is due to their polar and non-polar ends. Upon change, they might convert from ββ to more stable configuration of αβ and βα structures. The αβ that is called hopane is predominantly available in crude oil and some petroleum products. Hopanes are relatively involatile, resist biodegradation, geologically mature and relatively stable in the environment (Simoneit et al., 1988) however there are chemical characteristics and properties among hopanes that compounds with higher number of carbon (*e.g.* C_{35}) shows bigger resistibility against biodegradation than lower numbered such as C_{31} (Frontera et al, 2002) (Fig. 5).

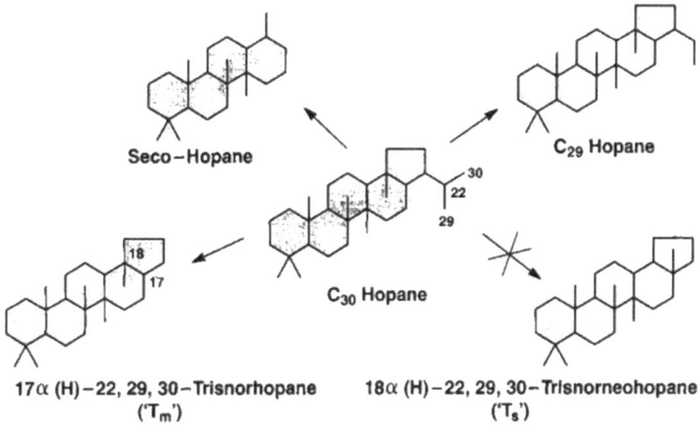

Fig. 5. Hopane chemical structures.

Homohopanes are the name of hopane series in which the number of carbon arises by thirty (n=30). The relatively more abundant homohopanes (C_{31}-C_{35}) are showed in less oil contaminated sites with significant loss of C_{30} (Colombo et al, 2005). In environmental samples from the biomass burning, hopane appears together with moretanes with abundance of C_{27} and C_{31} (Standley and Semoneit, 1987) that overlaps in common peaks (Omar et al., 2006).

Hopane is found in mineral oil and coal based fuel and lubricants (Kapalan et al, 2001). An unusually high proportion of the C_{29} hopane is often associated with oil derived from carbonate source rock oil which includes most of those from the Middle East however existence of oleanane is indicative of Southeast Asian oil.

9. New dimension of biomarkers; A reliability test over hopane

PAHs and hopane compounds in the study area were identified by comparing chromatograms of samples and standard solutions in the GC-MS. Chromatograms were compared in their retention time, surface area and mass spectra. Some of samples from the core intervals in various locations presented an absence or irregularity of appearances of peaks (representing compounds) in the chromatograms. Since some of Hopane compounds were missing, the identification of sources was difficult. We noticed that hopanes with carbon numbers from 27 to 30 (single peaks) are the most missing or depleting compounds among others. However hopanes with carbon numbers from 31 to 35 that appears in twin peaks (S and R stereochemistry) are dominant showing an unchanged structure. In this report, samples mostly from offshore locations such as Malacca and Klang showed this phenomenon. These samples are observed to be identical in their source and origin of production. They are highly pyrogenic appearing in extremely depleted MP/P and L/H ratios. Meantime, it is observed that hopane chromatograms appearing shorten or fade up in single peaks (C_{27}-C_{30}). Twin peaks however showed more resistance than single peaks, the depletion was also observed among them.

The scenarios are different among samples with petrogenic sources such as crankcase oil, spilled oil and lubricating oil. They show taller and sharper single peaks together with complete twin peaks representing C_{31} to C_{35}. We believe that the stereochemistry of twin peaks provides resistibility against temperature increase rather than single peaks. Thus, it is criticized that high temperature especially in combustion process of petroleum in internal engines may cause destruction on compound structure appear as demolished or depleted peaks in chromatograms (Peters et al 1992; Colombo et al, 2005). Hence, application of some compound and isomer pair ratios of hopane are failed to assist source identification of PAHs.

The correlation was applied for statistical comparison between MP/P and other ratios such as C_{31}-C_{35}/C_{30}, C_{29}/C_{30}, Tm/Ts and C_{31}-C_{35}/C_{29}. Increasing of hopane indices like C_{31}-C_{35}/C_{30} and C_{31}-C_{35}/C_{29} with depletion of MP/P (combusted) may suggest demolishing of chemical structures in C_{30}, C_{29} and MP compounds during the high temperature combustion (Peters et al, 1992).

Therefore in an environmental sample such as sediment, theoretically a decrease in MP/P values renders high temperature in combustion process. Here, the theory criticizes the possible changes on chemical structure of single peak hopane compounds with carbon numbers ranging from C_{27} to C_{30}.

A positive correlation were observed between Offshore locations such as Malacca and Klang together with near shore station such as Klang City demonstrate combustion of petroleum where several and average MP/P appear to be lower than unity. Here there are negative correlation between the MP/P and hopane indices of C_{31}-C_{35}/C_{30} and C_{31}-C_{35}/C_{29}. These correlation values indicate that combustion results lower values of either C_{29} or C_{30}. (Takada et al., 1990; Prahl and Carpenter, 1983; Pruel and Guinn, 1987; Garrigues et al., 1995).

10. Natural vs. anthropogenic PAHs

PAHs are known as anthropogenic and/or natural compounds, based on their sources of production (Simoneit and Mazurek, 1982; National Academy of Science, 2002). Natural process is called a procedure that bacterial and algae are involved. This process results in-situ production of PAHs that produce limited concentrations (Hites et al., 1977; Prahl and Carpenter, 1979). Anthropogenic processes usually produce greater concentration of PAH in the environment. They include combustion of organic matter such as plant and/or oil and direct release of oil and its derivatives into the environment (Neff, 1979). Thus natural in-situ PAH generation in sedimentary environment is negligible in the total concentration of anthropogenic amount. There are limited locations around the world that produce natural based PAHs. These reports are mostly from Brazilian tropical forest where scientists report appearance of Phenanthrene, Naphthalene and Perylene in remote and virgin locations (Wilcke et al, 2003, 2004).

Likewise reports from tropical forest of Brazil, we expect to have a natural contribution of some specific polycyclic aromatic hydrocarbons such as Phenanthrene and Perylene in Peninsular Malaysia. As it has been discussed earlier, it is too hard to differentiate specific compounds as background level from nature from those as anthropogenic input using ordinary instrumentations. The application of hopane assisted us to technically differentiate those natural from anthropogenic individual PAH. Dated sedimentary intervals along the core have shown deposited PAH in which represents an era before the oil exploration and usage contain limited but detectable concentrations of Phenanthrene. The same signature has been frequently found in dated ancient sediment from Malacca and Tebrau in which intervals represent an era of 17th century. This research expects a constant input of natural hydrocarbons into the study area.

11. High fluxes; Climate contribution to distribution of PAHs

Malaysia is located near the equator where the weather is characterized as hot and humid with constant daylight time of around 12 hours and heavy rainfall. This cause abundance of plants and thus increase in available organic matter in the environment. Daily heavy rainfall basically washes away organic material such as total organic carbon that is associated with PAHs from the city and land surface into the water. Since the media for PAHs transport is always available in the environment, it is expected to record the highest existed PAHs in the sedimentary environment.

Malaysia has been experiencing a rapid development in modernization, transport, urbanization and industries starting 1950's. Hence, it is hypothesized that due to the massive land development of the post-independence, the marine environment of Malaysia such as estuaries and coastal water should receive a considerable amount of TOC via rainfall and drainage runoff. Organic compounds including PAHs adhere to organic contents and are therefore able to travel over distances. In almost all stations of the study area, the concentrations of TOC were found to be very high as compared to other areas in the world.

There are basically four phases in cores from near shore stations namely pre-war (Pre-WWII), war-independency era, rapid development and finally modern input. The first phase belongs to era represents sediments with natural PAHs input or minimum anthropogenic input from pyrolysis of organic matters. The second phase intervals represent deposition during WWII and pre-independence. The third phase represents post-independence and rapid development (1956-1990) that shows the highest PAHs with oil signature. The last phase represents mostly sudden drop in PAHs in concentration however the sources are remained same as phase two and three (Fig. 6). The samples from the offshore unlikely have shown different results from the near shore cores. The Offshore cores are more erratic and expected to be derived presumably by the input from atmospheric movement. The results from the Offshore core show that MP/P ratios of 1 (in average) suggesting that the source of the PAHs were pyrogenic originated from street and urban dust and transported with atmospheric movements (Takada et al., 1990; 1991).

The near shore stations, the MP/P values indicated highly matched identified source comparing offshore cores. Since offshore PAHs are mainly derived by atmosphere, near shore locations receive via street run off, canals and drainage systems due to climate condition and rainfall.

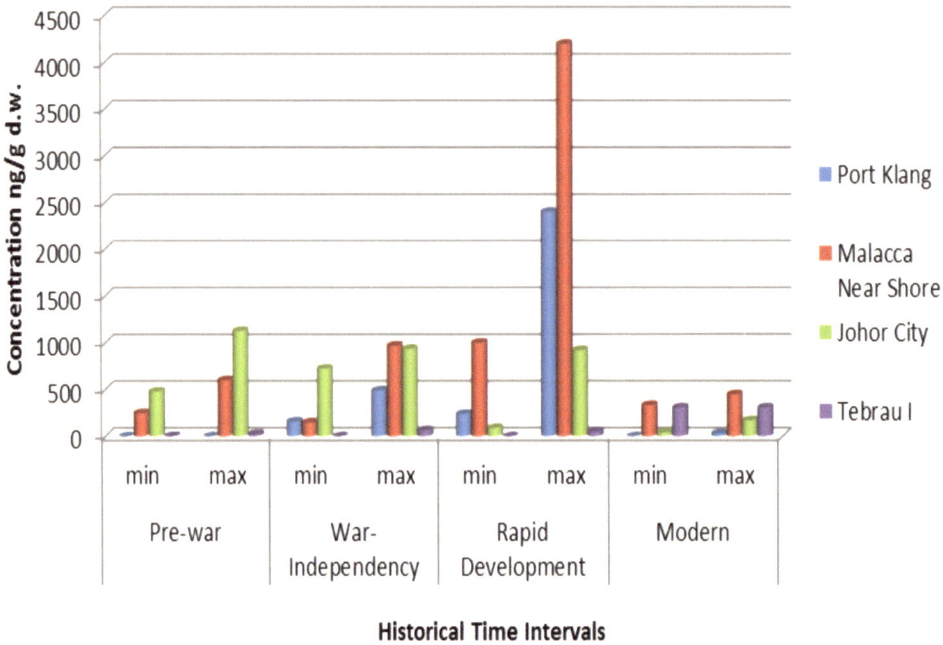

Fig. 6. The concentration of Polycyclic Aromatic Hydrocarbon (ng/g d.w.) during four identical time period in Malaysian history in selected study area.

12. References

Aizenshtat, Z., 1973. Perylene and its geochemical significance. *Geochimica et Cosmochimica Acta* 37: 559-567.

Baker, J. E., Capel P.D. and Eisenreich S.J., 1986. Influence of colloids on sediment-water partition coefficients of polychlorobiphenyl congeners in natural waters. *Environmental Science and Technology*, 20: 1136-1143.

Barra, R., P. Popp, R. Quiroz, H. Treutler, A. Araneda, C. Bauer, R. Urrutia, 2006. Polycyclic aromatic hydrocarbons fluxes during the past 50 years observed in dated sediment cores from Andean mountain lakes in central south Chile. *Ecotoxicology and Environmental Safety*. 63: 52-60.

Barrick, R. C. and F. G. Prahl, 1987. Hydrocarbon Geochemistry of the Puget Sound Region - III. Polycyclic Aromatic Hydrocarbons in Sediments. *Estuarine, Coastal and Shelf Science* 25: 175-191.

Bates, T. S., Hamilton S. E. and Cline J.D., 1984. Vertical transport and sedimentation of hydrocarbons in the central main basin of Puget Sound, Washington. *Environmental Science and Technology*, 18: 299-305.

Bauer, J.E. and D.G. Capone, 1985. Degradation and mineralization of the polycyclic aromatic hydrocarbons anthracene and naphthalene in intertidal marine sediments. *Applied Environmental Microbiology*, 50: 81-90.

Baumard, P., H. Budzinski, P. Garrigues, T. Burgeot, X. Michel, J. Bellocq (1999). Polycyclic aromatic hydrocarbons (PAH) burden of mussels (*Mytilus* sp.) in different marine environments in relation with sediment PAH contamination, and bioavailability. *Marine Environmental Research* 47: 415-439.

Baumard, P., Budzinski, H., Garrigues, P., Sorbe, J. C., Burgeot, T. and Bellocq, J., 1998. Concentrations of PAHs (Polycyclic Aromatic Hydrocarbons) in various marine organisms in relation to those in sediments and to trophic level. *Marine Pollution Bulletin*, 36: 951-960

Blumer M. 1976. Polycyclic aromatic compounds in nature. *Science*, 234: 34-45.

Boonyatumanond, R., G. Wattayakorn, A. Amano, Y. Inouchi, H. Takada, 2007. Reconstruction of pollution history of organic contaminants in the upper Gulf of Thailand by using sediment cores: First report from Tropical Asia Core (TACO) project. *Marine pollution bulletin* 54: 554-565.

Boonyatumanond R., Wattayakom, G., Togo A. and Takada, H., 2006. Distribution and origins of Polycyclic Aromatic Hydrocarbons in estuarine, rivers and marine sediments in Thailand. *Marine Pollution Bulletin*, 52: 942-956

Bouloubassi, I. and A. Saliot (1993). Dissolved, Particulate and Sedimentary Naturally Derived Polycyclic Aromatic-hydrocarbons in a Coastal Environment - Geochemical Significance. *Marine Chemistry* 42: 127-143.

Budzinski, H., I. Jones, J. Bellocq, C. Pierrard, P. Garrigues, 1997. Evaluation of sediment contamination by polycyclic aromatic hydrocarbons in the sediment of the Georges River estuary. *Marine chemistry*, 58: 85-97.

Chandru, K., M. P. Zakaria, S. Anita, A. Shahbazi, M. Sakari, P.S. Bahry, C. A. R. Mohamed (2008). Characterization of alkanes, hopanes, and polycyclic aromatic hydrocarbons (PAHs) in tar-balls collected from the east coast of peninsular Malaysia. *Marine Pollution Bulletin*, 56: 950-962.

Christensen, E. R., X. Zhang, 1993. Sources of polycyclic aromatic hydrocarbons to Lake Michigan determined from sedimentary records. *Environmental Science and Technology*, 27: 139-146.

Colombo, J.C., Cappelletti, N., Lasci, J., Migoya, M.C., Speranza, E. and Skorupka, C.N. (2005). Sources, Vertical Fluxes and Accumulation of Aliphatic Hydrocarbons in Coastal Sediments of the Rio de la Plata Estuary, Argentina. *Environmental Science and Technology*, 39, 8227-8234.

Cranwell P. A. and V. K. Koul, 1989. Sedimentary Record of Polycyclic Aromatic and aliphatic Hydrocarbons in the Windermere Catchment. *Water Research*, 23: 275-283

Curtosi, A. E. Pelletier, C. L. Vodopivez, W. P. MacCormack, 2007. Polycyclic aromatic hydrocarbons in soil and surface marine sediment near Jubany Station Antarctica. Role of permafrost as a low-permeability barrier. *Science of the Total Environment* 383: 193–204

Farrington, J. W., E. D. Goldberg, R. W. Risebrough, J. H. Martin, V. T. Bowen 1983. "Mussel Watch" 1976-1978: An overview of the trace-metal, DDE, PCB, Hydrocarbons, and artificial radionuclide data. *Environmental Science and Technology*. 17: 490-496.

Fernandez, P., Rose, N.L., Vilanova, R.M., Grimalt, J.O., 2002. Spatial and temporal comparison of polycyclic aromatic hydrocarbons and spheroidal carbonaceous particles in remote European lakes. *Water, Air and Soil Pollution: Focus* 2: 261-274.

Frontera-Suana, R., Bost, F. D., McDonald, T. J. and Morris, P. J., (2002). Aerobic Biodegradation of Hopanes and other Biomarkers by Crude Oil-Degrading Enrichment Culture. *Environmental Science and Technology*, 36, 4585-4592.

Gardner, W.S., R.F. Lee, K.R. Tenore and L.W. Smith, 1979. Degradation of selected polycyclic aromatic hydrocarbons in coastal sediments: importance of microbes and polychaete worms. *Water, Air, Soil Pollution*, 11: 339-347.

Garrigues, P., Budzinski, H., Manitz, M. P. and Wise, S. A. 1995. Pyrolitic and petrogenic input in recent sediments: A definitive signature through Phenantherene and Chrysene compounds distribution. *Journal of Polycyclic Aromatic Compounds* 7: 175-184.

Gevao, B., J. Hamilton-Taylor, C. Murdoch, K. C. Jones, M. Kelly, B. J. Tabner, 1997. Depositional time trends and remobilization of PCBs in lake sediments. *Environmental Science and Technology*, 31: 3274-3280.

Graef, W. and H. Diehl, 1966. The natural Normal Levels of Carcinogenic PCAH and the Reasons therefore. *Arch. Hyg. Bakteriol.* 150: 49-59

Gschwend, P. M. and R. A. Hites, R. A., 1981. Fluxes of polycyclic aromatic hydrocarbons to marine lacustrine sediments in the northeastern United States. *Geochimica et Cosmochimica Acta*, 45: 2359-2367.

Gustafsson, O., F. Haghseta, C. Chan, J. Macfarlane, P. Gschwend (1997). Quantification of The Dilute Sedimentary Soot Phase: Implications for PAH Speciation and Bioavailability. *Environmental Science and Technology* 31(1): 203-209.

Hancock J. L., H. G. Applegate and J. D. Dodd, 1970. Polynuclear Aromatic Hydrocarbons on Leaves. *Atmospheric Environment*. 4: 363-370.

Hartmann P. C., J. G. Quinn, R. W. Carins, J. W. King, 2005. Depositional History of Organic Contaminants in Narragansett Bay, Rhode Island, USA. *Marine Pollution Bulletin* 50 (4): 388-395.

Hase A. and Hites R.A. 1976. Identification and Analysis of Organic Pollutants in Water. *Geochimica et Cosmochimia Acta* 40: 1141.

Helfrich, J. and D. E. Armstrong, 1986. Polycyclic aromatic hydrocarbons in sediment of Lake Michigan. *Journal of Great Lake Research*, 12: 192-199.

Hites, R. A., R. E. Laflamme, J. W. Farrington, 1977. Sedimentary polycyclic aromatic hydrocarbons: the historical record. *Science*, 198: 829-831.

Hostettler, F. D., W. E. Pereira, K. A. Kvenvolden, A. Geen, S. N. Luoma, C. C. Fuller, R. Anima, 1999. A record of hydrocarbon input to San Francisco Bay as traced by biomarker profiles in surface sediment and sediment cores. *Marine Chemistry* 64:115–127

Ikenaka, Y., H. Eun, E. Watanabe, F. Kumon, Y. Miyabara, 2005. Estimation of sources and inflow of dioxin and polycyclic aromatic hydrocarbons from the sediment core of Lake Suwa, Japan. *Environmental Pollution*, 138: 529-537.

Jones, D.M., S.J. Rowland, A.G. Douglas and S. Howells, 1986. An examination of the fate of Nigerian crude oil in surface sediments of the Humber Estuary by Gas Chromatography and Gas Chromatography-Mass Spectrometry. *International Journal of Environmental Analytical Chemistry*, 24: 227 247.

Kamens, R.M., Guo, Z., Fulcher, J.N., Bell, D.A., 1988. Influence of humidity, sunlight, and temperature on the daytime decay of polycyclic aromatic hydrocarbons on atmospheric soot particles. *Environmental Science and Technology*, 22: 103-108.

Kamens, R.M., Fulcher, J.N., Guo, Z., 1986. Effects of temperature on wood soot PAH decay in atmospheres with sunlight and low NOx. *Atmosphere Environment*, 20: 1579-1587.

Kapalan, I. R.; Lu, S.-T.; Alimi, H. M.; MacMurphey, J. (2001). Fingerprinting of high boiling hydrocarbon fuels, asphalts and lubricants. *Environ. Forensics, 2*, 231-248.

Knight C. V., Graham M. S. and Neal B. S., 1983. Polynuclear Aromatic Hydrocarbons: Formation, Metabolism and Measurement *In Polynuclear Aromatic Hydrocarbons and associated organic emissions for catalytic and noncatalytic wood heaters.* (Edited by Cooke M. and Dennis A. J.), 689-708. Battelle Press, Columbus, Ohio.

Laflamme, R.E. and R.A. Hites, 1978. The global distribution of polycyclic aromatic hydrocarbons in recent sediments. *Geochimica Cosmochima Acta,* 42: 289-303.

Lee, M. L., G. P. Prado, J. B. Howard, R. A. Hites, 1977. Source identification of urban airborne polycyclic aromatic hydrocarbons by gas chromatography mass spectrometry and high resolution mass spectrometry. *Biomedical Mass Spectrometry*, 4: 182-186.

Liu, G. Q., G. Zhang, X. D. Li, J. Li, X. Z. Peng, S. H. Qi, 2005. Sedimentary record of polycyclic aromatic hydrocarbons in a sediment core from the Pearl River estuary, South China. *Marine Pollution Bulettin,* 51: 912-921.

Malaysian Marine Department, 2003. Annual Report.

Marr, L.C., Kirchstetter, T.W., Harley, R.A., Miguel, A.H., Hering, S.V., Hammond, S.K., 1999. *Environmental Science and Technology* 33:3091-3099.

Martel, L., M. J. Gagnon, R. Masse and A. Leclerc, 1987. The spatio-temporal variations and fluxes of polycyclic aromatic hydrocarbons in the sediment of the Saguenay Fjord, Quebec, Canada. *Water Research*, 21: 699-707.

Moriwaki, H., K. Katahiraa, O. Yamamotoa, J. Fukuyamaa, T. Kamiuraa, H. Yamazakib, S. Yoshikawac, 2005. Historical trends of polycyclic aromatic hydrocarbons in the reservoir sediment core at Osaka. *Atmospheric Environment,* 39: 1019–1025

Morris W. A., J.K. Versteeg a, C.H. Marvin b, B.E. McCarry b, N.A., Rukavina, 1994. Preliminary comparisons between magnetic susceptibility and polycyclic aromatic hydrocarbon content in sediments from Hamilton Harbour, western Lake Ontario. *The Science of the Total Environment* 152: 153-160

National Academy of Science (2002). *Oil in the Sea; input, fates and effects*. National Academy Press, Washington D.C. Press.

Neff, J. M. 1979. *Polycyclic Aromatic Hydrocarbon in the Aquatic Environment: Sources, Fates and Biological Effects*. Applied Science Publishers, London.

Nielsen, T., Jorgensen, H.E., Larsen, J.C., Poulsen, M. 1996. City air pollution of polycyclic aromatic hydrocarbons and other mutagens: occurrence, sources and health effects. *Science of the Total Environment* 189(190): 41-49.

Nielsen, T., 1988. The decay of benzo(a)pyrene and cyclopenteno(cd)pyrene in the atmosphere. *Atmospheric Environment* 22: 2249–2254.

Okuda, T., Kumata, H., Zakaria, M.P., Naroaka, H., Ishiwatari, R., and Takada, H. 2002. Source identification of Malaysian atmospheric polycyclic aromatic hydrocarbons nearby forest fires using molecular and isotopic compositions. *Atmospheric Environment* 36: 611-618.

Ou, S. M., J.H. Zheng, J.S. Zheng, B.J. Richardson and P.K.S. Lam (2004). Petroleum hydrocarbons and polycyclic aromatic hydrocarbons in the surficial sediments of Xiamen Harbour and Yuan Dan Lake, China, *Chemosphere* 56: 107–112.

Pereira, W. E., Hostettler, F. D., Luoma, S. N. Van Geen, A., Fuller, C. C., Anima, R. G., 1999. Sedimentary record of anthropogenic and biogenic Polycyclic Aromatic Hydrocarbons (PAHs) in San Francisco Bay, California. *Marine Chemistry* 64: 99-113.

Peters, K. E. and Moldowan, J. M. 1993. *The Biomarker Guide: Interpreting Molecular Fossils in Petroleum and Ancient Sediments.* Prentice Hall, Englewood Cliff, N.J. 363 p.

Peters, K.E., Scheuerman, G.L., Lee, C.Y., Moldowan, J.M., Reynolds, R.N. and Pena, M.M. (1992). Effects of refinery processes on biological markers. *Energy and Fuels*, 6, 560-577.

Prahl, F.G. and Carpenter, R., 1984. Hydrocarbons in Washington coastal sediments. *Estuarine, Coastal Shelf Science* 18: 703-720.

Prahl, F. G. and Carpenter, R. 1983. Polycyclic aromatic hydrocarbon (PAH)-phase associations in Washington coastal sediment. *Geochimica et Cosmochimica Acta*, 47(6): 1013-1023

Prahl F.G. and Carpenter R. 1979. The role of zooplankton fecal pellets in the sedimentation of PAHs in Dabob Bay, Washington. *Geochim Cosmochemica Acta*, 43: 1959-1968.

Pruel, R.J. and J. G. Quinn, 1988. Accumulation of polycyclic aromatic hydrocarbons in crankcase oil. *Environmental pollution*, 49: 89–97.

Omar, N.Y.M.J., T. C. Mon, N. A. Rahman, M. R. Abas. 2006. Distribution and health risk of polycyclic aromatic hydrocarbons (PAHs) in atmospheric aerosols of Kuala Lumpur, Malaysia. *Science of the Total Environment* 369: 76-81

Quiroz R., P. Popp, R. Urrutia, C. Bauer, A. Araneda, H. C. Treutler, R. Barra, 2005. PAH fluxes in the Laja Lake of south central Chile Andes over the last 50 years: Evidence from a dated sediment core. *Science of the Total Environment*, 349: 150–160

Rachdawong, P., E. R. Christensen and J. F. Carls, 1998. Historical PAH Fluxes to Lake Michigan Sediments Determined by Factor Analysis. *Water Research*, 32 (8): 2422-2430.

Readman, J. W., R.F.C. Mantoura and M. M. Rhead, 1987. A record of the polycyclic aromatic hydrocarbons (PAHs) pollution obtained from accreting sediments of the Tamar Estuary, UK; Evidences for non-equilibrium behavior of PAH. *The Science of the Total Environment*, 66: 73-94

Readman, J.W., R.F.C. Mantoura, M.M. Rhead and L. Brown, 1982. Aquatic distribution and heterotrophic degradation of polycyclic aromatic hydrocarbons in the Tamar Estuary. *Estuarine Coastal Shelf Science*, 14: 36-38.

Rose, N. L. and B. Rippey, 2002. The historical record of PAH, PCB, trace metal and fly-ash particle deposition at the remote lake in north-west Scotland. *Environmental Pollution*, 117: 121-132.

Sakari, M., Zakaria, M. P., Lajis, N., Mohamed, C. A. R., Chandru, K., Shahpoury. 2011. Polycyclic Aromatic Hydrocarbons and Hopane in Malacca Coastal Water: 130 Years of Evidence for their Land-based Sources. *Environ. Forensic.* 12: 63-78.

Sakari, M., Zakaria, M. P., Lajis, N., Mohamed, C. A. R., Chandru, K., Shahpoury, P., Shahbazi, A. and Anita, S. 2010a. Historical profiles of Polycyclic Aromatic Hydrocarbons (PAHs), sources and origins in dated sediment cores from Port Klang, Straits of Malacca, Malaysia. *Coast. Mar. Sci.* 34: (1): 140-155.

Sakari, M., Zakaria, M. P., Lajis, N., Mohamed, C. A. R., Chandru, K., Shahpoury, P., Mokhtar, M. and Shahbazi, A. 2010b. Urban vs. Marine Based Oil Pollution in the Strait of Johor, Malaysia: A Century Record. *Soil & Sed. Cont.* 19: 644-666.

Sakari, M 2009. Characterization, Concentration and Depositional History of Polycyclic Aromatic Hydrocarbons and Hopane in Selected Locations of Peninsular Malaysia. PhD dissertation. Universiti Putra Malaysia, Sri Serdang, Selangor, Malaysia.

Simoneit, B. R. T., Cox, R. E. and Standley, L. J. (1988). Organic Matter of the Troposphere-IV. Lipids in Harmattan aerosols of Nigeria. *Atmospheric Environment*, 22, 983-1004.

Simoneit, B.R.T. and Mazurek, M.A., 1982. Organic matter of the troposphere - II. Natural background of biogenic lipid matter in aerosols over the rural western United States. *Atmospheric Environment*.16: 2139-2159.

Standley, L. J. and Semoneit, B. R. T. (1987). Characterization of extractable plant wax, resin, and thermally matured components in smoke particles from prescribed burns. *Environmental Science and Technology* 21, 163-169.

Steinhauer, S. S. and P. D. Boehm, 1992. The composition and distribution of saturated and aromatic hydrocarbons in near shore sediments, river sediments and coastal peat of the Alaskan Beaufort Sea: Implications for detecting anthropogenic hydrocarbon inputs. *Marine Environmental Research* 33: 223-253.

Su, M., E. R. Christensen, J. F. Karls, 1998. Determination of PAHs sources in dated sediments from Green Bay, Wisconsin, by a chemical mass balance model. *Environmental Pollution*, 98: 411-419.

Takada, H., Onda, T., Ogura, N. 1990. Determination of polycyclic aromatic hydrocarbons in urban street dusts and their source materials by capillary gas chromatography. *Environmental Science and Technology* 24: 1179-1186.

Takada, H., Onda, T., Harada, M., Ogura, N. 1991. Distribution and sources of polycyclic aromatic hydrocarbons (PAHs) in street dust from Tokyo metropolitan area. *Science of the Total Environment* 107: 45-69.

Tan, Y. L. & Heit, M. 1981 Biogenic and abiogenic polynuclear aromatic hydrocarbons in sediments from two remote Adirondack Lakes. *Geochimica et Cosmochimica Acta* 45: 2267–2279.

Taylor, P. N. and J. N. Lester, 1995. Polynuclear aromatic hydrocarbons in a River Thames sediment core. *Environmental Technology*, 16: 1155-1163.

Tolosa, I., de Mora, S., Sheikholeslami, M.R., Villeneuve, J.P., Bartocci, J., Cattini, C., 2004. Aliphatic and aromatic hydrocarbons in coastal Caspian Sea sediments. *Marine Pollution Bulletin* 48: 44-60.

Tuominen, J., H. Pyysalo, J. Laurikko, T. Nurmela, 1987. Application of GLC-selected ion monitoring (SIM)-technique in analyzing Polycyclic Organic Compounds in Vehicle Emissions. *The Science of Total Environment* 59: 207-210.

Volkman, J. K., A. T. Revill, A. P. Murray (1997). Applications of Biomarkers for Identifying Sources of Natural and Pollutant Hydrocarbons in Aquatic Environments. In *Molecular Markers in Environmental Geochemistry*, ed. R. P. Eganhouse, American Chemical Society Press, Washington, D.C. pp. 279-313.

Wakeham, S. G., C. Schaffner, W. Giger (1980). Polycyclic aromatic hydrocarbons in recent lake sediments - I. Compounds having anthropogenic origin. *Geochimica et Cosmochimica Acta* 44: 403-413.

Wakeham, S. G., C. Schaffner, W. Giger, 1979. Polycyclic aromatic hydrocarbons in recent lake sediments- I. compounds having anthropogenic origins. *Environmental Science and Techonology*, 403-413.

Wang, Z., Fingas, M., 2005. Oil and petroleum product fingerprinting analysis by gas chromatographic techniques. In: Nollet, L. (Ed.), *Chromatographic Analysis of the Environment*, third ed. CRC Press, New York.

Wilcke W., Amelung W, Krauss M., Mrtius C, Bandeira A and Garcia MVB. 2003. Polycyclic aromatic hydrocarbon (PAH) patterns in climatically different ecological zones of Brazil. Org Geochem 34: 1405–1417.

Wilcke W., Krauss M., Lilienfein J. and Amelung W. 2004. Polycyclic aromatic hydrocarbon storage in a typical Cerrado of the Brazilian savanna. J Environ Qual 33: 946–955

Witt, G. and E. Trost, 1999. Polycyclic aromatic hydrocarbons (PAHs) in sediments of the Baltic Sea and of the German coastal waters. *Chemosphere*, 38: 1603-1614.

Wong CS., G. Snders, DR. Engstrom, DT. Long, DL. Swackhamer, SJ. Eisenreich, 1995. Accumulation, Inventory and Diagenesis of Chlorinated Hydrocarbons in Lake Ontario Sediments. *Environmental Science and Technology* 29: 2661-2672.

Yim, U. H., Hong, S. H., Shim, W. J., Oh, J. R., Chang, M., 2005. Spatio-temporal distribution and characteristics of PAHs in sediments from Masan Bay, Korea. *Marine Pollution Bulletin* 50: 319-326.

Youngblood, W. W. and Blumer, M. 1975. Polycyclic aromatic hydrocarbons in the environment: homologous series in soils and recent marine sediments. *Geochimica et Cosmochimica Acta*, 39: 1303-1314.

Yunker, M.B., Macdonald, R.W., Vingarzan, R., Mitchell, R.H., Goyette, D., Sylvestre, S., 2002. PAHs in the Fraser River basin: a critical appraisal of PAH ratios as indicators of PAH source and composition. *Organic Geochemistry,* 33: 489-515.

Yunker, M. B., R. W. MacDonald, D. Goyette, D. W. Paton, B. R. Fowler, D. Sullivan, J. Boyd (1999). Natural and Anthropogenic Inputs of Hydrocarbons to the Strait of Georgia. *The Science of the Total Environment* 225(3): 181-209.

Yunker, M. B. and Macdonald R. W., 1995. Composition and Origin of Polycyclic Aromatic Hydrocarbons in the Mackenzie River and on the Beaufort Sea Shelf. *Arctic*, 48(2): 118-129.

Zakaria, M. P., Takada, H., Tsutsumi, S., 1999. American Chemical Society (ACS) national meeting. Division of Environmental Chemistry (J). 39(2): 6.

Zakaria M.P., Horinouchi A., Tsutsumi S., Takada H., Tanabe S. and Ismail A. 2000. Oil Pollution in the Strait of Malacca, Malaysia: Application of Molecular Markers for Source Identification. *Environmental Science and Technology* 34: 1189-1196

Zakaria, M. P., Okuda, T., Takada, H., 2001. Polycyclic Aromatic Hydrocarbons (PAHs) and Hopanes in stranded tar-balls on the coasts of Peninsular Malaysia: applications of biomarkers for identifying sources of oil pollution. *Marine Pollution Bulletin* 42(12): 1357-1366.

Zakaria M. P., Takada H., Kumata H., Nakada N., Ohno K., Mihoko Y. 2002. Distribution of Polycyclic Aromatic Hydrocarbons (PAHs) in rivers and estuaries in Malaysia: widespread Input of petrogenic hydrocarbons. *Environmental Science and Technology* 36:1907-1918.

Zakaria, M. P., Takada H. 2003. *Petroleum hydrocarbon pollution: A closer look at the Malaysian legislations on marine environment*. Paper presented at First joint seminar on oceanography, NRCT-JSPS. Chien Mai, Thailand. December 2003.

Zakaria M.P. and Mahat A. A. (2006). Distribution of Polycyclic Aromatic Hydrocarbon (PAHs) in Sediments in the Langat Estuary. *Coastal Marine Science Journal*. 30(1).

Zhang, X., E. R. Christensen, M. F. Gin, 1993. Polycyclic aromatic hydrocarbons in dated sediments from Green Bay and Lake Michigan. *Estuaries*, 16: 638-652.

7

The Mass Distribution of Particle-Bound PAH Among Aerosol Fractions: A Case-Study of an Urban Area in Poland

Wioletta Rogula-Kozłowska[1], Barbara Kozielska[2], Barbara Błaszczak[1] and Krzysztof Klejnowski[1]
[1]*Institute of Environmental Engineering of the Polish Academy of Sciences, Zabrze,*
[2]*Faculty of Energy and Environmental Engineering, Department of Air Protection, Silesian University of Technology, Gliwice, Poland*

1. Introduction

Zabrze is one of the fourteen Silesian cities that form together the Silesian Agglomeration (Fig. 1-4). The Silesian Agglomeration lies in the center of the Silesia Province, occupies 1230 km^2, its population is about 2.1 million (1691 inhabitants per one square kilometer). It is one of the most urbanized and industrialized regions of Central Europe. Such a dense concentration of people on such a heavily urbanized and industrialized area is unique in Europe. About 50% of the Silesia Province gross product and 7% of the gross domestic product come from the Silesian Agglomeration. Six European capital cities, Berlin, Prague, Wien, Bratislava, Budapest and Warsaw, lie within 600 km from Katowice, the capital city of the Agglomeration. The main transport routes linking Poland with Western Europe run through it in all directions.

From the air protection point of view, the Silesian Agglomeration is one of the most interesting regions both in Poland and in Europe. Hard coal of better quality, mined in the western part of the Agglomeration since the 18th century, has been processed into coke and gas in five cities of the Agglomeration. Poorer quality coal, from the eastern part of the Agglomeration, is burnt in several great power stations, smaller power and heating plants, local heating plants and in domestic stoves. Almost all branches of industry, such as electrical, chemical, glass-making, textile, clothing, and ceramic industries, ferrous and non-ferrous metallurgy, machine-building, hard coal mining and coking very actively have been deteriorating the natural environment for about 200 years.

Nevertheless, three recent decades of economical changes forced in the Silesian Agglomeration the greatest in Poland drop of industrial air pollution (in Zabrze, yearly dust fall exceeded 2100 g m^{-2} in the 70s, oscillated between 700 and 800 g m^{-2} in the 80s, was less than 350 g m^{-2} after 1995). The concentration of the ambient particulate matter (PM) dropped significantly (Fig. 1). The greatest drop of the PM concentrations occurred between 1985 and 2000, when the total industrial emission of air pollutants drastically decreased in the effect of the transformation of the whole national industry and closing or restructuring of many great plants in the 80s. In 1990, the list of 80 Polish plants exerting the greatest impact on the environment was announced. More than 20 of them were located in the Katowice Region whose the Silesian Agglomeration was a sub-region. The actions to lower the emission from these plants and limiting the output of heavy industry decreased the emission of dust and its gaseous precursors (Central Statistical Office of Poland [CSO], 1976-2001) and halved the ambient concentrations of dust at measuring points (at the beginning of the 80s, in Dąbrowa Górnicza the PM concentration was almost 0.5 mg m^{-3}). During the 90s, the PM concentrations continued to drop and since 2000 the yearly PM concentrations have remained constant at the approximate level between 65-110 µg m^{-3} depending on a measuring site. The industrial emission of PM and its precursors (SO_2, NO_x) has also remained almost stable (Central Statistical Office of Poland [CSO], 2001-2011).

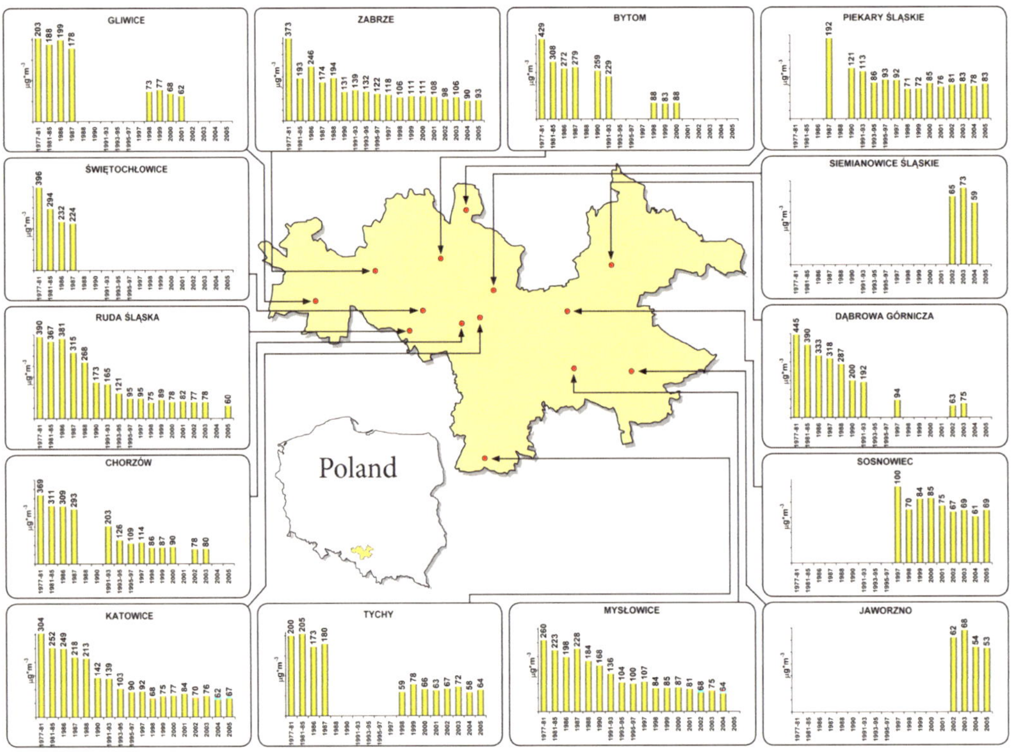

Fig. 1. The total PM concentrations (µg m^{-3}) in 14 cities of the Silesian Agglomeration in 1977-2005 (SSI, 1980-2006)

The concentrations of PM_{10}[1], although slightly lower than in the 90s, are still high in the Silesian Agglomeration and have also been stabilized in each of the cities since 2000 (Fig. 2).

[1]For a number $d>0$, PM_d is the fraction of the particles that have the aerodynamic diameter not greater than d. For $0<c<d$, $PM_{c\text{-}d}$ denotes the fraction of particles with the diameters between c and d. We have:
$PM_{2.5}$–particles with the aerodynamic diameter not greater than 2.5 µm (fine particles)
$PM_{2.5\text{-}10}$–particles with the aerodynamic diameter between 2.5 and 10 µm (coarse particles)
PM_{10}–particles with the aerodynamic diameter not greater than 10 µm ($PM_{2.5}$ and $PM_{2.5\text{-}10}$ together)

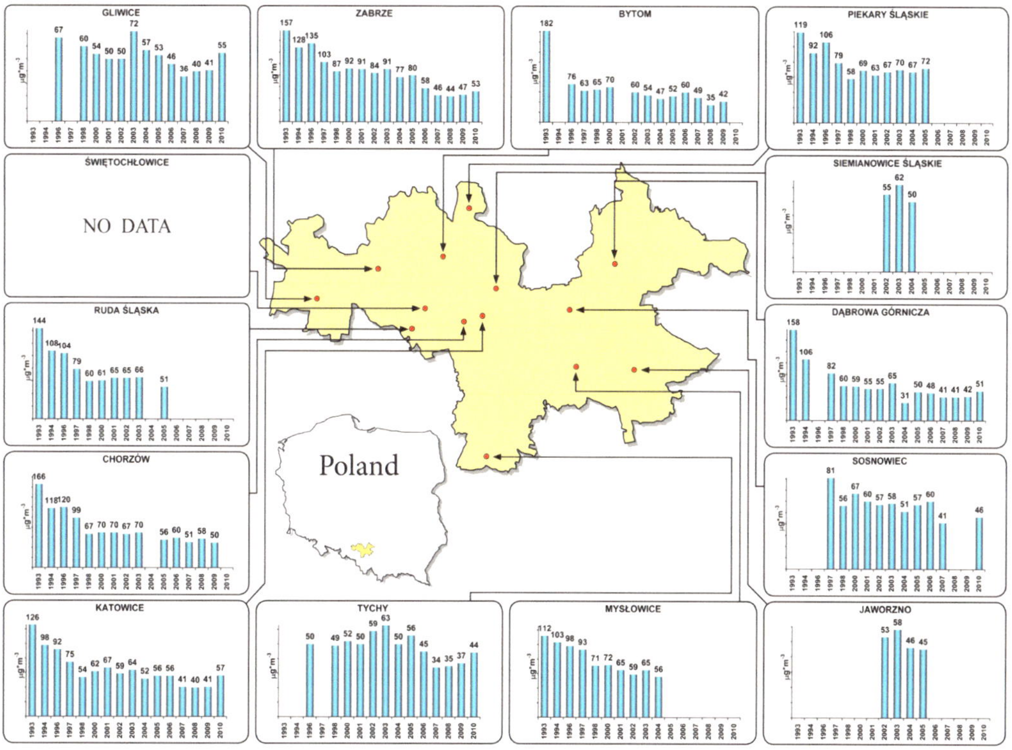

Fig. 2. The PM_{10} concentrations (µg m^{-3}) in 14 cities of the Silesian Agglomeration in 1993-2010 (SSI, 1980-2006; The Provincial Inspector for Environmental Protection in Katowice [PIEP], 2002-2011)

There are hundreds of organic compounds to be found in PM, among them more than one hundred polycyclic aromatic hydrocarbons (PAH). Since PAH tend to have low vapor pressures, they are usually adsorbed onto PM in the atmosphere. The vapor pressure of a PAH is inversely proportional to the number of rings it contains. As a result, the larger molecular weight PAH (≥ 4) are mostly adsorbed onto PM in atmospheric samples, while the lower molecular weight PAH can be found both free in the atmosphere and bound to particles. In some conditions, from 72% to even > 98% of four-, six-, and seven-ring ambient PAH are bound onto particles of PM_3 and PM_7 (Sheu et al., 1997). Seventeen PAH: acenaphtene (Acy), acenaphthylene (Ace), anthracene (An), benzo(a)anthracene (BaA), benzo(a)pyrene (BaP), benzo(e)pyrene (BeP), benzo(b)fluoranthene (BbF), benzo(j)fluoranthene (BjF), benzo(k)fluoranthene (BkF), benzo(g,h,i)perylene (BghiP), chrysene (Ch), dibenzo(a,h)anthracene (DBA), fluoranthene (Fl), fluorene (F), phenanthrene (Ph), pyrene (Py) and indeno(1,2,3-cd)pyrene (IP) are most often investigated. The most hazardous property of air pollutants is their carcinogenicity. Seven PAH: BaA, BaP, BbF, BkF, Ch, DBA and IP the USEPA classified as probable human carcinogens (http://www.epa.gov/).

The most important natural sources of PAH are forest fires and eruptions of volcanoes. However, amounts of the natural ambient PAH are small compared to the amounts of the anthropogenic PAH. The majority of anthropogenic PAH come from incomplete combustion of fossil fuels or organic matter (Sienra et al., 2005; Zou et al., 2003). In urban air this includes PAH combustion in car engines and residential heating (Harrison et al., 1996; Manoli et al., 2004; Kristensson et al., 2004).

Since 1977, in all cities of the Silesian Agglomeration, the State Sanitary Inspection (SSI, Department in Katowice) has measured ambient concentrations of PM-bound BaP continuously and 8 other PAH periodically (Fl, BaA, Ch, BbF, BkF, DBA, BghiP, IP; several years in each city). The total ambient concentration of these nine PAH drastically decreased in the period 1977–2005 (Fig. 3). The concentrations of PM-bound BaP dropped rapidly in all cities of the Agglomeration in 1980–1990 (Fig. 4). In each city the vast yearly concentrations of BaP from 1977–1981, reaching 500 ng m^{-3} in Ruda Śląska, decreased several times during this decade. During the 90s, the concentrations of BaP dropped to less than 20 ng m^{-3} in each city of the Agglomeration. These concentrations have been still very high and hazardous to humans and the concentrations of PM_{10}-bound BaP, measured since 2001, were high (Fig. 4), like the concentrations of PM-bound one.

In general, the ambient particles with the aerodynamic diameter greater than 10 μm are not inhalable. Therefore, from the sanitary point of view, the concentrations of PM_{10} should be the measure of the hazard from ambient particles. In all cities of the Agglomeration, the yearly average concentrations of PM_{10} and PM_{10}-bound BaP exceed their permissible levels (40 μg m^{-3} and 1 ng m^{-3}, respectively, Fig. 2). The limit of 40 μg m^{-3} on the yearly average concentration of PM_{10} has been in effect in majority of the European countries for about 20 years (in Poland since 1998). Although the air quality has been improving for the latest thirty years, there is no city in the Silesian Agglomeration where this standard was not yearly exceeded in the period 1993-2010 (Fig. 2). Even the yearly average $PM_{2.5}$ concentrations, measured continuously since 2001 in Zabrze, exceed 25 μg m^{-3}, the standard for the $PM_{2.5}$ concentrations, every year (Rogula-Kozłowska et al., 2010; Klejnowski et al., 2007a, 2007b, 2009; PIEP, 2002-2011). Although the spectacular reduction of the industrial emissions, especially of coarse dust, caused significant drop of the concentrations of PM and PM-bound BaP the PM_{10} concentrations decreased only a little (Fig. 2).

The poor air quality conditions in the cities of the Agglomeration are due to bad spatial arrangement of industrial and urban infrastructure. The industrial objects are interspersed with living quarters. The differences in the level of industrialization, urbanization and land use cause spatial non-homogeneity of the air pollution from industrial, municipal or vehicular sources even over small areas. The area of the Agglomeration is also affected by periodically occurring episodes of very high concentrations of air pollutants (especially of PM_{10} in winter in city centers), which inflate the yearly PM concentrations.

The inventory of the Silesian air pollution sources from 2006 contains 107 business entities emitting PM_{10} and BaP within the Silesian Agglomeration (Air protection program [APP], 2010). In 2006, they emitted 4.9 Gg of PM_{10} and 1.3 Mg of benzo(a)pyrene into the air. In average, this industrial emission was 44% and 28% of, respectively, the total PM_{10} and BaP emissions in the Agglomeration. The municipal and household emission[2], mainly from local heating plants and domestic stoves, have even greater share in the PM_{10} and the BaP total

[2]Heating plants and furnaces fed with solid fuel (mainly coal) are main sources of PM_{10} and BaP. They include small (local) heating plants and domestic stoves. In the Silesian Agglomeration, the estimated average of the household heat demands covered by solid fuel combustion in domestic stoves is still greater than 34%. The emissions of PM_{10} and PM_{10}-bound BaP from the solid fuel furnaces are more than 94% of their totals in the surface emission. It is due to bad technical condition and age of the heating plants and stoves and also to the poor quality of the combusted coal.

concentrations. In 2006, it was 5.3 Gg and 3.2 Mg, what made 47% and 71% of the total PM_{10} and BaP emissions. This contribution is even greater in a heating season – in winter.

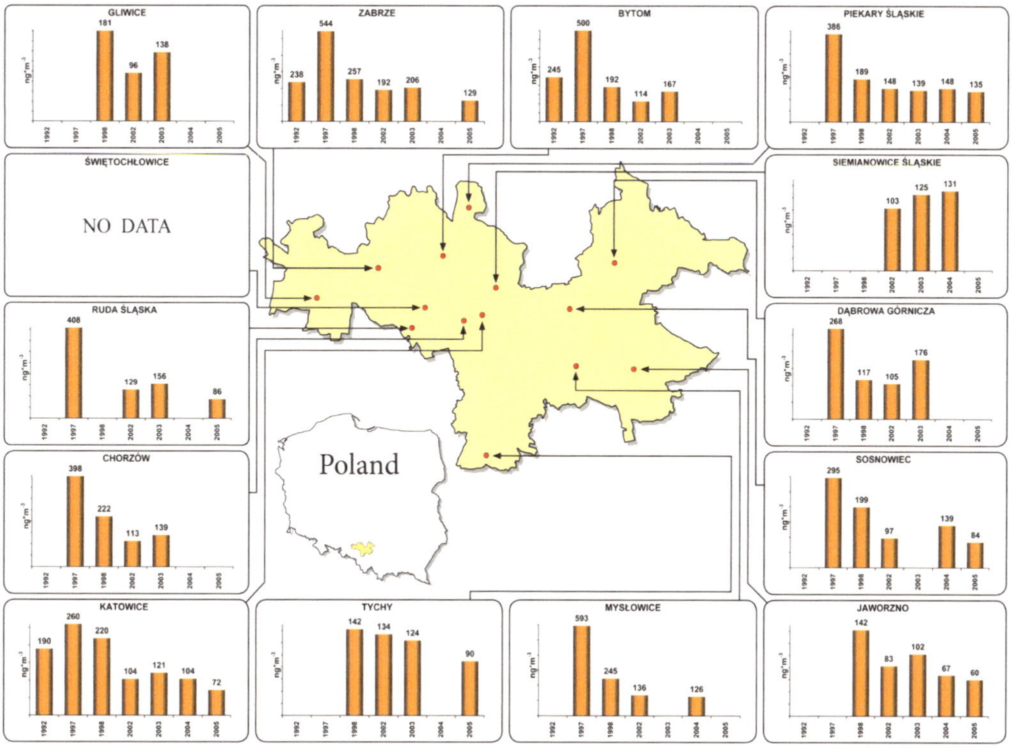

Fig. 3. The concentrations of the sum of 9 PM-bound PAH (Fl, BaA, Ch, BbF, BkF, BaP, DBA, BghiP, IP) in 14 cities of the Silesian Agglomeration in 1992-2005 (SSI, 1993-2006)

The elevated autumn and winter PAH concentrations in the Agglomeration my be linked with the emission from municipal sources. The seasonal dependence of the PM-bound PAH concentrations was observed in three Silesian cities: Katowice, Sosnowiec and Zawiercie in 2008. The seasonal PAH concentrations were 5.1-18.6 ng m^{-3} in spring, 5.4-7.6 ng m^{-3} in summer, 6.8-24.0 ng m^{-3} in autumn and 26.2-61.3 ng m^{-3} in winter (Zaciera et al., 2010).

Although the reduction of the industrial emission improved the air quality in the Silesian Agglomeration the hazard from PM_{10} (especially from the smallest particles) might grew owing to growth of vehicular emission. For the last two decades, the Poles have imported more than 920000 second-hand cars yearly. Majority of the cars were older than 10 years, they did not meet the emission standards and had high fuel-consumption.

In Gliwice, from April to June 2003, in a trafficked street canyon (1400 vehicles per hour), the average PM_{10} concentration was 94.0 µg m^{-3} and was higher by 40.0 µg m^{-3} than the one measured 100 m apart. The average concentration of total PAH, equal to 191.56 ng m^{-3}, was 1.5 times greater than the background concentration (Grynkiewicz-Bylina et al., 2005). In Zabrze, in summer 2005, the average concentration of the total PM_{10}- and the total $PM_{2.5}$-bound PAH at crossroads were 65.6 ng m^{-3} and 44.4 ng m^{-3}, and were 1.9 and 3.4 times greater than the background concentrations, respectively (Ćwiklak et al., 2009). In Bytom, on the turn of February and March 2007, the average concentrations of the total vehicular $PM_{2.5}$- and the total vehicular PM_{10}-bound PAH were between 56.2 and 73.4 ng m^{-3} and 75.1 and 91.0 ng m^{-3}, respectively. The significant in this city influence of the industrial emission and the emission from low sources was excluded by proper location of the measuring points (Kozielska et al., 2009).

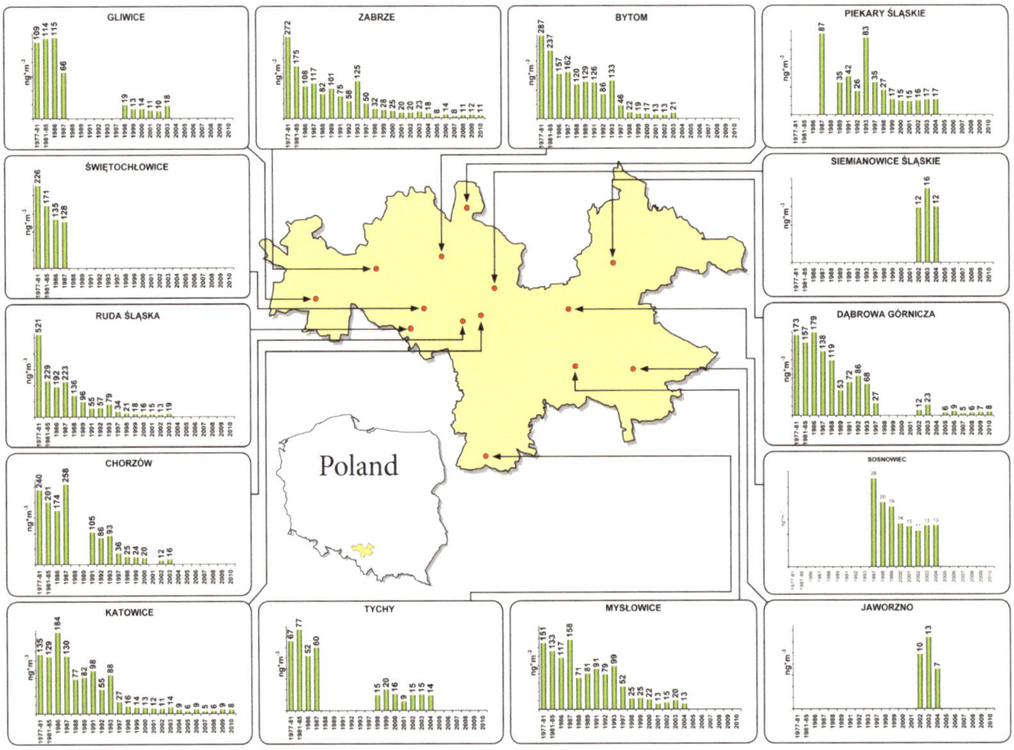

Fig. 4. The concentrations of PM-bound BaP in 1977-2001 and PM_{10}-bound BaP in 2002-2010 in 14 cities of the Silesian Agglomeration (SSI, 1982-2006; PIEP, 2005-2011)

The hazard from the air pollution in the neighborhood of trafficked roads is also elevated by the elevated toxicity of traffic PM_{10} containing allergens and carcinogens, also PAH (Kinney et al., 2000; Pakkanen et al., 2003; Vogt et al., 2003).

Now, in the Silesian Agglomeration, when the number of industrial and of the vast fugitive emission sources of PM (mine heaps, dumps) decreased, main PM and PAH sources are domestic furnaces, heat and power plants and road traffic. In the Agglomeration, the industrial and the municipal sources contribute to the air pollution much more than such sources in the Western Europe countries (EMEP, 2009), and the role of vehicular emission grows. As a result, the proportion of fine particles in PM increases. Ambient fine particles have relatively great surface to adsorb PAH (Ravindra et al., 2001; Sheu et al., 1997), so the fine dust is richer in PAH than the coarse one. Therefore, investigation of the particle size distribution of PM and the PAH content of PM fractions is crucial to abate the adverse effects of PM.

The work presents the method and the results of the investigations of the mass distribution of some PM-bound PAH among the PM fractions in the Silesian Agglomeration. The investigated PAH, three-ring: Acy, Ace, F, Ph, An, four-ring: Fl, Py, BaA, Ch, five-ring: BbF, BkF, BaP, DBA and six-ring: BghiP, and IP, are on the USEPA CWA list of the priority pollutants.

2. Methods

The site in Zabrze, selected for the experiment, is representative of the air pollution conditions typical in the Silesian Agglomeration - by the Directive 2008/50/EC definition, it is an urban background measuring point (Directive, 2008). The effects of the industrial and the municipal emissions on living quarters of the Agglomeration are represented and may be observed here very well.

Ambient dust was sampled with the use of a thirteen stage DEKATI low pressure impactor (DLPI), which collects thirteen PM fractions onto thirteen separate substrate filters (Table 1). The principle of DLPI operating may be found in (Klejnowski et al., 2010).

There were two periods of sampling: from 7 May to 2 August (summer) and from 26 October to 27 December (winter) 2007. One sample-taking lasted about one week. Seven such sample-takings were done in winter and nine in summer, the measurements covered the sampling periods in 98 and 92%, respectively, and whole summer (2nd and 3rd quarter of 2007) and winter (1st and 4th quarter of 2007) in 27% and 45%.

The mass of the dust collected on aluminum substrates was determined by weighing the substrates before and after exposure on a Mettler Toledo micro-balance. Before weighing, the substrates were kept in the weighing room for 48 hours (temperature 20 ± 2^0C, relative air humidity $48\pm5\%$). The concentrations of the fractions of PM were computed from the volume of air passed through the impactor and the masses of the dust collected on its stages. All the samples (substrates), till analyzing, were kept in a refrigerator in tight and lightproof containers.

The winter and summer samples were developed separately. For each of the thirteen PM fractions, all its samples (7 in winter and 9 in summer samples per fraction) were extracted together in an ultrasonic bath in dichloromethane (CH_2Cl_2). The extract was percolated, washed and dried by evaporating in the helium atmosphere. The dry residue was diluted in propane-2 ($CH_3CH(OH)CH_3$) and distilled water was added to receive the proportion 15/85 (v/v) of propanol-2 to water. For selective purification, the resulting samples were solidified (SPE) by extracting in columns filled with octadecylsilane (C_{18}, Supelco). PAH were eluted with the use of dichloromethane (CH_2Cl_2). The extract of a PAH fraction was condensed in the helium atmosphere to the volume of 0.5 cm^3. The samples were analysed on a Perkin Elmer Clarus 500 gas chromatograph with a Flame Ionization Detector (FID). An RTX-5 Restek capillary 30 m x 0.32 mm x 0.25 μm column was used to separate the sample components. The flow of the carrier gas, helium, was 1.5 $cm^3\,min^{-1}$. Calibration curves for 15 PAH standards were used in quantitative determinations. The linear correlation of the surfaces of the peaks with the PAH concentrations was checked in the concentration range 1 - 4 ng μl^{-1}. The correlation coefficient ranged from 0.90 to 0.97. The time of the whole analysis was 40 min. FID was provided with hydrogen (45 $cm^3\,min^{-1}$) and air (450 $cm^3\,min^{-1}$). The recoveries of PAH were determined using a standard containing the 15 PAH. They ranged from 85% to 93%.

3. Discussion of the results

3.1 Concentrations of PM- and PM-bound PAH. Origin of PAH in Zabrze

The winter PM_{10} concentrations exceeded 46 µg m^{-3}, the summer ones reached almost 19 µg m^{-3} (Table 1). Such a difference is due to very high emission of PM from combustion of fossil fuels in winter, specific of the Silesian Agglomeration (Rogula-Kozłowska et al., 2008; Pastuszka et al., 2010). The average concentrations of PM_{10} and PM in the experimental period (33.5 µg m^{-3} and 32.5 µg m^{-3}) were lower than the yearly concentrations of PM_{10} in 2007 and PM in 2005 (Fig. 1 and 2). The explanation may by that the former were sampled with DLPI at about 5 m above the ground level during about 16 weeks, missing a part of winter when in Zabrze the highest PM concentrations occur (Klejnowski et al., 2007a, 2009), the latter were measured by SSI at the height 2.5 m during the whole year.

Fraction, µm		PM, µg m^{-3}	PAH, ng m^{-3} Acy	Ace	F	Ph	An	Fl	Py	BaA	Ch	BbF	BkF	BaP	DBA	BghiP	IP	∑PAH
0.03-0.06	W	0.4	nd	nd	0.14	0.14	nd	0.03	0.04	nd	0.08	0.07	0.08	0.05	nd	nd	nd	0.62
	S	0.09	0.01	0.02	0.02	0.05	nd	0.04	0.02	nd	0.11	0.14	0.14	0.04	0.01	0.05	0.06	0.7
0.06-0.108	W	0.59	nd	nd	0.11	0.1	nd	0.04	0.06	nd	nd	0.07	0.11	nd	nd	nd	nd	0.49
	S	0.3	0.02	0.03	0.08	0.14	0.09	0.02	0.02	nd	0.05	0.1	0.07	0.04	nd	nd	nd	0.66
0.108-0.17	W	1.54	nd	nd	0.16	0.16	0.05	0.07	0.07	0.04	0.09	0.22	0.18	0.15	nd	nd	nd	1.19
	S	0.64	0.01	0.03	0.14	0.17	0.03	0.03	0.09	nd	0.06	0.15	0.06	0.03	0.01	0.01	0.01	0.84
0.17-0.26	W	6.57	0	0.05	0.12	0.25	0.03	1.29	1.4	1.62	1.83	1.16	1.25	1.37	0.06	0.21	0.25	10.88
	S	1.25	0.01	0.01	0.09	0.16	0.1	0.02	0.04	0.01	0.01	0.04	0.02	0.01	nd	0.01	nd	0.52
0.26-0.40	W	8.19	nd	0.1	0.28	0.99	0.33	5.08	6.75	5.79	5.51	4.17	4.35	5.39	0.24	0.03	1.96	40.97
	S	2.08	nd	nd	0.04	0.1	0.01	0.02	0.02	0.01	0.03	0.02	0.04	0.02	nd	nd	nd	0.3
0.40-0.65	W	8.77	0.09	0.1	0.27	1.1	0.16	4.64	5.03	5.29	5.06	3.79	4.2	4.86	0.23	1.67	2.14	38.65
	S	3.09	nd	nd	0.03	0.1	0.02	0.02	0.02	0.01	0.04	0.01	0.02	0.02	nd	nd	nd	0.31
0.65-1.0	W	7.59	0.1	0.08	0.28	1.33	0.31	4.65	4.68	5.04	4.54	3.36	3.59	4.27	0.2	1.1	1.77	35.3
	S	2.2	nd	nd	0.01	0.02	nd	0.02	0.01	nd	0.05	0.01	0.03	0.01	nd	nd	nd	0.14
1.0-1.6	W	5	0.1	0.05	0.13	0.86	0.06	3.23	4.24	3.31	3.48	1.88	2.13	2.79	0.09	0.34	0.78	23.47
	S	1.67	0.01	nd	nd	0.02	nd	0.02	0.01	0.01	0.03	0.05	0.06	0.01	nd	nd	0.01	0.24
1.6-2.5	W	2.66	nd	nd	0.04	0.09	0.22	0.26	0.26	0.26	0.3	0.21	0.28	0.32	nd	0.02	0.08	2.34
	S	1.5	0.02	0.03	0.15	0.16	0.04	0.03	0.04	nd	nd	0.1	0.07	0.13	nd	nd	nd	0.77
2.5-4.4	W	2.19	nd	nd	0.1	0.17	0.12	0.04	0.05	0.05	nd	0.05	0.04	nd	nd	nd	nd	0.63
	S	2.15	nd	0.01	0.02	0.04	0.01	0.02	0.01	nd	0.09	0.04	0.02	0.11	nd	nd	nd	0.37
4.4-6.8	W	1.55	nd	nd	0.14	0.21	0.09	0.02	0.04	nd	0.08	0.11	0.11	0.08	nd	nd	nd	0.88
	S	1.99	0.02	0.04	0.09	0.06	0.02	0.03	0.03	nd	0.07	0.07	0.05	0.17	nd	nd	nd	0.62
6.8-10.0	W	1.24	nd	nd	0.14	0.13	0.04	0.02	0.03	nd	0.08	0.08	0.12	0.05	nd	nd	nd	0.69
	S	1.94	nd	0.01	0.02	0.02	0.01	0.02	0.06	nd	0.04	0.07	0.07	0.06	nd	nd	nd	0.39
> 10.0	W	2.28	nd	nd	0.15	0.19	0.09	0.04	0.04	nd	0.04	0.1	0.11	0.08	nd	nd	nd	0.85
	S	2.9	0.01	0.02	0.05	0.06	0.01	0.02	0.05	nd	0.08	0.02	0.03	0.13	nd	nd	nd	0.48

nd – not detected

Table 1. The concentrations of 13 PM fractions and the fraction-related PAH in Zabrze in winter (W) and summer (S) 2007

The winter concentration of total PAH (ΣPAH) in Zabrze in 2007 were only a little higher than the average concentration of the sum of 9 PAH (Σ9PAH) observed by SSI in 2005 (Table 2, Fig. 3). The summer concentrations of ΣPAH in Zabrze in 2007 were lower than 7 ng m^{-3}. The average concentrations of BaP, in 2007 (about 10 ng m^{-3} in the whole measuring period, 1 ng m^{-3} and 20 ng m^{-3} in summer and winter) very well agree with the BaP concentrations measured by SSI in the recent years (Fig. 4).

The winter concentrations of ΣPAH bound onto particular PM fractions in Zabrze in 2007 were from 0.5 to almost 41.0 ng m^{-3}. ΣPAH bound onto $PM_{0.26-0.40}$, $PM_{0.40-0.65}$, $PM_{0.65-1.0}$ had the greatest concentrations (Table 1) because of the high concentrations of Fl, Py, BaA, Ch, BbF, BkF, BaP.

The summer concentrations of ΣPAH bound onto PM fractions in Zabrze in 2007 were from 0.15 to 0.85 ng m^{-3}. ΣPAH bound onto $PM_{0.03-0.06}$, $PM_{0.06-0.108}$, $PM_{0.108-0.17}$, $PM_{0.17-0.26}$, $PM_{1.6-2.5}$ and $PM_{4.4-6.8}$ had higher concentrations than ΣPAH bound to other fractions.

In winter, the PM concentrations and the concentrations of majority the PM fractions were two or three times higher than in summer. For $PM_{0.03-0.06}$ it was 4.5, for $PM_{0.17-0.26}$ more than 5 and for $PM_{0.26-0.40}$ almost 4 times (Table 1). The $PM_{2.5-10}$ concentrations in summer were close to or higher than the $PM_{2.5-10}$ winter concentrations.

The masses of $PM_{2.5}$ and PM_1 were 89 and 72% in winter and 68 and 51% in summer of the mass of PM_{10}, respectively (Fig. 5). The concentrations of ΣPAH behaved similarly. In winter, the concentrations of $PM_{2.5}$ and PM_1-bound ΣPAH were 99 and 82% of the PM_{10}-bound ΣPAH concentrations; in summer it was 77 and 69%, respectively. Despite the disparities between the concentrations of PM_1-, $PM_{1-2.5}$ and $PM_{2.5-10}$-bound ΣPAH in both seasons (Fig. 5), the mass contributions of ΣPAH to PM_1, $PM_{1-2.5}$, and $PM_{2.5-10}$ was 0.02–0.04% in summer and 0.38, 0.34 and 0.04% in winter, respectively. In each season, PM_1 and $PM_{1-2.5}$ had almost equal PAH contents. It means that in each season, PM_1 and $PM_{1-2.5}$ as well as PM_1- and $PM_{1-2.5}$-bound PAH came from the same sources.

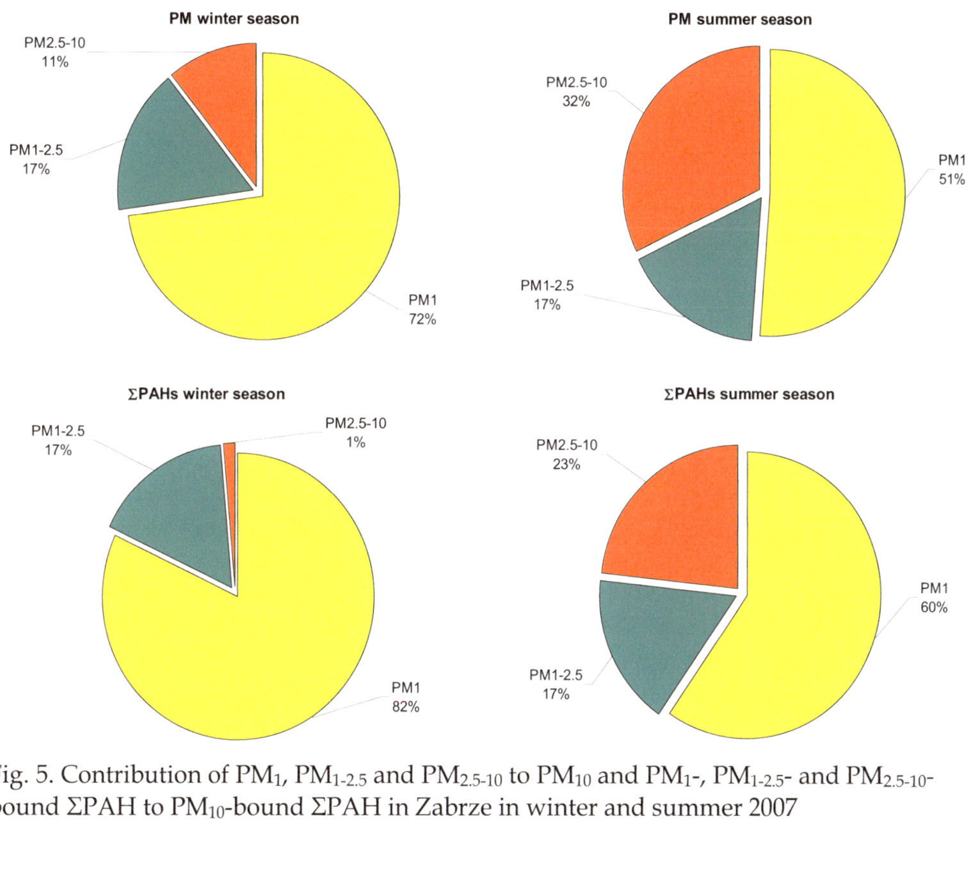

Fig. 5. Contribution of PM_1, $PM_{1-2.5}$ and $PM_{2.5-10}$ to PM_{10} and PM_1-, $PM_{1-2.5}$- and $PM_{2.5-10}$-bound ΣPAH to PM_{10}-bound ΣPAH in Zabrze in winter and summer 2007

The differences between the summer and the winter concentrations of the fraction-bound ΣPAH are greater than the differences between the concentrations of the PM fractions. The concentrations of ΣPAH bound onto $PM_{0.17-0.26}$, $PM_{0.26-0.40}$, $PM_{0.40-0.65}$, $PM_{0.65-1.0}$, $PM_{1.0-1.6}$ are greater from 21 to 251 times in winter than in summer. Noticeably, these fractions are usually formed by primary particles originating from combustion (Chow, 1995; Zhao et al., 2008; Wingfors et al., 2011).

For each of $PM_{0.06-0.108}$, $PM_{0.108-0.17}$, $PM_{0.17-0.26}$, $PM_{0.26-0.40}$, $PM_{0.40-0.65}$ and $PM_{1.6-2.5}$, among the fifteen PAH bound onto each of these fractions, Ph, one of the markers of emission from car engines (Harrison et al., 1996), had the highest summer concentrations. The summer concentrations of F, Py, An and Ch bound onto these fractions were also high. These PAH are also attributed to combustion of gasoline and oil in car engines (Chang et al., 2006; Miguel et al., 1998). In winter, Py, BaA, Ch, BbF and BkF had the greatest concentrations among the 15 PAH in each of these fractions. They belong to CPAH (Prahl & Carpenter, 1983), the nine so called combustion PAH (Fl, Py, BaA, BbF, BkF, BaP, BeP, IP and BghiP; Rogge et al., 1993a, 1993b; Kavouras et al., 1999; Bi et al., 2002; Manoli et al., 2002, 2004; Sienra et al., 2005). In winter, about 80% of the mass of these PAH was in PM_1. In summer, almost 100% of DBA and BghiP and about 80% of IP were in PM_1. Each of the remaining PAH was contained in particles greather than 1 μm at least in several dozen percent (Fig. 6).

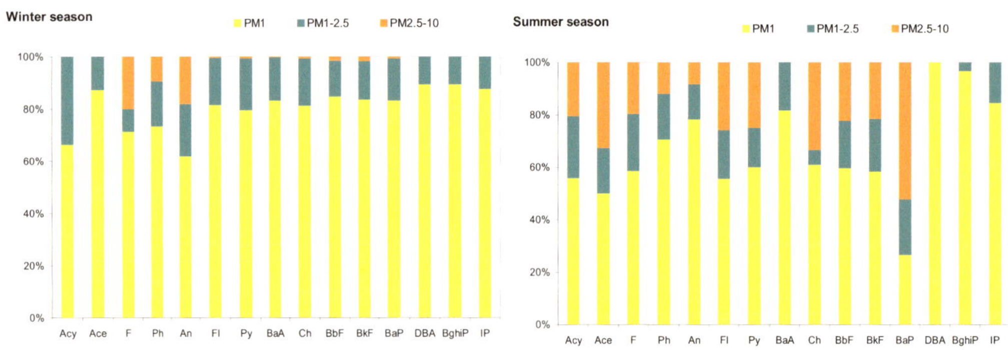

Fig. 6. Contribution of PM_1-, $PM_{1-2.5}$- and $PM_{2.5-10}$-bound PAH to PM_{10}-bound PAH in Zabrze in winter and summer 2007

In summer, the concentration of PM_1-bound BaP was a little more than 20% of the concentration of PM_{10}-bound BaP. The concentrations of BaP were greatest among the concentrations of $PM_{1-2.5}$- and $PM_{4.4-6.8}$-bound PAH.

The winter and the summer profiles of PM-bound PAH differ. In winter, five- and four-ring PAH were 87% of PM-bound ΣPAH. In summer, five- and four-ring PAH were only 58.5% of PM-bound ΣPAH. Three-ring PAH were 39.1% of PM-bound ΣPAH in summer, six-ring PAH were 6.6% and 2.4% of PM-bound ΣPAH in winter and summer, respectively (Table 1).

In winter, $PM_{0.26-0.40}$, $PM_{0.40-0.65}$, $PM_{0.65-1.0}$, $PM_{1.0-1.6}$ were richest in PAH (Fig. 7). $PM_{0.26-1.6}$ contained more than 88% of each: ΣPAH, six-, five- and four-ring PAH; for three-ring PAH it was over 67%. In summer, the PM fractions of PAH between the fractions was uniform except for six-ring PAH, whose contribution to $PM_{0.03-0.06}$ was over 73%.

The mass size distribution of PM is multimodal. Usually, PM is represented by three subdistributions (modes). They are called the nucleation, accumulation and coarse modes. The nucleation mode covers the mass distribution of the population of particles with diameters up to approximately 0.1 μm, the accumulation mode - the mass distribution of the particles with diameters in the interval 0.1-2 μm, and the coarse mode is for the particles with diameters greater than 2 μm (Willeke & Whitby 1975; Sverdrup & Whitby, 1977; Hinds, 1998). In practice, the particles in the nucleation mode weigh very little and the mass distribution density may have only two maxima, representing the accumulation and the coarse modes.

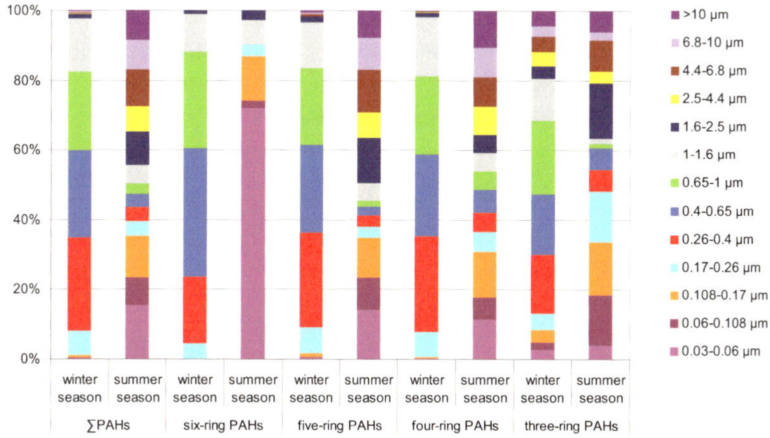

Fig. 7. Distributions of ΣPAH, six-ring PAH, five-ring PAH, four-ring-PAH and three-ring PAH among 13 PM fractions in Zabrze in winter and summer 2007

In Zabrze, in summer 2007, the PM mass distribution with respect to the particle aerodynamic diameter was bimodal. The probability density function had two maxima, one in the interval of particle diameters 0.4-0.65 μm (accumulation mode), and the second between 6.8 and 10 μm (coarse mode, Fig. 8). In winter, the mass distribution was unimodal, the density function had its only maximum between 0.26-0.4 μm.

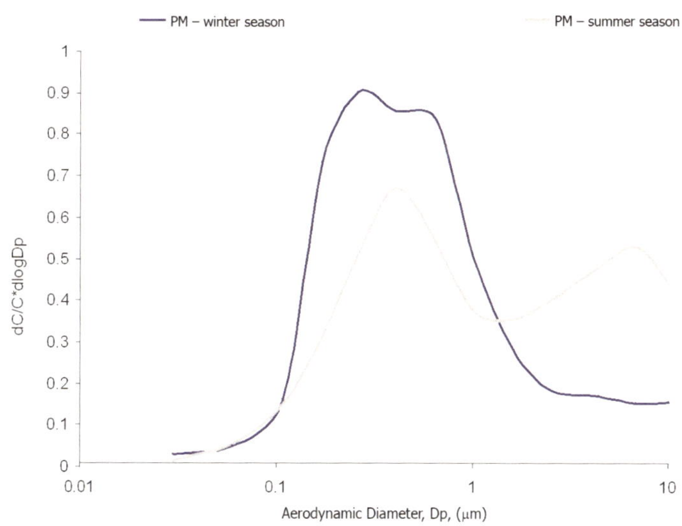

Fig. 8. Mass size distribution of PM relative to the aerodynamic diameter in Zabrze in winter and summer 2007

In winter, the PAH and ΣPAH mass distributions with respect to particle aerodynamic diameter of the particles they were adsorbed onto, except for the three-ring PAH, were bi- or trimodal, with one mode, like for PM, between 0.26 and 0.4 µm (Fig. 9 and 10). The second maximum occurred usually between 0.65 and 1.0 µm (except for BbF, BkF, BaP) and the third one between 4.4 and 6.8 µm or 6.8 and 10 µm (except for BaA, IP, DBA and BghiP). Three-ring PAH, in winter and in summer, were not detectable in some PM fractions (Table 1) because the lighter than Ph ambient species occur in the gaseous form (Guo et al., 2003; Fang et al., 2006; Akyüz & Çabuk, 2008). In summer, ΣPAH and, in general, four-, five-, and six-ring PAH (Fl, Py, Ch, BbF, BkF, BaP) had tri- or bimodal distributions. The distributions of IP and BghiP were bimodal, and of DBA—unimodal.

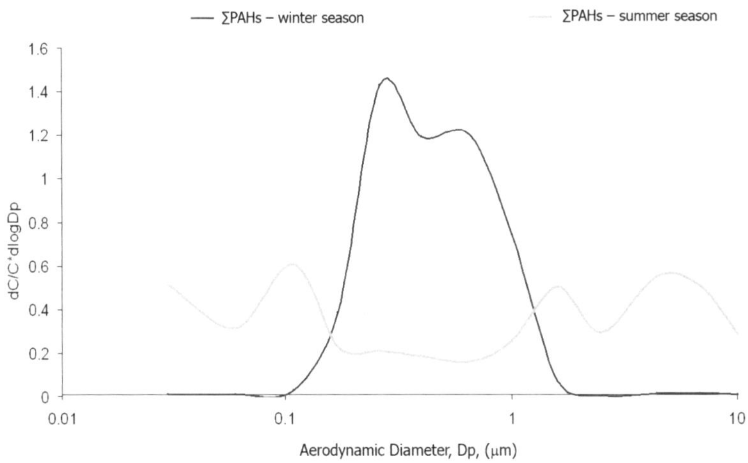

Fig. 9. Mass size distribution of PM-bound ∑PAH relative to the aerodynamic diameter of the particles PAH are adsorbed on in Zabrze in winter and summer 2007

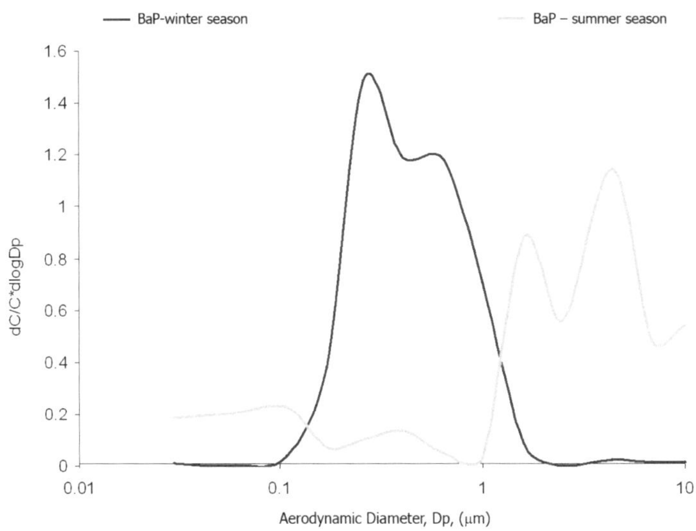

Fig. 10. Mass size distribution of PM-bound BaP relative to the aerodynamic diameter of the particles BaP is adsorbed on in Zabrze in winter and summer 2007

Like in winter, in summer too, majority of the probability functions of PAH and the function of ΣPAH had one maximum between 1.0 and 1.6 μm or 1.6 and 2.5 μm and the second between 4.4 and 6.8 μm or 6.8 and 10 μm (coarse particles).
The PAH diagnostic ratio (DR) is a proportion of ambient PAH concentrations. DR, in some way characterizes the origin of the involved PAH. In the present work, DR are used to determine the effect of the combustion sources in Zabrze on the concentrations of PAH bound to some PM fractions (PM_1, $PM_{2.5}$, PM_{10} and fractions where the majority of the probability functions of size distributions of PAH or ΣPAH concentrations assume their maximum). Some DR were taken from literature (Table 2).

Emission source	[BaA]/[BaP]	[Fl]/([Py]+[Fl])	[BaA]/([Ch]+[BaA])	[BbF]/[BkF]	[Ph]/([Ph]+[An])	*ΣCPAHs/ΣPAHs
Vehicular emissions		0.4 – 0.5 [2, 3, 4] 0.44a) [5]	0.2 – 0.35 [4]			0.41e) [5] 0.51f) [5]
Gasoline emissions	0.5 [1]	0.4 ± 0.08 [6] < 0.5 [7, 8]		1.1 – 1.5 [13]	0.50 [15]	
Diesel emissions	1.0 [1]	> 0.5 [7, 8] 0.6 – 0.7 [5]			0.65 [15]	
Used motor oil		0.36 ± 0.08 [6]				
Crude oil			0.16 ± 0.12 [6]		> 0.70 [6]	
Combustion (stationary sources)						0.78 ± 0.16g) [2]
Coal combustion		> 0.5 [3, 4] 0.57 [10]	0.46 [10] 0.50b) [11] 0.17 – 0.36 [12]	3.5 – 3.9 [14]	0.76 [15]	
Wood combustion	1.0 [1]	> 0.5 [3, 4] 0.51 [9,10]	0.40 ± 0.09 [6]	0.8 – 1.1 [14]		
Natural gas combustion		0.49 [9,10]				
Other sources				2.5 – 2.9c) [14]	> 0.7d) [16]	

a) Vehicular emissions - converter equipped automobiles; b) Coal combustion for domestic heating; c) Smelters; d) Associated with lubricant oil and fossil fuels; e) Non-catalyst automobiles; f) Catalyst-equipped automobiles; g) General dominance of combustion sources
[1]: (Li & Kamens, 1993); [2]: (Kavouras et al., 1999); [3]: (Zencak et al., 2007); [4]: (Yunker et al., 2002); [5]: (Rogge et al., 1993a, 1993b); [6]: (Sicre et al., 1987); [7]: (Ravindra et al., 2006); [8]: (Ravindra et al. 2008); [9]: (Galarneau, 2008); [10]: (Tang et al., 2005); [11]: (Gschwend & Hites, 1981); [12]: (Dickhut et al., 2000); [13]: (Masclet et al., 1987); [14]: (Khalili et al., 1995); [15]: (Alves et al., 2001)
*ΣCPAHs/ΣPAHs=([Fl]+[Py]+[BaA]+[BbF]+[BkF]+[BaP]+[BeP]+[IP]+[BghiP])/([Acy]+[Ace]+[F]+[Ph]+[An]+[Fl]+[Py]+[BaA]+[Ch]+[BbF]+[BkF]+[BaP]+[DBA]+[BghiP]+[IP])

Table 2. PAH diagnostic ratios (DR) from literature

The effect of stationary combustion sources on the ambient PAH concentrations is characterized by the proportion ΣCPAH/ΣPAH of the PM-bound ΣCPAH and ΣPAH concentrations (Rogge et al., 1993a, 1993b; Kavouras et al., 1999; Manoli et al., 2004; Sienra et al., 2005). In Zabrze, ΣCPAH/ΣPAH indicate that in winter $PM_{0.26-0.4}$-, $PM_{0.65-1.0}$- and $PM_{1.6-2.5}$-bound PAH came from stationary combustion sources (Table 2 and 3). In winter, these fractions contained half of the mass of the PM-bound PAH and, not surprisingly, the winter PM_1-, $PM_{2.5}$- and PM_{10}-bound PAH should in general come from stationary combustion.

Probably, $PM_{6.8\text{-}10}$-bound PAH came in part from vehicular sources (ΣCPAH/ΣPAH=0.56). In summer, PM_1- and $PM_{2.5}$-bound PAH might also have come from vehicular sources (ΣCPAH/ΣPAH was lower than in winter) but the summer concentrations of $PM_{1.6\text{-}2.5}$-, $PM_{6.8\text{-}10}$- and PM_{10}-bound PAH were affected by stationary combustion sources as well. However, in summer, the degradation processes (oxidation, photochemical reactions) and in winter the variability of the parameters of combustion processes and the variety of used fuels limit the reliability of DR making the source apportionment uncertain (Kavouras et al., 1999; Hong et al., 2007). Nevertheless, other DR confirm these findings. [Ph]/([Ph]+[An]) indicates PAH from hard coal combustion and, in winter, from unidentified sources and crude oil (traffic). [BaA]/([BaA]+[Ch]) indicates the effect of crude oil combustion on the PAH concentrations also in summer (0.10, 0.11 and 0.08 for PM_1, $PM_{2.5}$ and PM_{10}, respectively). For $PM_{0.26\text{-}0.4}$-bound PAH, [BaA]/([BaA]+[Ch]) is 0.27 indicating car exhaust (Tables 2 and 3). The winter [BaA]/([BaA]+[Ch]) for $PM_{1.6\text{-}2.5}$-bound PAH is 0.46, for $PM_{1.6\text{-}2.5}$-bound PAH—0.53, they are higher than in summer and prove that hard coal combustion was the main source of ambient PAH in Zabrze.

The effect of vehicular sources and wood combustion on the ambient PAH concentrations is reflected by [BbF]/[BkF]. In summer [BbF]/[BkF] were slightly higher than in winter for majority of the PM fractions (Table 3). [BbF]/[BkF] allows to apportion PM_1-, $PM_{2.5}$- and PM_{10}-bound PAH to traffic in summer. In winter, except for $PM_{6.8\text{-}10}$, [BbF]/[BkF] were between 0.8 and 1.1. It indicates the contribution of wood combustion to the winter PAH concentrations. The values of [BaA]/[BaP] also indicate the contribution of wood combustion to the winter concentration of PAH in Zabrze (Tables 2 and 3).

Fraction, μm	[BaA]/[BaP]		[Fl]/([Py]+[Fl])		[BaA]/([Ch]+[BaA])		[BbF]/[BkF]		[Ph]/([Ph]+[An])		ΣCPAHs/ΣPAHs	
	S	W	S	W	S	W	S	W	S	W	S	W
$PM_{0.26\text{-}0.4}$	0.75	1.07	0.51	0.43	0.27	0.51	0.59	0.96	0.91	0.75	0.46	0.83
$PM_{0.65\text{-}1.0}$	-	1.18	0.58	0.50	-	0.53	0.34	0.94	1.00	0.81	0.50	0.82
$PM_{1.6\text{-}2.5}$	0.02	0.81	0.40	0.50	1.00	0.46	1.44	0.77	0.79	0.29	0.65	0.73
$PM_{6.8\text{-}10}$	-	-	0.31	0.43	-	-	1.03	0.67	0.82	0.77	0.81	0.56
$PM_{1.0}$	0.23	1.11	0.42	0.47	0.10	0.51	1.29	0.93	0.75	0.82	0.54	0.82
$PM_{2.5}$	0.16	1.11	0.43	0.46	0.11	0.51	1.25	0.92	0.76	0.81	0.57	0.82
PM_{10}	0.08	1.11	0.44	0.46	0.08	0.50	1.26	0.92	0.77	0.80	0.60	0.81

Table 3. PAH diagnostic ratios (DR) in summer (S) and winter (W) 2007 in Zabrze

In winter, [Fl]/([Fl]+[Py]) were close to the values characteristic of PAH coming from wood or natural gas combustion (Galarneau, 2008). In general, [Fl]/([Fl]+[Py]) were lower in summer than in winter. [Fl]/([Fl]+[Py]), oscillating about 0.4, suggest the effect of car exhaust. Relatively high summer [Fl]/([Fl]+[Py]) for $PM_{0.26\text{-}0.4}$ and $PM_{0.65\text{-}1.0}$ suggest diesel car exhaust as the source of PAH in these fractions (Rogge et al., 1993a, 1993b).

3.2 Comparison of the concentrations of PM- and PM-bound PAH in Zabrze and selected sites in the world

The ambient concentrations of PAH bound to various PM fractions, especially to $PM_{2.5}$ and PM_{10}, are investigated at many sites in the world (Kavouras et al., 1999; Odabasi et al., 1999; Panther et al., 1999; Takeshi & Takashi, 2004; Mantis et al., 2005). Most often, the investigations cover short periods, results of long-term experiments are not abundant. The direct comparison of the concentrations of PM-bound PAH from various urban areas should be done cautiously. The data may be affected by the method of sampling and the technique of determination of PAH in the dust samples (Takeshi & Takashi, 2004). The local conditions, meteorological (air temperature and relative air humidity, direction and velocity of wind, precipitation) and other, neighborhood of the sampling point, a season of a year are also important (Wang X.H. et al., 2007; Evagelopoulos et.al., 2010).

Table 4 is a collation of the concentrations of the sum of 15 PAH (ΣPAH) and PM_1-, $PM_{2.5}$- and PM_{10}-bound BaP in Zabrze in 2007 and in various places in the world, mainly in Europe. BaP was selected because it is used as an air pollution indicator. The data cover the period from 2005 to 2009. In the last column of Table 4, in parentheses, the number of analyzed compounds is given. The concentration of ΣPAH was computed by summing up the concentrations of these PAH that were investigated in Zabrze. The table allows for rough evaluation of differences in the concentrations of BaP and ΣPAH bound to several PM fractions in various countries.

The ranges of the concentrations of BaP and ΣPAH are wide because of the causes mentioned above and different numbers of PAH in ΣPAH. In general, the lowest concentrations (lower than 0.1 ng m^{-3} for BaP and than 2 ng m^{-3} for ΣPAH) occurred in clean areas (not affected by vehicular or industrial sources), such as Virolahti (regional background station) in Finland (Makkonen et al., 2010) and Chréa National Park (great forest area) in Algeria (Ladji et al., 2009). In most places the BaP concentration did not exceed 1 ng m^{-3}, the limit established by the European Commission. The exceptions were the urban-traffic site in Oporto (Portugal; Slezakova et al., 2011), road tunnel in Marseille (France; El Haddad et al., 2009), urban background station in Flanders (Belgium; Vercauteren et al., 2011), measuring point in Zonguldak (Turkey; Akyüz & Çabuk, 2008) and urban background station and crossroads in Zabrze (Ćwiklak et al., 2009).

Majority of ambient PAH, especially in urbanized areas, are anthropogenic (Kulkarni & Venkataraman, 2000; Hien et al., 2007; Wang X.H. et al., 2007), and come mainly from combustion of fossil fuels, wastes or biomass and also from industry and road traffic. Most of the greatest values of BaP concentrations shown in Table 4, reaching 50 ng m^{-3}, come from sites located in industrialized and densely populated areas in Asiatic countries, such as Fushun (residential-commercial site, Kong et al., 2010), Beijing (campus site, Wang H. et al., 2009), Guiyu (electronic waste recycling site, Deng et al., 2006) in the People's Republic of China or Chennai in the Republic of India (Mohanraj et al., 2011). The concentrations of ΣPAH in these regions were also high, some times, like in Funshun, China, where the ΣPAH concentration reached 1.9 μg m^{-3}, many times higher than elsewhere in the world. In Europe, the highest BaP and ΣPAH concentrations were at traffic station in Sweden and at traffic and urban sites in Zabrze (Table 4).

BaP and ΣPAH tend to accumulate in the finest particles of PM (Table 4). Like in Zabrze, this tendency may be observed in the road tunnel in Lisbon, Portugal (Oliveira et al., 2011) and at the urban background and traffic stations in Los Angeles, USA (Phuleria et al., 2007).

Location		Sampling period	Fraction	Concentration ($ng\ m^{-3}$)	
				BaP [a)]	∑PAH [b)]
Zabrze (Poland) [1)]		summer 2007	PM_1	0.17	3.48 (15)
			$PM_{2.5}$	0.31	4.49 (15)
			PM_{10}	0.65	5.86 (15)
		winter 2007	PM_1	16.08	128.10 (15)
			$PM_{2.5}$	19.19	153.10 (15)
			PM_{10}	19.32	156.11 (15)
Zabrze (Poland) [2)]	urban background	summer 2005	$PM_{2.5}$	*0.90*	12.80 (15)
			PM_{10}	*1.20*	21.60 (15)
	crossroads		$PM_{2.5}$	*2.10*	43.40 (15)
			PM_{10}	*4.00*	63.80 (15)
Flanders (Belgium) [3)]	urban background	Oct 2006 - Mar 2007	PM_{10}	1.18	82.24 (15)
		Apr - Sep 2007	PM_{10}	0.51	74.68 (15)
	rural background	Oct 2006 - Mar 2007	PM_{10}	0.81	45.90 (15)
		Apr - Sep 2007	PM_{10}	0.41	32.23 (15)
Virolahti (Finland), regional background [4)]		summer 2006	PM_1	< 0.08	1.77 (13)
			$PM_{2.5}$	< 0.08	1.77 (13)
			PM_{10}	< 0.08	1.78 (13)
		winter 2006	PM_1	0.52	7.54 (13)
			$PM_{2.5}$	0.69	14.53 (13)
			PM_{10}	0.73	13.9 (13)
Marseille (France), tunnel [5)]		winter 2008	$PM_{2.5}$	2.42	74.67 (8)
			PM_{10}	6.73	108.39 (8)
Toulouse (France) [6)]	traffic	Apr 2006	PM_{10}	*0.10*	18.20 (15)
	urban		PM_{10}	*0.15*	11.45 (15)
	industrial		PM_{10}	*0.80*	20.40 (15)
Thessaloniki (Greece), kerbside [7)]	street level	Jan/Feb 2006	$PM_{<0.95}$	0.47	11.94 (13)
		Jan/Feb 2006	$PM_{0.95-1.5}$	0.10	2.36 (13)
		Jan/Feb 2006	$PM_{1.5-3}$	0.04	1.41 (13)
		Jan/Feb 2006	$PM_{3-7.5}$	0.02	0.98 (13)
		Jan/Feb 2006	$PM_{>7.5}$	0.02	0.80 (13)
	rooftop level	Jan/Feb 2006	$PM_{<0.95}$	0.45	11.54 (13)
		Jan/Feb 2006	$PM_{0.95-1.5}$	0.10	2.45 (13)
		Jan/Feb 2006	$PM_{1.5-3}$	0.03	1.06 (13)
		Jan/Feb 2006	$PM_{3-7.5}$	0.02	0.79 (13)
		Jan/Feb 2006	$PM_{>7.5}$	0.02	0.63 (13)
Athens (Greece), indoor samples [8)]	smoking area	16-27 Jul 2007	PM_1	0.10	0.80 (15)
			$PM_{2.5}$	0.14	1.11 (15)
			PM_4	1.73	21.33 (15)
	no-smoking area		PM_1	0.03	0.44 (15)
			$PM_{2.5}$	0.05	0.65 (15)
			PM_4	0.00	5.13 (15)
	passive smokers\` area		PM_4	3.58	29.04 (15)
Kozani (Greece), urban area surrounded by opencast coal mining [9)]		Dec 2005 – Oct 2006	$PM_{2.5}$	0.38	4.77 (15)
			PM_{10}	0.11	1.46 (15)
Rome (Italy), downtown [10)]		Apr - Jul 2007	PM_1	-	2.10 (14)
			$PM_{2.5}$	-	2.29 (14)
			PM_{10}	-	2.37 (14)

		Oct 2007 - Feb 2008	PM_1	-	6.70 (14)
			$PM_{2.5}$	-	7.77 (14)
			PM_{10}	-	7.98 (14)
Lisbon (Portugal), roadway tunnel [11]		Oct 2008	$PM_{0.49}$	6.98	113.80 (15)
			$PM_{0.49-0.95}$	0.35	14.70 (15)
			$PM_{0.95-2.5}$	0.17	7.33 (15)
			$PM_{2.5-10}$	0.19	2.66 (15)
Oporto Metropolitan Area (Portugal), urban-traffic [12]		Dec 2008	PM_{10}	2.02	20.67 (15)
			$PM_{2.5}$	1.88	18.96 (15)
Umeä (Sweden), traffic [13]		autumn 2009	$PM_{2.5}$	25.00	196.45 (12)
Zonguldak (Turkey), industrial city [14]		summer 2007	$PM_{2.5}$	0.40	5.70 (14)
		winter 2007	$PM_{2.5}$	15.70	152.70 (14)
		summer 2007	$PM_{2.5-10}$	0.20	1.60 (14)
		winter 2007	$PM_{2.5-10}$	0.70	10.50 (14)
Kabul (Afganistan), urban [13]		autumn 2009	$PM_{2.5}$	6.70	55.97 (12)
Mazar-eSharif (Afganistan), urban [13]			$PM_{2.5}$	0.09	2.19 (12)
Beijing (China); campus site [15]		summer 2005-2007	$PM_{2.5}$	1.19	39.28 (15)
			$PM_{2.5-10}$	1.14	24.87 (15)
		winter 2005-2007	$PM_{2.5}$	19.82	360.71 (15)
			$PM_{2.5-10}$	5.09	102.00 (15)
Guiyu (China), electronic waste recycling site [16]		AugI - Sep 2004	TSP	15.40	144.85 (15)
			$PM_{2.5}$	8.85	99.26 (15)
Xiamen (China) [17]	industrial-traffic intersection site	summer 2005	PM_{10}	0.10	5.24 (15)
		autumn 2005	PM_{10}	1.40	37.10 (15)
	residential site	summer 2005	PM_{10}	0.00	1.69 (15)
		autumn 2005	PM_{10}	1.50	26.60 (15)
Fushun (China) [18]	urban background	2004-2005	$PM_{2.5}$	10.71	261.82 (13)
			$PM_{2.5-10}$	1.98	72.44 (13)
	residential-commercial		$PM_{2.5}$	48.44	1899.36 (13)
			$PM_{2.5-10}$	3.44	166.92 (13)
Jinzhou (China) [18]	urban background		$PM_{2.5}$	13.61	190.86 (13)
			$PM_{2.5-10}$	1.13	20.78 (13)
	residential		$PM_{2.5}$	6.61	106.94 (13)
			$PM_{2.5-10}$	0.77	16.40 (13)
Chennai City (India) [19]	urban-residential	Dec 2009 – Feb 2010	$PM_{2.5}$	6.50	365.30 (11)
		Apr - Aug 2009	$PM_{2.5}$	7.40	326.90 (11)
	industrial/traffic site	Dec 2009 – Feb 2010	$PM_{2.5}$	8.10	681.80 (11)
		Apr - Aug 2009	$PM_{2.5}$	10.60	456.60 (11)
	urban-commercial	Dec 2009 – Feb 2010	$PM_{2.5}$	16.20	448.00 (11)
		Apr - Aug 2009	$PM_{2.5}$	2.50	313.90 (11)
Ho Chi Minh (Vietnam), roadside [20]		dry season, Jan-Feb 2005	TSP	*2.00*	39.80 (9)
		rainy season, Jul 2005	TSP	*5.70*	58.70 (9)

Boumerdes (Algeria), urban traffic congestion [21]		PM_{1}	0.11	1.53 (14)
		PM_{1-10}	0.02	0.33 (14)
Rouiba-Réghaia (Algeria), industrial zone [21]	Oct 2006	PM_{1}	0.30	2.70 (14)
		PM_{1-10}	0.04	0.59 (14)
Chréa National Park (Algeria), forest ecosystem [21]		PM_{1}	0.02	0.28 (14)
		PM_{1-10}	0.002	0.10 (14)
Golden (British Columbia, Canada), residential [22]	spring 2006	$PM_{2.5}$	0.14	1.76 (15)
	winter 2007	$PM_{2.5}$	2.67	31.39 (15)
Long Beach, California (USA), coastal city [23]	Jan - Mar 2005	$PM_{0.25}$	0.17	0.94 (10)
		$PM_{0.25-2.5}$	0.03	0.43 (10)
		$PM_{2.5-10}$	0.00	0.01 (10)
Los Angeles (USA) [24], background	Jan 2005	$PM_{0.108-2.5}$	0.03	0.31 (9)
		$PM_{>2.5}$	0.16	1.45 (9)
Los Angeles (USA) [24], freeway		$PM_{0.108-2.5}$	0.04	0.38 (9)
		$PM_{>2.5}$	0.17	1.82 (9)

[a] Underlined italics mark the values read from a chart; [b] The number of PAH taken to compute ΣPAH concentration is in parentheses

[1]: (this study); [2]: (Ćwiklak et al., 2009); [3]: (Vercauteren et al., 2011); [4]: (Makkonen et al., 2010); [5]: (El Haddad et al., 2009); [6]: (Dejean et al., 2009); [7]: (Chrysikou et al., 2009); [8]: (Saraga et al., 2010); [9]: (Evagelopoulos et al., 2010); [10]: (Di Filippo et al., 2010); [11]: (Oliveira et al., 2011); [12]: (Slezakova et al., 2011); [13]: (Wingfors et al., 2011); [14]: (Akyuz & Cabuk, 2008); [15]: (Wang et al., 2009); [16]: (Deng et al., 2006); [17]: (Hong et al., 2007); [18]: (Kong et al., 2010); [19]: (Mohanraj et al., 2011); [20]: (Hien et al., 2007); [21]: (Ladji et al., 2009); [22]: (Ding et al., 2009); [23]: (Krudysz et al., 2009); [24]: (Phuleria et al., 2007)

Table 4. Comparison of the PM and PM-bound PAH concentrations at various sites in the world

Like in Zabrze in 2007, at some other sites in the world the BaP and ΣPAH concentrations depend on a season of a year and are higher in cold seasons. One of the most obvious causes of the seasonal variability of the PAH concentrations is home heating, central or individual, that is important, if not the most important, source of air pollutants in winter. Moreover, the meteorological conditions in winter (shallow mixing layer) are favorable for local occurrences of high air pollution, like in Zabrze, Zonguldak (Turkey) and in Golden (Canada). Instead, in summer, higher air temperatures and solar radiation intensify desorption of PAH from PM particles and their photochemical decomposition (Odabasi et al., 1999; Hong et al., 2007). Therefore, lower in summer than in winter concentrations of PM-bound PAH may also be due to releasing of PAH from PM particles. To prove it in Zabrze, the gas phase of ambient PAH would have to be investigated. In summer, more favorable conditions for dispersion and dilution of air pollutants (Mantis et al., 2005) and washing out of particles (with adsorbed PAH) by precipitation occur.

The concentrations of BaP and ΣPAH in Zabrze in summer 2007 were lower than in 2005 (Table 4). It may be due to the differences in the sampling periods (long sampling periods in 2007, 24-hour sampling in 2005), in the method of sampling (cascade impactor in 2007, manual sampler with a separating head in 2005) and in the method of PAH determination (combining of extracts for a season and GC-FID in 2007, averaging of diurnal concentrations in a season and GC-MS in 2005) (Ćwiklak et al., 2009).

Nevertheless, the concentrations at two measuring points in Zabrze in summer 2005, several times lower than in winter 2007, suggest that the municipal emission in Silesian Agglomeration may be much greater problem than the traffic emission. Instead, in other European cities, the greatest problem is the traffic emission. The ΣPAH concentrations in Zabrze in 2007 were higher than in other European cities except for the traffic sites (road tunnels) in Marseille (France) and Lisbon (Portugal).

3.3 Health hazard from PAH in Zabrze

Ambient PAH endanger human health by their mutageneity and carcinogenicity. Their strong adverse biological effect is documented by numerous works (Grimmer et al., 1986; White, 2002; Yan et al., 2004).

The risk from the exposure to particular PAH is expressed in terms of the most cancerogenous PAH, BaP, as the toxicity equivalence factor (TEF). The carcinogenicity of a combination of PAH, the BaP equivalence (BEQ), is computed as the linear combination of the concentrations of PAH entering the PAH combination and their TEF (Nisbet & LaGoy, 1992). BEQ for the 15 PAH discussed in this paper is:

$$\begin{aligned} BEQ = &[Acy] \times 0.001 + [Ace] \times 0.001 + [F] \times 0.001 + [Ph] \times 0.001 + [An] \times 0.01 + \\ &+[Fl] \times 0.001 + [Py] \times 0.001 + [BaA] \times 0.1 + [Ch] \times 0.01 + [BbF] \times 0.1 + [BkF] \times 0.1 + \\ &+[BaP] \times 1 + [DBA] \times 5 + [BghiP] \times 0.1 + [IP] \times 0.1 \end{aligned} \quad (1)$$

In winter, BEQ for $PM_{0.17\text{-}0.26}$-, $PM_{0.26\text{-}0.40}$-, $PM_{0.40\text{-}0.65}$-, $PM_{0.65\text{-}1.0}$-, and $PM_{1.0\text{-}1.6}$-bound PAH were 2.85 ng m^{-3}, 7.09 ng m^{-3}, 14.65 ng m^{-3}, 11.07 ng m^{-3} and 5.28 ng m^{-3}, respectively (Table 5). The five fractions contain 78% of the mass of PM_{10} and 96% of PM_{10}-bound ΣPAH (Table 1). The summer BEQ of PAH in these fractions do not exceed 1 ng m^{-3}. BEQ for the remaining eight fractions do not exceed 1 ng m^{-3} in both seasons. PAH in the seven fractions contained in $PM_{0.108\text{-}2.5}$ have BEQ higher, often several times, in winter than in summer.

BEQ for PAH in $PM_{2.5}$ and PM_{10} in summer 2007 in Zabrze were 0.79 ng m^{-3} and 1.16 ng m^{-3}, respectively. They were two or almost three times lower than Ćwiklak et al. (2009) determined in summer 2005 at the urban background site in Zabrze. They are comparable with other, foreign, values (Xiamen: 0.85 ng m^{-3} and 0.92 ng m^{-3} for PM10 and PM2.5 bound PAH; Hong et al., 2007). Instead, the winter BEQ for PAH in PM2.5 and PM10 are very high, 41.72 ng m^{-3} and 41.91 ng m^{-3}, and they are higher than BEQ computed for PAH in these fractions in areas with very high PAH concentrations, such as in Shanghai (BEQ equal to 15.77 ng m^{-3}; Guo et al., 2004) or in some Japanese cities (BEQ about 2 ng m^{-3}; Takeshi & Takashi, 2004).

The closer $\Sigma PAH_{carc}/\Sigma PAH$ to 1 are the more hazardous to humans ambient PAH are (Hong et al., 2007). In Zabrze, in summer, the values of $\Sigma PAH_{carc}/\Sigma PAH$ were between 0.17 and 0.74 (Table 5). They are dispersed owing to differentiation of the PAH profiles (Table 1). The winter $\Sigma PAH_{carc}/\Sigma PAH$ are close to or higher than 0.5. In general, in Zabrze, the values of $\Sigma PAH_{carc}/\Sigma PAH$ were high, much higher than ones determined for other urban areas (Bourotte et al., 2005; Sienra et al., 2005). They are comparable with the values received by Chen et al. (2004) for dust emitted from coal combustion.

Fraction		BEQ, ng m^{-3}	*$\sum PAH_{carc}/\sum PAH$
$PM_{0.03\text{-}0.06}$	W	0.07	0.44
	S	0.32	0.70
$PM_{0.06\text{-}0.108}$	W	0.02	0.37
	S	0.06	0.39
$PM_{0.108\text{-}0.17}$	W	0.20	0.57
	S	0.10	0.39
$PM_{0.17\text{-}0.26}$	W	2.85	0.69
	S	0.07	0.17
$PM_{0.26\text{-}0.40}$	W	7.09	0.67
	S	0.03	0.39
$PM_{0.40\text{-}0.65}$	W	14.65	0.66
	S	0.02	0.34
$PM_{0.65\text{-}1.0}$	W	11.07	0.65
	S	0.01	0.63
$PM_{1.0\text{-}1.6}$	W	5.28	0.62
	S	0.02	0.74
$PM_{1.6\text{-}2.5}$	W	0.50	0.62
	S	0.15	0.39
$PM_{2.5\text{-}4.4}$	W	0.02	0.23
	S	0.12	0.70
$PM_{4.4\text{-}6.8}$	W	0.10	0.43
	S	0.18	0.55
$PM_{6.8\text{-}10.0}$	W	0.07	0.48
	S	0.07	0.63
$PM_{>10.0}$	W	0.10	0.39
	S	0.14	0.54
PM_1	W	35.93	0.66
	S	0.62	0.43
$PM_{1\text{-}2.5}$	W	5.78	0.62
	S	0.17	0.47
$PM_{2.5\text{-}10}$	W	0.19	0.39
	S	0.37	0.61
$PM_{2.5}$	W	41.72	0.65
	S	0.79	0.44
PM_{10}	W	41.91	0.65
	S	1.16	0.48

*$PAH_{carc}/\Sigma PAH$=([BaA]+[BaP]+[BbF]+[BkF]+[Ch]+[DBA]+[IP])/([Acy]+[Ace]+[F]+[Ph]+[An]+[Fl]+[Py]+[BaA]+[Ch]+[BbF]+[BkF]+[BaP]+[DBA]+[BghiP]+[IP])

Table 5. BEQ and $\sum PAH_{carc}/\sum PAH$ for PM-bound PAH In the PM fractions in Zabrze in winter (W) and summer (S) 2007

4. Conclusion

In Zabrze, the winter PM-bound BaP concentrations are 19 times greater than the limit for the yearly average BaP concentrations (1 ng m^{-3}). Both PM-bound BaP and Σ15PAH concentrations are much greater than the concentrations in other European cities. Despite the general improvement of the air quality in the Silesian Agglomeration during the last thirty years (decrease of the concentrations of ambient coarse particles and PAH related with this fraction), the concentrations of PM-bound BaP and Σ15PAH are high. On the local scale, in winter, the most important sources of fine particles and particle-bound PAH are municipal sources (hard coal, wood and garbage combustion), and/or electric power and heat production from coal; in summer it is vehicular emission. At the urban background measuring point in Zabrze, the vast differences in the seasonal ambient concentrations of PAH and in PAH profiles refute the supposition on industry affecting mostly the air quality in the Silesian Agglomeration. However, DR applied in determination of probable PAH sources are not reliable and their application may give contradictive results (Simcik et al., 1999; Sienra et al., 2005; Evangelopoulos et al., 2010; Dvorská et al., 2011). The exact apportionment of PAH to sources needs measuring of the diurnal PM-bound PAH concentrations and some statistical reasoning must be done (e.g. multivariate factor analysis to apportion combinations of PAH to sources). The data base containing the intervals of DR determined for the specific conditions in the Silesian Agglomeration in vicinities of real PAH sources would appear very helpful. Also, the measurements of ambient gaseous PAH concentrations are necessary.

The high ambient PAH concentrations and the high five- and six-ring PAH content of total PAH are hazardous to the Zabrze population, especially in winter. The concentrations of the carcinogenic PAH was never lower than 50% of the ΣPAH concentration in winter, and BEQ for PM_1, $PM_{2.5}$ and PM_{10} were 35.93 ng m^{-3}, 41.72 ng m^{-3} and 41.91 ng m^{-3}, respectively.

In winter, all the four-, five- and six-ring PAH and ΣPAH had two- or trimodal distributions with one maximum between 0.26 and 0.4 µm, the second usually between 0.65 and 1.0 µm and the third between 4.4 and 10 µm. The greatest BEQ and $\Sigma PAH_{carc}/\Sigma PAH$ found for $PM_{0.17-1.6}$ suggest elevated toxicity of very fine particles, which are the core mass of PM, in Zabrze in winter. In summer, the distribution of ΣPAH and particular PAH with respect to the aerodynamic diameter of particles they are bound to, the values of BEQ and $\Sigma PAH_{carc}/\Sigma PAH$ were similar to those from other sites in the world (Chen et. al., 2004; Akyüz & Çabuk, 2008). The contribution of PM_1- and $PM_{2.5}$-bound ΣPAH to ΣPAH was in Zabrze, like elsewhere (Chrysikou et al., 2009; Kong et al., 2010; Makkonen et al., 2010; Oliveira et al., 2011), very high (99 and 82% in winter and 77 in 69% summer, respectively).

5. Acknowledgment

The work was partially supported by grant No. N N523 421037 from the Polish Ministry of Science and Higher Education.

6. References

Air protection program for areas of Silesia, which were found oversize levels of substances in the air. Appendix to Resolution No. III/52/15/2010 Silesian Provincial Assembly of 16 June 2010, Katowice, 2010 (in polish)

Akyüz, M., Çabuk, H. (2008): Particle-associated polycyclic aromatic hydrocarbons in the atmospheric environment of Zonguldak, Turkey, *Science of the Total Environment* 405 (1-3), pp. 62-70

Alves, C., Pio, C., Duarte, A. (2001). Composition of extractable organic matter of air particles from rural and urban Portuguese areas, *Atmospheric Environment*, 35 (32), pp. 5485-5496

Bi, X., Sheng, G., Peng, P.A., Zhang, Z., Fu, J. (2002). Extractable organic matter in PM10 from LiWan district of Guangzhou City, PR China, *Science of the Total Environment,* 300 (1-3), pp. 213-228

Bourotte, C., Forti, M.-C., Taniguchi, S., Bícego, M.C., Lotufo, P.A. (2005). A wintertime study of PAHs in fine and coarse aerosols in São Paulo city, Brazil, *Atmospheric Environment* 39 (21), pp. 3799-3811

Central Statistical Office of Poland. Environment 1975- 2000. Information and statistical studies, Warsaw 1976-2001

Central Statistical Office of Poland. Environment 2000- 2010. Information and statistical studies, Warsaw 2001-2011

Chang, K.-F., Fang, G.-C., Chen, J.-C., Wu, Y.-S. (2006). Atmospheric polycyclic aromatic hydrocarbons (PAHs) in Asia: A review from 1999 to 2004, *Environmental Pollution,* 142 (3), pp. 388-396

Chen, Y., Bi, X., Mai, B., Sheng, G., Fu, J. (2004). Emission characterization of particulate/gaseous phases and size association for polycyclic aromatic hydrocarbons from residential coal combustion, *Fuel,* 83 (7-8), pp. 781-79

Chow, J.C. (1995): Measurement methods to determine compliance with ambient air quality standards for suspended particles, *Journal of the Air and Waste Management Association,* 45 (5), pp. 320-382

Chrysikou, L.P., Gemenetzis, P.G., Samara, C.A. (2009). Wintertime size distribution of polycyclic aromatic hydrocarbons (PAHs), polychlorinated biphenyls (PCBs) and organochlorine pesticides (OCPs) in the urban environment: Street- vs rooftop-level measurements. *Atmospheric Environment,* 43 (2), pp. 290-300

Ćwiklak, K., Pastuszka, J.S., Rogula-Kozłowska W. (2009). Influence of traffic on particulate-matter polycyclic aromatic hydrocarbons in Urban atmosphere of Zabrze, Poland, *Polish Journal of Environmental Studies,* 18 (4), pp. 579-585

Dejean, S., Raynaud, C., Meybeck, M., Della Massa, J.-P., Simon, V. (2009): Polycyclic aromatic hydrocarbons (PAHs) in atmospheric urban area: monitoring on various types of sites. *Environmental Monitoring and Assessment,* 148 (1-4), pp. 27-37

Deng, W.J., Louie, P.K.K., Liu, W.K., Bi, X.H., Fu, J.M., Wong, M.H. (2006). Atmospheric levels and cytotoxicity of PAHs and heavy metals in TSP and PM2.5 at an electronic waste recycling site in southeast China, *Atmospheric Environment,* 40 (36), 6955

Dickhut, R.M., Canuel, E.A., Gustafson, K.E., Liu, K., Arzayus, K.M., Walker, S.E., Edgecombe, G., Gaylor, M.O., MacDonald, E.H. (2000). Automotive sources of carcinogenic polycyclic aromatic hydrocarbons associated with particulate matter in the Chesapeake Bay region, *Environmental Science and Technology,* 34 (21), pp. 4635-4640

Di Filippo, P., Riccardi, C., Pomata, D., Gariazzo, C., Buiarelli, F. (2010). Seasonal abundance of particle-phase organic pollutants in an urban/industrial atmosphere. *Water, Air, and Soil Pollution* 211 (1-4), pp. 231-250

Ding, L.C., Ke, F., Wang, D.K.W., Dann, T., Austin, C.C. (2009): A new direct thermal desorption-GC/MS method: Organic speciation of ambient particulate matter collected in Golden, BC, *Atmospheric Environment* 43 (32), pp. 4894-4902

Directive 2008/50/EC of the European Parliament and of the Council of 21 May 2008 on ambient air quality and cleaner air for Europe

Dvorská, A., Lammel, G., Klánová, J. (2011). Use of diagnostic ratios for studying source apportionment and reactivity of ambient polycyclic aromatic hydrocarbons over Central Europe, *Atmospheric Environment,* 45 (2), pp. 420-427

El Haddad, I., Marchand, N., Dron, J., Temime-Roussel, B., Quivet, E., Wortham, H., Jaffrezo, J.L., Baduel, C., Viosin, D., Besombes, J.L., Gille, G. (2009). Comprehensive primary particulate organic characterization of vehicular exhaust emissions in France, *Atmospheric Environment,* 43 (39), pp. 6190-6198

EMEP. *Transboundary Partuculate Matter in Europe*, Status report 4/2009

Evagelopoulos, V., Albanis, T.A., Asvesta, A., Zoras, S. (2010). Polycyclic aromatic hydrocarbons (PAHs) in fine and coarse particles, *Global Nest Journal,* 12 (1), pp. 63-70

Fang, G.-C., Wu, Y.-S., Chen, J.-C., Chang, C.-N., Ho, T.-T. (2006). Characteristic of polycyclic aromatic hydrocarbon concentrations and source identification for fine and coarse particulates at Taichung Harbor near Taiwan Strait during 2004–2005, *Science of the Total Environment,* 366 (2-3), pp. 729-738

Galarneau, E. (2008). Source specificity and atmospheric processing of airborne PAHs: Implications for source apportionment, *Atmospheric Environment*, 42 (35), pp. 8139-8149

Grimmer, G., Abel, U., Brune, H., Deutsch-Wenzel, R., Emura, M., Heinrich, U., Jacob, J., Kemena, A., Misfeld, J., Mohr, U. (1986). Evaluation of environmental carcinogens by carcinogen-specific test systems, *Experimental Pathology,* 29 (2), pp. 65-76

Grynkiewicz-Bylina, B., Rakwic, B., Pastuszka, J.S.(2005). Assessment of exposure to traffic-related aerosol and to particle-associated PAHs in Gliwice, Poland, *Polish Journal of Environmental Studies,*14 (1), pp. 117-123

Gschwend, P.M., Hites, R.A. (1981). Fluxes of polycyclic aromatic hydrocarbons to marine and lacustrine sediments in the northeastern United States, *Geochimica et Cosmochimica Acta,* 45 (12), pp. 2359-2367

Guo, H., Lee, S.C., Ho, K.F., Wang, X.M., Zou, S.C. (2003): Particle-associated polycyclic aromatic hydrocarbons in urban air of Hong Kong, *Atmospheric Environment,* 37 (38), pp. 5307-5317

Guo, H.L., Lu, C.G., Yu, Q., Chen, L.M. (2004). Pollution characteristics of polynuclear aromatic hydrocarbons on airborne particulate in Shanghai. Journal of Fudan University (Natural Science), 43, pp. 1107

Harrison R.M., Smith D.J.T., Luhana L., (1996). Source apportionment of atmospheric polycyclic aromatic hydrocarbons collected from an urban location in Birmingham. UK, *Environmental Science and Technology*, 30(3), pp. 825–832

Hien, T.T., Thanh, L.T., Kameda, T., Takenaka, N., Bandow, H. (2007). Distribution characteristics of polycyclic aromatic hydrocarbons with particle size in urban aerosols at the roadside in Ho Chi Minh City, Vietnam, *Atmospheric Environment,* 41 (8), pp. 1575-1586

Hinds, W.C. (1998): Aerosol technology. Properties, behaviour, and measurement of airborne particles. Second Edition. John Wiley & Sons, Inc. New York

Hong, H.S., Yin, H.L., Wang, X.H., Ye, C.X. (2007). Seasonal variation of PM10-bound PAHs in the atmosphere of Xiamen, China. *Atmospheric Research,* 85 (3-4), pp. 429-441

http://www.epa.gov/

Inspection for Environmental Protection, Silesian Voivodship Inspectorate for Environmental Protection in Katowice, The environmental status in the Silesia region in 1999-2000, Library of Environmental Monitoring, Katowice 2001 (WIOŚ, 2001) (in polish)

Kavouras, I.G., Lawrence, J., Koutrakis, P., Stephanou, E.G., Oyola, P. (1999). Measurement of particulate aliphatic and polynuclear aromatic hydrocarbons in Santiago de Chile: Source reconciliation and evaluation of sampling artifacts, *Atmospheric Environment,* 33 (30), pp. 4977-4986

Khalili, N.R., Scheff, P.A., Holsen, T.M. (1995). PAH source fingerprints for coke ovens, diesel and gasoline engines, highway tunnels, and wood combustion emissions, *Atmospheric Environment,* 29 (4), pp. 533-542

Kinney, P.L., Aggarwal, M., Northridge, M.E., Janssen, N.A.H., Shepard, P. (2000). Airborne concentrations of PM2.5 and diesel exhaust particles on Harlem sidewalks: A community-based pilot study, *Environmental Health Perspectives,* 108 (3), pp. 213-218

Klejnowski, K., Krasa, A., Rogula, W. (2007a). Seasonal variability of concentrations of total suspended particles (TSP) as well as PM10, PM2.5 and PM1 modes in Zabrze, Poland, *Archives of Environmental Protection,* 33 (3), pp. 15-27

Klejnowski, K., Talik, E., Pastuszka, J., Rogula, W., Krasa, A. (2007b). Chemical composition of surface layer of PM1, PM1-2.5, PM2.5-10, *Archives of Environmental Protection*, 33 (3), pp. 89-95

Klejnowski, K., Rogula-Kozłowska, W., Krasa, A., (2009). Structure of atmospheric aerosol in Upper Silesia (Poland) - Contribution of PM2.5 to PM10 in Zabrze, Katowice and Częstochowa in 2005-2007, *Archives of Environmental Protection*, 35 (2), pp. 3-13

Kong, S., Ding, X., Bai, Z., Han, B., Chen, L., Shi, J., Li, Z. (2010). A seasonal study of polycyclic aromatic hydrocarbons in PM2.5 and PM2.5-10 in five typical cities of Liaoning Province, China *Journal of Hazardous Materials* 183 (1-3), pp. 70-80

Kozielska, B., Rogula-Kozłowska, W., Pastuszka, J.S. (2009). Effect of road traffic concentration of PM2.5, PM10 and PAHs in zones of high and low municipal emission, *Polska Inżynieria Środowiska pięć lat po wstąpieniu do Unii Europejskiej, Monografie Komitetu Inżynierii Środowiska PAN*, 58(1), pp. 129-137, Lublin, ISBN 978-83-89293-81-7 (in polish)

Kristensson, A., Johansson, C., Westerholm, R., Swietlicki, E., Gidhagen, L., Wideqvist, U., Vesely, V. (2004). Real-world traffic emission factors of gases and particles measured in a road tunnel in Stockholm, Sweden, *Atmospheric Environment*, 38 (5), pp. 657-673

Krudysz, M.A., Dutton, S.J., Brinkman, G.L., Hannigan, M.P., Fine, P.M., Sioutas, C., Froines, J.R. (2009). Intra-community spatial variation of size-fractionated organic compounds in Long Beach, California. *Air Quality, Atmosphere and Health,* 2 (2), pp 69-88

Kulkarni, P., Venkataraman, C. (2000). Atmospheric polycyclic aromatic hydrocarbons in Mumbai, India. *Atmospheric Environment* 34 (17), pp. 2785-2790

Ladji, R., Yassaa, N., Balducci, C., Cecinato, A., Meklati, B.Y. (2009). Distribution of the solvent-extractable organic compounds in fine (PM1) and coarse (PM1-10) particles in urban, industrial and forest atmospheres of Northern Algeria, *Science of the Total Environment,* 408 (2), pp. 415-424

Li, C.K., Kamens, R.M., (1993). The use of polycyclic aromatic hydrocarbons as sources signatures in receptor modeling. *Atmospheric Environment,* 27A, pp. 523–532

Makkonen, U., Hellén, H., Anttila, P., Ferm, M. (2010): Size distribution and chemical composition of airborne particles in south-eastern Finland during different seasons and wildfire episodes in 2006, *Science of the Total Environment* 408, (3), pp. 644-651

Manoli, E., Voutsa, D., Samara, C. (2002). Chemical characterization and source identification/apportionment of fine and coarse air particles in Thessaloniki, Greece, *Atmospheric Environment*, 36 (6), pp. 949-961

Manoli, E., Kouras, A., Samara, C. (2004). Profile Analysis of Ambient and Source Emitted Particle-Bound Polycyclic Aromatic Hydrocarbons from Three Sites in Northern Greece, *Chemosphere,* 56 (9), pp. 867-878

Mantis, J., Chaloulakou, A., Samara, C. (2005). PM10-bound polycyclic aromatic hydrocarbons (PAHs) in the Greater Area of Athens, Greece, *Chemosphere* 59 (5), pp. 593-604

Masclet, P., Bresson, M.A., Mouvier, G. (1987). Polycyclic aromatic hydrocarbons emitted by power stations, and influence of combustion conditions, *Fuel,* 66 (4), pp. 556-562

Miguel, A.H., Kirchstetter, T.W., Harley, R.A., Hering, S.V. (1998). On-road emissions of particulate polycyclic aromatic hydrocarbons and black carbon from gasoline and diesel vehicles, *Environmental Science and Technology,* 32 (4), pp. 450-455

Mohanraj, R., Solaraj, G., Dhanakumar, S. (2011). Fine particulate phase PAHs in ambient atmosphere of Chennai metropolitan city, India. *Environmental Science and Pollution Research,* 18 (5), pp. 764-771

Nisbet, I.C.T., LaGoy, P.K.: (1992) Toxic equivalency factors (TEFs) for polycyclic aromatic hydrocarbons (PAHs), *Regulatory Toxicology and Pharmacology,* 16 (3), pp. 290-300

Odabasi, M., Vardar, N., Sofuoglu, A., Tasdemir, Y., Holsen, T.M. (1999). Polycyclic aromatic hydrocarbons (PAHs) in Chicago air, *Science of the Total Environment,* 227 (1), pp. 57-67

Oliveira, C., Martins, N., Tavares, J., Pio, C., Cerqueira, M., Matos, M., Silva, H., Oliveira, C., Camões, F. (2011). Size distribution of polycyclic aromatic hydrocarbons in a roadway tunnel in Lisbon, Portugal, *Chemosphere,* 83 (11), pp. 1588-1596

Pakkanen, T.A., Kerminen, V.-M., Loukkola, K., Hillamo, R.E., Aarnio, P., Koskentalo, T., Maenhaut, W. (2003). Size distributions of mass and chemical components in street-level and rooftop PM1 particles in Helsinki, *Atmospheric Environment,* 37 (12), pp. 1673-1690

Panther, B.C., Hooper, M.A., Tapper, N.J. (1999). A comparison of air particulate matter and associated polycyclic aromatic hydrocarbons in some tropical and temperate urban Environments, *Atmospheric Environment,* 33 (24-25), pp. 4087-4099

Pastuszka, J.S., Rogula-Kozłowska, W., Zajusz-Zubek, E. (2010). Characterization of PM10 and PM2.5 and associated heavy metals at the crossroads and urban background site in Zabrze, Upper Silesia, Poland, during the smog episodes, *Environmental Monitoring and Assessment,* 168 (1-4), pp. 613-627

Phuleria, H.C., Sheesley, R.J., Schauer, J.J., Fine, P.M., Sioutas, C. (2007). Roadside measurements of size-segregated particulate organic compounds near gasoline and diesel-dominated freeways in Los Angeles, CA, *Atmospheric Environment,* 41 (22), pp. 4653-4671

Prahl, F.G., Carpenter, R. (1983). Polycyclic aromatic hydrocarbon (PAH)-phase associations in Washington coastal sediment, *Geochimica et Cosmochimica Acta,* 47(6), pp.1013–1023

Program ochrony powietrza dla stref województwa śląskiego, w których stwierdzone zostały ponadnormatywne poziomy substancji w powietrzu. Załącznik do uchwały Nr III/52/15/2010 Sejmiku Województwa Śląskiego z dnia 16 czerwca 2010 r. Katowice, 2010 r. (in polish)

Querol, X., Alastuey, A., Rodriguez, S., Plana, F., Mantilla, E., Ruiz, C.R. (2001). Monitoring of PM10 and PM2.5 around primary particulate anthropogenic emission sources, *Atmospheric Environment,* 35 (5), pp. 845–858

Ravindra, Mittal, A.K., Van Grieken, R. (2001). Health risk assessment of urban suspended particulate matter with special reference to polycyclic aromatic hydrocarbons: A review, *Reviews on Environmental Health,* 16 (3), pp. 169-189

Ravindra, K., Bencs, L., Wauters, E., De Hoog, J., Deutsch, F., Roekens, E., Bleux, N., Berghmans, P., Van Grieken, R. (2006). Seasonal and site-specific variation in vapour and aerosol phase PAHs over Flanders (Belgium) and their relation with anthropogenic activities, *Atmospheric Environment*, 40 (4), pp. 771-785

Ravindra, K., Sokhi, R., Van Grieken, R. (2008). Atmospheric polycyclic aromatic hydrocarbons: Source attribution, emission factors and regulation, *Atmospheric Environment*, 42 (13), pp. 2895-2921

Rogge, W.F., Hildemann, L.M., Mazurek, M.A., Cass, G.R., Simonelt, B.R.T. (1993a). Sources of fine organic aerosol. 3. Road dust, tire debris, and organometallic brake lining dust: Roads as sources and sinks, *Environmental Science and Technology*, 27 (9), pp. 1892-1904

Rogge, W.F., Hildemann, L.M., Mazurek, M.A., Cass, G.R., Simoneit, B.R.T. (1993b). Sources of fine organic aerosol. 2. Noncatalyst and catalyst-equipped automobiles and heavy-duty diesel trucks, *Environmental Science and Technology*, 27 (4), pp. 636-651

Rogula-Kozłowska, W., Pastuszka, J.S., Talik, E. (2008). Influence of vehicular traffic on concentration and particle surface composition of PM10 and PM2.5 in Zabrze, Poland, *Polish Journal of Environmental Studies*, 17 (4), pp. 539-548

Rogula-Kozłowska W., Klejnowski K., Krasa A., Szopa S.(2010). Concentration and elemental composition of atmospheric fine particles in Silesia Province, Poland. [In] Environmental Engineering III, Pawłowski, Dudzińska. & Pawłowski (eds.), Taylor & Francis Group, London, pp. 75-81

Saraga, D.E., Maggos, T.E., Sfetsos, A., Tolis, E.I., Andronopoulos, S., Bartzis, J.G., Vasilakos, C. (2010): PAHs sources contribution to the air quality of an office environment: Experimental results and receptor model (PMF) application. *Air Quality, Atmosphere and Health*, 3 (4), pp. 225-234

Sheu, H.-L., Lee, W.-J., Lin, S.J., Fang, G.-C., Chang, H.-C., You, W.-C. (1997). Particle-bound PAH content in ambient air, *Environmental Pollution*, 96 (3), pp. 369-382

Sicre, M.A., Marty, J.C., Saliot, A., Aparicio, X., (1987). Aliphatic and aromatic hydrocarbons in the Mediterranean aerosol. *International Journal of Environmental Analytical Chemistry*, 29, pp. 73–94

Sienra, M.del.R., Rosazza, N.G., Préndez, M. (2005). Polycyclic Aromatic Hydrocarbons and Their Molecular Diagnostic Ratios in Urban Atmospheric Respirable Particulate Matter. *Atmospheric Research*, 75 (4), pp. 267-281

Simcik, M.F., Eisenreich, S.J., Lioy, P.J. (1999). Source apportionment and source/sink relationships of PAHs in the coastal atmosphere of Chicago and Lake Michigan, *Atmospheric Environment*, 33 (30), pp. 5071-5079

Slezakova, K., Castro, D., Begonha, A., Delerue-Matos, C., Alvim-Ferraz, M.D.C., Morais, S., Pereira, M.D.C. (2011). Air pollution from traffic emissions in Oporto, Portugal: Health and environmental implications, *Microchemical Journal*, 99 (1), pp. 51-59

Smith, D.J.T., Harrison, R.M. (1998) Atmospheric Particles. Wiley, New York

State Sanitary Inspection, Department in Katowice. Atmospheric pollution in the Katowice Province in 1979. Katowice 1980, (for official use, in polish)

State Sanitary Inspection, Department in Katowice. Atmospheric pollution in the Katowice Province in the years 1983 and 1984. Katowice 1984/85, (for official use, in polish)

State Sanitary Inspection, Department in Katowice. Atmospheric pollution in the Katowice Province in the years 1985 - 1987. Katowice 1988, (in polish)

State Sanitary Inspection, Department in Katowice. Atmospheric pollution in the Katowice Province in the years 1988 - 1990, Katowice 1991, (in polish)

State Sanitary Inspection, Department in Katowice. Atmospheric pollution in the Katowice Province in the years 1991 - 1993, Katowice 1994, (in polish)

State Sanitary Inspection, Department in Katowice. Atmospheric pollution in the Katowice Province in the years 1996 - 1997, Katowice 1998, (in polish)

State Sanitary Inspection, Department in Katowice. Atmospheric pollution in the Katowice Province in the years 1997 - 1998 and in the former provinces Bielsko and Czestochowa in 1998, Katowice 1999, (in polish)

State Sanitary Inspection, Department in Katowice. Atmospheric pollution in the Silesian Province in the years 2001 - 2002, Katowice 2003, (in polish)

State Sanitary Inspection, Department in Katowice. Atmospheric pollution in the Silesian Province in the years 2002 - 2003, Katowice 2004, (in polish)

State Sanitary Inspection, Department in Katowice. Atmospheric pollution in the Silesian Province in the years 2003 - 2004, Katowice 2005, (in polish)

State Sanitary Inspection, Department in Katowice. Atmospheric pollution in the Silesian Province in 2005, Katowice 2006, (in polish)

Sverdrup, G.M., Whitby, K.T. (1977). Determination of submicron atmospheric aerosol size distributions by use of continuous analog sensors, *Environmental Science and Technology*, 11 (13), pp. 1171-1176

Takeshi, O., Takashi, A. (2004).Spatial distributions and profiles of atmospheric polycyclic aromatic hydrocarbons in two industrial cities in Japan, *Environmental Sciences and Technology*, 38 (1), pp. 49-55

Tang, N., Hattori, T., Taga, R., Igarashi, K., Yang, X., Tamura, K., Kakimoto, H., Mishukov, V.F., Toriba, A., Kizu, R., Hayakawa, K. (2005). Polycyclic aromatic hydrocarbons and nitropolycyclic aromatic hydrocarbons in urban air particulates and their relationship to emission sources in the Pan-Japan Sea countries, *Atmospheric Environment*, 39 (32), pp. 5817-5826

The Provincial Inspectorate for Environmental Protection, The environmental status in the Silesia region in 2001, Library of Environmental Monitoring, Katowice 2002, (in polish)

The Provincial Inspectorate for Environmental Protection. The environmental status in the Silesia region in 2002, Katowice 2003, (in polish)

The Provincial Inspectorate for Environmental Protection. The environmental status in the Silesia region in 2003, Katowice 2004, (in polish)

The Provincial Inspectorate for Environmental Protection. The environmental status in the Silesia region in 2004, Katowice 2005, (in polish)

The Provincial Inspectorate for Environmental Protection. The environmental status in the Silesia region in 2005, Katowice 2006, (in polish)

The Provincial Inspectorate for Environmental Protection. The environmental status in the Silesia region in 2006, Katowice 2007, (in polish)

The Provincial Inspectorate for Environmental Protection. The environmental status in the Silesia region in 2007, Katowice 2008, (in polish)

The Provincial Inspectorate for Environmental Protection. The environmental status in the Silesia region in 2008, Katowice 2009, (in polish)

The Provincial Inspectorate for Environmental Protection. The environmental status in the Silesia region in 2009, Katowice 2010, (in polish)

The Provincial Inspectorate for Environmental Protection. The environmental status in the Silesia region in 2010, Katowice 2011, (in polish)

Vercauteren, J., Matheeussen, C., Wauters, E., Roekens, E., van Grieken, R., Krata, A., Makarovska, Y., Maenhaut, W., Chi, X., Geypens, B. (2011). Chemkar PM10: An extensive look at the local differences in chemical composition of PM10 in Flanders, Belgium, *Atmospheric Environment,* 45 (1), pp. 108-116

Vogt, R., Kirchner, U., Scheer, V., Hinz, K.P., Trimborn, A., Spengler, B. (2003). Identification of diesel exhaust particles at an Autobahn, urban and rural location using single-particle mass spectrometry, *Journal of Aerosol Science,*34 (3), pp. 319-337

Wang, X.H., Ye, C.X., Yin, H.L., Zhuang, M.Z., Wu, S.P., Mu, J.L., Hong, H.S. (2007). Contamination of Polycyclic Aromatic Hydrocarbons Bound to PM10/PM2.5 in Xiamen, China, *Aerosol and Air Quality Research,* 7(2), pp.260–276

Wang, H., Zhou, Y., Zhuang, Y., Wang, X., Hao, Z. (2009). Characterization of PM2.5/PM2.5-10 and source tracking in the juncture belt between urban and rural areas of Beijing, *Chinese Science Bulletin,* 54 (14), pp. 2506-2515

Willeke, K., Whitby, K.T. (1975). Atmospheric aerosols: size distribution interpretation, *Journal of the Air Pollution Control Association,* 25 (5), pp. 529-534

Wingfors, H., Hägglund, L., Magnusson, R. (2011). Characterization of the size-distribution of aerosols and particle-bound content of oxygenated PAHs, PAHs, and n-alkanes in urban environments in Afghanistan, *Atmospheric Environment,* 45 (26), pp. 4360-4369

White, P.A. (2002). The genotoxicity of priority polycyclic aromatic hydrocarbons in complex mixtures, *Mutation Research - Genetic Toxicology and Environmental Mutagenesis,* 515 (1-2), pp. 85-98

Yan, J., Wang, L., Fu, P.P., Yu, H. (2004). Photomutagenicity of 16 polycyclic aromatic hydrocarbons from the US EPA priority pollutant list, *Mutation Research - Genetic Toxicology and Environmental Mutagenesis,* 557 (1), pp. 99-108

Yunker, M.B., Macdonald, R.W., Vingarzan, R., Mitchell, R.H., Goyette, D., Sylvestre, S. (2002). PAHs in the Fraser River basin: A critical appraisal of PAH ratios as indicators of PAH source and composition, *Organic Geochemistry,* 33 (4), pp. 489-515

Zaciera M., Kurek J., Dzwonek L., Feist B., Jędrzejczak A. (2011) Seasonal variability of PAHs and nitro-PAHs concentrations in total suspended particulate matter in ambient air of cities of silesian voivodeship, [In] Modern achievements in the protection ambient air, Musialik-Piotrowska A. & Rutkowski J.D., Polish Association of Sanitary Engineers and Technicians Wroclaw, pp. 401-406; (in polish)

Zencak, Z., Klanova, J., Holoubek, I., Gustafsson, Ö. (2007). Source apportionment of atmospheric PAHs in the western balkans by natural abundance radiocarbon analysis, *Environmental Science and Technology,* 41 (11), pp. 3850-3855

Zhao, Y., Wang, S., Duan, L., Lei, Y., Cao, P., Hao, J. (2008). Primary air pollutant emissions of coal-fired power plants in China: Current status and future prediction. *Atmospheric Environment,* 42 (36), pp. 8442-8452

Zou, L.Y., Zhang, W., Atkiston, S. (2003). The characterisation of polycyclic aromatic hydrocarbons emissions from burning of different firewood species in Australia, *Environmental Pollution,* 124 (2), pp. 283-289

8

Global Distillation in an Era of Climate Change

Ross Sadler and Des Connell
Griffith University
Australia

1. Introduction

When a chemical enters the natural environment it undergoes change by several processes. It can be transported by the movement of the sector of the environment it enters, for example ocean currents and atmospheric movements, to a different geographical location. In addition as it is transported it is usually diluted so that its concentration is reduced. Chemical degradation processes take place which result in the production of more polar and water soluble products and leaving residual initial chemical. As all of these processes occur there is a distribution of the chemical between the phases in the environment. These processes have been described in a series of papers and books by Mackay as reported in Mackay et al (2009).

The processes shown in Figure 1 illustrate the basic processes involved in the partitioning of a chemical into phases in the environment. All of the processes shown involve two phases and movement backwards and forwards of the chemical and thus can be characterised by a partition coefficient. The partition coefficient is the ratio of the chemical in the two phases at equilibrium and isoften represented by the symbol, K. Thus the air - water partition coefficient is represented by K_{AW} and is better known as the dimensionless Henrys Law constant (Shiu and Mackay, 1986), the fish - water partition coefficient is $K_{B,}$ (Connell, 1990) the sediment - water coefficient is K_D (Gobas and Maclean, 2003) and the other partition processes can be represented in a similar way. It is of interest to note that as a result of these partition processes a chemical can occur in very low concentrations in the atmosphere, low concentration in water but relatively high concentrations in fish and other aquatic biota.

Most of these partition coefficients can be calculated from partition values arrived at by laboratory measurements. For example the Henrys Law constant and the octanol - water partition coefficient, K_{OW} , can be measured in the laboratory. The K_{OW} can be used to calculate the fish - water partition coefficient, K_B value and also the K_D value and is extensively used to model chemical partitions in the environment. The octanol - air partition coefficient, K_{OA}, can be calculated or measured in the laboratory and can be used to evaluate the partitioning of a chemical into organisms as a result of concentrations in the atmosphere.

The Persistent Organic Pollutants (POPs) are a group of mainly chlorinated hydrocarbon insecticides and dioxins which are often reported to undergo global distillation (Fernandez and Grimalt, 2003). These substances undergo the partition processes illustrated diagrammatically in Figure 1. However if they are discharged to the environment in the warmer zones of the planet a proportion will partition into the atmospheric phase according to the partition coefficient at that temperature. Movement in the atmosphere through winds results in transport to different geographical regions which, with the polar zones, have a colder temperature. The partition coefficient values are displaced to favour the solid or liquid phase and not the atmospheric phase. Thus a transfer of chemical can occur from warmer to the polar zones of the planet by a process somewhat similar to distillation - global distillation. This can occur in a series of steps or hops (Ma, 2010).

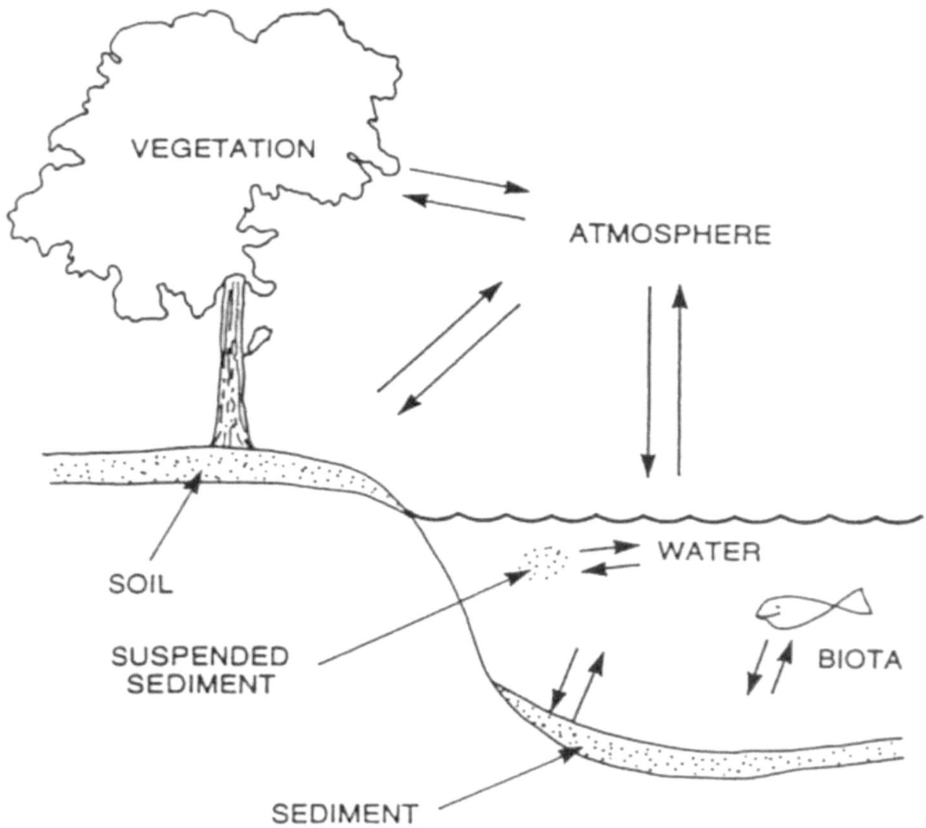

Fig. 1. Pathways for Movement of Persistent Organic Pollutants within the Environment

1.1 Physicochemical properties of POPS

There is no direct evidence that all of the compounds which are currently classified as POPs (see Table 1) exhibit the property of global distillation. However the POPs which have been identified as exhibiting global distillation share a set of properties which lead to their long range transport to the polar zones in the Arctic and are presumed (rather than demonstrated to also do the same in the Antarctic). Firstly all of these compounds are chlorohydrocarbons which means there is a limited range of covalent bonds present in them. These bonds are mainly C-C, C=C aromatic, C=C, C-H, C-Cl which are resistant to oxidation and hydrolysis - the main degradation processes in the environment. This leads to environmental persistence with a half life of between 2 to 10 years in soil and 0.25 to 2.5 years in air. Aqueous solubility is low at 1 to 7 g/m^3 with a correspondingly high solubility in lipid reflected in the K_{OW} values from about 10^2 to 10^6.

1.2 Climate change and long range transport of POPs

The effects of climate change on the global distillation process have been evaluated by several authors (e.g. Macdonald et al., (2003), Sadler et al., (2011)). Macdonald et al., (2003) pointed out that while the partition process can be successfully modelled, the impact of other environmental processes such as altered rainfall and changes to the particulate content of the atmosphere are difficult to assess. In the years since this review, a great deal more has been learned regarding the processes that underlie global distillation and also the effects and severity of climate change. For this reason, an updated review of the literature has been undertaken with a view to providing a prediction of long range transport of POPs for both the Northern and Southern Hemispheres that is relevant to the current state of knowledge.

2. Evidence for global distillation

2.1 Northern Hemisphere

Since the process was first proposed by Wania and Mackay (1995), numerous studies (notably in the Northern Hemisphere) have provided evidence that would be supportive of a global distillation model in respect of global transport of persistent organic pollutants. Polychlorinated biphenyls (PCB) have been amongst the most heavily studied of pollutants in this regard and reports of PCB migration include those in Northern Hemisphere soils (Meijer, et al., 2002), European high mountain lakes (Carera et al, 2002) levels observed in Norwegian mosses (Lead et al., 1996), deposition patterns in Canadian lakes (Muir et al., 1996) and Russian Lakes (McConnell *et al.*, 1996). Studies universally point to enhanced retention of the more chlorinated PCB congeners at lower latitudes, probably reflecting a more facile volatilization of the simpler PCBs and their subsequent migration to the poles. It is clear that although overall distribution patterns are consistent with the theory of global distribution and transport, a number of other local factors have to be taken into consideration when ascribing differences to global transport. Regression plots generally show considerable scatter with regard to the trend line, particularly with octa-PCBs (Meijer et al., 2002). This may account for the fact that other studies have found only partial evidence in support of the expected global transport process.

Reports of similar distribution patterns have also been produced in respect of global distribution of PCBs and organochlorine pesticides in other media. For example Kalantzi et al., (2001), noted that the distribution of PCB congeners was highest in European and North American butter and lowest in butter samples from the southern hemisphere. But the authors rightly point out that there are a number of compounding factors, such as relative efficiency of uptake from air/foliage, history of use in the area, etc. A number of other POPs included in the study showed maximal levels in areas where they are still in use. Trapping of PCBs and organochlorine pesticides in pine needles in Canada has also been ascribed to their delivery via global transport processes (Davidson et al., 2003).

Additional evidence of POP migration in accordance with the global transport/distillation model has come from measurement of levels in air over an extended period of time. There are now a number of networks dedicated to this type of surveillance. The underlying premise is that with a decrease in evolution of POPs following withdrawal of their use in many countries, levels in air should decrease with increasing time. For example, Sweetman and Jones (2000) reported a decrease of this kind with respect to PCBs. Their study, which centered around Hazelrigg, UK, noted a significant downward trend in PCB congener levels with time. Similar results were also obtained as regards levels of PCB congeners around the

Great Lakes region (Slmck et al., 1999). In contrast, the situation in the Arctic is more complex, as there will be an ongoing input of POPs from the continuing use in certain countries and also in terms of completion of global migration which commenced in other areas when the pesticides were routinely employed. The levels observed in Arctic air will principally result from global transport and also to a lesser extent, from revolatilization of POPs already in the Arctic environment. Hung et al., 2001 observed a rather mixed response in terms of trends for PCB congeners in Arctic air with time. There was generally a lack in decline of temporal trend, particularly as regards the more chlorinated congeners. The authors considered this to be basically supportive of a global transport model. Nevertheless, there were some exceptions, notably PCB 180, which whilst polychlorinated, showed a distinctive downward temporal trend. It was postulated that the heavier congeners may be more subject to differential removal processes by snow and particulate scavenging, whereas the lighter PBCs may be more prone to attack by hydroxyl radicals.

There have also been a number of reports of global transport in respect of a variety of pollutants apart from PCB congeners. Several of these have been included in the discussion above, but pollutants for which similar patterns have been observed include endosulfan, HCB and HCH (Carrera et al., 2002), lindane (Zhang et al., 2008) and phthalates (Xie et al., 2007). Significant attention has been paid to α- and γ-HCH (lindane) as well as endosulfan, because of their relatively recent use. In 2002, γ-HCH was shown to have reached phase equilibrium in the North Atlantic as has α-HCH, but the surface waters of the tropical and southern Atlantic were strongly undersaturated with γ-HCH, especially between 30°N and 20 °S. It must be noted that the state of the air-sea equilibria of α-HCH and γ-HCH are very different. The sporadic occurrence of γ-HCH in air makes it difficult to obtain representative results from transect cruises (Lakaschus et al., 2002). Weber et al., (2006) collated data for these pesticides from a number of sampling expeditions over the past decade. Although all could be detected in the Arctic environment, different patterns of distribution were observed for each substance with α-endosulfan predominating in the western Arctic and γ-HCH in the central Arctic. It was concluded that coastal sources may be important as regards γ-HCH, whereas air exchange is the major pathway of input as regards α-endosulfan.

Hargrave et al., 1997 examined the Arctic air-seawater fluxes of a number of POPs in the Canadian Arctic Archipelago during 1993. All of the pesticides showed a bimodal seasonal distribution of concentrations with maxima in February-May and July-August separated by minimum values in June. Organochlorine levels in air increased sharply during April and May and decreased during June, coincident with the onset of the open water period. Volatilization loses for HCH isomers during the open water period were estimated to have been small, whilst those for HCB and dieldrin were significant. These workers obtained evidence for significant deposition of toxaphenes, chlordanes and α-endosulfan during the open water period. The former USSR and Eastern Europe were probable emission sources for atmospheric contaminants.

An alternative approach to demonstrating the operation of a global distillation process would be the demonstration of summer vs. winter differences as regards deposition of POPs. In theory, higher levels of POPs would be expected to be transported during the summer, as a result of increased volatiliziation. Although data of this kind have been demonstrated for a number of European sites (cf. Carrera et al., 2002), it is impossible to separate the processes of revolatiliziation resulting from summer temperatures and increased use patterns at this same period. Diel variations in polybrominated diphenyl ethers and chlordanes in air have also been noted (Moeckel et al., 2008). The use of isotope tracers or more probably isotope ratios has also been advanced as a possible means of confirming the processes involved in long range transport of POPs (Dickhut et al., 2004).

There is also evidence that apart from global migration towards the poles, distillation phenomena can be invoked to explain the increasing concentrations of these pollutants with altitude. For example, Gallego et al., 2007 studied the distribution of PBDEs and PCBs in fish from high European mountain lakes (Pyranees and Tatra Mountains respectively). PCB levels in fish muscle were found to be significant in both study areas, with higher levels being recorded in the Tatra Mountains. This would be expected as the latter region experiences lower minima than the former. As regards the recorded levels of PBDEs, there was a good correlation between their levels and those of the PCBs in samples from the lakes in the Pyranees. The authors considered that this pointed to a predominance of temperature effects in this case, as would be predicted from the global distillation model. This relationship did not apply in the Tatra Mountains, probably reflecting the activity of other (unspecified) processes and also the fact that PBDEs are of later introduction than PCBs. One possibility would be the more recent and probably less restrained use of PCDEs during and immediately after the existence of the Eastern Bloc regimes (cf. Daly and Wania, 2005). Similar conclusions were reached by Demers et al., (2007), in a study of PCB and other organochlorine levels in trout from lakes in British Columbia and Alberta. Although the results obtained were in keeping with overall global distillation, this process alone could not explain the levels of contaminants found in the trout. The authors considered that both the feeding behavior of the fish and the octanol-water partition coefficient of the contaminant in question were also important factors in determining the observed distribution.

However, Wania and Westgate (2008) have pointed to several important differences between so-called mountain cold trapping and polar cold trapping. Although the same families of pollutants, in particular the PCBs, have been shown to concentrate in both high latitude and high altitude locations, a detailed comparison of studies shows that different chemicals concentrate in high latitudes than at high elevations. The chemicals that become enriched in mountains tend to be less volatile than those preferentially accumulating in polar regions by about two orders of magnitude. Wania and Westgate (2008) hypothesized that the temperature dependence of the precipitation scavenging efficiency of organic chemicals underlies mountain cold-trapping. In both polar and mountain cold trapping, temperature gradients and their impact on gas phase/condensed phase partitioning play a crucial role. Nevertheless, these processes are controlled by different mechanisms and affect different chemicals. In the case of polar cold-trapping the temperature dependence of partitioning between the Earth's surface and atmosphere is at the basis of the grass-hopper effect. In the case of mountain cold trapping the temperature dependence of partitioning between the various atmospheric components (gas phase vs. particles, rain droplets and snow flakes) is important. Franz and Eisenreich (1998) have pointed to the relative efficiency of snow as compared to rain for scavenging of PCBs and PAHs from the atmosphere, citing porosity differences as a major factor.

2.2 Southern Hemisphere

Evidence for operation of the process of global distillation in the Southern Hemisphere is generally more limited, albeit with at least some grounds for supporting the hypothesis (Corsolini et al., 2002; Noël et al., 2009). A major factor to be considered when comparing the two processes is the relative differences in landforms between the two hemispheres. Whereas an extensive landmass extends towards the poles in the northern hemisphere, the Southern Hemisphere consists of large expanses of open ocean, with only South America extending close to the Antarctic. Consequently, it is not surprising that the few studies of POPs distribution in the Southern Hemisphere have not provided much evidence of global migration. Studies of seawater are subject to many compounding effects and thus a North-South transect that included a large number of sampling sites off the coast of Africa, failed to show any evidence of increased PCB content at southern latitudes (Nizzetto *et al.*, 2008). The concentrations of HCHs in air and surface water of the Arctic have been shown to exceed those of the Antarctic by one to two orders of magnitude (Lakaschus *et al.*, 2002). Soil and sediment samples from James Ross Island were shown to contain low levels of PCBs, PAHs, p,p'-DDT, DDE, and DDD, with generally lower levels being detected in the sediments. A prevalence of low-mass PAHs, less chlorinated PCBs, and more volatile chemicals was taken to indicate that the long-range atmospheric transport from populated areas of Africa, South America, and Australia as the most probable contamination source for the solid matrices in James Ross Island (Klánová *et al.*, 2008).

Low levels of various organochlorine samples have been recorded in moss samples from high altitude southern hemisphere locations (Grimalt *et al.*, 2004) and also from the Antarctic (Focardi *et al.*, 1991). (Mosses are favoured for this type of observation because of their rather simple physiology compared to higher plants.). When compared to the levels of these pollutants found in similar locations within the northern hemisphere, the levels are small, with the Antarctic values being particularly low. As has already been explained, there are differences in the underlying principles of high altitude vs. polar deposition, even though there is some overall resemblance in the processes.

The general lack of supporting observations for operation of global distillation in the Southern Hemisphere is probably a reflection of a number of factors. As mentioned above, the Southern Hemisphere contains far larger expanses of ocean than does the Northern Hemisphere. There has to date been little evidence obtained to support the operation of global distillation in oceans, probably because of the susceptibility of deposited POPs to scavenging by water-borne particulates (notably those of phytoplankton origin), with subsequent benthic deposition (Dachs *et al.*, 2002).

Moreover, it must be conceded that the overall shape of continents in the Southern Hemisphere differs markedly from that of their Northern Hemisphere counterparts. The two land masses that would provide a terrestrial pathway for POPs from the equator towards the poles (viz. South America and Africa) are both triangular in shape. Given that global distillation works best in terrestrial environments (Dachs *et al.*, 2002), a persistent organic pollutant moving towards the Arctic in the Northern Hemisphere would have a relatively good chance of progressing via land. In contrast, movement southwards in Africa or South America would significantly increase the chance of a persistent organic pollutants' entering the marine environment and hence being lost to the global distillation process. And although the southern tip of South America does lie close to Antarctica, southward migration would still require that the pollutant traverse a significant tract of the Southern Ocean. In the case of Africa, the migration distance across water would be so significant, as to probably relegate the global distillation process to a minor role. Interestingly, Gioia *et al.*, 2008 noted that PCB levels in the North Atlantic appeared to be governed primarily by transport from source regions, whereas corresponding levels in the South Atlantic seemed to be driven by temperature changes via air-water exchange with the ocean. This would be in keeping with the overall observations above.

Finally, evidence of pollutant migration throughout the African and South American continents would probably be blurred by the ongoing/recent use of organochlorine pesticides in some of the countries.

The relative differences in pollutant transport by global distillation in the two hemispheres are highlighted in the accompanying figure (Figure 2).

2.3 Alternative explanations

Reference has already been made to the fact that global distillation may, in certain circumstances provide only a partial explanation for the observed long-range transport of pollutants (Dachs *et al.*, 2002). von Waldow *et al.*, 2010 proposed an alternative differential removal hypothesis, which proposes that fractionation results from different loss rates from the atmosphere, acting along a gradient of remoteness from emission sources. They successfully applied the associated model to explain the observed differences in PCB concentrations in European air, using data from an ongoing study in which transects from England to Norway are monitored. From their data analysis, the observed concentrations were better correlated with distance from the source than with temperature.

It must be emphasized that these data pertain only to a relatively limited transect within the Northern Hemisphere. Moreover, the authors point to possible anomalies in data collection using semipermeable membrane devices as a source of error in some calculations. The failure of this study to find evidence in support of global distillation does not negate the concept overall. Various other examples have been given above in which global distillation was shown to be of secondary importance to other processes. The two processes (viz. global distillation and differential removal) may be seen as competing and further discussion of their relative importance in terms of climate change is provided below.

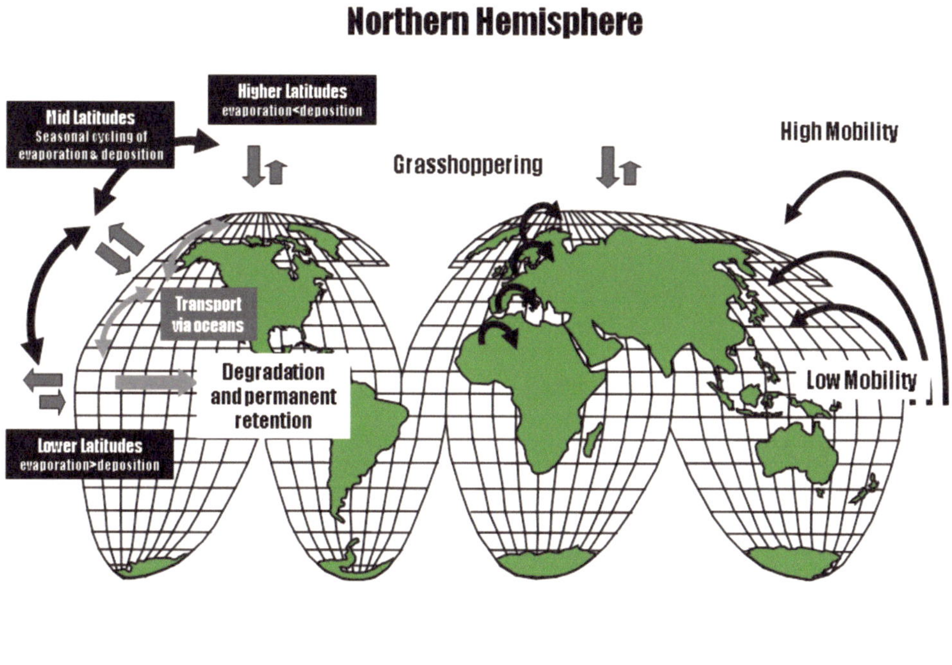

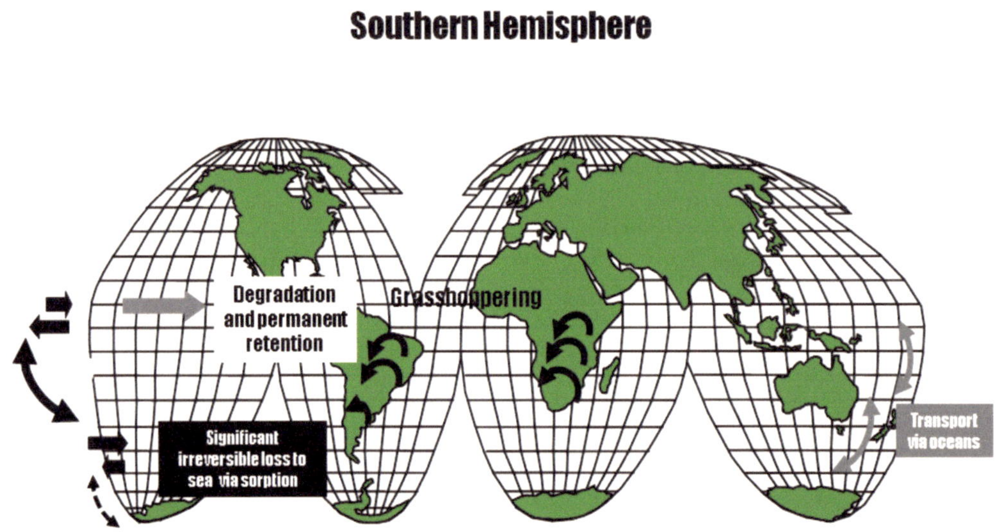

[1]Although multiple hops are involved in the global distillation process of POPs (Wania, 2003), only one is shown in these diagrams for simplicity.

Fig. 2. Global Distillation Processes in Northern and Southern Hemispheres[1]

3. Processes involved in global distillation

Wania (2003) divided pollutants into:

- 'Fliers' which are very volatile pollutants and unlikely to deposit even at the poles.
- 'Single hoppers' which are volatile enough to be carried all the way to the poles, where they will ultimately deposit by condensation.
- 'Multiple hoppers' which will be carried toward the poles by repeated evaporation and condensation cycles.
- 'Swimmers', which are non-volatile and will be transported via the oceans.

Of these four groups, it is the multiple hoppers that will be subject to transport by global distillation. This group includes the persistent organic pollutants, which are enumerated in the Table below (Table 1).

3.1 Basic underlying processes

The ability of a substance to undergo long range transport via global distillation will be governed by three factors:

1. Its ability to enter the air compartment by volatilization, if it is not emitted directly to air.
2. Its ability to remain stable during the transport process.
3. Its ability to deposit in the Arctic region.

Thus the main physicochemical parameters that govern global distillation will be the Henry's Law coefficient (H), the octanol-air partition coefficient (K_{OA}) and the air-water partition coefficient (K_{AW}).

Chemical	Log K_{OA}	Log k_{AW}	H
Aldrin	4.44	-2.23[11]	7.0669×10^{-3} atm-m^3/mole
Chlordane	9.21 (*cis*)[4] 9.16 (*trans*)[4]	2.31[12]	4.105×10^{-3} atm-m^3/mole
p,p'-Dichlorodiphenyltrichloroethane (DDT)	10.08[4]	-2.48[11]	2.232×10^{-3} atm-m^3/mole
Dieldrin	8.54	-3.56[11]	1.1099×10^{-4} atm-m^3/mole
Endrin	10.1[5]	-4.66[12]	3.08×10^{-4} atm-m^3/mole
Heptachlor	6.12[4]	-0.02	4.47×10^{-2} atm-m^3/mole
Hexachlorobenzene (HCB)	7.563[2]	-1.64[11]	3.61×10^{-2}
Mirex	10.7[12]	-5.42[12]	8.28×10^{-3} atm-m^3/mole
Polychlorinated biphenyls	7.01 -10.75[2]	0.349- (-5.08)[12]	3.3×10^{-4} - 5×10^{-5} atm-m^3/mole
Polychlorinated dibenzo-p-dioxins	8.564 – 11.66[2]	-5.02 (TCDD[12])	1.47×10^{-3} (TCDD)
Polychlorinated dibenzofurans	10.281[2]	(-2.511) – (-5.47)[12]	8.1×10^{-5} atm-m^3/mole
Toxaphene	7.6[12]	-1.16[12]	6.0×10^{-6} atm-m^3/mole
Alpha hexachlorocyclohexane	7.26[5]	-3.51[11]	6.9×10^{-6} atm-m^3/mole
Beta hexachlorocyclohexane	8.13[5]	-5.32[11]	2.35×10^{-9}
Chlordecone	8.92[12]	-6.69[6]	2.53×10^{-3}
Technical endosulfan and its related isomers	6.00 (α-endosulfan[5])	-2.72 (α-endosulfan[11])	1.06
Hexabromobiphenyl	13.6[12]	-4.62[6]	1.38×10^{-6} to 5.7×10^{-3}
Hexabromodiphenyl ether and heptabromodiphenyl ether (commercial octabromodiphenyl ether)	12.113 – 12.273[2]	-4.29[12]	1.88×10^{-7} – 7.48×10^{-8} atm-m^3/mole
Lindane	8.04[4]	-3.96[6]	6.1×10^{-5}
Pentachlorobenzene	6.539[2, 3]		7.03×10^{-4}
Perfluorooctane sulfonic acid, its salts and perfluorooctane sulfonyl fluoride	6.5 – 7.5[9]	$<2 \times 10^{-6}$ [10]	3.09×10^{-9} atm-m^3/mole
Tetrabromodiphenyl ether and pentabromodiphenyl ether (commercial pentabromodiphenyl ether)	10.04 (TBDE)[8] 11.2 (PBDE)	-3.67[6, 7]	1.18×10^{-6} – 3.54×10^{-6} atm-m^3/mole

[1]As defined by the Stockholm Convention (http://chm.pops.int/Home/tabid/2121/Default.aspx);
[2]Data from Li et al., 2006, all values for 20°C unless otherwise stated. [3]Data for 18.7°C.
[4]Data from Harner and Bidleman, 1998 (all data for 20°C). [5]Data from Shoeib and Harner, 2002 (all data for 25°C).[6]Data from Scheringer et al., 2006. [7]Data for BDE-99. [8]Data from Wania et al., 2002 (all data for 20°C). [9]Data from Shoeib et al., 2002 (all data for 20°C). [10]Data from POPRC, 2007. [11]Data for 25°C
[12]Calculated according to the method of Meylan and Howard (2005)

Table 1. Persistent Organic Pollutants[1]

Using the above modelling parameters, Wania (2003) developed an averaged global distribution model (the Globo-POP model) to predict both intermediate term (1 year ongoing emissions) and long term (10 year ongoing emissions) Arctic Contamination Potential (ACP). The ACP of most chemicals is sensitive to the temperature dependence of the partition coefficients, temperature, atmospheric mixing coefficients, and sea ice cover. The substances with significant ACP were in the range elevated ACP overlap in the range $6.5 < \log K_{OA} < 10$ and $-0.5 > \log K_{AW} > -3$, corresponding to chemicals that have significant ability to bioaccumulate. From the modelling, a parameter mACP, representing the potential for relative enrichment in the Arctic was developed.

$$mACP = [m_{T1} - m_{A1}]/m_{TG} \times 100\% \quad (1)$$

where m_{T1} and m_{A1} are the mass of chemical in all compartments and in the four atmospheric compartments of zone 1 (N-Polar) of the Globo-POP model, respectively. m_{TG} is the mass of chemicals in all model compartments. Wania (2006) later added a parameter for estimating absolute contamination of the Arctic environment (eACP):

$$eACP = [m_{T1} - m_{A1}]/e_{TG} \times 100\% \quad (2)$$

where e_{TG} is the mass of chemical emitted cumulatively to the global environment. m_{TG} will always be smaller than e_{TG} because of chemical losses by degradation and loss processes such as transfer to the deep sea, with subsequent loss to deeper sediment layers.

3.2 Atmospheric sorption

During transport, a persistent organic pollutant will always be subject to loss processes. These could include sorption onto particles with subsequent deposition, either by wet or dry deposition processes. Little is known of the relative importance of wet vs. dry deposition fluxes for persistent organic pollutants, although in a study of a number of sites in Canada, Yao et al. (2008) demonstrated that wet deposition was more important than dry deposition for most organochlorine pesticides. Clearly, there is a need for similar studies in other geographic regions. Moreover, it cannot be assumed that all scavenging will take place via rain, particularly in higher latitudes, where a significant amount of precipitation takes place via snowfall. Snow has been demonstrated to be a more efficient scavenger of airborne particulates than normal rainfall, probably largely because of the larger size and surface area of snowflakes. Franz and Eisenreich (1998) studied the scavenging of PCBs and PAHs by snow in Minnesota. Although anomalous, gas scavenging of the two pollutant classes was a relatively minor contributor to the overall snow removal flux. Particle scavenging ratios ranged from 5×10^4 to 5×10^7 for the snow events, as compared to 10^3 to 10^5 for rainfall events. Snow particle scavenging was slightly greater for the more volatile PCBs and PAHs. The authors hypothesized that lower molecular weight semi-volatile organic carbon compounds are associated with a different particle size spectrum than their less volatile counterparts.

The characteristics of air-particulate sorption for POPs are somewhat obscure and it is almost certain that the sorption characteristics of particulates from different sources will not be the same (cf. Gustafason and Dickhut, 1997). Once sorbed to a particle and carried to the earth by some form of deposition, a pollutant can be considered to have been removed from the global distillation process, at least temporarily (cf. Scheringer *et al.*, 2000). Such removal is particularly significant in the case of persistent organic pollutants, as a result of their ability to sorb to particulates. Removal of this kind does not preclude the pollutant from undergoing desorption and revolatilization at some later stage although it will cause a temporary disruption to the global transport process.

3.3 Atmospheric destruction

Atmospheric destruction of POPs is obviously another competing process and one with the potential to permanently remove the molecules from global transport processes. The mechanisms are generally less well understood than are those of destruction of volatile organics in air, but hydroxyl radicals, nitrate radicals, and ozone are believed to be involved (Boethling *et al.*, 2009). The potential for POPs to associate with atmospheric particles provides a further complication, although there is some evidence to suggest that sorbed POPs may still be subject to oxidation processes. In the case of atmospheric ozone and hydroxyl radical destruction, the half life ($t_{1/2}$)of a persistent organic pollutant is given by:

$$t_{1/2} = 0.693k_{Oxidant}[Oxidant] \quad (3)$$

where $k_{Oxidant}$ is the rate constant in units of cm^3 $molecule^{-1}$ s^{-1}, and [Oxidant] is the oxidant concentration in units of molecules (or radicals) cm^{-3}. On a global basis, the situation is complicated by the fact that the concentration of these oxidants vary from place to place and also exhibit seasonal as well as diel variation.

3.4 Aquatic sorption

When considering atmospheric volatilization and deposition processes, it is important to appreciate that the situation over water will differ from that over land. Over the oceans, the operative processes will be diffusive air-water exchange, as well as exchange involving wet and dry particulates. In terms of deposition, the latter terms will equate to wet and dry deposition. These processes will also be operative in the case of terrestrial systems but the situation will be complicated by air-soil exchange and air-vegetation exchange. Under normal circumstances in the oceans, air-water exchange is likely to be the dominant process (Dachs *et al.*, 2002). The air-water partition coefficient is related to the Henry's Law Coefficient by the following equation:

$$\frac{1}{k_{aw}} = \frac{1}{k_a H'} + \frac{1}{k_w} \quad (4)$$

where k_{AW} is the air-water partition coefficient, k_A and k_W are the POP mass transfer coefficients (m d^{-1}) in the air and water films, respectively and H′ is the temperature corrected and dimensionless Henry's law Constant. k_{AW} is known to be very sensitive to temperature, because of its dependence on the Henry's Law Constant, but it can also be affected by a number of other environmental variables such as wind speed. Therefore, the latitudinal variation of k_{AW} will also be a function of prevailing winds. The air-water flux for a persistent organic pollutant will be described by the equation:

$$F_{aw} = k_{aw} \left\{ \frac{C_a}{H'} - C_w \right\} \quad (5)$$

where C_A and C_W are the concentrations in the air and water phases respectively.

Once a persistent organic pollutant lands on the water surface, it is unlikely to dissolve because of its low inherent solubility. The pollutant will however be subject to processes of aquatic sorption by suspended particles, notably plankton. As with atmospheric deposition (considered above), the pollutant could theoretically later desorb and revolatilize. This is however less likely in the case of phytoplankron sorption, as the phytoplankton cell will eventually senesce and deposit in the benthic layer. Dachs *et al.*, (2002) modelled the phytoplankton uptake flux (F_{WP}) according to the equation:

$$F_{wp}=k_{wp}\left\{C_{w}-\frac{k_{dep}}{k_{up}}C_{P}\right\} \tag{6}$$

where k_{WP} is the water-phytoplankton mass transfer rate constant, k_{dep} and k_{up} are the depuration and uptake rate constants respectively and C_P is the concentration in the phytoplankton. The overall sinking flux for a POP, associated with phytoplankton (F_{sink}) could be calculated from:

$$\log F_{sink} = 1.8 \times F_{OC} \times C_p$$

$$\text{where } \log F_{OC} = 2.09 + 0.81 \log [\text{Chlorophyll}] \tag{7}$$

The authors suggested that the process would be of major significance in regions of ocean upwelling, where nutrient rich waters are brought to the surface, resulting in high primary productivity. In the overall transfer of POPs from the atmosphere to benthic sediment, the rate limiting processes will either be air-water transfer rate or the sinking rate. With POPs of log K_{OW} < 6.0, the sinking process will provide the rate determinant step in most cases, whereas for more lipophilic compounds, sinking fluxes are probably rate limiting at lower latitudes and air-water exchange at higher latitudes. These biogeochemical processes have the ability to disrupt global distillation processes over the oceans and as has already been mentioned, are probably important in terms of explaining the differences in global distillation between the northern and southern hemisphere. Supporting evidence in respect of the role of these plankton mediated biogeochemical controls in the Mediterranean Sea has recently been produced by Berrojalbiz *et al.*, (2011).

4. Climate change scenarios

Since 1990, considerable attention has been paid to climate change on an international basis, largely through the efforts of the Intergovernmental Panel on Climate Change (IPCC). To date, four assessment reports have been published, the most recent being in 2007 (IPCC, 2007a; IPCC 2007b; IPCC 2007c) and a fifth assessment report is expected to be published in due course. However, the topic of pollutant behaviour in response to climate change remains under-investigated. By far the most detailed studies of pollutant behaviour in response to climate change are those of MacDonald (MacDonald et al., 2002, MacDonald et al., 2003) which pertain to the Canadian Arctic. Although the majority of attention has been devoted to effects in the Northern Hemisphere, a recent review has considered possible climate change effects on pollutant behaviour in the Southern Hemisphere (Sadler et al., 2011). Further studies of the effects of selected climate change phenomena on behaviour of certain persistent organic pollutants have been carried out by some authors (Dalla Vale et al., 2007; Lamon et al., 2009, Ma and Cao, 2010).

When considering effects of a phenomenon such as climate change upon a complex process such as global transport, it is necessary to conduct the assessment with reference to as many variables s possible if a true picture is to be obtained. In a number of studies that have been conducted to date, there has been a tendency to consider temperature as the major driver associated with climate change. This is clearly not the case (cf. IPCC 2007a, IPCC 2007b). Climate change can be considered to consist of the following manifestations:

- Increases in land, water and air temperatures
- Increases in the intensity of extreme weather events
- Changes in salinization patterns within the oceans
- Rises in sea levels

In respect of temperature changes, it must be noted that although it is the most frequently discussed aspect of climate change, current predictions of global temperature rise are well below 10°C in 100 years. Most of these projections constitute relatively small changes and are unlikely on their own to have a major effect on pollutant behaviour. It is our view that more significant effects on pollutant behaviour (including ones mediated by temperature) will be occasioned by the increased intensity of extreme weather events. IPCC predictions point to significant changes in the intensity of heatwaves, tropical storms (whose range may well extend into areas currently free of such events). Such events have the potential to cause far greater changes in environmental parameters, notably temperature albeit over a shorter period. Hence, the modelling of climate change mediated effects becomes a far more complex process.

Another frequently neglected aspect of climate change is its potential to alter physical conditions leading to changes in both production environments and also receiving environments. This may be reflected in a number of ways, including patterns of use. Persistent organic pollutants such as DDT have been banned in many countries and hence their evolution rates declined significantly in the recent past. But the effects of warming have seen the occurrence of diseases such as Malaria increase in geographical scope in some countries (notably Africa). This has occasioned a reintroduction of these pesticides and hence a potential for increased evolution rates (cf. WHO 2009).

Equally, the warming effects associated with retreat of the polar ice caps and glaciers have major potential to alter the scenarios as regards persistent organic pollutant sorption. The air-ice interface has been recognized as a major site of sorption for these pollutants and loss of this interface will almost certainly lead to a remobilization of pollutants both from polar and high altitude regions. Thus, even if global/mountain transport processes continue to operate at their present or some altered rate, persistent organic pollutants delivered to these sites cannot be expected to remain in their traditional niches.

Intense extreme weather events also have the potential to cause a significant export of nutrients from land and offering the potential for increased primary productivity in the oceans (Sadler et al., 2011). As has already been pointed out above and will be discussed more fully below, this too has significant ability to affect global transport phenomena. Glacial melt can also make a significant similar contribution in this type of area, with the potential to also transport persistent organic pollutants released from the ice.

It follows from the foregoing remarks that a detailed consideration of the processes involved in global transport of pollutants is required if a meaningful assessment of climate change effects is to be made. Whilst temperature will be an important driver for all these processes, it or any other factor cannot be considered in isolation.

4.1 Effects on modelling parameters

As has already been outlined, the most simple parameters associated with global distillation modelling are the octanol-air partition coefficient, the air-water partition coefficient and the Henry's Law Constant. The octanol-air partition coefficient is known to vary log-linearly with temperature. Shoeib and Harner (2002) measured octanol-air partition coefficients for a range of persistent organic pollutants, over the range 278-308°K and plotted the response as log K_{OA} vs 1000/T (°K). From their data, they concluded that it was possible to calculate the variation in log K_{OA} from a simple regression equation. Using this equation, the variation in log K_{OA} for a 5°K and a 10°K temperature change as associated with global warming scenarios has been calculated for a number of persistent organic pollutants and is shown in Table 2.

POP	293°K (20°C)	298°K (25°C)	303°K (30°C)
HCB	7.55	7.38	7.23
α-HCH	7.80	7.61	7.43
γ-HCH	8.04	7.85	7.66
Heptachlor	7.84	7.64	7.45
Aldrin	8.29	8.08	7.87
trans-Chlordane	9.16	8.87	8.59
cis-Chlordane	9.21	8.92	8.63
α-Endosulfan	8.88	8.64	8.40
Dieldrin	9.11	8.90	8.69
Endrin	8.38	8.13	7.89
p, p-DDT	10.08	9.82	9.56

[1]Data calculated using the regression equation of Shoeib and Harner 2002.

Table 2. Variation in log K_{OA} values for Temperature Rises associated with Global Warming[1]

It is readily seen that the temperature rises associated with climate change (and the values given in the above table are on the extreme side of predictions) cause relatively small changes in the octanol-air partition coefficient. In general, rising temperatures will tend to decrease sorption of persistent organic pollutants on particulates. But it must be remembered that the changes referred to above are also well within the range of uncertainty associated with determinations of octanol-air partition coefficients and temperature.

It must also be remembered that the majority of observations, linking K_{OA} to particle sorption pertain to urban situations (cf. Radonić et al., 2011). The nature of the particles with which persistent organic pollutants will interact in non-urban situations may be entirely different, particularly in a climate change scenario and the subject may warrant some further specific consideration. Predictions of climate change scenarios generally include an increased frequency of intense dust storms and although currently, the majority of pollutants appear to exist in the gaseous phase, there have been contrary reports from some agricultural areas (cf. Yao et al., 2008). There is a need to obtain more information on the partitioning behaviour of pesticides with dust derived from agricultural soils, which will almost certainly form the majority of climate change associated dust.

Although it is possible to make theoretical estimates of log K_{AW} (cf. Table 1, Meylan and Howard, 2005), in the actual environment a number of parameters are known to affect the actual value of this parameter. Temperature plays an important role in affecting K_{AW}, but the parameter is also subject to other influences, notably wind speed and various diffusion/transport variables (cf. Dachs et al., 2002). Prediction of future wind speed changes with current state of computer power remains a difficult task and thus a subject of active research. While definite trend changes in severe cyclonic wind intensity and frequency have not yet established, preliminary research results suggest that significant alteration in cyclonic wind intensity and frequency are possible. In terms of non-cyclonic wind intensity, there appears good evidence to suggest significant increases, based upon modelling using the IPCC A1F1 scenario (temperature change +2.4-6.4°C, best estimate +4°C, over the final decade of the 21st century vs. the final decade of the 20th century) (CSIRO 2007).

In terms of equation 2, it can be shown that both k_A and k_w are sensitive to wind speed. The equation is clearly sensitive to changes in Henry's Law Coefficient, which is known to respond to temperature (see below). Obviously, climate change scenarios would be expected to involve changes of temperature. But even in the absence of any temperature effect, it can be shown that a 5% increase in wind speed (considered an upper best estimate by CSIRO (2007) for the next 50 years) can have a noticeable effect on the value of k_{AW}. Later in the 21st century, there may be greater increases in wind speed, depending on the forcings operative at the time.

That temperature has a major effect on the value of the Henry's Law Coefficient has been demonstrated both experimentally and by calculation (Staudinger and Roberts, 2001). The variation of Henry's Law Coefficient with temperature is given by:

$$H_t' = \left\{H_{293^\circ K}\right\}\left\{10^{-\frac{\Delta H^0}{R}\left[\frac{1}{T}-\frac{1}{293}\right]}\right\} \quad (8)$$

where H_T' is the value of H′ at a given temperature, T, ΔH^0 is the enthalpy of phase change ($J\ mol^{-1}$) and R is the universal gas constant (8.314 $J\ mol\ K^{-1}$).

The variation of H′ with temperature is not uniform for all environmental contaminants. On the basis of an extensive literature survey, Staudinger and Roberts (2001) estimated the average % rise in Henry's Law Constant/10°C to be 60% in the case of hydrocarbons, 90% in the case of miscellaneous substances and 140% in the case of chlorinated organic pesticides and PCBs. Unfortunately, insufficient data of this kind are currently available in the case of the more recent additions to the list of persistent organic pollutants. Kűhne et al., (2005) suggested that temperature variation of Henry's Law Coefficient is maximal for compounds that are polar or have significant hydrogen bond interaction capacity. In this case, it could be expected that the temperature variation of the more recent additions to the POPs register would be significant.

The effect of a 5°C temperature rise on the temperature corrected Henry's Law constant for a number of persistent organic pollutants is shown in Table 3. These values were derived using the USEPA's on-line calculator (USEPA 1996).

Because of its effects on modelling parameters, notably k_{AW}, the temperature effect on the Henry's Law Constant is probably the largest single effect that climate change is likely to exert. From the data presented above and Equation 4, it can be seen that the difference caused by even a 5°C rise of temperature would have a significant effect on the Henry's Law constant and also on dependent parameters such as k_{AW}. Although the temperature effect is the most prominent influence of all on the Henry's Law Constant, a number of other climate change factors have the ability to affect this parameter as well. These have recently been discussed (Sadler et al., 2011) and although their effects will probably be secondary to the temperature perturbations associated with climate change, they may be of significant in some instances. Of special relevance to climate change scenarios is the effect of suspended particles, which would tend to offset the temperature increases in the Henry's Law Constant. These will be further discussed below.

POP	H' at 20°C	H' at 25°C	Ratio of H' at 25°C/H' at 20°C (%)
HCB	0.0310	0.0540	174
α-HCH	0.000241	0.000433	237
γ-HCH	0.000319	0.000572	179
Heptachlor	0.0268	0.0446	166
Aldrin	0.00384	0.00695	181
α-Endosulfan	0.000261	0.000458	198
Dieldrin	0.000312	0.000617	198
Endrin	0.000165	0.000307	186
p, p-DDT	0.000139	0.000331	238

[1]Data obtained using on-line calculator (USEPA, 1996)

Table 3. Variation in H' values for a 5°C Temperature Rise associated with Global Warming[1]

4.2 Modelling studies relating to effects of climate change on global transport

To date, only a few specific studies relating climate change to effects on global transport have been published, although it is possible to make extrapolations from other published work. Lamon et al., (2009) examined the effects of climate change on global levels of PCBs. Using two IPCC scenarios, they were able to demonstrate that increased temperature would probably be the major driver as regards climate change effects on PCB transport. Higher temperatures were considered to drive increased primary and secondary volatilization emissions of PCBs and enhance transport from temperate regions to the Arctic. The largest relative increase in concentrations of both PCBs in air was predicted to occur in the high Arctic and the remote Pacific Ocean. Higher wind speeds were predicted to result in more efficient intercontinental transport of PCB congeners.

Ma and Cao (2010) developed a perturbed air-surface coupled to simulate and predict perturbations of POPs concentrations in various environmental media under given climate change scenarios. Their studies pertained to α- and γ-HCHs, HCB, PCB-28 and -153. The HCHs, HCB and OCB 153, showed strong perturbations as regards emissions under climate change scenarios. The largest perturbation was found in the soil-air system for all these chemicals with the exception of α-HCH, which showed a stronger perturbation in the water-air system. The study also included terms for degradation, as would be operative under climate change scenarios. As would be expected, climate change also has a significant effect on the degradative pathways for pollutants, potentially undergoing global transport. The scenarios investigated resulted in a maximum 12% increase of HCB and a 10% increase of α-HCH levels in air during the first year of disturbed conditions. Following the increases during the first several years, the perturbed atmospheric concentrations tend to result in decreased air levels due to degradation processes. It is clear from this study that simple modelling of global transport cannot be undertaken, without a proper consideration of all factors that might be affected by climate forcings. As has already been noted in this review, the so-called global distillation is only one pathway available for transport and fate of persistent organic pollutants. Its relative importance may well change under some climate change scenarios.

4.3 A possible role for phytoplankton

All of the above studies pertain to Northern Hemisphere situations. As has already been pointed out above, the role of deposition on the oceans provides an important sink for persistent organic pollutants and is likely to be operative particularly in the Southern Hemisphere (cf. Dachs et. al. 2002). Jurado et al., (2004) noted the importance of phytoplankton, especially in areas of high primary productivity as regards controlling levels of PCBs in the oceans. They stressed the need for further research, particularly as regards seasonal influences on phytoplankton populations and the associated biogeochemical cycles.
The effect of climate change on phytoplankton levels has been the subject of a number of papers in recent years (Hayes et al., 2005; Hallegraeff et al., 2009; Sadler et al., 2011). All of these studies have identified the need for intense ongoing monitoring. There is a general belief that the effects of climate change will see an increase in the frequency and intensity of blooms, an expansion of the geographical range and seasonal window for warm water species. The general concern expressed by these studies has been a possible increase in toxic algae and the attendant public health problems. But it is equally true that these projected increases could translate into increased sinking fluxes of persistent organic pollutants, deposited on the ocean surface.

In order to appreciate the magnitude of what phytoplankton blooms could achieve in terms of removal of persistent organic pollutants, it is necessary to consider the magnitude of the F_{sink} term (Equation 7). Dachs et al.,(2002) calculated sinking fluxes on the basis of chlorophyll levels in oceans, as recorded by satellite imaging. Although chlorophyll concentrations are a less than satisfactory indicator of algal biomass, under the circumstances, it must be conceded that this was probably the best available indicator.

Seasonal growth of algae represents what is probably the normal situation at present, but given the fact that blooms will become more frequent with climate change, it is important to examine the relative ability of such occurrences in terms of capturing persistent organic pollutants. For comparison, therefore, data on two *Trichodesmium* bloom events are included. This organism has been chosen because of its adaptive abilities (cf. Bell et al., 2005) which make it an ideal candidate to successfully proliferate under climate change conditions. As is shown in the table below, bloom situations could effectively remove four to six times as much persistent organic pollutant than the relatively static situation reported by Dachs et al., (2002). This means that the phytoplankton blooms which are expected to result from climate change, particularly in the Southern Hemisphere provide an important and hitherto neglected contribution to the overall observed global transport of pollutants.

The possibility also exists that in the Arctic Region, with the retreat of the Arctic ice cover, phytoplankton blooms will occur in the resulting ocean, providing a sink for released pollutants (see next section). There is already evidence of earlier algal blooms in the Arctic Region (Kahru et al., 2010) but although considerable effort has been devoted to effects of climate change on higher vertebrates in the Arctic, there remains a significant gap of knowledge with respect to dynamics of phytoplankton (Wassmann et al., 2011).

Scenario	Reference	No of Phytoplankton Cellsx $10^3 L^{-1}$	[Chlorophyll] ($mg\ m^{-3}$)	F_{OC}[1]
Ocean at 60-75 S	Dachs et al (2002)	-	0.1-18	21-1068
Trichodesmium bloom, Arabian Sea	Desa et al., 2005	690.9	126.75	6214
Trichodesmium bloom Moreton Bay, Queensland	Queensland Department of Environment and Resource Management, unpublished Data	500 (calculated from number of trichomes, assuming 100 cells per trichome)	87	4581

[1]log FOC = 2.09 + 0.81 log [Chlorophyll]

Table 4. Phytoplankton and Organic Carbon Sinking Fluxes

The end product of this type of pollutant sink will almost certainly be an increase in the levels of persistent organic pollutants in benthic communities. Results such as those obtained by Berrojalbiz et al., (2011) tend to highlight the importance of phytoplankton processes as sinks for persistent organic pollutants. With regard to climate change phenomena, a full discussion of causal factors for algal blooms is given elsewhere (Hayes et al., 2005; Hallegraeff et al., 2009; Sadler et al., 2011). It will be particularly important to pinpoint sites of upwellings. Upwellings are of two kinds:

1. Upwellings caused by trade winds.
2. Upwellings caused by extreme weather events such as cyclones (hurricanes, typhoons).

Both types have the potential to be affected by climate change. Upwellings caused by trade winds are known to decrease during El Niňo events and increase during La Niňa events. With the expected accentuation of these events as a result of climate change (IPCC 2007a) it could be expected that changes in the intensity and patterns of these upwellings would be observed. Similarly, the expected increase in intensity of cyclones (hurricanes, typhoons) with climate change will have an influence on the second type of upwelling. It is significant that some surveys of POPs in ocean transects have shown increased levels in waters off the west coast of Africa – a major site for ocean upwellings (Gioia et al., 2008).

4.4 Fate of transported pollutants

It is established that both Arctic cold trapping and mountain cold trapping lead to an accumulation of pollutants in the coldest regions, although the actual pattern of pollutant retention is different (Daly and Wania, 2005). To date however, there seems to have been a tacit assumption that climate change has the potential to alter these transport processes, but that any pollutants reaching the usual site of accumulation will be retained in the normal manner. Such a view is clearly untenable, particularly in respect of the Arctic, where significant ice melt has already taken place. It is known that the air-ice interface provides a particularly strong site for sorption of persistent organic pollutants (Hoff et al., 1995) and hence some examination of the fate of cold trapped pollutants is warranted. In fact, regions of the Arctic are particularly sensitive to changes in temperature as the Greenland Ice Sheet is normally only very close to freezing point (Archer and Rahmstorf, 2010).

Daly and Wania (2005) noted that concentrations of persistent organic pollutants in air and water increased during periods of melt activity. More recently, Stocker et al., (2007) modelled the effect of snow and ice on distribution of semivolatile organic compounds, with particular reference to long range transport. It was concluded that polar ice receives most chemicals from snow precipitation, although HCH and PBDEs are largely brought to the Arctic region in a particle-bound form. The polar ice caps serve to prevent air-water exchange of gaseous pollutants and the major removal pathway of pollutants in surface ice is via transport to lower layers of the ice cap. Because of its open nature, snow offers a greater potential for both diffusion and revolatilization than does ice in the polar regions. Therefore, the loss of ice from both polar and mountain regions has the potential to bring about a redistribution of pollutants already in this layer. It can be expected that the pollutants will largely move to the aquatic phase in the first instance, although more severe warming could lead to increases in the air phase as well. In the case of the Arctic, the pollutants released to the aquatic compartment will have a significantly higher potential to enter food webs. Pollutants entering mountain meltwater will be carried downstream and it is noteworthy that a number of major river-systems throughout the world have a glacial origin.

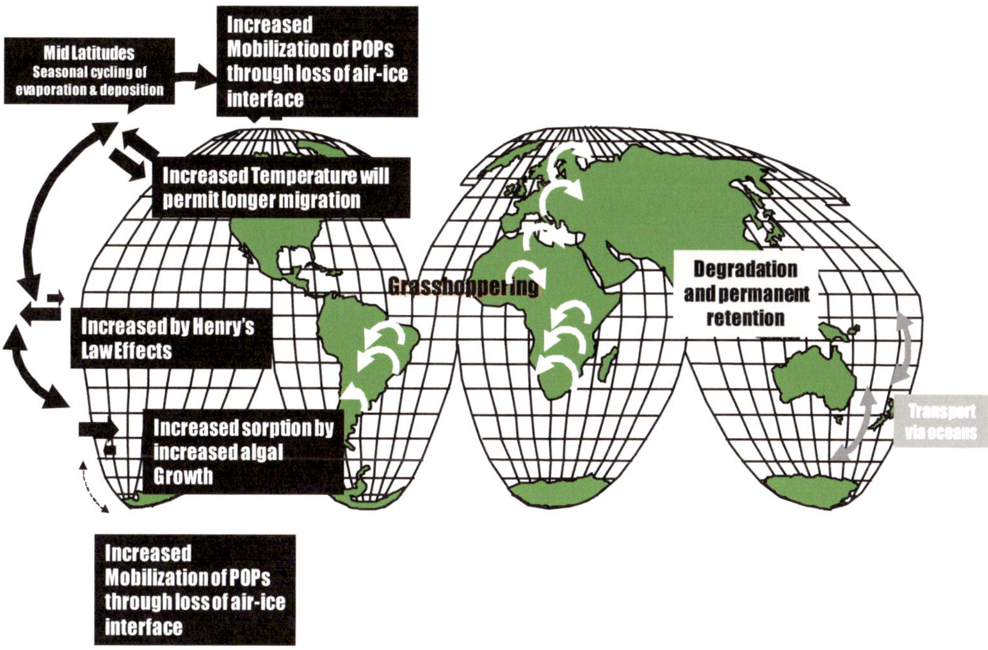

Fig. 3. Global Transport Processes for POPs as modified by Climate Change

5. Overall conclusions

The topic of global distillation has been reviewed with respect to both Northern and Southern Hemispheres. It appears that the global distillation phenomenon is best suited for operation in the Northern Hemisphere, although even here, it may be subject to significant competition from other processes. It is hypothesized that climate change will bring about significant interference with the operation of global distillation in both hemispheres. Although the increased temperatures associated with climate change have the potential to result in increased volatilization of persistent organic pollutants, the altered conditions may lead to increased losses of pollutants from the atmosphere through increased availability of hydroxyl radicals and other species active in the destruction of these substances in the atmosphere. There will be decreases in snow and rain scavenging in many areas, as a result of changes in precipitation patterns and a higher probability of particulate sorption in the atmosphere. The increase in phytoplankton blooms that is predicted to accompany climate change will provide a significant sink for persistent organic pollutants. Finally, the loss of ice at the poles will result in a redistribution of pollutants that have already been transport there, or those which subsequently arrive as a result of long range transport. The accompanying figure (Figure 3) provides a summary of the expected effects of climate change on global transport of persistent organic pollutants.

6. References

Archer, D., and Rahmstorf, S., 2010. *The Climate Crisis*. Cambridge University Press, ISBN 978-0-521-73255-0 UK.

Bell, P. R. F., Unwins, P. J. R., Elnmetri, I., Phillips, J. A., Fu, F-X. & Yago, A. J. E. (2005). Laboratory Culture Studies of *Trichodesmium* isolated from the Great Barrier Reef Lagoon, Australia. *Hydrobiologia*, Vol. 532, pp. 9-21, ISSN: 0018-8158

Berrojalbiz, N., Dachs J., Del Vento, S., Ojeda, M. J., Valle, M. C., Castro-Jiminéz, J., Mariani, G., Wollgast, J. & Hanke, G. (2011). Persistent Organic Pollutants in Mediterranean Seawater and Processes Affecting Their Accumulation in Plankton. *Environmental Science and Technology*, Vol. 45, pp. 4315-4222, ISSN 0013-936X.

Boethling, R., Fenner, K., Howard, P., Klečka, G., Madsen, T., Snape J. R. and Whelhan, M. J. (2009). Environmental Persistence of Organic Pollutants: Guidance for Development and Review of POP Risk Profiles. *Integrated Environmental Assessment and Management*, Vol. 5, pp. 539-556, ISSN 1551-3793.

Carrera, G., P. Fernandez, Grimalt, J.O., Ventura, M., Camarero, L., Catalan, J. Nickus, U., Thies H. & Psenner, R., 2002. Atmospheric Deposition of Organochlorine Compounds to Remote High Mountain Lakes of Europe. *Environmental Science and Technology*, Vol 36, pp. 2581-2588, ISSN 0013-936X.

Connell, D.W. (1990). *Bioaccumulation of Xenobiotic Compounds*. CRC Press, ISBN: 0849348102, Boca Raton, FL, USA.

Corsolini, S., Kannan, K. Imagawa, T. Focardi, S. & Geisy, J.P. (2002). Polychloronaphthalenes and Other Dioxin-like Compounds in Arctic and AntArctic Marine Food Webs. *Environmental Science and Technology*, Vol. 36, pp. 4229-4237, ISSN 0013-936X.

CSIRO (2007). Climate Change in Australia. ISBN 9781921232947. CSIRO, Canberra.

Dachs, J., Lohmann,R., Ockenden, W., Méjanelle, L., Eisenreich, S.J & Jones, K.C. (2002). Oceanic Biogeochemical Controls on Global Dynamics of Persistent Organic Pollutants. *Environmental Science and Technology*, Vol. 36, pp. 2851-2858, ISSN 0013-936X.

Dalla Vale, M., Codato, E. & Marcomini. A.(2007). Climate change influence on POPs distribution and fate: A case study. *Chemosphere*, 67, 1287-1295, ISSN: 0045-6535.

Daly, G. L. & Wania, F. (2005). Organic Contaminants in Mountains. *Environmental Science and Technology*, Vol. 39, pp. 385-398, ISSN 0013-936X.

Davidson D., Wilkinson, A.C., Blais, J.M., Kempe, L., McDonald, K.M. & Schindler, D.W. (2003). Orographic Cold-Trapping of Persistent Organic Pollutants by Vegetation in Mountains of Western Canada. *Environmental Science and Technology*, Vol. 37, pp. 209-215, ISSN 0013-936X.

Demers, M.J. Kelly, E.N. Blais, J.M. Pick, F.R. St Louis, V.L. & Schindler, D.W. (2007). Organochlorine Compounds in Trout from Lakes over a 1600 Meter Elevation Gradient in the Canadian Rocky Mountains. *Environmental Science and Technology*, Vol. 41, pp. 2723-2729, ISSN 0013-936X.

Desa, E., Suresh, T., Matondkar, S .G. P, Desa, E., Goes, J., Mascarenhas, A., Parab, S.G., Shaikh, N. & Fernandes, C. E. G. (2005). Detection of *Trichodesmium* Bloom Patches along the Eastern Arabian Sea by IRS/P4 OCM Ocean Colour Sensor and by in-site Measurements. *Indian Journal of Marine Sciences*, Vol. 34, pp. 374-386 ISSN: 0379-5136.

Dickhut, R.M., Padma, T.V. & Cincinelli, A. (2004). Fractionation of Stable Isotope-Labeled Organic Pollutants as a Potential Tracer of Atmospheric Transport Processes. *Environmental Science and Technology*, Vol.38, pp. 3871-3876, ISSN 0013-936X.

Fernandez, P. & Grimalt, J.O. (2003). On the global distribution of persistent organic pollutants. *Chimea*, Vol. 57, pp. 514 - 521, ISSN: 1080-6059

Focardi, S., Gaggi, C., Chemello, G. & Bacci, E. (1991). Organochlorine residues in moss and lichen samples from two AntArctic areas. *Polar Record*, Vol. 27, pp. 241-244, ISSN: 0032-2474.

Franz, T.P. & Eisenreich, S.J. (1998). Snow Scavenging of Polychlorinated Biphenyls and Polycyclic Aromatic Hydrocarbons in Minnesota. *Environmental Science and Technology*, Vol. 32, pp. 1771-1778, ISSN 0013-936X.

Gallego, E., Grimalt, J.O., Bartrons, M.,. Lopez, J.F., Camarero, L., Catalan, J., Stuchlik, E. & Battarbee, R. (2007). Altitudinal Gradients of PBDEs and PCBs in Fish from European High Mountain Lakes. *Environmental Science and Technology*, Vol. 41 pp., 2196-2201, ISSN 0013-936X.

Gobas, F. & Maclean, L.G. (2003). Sediment- water distribution of organic contaminants in aquatic ecosystems - the role of organic carbon mineralization. *Environmental. Science and Technology*, Vol. 37, pp, 735- 741, ISSN 0013-936X.

Grimalt, J.O., Borghini, F., Sanchez-Hernandez, J.C., Barra, R., Torres Garcia, C. J. & Focardi, S. (2004). Temperature Dependence of the Distribution of Organochlorine Compounds in the Mosses of the Andean Mountains. *Environmental Science and Technology*,Vol. 42, pp. 1416-1422, ISSN 0013-936X.

Gioia, R., Nizzetta, L., Lohmann, R., Dachs, J., Temme, C. & Jones, K.C. (2008). Polychlorinated Biphenyls (PCBs) in Air and Seawater of the Atlantic Ocean: Sources, Trends and Processes. *Environmental Science and Technology* Vol. 38, pp. 5386-5392, ISSN 0013-936X.

Gustafason, K.E. & Dickhut, R.M. (1997). Response to Comment on "Particle/Gas Concentrations and Distributions of PAHs in the Atmosphere of Southern Chesapeake Bay". *Environmental Science and Technology*, Vol. 31, pp. 3738-3739, ISSN 0013-936X.

Hallegraeff G., Beardall, J., Brett, S., Doblin M., Hosja, W., de Salas, M. and Thompson, P. Marine Climate Change in Australia. Impacts and Adaptation Responses. 2009 Report Card. Phytoplankton. In A Marine Climate Change Impacts and Adaptation Report Card for Australia 2009 (Eds. E.S. Poloczanska, A.J. Hobday and A.J. Richardson), NCCARF Publication 05/09, ISBN 978-1-921609-03-9, Gold Coast, Australia.

Hargrave, B. T., Barrie, L.A., Bidleman, T.F. & Welch, H.E. (1997). Seasonality in Exchange of Organochlorines between Arctic Air and Seawater. *Environmental Science and Technology*, Vol. 31, pp. 3258-3266, ISSN 0013-936X.

Harner, T & Bidleman, T.F. (1998). Measurement of octanol-air partition coefficients for polycyclic aromatic hydrocarbons and polychlorinated naphthalenes. *Journal of Chemical and Engineering Data*, Vol. 43, pp. 40-46, ISSN 0021-9568.

Hayes G. C., Richardson, A. J. & Robinson, C (2005). Climate Change and Marine Phytoplankton. *Trends in Ecology and Evolution*, Vol. 20, pp. 337-344, ISSN: 0169-5347.

Hoff, J. T., Wania, F., Mackay, D. And Gilham, R. (1995). Sorption of Nonpolar Organic Vapors by Ice and Snow. *Environmental Science and Technology*, Vol. 29, pp. 1982-1989, ISSN 0013-936X.

Hung, H., Halsall, C.J., Blanchard, P., Li, H.H., Fellin, P., Stern, G. & D.B. Rosenberg, D.B. (2001). Are PCBs in the Canadian Arctic Atmosphere Declining? Evidence from 5 Years of Monitoring. *Environmental Science and Technology*, Vol. 35, pp. 1303-1311, ISSN 0013-936X.

IPCC 2007a. Intergovernmental Panel on Climate Change (IPCC). 2007. Fourth assessment report. Climate Change 2007 - The Physical Science Basis. Cambridge, ISBN 978 0521 88009-1, Cambridge University Press.

IPCC 2007b. Intergovernmental Panel on Climate Change (IPCC). 2007. Fourth assessment report. Climate Change 2007 -Impacts, Adaptation and Vulnerability. Cambridge, ISBN 978 0521 88010-7, Cambridge University Press.

IPCC 2007c. Intergovernmental Panel on Climate Change (IPCC). 2007. Fourth assessment report. Climate Change 2007 - Mitigation of Climate Change. Cambridge, 978 0521 88011-4, Cambridge University Press.

Kahru, M., Brotas, V., Manzano-Sarabia, M., and Mitchell, B. G. (2010). Are Phytoplankton Blooms occurring earlier in the Arctic? *Global Change Biology*, Vol. 17, pp. 1733-1739, ISSN 1365-2486.

Kalantzi, O.I., Alcock, R.E., Johnston, P.A., Santillo, D., Stringer, R.L.,. Thomas, G.O. & Jones, K.C. (2001). The Global Distribution of PCBs and Organochlorine Pesticides in Butter. *Environmental Science and Technology*, Vol. 35, pp. 1013-1018, ISSN 0013-936X.

Klánová, J., Matykiewiczová, N., Máčka, Z., Prošek, P., Láska, K. & Klán, P. (2008). Persistent Organic Pollutants in Soils and Sediments from James Ross Island, Antarctica. *Environmental Pollution*, Vol. 152, pp. 416-423, ISSN 0269-7491.

Kűhne, R., Ebert, R-U. & Schűűrmann, G. (2005). Prediction of the Temperature Dependency of Henry's Law Constant from Chemical Structure. *Environmental Science and Technology*, Vol. 39, pp. 6705-6711, ISSN 0013-936X.

Jurado, E., Lohmann, R.,. Meijer, S., Jones, K.C. and Dachs, J. (2004). Latitudinal and Seasonal Capacity of the Surface Oceans as a Reservoir of Polychlorinated Biphenyls. *Environmental Pollution* 128, 149-162. ISSN 0269-7491.

Lakaschus, S., Weber, K., Wania, F., Bruhn, R. & Schrems, O. (2002). The Air-Sea Equilibrium and TimeTrend of Hexachlorocyclohexanes in the Atlantic Ocean between theArctic and Antarctica. Environmental Science and Technology 36: 138-145, ISSN 0013-936X.

Lamon, L., von Waldow, H., Macleod, M., Scheringer, M., Marcomini, A. & Hungerbühler, K. (2009). Modeling the Global Levels and Distribution of Polychlorinated Biphenyls in Air under a Climate Change Scenario. *Environmental Science and Technology*, Vol. 43, pp. 5818-5824, ISSN 0013-936X.

Lead, W.A., Steinnes, E. & Jones, K.C. (2002). Atmospheric Deposition of PCBs to Moss (*Hylocomium splendens*) in Norway between 1977 and 1990. *Environmental Science and Technology* 30: 524-530, ISSN 0013-936X.

Li, X., Chen, J., Zhang, L., Quiao X. & Huang, L. (2006). The Fragment Constant Method for Predicting Octanol-Air PartitionCoefficients of Persistent Organic Pollutants at Different Temperatures. *Journal of Physical Chemistry* (Reference Data), Vol. 35, pp. 1366 - 1384, ISSN: 1520-6106.

Ma, J. (2010). Atmospheric transport of persistent semivolatile organic chemicals to the Arctic and cold condensation in the mid-troposphere - Part 1: 2-D modelling in mean atmosphere. *Atmospheric Chemistry and Physics*, Vol. 10, pp. 7303- 7314, ISSN ISSN 1680-7316.

Ma, J. & Cao, Z. (2010). Quantifying the Perturbations of Persistent Organic Pollutants Induced by Climate Change. *Environmental Science and Technology*, Vol. 44, pp. 8657-8573, ISSN 0013-936X.

Macdonald, R.W., Mackay, D. & Hickie, B. (2002). Contaminant amplification in the environment: Revealing the fundamental mechanisms. *Environmental Science and Technology*, Vol. 36, pp.: 457A–462A, ISSN 0013-936X.

Macdonald, R.W., Mackay, D., Li, Y-F & Hickie, B. (2003). How will global climate change affect risks from long range transport of persistent organic pollutants? *Ecological Risk Assessment*, Vol. 9, pp. 643-660, ISSN 1811-0231.

Mackay, D., Arnot, J.A., Webster, E. and Reid, L. (2009). The Evolution and Future of Environmental Fugacity Models. In *Ecotoxicology Modelling* . Devillers. J. Ed. Springer, pp 355 - 375, ISBN: 1441901965, Berlin, Germany.

McConnell, L., Kucklick, J.R., Bidleman, T.F., Ivanov, G.P. & Chernyak, S.M. (1996). Air-Water Gas Exchange of Organochlorine Compounds in Lake Baikal, Russia. *Environmental Science and Technology*, Vol. 30, pp. 2975-2983, ISSN 0013-936X.

Meijer, S.N., Steinnes, E., Ockenden, W.A. & Jones, K.C. (2002). Influence of Environmental Variables on the Spatial Distribution of PCBs in Norwegian and U.K. Soils: Implications for Global Cycling. *Environmental Science and Technology*, Vol. 36, pp. 2146-2153, ISSN 0013-936X.

Meylan, W.M & Howard, P.H. (2005). Estimating Octanol–Air Partition Coefficients with Octanol–Water Partition Coefficients and Henry's Law Constants. *Chemosphere*, Vol. 61, pp. 640-644, ISSN: 0045-6535.

Moeckel, C., MacLeod, M., Hungerbűhler, K & Jones, K.C. (2008). Measurement and Modeling of Diel Variability of Polybrominated Diphenyl Ethers and Chlordanes in Air. *Environmental Science and Technology*, Vol. 42, pp. 3219–3225, ISSN 0013-936X.

Muir, D.C.G., Omelchenko, A., Grift, N.P., Savoie, D.A., Lockhardt, W.L., Wilkinson, P. & Brunskill, G.J. (1996). Spatial Trends and Historical Deposition of Polychlorinated Biphenyls in Canadian Midlatitude and Arctic Lake Sediments. *Environmental Science and Technology*, Vol. 30, pp. 3609–3617, ISSN 0013-936X.

Nizzetto, L., Lohmann, R, Gioia, G., Jahnke, A., Temme, C., Dachs, J., Herckes, P., Diguardo, A. & Jones, K.C. (2008). PAHs in Seawater along a North-South Atlantic Transect: Trends, Processes and Possible Sources. *Environmental Science and Technology*, Vol. 42, pp. 1580 - 1585, ISSN 0013-936X.

Noël, M., Barrett-Lennard, L., Guinet, C. Dangerfield, N. & Ross, P. S. (2009). Persistent organic pollutants (POPs) in killer whales (*Orcinus orca*) from the Crozet Archipelago, Southern Indian Ocean. *Marine Environmental Research*, Vol. 68, pp 196-202, ISSN: 0141-1136.

Radonić, J., Miloradov, M. V., Sekulić, M. T., Kiruski, J., Djogjo, M. & Milanovanović, D. (2011). The Octanol–Air Partition Coefficient, K_{OA}, as a Predictor of Gas–Particle Partitioning of Polycyclic Aromatic Hydrocarbons and Polychlorinated Biphenyls at Industrial and Urban Sites. *Journal of the Serbian Chemical Society*, Vol. 76, pp. 1-12, ISSN 0352-5139.

Sadler R., Gabric, A., Shaw, G., Shaw, E. & Connell, D. (2011). An opinion on the distribution and behaviour of chemicals in response to climate change, with particular reference to the Asia-Pacific Region. *Toxicological and Environmental Chemistry*, Vol. 93, pp. 1-29, ISSN: 0277-2248.

Scheringer, M., Wegman, F., Fenner, K. & Hungerbűhler, K. (2000). Investigation of the Cold Condensation of Persistent Organic Pollutants with a Global Multimedia Fate Model. *Environmental Science and Technology*, Vol. 34, pp. 1842-1850, ISSN 0013-936X.

Scheringer, M., M. McLeod, and F. Wegmann, 2006. Analysis of Four Current POP Candidates with the OECD Pov and LRTP Screening Tool. http://www.sust-chem.ethz.ch/docs/POP_Candidates_OECD_Tool.pdf

Shiu, W.Y. & Mackay, D. (1986). A critical review of aqueous solubilities, vapour pressures. Henrys Law constants and octanol - water partition coefficients of polychlorinated biphenyls. *Journal of Physical Chemistry. Reference Data*, Vol. 15, pp. 911 - 929, ISSN 1520-6106.

Shoeib, M. & Harner,T. (2002). Using Measured Octanol-Air Partition Coefficients to Explain Environmental Partitioning of Organochlorine Pesticides. *Environmental Toxicology and Chemistry*, Vol. 21, pp. 984-990, ISSN 1552-8618.

Shoeib, M., T. Harner and K. Kannan, 2002. Octanol-Air Partition Coefficients and Indoor Air Measurements of PFOS and Precursor Compounds. SETAC Abstracts P212. http://abstracts.co.allenpress.com/pweb/setac2002/document/20575

Simcik, M., Basu, I., Sweet, C. & Hites, R. A., (1999). Temperature Dependence and Temporal Trends of Polychlorinated Biphenyl Congeners in the Great Lakes Atmosphere. *Environmental Science and Technology*, Vol. 33, pp. 1991-1995, ISSN 0013-936X.

Staudinger, J. & Roberts, P.V. (2001). A Critical Compilation of Henry's Law Constant Temperature Dependence Relations for Organic Compounds in Dilute Aqueous Solutions. *Chemosphere*, Vol. 44, pp. 561-576, ISSN: 0045-6535.

Stocker, J., Scheringer, M., Wegman, F., & Hungerbűhler K. (2007). Modeling the Effect of Snow and Ice on the Global Environmental Fate and Long-Range Transport Potential of Semivolatile Organic Compounds. *Environmental Science and Technology*, Vol. 41, pp. 6192-6198, ISSN 0013-936X.

Sweetman, A.J. & Jones, K.C. (2000). Modeling Historical Emissions and Environmental Fate of PCBs in the United Kingdom. *Environmental Science and Technology*, Vol. 34, pp. 863-869, ISSN 0013-936X.

United States Environmental Protection Agency (USEPA) (1996). On-line Tools for Site Assessment Calculation. Estimated Henry's Law Constants. http://www.epa.gov/athens/learn2model/part-two/onsite/esthenry.html

Von Waldow, H., MacLeod, M., Jones, K. Scheringer, M. & Hungerbühler, K. (2010). Remoteness from Emission Sources Explains the Fractionation Pattern of Polychlorinated Biphenyls in the Northern Hemisphere. *Environmental Science and Technology*, Vol. 44, pp. 6183–6188, ISSN 0013-936X.

Wania, F., (2003). Assessing the Potential of Persistent Organic Chemicals for Long-Range Transport and Accumulation in Polar Regions. *Environmental Science and Technology*, Vol. 37, pp. 1344-1351, ISSN 0013-936X.

Wania F. (2006). Potential of degradable organic chemicals for absolute and relative enrichment in the Arctic. *Environmental Science and Technology*, Vol 40, pp. 569–577, ISSN 0013-936X.

Wania, F., & Mackay, D. (1995). A Global Distribution Model for Persistent Organic Chemicals. *Science of the Total Environment*, Vol. 160/161, pp. 211-232, ISSN: 0048-9697

Wania, F., & Mackay. (1996). Tracking the distribution of persistent organic pollutants. *Environmental Science and Technology* Vol. 30, pp. 390A-396A, ISSN 0013-936X.

Wania F. & Westgate, J.N. (2008). On the Mechanism of Mountain Cold-Trapping of Organic Chemicals. *Environmental Science and Technology*, Vol. 42, pp. 9092-9098, ISSN 0013-936X.

Wania, F., Lei, Y.D. & T. Harner, T. (2002). Estimating Octanol-Air Partition Coefficients of Nonpolar Semivolatile Organic Compounds from Gas Chromatographic Retention Times. *Analytical Chemistry* Vol. 74, pp. 3476-3483, ISSN 0003-2700.

Wassmann, P., Duarte, C. M., Agusti, S. & Sejr, M. K. (2011). Footprints of Climate Change in the Arctic Marine Ecosystem. *Global Change Biology*, Vol. 17, pp. 1325-1249, ISSN 1365-2486.

Weber, I., Halsall, C.J., Muir, D.C.G., Texiera, C., Burniston, D.A., Strachan, W.M.J., Hung, H., Mackay, N., Arnold, D. & Kylin, H. (2006). Endosulfan and γ-HCH in the Arctic: An Assessment of Surface Seawater Concentrations and Air-Sea Exchange. *Environmental Science and Technology*, Vol. 40, pp. 7570-7576, ISSN 0013-936X.

WHO (World Health Organization) (2009). Protecting health from climate change. World Health Organization, ISBN: 978 92 4 159888 0, Geneva, Switzerland.

Xie, Z., Erbinghaus, R., Temme, C., Lohmann, R., Caba, A. & Ruck, W.W. (2007). Occurrence and Air-Sea Exchange of Phthalates in the Arctic. *Environmental Science and Technology*, Vol. 41, pp. 4555-4560, ISSN 0013-936X.

Yao, Y., Harner, T., Blanchard, P., Tuduri, L., Waite, D., Poissant, L., Murphy, C., Belzer, W., Aulagnier, F. &. Sverko, E.(2008). Pesticides in the Atmosphere Across Canadian Agricultural Regions. *Environmental Science and Technology*, Vol. 42, pp. 5931-5937, ISSN 0013-936X.

Zhang, L., Ma, J., Venkatesh, S., Li, J-F. & Cheung. P. (2008). Modelling of Episodic Intercontinental Long-Range Transport of Lindane. *Environmental Science and Technology*, Vol. 42, pp. 8791-8797, ISSN 0013-936X.

9

Rapid Detection and Recognition of Organic Pollutants at Trace Levels by Surface-Enhanced Raman Scattering

Zhengjun Zhang, Qin Zhou and Xian Zhang
Tsinghua University
P. R. China

1. Introduction

Organic pollutants are harmful even at trace level in the environment, and they are difficult to detect at that concentration. Our work provides a rapid and sensitive method – surface-enhanced Raman scattering – to detect and distinguish isomers of organic pollutants.

2. Application of surface-enhanced Raman scattering to organic pollutant detection

In the modern world, environmental problems have attracted more and more attention, for environmental pollutants are extremely harmful to human beings' health. Environmental pollutants, such as persistent organic pollutants, are widely separated in the environment and difficult to detect at trace level.

Within persistent organic pollutants, polychlorinated biphenyls (PCBs), due to their excellent dielectric properties, had been widely used since the 1920s in transformers, heat transfers, capacitors, etc., and had polluted nearly everywhere in the world.[1] In recent years, however, they have been found to be very harmful to human beings. They may cause serious diseases, such as cancers and gene distortion, when exceeding the critical dose in human bodies, and more seriously, PCBs can be accumulated in plants and animals from the environment and yield higher doses in human bodies, making PCBs very dangerous to human beings even in trace amounts.[1-3] Therefore, the detection of PCBs in trace amounts is crucial. Currently, the mostly applied detection technique for PCBs is the combination of high-resolution gas chromatography and mass spectrometry. It requires, however, very sophisticated devices, standard samples, complicated pretreatments of samples, favourable experimental environments and experienced operators.[4-7] Thus, new methods are demanded especially for the rapid detection of trace amounts of PCBs.

Surface-enhanced Raman scattering (SERS) has been proven to be an effective way to detect some organics.[8] With the great progress of nanoscale technology in recent years, SERS has attracted enormous attention due to its excellent performance and potential applications in the detection of molecules in trace amounts, even single molecule detection. For instance, using Ag nanorods as SERS substrates Rhodamine 6G with concentration of

10^{-14} M (dissolved in water) was detected;[9] with the alumina-modified AgFON substrates, bacillus subtilis spores were detected to 10^{-14} M;[10, 11] Vo-Dinh reported even the detection of specific nucleic acid sequences by the SERS technique.[12-14] In spite of the numerous studies on the application as a chemical and biological sensor,[15-17] the SERS technique has not yet been employed to detect PCBs as they are hardly dissolved in water.
While lots of researchers investigate SERS in the detection of biological and medical molecules,[12-17] SERS has also proved an excellent method in environment pollutants' detection, such as trace amounts of polychlorinated biphenyls (PCBs).[45, 47]

3. Fingerprint character of SERS: Understanding and simulating Raman spectrum of organic pollutants

SERS is an excellent method to detect and recognize trace amounts of organic pollutants as organic pollutants have different Raman spectrums due to their different molecule vibration modes, even when they have similar physical and chemical properties. These different Raman spectrums have peaks with different Raman shifts and different peak heights. These peaks are sharp and unique, which show a fingerprint character of SERS spectra and make SERS spectra easily distinguishable from each other.
The Raman spectrum shows detailed structure information of organic pollutants. Therefore, one can detect and recognize organic pollutants via the SERS spectrum even at trace level, just like one can recognize crystal structures via X-ray diffraction.
The relationship between Raman spectrum and molecule vibration modes can be analyzed by density functional theory. We performed a simulation using the Gaussian 03 programme package with the density functional theory. The simulations were carried out with the Becke's three-parameter hybrid method using the Lee-yang-Parr correlation functional (B3LYP) and the LANL2DZ basis set.[46] The Gaussian View was used to input investigated compounds data visually.

4. Fabrication high sensitive silver nanorods SERS substrates

The detection sensitivity of SERS depends considerably on the surface property of the SERS substrate. High aspect ratio, nanostructured Ag, Au, Cu substrates are proved to be good SERS substrates. For instance, using ordered arrays of gold particles prepared through a porous alumina template as the SERS substrate, Rhodamine 6G (R6G) molecules were detected to a concentration limit of 10^{-12} M; arrays of silicon nanorods coated with thin films of Ag served as good SERS substrates for R6G molecule detection, etc.[12, 16] Thus the preparation of SERS substrates with preferred surface property is of great importance. There are several methods to prepare these kinds of SERS substrates and in this chapter we take glancing angle deposition as an example.
Glancing angle deposition (GLAD) technique is a simple but powerful means of producing thin films with pre-designed nanostructures, such as nanopillars, slanted posts, zigzag columns and spirals. Silver nanorod arrays prepared by GLAD are excellent SERS substrates.
In addition, the SERS properties are related to the optical properties of the nanorod arrays. Both SERS properties and optical properties depend on the structure of the nanorods, such as the shape, length, separation, tilting angle and so on, which can be tuned by the deposition conditions.

4.1 Fabrication of sensitive SERS substrates by GLAD

The detection sensitivity of the SERS technique depends greatly on the surface property of the SERS substrate.[40, 41] Among the approaches so far available to prepare nanostructured materials, the glancing angle deposition (GLAD) technique is a simple but powerful means of producing thin films with pre-designed nanostructures,[42, 43] such as nanopillars, slanted posts, zigzag columns, spirals,[18] [19] etc. [20-23] For example, arrays of Ag nanorods were found to be good SERS substrates for the detection of trans-1,2-bis(4-pyridyl)ethane molecules, with a SERS enhancement factor greater than 10^8.[16] It is therefore of great interest to investigate the growth of metal nanostructures by the GLAD technique.[44]

Pristine Si wafers with (001) orientation were used as substrates. These were supersonically cleaned in acetone, ethanol and de-ionized water baths in sequence, and were fixed on the GLAD substrate in an e-beam deposition system. The system was pumped down to a vacuum level of 3×10^{-5} Pa and then the thin Ag film was deposited on the substrate with a depositing rate of 0.5 nm/s, with the thickness monitored by a quartz crystal microbalance. To produce films of aligned Ag nanorods, the incident beam of Ag flux was set at ~ 85 ^{o} from the normal of the silicon substrate, at different substrate temperatures. The morphology and structure of the thin Ag films was characterized by scanning electron microscope (SEM), transmission electron microscope (TEM) and high-resolution TEM, selected area diffraction (SAD) and X-ray diffraction (XRD), respectively. The performance of the nanostructured Ag films as SERS substrates was evaluated with a micro-Raman spectrometer using R6G as the model molecule.

It is well known that the major factors influencing the growth morphology of the films by GLAD are the incident direction of the depositing beam flux, the temperature and the movement of the substrate, and the deposition rate, etc. When fixing the incident Ag flux at ~85 ^{o} from the normal of the substrate and the deposition rate at ~0.5 nm/s, the growth morphology of the Ag films was greatly dependent on the temperature and movement of the substrate. Figure 1 shows the growth morphology of thin Ag films versus the temperature and movement of the substrate. The SEM micrographs were taken by a FEI SEM (QUANTA 200FEG) working at 20 kV.

Figure 1(a) and (b) shows typical SEM images of the surface morphology of thin Ag films deposited at 120 ^{o}C, without substrate rotation and with substrate rotation at a speed of 0.2 rpm, respectively. One sees from the images that at this temperature, Ag nanorods formed in two films with a length of 500 nm, yet they were not well separated - most nanorods were joined together. A major difference between the two is the growth direction of the joined nanorods, i.e. without rotation the nanorods grew at a glancing angle on the substrate, while with substrate rotation the nanorods grew vertically aligned. Another difference noticeable is the size of the nanorods, i.e. nanorods grown with substrate rotation have a slightly larger diameter.

Figure 1(c) and (d) shows respectively the surface morphology of thin Ag films deposited at -40 ^{o}C, without substrate rotation and with rotation at a speed of 0.2 rpm. Comparing with figures 1(a) and (b), it can be seen that the decrease in the deposition temperature led to the separation of Ag nanorods in the two films, while the rotation of the substrate also determined the growth direction and diameter of the nanorods, as observed from figures 1(a) and 1(b). The Ag nanorods grown at this temperature are 20-30 nm in diameter, ~ 800 nm in length and are well separated. Therefore, through adjusting the temperature and movement of the substrate one can grow well separated and aligned Ag nanorods on planar silicon substrates.

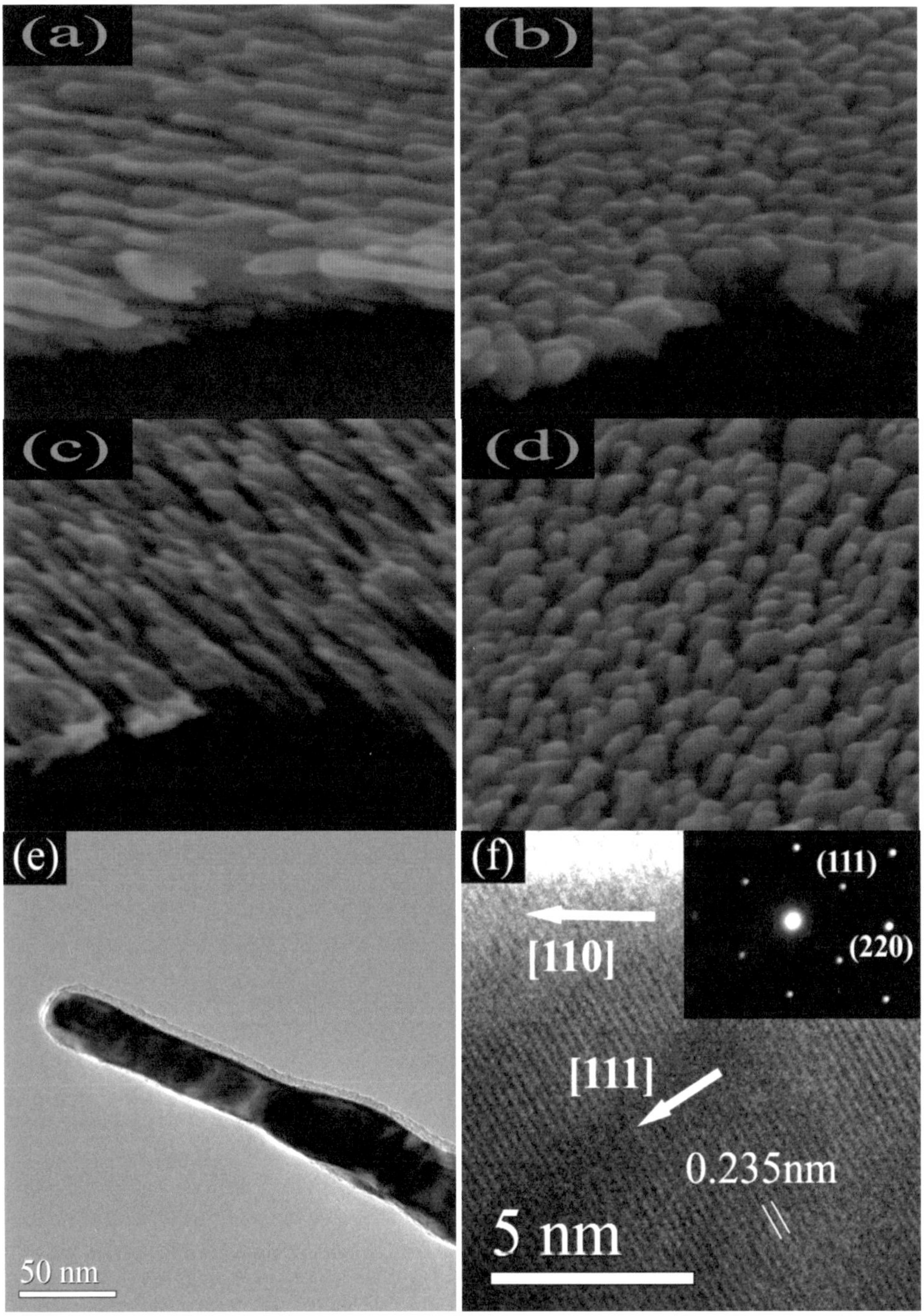

Fig. 1. Growth morphology of thin Ag films by GLAD at various conditions. (a) at 120 °C without substrate rotation; (b) at 120 °C and substrate rotation at 0.2 rpm; (c) at -40 °C without substrate rotation; and (d) at -40 °C and substrate rotation at 0.2 rpm. (e) and (f) shows respectively a bright-field TEM and a HRTEM image of the nanorods shown by figure 1(c); inset of (f) is the corresponding SAD pattern.

Figure 1(e) and 1(f) shows respectively a bright-field TEM and a HRTEM image of Ag nanorods shown by figure 1(c); inset of figure 1(f) is the corresponding SAD pattern. The images and the SAD pattern were taken with a JEM-2011F working at 200 kV. One sees from the figures that the Ag nanorod is ~ 30 nm in diameter and its micro-structure is single crystalline. By indexing the SAD pattern it is noticed that during the growth process the {111} plane of the nanorod was parallel to the substrate surface, with its axis along the <110> direction. This was confirmed by XRD analysis. Figure 2 shows a XRD pattern of the Ag nanorods shown by figure 1(c). The pattern was taken with a Rigaku X-ray diffractometer using the Cu k□ line, working at the θ -2 θ coupled scan mode. From the figure, a very strong (111) texture is observed, indicating that the {111} plane of the Ag nanorods was parallel to the substrate surface. These suggest that one can produce arrays of aligned, single crystalline Ag nanorods by the GLAD technique even at a low substrate temperature, i.e. -40 °C.

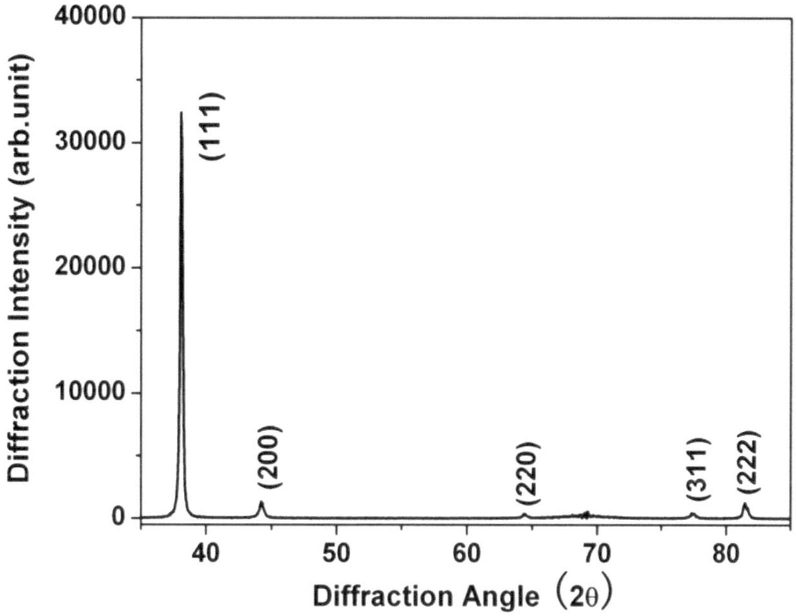

Fig. 2. A XRD pattern of the Ag film consisting of well separated, single crystalline nanorods shown by figure 1(c).

By using Rhodamine 6G as the model molecule, the performance of thin Ag films shown by figures 1(a)-(d) is examined as the SERS substrates. These samples were dipped in a 1×10^{-6} mol/L solution of R6G in water for 30 minutes and dried with a continuous gentle nitrogen blow. Figure 3(a) and 3(b) show Raman spectra of R6G obtained on the four nanostructured Ag films by a Reinshaw 100 Raman spectrometer using a 514 nm Ar^+ laser as the excitation source. It is observed that with the thin Ag films as the SERS substrate, all spectra exhibit clearly the characteristic peaks of R6G molecules, at 612, 774, 1180, 1311, 1361, 1511, 1575 and 1648 cm^{-1}, respectively.[12] However, the intensity of the Raman peaks was dependent on the morphology of the films. It is noticed that on Ag films consisting of well separated nanorods, see figures 3(b), the Raman peaks of R6G are much stronger than those on films of joined nanorods, see figures 3(a). This suggests that arrays of aligned but well separated Ag nanorods represent excellent SERS performance.

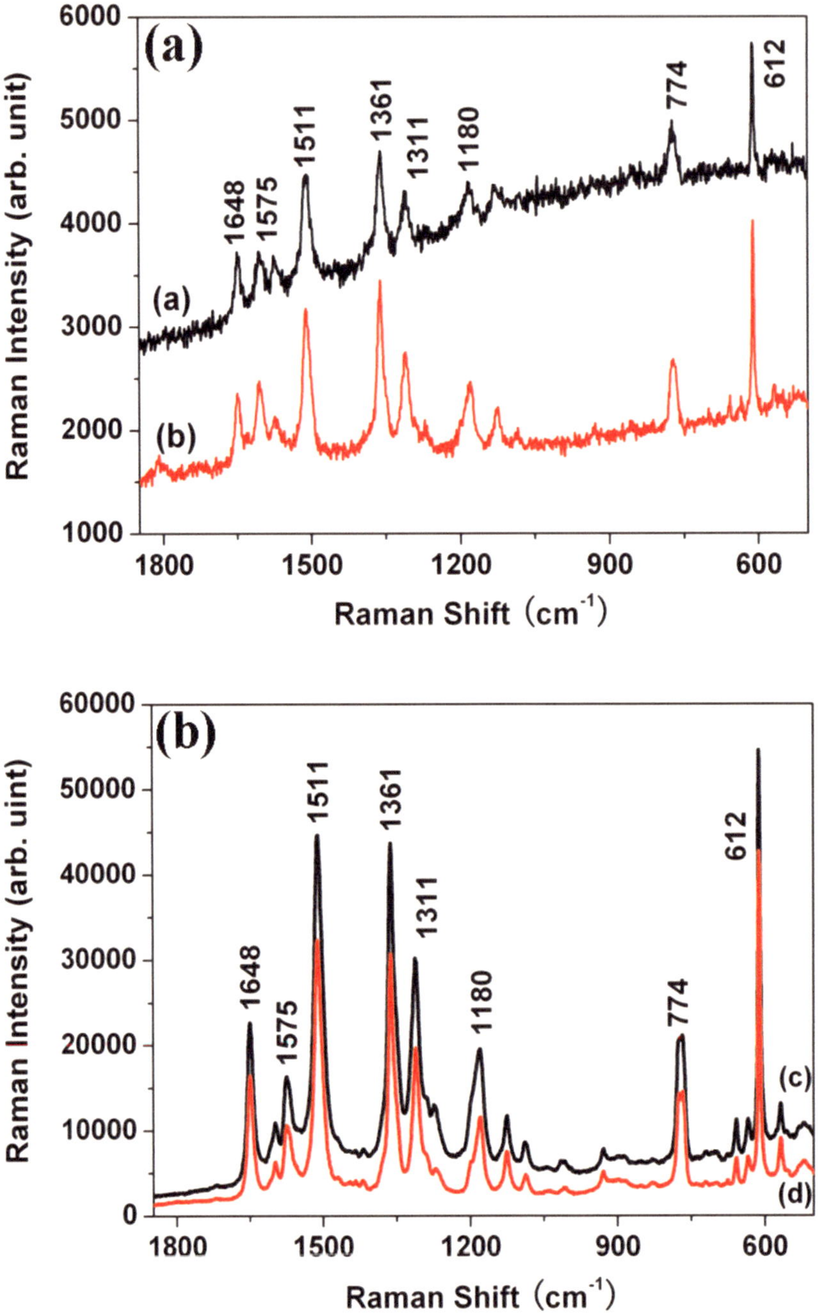

Fig. 3. Raman spectra of R6G on thin Ag films consisting of (a) joined nanorods shown by figures 1(a) (black line) and 1(b) (grey line); and (b) separated Ag nanorods shown by figures 1(c) (black line) and 1(d) (grey line), respectively, at a concentration of $1x10^{-6}$ mol/L.

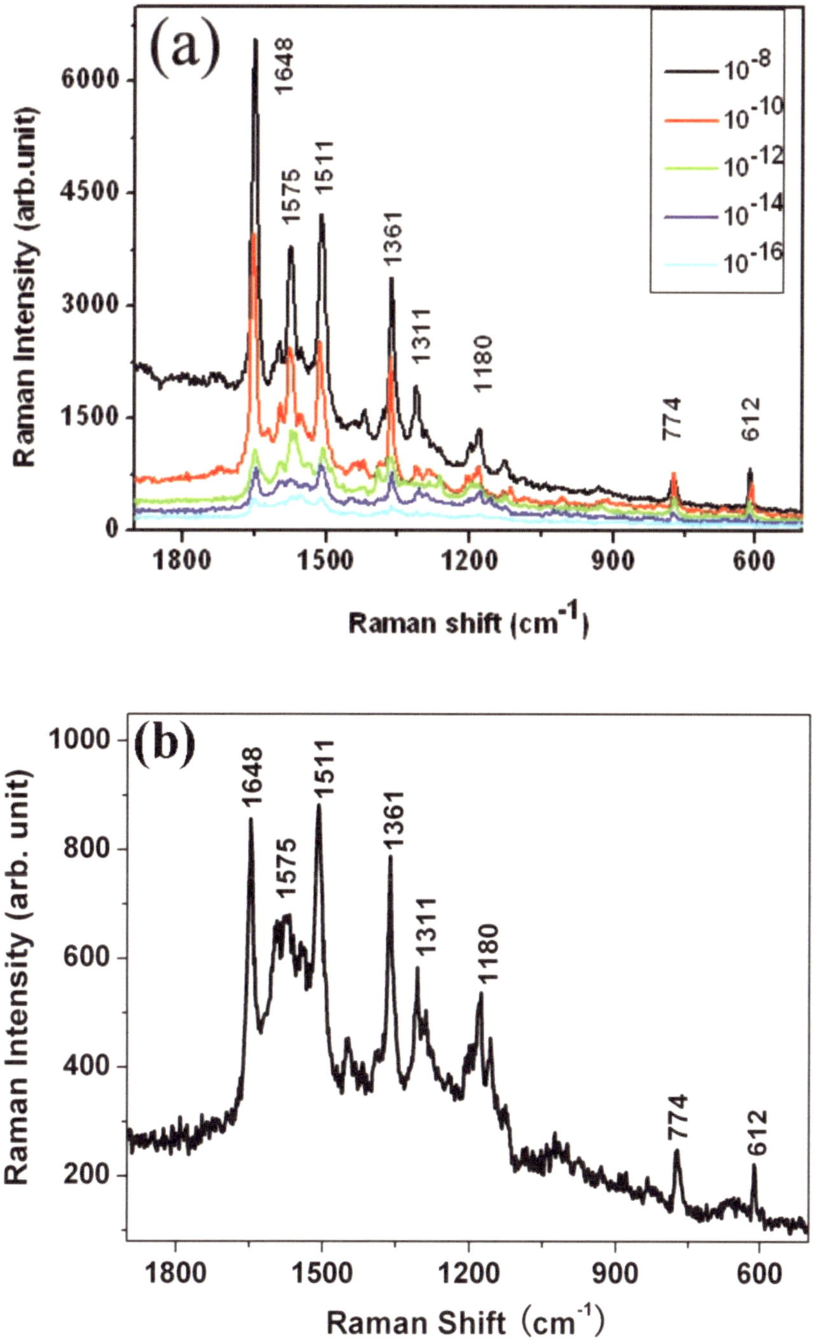

Fig. 4. (a) Raman spectra of R6G at concentrations ranging from $1x10^{-8}$ to $1x10^{-16}$ mol/L; and (b) the Raman spectrum of R6G at a concentration of $1x10^{-14}$ mol/L, on the thin Ag film consisting of well separated, single crystalline Ag nanorods.

Using arrays of aligned Ag nanorods shown by figures 1(c) and 1(d) as SERS substrates, we examined the detection limit of R6G molecules in water by the SERS technique. Figure 4(a) shows Raman spectra of R6G obtained on Ag nanorods shown by figure 1(c), as a function of the concentration of R6G in water ranging from $1x10^{-8}$ to $1x10^{-16}$ mol/L. Similar results were also obtained for Ag nanorods shown by figure 1(d). The Raman spectra were obtained by one scan with an accumulation time of 10 s, at a laser power of 1 % to avoid decomposition of R6G. It is found that characteristic peaks of R6G were all observed at all concentrations. To clearly show this, we plot the Raman spectrum at 10^{-14} mol/L in figure 4(b). It is noticed that although the intensity of the peaks is almost two orders lower than that at 10^{-6} mol/L, the spectrum contains the clear characteristic peaks of R6G.[12] These suggest that Ag films consisting of aligned and well separated Ag nanorods with single crystalline could serve as excellent SERS substrate for trace amount detection of R6G molecules. However, in the Raman spectrum at 10^{-16} mol/L in figure 4(a), some of the peaks of R6G disappear. That suggests the concentration limit of this method is 10^{-14}mol/L in the authors' work.[9]

4.2 Enhancing the sensitivity of SERS substrates via underlayer films

Although the Ag nanorod arrays present sensitive SERS performance, it is still necessary to enable the substrate to detect organic pollutants at trace amount with adequate sensitivity. There are several ways to promote the sensitivity of Ag nanorods as SERS substrates.

Much effort has been devoted to achieving highly sensitive SERS substrates. In particular, multilayer structures can improve SERS enhancement, such as "sandwich" structures with silver oxide or carbon inside and Ag or Au as both underlayer and overlayer.[24-28] Other researchers found that multilayer structures of Ag/Au nanostructures on the smooth metallic underlayer exhibited better SERS sensitivity compared to those without metallic underlayer (EF = 5×10^8).[29-31] However, the factor that governs the enhancement for multilayer structures is not very clear. Recently, Misra *et al.* obtained remarkably high SERS sensitivity using a micro-cavity with a radius of several micrometers.[32] Shoute *et al.* obtained high SERS signals (EF = 6×10^6) for molecules adsorbed on the silver island films supported by thermally oxidized silicon wafers and declared that the additional enhancement was due to the optical interference effect.[33] All the above experiments and those conducted by Driskell *et al.* suggested that the underlayer reflectivity could play an important role in the multilayer SERS substrates.[29]

We have investigated in detail the relationship of underlayer reflectivity and the SERS enhancement of Ag nanorod substrates prepared by oblique angle deposition. We use thin films of different materials with different thicknesses as underlayers to modulate the reflectivity systematically. With the coating of the same Ag nanorods, we find that the SERS intensity increases linearly with the underlayer reflectivity. This conclusion can be explained by a modified Greenler's model we recently developed.[34]

To change the reflectivity of the underlayer films, one can vary the dielectric constant and the thickness of the films systematically. We proposed to use Ag, Al, Si and Ti films, since they have different dielectric constants and can be fabricated easily. With a transfer matrix method, we can calculate the reflectivity of those films.[35, 36] Figure 5(a) shows the calculated reflectivity spectra of 100 nm Ag, Al, Si, and Ti films. In general, the reflectivity, $R_{Ag} > R_{Al} > R_{Ti} > R_{Si}$, except that at $\lambda \sim 600$ nm where the Si film has a large constructive interference. Figure 5(b) plots the film thickness dependent reflectivity for Ag, Al, Si and Ti at a fixed wavelength $\lambda_0 = 785$ nm. The reflectivity of Ag, Al, Ti, e.g. metals, increases monotonically with the film thickness *d*. The reflectivity *R* of Ag, Al and Ti thin films increases sharply when $d < 100$ nm and almost remains unchanged when $100 \text{ nm} \leq d \leq 400$ nm, while R_{Si} shows an oscillative behaviour due to the interference effect of a dielectric layer.

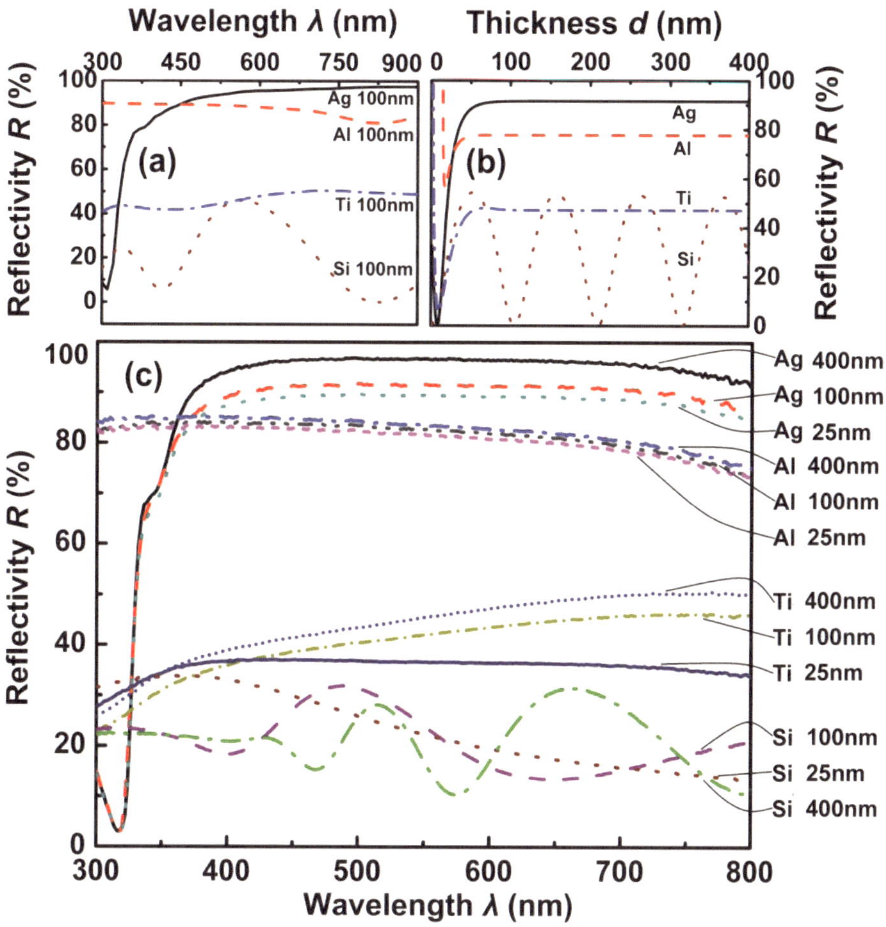

Fig. 5. (a) Calculated reflectivity *R* of thin Ag, Al, Si and Ti films at different wavelengths λ with film thickness of 100 nm; b) calculated reflectivity *R* of thin Ag, Al, Si and Ti films with different thicknesses *d* at $\lambda_0 = 785$ nm; c) experimentally obtained reflectivity spectra of thin Ag, Al, Si and Ti films with different thicknesses.

We deposited thin Ag, Al, Si and Ti films, all with thickness *d* = 25, 100, and 400 nm, respectively, to achieve different reflectivity. All depositions were carried out in a custom-designed electron-beam deposition system.[16] Before the deposition, the glass slide substrates were cleaned by piranha solution ($H_2SO_4 : H_2O_2 = 4:1$ in volume). The pellets of source materials, Ag, Al, Ti, with 99.99% purity, were purchased from Kurt J. Lesker Company, and Si with 99.9999% purity was purchased from Alfa Aesar Company. The film thickness was monitored *in situ* by a quartz crystal microbalance (QCM) facing toward the vapour source. After the deposition, the reflectivity of the deposited thin films was measured by an Ultraviolet-Visible Spectrophotometer (UV-Vis) double beam spectrophotometer with an integrating sphere (Shimadzu UV-Vis 2450). Figure 1(c) shows the reflectivity spectra of the twelve thin films obtained. The shapes of the reflection spectra are qualitatively consistent with those predicted by the calculations, as shown in Figs. 1(a) and (b). At the same wavelength, in general, $R_{Ag} > R_{Al} > R_{Ti} > R_{Si}$. In the visible wavelength region, the reflectivity of Ag, Al and Ti increases with the thickness d, while Si demonstrates an oscillating behaviour.

The twelve deposited planar thin film samples were then loaded into another custom-designed electron-beam evaporation system for Ag nanorod deposition through the so-called oblique angle deposition (OAD).[16, 29, 37] In this deposition, the background pressure was 1×10^{-7} Torr and the substrate holder was rotated so that the deposition flux was incident onto the thin films with an angle $\theta = 86^{\circ}$ with respect to the surface normal of the substrate holder. The Ag nanorod arrays were formed through a self-shadowing effect.[16, 29, 37] During the deposition, the Ag deposition rate was monitored by a QCM directly facing the incident vapour. The deposition rate was fixed at 0.3 nm/s and the deposition ended when the QCM read 2000 nm (our optimized condition).

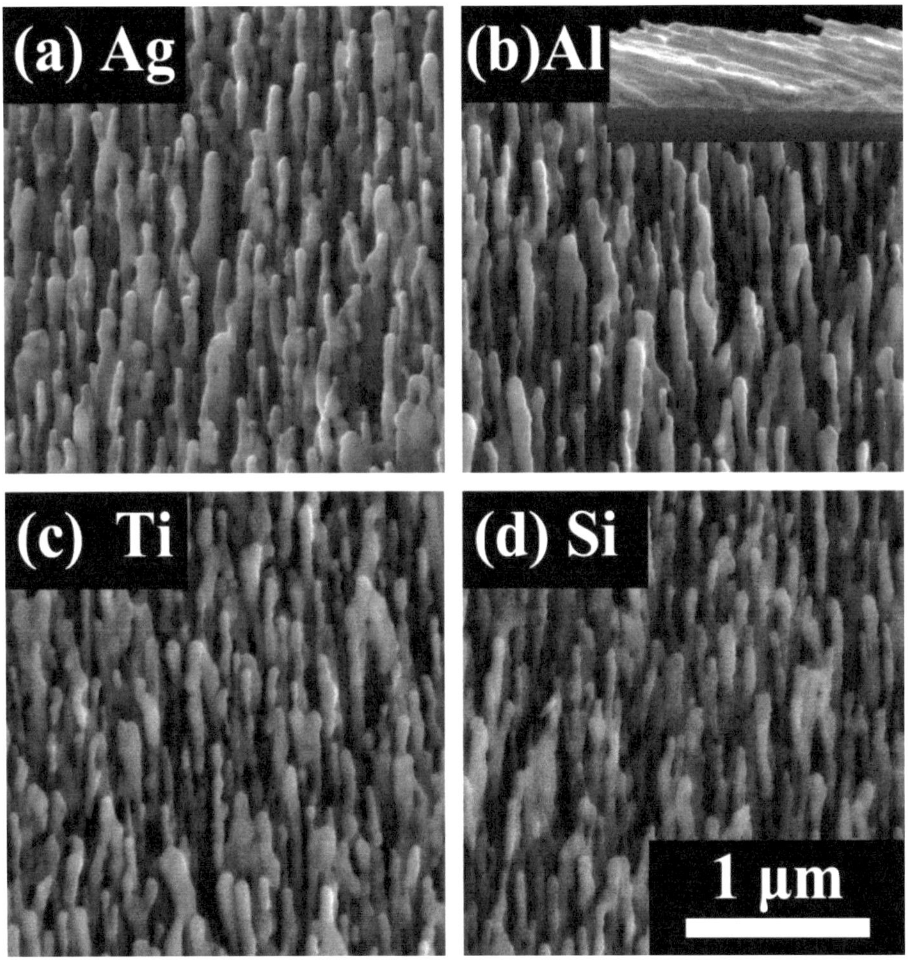

Fig. 6. Representative SEM images of Ag nanorod arrays on 100 nm underlayer thin films with different materials: (a) Ag; (b) Al; (c) Ti; (d) Si. All the figures have the same scale bar.

The morphologies of the Ag nanorod arrays on different thin film substrates were characterized by a scanning electron microscope (SEM, FEI Inspect F). The typical top-view SEM images are shown in Fig. 6 and they all look very similar. From the cross-section and top-view SEM images, the length *L*, diameter *D* and separation *S* of these Ag nanorods on different planar thin films are obtained statistically: L_{Ag} = 940 ± 70 nm, D_{Ag} = 90 ± 10 nm, S_{Ag} = 140 ± 30 nm; L_{Al} = 950 ± 50 nm, D_{Al} = 90 ± 10 nm, S_{Al} = 140 ± 30 nm; L_{Si} = 900 ± 50 nm, D_{Si} = 80 ± 10 nm, S_{Si} = 130 ± 20 nm; and L_{Ti} = 930 ± 60 nm, D_{Ti} = 90 ± 10 nm, S_{Ti} = 130 ± 20 nm, respectively. The Ag nanorod tilting angles β were measured to be about 73°with respect to substrate normal, which are consistent with our previous results.[16, 29, 37] These structure parameters are very close to one another, implying that the Ag nanorod arrays deposited on different thin film substrates are statistically the same. The SERS response of these Ag nanorod substrates were evaluated under identical conditions: A 2 μL droplet of a Raman probe molecule, trans-1, 2- bis (4-pyridyl) ethylene (BPE) with a concentration of 10^{-5} M, was uniformly dispersed onto the Ag nanorod substrates. The SERS spectra were recorded by the HRC-10HT Raman Analyzer from Enwave Optronics Inc., with an excitation wavelength of λ_0 = 785 nm, a power of 30 mW and an accumulation time of 10 s.

Figure 7(a) shows the representative BPE SERS spectra obtained at λ_0 = 785 nm from the Ag nanorod arrays on thin Ag, Al, Si and Ti film underlayers (thickness *d* = 100 nm). Each spectrum is an average of at least 15 different spectra taken at different spots on the substrates. All of them show the three main Raman bands of BPE, $\Delta\nu$ = 1639, 1610, and 1200 cm^{-1}, which can be assigned to the C=C stretching mode, aromatic ring stretching mode and in-plane ring mode, respectively.[38] The SERS intensity of the Ag nanorods grown on thin Ag film are higher than others and the SERS intensity of the Ag nanorods on Al film is larger than that on Ti film. The Ag nanorods on Si film show the smallest SERS intensity. According to Fig. 5(c), this seems to follow a trend: the larger the underlayer reflectivity, the larger the SERS intensity. To quantitatively compare the SERS response of these substrates, the Raman peak intensity I_{1200} at $\Delta\nu$ = 1200 cm^{-1} is analyzed.

Figure 7(b) plots the SERS intensity I_{1200} versus the reflectivity *R* of the underlayer thin films at λ_0 = 785 nm. The error bar for the Raman intensity is the standard deviation from 15 or more measurements from multiple sampling spots on the same substrates and the error bar for the reflectivity data is calculated from multiple reflectivity measurements at λ_0 = 785 nm. In Fig. 7(b), the SERS intensity and reflectivity follow a linear relationship: when the reflectivity of the underlayer increases, the SERS enhancement factor increases. This linear relationship of the underlayer reflectivity and SERS intensity can be explained by a modified Greenler's model developed by Liu *et al.*[34] Greenler's model is proposed through classical electrodynamics to explain the effects of the incident angles and polarization, and the collecting angle on the Raman scattering from a molecule adsorbed on a planar surface.[32] The modified Greenler's model extended the Greenler's model from a planar surface to Ag nanorod substrates and considered the effect of the underlying substrate.[34] The main point of the modified Greenler's model is to consider the conditions of both the incident and scattering fields near the molecule absorbed on a nanorod to calculate the enhancement.

Assuming that relative Raman intensity η is the ratio of the total Raman scattering power to incident light power, according to the modified Greenler's model, η excited by an unpolarized light can be explicitly expressed as[34]

$$\eta = \langle E^2_{\mathrm{Raman}} \rangle / \langle E^2_{\mathrm{incident}} \rangle$$
$$= \frac{1}{2}\{[1 + R_p + n_2^4 R'_p \cos\delta_p \cos 2(\varphi - \beta) + 2n_2^2 R'^{\frac{1}{2}}_p \cos(\delta'_p + 2\pi\Delta/\lambda)\sin 2\beta$$
$$+2n_2^2 R_p^{\frac{1}{2}} R'^{\frac{1}{2}}_p \sin 2\varphi \cos(\delta'_p + 2\pi\Delta/\lambda - \delta_p)](1 + R_p + 2\sqrt{R_p}\cos\delta_p)\cos^2(\varphi - \beta)$$
$$+[1 + n_2^4 R'_s + 2n_2^2 R'^{\frac{1}{2}}_s \cos(\delta'_s + 2\pi\Delta/\lambda)]\}$$

where R_p and R_s are the reflectivity of *p*- and *s*-polarized lights by the Ag nanorod surface, and R'_p and R'_s are the reflectivity of *p*- and *s*-polarized components by the underlayer thin film; n_2 is complex refractive index of Ag, and $n_2 = 0.03 + 5.242i$ (for λ_0 = 785 nm); φ is the light incident angle, and β is the Ag nanorod tilting angle; $\Delta = d\,(1 + \cos 2\varphi) / \cos\varphi$, where d is the thickness of Ag nanorod layer; δ_p, δ_s, δ'_p, and δ'_s are the reflectivity phase shifts of *p*- and *s*-polarization E-fields from Ag nanorods and underlayer thin film, defined as

$$\delta_p = \tan^{-1}[\mathrm{Im}(r_p)/\mathrm{Re}(r_p)],\ \delta_s = \tan^{-1}[\mathrm{Im}(r_s)/\mathrm{Re}(r_s)],$$

$$\delta'_p = \tan^{-1}[\mathrm{Im}(r'_p)/\mathrm{Re}(r'_p)],\ \delta'_s = \tan^{-1}[\mathrm{Im}(r'_s)/\mathrm{Re}(r'_s)]$$

By setting the light incident angle $\varphi = 0^o$, the Ag nanorod tilting angle $\beta = 73^o$, the thickness of Ag layer d = 300 nm, the relative Raman intensity η as a function of the underlayer reflectivity R at λ_0 = 785 nm is calculated and plotted in Fig. 7(c). It shows that the η indeed increases linearly with R, which is in very good agreement with our experimental data shown in Fig. 7(b). Therefore, the underlayer reflectivity is one significant parameter to consider for improving the SERS response of multilayer substrates.

Both our experiments and the modified Greenler's model demonstrate that the higher the underlayer reflectivity, the higher the SERS intensity for the Ag nanorod based SERS substrates. Accordingly, in order to further improve the SERS response of the Ag nanorod substrates, one can further increase the reflectivity of the underlayers through a proper surface coating such as multilayer dielectric coating.[39]

5. Detecting trace amount PCBs by SERS method

With the highly sensitive SERS substrates described before, one can detect trace amount organic molecules by the SERS method.

SERS is extremely sensitive in water solutions, for water does not have any Raman peaks. When detecting organic pollutants in nonaqueous systems, we use volatile organic solvents, such as acetone, to dilute pollutants. As the organic solvent shows high Raman background, we need to make the solvent volatilizated completely before SERS measurement.

The powders of 2, 3, 3′, 4, 4′-pentachlorinated biphenyl used in this study were commercially available from the AccuStandard Company. Since there is no Raman data of 2, 3, 3′, 4, 4′- pentachlorinated biphenyl reported, we first measured its Raman spectrum and that of acetone, for comparison, see Figure 8(a). To clearly show most characteristic peaks of 2, 3, 3′, 4, 4′-pentachlorinated biphenyl, the Raman spectrum was plotted in two regions of 300 to 1000 cm^{-1} and 1000 to 1700 cm^{-1} respectively, see Figures 11(b) and 11(c). From the figures one sees that the strongest peaks are located at 342, 395, 436, 465, 495, 507, 517, 598, 679, 731, 833, 891, 1032, 1136, 1179, 1254, 1294, 1573, and 1591 cm^{-1}, respectively; while for acetone the characteristic peaks are at 530, 786, 1065, 1220, 1428 and 1709 cm^{-1}, respectively. It is suggested that 2, 3, 3′, 4, 4′-pentachlorinated biphenyl is distinguishable from acetone and that acetone can be used as the solvent for the SERS measurements, as 2, 3, 3′, 4, 4′-pentachlorinated biphenyl is not soluble in water.

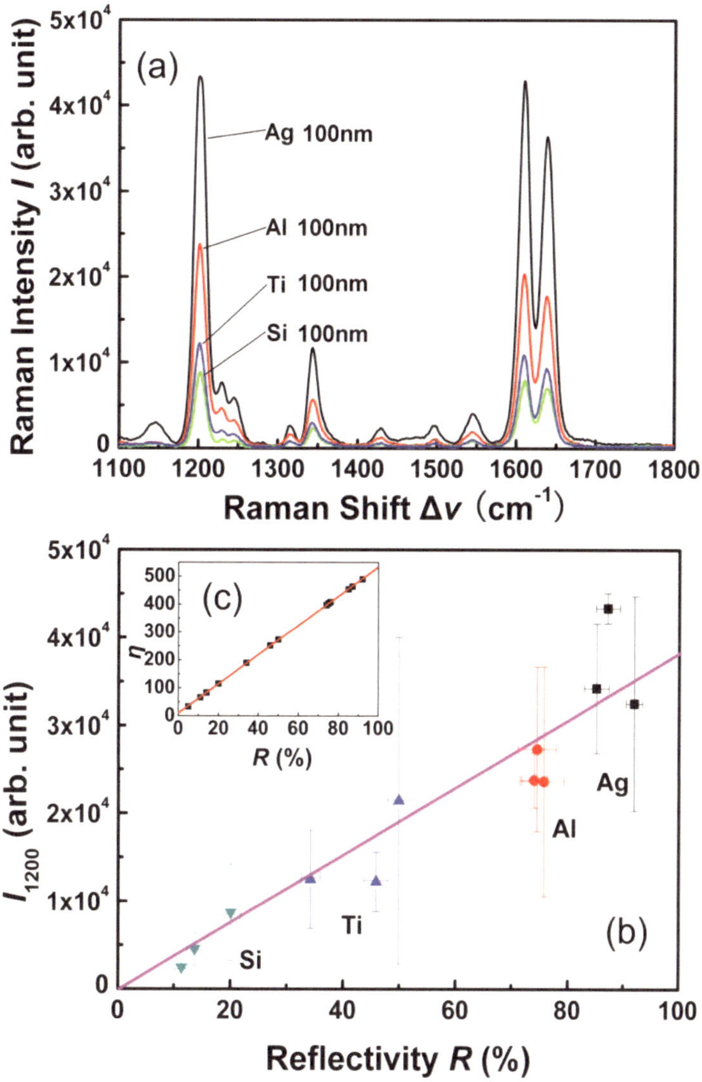

Fig. 7. (a) BPE SERS spectra obtained from Ag nanorod arrays deposited on 100 nm thin Ag, Al, Si and Ti film underlayers; (b) the plot of experimental Raman intensity as a function of underlayer reflectivity at $\lambda_0 = 785$ nm. Different symbol groups represent different kinds of substrates. (c) The plot of the enhanced Raman intensity ratio η as a function of underlayer reflectivity calculated by the modified Greenler's model.

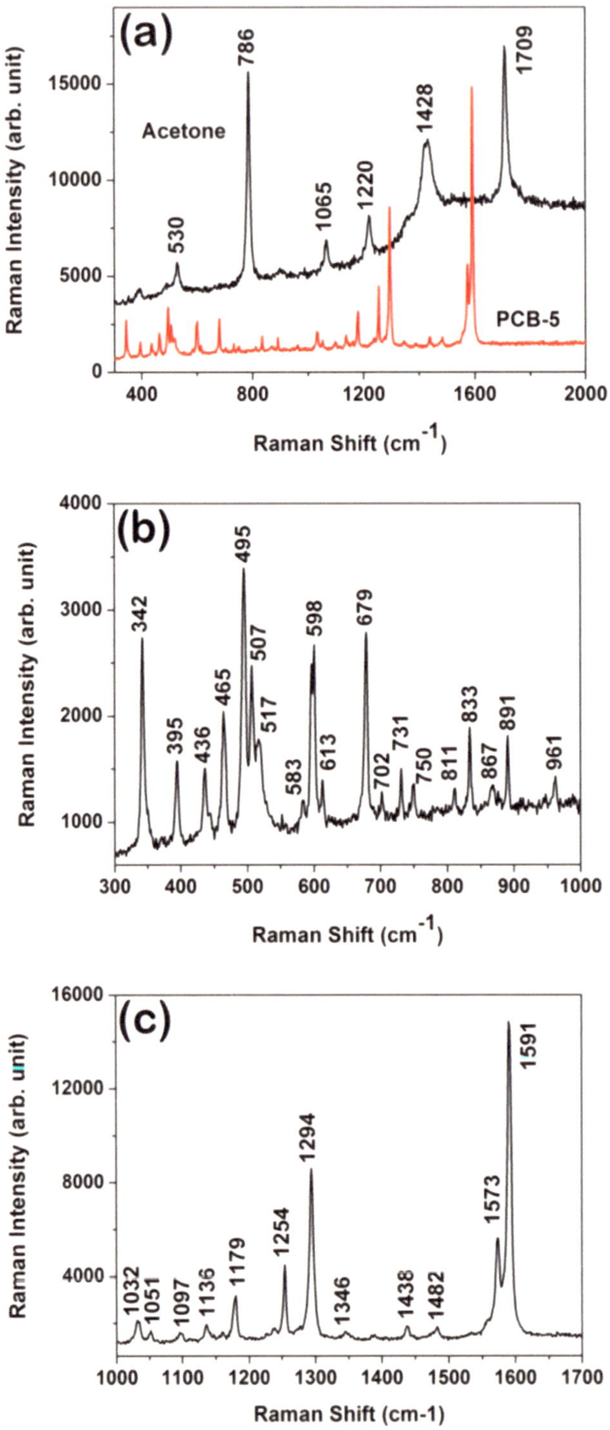

Fig. 8. (a) Comparison of Raman spectra of PCB-5 powders and acetone; (b) and (c) show details of the Raman spectrum of PCB-5 powders.

Because the SERS sensitivity is also dependent on the sample treatment, we employed in this study a very simple method to prepare SERS samples, i.e. dropping a small volume (~ 0.5 uL) of solutions of 2, 3, 3′, 4, 4′-pentachlorinated biphenyl in acetone on Ag nanorods using a top single channel pipettor and then blowing away the acetone with a continuous, gentle nitrogen blow. Figure 9(a) shows the Raman spectra of 2, 3, 3′, 4, 4′-pentachlorinated biphenyl dissolved in acetone at concentrations of 10^{-4} to 10^{-10} mol/L, respectively. The accumulation time of each Raman spectrum was 50 seconds and we used only 1% laser power to avoid changing of pentachlorinated biphenyl. When the small volume (~ 0.5 uL) of solutions of 2, 3, 3′, 4, 4′-pentachlorinated biphenyl was dropped on Ag nanorods, it became a circular spot with diameter of about 4 mm. The Raman spectrum was accumulated from a 2 um diameter circular area on the substrates. Therefore, for the solution at concentration of 10^{-10} mol/L, only about ten 2, 3, 3′, 4, 4′-pentachlorinated biphenyl molecules ($2*10^{-23}$ mol) would be accumulated in SERS; if that was at concentration of 10^{-8} mol/L, about 1000 molecules ($2*10^{-21}$ mol) would be accumulated and so on.

One sees that the Raman peaks of 10^{-4} mol/L PCB-5 solution located at 342, 495, 598, 679, 1032, 1136, 1179, 1254, 1294, 1573 and 1591 cm^{-1} march the Raman peaks of powder PCB-5 very well, this is quite different from the characteristic peaks of acetone. The peak around 1390 cm^{-1} represents disordered and amorphous carbon on the substrates. Figure 9(b) shows the SERS spectra of 10^{-8} mol/L PCB-5. Peaks located at 495, 1032, 1294, 1573 and 1591 cm^{-1} can march the Raman peaks of powder PCB-5. It indicates that the peaks shown in Figure 9(a) and (b) are the characteristic peaks of dissolved PCB-5, and PCB-5 with a concentration of 10^{-8} mol/L can be detected by the SERS method in the authors' work.

Large scale arrays of aligned and well separated single crystalline Ag nanorods on planar silicon substrate can be fabricated by GLAD method and these Ag films can be used as SERS substrates. With these substrates 2, 3, 3′, 4, 4′- PCB-5 molecules were detected even at a concentration of 10^{-8} mol/L by the SERS method, which indicates that trace amount of PCBs can be detected by the SERS method with Ag nanorods as SERS substrates.[45]

6. Rapid recognition of isomers and homologues of PCBs at trace levels by SERS

Detecting trace amount PCBs by the SERS method is introduced in section 4, but simple detection is not enough for organic pollutant detection, one also needs to distinguish kinds of organic pollutants from each other.

Furthermore, isomers and homologues of organic pollutants are hard to distinguish - especially in trace amounts - due to the similarities in their physical and chemical properties. The SERS method with silver nanorods as a substrate can be used to identify the Raman characteristics of isomers of monochlorobiphenyls and recognize these compounds, even at trace levels.

The Raman spectra of biphenyl, 2-, 3- and 4-chlorobiphenyls were measured by a Renishaw Raman 100 spectrometer using a 633 nm He-Ne laser as the excitation source at room temperature. Powders of these compounds are commercially available from the AccuStandard Company. Simulation of these Raman spectra was performed using the Gaussian 03 programme package with the density functional theory, to better understand the vibrational modes observed and figure out fingerprints of these compounds. For the SERS measurements, powders of chlorobiphenyl were dissolved in acetone to concentrations from 10^{-4} to 10^{-10} mol/L. The substrates were Ag nanorods prepared by electron beam deposition. The deposition of the Ag nanorods was described as aforementioned. A small volume of the solutions (~ 0.5 µL) was dropped on the surface of Ag nanorods and acetone was blown away using a nitrogen flow.

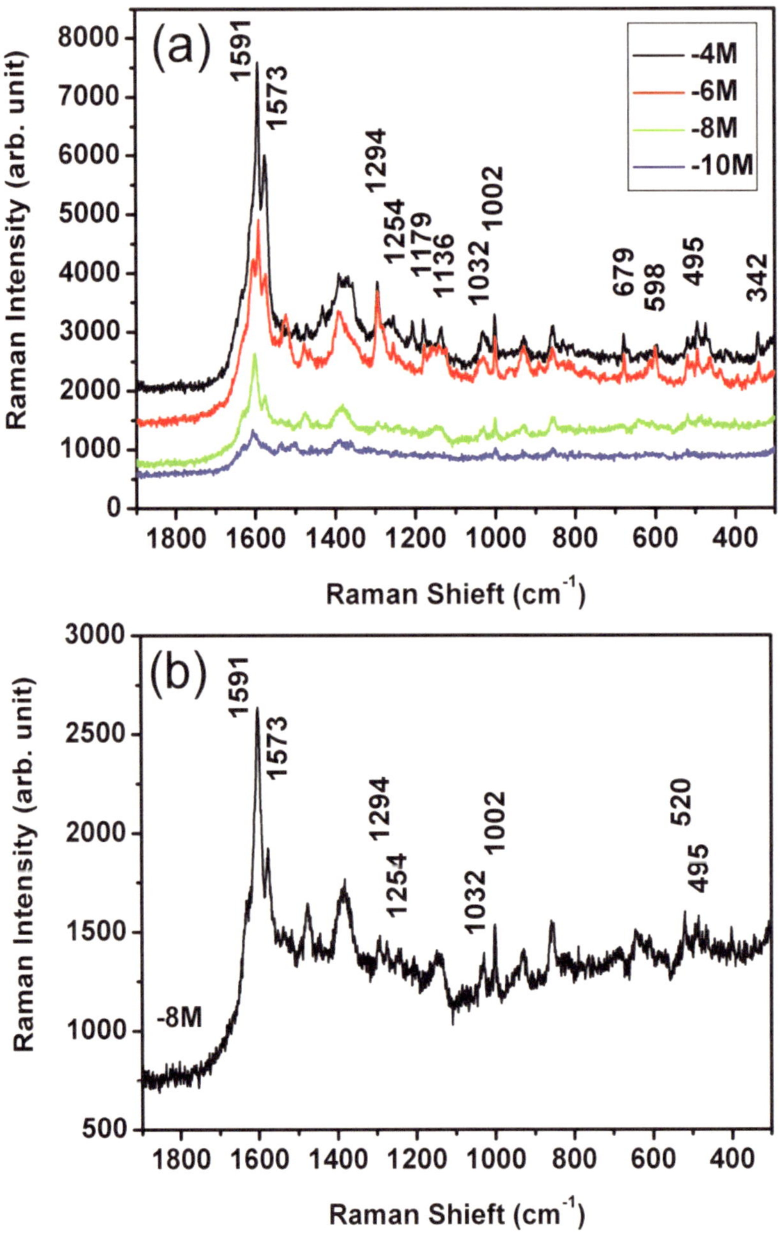

Fig. 9. (a) SERS spectra of PCB-5 dissolved in acetone with various concentrations; (b) SERS spectrum of PCB-5 in acetone at a concentration of 10^{-8} mol/L.

Figures 10 (a), (b), (c) and (d) show the measured Raman spectrum of biphenyl, 2-, 3- and 4-chlorobiphenyl, respectively. One sees that the four derivatives have strong peaks at ~ 3065, 1600, 1280, 1030 and 1000 cm^{-1}, demonstrating the common feature of biphenyl and its derivatives. One may also notice the differences among the Raman spectra of the four derivatives. For example, (1) biphenyl, 3- and 4-chlorobiphenyl have strong Raman peaks around 1276 cm^{-1}, while the peak for 2-chlorobiphenyl was at ~ 1297 cm^{-1}; (2) biphenyl has a strong peak at 738 cm^{-1}, 2- and 4-chlorobiphenyl have strong peaks at ~ 760 cm^{-1}, but the peak for 3-chlorobiphenyl was negligible; (3) both 2- and 3-chlorobiphenyl have strong peaks around ~ 680 cm^{-1}, while biphenyl and 4-chlorobiphenyl have no visible peak nearby; (4) only 2-chlorobiphenyl has a strong peak at ~ 432 cm^{-1}. The above features might be used to detect and distinguish biphenyl, 2-, 3 and 4-chlorobiphenyl.

To gain a clear understanding of these features, we performed simulations using the Gaussian 03 programme package with the density functional theory. The simulations were carried out with the Becke's three-parameter hybrid method using the Lee-yang-Parr correlation functional (B3LYP) and the LANL2DZ basis set.[46] The Gaussian View was used to input investigated compounds' data visually. The π bond length of the benzene ring was set to be 1.409 Å, the σ bond length between C and H atoms was set to be 1.088 Å and the σ bond length between C and Cl atoms was set to be 1.760 Å.

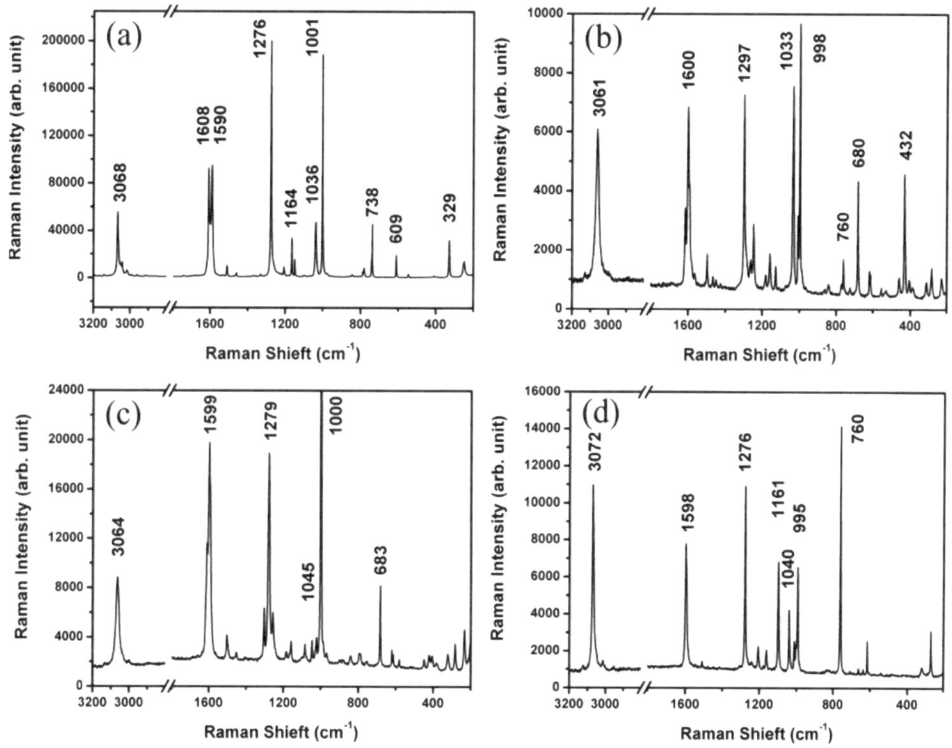

Fig. 10. Raman spectra of (a) biphenyl; (b) 2-chlorobiphenyl; (c) 3-chlorobiphenyl; and (d) 4-chlorobiphenyl, measured using powders commercially available from the AccuStandard Company.

Table I lists major vibrational modes of the four derivatives obtained by the above simulations. The common features of their Raman spectra at ~ 3065, 1600, 1280, 1030 and 1000 cm^{-1}, seen in Figure 10, can be attributed to the C-H stretching mode (~ 3100 cm^{-1}), the ring CCC stretching mode (~ 1650 cm^{-1}), the C-C bridge bond stretching mode (~ 1280 cm^{-1}), the C-H bending in-plane mode (1050 to 1100 cm^{-1}) and the CCC trigonal breathing mode(~ 1000 cm^{-1}), respectively.

Vibrational Model	Raman Shift /cm-1	Raman Int.of 2-chlorobiphenyl	Raman Int.of 3-chlorobiphenyl	Raman Int.of 4-chlorobiphenyl
CH stretching	3100	511	359	351
CCC stretching	1650	31	34	38
CC-bridge stretching	1280	50	50	62
CH bending in-plane	1050	46	22	15
trigonal breathing	1000	34	44	27
CCC bending in-plane (1-4 direction)	760	3	4	23
CCC bending in-plane (3-6 direction)	680	11	8	7
CCC bending in-plane (2-5 direction)	460	13	0	0

Table 1. Major simulated vibrational modes for 2-, 3- and 4-chlorobiphenyl.

One sees that due to the replacement of the H by Cl atom, the CCC bending (ring deformation) in-plane modes (~760 cm^{-1}, 680 cm^{-1} and 460 cm^{-1}) changed differently for 2-, 3- and 4-chlorobiphenyl. Figures 11 (a), (b) and (c) show respectively the strong CCC bending in-plane mode for 2-, 3- and 4-chlorobiphenyl. It is seen that for the 2-chlorobiphenyl, the 2-5 direction CCC bending at 460 cm^{-1} is the strongest one, the 3-6 direction CCC bending at 680 cm^{-1} has a similar intensity, while the 1-4 direction bending at 760 cm^{-1} is weak. For 3-chlorobiphenyl, the 2-5 direction bending mode is negligible, the 3-6 direction bending mode is strong, while the 1-4 direction bending mode is weak. For 4-chlorobiphenyl, the 2-5 direction bending mode is negligible, the 3-6 direction bending mode is weak, while the 1-4 direction bending mode is strong. These results are in agreement with the experimental measurements and suggest that these features in the CCC bending in-plane modes can be used to recognize the three homologues.

The substrate used in the SERS measurements was Ag nanorods prepared by the electron beam deposition technique. Powders of 2-, 3- and 4-chlorobiphenyl were dissolved in acetone and diluted into solutions with a concentration ranging from 10^{-4} to 10^{-10} mol/L. A small volume (0.5 μL) of these solutions was dropped on Ag nanorods and the acetone was blown away using a gentle nitrogen flow. Figures 12 (a), (b) and (c) show respectively the SERS spectra of the 2-, 3- and 4-chlorobiphenyl, at various concentrations. The accumulation time of each spectrum was fixed at 30 seconds per 100 cm^{-1} and we used only 10% laser power (0.47mW) to avoid radiation damage. From these figures we notice that the characteristic Raman peaks are all clearly observed for three homologues, even at a concentration of 10^{-8} mol/L, suggesting that the SERS technique is able to detect chlorobiphenyls even at such a low concentration.

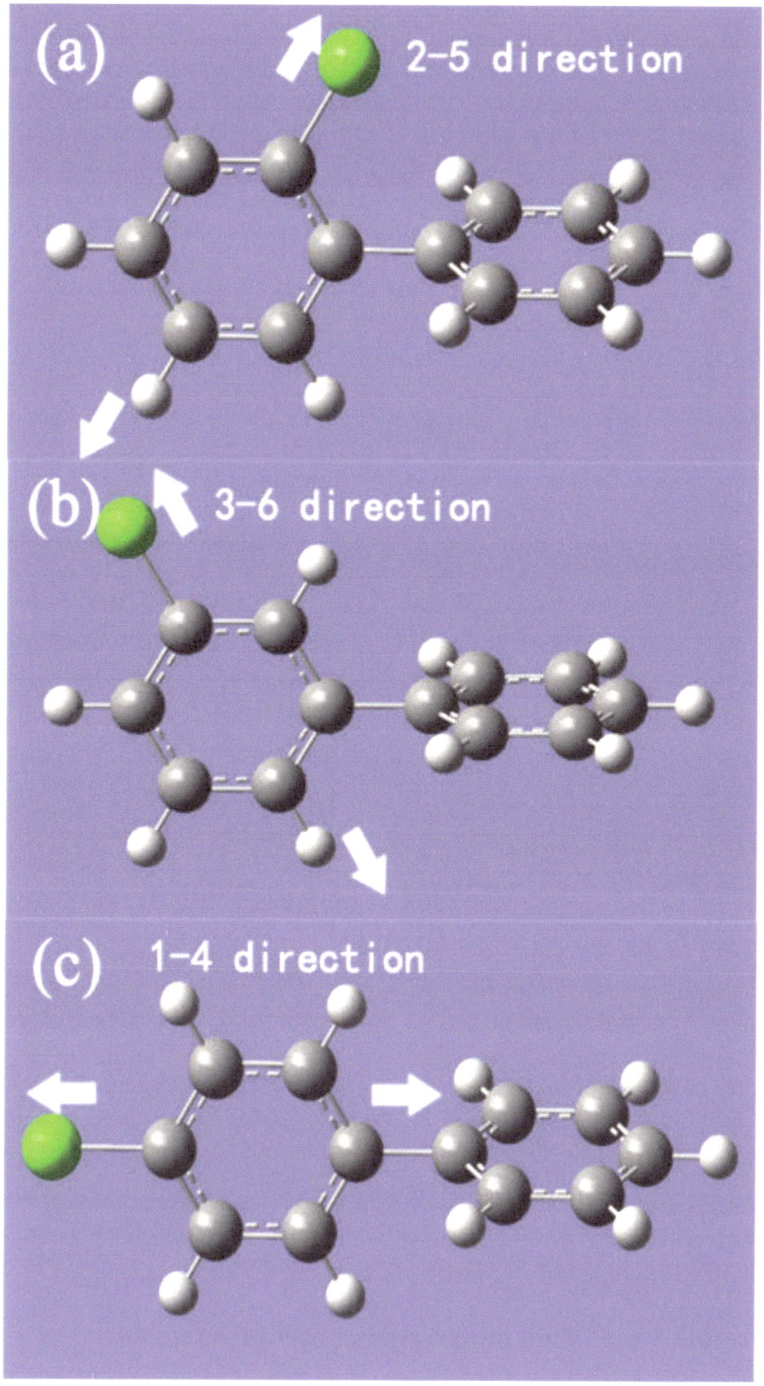

Fig. 11. The strong ring deformation in-plane modes for the three homologues of chlorobiphenyl. (a) 2-chlorobiphenyl; (b) 3-chlorobiphenyl; and (c) 4-chlorobiphenyl.

Figure 12 (d) compares SERS spectra of the three homologues at a concentration of 10^{-6} mol/L. One sees that at this concentration the three spectra show clearly the common features as mentioned above around 1600, 1280, 1030 and 1000 cm^{-1} (the 3100 cm^{-1} was not measured). The spectra also clearly show the characteristic Raman peaks for three homologues, i.e. the difference in the CCC bending (ring deformation) in-plane modes caused by the Cl atom replacement. These suggest that by using Ag nanorods as substrates, the SERS technique is capable of detecting chlorobiphenyls at trace amounts and is capable of recognizing the homologues at small concentrations.

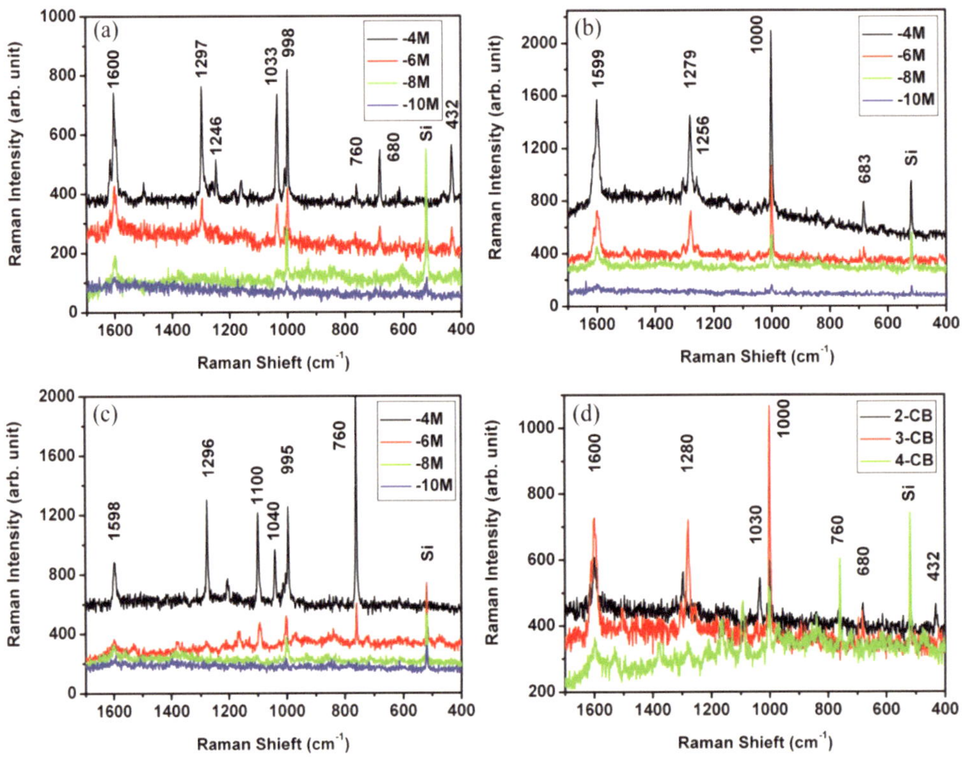

Fig. 12. SERS spectra of (a) 2-chlorobiphenyl; (b) 3-chlorobiphenyl; and (c) 4-chlorobiphenyl at concentrations from 10^{-4} to 10^{-10} mol/L in acetone. (d) Compares the SERS spectra of the three homologues at a concentration of 10^{-6} mol/L.

In summary, based on the understanding of Raman characteristics of these compounds, one can detect and recognize the homologues of chlorobiphenyls, even at the trace amount, by using the SERS technique with Ag nanorods as substrates.[17]

7. Detecting trace POPs in real environmental samples

In sections 4 and 5 we introduced the SERS method to detect and distinguish trace amount PCBs and their isomers and homologues. In those experiments, the PCBs are in acetone solutions, as fundamental study. In this section we introduce some examples in practical trace POPs detection.

7.1 Detecting trace PCBs in dry soil samples

The polluted soil samples were dried and made into small powers which were acquired from the Nanjing Institute of Soil (China). With a combination of the high-resolution gas chromatography and mass spectrometry techniques, sample I proved to contain about 5 μg/g PCBs and sample II proved to contain about 300 μg/g PCBs. 0.2 g soil sample I was put into 20 mL acetone and was agitated uniformly for about 5 minutes. This suspension was precipitated for 30 minutes and the transparent acetone solution in the upper layer was taken as solution sample A. 0.2 g soil sample I was put into 200 mL acetone and solution sample B was obtained through the aforementioned process. 0.2 g soil sample II was put into 20 mL acetone to obtain solution sample C and was put into 200 mL acetone to obtain solution sample D.

The Ag nanorods SERS substrates were put into the solution samples A, B, C and D, respectively. After 30 minutes, the Ag nanorods substrates were taken out of the solutions and the acetone on the substrates was blown away using a nitrogen flow. The Raman spectra of these substrates dipped into solution samples were measured by a Renishaw Raman 100 spectrometer using a 633 nm He-Ne laser as the excitation source at room temperature.

Figures 13 (a), (b), (c) and (d) show the measured Raman spectrum of the Ag substrates dipped into sample A, B, C and D, respectively. From Figures 13 (a), (b) and (c), one sees peaks at ~1600, 1280, 1240, 1150, 1030 and 1000 cm^{-1} clearly, demonstrating the common feature of PCBs. The peaks around 1590~1600cm^{-1} present benzene stretching vibration mode; the peak around 1280cm^{-1} presents CC bridge stretching vibration mode; the peak around 1030cm^{-1} presents CH bending in-plane mode; the peak around 1000cm^{-1} presents trigonal breathing vibration mode; and peaks around 1240~1250cm^{-1} and 1140~1200cm^{-1} present the vibration peaks induced by Cl substituent. These characteristic peaks suggest that PCBs in dry soil can be detected by the SERS method by dissolving into acetone. The most widely used PCBs are trichlorobiphenyls and pentachlorobiphenyls, we assumed that the molecular weight of the PCBs in the soil samples is 300, then the concentration of the PCBs acetone solution in solution sample A, B, C and D are about 10^{-5} mol/L, 10^{-5} mol/L, 10^{-6} mol/L, 10^{-7} mol/L, and 10^{-8} mol/L, respectively. Thus, with silver nanorod substrates, 5ug/g PCBs in dry soil samples can be detected by the SERS method.

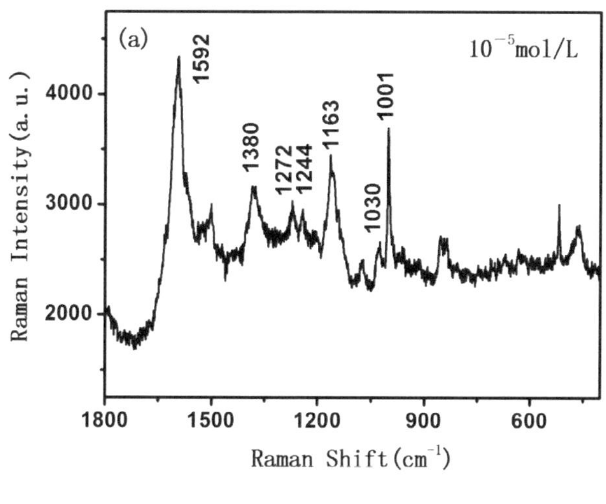

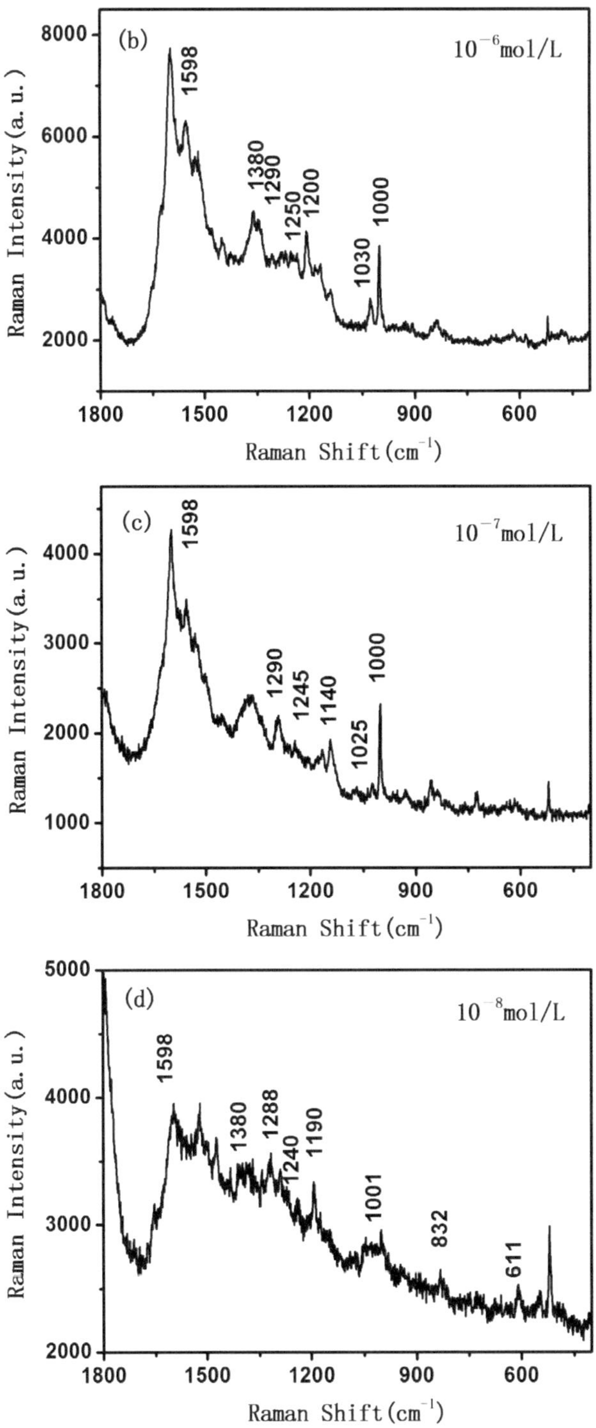

Fig. 13. SERS spectra of PCBs in dry soil samples after being treated by acetone: (a) ~10^{-5} mol/L; (b) ~10^{-6} mol/L; (c) ~10^{-7} mol/L; (d) ~10^{-8} mol/L.

7.2 Detecting PCBs in white spirit

PCBs in white spirit can also be detected by the SERS method with silver nanorod substrates. The concentration of PCBs in white spirit is about 10^{-4} mol/L. We put a drop of PCBs "polluted" white spirit on the silver nanorod substrates and made the white spirit volatilized away. Then, we found PCBs Raman signal with the SERS method described before. Figures 14 (a) and (b) show the SERS spectra of pure white spirit and white spirit with 10^{-4} mol/L PCBs, respectively. One can recognize characteristic Raman peaks of PCBs around 1590, 1290, 1240, 1030 and 1000 cm^{-1} in Figure 14 (b).

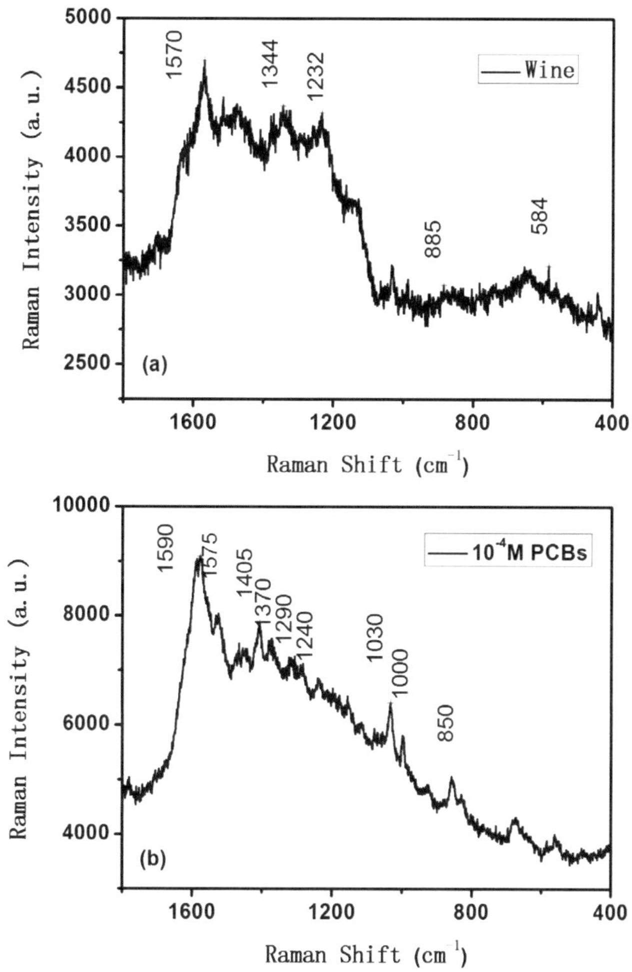

Fig. 14. SERS spectra of white spirit without and with PCBs.

7.3 Detecting melamine in milk

In the year 2009, milk produced by Sanlu Co. (China) was found to contain amounts of Melamine in much higher concentrations than usual. Milk with Melamine seems to contain more protein when detecting nitrogen concentration, but it is poisonous to children. With the SERS method with silver nanorods as substrates, we detected Melamine in milk. Figures 15 (a) and (b) show the SERS spectra of pure Melamine and milk with Melamine.

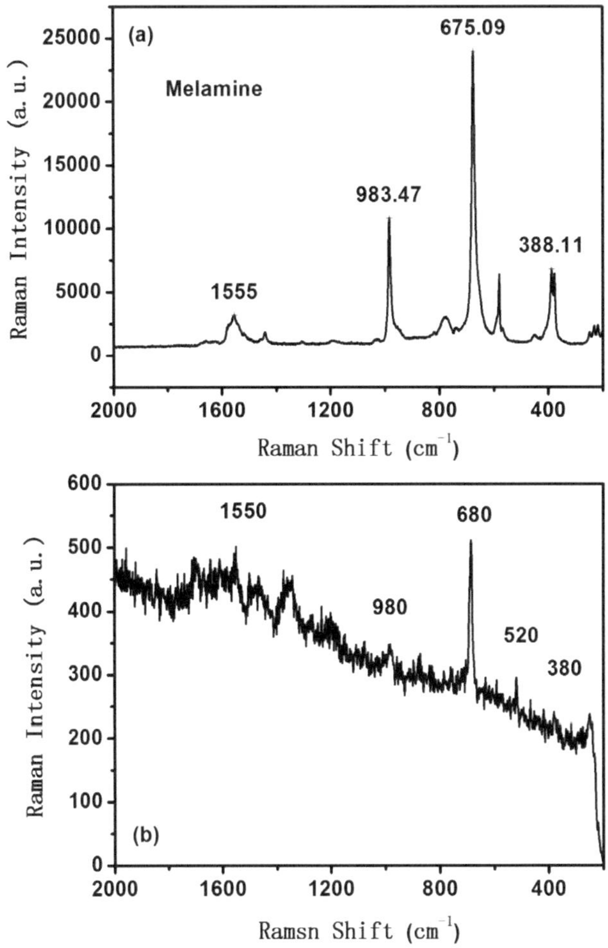

Fig. 15. SERS spectra of Melamine and trace Melamine in milk.

8. Summary

Although persistent organic pollutants such as PCBs are difficult to detect at trace amount, they can be detected and recognized rapidly via the SERS technique. Ag nanostructured SERS substrates prepared by the glancing angle deposition method are excellent at detection and their sensitivity can be further improved by tuning the thin underlayer films and Ag nanorod structures. With well designed and prepared Ag nanostructured SERS substrates, pentachlorinated biphenyl molecules are detected and isomers of chlorobiphenyl molecules are recognized at trace level using the SERS method. These series of studies provide a potential method for trace pollutant detection via nano technology.

9. References

[1] Ross G. The public health implications of polychlorinated biphenyls (pcbs) in the environment. *Ecotoxicology and Environmental Safety*, 2004, 59(3): 275~291.

[2] Ohtsubo Y, Kudo T, Tsuda M, et al. Strategies for bioremediation of polychlorinated biphenyls. *Applied Microbiology and Biotechnology*, 2004, 65(3): 250~258.

[3] Cicchetti D V, Kaufman A S, Sparrow S S. The relationship between prenatal and postnatal exposure to polychlorinated biphenyls (pcbs) and cognitive, neuropsychological, and behavioral deficits: a critical appraisal. *Psychology in the Schools*, 2004, 41(6): 589~624.

[4] Hong J E, Pyo H, Park S J, et al. Determination of hydroxy-pcbs in urine by gas chromatography/mass spectrometry with solid-phase extraction and derivatization. *Analytica Chimica Acta*, 2005, 531(2): 249~256.

[5] Namiesnik J, Zygmunt B. Selected concentration techniques for gas chromatographic analysis of environmental samples. *Chromatographia*, 2002, 56Suppl. S: S9~S18.

[6] Pitarch E, Serrano R, Lopez F J, et al. Rapid multiresidue determination of organochlorine and organophosphorus compounds in human serum by solid-phase extraction and gas chromatography coupled to tandem mass spectrometry. *Analytical and Bioanalytical Chemistry*, 2003, 376(2): 189~197.

[7] Barra R, Cisternas M, Suarez C, et al. Pcbs and hchs in a salt-marsh sediment record from south-central Chile: use of tsunami signatures and cs-137 fallout as temporal markers. *Chemosphere*, 2004, 55(7): 965~972.

[8] Moskovits M. Surface-enhanced spectroscopy. *Reviews of Modern Physics*, 1985, 57(3): 783~826.

[9] Zhou Q, Li Z C, Yang Y, et al. Arrays of aligned, single crystalline silver nanorods for trace amount detection. *Journal of Physics D-Applied Physics*, 2008, 41(15200715).

[10] Kudelski A. Analytical applications of raman spectroscopy. *Talanta*, 2008, 76(1): 1~8.

[11] Zhang X Y, Zhao J, Whitney A V, et al. Ultrastable substrates for surface-enhanced raman spectroscopy: al2o3 overlayers fabricated by atomic layer deposition yield improved anthrax biomarker detection. *Journal of the American Chemical Society*, 2006, 128(JA063876031): 10304~10309.

[12] Tan R Z, Agarwal A, Balasubramanian N, et al. 3d arrays of sers substrate for ultrasensitive molecular detection. *Sensors and Actuators a-Physical*, 2007, 139(1-2Sp. Iss. SI): 36~41.

[13] Isola N R, Stokes D L, Vo-Dinh T. Surface enhanced raman gene probe for hiv detection. *Analytical Chemistry*, 1998, 70(7): 1352~1356.

[14] Vodinh T, Houck K, Stokes D L. Surface-enhanced raman gene probes. Analytical Chemistry, 1994, 66(20): 3379~3383.

[15] Tripp R A, Dluhy R A, Zhao Y P. Novel nanostructures for sers biosensing. *Nano Today*, 2008, 3(3-4): 31~37.

[16] Chaney S B, Shanmukh S, Dluhy R A, et al. Aligned silver nanorod arrays produce high sensitivity surface-enhanced raman spectroscopy substrates. *Applied Physics Letters*, 2005, 87(0319083).

[17] Zhang Z Y, Zhao Y P. Tuning the optical absorption properties of ag nanorods by their topologic shapes: a discrete dipole approximation calculation. *Applied Physics Letters*, 2006, 89(0231102).
[18] Malac M, Egerton R F, Brett M J, et al. Fabrication of submicrometer regular arrays of pillars and helices. *Journal of Vacuum Science & Technology B*, 1999, 17(6): 2671～2674.
[19] Dick B, Brett M J, Smy T. Investigation of substrate rotation at glancing-incidence on thin-film morphology. *Journal of Vacuum Science & Technology B*, 2003, 21(6): 2569～2575.
[20] Dick B, Brett M J, Smy T J, et al. Periodic magnetic microstructures by glancing angle deposition. *Journal of Vacuum Science & Technology A-Vacuum Surface and Films*, 2000, 18(4Part 2): 1838～1844.
[21] Alouach H, Fujiwara H, Mankey G J. Magnetocrystalline anisotropy in glancing angle deposited permalloy nanowire arrays. *Journal of Vacuum Science & Technology A*, 2005, 23(4): 1046～1050.
[22] Singh J P, Tang F, Karabacak T, et al. Enhanced cold field emission from 100 oriented beta-w nanoemitters. *Journal of Vacuum Science & Technology B*, 2004, 22(3): 1048～1051.
[23] Hawkeye M M, Brett M J. Glancing angle deposition: fabrication, properties, and applications of micro- and nanostructured thin films. *Journal of Vacuum Science & Technology* A, 2007, 25(5): 1317～1335.
[24] Leverette C L, Shubert V A, Wade T L, et al. Development of a novel dual-layer thick ag substrate for surface-enhanced raman scattering (sers) of self-assembled monolayers. *Journal of Physical Chemistry B*, 2002, 106(34): 8747～8755.
[25] Li H G, Cullum B M. Dual layer and multilayer enhancements from silver film over nanostructured surface-enhanced raman substrates. *Applied Spectroscopy*, 2005, 59(4): 410～417.
[26] Li H G, Baum C E, Sun J, et al. Multilayer enhanced gold film over nanostructure surface-enhanced raman substrates. *Applied Spectroscopy*, 2006, 60(12): 1377～1385.
[27] Yang Y A, Bittner A M, Kern K. A new sers-active sandwich structure. *Journal of Solid State Electrochemistry*, 2007, 11(2): 150～154.
[28] Mulvaney S P, He L, Natan M J, et al. Three-layer substrates for surface-enhanced raman scattering: preparation and preliminary evaluation. *Journal of Raman Spectroscopy*, 2003, 34(2): 163～171.
[29] Driskell J D, Shanmukh S, Liu Y, et al. The use of aligned silver nanorod arrays prepared by oblique angle deposition as surface enhanced raman scattering substrates. *Journal of Physical Chemistry C*, 2008, 112(4): 895～901.
[30] Driskell J D, Lipert R J, Porter M D. Labeled gold nanoparticles immobilized at smooth metallic substrates: systematic investigation of surface plasmon resonance and surface-enhanced raman scattering. *Journal of Physical Chemistry B*, 2006, 110(35): 17444～17451.
[31] Addison C J, Brolo A G. Nanoparticle-containing structures as a substrate for surface-enhanced raman scattering. *Langmuir*, 2006, 22(21): 8696～8702.

[32] Misra A K, Sharma S K, Kamemoto L, et al. Novel micro-cavity substrates for improving the raman signal from submicrometer size materials. *Applied Spectroscopy*, 2009, 63(3): 373~377.

[33] Shoute L, Bergren A J, Mahmoud A M, et al. Optical interference effects in the design of substrates for surface-enhanced raman spectroscopy. *Applied Spectroscopy*, 2009, 63(2): 133~140.

[34] Liu Y J, Zhao Y P. Simple model for surface-enhanced raman scattering from tilted silver nanorod array substrates. *Physical Review B*, 2008, 78(0754367).

[35] Mitsas C L, Siapkas D I. Generalized matrix-method for analysis of coherent and incoherent reflectance and transmittance of multilayer structures with rough surfaces, interfaces, and finite substrates. *Applied Optics*, 1995, 34(10): 1678~1683.

[36] Fu J X, Park B, Zhao Y P. Nanorod-mediated surface plasmon resonance sensor based on effective medium theory. *Applied Optics*, 2009, 48(23): 4637~4649.

[37] Abell J L, Driskell J D, Dluhy R A, et al. Fabrication and characterization of a multiwell array sers chip with biological applications. *Biosensors & Bioelectronics*, 2009, 24(12): 3663~3670.

[38] Yang W H, Hulteen J, Schatz G C, et al. A surface-enhanced hyper-raman and surface-enhanced raman scattering study of trans-1,2-bis(4-pyridyl)ethylene adsorbed onto silver film over nanosphere electrodes. Vibrational assignments: experiment and theory. *Journal of Chemical Physics*, 1996, 104(11): 4313~4323.

[39] Zhou Q, Liu Y J, He Y P, et al. The effect of underlayer thin films on the surface-enhanced raman scattering response of ag nanorod substrates. *Applied Physics Letters*, 2010, 97(12190212).

[40] Michaels A M, Jiang J, Brus L. Ag nanocrystal junctions as the site for surface-enhanced raman scattering of single rhodamine 6g molecules. *Journal of Physical Chemistry B*, 2000, 104(50): 11965~11971.

[41] Mcfarland A D, Young M A, Dieringer J A, et al. Wavelength-scanned surface-enhanced raman excitation spectroscopy. *Journal of Physical Chemistry B*, 2005, 109(22): 11279~11285.

[42] Qin L D, Zou S L, Xue C, et al. Designing, fabricating, and imaging raman hot spots. *Proceedings of the National Academy of Sciences of the United States of America*, 2006, 103(36): 13300~13303.

[43] Liu Y J, Zhang Z Y, Zhao Q, et al. Surface enhanced raman scattering from an ag nanorod array substrate: the site dependent enhancement and layer absorbance effect. *Journal of Physical Chemistry C*, 2009, 113(22): 9664~9669.

[44] Gansel J K, Thiel M, Rill M S, et al. Gold helix photonic metamaterial as broadband circular polarizer. *Science*, 2009, 325(5947): 1513~1515.

[45] Zhou Q, Yang Y, Ni J, et al. Rapid detection of 2, 3, 3′, 4, 4′-pentachlorinated biphenyls by silver nanorods-enhanced raman spectroscopy. *Physica E-Low-Dimensional Systems & Nanostructures*, 2010, 42(5): 1717~1720.

[46] Fleming G D, Golsio I, Aracena A, et al. Theoretical surface-enhanced raman spectra study of substituted benzenes i. Density functional theoretical sers modelling of benzene and benzonitrile. *Spectrochimica Acta Part a-Molecular and Biomolecular Spectroscopy*, 2008, 71(3): 1049~1055.

[47] Zhou Q, Yang Y, Ni J E, et al. Rapid recognition of isomers of monochlorobiphenyls at trace levels by surface-enhanced raman scattering using ag nanorods as a substrate. Nano Research, 2010, 3(6): 423~428.

Part 3

Methods of Decontaminating the Environment from Organic Pollutants

10

Fenton´s Process for the Treatment of Mixed Waste Chemicals

Cláudia Telles Benatti[1] and Célia Regina Granhen Tavares[2]
[1]*Faculdade Ingá - UNINGÁ,*
[2]*Universidade Estadual de Maringá - UEM,*
Brazil

1. Introduction

In recent years, with an increase in the stringent water quality regulations due to environmental concerns, extensive research has focused on upgrading current water treatment technologies and developing more economical processes that can effectively deal with toxic and biologically refractory organic contaminants in wastewater. In this context, in order to avoid or mitigate the possible adverse health, safety, and environmental impacts, to grantee compliance with federal, state, and local environmental laws, or only to set an example to students, many institutions of higher education have supported researches that aim to establish a treatment process for practical and economic disposal of waste chemicals.

Waste chemicals in academic laboratories are by-products of research, teaching and testing activities. Waste chemicals from academic research laboratories can be considered one of the most polluting wastewaters and they pose more problems for the treatment and subsequent adequate disposal, due to their unique characteristics. These wastes are generated by laboratory operations, such as chemical analysis and research activities, including chemical and biological treatment experiments on a wide range of synthetic and natural wastewaters, and may include an abundance of unused laboratory reagents. Thus, they may present a great diversity of composition and volume, including refractory organics, toxic compounds and heavy metals, and may offer potential hazards to both health and environment.

The ultimate destination of waste is usually a treatment, storage, and disposal facility (National Research Council, 1995). The treatment of waste chemicals is typically via chemical action, such as neutralization, precipitation and reduction to yield a less toxic waste. However, in most cases, the treatment product still cannot be safely disposed of in the sanitary sewer. Most generators also adopt the practice of land filling or direct incineration of hazardous wastes. In this scenario, the development of economical methods to achieve a high degree of wastewater treatment is highly desirable.

The development and application of several Advanced Oxidation Processes (AOPs) to destroy toxic and biologically refractory organic contaminants in aqueous solutions concentrated significant research in the field of environmental engineering during the last decades. Among AOPs, the Fenton's reagent is an interesting solution since it allows high depuration levels at room temperature and pressure conditions using innocuous and easy to handle reactants. The inorganic reactions involved in Fenton process are well established and the process has been used for the treatment of a variety of wastewaters. The high efficiency of this technique can be explained by the formation of strong hydroxyl radical (•OH) and oxidation of Fe^{2+} to Fe^{3+}. Both Fe^{2+} and Fe^{3+} ions are coagulants, so the Fenton process can, therefore, have dual function, namely oxidation and coagulation, in the treatment processes (Badawy & Ali, 2006). It is essential, though, to investigate and set the operating conditions that best suits the wastewater that are being treated in order to achieve high degradation efficiencies.

In previous research work, chemical oxidation using Fenton's reagent was tested as a treatment method for mixed waste chemicals and optimized using a response surface methodology (Benatti et al., 2006). As predicted, the process optimization leaded to high COD removal (92.3%), with a high efficiency in the removal of heavy metals as a side effect. However, its major disadvantage is the production of $Fe(OH)_3$ sludge that requires further separation and proper disposal, with the consequent increase in operation costs. To overcome the drawback, the knowledge of the treatment process residue characteristics is imperative to design a successful waste management plan that may guarantee the viability of the applied technology to the wastewater treatment.

Most studies focused on applying the Fenton process to wastewater treatment and do not take into account the generated residues. Using the previously optimized set of process variables, the effects of Fenton´s reagent treatment of mixed waste chemicals are thoroughly discussed in the current chapter. The whole process was analysed from an environmental perspective, where the focus was not only the treatment of the wastewater, but also the characterization for correct final destination of the solids originating from the treatment process. Furthermore, the process was applied to wastewaters generated in different periods in order to study the effect of wastewater composition on the process efficiency.

This chapter also discusses about the unique nature of waste chemicals, presents different applications of Fenton´s reagent on the treatment of a variety of wastewaters, and provides an insight into the Fenton´s reaction mechanisms and some background on their operations conditions.

2. Theoretical approach

Advanced oxidation processes (AOPs) are based on the generation of very reactive species such as hydroxyl radical (•OH), a nonspecific, strong oxidant which reacts with most organic and biological molecules at near diffusion-controlled rates (>10^9 M^{-1} s^{-1}) (Büyüksönmez et al., 1999). Common AOPs involve Fenton and Fenton "like" processes, ozonation, photochemical and electrochemical oxidation, photolysis with H_2O_2 and O_3, high voltage electrical discharge (corona) process, TiO_2 photocatalysis, radiolysis, wet oxidation, water solutions treatment by electronic beams or γ-beams and various combinations of these methods (Kušić et al., 2007). Among AOPs, Fenton's reagent has been used (either alone or in combination with other treatments) as a chemical process for the treatment of a wide range of wastewaters. Recent applications of Fenton´s reagent include the pre-treatment of olive mill wastewater (Lucas & Peres, 2009), the treatment of landfill leachate (Deng & Englehardt, 2006; Zhang et al., 2005), copper mine wastewater (Mahiroglu et al., 2009), water-based printing ink wastewater (Ma & Xia, 2009) and cosmetic wastewaters (Bautista et al., 2007), the degradation of pesticide (Li et al., 2009; Chen et al., 2007), antibiotic (Ay & Kargi, 2010; Elmolla & Chaudhuri, 2009), high-strength livestock wastewater (Lee & Shoda, 2008) and organic compounds of nuclear laundry water (Vilve et al., 2009), the oxidation of combined industrial and domestic wastewater (Badawy & Ali, 2006), the pre-oxidation of pharmaceutical wastewaters (Martínez et al., 2003), the treatment of water-based paint wastewater (Kurt et al., 2006) and cellulose bleaching effluents (Torrades et al., 2003), the degradation of the explosives 2,4,6-trinitrotoluene (TNT) and hexahydro-1,3,5-trinitro-1,3,5-triazine (RDX) after iron pre-treatment (Oh et al., 2003), and the treatment of different streams of textile wastewaters, such as the treatment of hot desizing wastewaters (Lin & Lo, 1997), and the treatment of dye wastewaters (Wang et al., 2008; Gulkaya et al., 2006).

Fenton´s reagent is also combined with biological process, as a pre-treatment to enhance the biodegradability of the recalcitrant compounds and lower the toxicity (Padoley et al., 2011, Mandal et al., 2010, Badawy et al., 2009) or as a post-treatment to improve the efficiency of the wastewater treatment (Ben et al., 2009, Yetilmezsoy & Sakar, 2008).

Fenton's reagent, which involves homogenous reaction and is environmentally acceptable (Bham & Chambers, 1997), is a system based on the generation of very reactive oxidizing free radicals, especially hydroxyl radicals, which have a stronger oxidation potential than ozone; 2.8 V for •OH and 2.07 V for ozone (Heredia et al., 2001). The Fenton's reactions at acidic pH lead to the production of ferric ion and of the hydroxyl radical (Garrido-Ramírez et al., 2010; Gallard & De Laat, 2001; Walling, 1975):

$$Fe^{2+} + H_2O_2 \rightarrow Fe^{3+} + OH^- + \bullet OH \quad E_a = 39.5\ kJ\ mol^{-1}\ k_1 = 76\ M^{-1}s^{-1} \tag{1}$$

$$Fe^{3+} + H_2O_2 \rightarrow Fe^{2+} + HO_2\bullet + H^+ \quad E_a = 126\ kJ\ mol^{-1}\ k_2 = 0.001\text{-}0.01\ M^{-1}s^{-1} \tag{2}$$

Hydroxyl radicals may be scavenged by reaction with another Fe^{2+} or with H_2O_2 (Torrades et al., 2003; Chamarro et al., 2001; Lu et al., 1999):

$$\bullet OH + Fe^{+2} \rightarrow HO^- + Fe^{+3} \tag{3}$$

$$\bullet OH + H_2O_2 \rightarrow HO_2 \bullet + H_2O \tag{4}$$

Hydroxyl radicals may react with organics starting a chain reaction (Bianco et al., 2011; Dercová et al., 1999):

$$\bullet OH + RH \rightarrow H_2O + R\bullet, \text{ RH = organic substrate} \tag{5}$$

$$R\bullet + O_2 \rightarrow ROO\bullet \rightarrow \text{products of degradation} \tag{6}$$

Ferrous ions and radicals are produced during the reactions as shown below (Lu et al., 1999):

$$H_2O_2 + Fe^{3+} \rightleftarrows H^+ + FeOOH^{2+} \tag{7}$$

$$FeOOH^{2+} \rightarrow HO_2 \bullet + Fe^{2+} \tag{8}$$

$$HO_2 \bullet + Fe^{2+} \rightarrow HO_2^- + Fe^{3+} \tag{9}$$

$$HO_2 \bullet + Fe^{3+} \rightarrow O_2 + Fe^{2+} + H^+ \tag{10}$$

The efficiency of this process depends on several variables, namely temperature, pH, hydrogen peroxide, ferrous ion concentration and treatment time. The oxidizing potential of the hydroxyl radical is pH dependent, and varies from E^0 =1.8 V at neutral pH to +2.7 V in acidic solutions (Buxton et al., 1988, as cited in El-Morsi et al., 2002). Operating pH of the system has been observed to significantly affect the degradation of pollutants (Benitez et al., 2001; Kang & Hwang, 2000; Nesheiwat & Swanson, 2000; Lin & Lo, 1997). The degree of oxidation of organics with Fenton's reagent is maximum when the pH lies in the interval pH 3-5 (Lunar et al., 2000; Lin & Lo, 1997). Hydrogen peroxide is most stable in the range of pH 3–4, but the decomposition rate is rapidly increased with increasing pH above 5. Thus, the acidic pH level around 3 is usually optimum for Fenton oxidation (Gogate & Pandit, 2004; Neyens & Baeyens, 2003). However, in order to achieve high performances, these experimental conditions must be optimized (Homem et al., 2010). Some environmental applications of Fenton's reagent involve reaction modifications, including the use of high concentrations of hydrogen peroxide, the substitution of different catalysts such as ferric iron and naturally occurring iron oxides, and the use of phosphate-buffered media and metal-chelating agents. These conditions, although not as stoichiometrically efficient as the standard Fenton's reactions, are often necessary to treat industrial waste streams and contaminants in soils and groundwater (Büyüksönmez et al., 1999).

3. Materials and methods

3.1 Materials

All chemicals employed in this study were analytical grade. All solutions were prepared in distilled-deionized water. Glassware used in metals determination was washed with detergent, rinsed with tap water, soaked with HNO_3 (~50% v/v) for 24 h, and rinsed with distilled-deionized water prior to drying.

Reagent grade H_2O_2 was standardized using iodometric titration (U.S. Peroxide, 2003) and used as purchased. A 1 M $FeSO_4.7H_2O$ stock solution was prepared and standardized (Pavan et al., 1992) just before Fenton's experiments. Solutions of NaOH and H_2SO_4 were used for pH adjustments.

3.2 Sample preparation

Chemical effluents generated during the laboratory operations over a period of 17 months were monitored, collected and stored in clearly marked containers. Later on these effluents were divided into two groups and then mixed to obtain two combined samples (sample 1 and 2), produced in different periods of time that were used in the experiments of Fenton's oxidation.

3.3 Fenton´s experimental procedure

The Fenton's oxidation experiments were carried out under optimal conditions established in previous work (Benatti et al., 2006): ratio [COD]:[H_2O_2] = 1:9; ratio [H_2O_2]:[Fe^{2+}] = 4.5:1 and pH 4.

Experiments were carried out in 250 mL beakers with a solution volume of 150 mL that consisted of laboratory effluent without solids separation. The effluent was continuously mixed (100 rpm) in a jar test apparatus at room temperature. Firstly the pH of the solution was adjusted to 4 with NaOH (~30% w/v), and a sample was withdrawn and centrifuged at 2500 rpm for 5 min for the separation of suspended solids and COD_0 was determined in supernatant. The required amount of reagents was determined according to COD_0 value. The amount of $FeSO_4$ was first added to the reaction mixture. The Fenton reaction was then initiated by sequential addition of the required amount of H_2O_2, in three steps of equal volume added at intervals of 20 min, to moderate the rise in temperature that occurs as the reaction proceeds and to minimize quenching of •OH. The pH adjustments were performed using H_2SO_4 or NaOH solutions before each reagent addition and then at each hour. The Fenton reaction time was initiated by the addition of the first required amount of H_2O_2 to the reaction mixture. After reactions were completed (4 h), precipitation of the oxidized iron as $Fe(OH)_3$ was performed by adjusting the pH to 8 and then about 15 h of clarification under quiescent conditions. Final samples of supernatant were taken for COD measurements and for determination of residual hydrogen peroxide. All experiments were performed in duplicate.

The wastewater was also characterized in terms of pH, apparent color, turbidity, total phenols, sulfate, sulfide, chloride, total phousphorous, nitrogen and metals content (Ag, Al, Ca, Cd, Co, Cr, Cu, Fe, Hg, K, Mg, Mn, Na, Ni, Pb and Zn) before and after the Fenton's oxidation experiments.

3.4 Fenton´s residues characterization

The Fenton´s reagent residues were characterized through leaching tests, chemical fractionation and analysis of X-ray diffraction.

Sludges obtained at the end of the Fenton's process were named for ease of notation as residues 1 and 2 according to their originating effluent samples, and each one was divided into two portions. The first portion was transferred to an ambar glass flask and preserved at 4 °C for leaching tests. The second one was filtered under vacuum, rinsed with distilled-deionized water to eliminate the excess of sodium hydroxide, and then dried at 105 °C. The dried solids were removed from the filter, grinded and homogenized in a porcelain mortar, and stored for chemical fractionation and analysis of X-ray diffraction.

The leaching tests were carried out according to the Brazilian methodology (ABNT, 1987a). In these tests, 100 g of sludge was filtered through 0.45 μm cellulose ester membranes, the liquid phase was stored at 4 °C and the membranes containing the solid phases were transferred to 500 mL glass beakers. Distilled-deionized water (1:16) was added along with a sufficient quantity of acetic acid (0.5 N) to adjust the pH to 5.0±0.2. The suspension was stirred for 24 h. A pre-calculated amount of distilled-deionized water was then added to the suspension, the separation of phases was performed by filtering in 0.45 μm cellulose ester membrane, and the obtained solution was mixed with the stored liquid phase initially obtained. The mixed liquor was then preserved at 4 °C for metals determination. After analysis, the residues were classified according to ABNT-10004 (ABNT, 1987b) with regard to the maximum limits obtained in the extracts.

The chemical fractionation of solid residues was carried out by use of the following sequential dissolution procedure:

Step 1. Exchangeable ions were removed at room temperature for 1 h with magnesium chloride solution (1M $MgCl_2$, pH 7.0) and continuous agitation in Erlenmeyer flasks, using 1 g of dried precipitate and 8 mL of extractant, according to the methodology of Tessier et al. (1979).

Step 2. Amorphous material was removed by acid ammonium oxalate (pH 3.0) in a 4 h extraction in Kjeldahl flasks kept in the dark and at room temperature, using 1 g of dried precipitate and 100 mL of extractant, according to the methodology proposed by Camargo et al. (1986).

Step 3. Free Fe-oxides were removed from 0.5 g of dried samples by Na–dithionite–citrate–bicarbonate system in Kjeldahl flasks according to the methodology of Mehra & Jackson (1960).

Step 4. The residues remaining from Step 3 as well as the original samples were digested with a 2:1 mixture of nitric and perchloric acids.

Between each successive extraction, separation was performed by filtering the suspension through a 0.45 μm cellulose ester membrane. During filtration the residues were washed with distilled-deionized water. The liquid phase was transferred to volumetric flasks, diluted to mark, transferred to amber glass flasks, acidified to pH< 2 with concentrated nitric acid and preserved at 4 ◦C for metal determinations. The solids were transferred with a porcelain spoon and rinses of distilled-deionized water to a tare porcelain crucible and dried at 105 ◦C, cooled under vacuum in a desiccator, and weighed for weigh loss determinations. The solids were then crushed, manually homogenized using a porcelain spoon, dried again at 105 ◦C for moisture removal, and stored in a desiccator until ready for use in the next extraction step. These procedures were performed in three or more replicates.

The original dried solids as well as the solids from each extraction phase were analyzed by X-ray diffraction (XRD) using a Shimadzu D6000 diffractometer (Cu Kα radiation and λ = 1.54178 Å; scanning speed 2° 2θ min^{-1} for the ranges 5–70° 2θ). The mineralogical phases identification in generated X-ray diffractograms was performed by the position and intensity of diffraction planes. The amorphous phase diffractograms were determined by DXRD according to Schulze (1994), by subtraction of diffractograms intensity of original samples and of residues obtained after amorphous phase extraction.

3.5 Analytical methods

pH was measured with a Digimed-DM-20 pH meter calibrated with pH 4.01 and 6.86 Digimed standard buffers. The analyses of COD, settable solids, total phosphorous and soluble sulfate (turbidimetric method) were performed in accordance with standard methods (APHA-AWWA-WEF, 1998). Residual hydrogen peroxide was determined by the iodometric titration method (U.S. Peroxide, 2003). Total phenols were measured according to the colorimetric method of Folin-Ciocalteu reagent (Scalbert et al., 1989). Real and apparent color, turbidity, and sulfide were measured spectrophotometrically (spectrophotometer DR/2010, HACH, Loveland, CO) using the APHA platinum–cobalt standard method, the attenuated radiation method (direct reading), and the HACH sulfide test, respectively. Chloride was determined by the silver nitrate titration method (APHA-AWWA-WEF, 1998). Nitrogen (Kjeldahl) was measured according to Adolfo Lutz Institute Analytical Norms (Normas Analíticas do Instituto Adolfo Lutz, 1985). Metallic elements (Ag, Al, Ca, Cd, Co, Cr, Cu, Fe, Hg, K, Mg, Mn, Na, Ni, Pb and Zn) were determined in extracts by atomic absorption spectroscopy (Varian SpectrAA – 10 Plus).

The COD measured in the samples taken from Fenton's reactor was converted according to Eq. (11) to prevent the interference of H_2O_2 on COD analysis (Talini & Anderson, 1992).

$$COD = COD_M - R_p \times 0.25 \qquad (11)$$

where COD, COD value in the sample (mg O_2/L); COD_M, measured COD (mg O_2/L); R_p, residual hydrogen peroxide in the sample (mg/L).

The percent of COD removal was then determined through the following equation:

$$\eta(\%) = (\frac{COD_0 - COD_E}{COD_0}) * 100 \tag{12}$$

where η, percentage of COD removal; COD_0, measured COD in supernatant before oxidation (mg O_2/L); COD_E, COD value in clarified supernatant after precipitation (mg O_2/L).

4. Results and discussion

Throughout the monitoring period waste chemicals were generated as a consequence of chemical analyses and other research activities. Part of these residues is originated from chemical oxygen demand (COD), total phenols, nitrogen, protein, phosphate and sulfide determinations. Another significant part is constituted by diluted metal solutions containing Ag, Al, Ca, Cd, Co, Cr, Cu, Fe, Hg, K, Mg, Mn, Na, Ni, Pb and Zn, solutions standardization, and an abundance of unused laboratory reagents. As a consequence, the samples could be described as a quite complex wastewater which comprises different chemical species in dissolved, colloidal and particulate form. The main characteristics of the raw laboratory wastewater are presented in Table 1. Results in Table 1 show that the wastewater has high organic load (COD up to 2.7 g/L) and extremely low pH (<1.0). All raw samples contained an amount of brown solids presented as settable solids.

Parameter	Unity	Sample 1	Sample 2
pH		< 1	< 1
Apparent color	Pt/Co	12700	10250
Real color	Pt/Co	7150	1020
Turbidity	NTU	4010	2960
Total COD	mg O_2/L	2345	2676
Total phenols	mg/L	58.6	37.9
Settable solids	mL/L	4.5	4.0

Table 1. Characteristics of raw waste chemicals

Table 2 presents the results of chemical oxidation by Fenton´s reagent for both samples under optimum operation conditions. The characterization of the wastewater before and after Fenton's treatment is presented in Table 3.

Sample	Initial	End of the oxidation stage		End of the precipitation stage			
	COD mg O_2/L	COD mg O_2/L	% COD removal	residual H_2O_2 mg/L	Catalytic sludge mL	COD mg O_2/L	% COD removal
1	898	232	74,1	13,4	40	93	89,7
2	769	176	77,1	30,0	40	166	78,5

Table 2. Results of the waste chemical treatment by Fenton´s reagent.

Parameter	Unity	Sample 1			Sample 2		
		Raw supernatant	Initial effluent[a]	Final effluent at optimized conditions	Raw supernatant	Initial effluent[a]	Final effluent at optimized conditions
pH		<1	4	8	<1	4	8
Apparent color	Pt/Co	308	2205	137	503	1615	352
Turbidity	NTU	N.D.	42	10	N.D.	5	6
Soluble COD	mg O_2/L	1145	898	93	2576	769	166
Total phenols	mg/L	27.3	27.0	N.D.	39.8	35.9	N.D.
Sulfate	g/L	263	142	152	296	151	164
Sulfide	mg/L	0.04	0.07	0.01	0.05	0.12	0.01
Chloride[b]	mg/L	638	638	567	6523	6523	5247
Phosphorous	mg/L	203.8	90.5	0.1	394.2	233.2	0.4
Nitrogen	mg/L	1434	1038	566	38	38	N.D.
Metals:							
Ag	mg/L	1.1	1.1	0.8	1.1	1.1	1.0
Al	mg/L	18.8	5.2	1.4	6.8	6.4	4.5
Ca	mg/L	59.1	39.0	30.4	37.7	26.3	21.4
Cd	mg/L	0.3	0.3	0.3	6.9	2.3	0.5
Co	mg/L	1.2	0.8	0.8	1.3	0.8	1.0
Cr	mg/L	301.4	155.2	1.0	541.6	266.9	12.1
Cu	mg/L	1.7	0.3	0.6	4.5	2.0	0.3
Fe	mg/L	131.9	29.6	4.4	114.0	38.7	1.7
Hg	mg/L	1815.6	148.4	75.1	2769.3	70.7	99.8
K	mg/L	405.2	278.7	187.4	1084.8	646.4	645.3
Mg	mg/L	3.5	2.6	1.7	10.6	8.6	3.0
Mn	mg/L	0.7	0.6	0.5	0.8	0.4	0.1
Na	g/L	10.2	73.9	77.9	9.1	98.5	101.7
Ni	mg/L	0.9	1.2	1.1	1.1	1.3	1.1
Pb	mg/L	1.9	1.3	1.1	1.6	1.3	1.1
Zn	mg/L	0.9	0.5	N.D.	0.1	N.D.	N.D.

[a] Sample after pH adjustment to 4 with NaOH (30% w/v).
[b] Chloride content was determined at pH 8.

Table 3. Characteristics of the laboratory wastewater supernatant before (raw and after pH adjustment) and after oxidation at optimized conditions (N.D.: not detected).

One set of experiments was conducted as control experiments (without any addition of hydrogen peroxide or ferrous sulfate). The simple pH adjustment of the effluent to 8 did not contribute for significant COD or phenol removal. Though it was capable of minimizing the toxic metal content in solution as metal hydroxide precipitation (data not shown), it presented an undesirable effect. The raise of pH of the mixed waste chemicals to high values (pH=8) resulted, under temperatures below 20 °C, in a high instable solution, with the formation of a crystal solid phase simply by its manipulation. This is probably due to the formation of several inorganic substances, and for such substances solubility decreases with decreasing temperature.

Thus, Fenton's oxidation was conducted on wastewater without a previous precipitation of the metals. This process configuration favours the elimination of a pH adjustment step since the laboratory wastewater is highly acidic (see Tables 1 and 3) and the Fenton's reaction is also conducted under acidic conditions, with the raise of pH to 8 at the end of the process. Fenton's oxidation conducted under different conditions suggests that the presence of others metals do not interfere in the efficiency of oxidation of organic compounds by the Fenton's reagent.

Bidga (1995) describes the Fenton method as a process divided into four stages. First, pH is adjusted to low acidity, at pH value of 3-5. In this study, it was adjusted to optimum pH (pH=4). Then, main oxidation reaction takes place. The wastewater is then neutralized at pH of 7-8 and finally precipitation occurs. In this study pH=8 was adopted in order to favour the precipitation of metal species as hydroxides. Wastewater characteristics at different process stages are presented in Table 2.

During the Fenton oxidation process and, mainly, during the pH adjustment to 8, a large amount of flocks of various sizes in the wastewater were observed. According to Walling & Kato (1971), the small flocs were ferric hydroxo complexes formed by complex chain reactions of ferrous and hydroxide ions. After a period of natural sedimentation (about 15 hours), all flocs settled out in wastewater forming a catalytic sludge, whose volume is also presented in Table 2.

The supernatant separated by decantation revealed that the Fenton´s reagent oxidation was efficient in degrading organic matter in both samples, reaching 89.7 and 78.5% COD removal in samples 1 and 2, respectively. Moreover, the total phenols presence was not detected after the application of Fenton´s reagent. Moreover, the treated liquor under the optimized conditions showed Fe concentrations up to 4.4 mg/L. Thus, the Fe concentration could be kept below 15 mg/L, which is the maximum value for Fe concentration in disposed effluents imposed by CONAMA (Brazilian National Environmental Council) standards (CONAMA, 2008).

It is worth mentioning that great amount of NaOH were necessary to raise the pH from below 1 to 2. From this point on, the pH showed to be more sensitive to the sequential addition of NaOH solution. As a consequence, the effluent presented a high sodium concentration that increases with the raise of pH (see Table 3).

Regarding the sulfate content, the raw wastewater at pH 4 presented an initial sulfate concentration of 142 and 151 g/L for samples 1 and 2, respectively (see Table 3). Because sulfuric acid and sodium hydroxide solutions were used for pH adjustments, and ferrous sulfate was used as a catalyst in the Fenton's process, further amounts of sulfate resulted from the wastewater oxidation. Thus, the sulfate concentration still remained extremely high at the end of the process, reaching 152 and 164 g/L for samples 1 and 2, respectively. Although the Brazilian legislation does not directly limit its concentration for effluent discharge, the CONAMA Resolution No. 357 (CONAMA/2005) states that effluents must not give to the receiving waters characteristics different from those used in their classification. This resolution does not include the sulfate concentration as a parameter to be monitored, probably due to the fact that the damage caused by sulfate emissions is not direct, since sulfate is a chemically inert, non-volatile, and non-toxic compound. However, high sulfate concentrations can unbalance the natural sulfur cycle (Silva et al., 2002; Lens et al., 1998). The accumulation of sulfate rich sediments in lakes, rivers and seas may cause the release of toxic sulfides that can provoke damages to the environment (Ghigliazza et al.,

2000). In addition, the release of sulfate-rich wastewaters in sewage systems may cause the inhibition or even the collapse of the biological treatment system. Thus, a post treatment system is required to bring the sulfate levels of the treated wastewater to values that are less than the maximum allowable limit set by the local regulatory authority (1000 mg/L) in order to allow its discharged directly to the municipal biological treatment facilities.

Regarding the metal content, the Fenton's reagent process presented as a side effect the removal of certain elements, such as chromium (up to 99.4%) and aluminum (up to 73.1%).

Fu et al. (2009) studied the removal of heavy metal ions in metal–EDTA complexes by Fenton and Fenton-like reaction followed by hydroxide precipitation, and achieved high removal efficiencies while conventional technologies, such as hydroxide, sulfide and dithiocarbamate-type precipitants could hardly work for it. Although Fenton-like process presented higher efficiency than Fenton process, at optimal operation conditions ($[H_2O_2]_0$ = 141 mM, $[Fe^{2+}]_0$ = 1.0 mM, $[Fe^{3+}]_0$ = 1.0 mM, initial pH 3.0 and precipitation pH 11.0), the removal efficiency of Ni(II) were above 92% for the two systems.

Mahiroglu et al. (2009) investigated the treatability of combined acid mine drainage (AMD)—Flotation circuit effluents from copper mine via Fenton's process and pointed out that heavy metals in the AMD could also be reduced to very low levels via Fenton reactions by taking part in oxidation steps.

Despite the results, the heavy metals content in the treated liquor still exceeded the maximum allowable limits for effluent discharge according to the CONAMA Resolution No. 397 (CONAMA/2008), especially for mercury, lead, chromium and silver, whose limit for disposal are 0.01, 0.5, 0.5 and 0.1 mg/L, respectively This stresses the need for an additional treatment for the removal of the remaining heavy metals prior to discharge to the environment.

Finally, the application of the Fenton's reagent in the destruction of organic compounds in mixed waste chemicals generated a dark brown slurry phase mainly constituted by heavy metals-Fe(III)-iron sludge, with a formation of 267 mL of sludge per liter of oxidized wastewater. This slurry phase presented a percentage of total suspended solids between 2.4 and 2.5% (w/w).

Thus, the overall result of the waste chemicals treatment by Fenton´s Reagent is the production of an aqueous solution with a substantially lower total carbonaceous load. However, the treated liquor still presented levels of heavy metals and sulfate that were too high to meet discharge standards. Thus, the Fenton's reagent cannot be applied as a stand-alone treatment option, but it can be used in combination with other treatment techniques. Finally, it is noteworthy that a potentially hazardous solid residue is obtained as a by-product of the Fenton's treatment. Hence, it is important to characterize this material for proper waste management, satisfying environmental and health related criteria.

As a first approach to characterize the solids originating from the Fenton´s process, this study focused on identifying the metals present in the residues, on their crystal structure, on the specific chemical bond in which electrons are delocalized and mobile, and on their magnetic properties. As a second approach, and in order to complement the physical characterization of the residues, leaching tests were also performed in order to obtain correct predictions of elements possible mobilization process in the environment.

The results of metals determination in these residues (Table 4) showed that the predominant metals are silver, chromium, mercury and iron. The main source of the first three ones is COD analysis. Iron is mainly a result of the Fenton's process, once it is used as a catalyst of the reaction and undergoes precipitation after oxidation. Regarding cadmium, the results of effluent characterization (Table 3) showed that sample 1 presented an initial concentration of 0.3 mg/L, which was kept constant after the equalization and oxidation stages of effluent treatment. In consequence, this element was not detected in the originating residue from the Fenton's process (Residue 1).

Metal	Residue 1	Residue 2
Ag	30157.1	38678.3
Al	2184.5	362.0
Ca	1827.7	1839.4
Cd	ND	480.0
Co	61.2	50.2
Cr	20747.2	28395.0
Cu	210.0	538.8
Fe	400065.2	324171.4
Hg	70430.1	96926.7
Mn	221.7	282.6
Ni	49.8	36.4
Pb	57.2	11.3
Zn	96.3	50.1

Table 4. Total element determinations (mg/kg) in samples (N.D.: not detected).

Fig. 1 shows the X-ray diffractograms obtained for the studied residues. This figure shows that the amorphous phase is predominant in the material. Besides the amorphous material identified as 2-line ferrihydrite (Cornell & Schuwertmann, 1996), the presence of the crystalline phases HgCl (calomel) and AgCl (chlorargyrite) was identified in both residues. The three solid phases detected by X-ray diffraction are also the ones that present the highest concentration considering the total metal content presented in Table 3. Although the studied residues were produced by the oxidation of mixed laboratory wastewaters generated in different periods, they present similar characteristics.

The knowledge of the total concentration of metals in residues of Fenton´s process is, however, not enough to evaluate their environmental impact. Addressing the chemical form of the element instead of the total trace element concentration renders the information gained through careful analysis much more valuable (Cornelis & Nordberg, 2007). The mobility of trace metals, as well as their bioavailability and related ecotoxicity to plants, depends strongly on their specific chemical forms or ways of binding (Quevauviller et al., 1997). Consequently, it is necessary to determine the solubility of these metals, which depends on the way of their association in residue.

Generally speaking, the sequential extraction of trace elements adopted in this study uses a series of chemical reagents, each time stronger, under specific conditions, to dissolve one or more specific phases from the solid sample, while preserving others. In the extract it is possible to speciate elements solubilised during the dissolution. By studying the distribution of the metals between the different phases, their contamination risks can be ascertained, because it is possible to divide a specific metal into fractions of increasing stability.

Thus, in order to study the distribution of the metals in residues and the phases to which they are bound, a sequential extraction procedure was designed based on the characteristics of the Fenton's process as well as on the analysis of the X-ray diffractograms presented in Fig. 1. In this procedure, the total metal content is divided into four fractions of increasing stability: exchangeable (Fraction 1), amorphous iron oxide (Fraction 2), crystalline iron oxide (Fraction 3), and inert or residual (Fraction 4). The solubility of the metals in the residues can be associated with their extraction, decreasing in the order of extraction sequence. In this sense, samples with a higher metal content in Fraction 1 will be potentially more dangerous than those that present a lower content in this fraction.

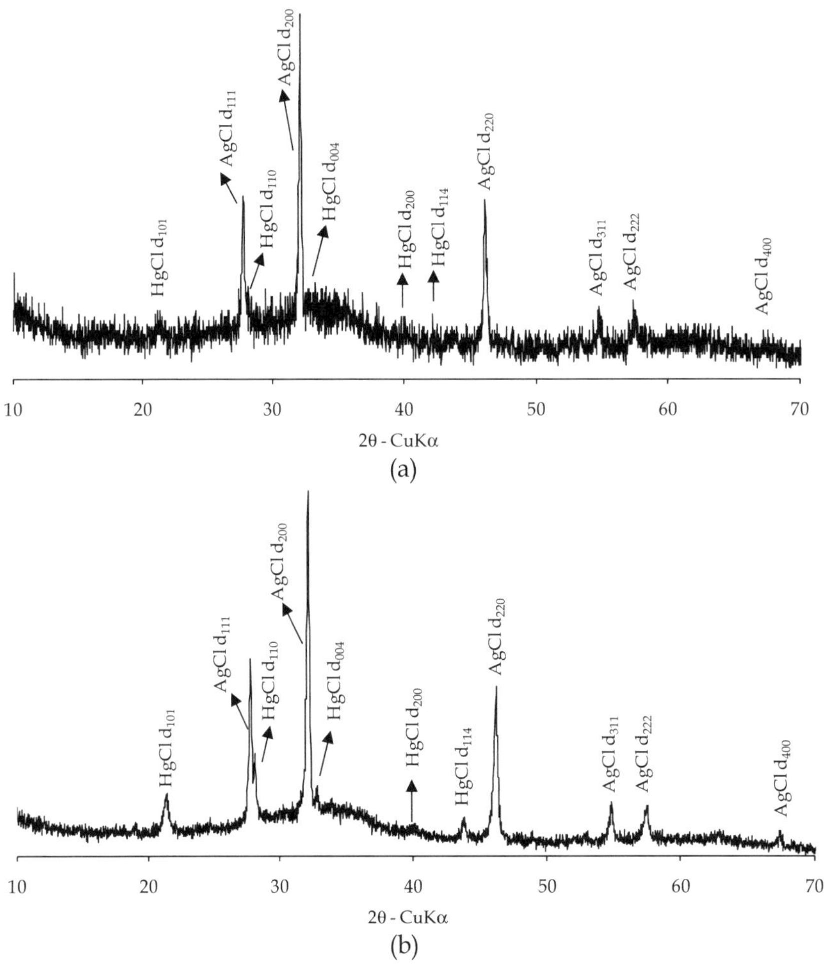

Fig. 1. X-ray diffractograms of the residue of Fenton's process: (a) Residue 1 and (b) Residue 2.

Metals extracted in Fraction 1 (exchangeable) correspond to those weakly absorbed, in particular to those retained in the residue surface with a weak electrostatic force. Changes in water ionic composition greatly affect the sorption-desorption processes of this fraction. Fraction 2 contains metals bound to amorphous iron oxide and corresponds to the reactive part of the iron compounds of the residues. Fraction 3 contains metals bound to crystalline iron oxide. This extraction favours the preferential orientation of remaining metals when submitted to X-ray diffractometry. Fraction 4 (inert or residual) contains primary and secondary minerals, mostly silicates, titanium and aluminum oxides, which can retain metals in their crystalline structure, removed from the laboratory wastewaters. These minerals are not expected to be released in a reasonable space of time in nature's normal conditions. Metals of this fraction are chemically stable and biologically inactive.

Fig. 2 shows the weight loss of the sample after each stage of the sequential extraction method adopted. As can be seen, the last two phases of the sequential extraction (crystalline iron oxide and residual), the less reactive fraction of the residues, represent only a small percentage in mass, about 11.2 and 16.6% for Residues 1 and 2, respectively. In fact, the residues originating from the Fenton's process is mainly constituted of amorphous material (over 80%) and most metals are co-precipitated to this fraction.

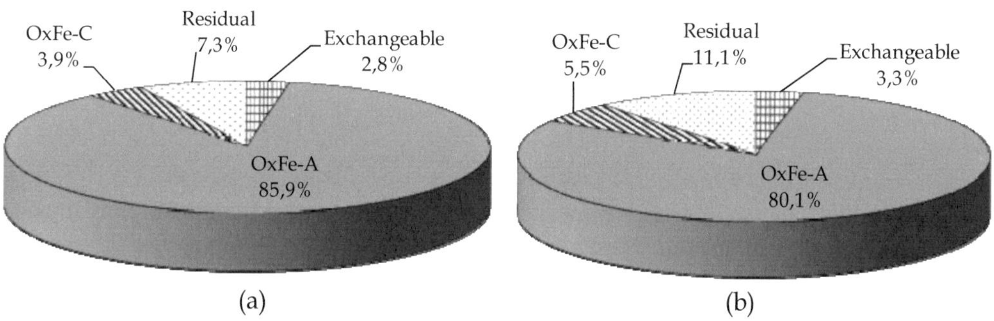

Fig. 2. Mass fractionation (% in mass) of Residues 1 (a) and 2 (b). Exchangeable, amorphous iron oxide (OxFe-A), crystalline iron oxide (OxFe-C), and final residual.

Table 5 presents the results of metals concentration in the different fractions of the studied residues. Fig. 3 shows the metal partitioning, in percentage of the total amount, found in the residues fractions. As can be seen, the sum of the four fractions (Table 5) is reasonably similar to the total contents obtained after digestion of the original samples (Table 4), indicating that no significant loss occurred during the sequential dissolution procedures, with recoveries of 96–100% in all cases. Thus, it is reasonable to assume that the dissolution procedure adopted was reliable for metals partitioning, increasing the confidence of the obtained data.

Several elements were detected in Fraction 1 (exchangeable), such as silver (0.2%), chromium (0.1%), copper (3.8%), iron (0.004%) and mercury (20.5%) of Residue 1; silver (0.1%), cadmium (5.2%), chromium (1.1%), copper (1.1%), iron (0.01%), mercury (31.6%) and zinc (0.2%) of Residue 2. These metals were associated to soluble salts as chlorites or sulfates, or simply absorbed to the residues surface.

Generally, both studied residues presented a similar distribution of metals at the different fractions obtained by the sequential dissolution procedure employed. Fig. 3 shows the chemical fractionation of each residue. With regard to silver, this element was mainly associated to the residual fraction of both residues (over 97% extraction). Elements like aluminum, calcium, cobalt, chromium, iron, manganese, nickel and zinc were predominantly found in the amorphous phase of both residues. In the case of cadmium, which was detected only in Residue 2, 86.9% of the total was found in the amorphous phase and only 6.8% was found in the most stable phases, which means in the crystalline iron oxide and residual phases. With regard to mercury, besides the exchangeable phase previously mentioned, 21.7 and 14.8% were found in the amorphous phase, 0.2 and 0.02% in the crystalline fraction and 53.5 and 49.8% in final residual, in Residues 1 and 2, respectively. In Residue 1, lead was found predominantly in the amorphous phase, representing 73.1% of the total content, and the remaining content was distributed between the crystalline (7.4%) and residual (19.3%) fractions. In Residue 2, though, lead was mainly associated with the residual fraction (99.7%); its availability was therefore low since most of it was in the insoluble form.

Sample	Metal	Exchangeable	OxFe-A	OxFe-C	Residual	Sum	% Recovery
Residue 1	Ag	75.0	35.1	107.0	29344.0	29561.1	98.0
	Al	N.D.	2051.4	68.6	56.4	2176.4	99.6
	Ca	N.D.	1739.3	16.6	64.6	1820.5	99.6
	Cd	N.D.	N.D.	N.D.	N.D.	0.0	-
	Co	N.D.	57	2.6	1.5	61.1	99.9
	Cr	28.5	19195.5	1122.5	318.0	20664.5	99.6
	Cu	7.9	190.9	0.3	6.2	205.3	97.7
	Fe	15.7	370959.6	19230.1	8634.0	398839.4	99.7
	Hg	14450.1	15263.0	140.0	37688.3	67541.4	95.9
	Mn	N.D.	210.2	6.5	4.0	220.7	99.5
	Ni	N.D.	45.6	1.3	1.2	48.1	96.7
	Pb	N.D.	41.9	4.3	11.1	57.3	99.9
	Zn	N.D.	75.4	N.D.	20.2	95.6	99.3
Residue 2	Ag	9.7	4.9	12.4	38571.0	38598.0	99.8
	Al	N.D.	323.8	5.8	29.1	358.7	99.1
	Ca	N.D.	1467.2	287.9	82.9	1838.0	99.9
	Cd	25.1	417.0	12.0	20.3	474.4	98.8
	Co	N.D.	42.5	5.3	0.7	48.5	96.7
	Cr	310.1	26745.1	725.6	145.6	27926.4	98.3
	Cu	5.6	529.6	0.1	3.2	538.5	100.0
	Fe	19.9	315243.4	4616.7	2934.5	322814.5	99.6
	Hg	30648.7	14296.5	23.6	48263.1	93231.9	96.2
	Mn	N.D.	273.1	2.4	0.4	275.9	97.7
	Ni	N.D.	33.7	1.3	0.8	35.8	98.5
	Pb	N.D.	N.D.	N.D.	11.2	11.2	99.7
	Zn	0.1	40.8	N.D.	7.9	48.8	97.3

Table 5. Metals concentration (mg/kg) in the different fractions (exchangeable, amorphous iron oxide (OxFe-A), crystalline iron oxide (OxFe-C), and final residual) of the studied residues (N.D.: not detected).

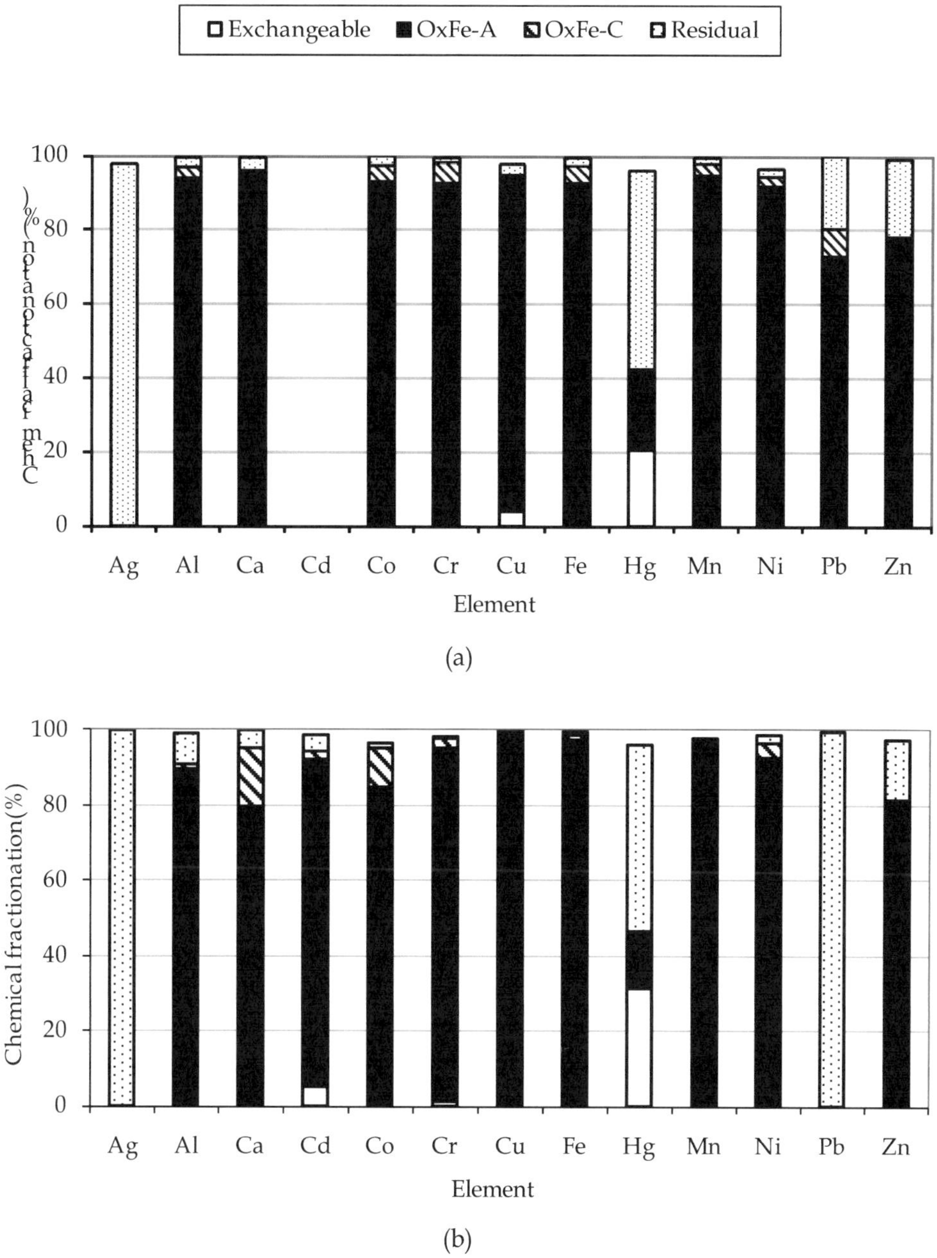

Fig. 3. Chemical fractionation of solids originating from the Fenton's process: (a) Residue 1 and (b) Residue 2. Exchangeable, amorphous iron oxide (OxFe-A), crystalline iron oxide (OxFe-C), and final residual.

Figs. 4 and 5 show the X-ray diffractograms of the studied residues. Fig. 6 shows the differential X-ray diffractograms for the amorphous phase of Residues 1 and 2.

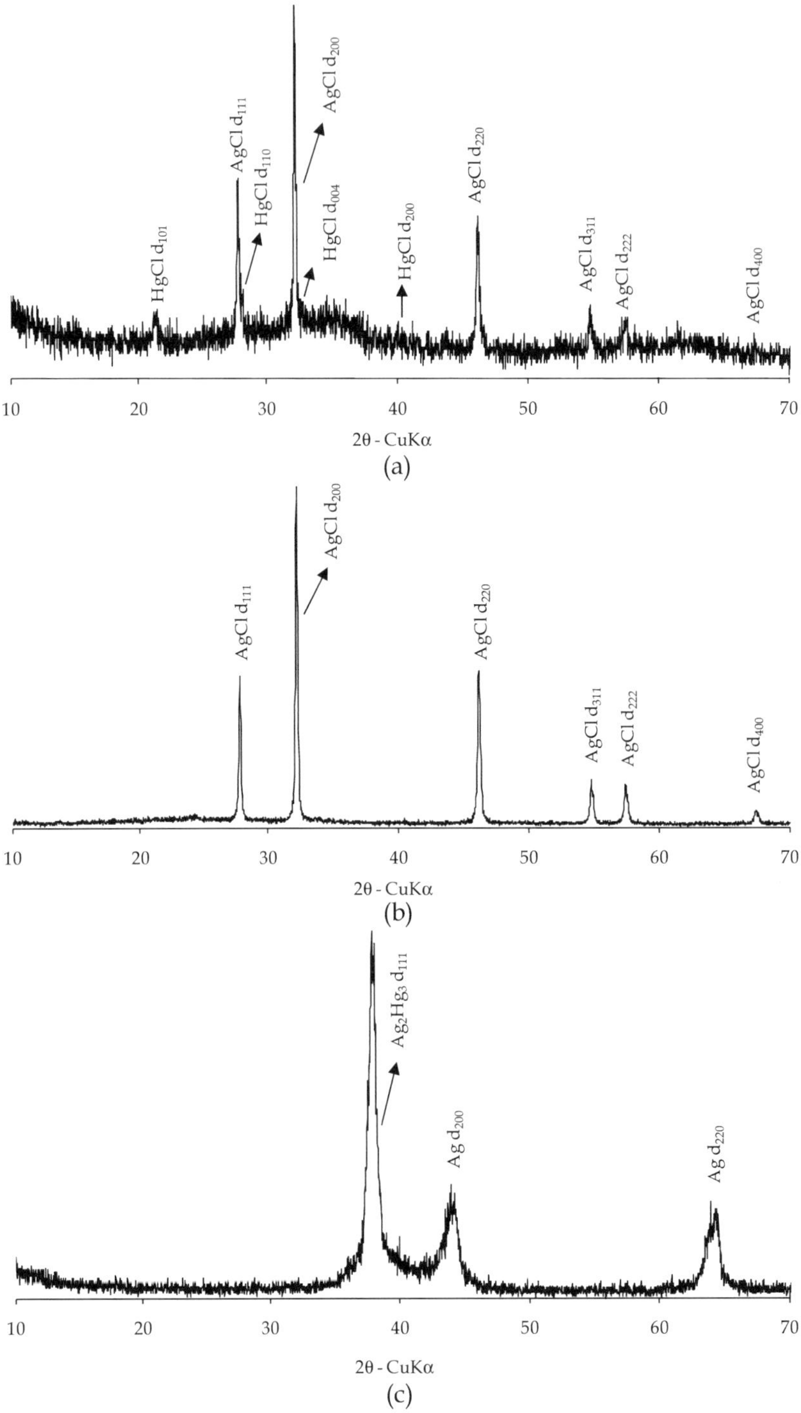

Fig. 4. X-ray diffractograms for the different fractions of Residue 1. (a) After exchangeable extraction; (b) after amorphous iron oxide extraction and (c) after crystalline iron oxide extraction.

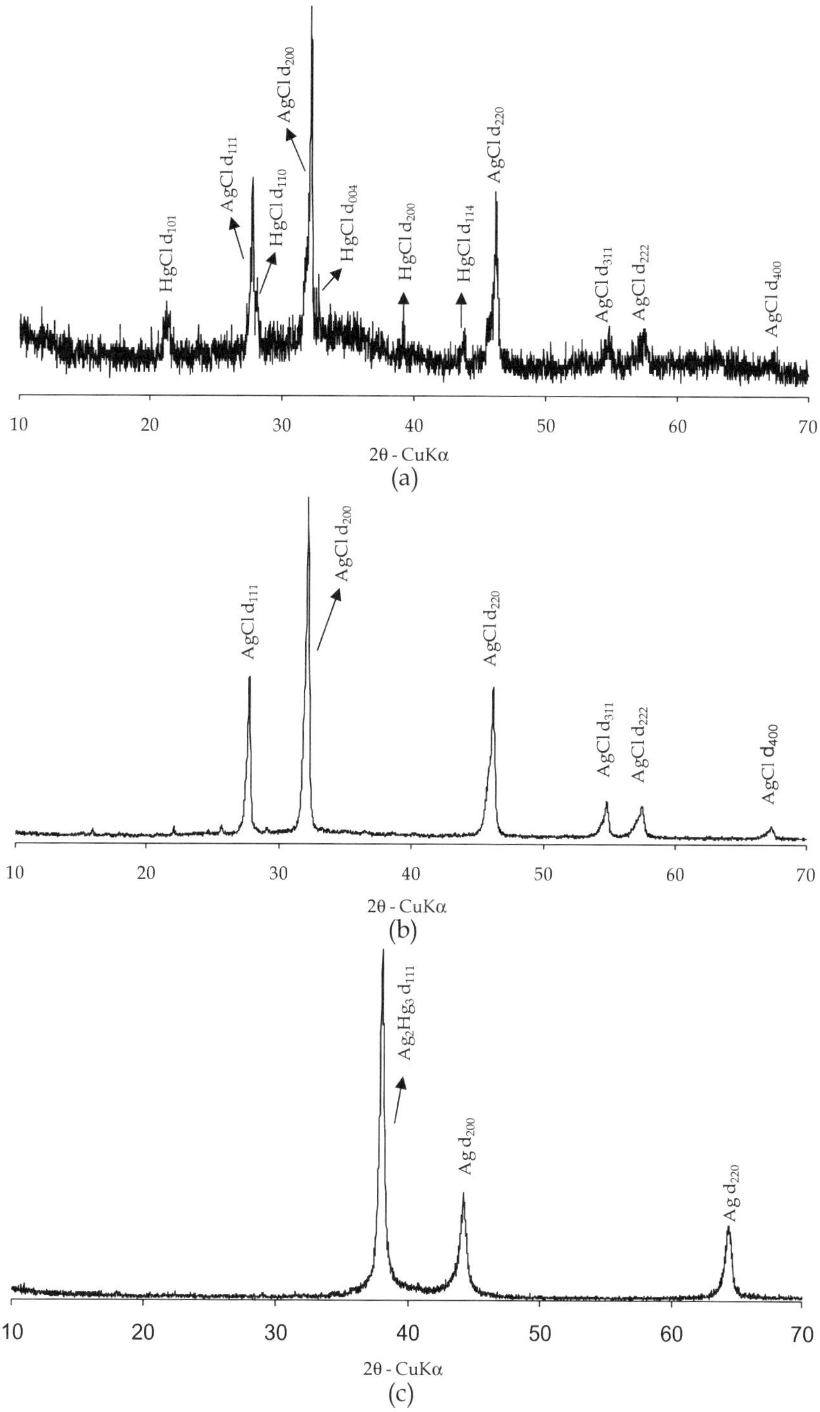

Fig. 5. X-ray diffractograms for the different fractions of Residue 2. (a) After exchangeable extraction; (b) after amorphous iron oxide extraction and (c) after crystalline iron oxide extraction.

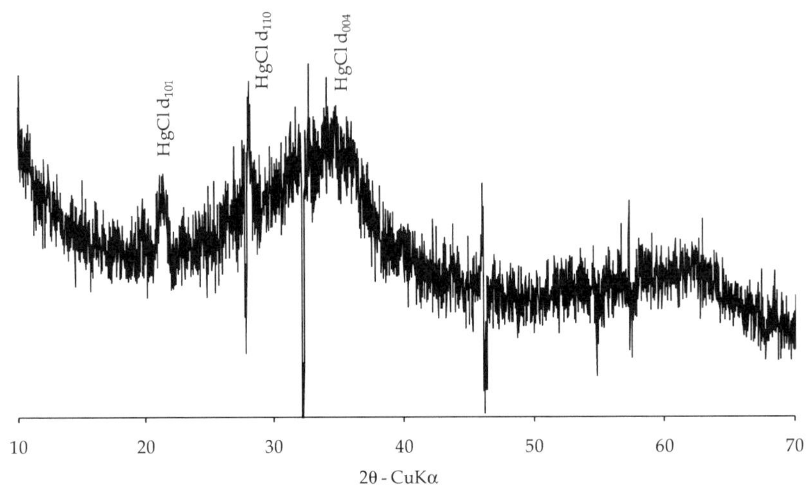

(a)

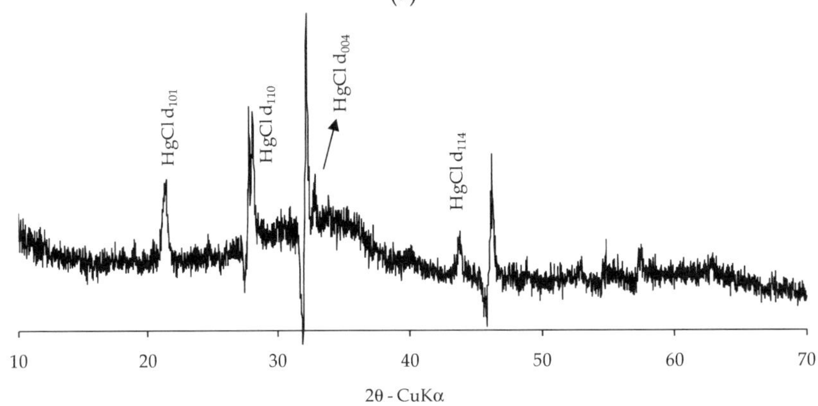

(b)

Fig. 6. Differential X-ray diffractograms for the amorphous phase and in the presence of HgCl: (a) Residue 1 and (b) Residue 2.

The position and intensity of diffraction planes indicate that there are no differences between the two solid samples. Therefore, although the wastewater samples were generated in distinct periods of time, their differences in chemical composition did not affect the mineralogical phases present in the solids resulting from their treatment by Fenton's reactions. From the interpretation of the main diffraction planes observed in the diffractograms, it was possible to identify the most likely compounds related to them. The solid residues originated from the Fenton's process presented a typical profile of material constituted predominantly by amorphous phase (2-line ferrihydrite (Schwertmann et al., 1982)), and silver and mercury compounds with well-defined crystalline structure were identified as chlorargyrite and calomel, respectively. The residual or inert fraction of the residues presented characteristics of a metallic league, formed mostly by elementary silver and mercury with minor amounts of other metals (Table 5). The Fenton's residues were originally brown, and did not exhibit detectable magnetic properties by the presence of a magnetic field, and a black coloured residual was obtained at the end of the sequential extraction procedure. The diffractograms in Fig. 6 present the amorphous phase of the studied residues with the presence of calomel, obtained by the subtraction of difractograms intensities of original samples (Fig. 1a and b) and of residues obtained after amorphous phase extraction (Figs. 4b and 5b). Therefore, differently to that observed for soils, acid ammonium oxalate extracted a portion of poorly crystalline crystals of calomel, besides most iron associated to 2-line ferrihydrite. The use of dithionite at the crystalline iron extraction was capable of reducing remaining mercury and silver of the selective dissolution procedure and formed a metallic league that presents diffraction planes of Ag and Ag_2Hg_3.

From the results of the selective dissolution, the solids originated from the Fenton's process are not stable, being composed fundamentally by high toxicity metals like chromium and mercury. Hence, the results suggest that leaching of metals into the environment can occur even under mild environmental conditions.

Finally, Table 6 presents the results of metal leachability for both studied residues determined by ABNT-NBR 10005 method (ABNT, 1987a). For comparative purposes, the same table shows the maximum levels of heavy metals allowed in Brazilian non-hazardous wastes (ABNT, 1987b). Based on the obtained chemical analytical results and on the classificatory procedure proposed by ABNT-NBR 10004 (ABNT, 1987b), the solid residues from the Fenton's process can be classified as Class I—hazardous. Both residues are characterized as toxic TL (leaking test). Residue 1 received the identification codes D009 and D011, due to its mean metal concentration above the maxima permitted for total chromium and mercury, and Residue 2, D007, D009 and D011 due to its cadmium, total chromium and mercury concentration, respectively.

Metals	Element code	Residue 1	Residue 2	Limits*
Cd	D007	N.D.	0.8	0.5
Pb	D008	N.D.	N.D.	5.0
Cr	D009	13.6	18.4	5.0
Hg	D011	49.8	243.1	0.1
Ag	D012	0.2	0.2	5.0

* Identification codes and maximum established limits according to ABNT-NBR 10004 (ABNT, 1987b).

Table 6. Results of metals leachability (mg/L) for both studied residues (N.D.: not detected).

Thus, the obtained results indicate a great potential for soil, surface and underground waters contamination by heavy metals (chromium, cadmium and mercury), if the residues originated from the Fenton's process are disposed of improperly. A stabilization process of the residues is necessary prior to their disposal in the environment. Furthermore, the results indicated a high potential of silver and mercury recovery, which represent a large part of the studied residues.

5. Conclusion

Chemical oxidation using Fenton's reagent under optimum conditions has proven to be a viable alternative to the oxidative destruction of organic pollutants in mixed waste chemicals, with a COD removal of 89.7 e 78.5% in samples 1 and 2, respectively. Moreover, total phenols were not detected in the treated liquor. However, the reported results also

indicated that no single technology could be applied to mixed waste chemicals as a stand-alone treatment option once the concentration of certain inorganic constituents, such as heavy metals and sulfate, still remained high for effluent discharge.
Regarding to the Fenton's residues, they were classified as hazardous according to Brazilian waste regulations. The application of the sequential dissolution procedure indicated that the metals in the Fenton's residues are mainly constituted of amorphous material (over 80%). Furthermore, the reactive fractions of the residues (exchangeable and amorphous iron oxide fractions) retain most of remaining metals. Therefore, the Fenton's residues present great potential for environmental contamination, and require an administration system and control of their final disposal. However, Fenton's residues present great potential for silver and mercury recovery, since these elements represent a great portion of the studied residues.

6. Acknowledgment

The support provided by CAPES and by the State University of Maringá is gratefully acknowledged.

7. References

ABNT - Associação Brasileira de Normas Técnicas. (1987a). Lixiviação de Resíduos - Procedimento, NBR 10005, Rio de Janeiro, R.J.

ABNT - Associação Brasileira de Normas Técnicas (1987b). Resíduos Sólidos-Classificação, NBR 10004, Rio de Janeiro, R.J.

APHA-AWWA-WEF—American Public Health Association, American Water Works Association, Water Environment Federation. (1998). *Standard Methods for the Examination of Water and Wastewater* (20th ed), ISBN 0875530478, Washington, DC.

Ay, F., Kargi, F. (2010). Advanced oxidation of amoxicillin by Fenton's reagent treatment. *Journal of Hazardous Materials*, Vol. 179, pp. 622-627, ISSN 0304-3894.

Badawy, M., Wahaab, R.A., El-Kalliny, A.S. (2009). Fenton-biological treatment processes for the removal of some pharmaceuticals from industrial wastewater. *Journal of Hazardous Materials*, Vol. 167, pp. 567-574, ISSN 0304-3894.

Badawy, M.I.; Ali, M.E.M. (2006). Fenton's peroxidation and coagulation processes for the treatment of combined industrial and domestic wastewater. *Journal of Hazardous Materials*, Vol. B136, pp. 961-966, ISSN 0304-3894.

Bautista, P., Mohedano, A.F., Gilarranz, M.A., Casas, J.A., Rodriguez, J.J. (2007). Application of Fenton oxidation to cosmetic wastewaters treatment. *Journal of Hazardous Materials*, Vol. 143, pp. 128-134, ISSN 0304-3894.

Ben, W., Qiang, Z., Pan, X., Chen, M. (2009). Removal of veterinary antibiotics from sequencing batch reactor (SBR) pretreated swine wastewater by Fenton's reagent. *Water Research*, Vol. 43, pp. 4392-4402, ISSN 0043-1354.

Benatti, C.T.; Tavares, C.R.G.; Guedes, T.A. (2006). Optimization of Fenton's oxidation of chemical laboratory wastewaters using the response surface methodology. *Journal of Environmental Management*, Vol. 80, pp. 66-74, ISSN 0301-4797.

Benitez, F.J., Acero, J.L., Real, F.J., Rubio, F.J., Leal A.I. (2001). The Role of Hydroxyl Radicals for the Decomposition of p-hydroxy Phenylacetic Acid in Aqueous Solutions. *Water Research*, Vol. 35, No. 5, pp. 1338-1343, ISSN 0043-1354.

Bham, A.A., Chambers, R.P. (1997). Degradation of high molecular weight chlorinated aromatics and aliphatics in bleach plant effluent by Fenton's Reagent. *Advances in Environmental Research*, Vol. 1, pp. 135–143, ISSN 1093-7927.

Bianco, B., Michelis, I., Vegliò, F. (2011). Fenton treatment of complex industrial wastewater: Optimization of process conditions by surface response method. *Journal of Hazardous Materials*, Vol. 186, pp. 1733–1738, ISSN 0304-3894.

Bigda, R.J. (1995). Consider Fenton's Chemistry for Wastewater Treatment. *Chemical Engineering Progress*, December, pp. 62–66, ISSN 0360-7275.

Büyüksönmez, F., Hess, T.F., Crawford, R.L., Paszczynski, A., Watts, R.J. (1999). Optimization of Simultaneous Chemical and Biological Mineralization of Perchloroethylene. *Applied and Environmental Microbiology*, Vol. 65, No. 6 (Jun), pp. 2784-2788, ISSN 0099-2240.

Camargo, O.A. ; Moniz, A.C. ; Jorge, J.A. ; Valadares, J.M.A.S. (1986). *Métodos de análise química, Mineralógica e física de solos do Instituto Agronômico de Campinas*, Boletim Técnico no. 106, Instituto Agronômico, ISBN 8585564, Campinas, S.P.

Chamarro, E., Marco, A., Esplugas, S., 2001. Use of Fenton reagent to improve organic chemical biodegradability. *Water Research*, Vol. 35, No. 4, pp. 1047–1051, ISSN 0043-1354.

Chen, S., Sun, D., Chung, J.S. (2007). Treatment of pesticide wastewater by moving-bed biofilm reactor combined with Fenton-coagulation pre-treatment. *Journal of Hazardous Materials*, Vol. 144, pp. 577–584, ISSN 0304-3894.

CONAMA - Conselho Nacional do Meio Ambiente. (2005). Resolution No. 357, March 17, 2005, SãoPaulo, Brazil.

CONAMA - Conselho Nacional do Meio Ambiente. (2008). Resolution No. 397, April 03, 2008, SãoPaulo, Brazil.

Cornelis, R., Nordberg, M. (2007). General Chemistry, Sampling, Analytical Methods and Speciation, In: *Handbook on the toxicology of metals*, Nordberg, G.F. (Eds.), pp. 11-38, Academic Press, ISBN 9780123694133, USA.

Cornell, R.M., Schuwertmann, U. (1996). *The Iron Oxides: Structure, Properties, Reactions, Occurrence and Uses*. VCH Publishers, ISBN 9783527302741, Weinheim, Germany.

Deng, Y., Englehardt, J.D. (2006). Treatment of landfill leachate by the Fenton process. *Water Research*, Vol. 40, pp. 3683–3694, ISSN 0043-1354.

Dercová, K., Vrana, B., Tandlich, R., Šubová, L. (1999). Fenton's type reaction and chemical pretreatment of PCBs. *Chemosphere*, Vol. 39, No. 15, pp. 2621–2628, ISSN 0045-6535.

Elmolla, E., Chaudhuri, M. (2009). Optimization of Fenton process for treatment of amoxicillin, ampicillin and cloxacillin antibiotics in aqueous solution. *Journal of Hazardous Materials*, Vol. 170, pp. 666–672, ISSN 0304-3894.

El-Morsi, T.M., Emara, M.M., Abd El Bary, H.M.H., Abd-El-Aziz, A.S., Friesen, K.J. (2002). Homogeneous degradation of 1,2,9,10-tetrachlorodecane in aqueous solutions using hydrogen peroxide, iron and UV light. *Chemosphere*, Vol. 47, pp. 343–348, ISSN 0045-6535.

Fu, F., Wang, Q., Tang, B. (2009). Fenton and Fenton-like reaction followed by hydroxide precipitation in the removal of Ni(II) from NiEDTA wastewater: A comparative study. *Chemical Engineering Journal*, Vol. 155, pp. 769–774, ISSN 1385-8947.

Gallard, H., De Laat, J. (2001). Kinetics of oxidation of chlorobenzenes and phenyl-ureas by Fe(II)/H_2O_2 and Fe(III)/H_2O_2. Evidence of reduction and oxidation reactions of intermediates by Fe(II) or Fe(III). *Chemosphere*, Vol. 42, pp. 405–413, ISSN 0045-6535.

Garrido-Ramírez, E.G., Theng, B.K.G, Mora, M.L. (2010). Clays and oxide minerals as catalysts and nanocatalysts in Fenton-like reactions – A review. *Applied Clay Science*, Vol. 47, pp. 182–192, ISSN 0169-1317.

Ghigliazza, R., Lodi, A., Rovatti, M. (2000). Kinetic and process considerations on biological reduction of soluble and scarcely soluble sulfates. *Resources, Conservation and Recycling*, Vol. 29, pp. 181-194, ISSN 0921-3449.

Gogate, P.R., Pandit, A.B. (2004). A Review of Imperative Technologies for Wastewater Treatment I: Oxidation Technologies at Ambient Conditions. *Advances in Environmental Research*, v.8, pp. 501-551, ISSN 1093-7927.

Gulkaya, I., Surucu, G.A., Dilek, F.B. (2006). Importance of H_2O_2/Fe^{2+} ratio in Fenton's treatment of a carpet dyeing wastewater. *Journal of Hazardous Materials*, Vol. B136, pp. 763–769, ISSN 0304-3894.

Heredia, J.B., Terregrosa, J., Dominguez, J.R., Peres, J.A. (2001). Kinetic model for phenolic compound oxidation by Fenton's reagent. *Chemosphere*, Vol. 45, pp. 85–90, ISSN 0045-6535.

Homem, V., Alves, A., Santos, L. (2010). Amoxicillin degradation at ppb levels by Fenton's oxidation using design of experiments. *Science of the Total Environment*, Vol. 408, pp. 6272–6280, ISSN 0048-9697.

Kang, Y.W., Hwang, K.Y. (2000). Effects of Reaction Conditions on the Efficiency in the Fenton Process. *Water Research*, Vol. 34, No. 10, pp. 2786-2790, ISSN 0043-1354.

Kurt, U., Avsar, Y., Gonullu, M.T. (2006). Treatability of water-based paint wastewater with Fenton process in different reactor types. *Chemosphere*, Vol. 64, pp. 1536–1540, ISSN 0045-6535.

Kušić, H., Božić, A.L., Koprivanac, N. (2007). Fenton type processes for minimization of organic content in coloured wastewaters: Part I: Processes optimization. *Dyes and Pigments*, Vol. 74, pp. 380-387, ISSN 0143-7208.

Lee, H., Shoda, M. (2008). Removal of COD and colour from livestock wastewater by the Fenton method. *Journal of Hazardous Materials*, Vol. 153, pp. 1314–1319, ISSN 0304-3894.

Lens, P.N.L., Visser, A., Janssen, A.J.H., Hulshoff Pol, L.W., Lettinga, G. (1998). Biotechnological treatment of sulfate-rich wastewaters. *Critical Reviews in Environmental Science and Technology*, Vol. 28, No. 1, pp. 41-88, ISSN 1064-3389.

Li, R., Yang, C., Chen, H., Zeng, G., Yu, G., Guo, J. (2009). Removal of triazophos pesticide from wastewater with Fenton reagent. *Journal of Hazardous Materials*, Vol. 167, pp. 1028–1032, ISSN 0304-3894.

Lin, S.H., Lo, C.C. (1997). Fenton Process for Treatment of Desizing Wastewater. *Water Research*, Vol. 31, No. 8, pp. 2050-2056, ISSN 0043-1354.

Lu, M.C., Chen, J.N., Chang, C.P. (1999). Oxidation of dichlorvos with hydrogen peroxide using ferrous ion as catalyst. *Journal of Hazardous Materials*, Vol. B65, pp. 277–288, ISSN 0304-3894.

Lucas, M.S., Peres, J.A. (2009). Removal of COD from olive mill wastewater by Fenton's reagent: Kinetic study. *Journal of Hazardous Materials*, Vol. 168, pp. 1253–1259, ISSN 0304-3894.

Lunar, L., Sicilia, D., Rubio, S., Rez-Bendito, D., Nickel, U. (2000). Degradation of Photographic Developers by Fenton's Reagent: Condition Optimization and Kinetics for Metol Oxidation. *Water Research,* Vol. 34, No. 6, pp. 1791-1802, ISSN 0043-1354.

Ma X.J., Xia, H.L. (2009). Treatment of water-based printing ink wastewater by Fenton process combined with coagulation. *Journal of Hazardous Materials,* Vol. 162, pp. 386–390, ISSN 0304-3894.

Mahiroglu, A., Tarlan-Yel, E., Sevimli, M.F. (2009). Treatment of combined acid mine drainage (AMD)—Flotation circuit effluents from copper mine via Fenton's process. *Journal of Hazardous Materials,* Vol. 166, pp. 782–787, ISSN 0304-3894.

Mandal, T., Dasgupta, D., Mandal, S., Datta, S. (2010) Treatment of leather industry wastewater by aerobic biological and Fenton oxidation process. *Journal of Hazardous Materials,* Vol. 180, pp. 204–211, ISSN 0304-3894.

Martínez, N.S.S., Fernández, J.F., Segura, X.F., Ferrer, A.S. (2003). Preoxidation of an extremely polluted industrial wastewater by the Fenton's reagent. *Journal of Hazardous Materials,* Vol. B101, pp. 315–322, ISSN 0304-3894.

Mehra, O.P.; Jackson, M. (1960). Iron oxide removal from soils and clays by a dithionite-citrate system buffered with sodium bicarbonate, *Proceedings of the 7th National Clay Conference,* Pergamon Press, New York, pp. 317–327.

National Research Council, 1995. Prudent Practices in the Laboratory Handling and Disposal of Chemicals. National Academy Press, Washington, DC. ISBN 0309052297.

Nesheiwat, F.K., Swanson, A.G. (2000). Clean Contaminated Sites using Fenton's Reagent. *Chemical Engineering Progress,* Vol. 96, No. 4, pp. 61-66, ISSN 0360-7275.

Neyens, E., Baeyens, J. (2003). A review of classic Fenton's peroxidation as an advanced oxidation technique. *Journal of Hazardous Materials, Vol.* 98(B), pp. 33–50, ISSN 0304-3894.

Normas Analíticas do Instituto Adolfo Lutz. (1985). Métodos químicos e físicos para análises de alimentos (3rd ed). Editoração Débora D. Estrella Rebocho,São Paulo, BR.

Oh, S.Y., Chiu, P.C., Kim, B.J., Cha, D.K. (2003). Enhancing Fenton oxidation of TNT and RDX through pretreatment with zero-valent iron. *Water Research,* Vol. 37, pp. 4275–4283, ISSN 0043-1354.

Padoley, K.V., Mudliar, S.N. , Banerjee, S.K., Deshmukh, S.C., Pandey, R.A. (2011). Fenton oxidation: A pretreatment option for improved biological treatment of pyridine and 3-cyanopyridine plant wastewater. *Chemical Engineering Journal,* Vol. 166, pp. 1–9, ISSN 1385-8947.

Pavan, M.A., Bloch, M.F., Zempulski, H.C., Miyakawa, M., Zocoler, D.C. (1992). *Manual de análise química de solo e controle de qualidade.* Instituto Agronômico do Paraná e IAPAR, Londrina, BR.

Quevauviller, Ph. Rauret, G. López-Sánchez, J.-F. Rubio, R. Ure, A. Muntau, H. (1997). Certification of trace metal extractible contents in a sediment reference material (CRM 601) following a three-step sequential extraction procedure. *Science of the Total Environment,* Vol. 205, pp. 223–234, ISSN 0048-9697.

Scalbert, A.; Monties, B.; Janin, G. (1989). Tannins in wood: comparison of different estimation methods. *Journal of Agricultural and Food Chemistry,* Vol. 37, pp. 1324–1329, ISSN 0021-8561.

Schulze, D.G. (1994). Differential X-ray diffraction analysis of soil minerals, In: *Quantitative Methods in Soil Mineralogy*, J.E. Amonette, L.W. Zelazny (Eds.), pp. 412–429, Soil Sci. Soc. Amer. Misc. Publ, ISBN 0891188061, Madison.

Schwertmann, U. Schulze, D.G. Murad, E. (1982). Identification of ferrihydrite in soils by dissolution kinetics, differential X-ray diffraction and Mössbauer spectroscopy. *Soil Science Society of America Journal*, Vol. 46, pp. 869–875, ISSN 0361-5995.

Silva, A.J., Varesche, M.B., Foresti, E., Zaiat, M. (2002). Sulphate removal from industrial wastewater using a packed-bed anaerobic reactor. *Process Biochemistry*, Vol. 37, pp. 927-935, ISSN 0032-9592.

Talini, I., Anderson, G.K. (1992). Interference of Hydrogen Peroxide on the Standard COD Test. *Water Research*, Vol. 26, No. 1, pp. 107-110, ISSN 0043-1354.

Tessier, A.; Campbell, P.G.C.; Bisson, M. (1979). Sequential extraction procedure for the speciation of particulate trace metals. *Analytical Chemistry*, Vol. 51, pp. 844–851, ISSN 0003-2700.

Torrades, F., Pérez, M., Mansilla, H.D., Peral, J. (2003). Experimental design of Fenton and photo-Fenton reactions for the treatment of cellulose bleaching effluents. *Chemosphere*, Vol. 53, pp. 1211–1220, ISSN 0045-6535.

U.S. Peroxide (2003). Methods for residual peroxide determination: iodometric titration. Available from: <http://www.h2o2.com/intro/analytical.html>.

Vilve, M., Hirvonen, A., Sillanpää, M. (2009). Effects of reaction conditions on nuclear laundry water treatment in Fenton process. *Journal of Hazardous Materials*, Vol. 164, pp. 1468–1473, ISSN 0304-3894.

Walling, C., 1975. Fenton's reagent revisited. *Accounts of Chemical Research*, Vol. 8, pp. 125–131, ISSN 0001-4842.

Walling, C., Kato, S. (1971). The oxidation of alcohols by Fenton's reagent: the effect of copper ion. *Journal of American Chemical Society*, Vol. 93, pp. 4275–4281, ISSN 0002-7863.

Wang, X., Zenga, G.,Zhua, J. (2008). Treatment of jean-wash wastewater by combined coagulation, hydrolysis/acidification and Fenton oxidation. *Journal of Hazardous Materials*, Vol. 153, pp. 810–816, ISSN 0304-3894.

Yetilmezsoy, K., Sakar, S. (2008). Improvement of COD and color removal from UASB treated poultry manure wastewater using Fenton's oxidation. *Journal of Hazardous Materials*, Vol. 151, pp. 547–558, ISSN 0304-3894.

Zhang, H., Choi, H.J., Huang, C.P. (2005). Optimization of Fenton process for the treatment of landfill leachate. *Journal of Hazardous Materials*, Vol. B125, pp. 166–174, ISSN 0304-3894.

11

Fundamental Mechanistic Studies of the Photo-Fenton Reaction for the Degradation of Organic Pollutants

Amilcar Machulek Jr.[1], Frank H. Quina[2], Fabio Gozzi[1], Volnir O. Silva[2], Leidi C. Friedrich[2] and José E. F. Moraes[3]
[1]*Universidade Federal de Mato Grosso do Sul, Departamento de Química - UFMS,*
[2]*Universidade de São Paulo, Instituto de Química and NAP-PhotoTech - USP,*
[3]*Universidade Federal de São Paulo, Escola Paulista de Engenharia - UNIFESP,*
Brazil

1. Introduction

Very few regions of the planet possess abundant fresh water and access to adequate fresh water resources can be expected to worsen as a result of population growth and industrial demands for water. Liquid effluents containing toxic substances are generated by a variety of chemistry-related industrial processes, as well as by a number of common household or agricultural applications. The inadequate management of these residues can cause contamination of the soil and of subterranean and surface water sources.

In general, the recovery of industrial effluents containing low levels of organic substances by conventional treatments is not economically viable. Thus, for example, removal of the pollutant by adsorption onto active charcoal, while often efficient, requires subsequent recovery or incineration of the charcoal and merely transfers the pollutant from one phase to another (Matthews, 1992). Substances that are biocides or that are non-biodegradable represent a particular threat to the environment and prevent the use of conventional biological treatments. Social and legal demands for environmental safety increasingly require that effluents discharged into the environment have minimal impact on human health, natural resources and the biosphere. These demands have fueled increasing research into the development of new, more effective and economically viable methods for pollution control and prevention. When applied to the degradation of pollutants, these reactions are usually grouped together under the designations of Advanced Oxidation Processes (AOP) or Advanced Oxidation Technologies (AOT).

2. The principal advanced oxidation processes (AOP)

Advanced Oxidation Processes typically employ chemical oxidizing agents in the presence of an appropriate catalyst and/or ultraviolet light (Legrini, et al., 1993; Sonntag, 2008; Matilainen & Sillanpää, 2010) to oxidize or degrade the pollutant of interest. AOPs have been employed for the degradation of a variety of organic pollutants, such as aliphatic and aromatic hydrocarbons, halocarbons, phenols, ethers, ketones, etc. Examples of the major types of AOPs that have been proposed in the literature include (oxidant [catalyst, when present]/light): H_2O_2/UV (Gryglik et al., 2010; Ho & Bolton, 1998); O_3/UV (Esplugas et al., 1994; Machulek et al., 2009a); O_3-H_2O_2/UV (Yue, 1993); [TiO_2]/UV (Gaya & Abdullah, 2008; Henderson, 2011; Jenks, 2005; Matthews, 1992); Fe(III)/[TiO_2]/UV-Vis (Domínguez et al., 1998); direct photolysis of water with vacuum UV (Gonzalez et al., 2004); Fenton reaction or H_2O_2-Fe(II) (Dao & Laat, 2011; Haddou et al., 2010; Kwon et al., 1999; Pignatello et al., 2006; Pontes et al., 2010); and the photo-Fenton reaction or H_2O_2 [Fe(II)/Fe(III)]/UV (Benitez et al., 2011; Huston & Pignatello, 1999; Kim & Vogelpohl, 1998; Kiwi et al., 1994; Machulek et al., 2007; Martyanov et al., 1997; Nichela et al., 2010; Pignatello et al., 2006; Ruppert et al., 1993).

In most AOP, the objective is to use systems that produce the hydroxyl radical, $HO^•$, or another species of similar reactivity such as the sulfate radical anion ($SO_4^{•-}$). The hydroxyl radical is one of the most reactive species known in aqueous solution, surpassed only by fluorine atoms, and reacts with the majority of organic substances with little or no selectivity and at rates often approaching the diffusion-controlled limit (unit reaction efficiency per encounter). The principal modes of reaction of $HO^•$ with organic compounds include hydrogen abstraction from aliphatic carbon, addition to double bonds and aromatic rings, and electron transfer (Bauer & Fallmann, 1997). These reactions generate organic radicals as transient intermediates, which then undergo further reactions, eventually resulting in final products (P_{oxid}) corresponding to the net oxidative degradation of the starting molecule (R). Overall, a typical AOP process can be can be generically represented as:

$$HO^• + R \rightarrow \rightarrow \rightarrow \rightarrow \rightarrow P_{oxid} \quad (1)$$

Evidently, however, this representation belies the underlying complexity of the intermediate steps that occur on the pathway(s) from R to P_{oxid}.

From an environmental standpoint, one must be sure that the degradation of the initial pollutant does not produce intermediate products that are as toxic as or more toxic than the initial pollutant that one wishes to degrade. Thus, it is absolutely essential that the progress of the degradation and the final degradation products be adequately characterized. Since the degradation of organic pollutants by the hydroxyl radical is typically stepwise, even a relatively simple molecule like 4-chlorophenol can give rise to a plethora of intermediates, with from six to two carbon atoms, in various states oxidation (Li et al., 1999a, 1999b; Jenks, 2005) before converging to the final mineralization products, CO_2, water and HCl. Identification and quantification of all of the intermediates and determination of the kinetics and mechanisms of the individual reactions represent formidable tasks. For actual industrial effluents or wastewaters, which are often complex mixtures of pollutants, the intricacy of the degradation reactions can be enormous and one or more of the components or intermediates may be resistant to degradation and accumulate in the system. Not surprisingly, basic research into degradation pathways and mechanisms is still in its infancy for most of the AOP.

Photochemical and photocatalytic processes have enormous potential for becoming viable alternatives to conventional chemical AOP for the treatment of polluted waters and effluents. Currently available photochemical technology permits the conversion of organic pollutants having a wide range of chemical structures into substances that are less toxic and/or more readily biodegradable. In favorable cases, they can cause total decomposition of the organic constituents of the pollutant, generally referred to as "total mineralization" (complete oxidation to carbon dioxide and water, plus inorganic salts of all heteroatoms other than oxygen). Light of wavelengths in the range of 250-400 nm, corresponding to the ultraviolet (UV) region of the spectrum, is most commonly used in photochemical degradation processes (Braslavsky et al., 2011). Since ultraviolet light is a natural component of solar radiation, the sun provides a low-cost, environmentally friendly, renewable source of ultraviolet photons in photochemical processes. Thus, the use of solar photochemical reactors is an extremely interesting, cost-effective option for treatment of effluents in many of the tropical and sub-tropical regions of the planet. In areas with marginal or inadequate solar radiation intensity, conventional photochemical reactors fitted with ultraviolet lamps or hybrid UV lamp/solar photoreactors can be employed. In addition to light, the common AOPs use low to moderate concentrations of environmentally compatible chemical reagents and are capable, in favorable cases, of complete mineralization of the organic constituents of aqueous effluents. Although the initial or primary photochemical steps of the reactions employed to treat effluents can be conceptually rather simple, a large number of subsequent quite complex chemical steps are often involved in the overall degradation process.

Of the AOPs that have been proposed thus far, the Fenton reaction and the light-accelerated Fenton reaction, commonly known as the photo-Fenton reaction, appear to be the most promising for practical industrial applications (Pignatello et al., 2006). In the remainder of this chapter, we shall focus our attention on fundamental mechanistic details of the Fenton and photo-Fenton reactions, information that is essential for the adequate design and control of photo-Fenton processes for the degradation of organic pollutants.

3. The Fenton reaction

Over a century ago, Fenton (Fenton, 1894) demonstrated that a mixture of H_2O_2 and Fe(II) in acidic medium had very powerful oxidizing properties. Although the precise mechanism of this reaction, now known as the Fenton reaction, is still the subject of some discussion (Bossmann et al., 1998; Pignatello et al., 1999, 2006), it is generally assumed to be an important chemical source of hydroxyl radicals. The classical mechanism is a simple redox reaction in which Fe(II) is oxidized to Fe(III) and H_2O_2 is reduced to hydroxide ion and the hydroxyl radical:

$$Fe^{2+}{}_{aq} + H_2O_2 \rightarrow Fe^{3+}{}_{aq} + HO^{\bullet} + OH^{-} \qquad (2)$$

For the degradation of organic molecules, the optimum pH for the Fenton reaction is typically in the range of pH 3-4 and the optimum mass ratio of catalyst (as iron) to hydrogen peroxide is 1.5 (Bigda, 1995).

In the conventional Fenton reaction, carried out in the absence of light, the ferric ion produced in reaction (2) can be reduced back to ferrous ion by a second molecule of hydrogen peroxide:

$$Fe^{3+}{}_{aq} + H_2O_2 + H_2O \rightarrow Fe^{2+}{}_{aq} + H_3O^{+} + HO_2{}^{\bullet -} \qquad (3)$$

However, this thermal reduction (reaction 3) is much slower than the initial step (reaction 2). Thus, although chemically very efficient for the removal of organic pollutants, the Fenton reaction slows down appreciably after the initial conversion of Fe(II) to Fe(III) and may require the addition of relatively large amounts of Fe(II) in order to degrade the pollutant of interest. Another important limitation of the Fenton reaction is the formation of recalcitrant intermediates that inhibit the complete mineralization. Particularly noteworthy is the formation of oxalic acid, $H_2C_2O_4$, a poisonous and persistent oxidation product of many degradation reactions. Since Fenton reactions are typically run at an initial pH of about 3 and oxalic acid is a relatively strong acid (with a first pK_a of 1.4), the accumulation of oxalic acid causes further acidification of the reaction mixture (to ca. pH 2) as the reaction proceeds. In addition, Fe(III) is very efficiently chelated by the oxalate anion. This prevents the reduction of Fe(III) back to Fe(II) and hence the complete mineralization of the organic matter. A still unresolved question in the mechanism of the Fenton degradation of organic material is the relative importance of other potential oxidizing species besides the hydroxyl radical. The stoichiometry of Fenton degradation reactions is complex and, in addition to Fe^{2+}/Fe^{3+} and hydrogen peroxide, can involve the participation of the hydroperoxyl radical, $HOO^•$, iron(IV) or ferryl, FeO^{2+}, dissolved molecular oxygen, organic hydroperoxides, and other intermediates formed during the degradation. Despite these potential limitations, the conventional Fenton reaction has been widely used for the treatment of effluents (Benitez et al., 1999; Pignatello et al., 2006).

Any reaction or process that enhances the rate of conversion of Fe(III) back to Fe(II) will in principle accelerate the rate of the Fenton reaction. In the electro-Fenton reaction, this is accomplished electrochemically. Chemically, the Fenton reaction can be efficiently catalyzed by certain types of organic molecules, especially benzoquinones or dihydroxybenzene (DHB) derivatives. The catalytic influence of DHBs on the Fenton reaction was originally reported by Hamilton and coworkers (Hamilton et al., 1966a, 1966b). Because DHBs such as catechol or hydroquinone and their analogs are common initial intermediates in the degradation of aromatic molecules, their presence in the reaction medium can result in efficient DHB-catalyzed redox cycling of Fe^{3+} back o Fe^{2+} (Chen & Pignatello, 1997; Nogueira et al., 2005). Indeed, in several cases, it has been shown that the addition of catechol or catechol derivatives can enhance the rate and the overall mineralization efficiency of Fenton reactions (Aguiar et al., 2007; Zanta et al., 2010). In addition, DHB-catalyzed redox cycling of iron may be important in fungal degradation of lignin and several possible mechanisms have been suggested (Aguiar et al., 2007). The third method of accelerating the Fenton reaction is via irradiation with ultraviolet light, generally known as the photo-assisted Fenton or photo-Fenton reaction, which will be the primary focus of the remainder of this chapter.

4. The Photo-Fenton reaction

4.1 General features of the Photo-Fenton reaction

About two decades ago, it was found that the irradiation of Fenton reaction systems with UV/Visible light strongly accelerated the rate of degradation of a variety of pollutants (Huston & Pignatello, 1999; Ruppert et al., 1993). This behavior upon irradiation is due principally to the photochemical reduction of Fe(III) back to Fe(II), for which the net reaction can be written as:

$$Fe^{3+}_{aq} + H_2O \xrightarrow{h\upsilon} Fe^{2+}_{aq} + HO^{\bullet} + H^+ \quad (4)$$

More detailed studies of the pH dependence of the photo-Fenton reaction have shown that the optimum pH range is ca. pH 3. The reason for this pH dependence becomes clear when one examines the speciation of Fe(III) as a function of pH (Figure 1) and the absorption spectra of the relevant Fe(III) species (Figure 2). At pH < 2, the dominant species is hexaquoiron(III), $Fe(H_2O)_6^{3+}$ [or simply Fe^{3+} for convenience], which absorbs weakly in the ultraviolet above 300 nm. At pH > 3, freshly prepared solutions of Fe(III) are supersaturated with respect to formation of colloidal iron hydroxide, $Fe(OH)_3$ and prone to precipitation of hydrated iron oxides upon standing for a prolonged period. At pH 3, however, the predominant Fe(III) species present in aqueous solution is $Fe(H_2O)_5(OH)^{2+}$ [or simply $Fe(OH)^{2+}$], which absorbs throughout much of the ultraviolet spectral region (Martyanov et al., 1997).

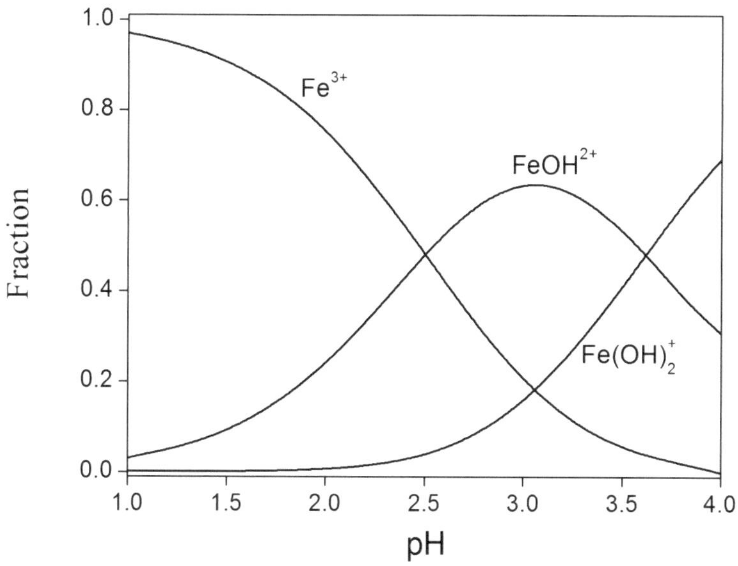

Fig. 1. Speciation of 0.5 mM Fe(III) between pH 1 and 4 at 25°C and an ionic strength of 0.1. Mole fractions of each species were calculated with the public domain program Hydra (Puigdomenech, 2010), employing the equilibrium constants for complexation supplied with the program (excluding insoluble iron species).

Studies of the photochemistry of $Fe(OH)^{2+}$ have shown (Pozdnyakov et al., 2000) that $Fe(OH)^{2+}$ undergoes a relatively efficient photoreaction to produce Fe(II) and the hydroxyl radical:

$$Fe(OH)^{2+} \xrightarrow{h\upsilon} Fe^{2+} + HO^{\bullet} \qquad (5)$$

Thus, irradiation of the Fenton reaction not only regenerates Fe(II), the crucial catalytic species in the Fenton reaction (reaction 2), but also produces an additional hydroxyl radical, the species responsible for the degradation of organic material. As a consequence of these two effects, the photo-Fenton process is faster than the conventional thermal Fenton process. Moreover, since Fe(II) is regenerated by light with decomposition of water (equations 4-5) rather than H_2O_2 (reaction 3), the photo-Fenton process consumes less H_2O_2 and requires only catalytic amounts of Fe(II).

The photo-Fenton reaction has several operational and environmental advantages. The classes of organic compounds that are susceptible to photodegradation via the Fenton reaction are rather well known (Bigda, 1995). The photo-Fenton process produces no new pollutants and requires only small quantities of iron salt. At the end of the reaction, if necessary, the residual Fe(III) can be precipitated as iron hydroxide by increasing the pH. Any residual hydrogen peroxide that is not consumed in the process will spontaneously decompose into water and molecular oxygen and is thus a "clean" reagent in itself. These features make homogeneous photo-Fenton based AOPs the leading candidate for cost-efficient, environmental friendly treatment of industrial effluents on a small to moderate scale (Pignatello et al., 2006). An early example of an industrial-scale application of the photo-Fenton process was the decontamination of 500 L batches of an industrial effluent containing 2,4-dimethylaniline in a photochemical reactor fitted with a 10 kW medium pressure mercury lamp (Oliveros et al., 1997).

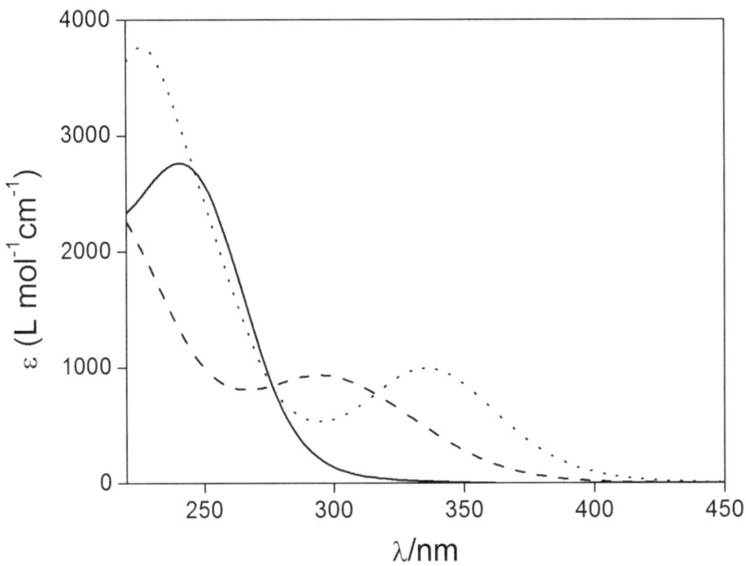

(a) Solid curve: pH 1.5 with perchloric acid, 90% $Fe(H_2O)_6{}^{3+}$ or Fe^{3+}, see Figure 1;
(b) Dashed Curve: pH 2.5 with perchloric acid, ca. 50:50 $FeOH^{2+}$:Fe^{3+}, see Figure 1;
(c) Dotted Curve: pH 1.5 plus added 500 mmol L^{-1} NaCl, 30% $FeCl^{2+}$ and 65% $FeCl_2{}^{+}$.

Fig. 2. Absorption spectra of 0.43 mM Fe(III) perchlorate under three typical conditions, expressed as the apparent extinction coefficient, ε, defined as the observed absorbance divided by the total Fe(III) concentration.

4.2 Inhibition of the Photo-Fenton reaction by chloride ion

Our interest in the fundamental mechanistic aspects of the photo-Fenton reaction arose from the necessity to optimize the degradation of organic material in effluents or wastewaters containing high concentrations of chloride ion. With such effluents, such as petroleum wastewaters from offshore marine environments or residues from organochloride pesticide production, the photo-Fenton reaction is strongly inhibited. Thus, although the photo-Fenton reaction results in essentially complete mineralization of phenol in the absence of chloride ion (Figure 3, curve e), in the presence of the chloride ion the mineralization process (indicated by the reduction in the amount of total organic carbon, TOC) stops after only partial decomposition of the organic material (Figure 3, curve c).

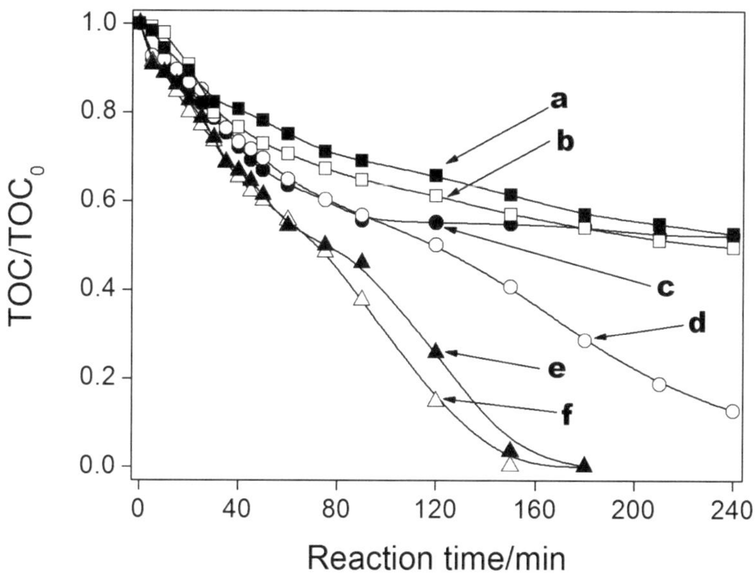

Fig. 3. Degradation of 12 mM phenol in the presence of 200 mM H_2O_2 and 0.5 mM Fe^{2+}. Fenton oxidation at $pH_{initial}$ 3.0 in the presence (curve a, ■) and absence (curve b, □) of 0.5 M NaCl. Photo-Fenton degradation (400 W medium pressure Hg vapor lamp - incident photon intensities were 1.2 x 10^{-4} Ein/s) at $pH_{initial}$ 3.0 with no pH control in the absence (curve e, ▲) or presence (curve c, •) of 0.5 M NaCl; or with pH 3.0 maintained throughout the reaction in the absence (curve f, Δ) or presence of 0.5 M NaCl (curve d, ○). For details, see (Machulek et al., 2007).

In the presence of chloride ion, the extent of mineralization in the photo-Fenton reaction (Figure 3, curve c) is comparable to that observed in the analogous thermal Fenton reaction carried out in the dark in the presence or absence of chloride ion (Figure 3, curves a and b). Although we (Moraes et al., 2004a, 2004b) and others (De Laat & Le, 2006; Kiwi et al., 2000; Maciel et al., 2004; Pignatello, 1992) had ascribed this inhibition to the preferential formation of the less-reactive $Cl_2^{\bullet-}$ radical anion instead of the desired hydroxyl radical, optimization of photo-Fenton reactions required a fuller understanding of the mechanistic details of the inhibition. For this purpose, we (Machulek et al., 2006) used nanosecond laser flash photolysis to investigate the influence of added chloride ion on the photocatalytic step that converts Fe(III) back to Fe(II) (equation 5), deliberately omitting H_2O_2 from the reaction mixture to prevent the thermal Fenton reaction. Although direct spectroscopic detection of the hydroxyl radical has proved elusive (Marin et al., 2011), the $Cl_2^{\bullet-}$ radical anion, which absorbs at 340 nm (ε_{340nm} = 8000 $M^{-1}cm^{-1}$), can be readily detected upon excitation of aqueous solutions of iron(III) at acidic pH in the presence of added sodium chloride at 355 nm with the third harmonic of a Nd-YAG laser. Differential absorption spectra (Figure 4) and kinetic traces (insert, Figure 4) showed fast formation of the $Cl_2^{\bullet-}$ radical anion, within the lifetime of the laser pulse (5 ns), and its subsequent decay via mixed first and second order kinetics. The net decrease in absorption at longer times, relative to that prior to the laser pulse, reflects the conversion of Fe(III) to Fe(II), which does not absorb in this spectral region.

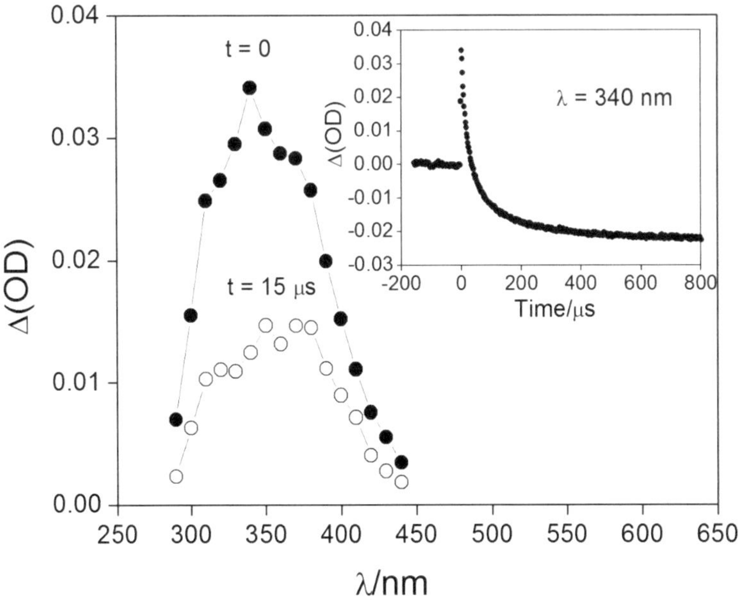

Fig. 4. Laser flash photolysis at 355 nm of 0.50 mM Fe(III) in the presence of 0.5 M NaCl at pH 1. Transient absorption spectra of $Cl_2^{\bullet -}$ immediately (•) and 15 μs (○) after the laser pulse. The insert shows the kinetics of $Cl_2^{\bullet -}$ disappearance monitored at λ= 340 nm.

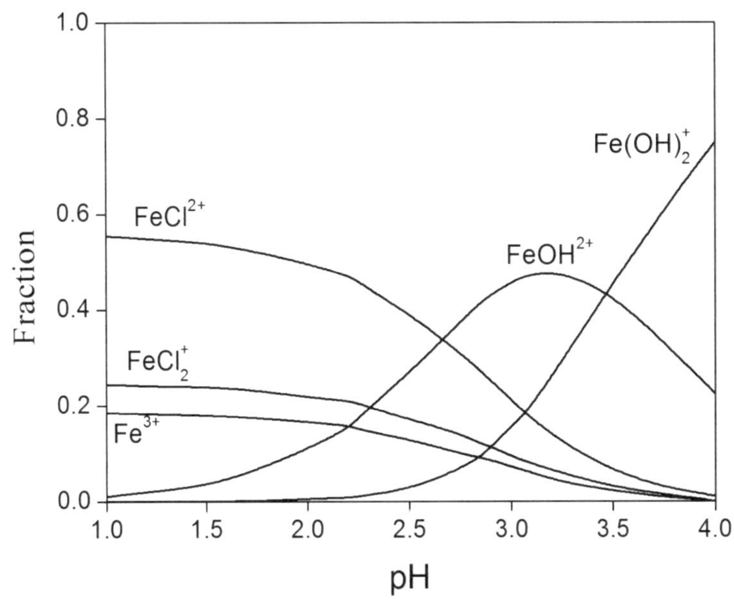

Fig. 5. Speciation of 0.5 mM Fe(III) between pH 1 and 4 at 25°C and an ionic strength of 0.2 in the presence of 0.1 M NaCl. Mole fractions of each species were calculated with the public domain program Hydra (Puigdomenech, 2010), employing the equilibrium constants for complexation supplied with the program (excluding insoluble iron species).

Laser flash photolysis data for the production and decay of $Cl_2^{\bullet-}$ were then obtained over a wide range of pH and concentration of Fe(III) and chloride ion and these data used to develop an explicit mechanistic model for the initial photoinduced processes involved in the photo-Fenton reaction in the presence of chloride ion (Machulek et al., 2006). As indicated in Figure 5, in the presence of chloride ion, the dominant species present at pH < 2.5 is no longer Fe^{3+}, but rather $FeCl^{2+}$ and $FeCl_2^+$ (coordinated waters being omitted).
Moreover, $FeCl^{2+}$ not only absorbs further out in the ultraviolet than $Fe(OH)^{2+}$ (Figure 2), but it also undergoes photolysis (equation 6) with a quantum yield higher than that of $Fe(OH)^{2+}$ (equation 5).

$$Fe(Cl)^{2+} \xrightarrow{h\upsilon} Fe^{2+} + Cl^{\bullet} \quad (6)$$

The photoproduced chlorine atoms rapidly react with chloride ions to form $Cl_2^{\bullet-}$ (equation 7), the species actually detected in the laser flash photolysis experiments:

$$Cl^{\bullet} + Cl^{-} \rightarrow Cl_2^{\bullet-} \quad (7)$$

and whose second order decay component is due to the disproportionation reaction:

$$Cl_2^{\bullet-} + Cl_2^{\bullet-} \rightarrow 2\,Cl^{-} + Cl_2 \quad (8)$$

The model employed to fit the experimentally observed decay curves for $Cl_2^{\bullet-}$ took into account the initial photochemical formation of the hydroxyl radical from $Fe(OH)^{2+}$ (equation 5) and of chlorine atoms from $FeCl^{2+}$ (equation 6) and the subsequent reactions of these via the set of elementary steps outlined in Table 1 (Machulek et al., 2006).

No.	Reaction	k (s^{-1} or $mol^{-1}\,L\,s^{-1}$)
Speciation Equilibria (I = 1)		
1	$Fe^{3+} + Cl^{-} \rightleftarrows FeCl^{2+}$	$k_1 = 4.79 \times 10^{10}$, $k_{-1} = 1 \times 10^{10}$
2	$Fe^{3+} + 2\,Cl^{-} \rightleftarrows FeCl_2^{+}$	$k_2 = 6.31 \times 10^{10}$, $k_{-2} = 1 \times 10^{10}$
Reactions of Iron Species		
3	$Fe^{2+} + Cl^{\bullet} \rightarrow Fe^{3+} + Cl^{-}$	$k_3 = 5.9 \times 10^9$
4	$Fe^{2+} + Cl_2^{\bullet-} \rightarrow Fe^{3+} + 2Cl^{-}$	$k_4 = 5 \times 10^6$
Reactions of Reactive Chlorine Species		
5	$Cl^{\bullet} + Cl^{-} \rightleftarrows Cl_2^{\bullet-}$	$k_5 = 7.8 \times 10^9$, $k_{-5} = 5.7 \times 10^4$
6	$Cl^{\bullet} + H_2O_2 \rightarrow HO_2^{\bullet} + Cl^{-} + H^{+}$	$k_6 = 1 \times 10^9$
7	$Cl_2^{\bullet-} + Cl_2^{\bullet-} \rightarrow 2Cl^{-} + Cl_2$	$k_7 = 2.8 \times 10^9$
8	$Cl_2^{\bullet-} + H_2O_2 \rightarrow HO_2^{\bullet} + 2Cl^{-} + H^{+}$	$k_8 = 1.4 \times 10^5$
9	$Cl_2^{\bullet-} + HO_2^{\bullet} \rightarrow 2Cl^{-} + H^{+} + O_2$	$k_9 = 3.1 \times 10^9$
10	$Cl^{-} + HO^{\bullet} \rightleftarrows HOCl^{\bullet-}$	$k_{10} = 4.2 \times 10^9$, $k_{-10} = 6.1 \times 10^9$
11	$HOCl^{\bullet-} + H^{+} \rightleftarrows H_2O + Cl^{\bullet}$	$k_{11} = 2.4 \times 10^{10}$, $k_{-11} = 1.8 \times 10^5$
12	$Cl_2^{\bullet-} + Cl^{\bullet} \rightarrow Cl^{-} + Cl_2$	$k_{12} = 1.4 \times 10^9$

Table 1. Set of reactions required to simulate the kinetics for production and decay of $Cl_2^{\bullet-}$ upon laser flash photolysis of Fe^{3+} in the presence of chloride ion over a wide range of pH and Fe^{3+} and NaCl concentrations.

The initial concentrations of $HO^{\bullet}$ and $C1^{\bullet}$ produced at t = 0 by the laser pulse were calculated from the relationship (equation 9):

$$\left[X^{\bullet}\right]_{t=0} = \left(I_{Laser}\right)\left(\frac{A_{Fe(X)}}{A_T}\right)\left(1-10^{A_T}\right)\left(\Phi_X\right) \tag{9}$$

where X = OH or Cl and Φ_{HO} = 0.21 and Φ_{Cl} = 0.47 are, respectively, the quantum yields for production of Fe(I1) from $Fe(OH)^{2+}$ or $Fe(Cl)^{2+}$ at 355 nm. $A_{Fe(X)}$ is the absorbance of either $Fe(OH)^{2+}$ or $Fe(Cl)^{2+}$ and $\mathbf{A_T}$ the total absorbance of the solution, both at the laser excitation wavelength of 355 nm (in an effective optical path length of 0.5 cm). The ratio $A_{Fe(X)}/A_T$ was calculated from the initial concentrations of the species $Fe(Cl)^{2+}$, $Fe(Cl)_2^+$, Fe^{3+}, $Fe(OH)^{2+}$, $Fe(OH)_2^+$ and $[Fe_2(H_2O)_4(OH)_2]^{4+}$, together with molar absorption coefficients at 355 nm taken from the literature absorption spectra (Byrne & Kester, 1978, 1981). For each experimental condition (pH, $[Cl^-]$, [Fe(III)]), the relative concentrations of the iron(III) species were calculated with the public domain speciation program Hydra (Puigdomenech, 2010), employing the set of standard equilibrium constants for complexation supplied with the program. The incident laser intensity (I_{laser}) was estimated by fitting the transient absorbance of $Cl_2^{\bullet-}$ at time zero at pH 1.0 in the presence of 0.5 M NaCl, where competitive photolysis of $Fe(OH)^{2+}$ is negligible. The concentrations of $HO^{\bullet}$ and $Cl^{\bullet}$ at t = 0 calculated from equation 9 served as the initial conditions for numerical solution of the set of reactions and rate constants listed in Table 1. This kinetic model provided a quantitative fit of the observed transient decay curves (like that shown in the inset of Figure 4) over the entire range of pH and concentrations of chloride ion and Fe(III) investigated.

Having established the basic mechanistic scheme for the photoinduced steps of the photo-Fenton reaction, one can then use it to infer the course of a typical photo-Fenton degradation in the presence of chloride ion. At the beginning of the photo-Fenton reaction, when the pH is still ca. 3.0, the concentration of $Fe(OH)^{2+}$ exceeds that of $FeCl^{2+}$ or $FeCl_2^+$, even in the presence of relatively high concentrations of chloride ion (Figure 5). The pH-dependent scavenging of $HO^{\bullet}$ by chloride ion (equations 10-11):

$$HO^{\bullet} + Cl^- \rightleftarrows HOCl^{\bullet-} \tag{10}$$

$$H^+ + HOCl^{\bullet-} \rightleftarrows H_2OCl^{\bullet} \rightarrow H_2O + Cl^{\bullet} \tag{11}$$

is also still a relatively inefficient process, so that the photochemical formation of $HO^{\bullet}$ should predominate. However, as the photo-Fenton process proceeds, partial degradation of the organic material causes the pH of the medium to fall to ca. pH 2.0, where photolysis of $FeCl^{2+}$ dominates over photolysis of $Fe(OH)^{2+}$ and where the chloride ion efficiently converts any $HO^{\bullet}$ formed in the system into the intrinsically much less reactive $Cl_2^{\bullet-}$ radical anion (Bacardit et al., 2007; Buxton et al., 1999; De Laat et al., 2004; George & Chovelon, 2002; Kiwi et al., 2000; McElroy, 1990; Moraes, et al., 2004a; Nadtochenko & Kiwi, 1998; Pignatello, 1992; Soler et al., 2009; Truong et al., 2004; Yu, 2004; Yu et al., 2004; Yu & Barker, 2003a, 2003b; Zapata et al., 2009). As a result, virtually complete inhibition of the photo-Fenton degradation of typical organic substrates (Kiwi & Nadtochenko, 2000; Machulek et al., 2007; Moraes et al., 2004a, 2004b; Pignatello, 1992) will occur at moderate chloride ion concentrations [>0.03 ***M*** NaCl for aliphatic hydrocarbons (Moraes et al., 2004a); >0.2 ***M*** NaCl for phenols] when the pH of the medium reaches pH 2.0 or below. On the other hand, this sequence of events clearly indicates that it should be possible to circumvent the inhibition by chloride ion by simply maintaining the pH at or slightly above pH 3 throughout the degradation process (see Figure 5). This is nicely illustrated by curve d in Figure 3, which shows that pH control does indeed permit nearly complete mineralization of phenol in the presence of chloride ion (Machulek et al., 2006, 2007). Although pH control also enhances the rate of the photo-Fenton reaction in the absence of chloride ion by optimizing the concentration of $Fe(OH)^{2+}$ (see Figure 1), the effect is much less dramatic (compare curves e and f in Figure 3).

Although the mechanism outlined in Table 1 nicely rationalizes the kinetics of formation and decay of $Cl_2^{\bullet-}$ on a fast time scale, it does not permit prediction of the accumulation of Fe(II) in the same system in the absence of H_2O_2 under steady-state irradiation. For this purpose, one must include additional kinetic steps that are too slow to be relevant on the time scale of the laser flash photolysis experiments. Experimentally (Machulek et al., 2009b) it is found that the concentration of Fe^{2+} eventually reaches a plateau value at long times, more quickly in the presence of chloride ion than in its absence (Figure 6). Simulation of this behavior required the inclusion of additional kinetic steps, in particular for the back reactions that result in reoxidation of Fe^{2+}. These additional kinetic steps are listed in Table 2.

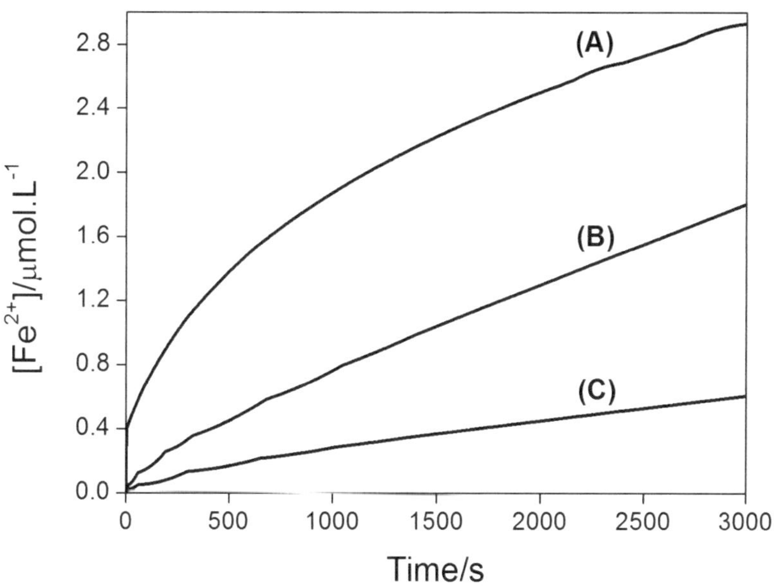

Fig. 6. The accumulation of ferrous ion during irradiation (350 nm) of 1.0 mM Fe(III) at: (A) pH 1.0 in the presence of 0.75 mol L^{-1} chloride ion; (B) at pH 3 in the absence of chloride ion; and (C) at pH 3 in the presence of 1.8 mmol L^{-1} sulfate ions.

No.	Reaction	k (s^{-1} or $mol^{-1}\,L\,s^{-1}$)
Speciation Equilibria (I = 1)		
1	$Fe^{3+} + H_2O \rightleftarrows FeOH^{2+} + H^+$	$k_1 = 1.91 \times 10^7$, $k_{-1} = 1 \times 10^{10}$
2	$Fe^{3+} + 2H_2O \rightleftarrows Fe(OH)_2^+ + 2H^+$	$k_2 = 3.39 \times 10^3$, $k_{-2} = 1 \times 10^{10}$
3	$2Fe^{3+} + 2H_2O \rightleftarrows [Fe_2(OH)_2]^{4+} + 2H^+$	$k_3 = 1.12 \times 10^7$, $k_{-3} = 1 \times 10^{10}$
Reactions of Reactive Oxygen Radicals		
4	$HO^\bullet + HO^\bullet \rightarrow H_2O_2$	$k_4 = 6.0 \times 10^9$
5	$HO^\bullet + H_2O_2 \rightarrow H_2O + HO_2^\bullet$	$k_5 = 2.7 \times 10^7$
6	$HO_2^\bullet + HO_2^\bullet \rightarrow H_2O_2 + O_2$	$k_6 = 8.3 \times 10^5$
7	$HO_2^\bullet + H_2O_2 \rightarrow HO^\bullet + O_2 + H_2O$	$k_7 = 0.5$
Reactions of Iron Species		
8	$Fe^{2+} + HO^\bullet \rightarrow Fe^{3+} + OH^-$	$k_8 = 4.3 \times 10^8$
9	$Fe^{2+} + H_2O_2 \rightarrow Fe^{3+} + HO^\bullet + OH^-$	$k_9 = 63$
10	$Fe^{3+} + HO_2^\bullet \rightarrow Fe^{2+} + O_2 + H^+$	$k_{10} = 1 \times 10^6$
11	$Fe^{3+} + H_2O_2 \rightarrow Fe^{2+} + HO_2^\bullet + H^+$	$k_{11} = 0.01$

Table 2. Set of additional ground state reactions required to fit the accumulation of Fe^{2+} during the steady-state irradiation of Fe^{3+} in the presence and absence of chloride ion (Machulek et al., 2009b)

4.3 Inhibition of the Photo-Fenton reaction by sulfate ion

Also indicated in Figure 6 is the inhibition of the formation of Fe^{2+} in the presence of a relatively low concentration of added sulfate ions, based on the rate constants for complexation of Fe(III) by sulfate in Table 3. Sulfate ion is often present in photo-Fenton reactions as a result of the addition of iron in the form of readily available sulfate salts (De Laat & Le, 2005). Sulfate ion complexes strongly with Fe(III) over a wide pH range (Figure 7) and the quantum yield of production of Fe^{2+} plus a sulfate anion radical from photolysis of $Fe(SO_4)^+$ is only about 0.05 (Benkelberg & Warneck, 1995). Iron phosphate is even more photoinert than the iron sulfate complex (Benkelberg & Warneck, 1995; Lee et al., 2003) and should be an even more powerful inhibitor of the photo-Fenton reaction.

4.4 Catalysis of the Photo-Fenton reaction by complexation of Fe(III)

The efficiency of the photo-Fenton process can be further enhanced by using organic carboxylic acids to complex Fe(III) (Pignatello et al., 2006). A particularly important example is provided by oxalic acid. Thus, unlike the thermal Fenton reaction, in which oxalic acid is a recalcitrant intermediate, in the photo-Fenton reaction it can act as a catalyst. Thus, ferrioxalate complexes can absorb light as far out as 570 nm, i.e., well into the visible region of the spectrum. Moreover, upon irradiation, they decompose efficiently (quantum yields of the order of unity) to Fe(II) and CO_2. The net result is that, in the presence of oxalate, the photo-Fenton reaction is intrinsically more efficient, can be induced by a wider range of wavelengths of light, and results in the mineralization of the oxalate ion. Thus, for example, in a municipal water treatment system, Kim & Vogelpohl (1998) found that, with UV irradiation, the photo-Fenton process was at least 30% more energy efficient in the presence of oxalate than in its absence. Clearly, the sensitivity of the ferrioxalate-catalyzed photo-Fenton process to both UV and visible light makes it particularly attractive for applications in which the sun is employed as the radiation source (Silva et al., 2010; Trovó & Nogueira, 2011).

No.	Reaction	k (s^{-1} or mol^{-1} L s^{-1})
Speciation Equilibria (I = 1)		
1	$Fe^{3+} + SO_4^{2-} \rightleftarrows FeSO_4^{+}$	$k_1 = 2.09 \times 10^{12}$, $k_{-1} = 1 \times 10^{10}$ (I=1)
2	$Fe^{3+} + 2SO_4^{2-} \rightleftarrows Fe(SO_4)_2^{-}$	$k_2 = 1.95 \times 10^{13}$, $k_{-2} = 1 \times 10^{10}$ (I=1)
3	$Fe^{2+} + SO_4^{2-} \rightleftarrows FeSO_4$	$k_3 = 1.55 \times 10^{11}$, $k_{-3} = 1 \times 10^{10}$ (I=1)
Reactions of Sulfate Ions		
4	$H^+ + SO_4^{2-} \rightleftarrows HSO_4^-$	$k_4 = 2.8 \times 10^{11}$, $k_{-4} = 1 \times 10^{10}$ (I=1)
5	$HSO_4^- + HO^\bullet \rightarrow SO_4^{\bullet-} + H_2O$	$k_5 = 3.5 \times 10^5$
6	$SO_4^{\bullet-} + H_2O \rightarrow H^+ + SO_4^{2-} + HO^\bullet$	$k_6 = 6.6 \times 10^2$
7	$SO_4^{\bullet-} + OH^- \rightarrow SO_4^{2-} + HO^\bullet$	$k_7 = 1.4 \times 10^7$
8	$SO_4^{\bullet-} + H_2O_2 \rightarrow SO_4^{2-} + H^+ + HO_2^\bullet$	$k_8 = 1.2 \times 10^7$
9	$SO_4^{\bullet-} + HO_2^\bullet \rightarrow SO_4^{2-} + H^+ + O_2$	$k_9 = 3.5 \times 10^9$
10	$SO_4^{\bullet-} + SO_4^{\bullet-} \rightarrow S_2O_8^{2-}$	$k_{10} = 2.7 \times 10^8$
Reactions of Iron Species		
11	$Fe^{2+} + SO_4^{\bullet-} \rightarrow Fe^{3+} + SO_4^{2-}$	$k_{11} = 3.0 \times 10^8$

Table 3. Set of additional ground state reactions required to fit the accumulation of Fe^{2+} during the steady-state irradiation of Fe^{3+} in the presence of sulfate ion (Machulek et al., 2009b).

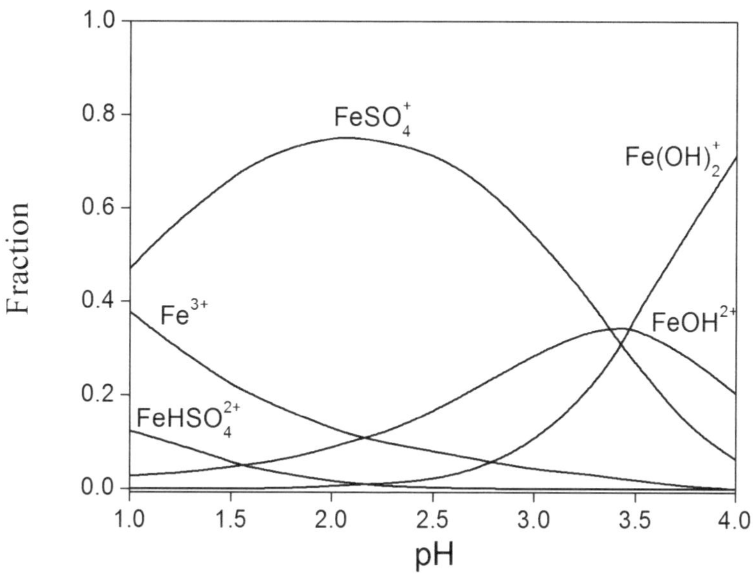

Fig. 7. Speciation of 1.0 mM Fe(III) between pH 1 and 4 at 25°C and an ionic strength of 0.2 in the presence of 1.8 mM sulfate. Mole fractions of each species were calculated with the public domain program Hydra (Puigdomenech, 2010), employing the equilibrium constants for complexation supplied with the program (excluding the insoluble iron species).

4.5 Concurrent Photo-Fenton and Fenton reactions

A final point that should be emphasized is that, under typical photo-Fenton reaction conditions, both the thermal Fenton reaction and the photo-Fenton reactions occur concurrently in the reaction mixture. Light is almost always the limiting reagent in photochemical reactions and photo-Fenton reactions are often conducted in reactors in which the solution is recirculated through the photoreactor from an external reservoir. Thus, at any given moment, the photo-Fenton reaction will be operative only in the portion of the total reaction mixture that is contained in the photoreactor *per se*, while the thermal Fenton reaction can proceed throughout the entire reaction volume. Indeed, comparison of the degradation of phenol under various conditions (Figure 3) strongly suggests that the initial phase of the reaction is dominated by the thermal reaction, with the inflection at about 80 min where the curves diverge indicating the point at which the photo-Fenton reaction becomes dominant.

Particularly interesting is the apparent insensitivity of the thermal Fenton degradation of phenol to the presence of added chloride ion, as indicated by the similarity of curves a and b of Figure 3. At the initial pH of 3 of the reaction mixture, the hydroxyl radical is inefficiently scavenged by chloride and produces, as the initial products of phenol degradation, the DHBs catechol and hydroquinone. Thus, unlike the photo-Fenton reaction, which depends on the intermediacy of the hydroxyl radical, the thermal degradation of phenol shifts to the DHB-catalyzed reaction mentioned above for the Fenton reaction until all of the DHBs have themselves been degraded (Chen and Pignatello, 1997). At that point, most of the iron(III) is rendered redox inert by complexation with oxalic acid and/or other recalcitrant aliphatic acids and the Fenton reaction ceases (Zanta et al., 2010).

5. Areas for future development

Despite significant advances in the last decade, the widespread successful application of photochemical technology for the treatment and decontamination of industrial residues and wastewaters is still not a reality. Research chemists and process engineers must work together to solve several important practical problems that limit the technological applications of the photo-Fenton reaction on a large scale (Braun et al., 1991). Since the overall effective quantum efficiency can be relatively low, long irradiation times must be employed in order to achieve total mineralization of the organic components of effluents. Electrical power consumption can then easily represent 60% of the total operating expense of a lamp-based photochemical reactor. Additional costs associated with non-solar photochemical reactors include cooling, maintenance and depreciation of the lamps and UV radiation shielding and protection for workers. One of the obvious economic benefits of the use of solar photochemical reactors is the elimination of costs associated with installation and maintenance of lamps and with electrical power consumption (Moraes et al., 2004b; Nascimento et al., 2007). In parallel with the development of solar reactors or hybrid lamp/solar reactors, an important area is the development of photocatalysts that operate effectively under visible or solar irradiation (Anpo, 2000).

Particularly in larger scale reactors, operational problems such as the formation of deposits on the reactor walls that block the incident radiation must be addressed. Optimization strategies (Braun et al., 1991, 1993; Cassano et al., 1995; Nascimento et al., 1994; Oliveros et al., 1998) must be applied to the design of reactors that homogenize the incidence of light on the reactor walls and maximize the total amount of light absorbed by the system. Reactor design must also take into account the strong interdependence between light absorption, mass transport and reaction kinetics in solutions with high optical densities and optimize the balance between the concurrent photochemical and thermal contributions to the Fenton reaction. Although the relatively long irradiation times required for the complete mineralization of organic contaminants tend to compromise the economic viability of stand-alone photo-Fenton processes, they can be viable as a pre-treatment to reduce the toxicity of the effluent to levels compatible with other remediation technologies such as conventional biological treatment. Hence, the design of photo-Fenton reactors should also take into account the necessity of facile integration with other effluent decontamination technologies. A final point that has only recently been considered in a more systematic context is the intensification of Fenton processes by increasing the temperature (Zazo et al., 2011), particularly attractive in the case of photo-Fenton reactions where a substantial part of the absorbed light energy is dissipated as heat in the solution.

6. Conclusions

Of the currently known advanced oxidation processes, the Fenton reaction and the photo-Fenton reaction appear to be the most promising for practical industrial applications on a moderate scale. The fundamental photomechanistic aspects of the photo-Fenton reaction have been clarified by studies of the photochemistry of Fe(IIII) in the absence of hydrogen peroxide in order to avoid the complications due to the competing Fenton reaction. The Fe(III) species primarily responsible for the initial photochemical generation of Fe(II) plus a hydroxyl radical is $Fe(OH)^{2+}$. Inhibition of the photo-Fenton by ions such as chloride or sulfate can be understood by considering the competitive pH-dependent complexation of Fe(III) by these ions, the competitive absorption and photochemistry of these complexes and/or the subsequent chemical reactions that can convert the hydroxyl radical into less reactive species. Adequate mechanisms, together with the relevant rate constants, are now available for the inhibition of the net photoconversion of Fe(III) to Fe(II) by both chloride and sulfate, allowing quantitative or semi-quantitative simulation of the experimentally observed inhibitory effects. As predicted by the mechanistic models, the inhibitory effect of chloride ion on real photo-Fenton degradations can be readily circumvented by simply maintaining the medium pH at ca. pH 3 throughout the reaction. Areas where additional progress would be welcome include an understanding of the role of temperature effects on the Fenton and photo-Fenton reactions and new strategies for further enhancing the efficiency of the photo-Fenton reaction. Improvements in photoreactor design are necessary in order to optimize the contribution of the photoinduced processes relative to the concurrent thermal Fenton reaction.

7. Acknowledgment

The authors acknowledge the Brazilian funding agencies CAPES and CNPq for financial and fellowship support. F.H.Q. is associated with NAP-PhotoTech, the USP Research Consortium for Photochemical Technology, and INCT-Catalysis. A.M. Jr. and J.E.F. Moraes are affiliated with INCT-EMA. A.M.Jr. thanks FUNDECT for financial support of the work at UFMS.

8. References

Aguiar, A.; Ferraz, A.; Contreras, D. & Rodríguez, J. (2007). Mecanismo e aplicação da reação de Fenton assistida por compostos fenólicos redutores de ferro. *Química Nova*, Vol. 30, No. 3, pp. 623-628, ISSN 0100-4042

Anpo, M. (2000). Utilisation of TiO_2 photocatalysts in green chemistry. *Pure and Applied Chemistry*, Vol. 72, No. 7, pp. 1265-1270, ISSN 0033-4545

Bacardit, J.; Stötzner, J.; Chamarro, E. & Esplugas, S. (2007). Effect of salinity on the photo-Fenton process. *Industrial & Engineering Chemistry Research*, Vol. 46, No. 23, pp. 7615-7619, ISSN 0888-5885

Bauer, R. & Fallmann, H. (1997). The photo-Fenton oxidation - a cheap and efficient wastewater treatment method. *Research on Chemical Intermediates*, Vol. 23, No. 4, pp. 341-354, ISSN 0922-6168

Benitez, F.J.; Beltran-Heredia, J.; Acero, J.L. & Rubio, F.J. (1999). Chemical decomposition of 2,4,6-triclorophenol by ozone, Fenton's reagent, and UV radiation. *Industrial & Engineering Chemistry Research*, Vol. 38, No. 4, pp. 1341-1349, ISSN 0888-5885

Benitez, F.J.; Acero, J.L.; Real, F.J.; Roldan, G. & Casas, F. (2011). Comparison of different chemical oxidation treatments for the removal of selected pharmaceuticals in water matrices. *Chemical Engineering Journal*, Vol. 168, No. 3, pp. 1149-1156, ISSN 1385-8947

Benkelberg, H. J. & Warneck, P. (1995). Photodecomposition of iron(III) hydroxo and sulfate complexes in aqueous solution: Wavelength dependence of OH and SO_4^- quantum yields. *Journal of Physical Chemistry*, Vol. 99, No. 14, pp. 5214-5221, ISSN 0022-3654

Bigda, R. (1995). Consider Fenton's chemistry for wastewater treatment. *Chemical Engineering Progress*, (Dec.), Vol. 91, No. 12, pp. 62-66, ISSN 03607275

Bolton, J. R. (1999). Light Compendium - Ultraviolet Principles and Applications. *Inter-American Photochemical Society Newsletter*, Vol. 22, No. 2, pp. 20-61

Bossmann, S.H.; Oliveros, E.; Göb, S.; Siegwart, S.; Dahlen, E.P.; Payawan, L.; Straub, M.; Wörner, M. & Braun, A.M. (1998). New evidence against hydroxyl radicals as reactive intermediates in the thermal and photochemically enhanced Fenton Reactions. *Journal of Physical Chemistry A*, Vol. 102, No. 28, pp. 5542-5550, ISSN 1089-5639

Braslavsky, S.E.; Braun, A.M.; Cassano, A.E.; Emeline, A.V.; Litter, M.I.; Palmisano, L.; Parmon, V.N. & Serpone, N. (2011). Glossary of terms used in photocatalysis and radiation catalysis (IUPAC recommendations 2011). *Pure and Applied Chemistry*, Vol. 83, No. 4, pp. 931-1014, ISSN 0033-4545

Braun, A.M.; Maurette, M.T. & Oliveros, E. (1991). *Photochemical Technology*. John Wiley & Sons, ISBN-13 978-0471926528, New York, USA

Braun, A.M.; Jakob, L.; Oliveros, E. & Nascimento, C.A.O. (1993). Up-scaling photochemical reactions, *In: Advances in Photochemistry*, Volman, D.H.; Hammond, G.S. & Neckers, D.C., Vol. 18, pp. 235-313, ISBN 9780471591337, John Wiley & Sons, USA

Buxton, G.V., Bydder, M. & Salmon, G.A. (1999). The reactivity of chlorine atoms in aqueous solution—Part II. The equilibrium $SO_4^{\bullet -} + Cl^- \rightarrow Cl^\bullet + SO_4^{2-}$. *Physical Chemistry Chemical Physics*, Vol. 1, No. 2, pp. 269–273, ISSN 1463-9076

Byrne, R.H. & Kester, D.R. (1978). Ultraviolet spectroscopic study of ferric hydroxide complexation. *Journal of Solution Chemistry,* Vol. 7, No. 5, pp. 373–383, ISSN 0095-9782

Byrne, R.H. & Kester, D.R. (1981). Ultraviolet spectroscopic study of ferric equilibria at high chloride concentration. *Journal of Solution Chemistry,* Vol. 10, No. 1, pp. 51–67, ISSN 0095-9782

Cassano, A.; Martín, C.A.; Brandi, R.J. & Alfano, O.M. (1995). Photoreactor analysis and design: fundamentals and applications. *Industrial & Engineering Chemistry Research,* Vol. 34, No. 7, pp. 2155-2201, ISSN 0888-5885

Chen, R. & Pignatello, J.J. (1997). Role of quinone intermediates as electron shuttles in Fenton and photoassisted Fenton oxidations of aromatic compouds. *Environmental Science & Technology,* Vol. 31, No. 8, pp. 2399-2406, ISSN 0013-936X

Dao, Y.H. & Laat, J. (2011). Hydroxyl radical involvement in the decomposition of hydrogen peroxide by ferrous and ferric-nitrilotriacetate complexes at neutral pH. *Water Research,* Vol. 45, No. 11, pp. 3309-3317, ISSN 0043-1354

De Laat, J.; Le, G.T. & Legube, B. (2004). A comparative study of the effects of chloride, sulfate and nitrate ions on the rates of decomposition of H_2O_2 and organic compounds by Fe(II)/H_2O_2 and Fe(III)/H_2O_2. *Chemosphere,* Vol. 55, No. 5, pp. 715-723, ISSN 0045-6535

De Laat, J. & Le, T. G. (2005). Kinetics and Modeling of the Fe(III)/H_2O_2 System in the Presence of Sulfate in Acidic Aqueous Solutions. *Environmental Science & Technology,* Vol. 39, No. 6, pp. 1811-1818, ISSN 0013-936X

De Laat, J. & Le, T. G. (2006). Effects of chloride ions on the iron(III)-catalyzed decomposition of hydrogen peroxide and on the efficiency of the Fenton-like oxidation process. *Applied Catalysis, B: Environmetal,* Vol. 66, No. 1-2, pp. 137-146, ISSN 0926-3373

Domínguez, C.; García, J.; Pedraz, M.A.; Torres, A. & Galán, M.A. (1998). Photocatalytic oxidation of organic pollutants in water. *Catalysis Today,* Vol. 40, No. 1, pp. 85-101, ISSN 0920-5861

Esplugas, S.; Yue, P.L. & Pervez, M.I. (1994). Degradation of 4-chlorophenol by photolytic oxidation. *Water Research,* Vol. 28, No. 6, pp. 1323-1328, ISSN 0043-1354

Fenton, H.J.H. (1894). Oxidation of tartaric acid in presence of iron. *Journal of the Chemical Society,* Vol. 65, pp. 899-901, ISSN 0368-1769

Gaya, U.I. & Abdullah, A.H. (2008). Heterogeneous photocatalytic degradation of organic contaminants over titanium dioxide: A review of fundamentals, progress and problems. *Journal of Photochemistry and Photobiology C: Photochemistry Reviews,* Vol. 9, No. 1, pp. 1-12, ISSN 1389-5567

George, C. & Chovelon, J.M. (2002). A laser flash photolysis study of the decay of $SO_4^{\cdot-}$ and $Cl_2^{\cdot-}$ radical anions in the presence of Cl^- in aqueous solution. *Chemosphere,* Vol. 47, No. 4, pp. 385–393, ISSN 0045-6535

Gonzalez, M.C.; Oliveros, E.; Wörner, M. & Braun, A.M. (2004). Vacuum-ultraviolet photlolysis of aqueous reaction systems. *Journal of Photochemistry and Photobiology C: Photochemistry Reviews,* Vol. 5, No. 3, pp. 225-246, ISSN 1389-5567

Gryglik, D.; Olak, M. & Miller, J.S. (2010). Photodegradation kinetics of androgenic steroids boldenone and trenbolone in aqueous solutions. *Journal of Photochemistry and Photobiology A: Chemistry*, Vol. 212, No. 1, pp. 14-19, ISSN 1010-6030

Haddou, M.; Benoit-Marquié, F.; Maurette, M.T. & Oliveros, E. (2010). Oxidative degradation of 2,4-dihydroxybenzoic acid by Fenton and photo-Fenton process: Kinetics, mechanisms, and evidence for the substitution of H_2O_2 by O_2. *Helvetica Chimica Acta,* Vol. 93, No. 6, pp. 1067-1080, ISSN 0018-019X

Hamilton, G.A.; Friedman, J.P. & Campbell, P.M. (1966a). The hydroxylation of anisole by hydrogen peroxide in the presence of catalytic amounts of ferric ion and catechol. Scope, requirements and kinetic studies. *Journal of the American Chemical Society,* Vol. 88, No. 22, pp. 5266–5268, ISSN 0002-7863

Hamilton, G.A.; Hanifin Jr., J.W. & Friedman, J.P. (1966b). The hydroxylation of anisole by hydrogen peroxide in the presence of catalytic amounts of ferric ion and catechol. Product studies, mechanism, and relation to some enzymic reactions. *Journal of the American Chemical Society,* Vol. 88, No. 22, pp. 5269–5272, ISSN 0002-7863

Henderson, M.A (2011). A surface science perspective on TiO_2 photocatalysis. *Surface Science Reports*, Vol. 66, No. 6-7, pp. 185-297, ISSN 0167-5729

Ho, T.L. & Bolton, J.R. (1998). Toxicity changes during the UV treatment of pentachlorophenol in dilute aqueous solution. *Water Research,* Vol. 32, No. 2, pp. 489-497, ISSN 0043-1354

Huston, P.L. & Pignatello, J.J. (1999). Degradation of selected pesticide active ingredientes and commercial formulations in water by the photoassisted Fenton reaction. *Water Research*, Vol. 33, No. 5, pp. 1238-1246, ISSN 0043-1354

Jenks, W.S. (2005). The Organic Chemistry of TiO_2 Photocatalysis of Aromatic Hydrocarbons, In: *Environmental Catalysis*, Vicki H. Grassian, 307-346, CRC, ISBN 1-57444-462-X, New York, USA

Kim, S.M. & Vogelpohl, A. (1998). Degradation of organic pollutants by the photo-Fenton process. *Chemical Engineering Technology,* Vol. 21, No. 2, pp. 187-191, ISSN 0930-7516

Kiwi, J.; Pulgarin, C. & Peringer, P. (1994). Effect of Fenton and photo-Fenton reactions on the degradation and biodegradability of 2-nitrophenols and 4-nitrophenols in water treatment. *Applied Catalysis, B: Environmetal,* Vol. 3, No. 4, pp. 335-350, ISSN 0926-3373

Kiwi, J., Lopez, A. & Nadtochenko, V. (2000). Mechanism and kinetics of the OH-radical intervention during Fenton oxidation in the presence of a significant amount of radical scavenger (Cl^-). *Environmental Science & Technology,* Vol. 34, No. 11, pp. 2162-2168, ISSN 0013-936X

Kwon, B.G.; Lee, D.S.; Kang, N. & Yoon, J. (1999). Characteristics of *p*-chlorophenol oxidation by Fenton's reagent. *Water Research,* Vol. 33, No. 9, pp. 2110-2118, ISSN 0043-1354

Lee, Y., J. Jeong, C. Lee, S. Kim and J. Yoon (2003) Influence of various reaction parameters on 2,4-D removal in photo/ferrioxalate/H_2O_2 process. Chemosphere, Vol. 51, No. 9, pp. 901-912, ISSN 0045-6535

Legrini, O.; Oliveros, E. & Braun, A.M. (1993). Photochemical processes for water treatment. *Chemical Reviews,* Vol. 93, No. 2, pp. 671-698, ISSN 0009-2665

Li, X.; Cubbage, J. W. & Jenks, W. S. (1999a). Photocatalytic degradation of 4-chlorophenol. 1. The 4-chlorocatechol pathway. *Journal of Organic Chemistry*, Vol. 64, No. 23, pp. 8525-8536, ISSN 0022-3263

Li, X.; Cubbage, J. W.; Tetzlaff, T. A. & Jenks, W. S., (1999b). Photocatalytic degradation of 4-chlorophenol. 1. The hydroquinone pathway. *Journal of Organic Chemistry*, Vol. 64, No. 23, pp. 8509-8524, ISSN 0022-3263

Machulek Jr., A.; Vautier-Giongo, C.; Moraes, J. E. F.; Nascimento, C. A. O. & Quina, F.H. (2006). Laser flash photolysis study of the photocatalytic step of the photo-Fenton reaction in saline solution. *Photochemestry and Photobiology*, Vol. 82, No. 1, pp. 208-212, ISSN 0031-8655

Machulek Jr., A.; Moraes, J.E.; Vautier-Giongo, C.; Silverio, C.A.; Friedrich, L.C.; Nascimento, C.A.O.; Gonzalez, M.C. & Quina, F.H. (2007). Abatement of the inhibitory effect of chloride anions in the photo-Fenton process. *Environmental Science & Technology*, Vol. 41, No. 24, pp. 8459-8463, ISSN 0013-936X

Machulek Jr., A.; Gogritcchiani, E.; Moraes, J.E.F.; Quina, F.H.; Braun, A.M. & Oliveros, E. (2009a). Kinetic and mechanistic investigation of the ozonolysis of 2,4-xylidine (2,4-dimethyl-aniline) in acid aqueous solution. *Separation and Purification Technology*, Vol. 67, No. 2, pp. 141-148, ISSN 1383-5866

Machulek Jr., A.; Moraes, J.E.F.; Okano, L.T.; Silvério, C.A. & Quina, F.H. (2009b). Photolysis of ferric íon in the presence of sulfate or chloride íons: implications for the photo-Fenton process. *Photochemical & Photobiological Sciences*, Vol. 8, No. 7, pp. 985-991, ISSN 1474-905X

Maciel, R.; Sant'Anna Jr., G. L. & Dezotti, M. (2004). Phenol removal from high salinity effluents using Fenton's reagent and photo-Fenton reactions. *Chemosphere*, Vol. 57, No. 7, pp. 711-719, ISSN 0045-6535

Marin, M.L.; Lhiaubet-Vallet, V., Santos-Juanes, L.; Soler, J.; Gomis, J.; Arques, A.; Amat, A. M. & Miranda, M.A. (2011). A photophysical approach to investigate the photooxidation mechanism of pesticides: Hydroxyl radical versus electron transfer. *Applied Catalysis, B: Environmetal*, Vol. 103, No. 1-2, pp. 48-53, ISSN 0926-3373

Martyanov, I.N.; Savinov, E.N. & Parmon, V.N. (1997). A comparative study of efficiency of photooxidation of organic contaminants in water solutions in various photochemical and photocatalytic systems. 1. Phenol photoxidation promoted by hydrogen peroxide in a flow reactor. *Journal of Photochemistry and Photobiology A: Chemistry*, Vol. 107, No. 1-3, pp. 227-231, ISSN 1010-6030

Matilainen, A. & Sillanpää, M. (2010). Removal of natural organic matter from drinking water by advanced oxidation processes. *Chemosphere*, Vol. 80, No. 4, pp. 351-365, ISSN 0045-6535

Matthews, R.W. (1992). Photocatalytic oxidation of organic contaminants in water: An aid to environmental preservation. *Pure and Applied Chemistry*, Vol. 64, No. 9, pp. 1285-1290, ISSN 0033-4545

McElroy, W.J. (1990). A laser photolysis study of the reaction of $SO4^{\bullet-}$ with Cl^- and subsequent decay of Cl_2^- in aqueous solution. *Journal of Physical Chemistry*, Vol. 94, No. 6, pp. 2435–2441, ISSN 0022-3654

Moraes, J.E.F.; Quina, F.H.; Nascimento, C.A.O.; Silva, D.N. & Chiavone-Filho, O. (2004a). Treatment of saline wastewater contaminated with hydrocarbons by the photo-Fenton process. *Environmental Science & Technology,* Vol. 38, No. 4, pp. 1183-1187, ISSN 0013-936X

Moraes, J.E.F.; Silva, D.N.; Quina, F.H.; Chiavone-Filho, O. & Nascimento, C.A.O. (2004b). Utilization of solar energy in the photodegradation of gasoline in water and of oil-field-produced water. *Environmental Science & Technology,* Vol. 38, No. 13, pp. 3746-3751, ISSN 0013-936X

Nadtochenko, V.A. & Kiwi, J. (1998). Photolysis of $Fe(OH)^{2+}$ and $Fe(Cl)^{2+}$ in aqueous solution. Photodissociation kinetics and quantum yields. *Inorganic Chemistry,* Vol. 37, No. 20, pp. 5233-5238, ISSN 0020-1669

Nascimento, C.A.O.; Teixeira, A.C.S.C.; Guardani, R.; Quina, F.H.; Chiavone-Filho, O. & Braun, A.M. (2007). Industrial wastewater treatment by photochemical processes based on solar energy. *Journal of Solar Energy Engineering*, Vol. 129, No. 1, pp. 45-52, ISSN 0199-6231

Nascimento, C.A.O.; Oliveros, E. & Braun, A.M. (1994). Neural-network modeling of photochemical processes. *Chemical Engineering Process,* Vol. 33, No. 5, pp. 319-324, ISSN 0255-2701

Nichela, D.; Haddou, M.; Benoit-Marquié, F.; Maurette, M.T.; Oliveros, E. & Einschlag, F.S.G. (2010). Degradation kinetics of hydroxy and hydroxynitro derivates of benzoic acid by Fenton-like and photo-Fenton techniques: A comparative study. *Applied Catalysis, B: Environmetal,* Vol. 98, No. 3-4, pp. 171-179, ISSN 0926-3373

Nogueira, R.F.P.; Silva, M.R.A. & Trovó, A.G. (2005). Influence of iron source on the solar photo-Fenton degradation of different classes of organic compounds. *Solar Energy,* Vol. 79, No. 4, pp. 384-392 , ISSN 0038-092X

Oliveros, E.; Benoit-Marquie, F.; Puech-Costes, E.; Maurette, M.T. & Nascimento, C.A.O. (1998). Neural network modeling of the photocatalytic degradation of 2,4-dihydroxybenzoic acid in aqueous solution. *Analusis,* Vol. 26, No. 8, pp. 326-332, ISSN 0365-4877

Oliveros, E.; Legrini, O.; Hohl, M.; Müller, T. & Braun, A.M. (1997). Industrial waste water treatment: Large scale development of a light-enhanced Fenton reaction. *Chemical Engineering and Processing,* Vol. 36, No. 5, pp. 397-405, ISSN 0255-2701

Pignatello, J.J. (1992). Dark and photoassisted Fe^{3+}- catalysed degradation of chlorophenoxy herbicides by hydrogen-peroxide. *Environmental Science & Technology,* Vol. 26, No. 5, pp. 944-951, ISSN 0013-936X

Pignatello, J.J.; Liu, D. & Huston, P. (1999). Evidence for an additional oxidant in photoassisted Fenton reaction. *Environmental Science & Technology,* Vol. 33, No. 11, pp. 1832-1839, ISSN 0013-936X

Pignatello, J.J; Oliveros, E. & Mackay, A. (2006). Advanced oxidation processes for organic contaminant destruction based on the Fenton reaction and related chemistry. *Critical Reviews in Environmental Science & Technology,* Vol. 36, No. 1, pp. 1-84. Errata. (2007). *Critical Reviews in Environmental Science & Technology,* Vol. 37, No. 3, pp. 273-275, ISSN 1064-3389

Pontes, R.F.F.; Moraes, J.E.F; Machulek, A. & Pinto, J.M. (2010). A mechanistic kinetic model for phenol degradation by the Fenton process. *Journal of Hazardous Materials,* Vol. 176, No. 1-3, pp. 402-413, ISSN 0304-3894

Pozdnyakov. I. P.; Glebov, E. M.; Plyusnin, V. F.; Grivin, V. P.; Ivanov, Y. V.; Vorobyev, D. Y. & Bazhin, N. M. (2000). Mechanism of $Fe(OH)^{2+}_{(aq)}$ photolysis in aqueous solution. *Pure and Applied Chemistry,* Vol. 72, No. 11, pp. 2187-2197, ISSN 0033-4545

Puigdomenech, I. (April 2010). Chemical Equilibrium Diagrams, 14.07.2011, Available from http://www.kemi.kth.se/medusa

Ruppert, G.; Bauer, R. & Heisler, G. (1993). The photo-Fenton reaction - an effective photochemical wastewater treatment process. *Journal of Photochemistry and Photobiology A: Chemistry,* Vol. 73, No. 1, pp. 75-78, ISSN 1010-6030

Silva, M.R.A.; Vilegas, W.; Zanoni, M.V.B. & Nogueira, R.F.P. (2010). Photo-Fenton degradation of the herbicide tebuthiuron under solar irradiation: Iron complexation and initial intermediates. *Water Research,* Vol. 44, No. 12, pp. 3745-3753, ISSN 0043-1354

Soler, J.; García-Ripoll, A.; Hayek, N.; Miró, P.; Vicente, R.; Arques, A. & Amat, A.M. (2009). Effect of inorganic íons on the solar detoxification of water polluted with pesticides. *Water Research,* Vol. 43, No. 18, pp. 4441-4450, ISSN 0043-1354

Sonntag, C. von. (2008). Advanced oxidation processes: mechanistic aspects. *Water Science & Technology,* Vol. 58, No. 5, pp. 1015-1021, ISSN 0273-1223

Trovó, A.G. & Nogueira, R.F.P. (2011). Diclofenac abatement using modified solar photo-Fenton process with ammonium iron(III) citrate. *Journal of the Brazilian Chemical Society,* Vol. 22, No. 6, pp. 1033-1039, ISSN 0103-5053

Truong, G.L.; De Laat, J. & Legube, B. (2004). Effects of chloride and sulfate on the rate of oxidation of ferrous ion by H_2O_2. *Water Research,* Vol. 38, No. 9, pp. 2384-2394, ISSN 0043-1354

Yu, X.-Y. & Barker, J. R. (2003a). Hydrogen Peroxide Photolysis in Acidic Aqueous Solutions Containing Chloride Ions. I.Chemical Mechanism. *Journal of Physical Chemistry A,* Vol. 107, No. 9, pp. 1313-1324, ISSN 1089-5639

Yu, X.-Y. & Barker, J. R. (2003b). Hydrogen Peroxide Photolysis in Acidic Aqueous Solutions Containing Chloride Ions. II.Quantum Yield of $HO^{\bullet}$ (Aq) Radicals. *Journal of Physical Chemistry A,* Vol. 107, No. 9, pp. 1325-1332, ISSN 1089-5639

Yu, X.-Y. (2004). Critical Evolution of Rate Constants and Equilibrium Constants of Hydrogen Peroxide Photolysis in Acidic Aqueous Solutions containing Chloride Ions. *Journal of Physical and Chemical Reference Data,* Vol. 33, No. 3, pp. 747-763, ISSN 0047-2689

Yu, X.Y., Bao, Z-C. & Barker, J.R. (2004). Free radical reactions involving $Cl^{\bullet}$, $Cl2^{\bullet -}$ and $SO4^{\bullet -}$ in the 248 nm photolysis of aqueous solutions containing $S_2O_8^{2-}$ and Cl^{-}. *Journal of Physical Chemistry A,* Vol. 108, No. 2, pp. 295–308, ISSN 1089-5639

Yue, P.L. (1993). Modelling of kinetics and reactor for water purification by photooxidation. *Chemical Engineering Science,* Vol. 48, No. 1, pp. 1-11, ISSN 0009-2509

Zanta, C.L.P.S., Friedrich, L.C., Machulek Jr., A, Higa, K.M. & Quina, F.H. (2010). Surfactant degradation by a catechol-driven Fenton reaction. *Journal of Hazardous Materials*, Vol. 178, No. 1-3, pp. 258-263, ISSN 0304-3894

Zapata, A.; Oller, I.; Bizani, E.; Sánches-Pérez, J.A.; Maldonado, M.I. & Malato, S. (2009). Evaluation of operational parameters involved in solar photo-Fenton degradation of a commercial pesticide mixture. *Catalysis Today*, Vol. 144, No. 1-2, pp. 94-99, ISSN 0920-5861

Zazo, J.A.; Pliego, G.; Blasco, S.; Casas, J.A. & Rodriguez, J.J (2011). Intensification of the Fenton process by increasing the temperature. *Industrial & Engineering Chemistry Research*, Vol. 50, No. 2, pp. 866-870, ISSN 0888-5885

12

Photocatalytic Degradation of Organic Pollutants: Mechanisms and Kinetics

Malik Mohibbul Haque[1], Detlef Bahnemann[2] and Mohammad Muneer[1*]
[1]*Department of Chemistry, Aligarh Muslim University,*
[2]*Institut fuer Technische Chemie, Leibniz Universität Hannover,*
[1]*India*
[2]*Germany*

1. Introduction

A wide variety of organic pollutants are introduced into the water system from various sources such as industrial effluents, agricultural runoff and chemical spills (Muszkat et al., 1994; Cohen et al., 1986). Their toxicity, stability to natural decomposition and persistence in the environment has been the cause of much concern to the societies and regulation authorities around the world (Dowd et al., 1998).

Development of appropriate methods for the degradation of contaminated drinking, ground, surface waters, wastewaters containing toxic or nonbiodegradable compounds is necessary. Among many processes proposed and/or being developed for the destruction of the organic contaminants, biodegradation has received the greatest attention. However, many organic chemicals, especially which are toxic or refractory, are not amendable to microbial degradation. Researcher showed their interest and started the intensive studies on heterogeneous photocatalysis, after the discovery of the photo-induced splitting of water on TiO_2 electrodes (Fujishima and Honda, 1972).

Semiconductor particles have been found to act as heterogeneous photocatalysts in a number of environmentally important reactions (Blake, 2001; Pirkanniemi, & Sillanpää, 2002; Gaya & Abdullah, 2008). Materials such as colloidal TiO_2 and CdS have been found to be efficient in laboratory-scale pollution abatement systems (Barni et al., 1995; Bellobono et al., 1994; Legrini et al., 1993; Mills & Hunte, 1997; Halmann, 1996), reducing both organic [e.g. halogenocarbons (Gupta & Tanaka, 1995; Martin et al., 1994; Read et al., 1996;), benzene derivatives (Blanco et al., 1996; Mao et al., 1996) detergents (Rao & Dube, 1996), PCB's (Huang et al., 1996), pesticides (Gianturco et al., 1997; Minero et al., 1996; Lobedank et al., 1997; Haque & Muneer 2003; Muneer & Bahnemann, 2002), explosives (Schmelling et al., 1996), dyes (Vinodgopal et al., 1996), cyanobacterial toxins (Liu et al., 2002)] and inorganic [e.g. N_2 (Ranjit et al., 1996), NO_3^- and NO_2^- (Mills et al., 1994; Ranjit et al., 1995; Kosanic & Topalov, 1990; Pollema et al., 1992), cyanides (Mihaylov et al., 1993; Frank & Bard 1977), thiocyanates (Draper & Fox, 1990), cyanates (Bravo et al., 1994), bromates (Mills et al., 1996) etc.] pollutants/impurities to harmless species. Semiconductor photocatalysts have been shown to be useful as carbon dioxide (Irvine et al., 1990) and nitrogen (Khan & Rao, 1991) fixatives and for the decomposition of O_3 (Ohtani et al., 1992), destruction of micro-organisms such as bacteria (Matsunaga & Okochi, 1995; Zhang et al., 1994; Dunlop et al.,

2002), viruses (Lee et al., 1997), for the killing of malignant cancer-cells (Kubota et al., 1994), for the photo splitting of water (Pleskov & Krotova, 1993; Grätzel, 1981), for the cleanup of oil spills (Gerisher & Heller, 1992; Berry & Mueller, 1994), and for the control of the quality of meat freshness in food industry (Funazaki et al., 1995). Photocatalytic semiconductor films have been studied for the purposes of laboratory scale up of some of the above applications (Dunlop et al., 2002; Byrne et al., 2002) and, in the case of TiO_2 films, for their photo-induced wettability properties (Wang et al., 1998, 1999; Sakai et al., 1998; Yu et al., 2002). This latter phenomenon, termed superhydrophilicity, is being explored for applications in the development of self-cleaning and antifogging surfaces.
Our research group at the Department of Chemistry, Aligarh Muslim University, Aligarh, India in collaboration with the Institut fuer Technische Chemie, Leibniz Universität Hannover, Hannover, Germany, have been actively involved in studying the photocatalytic degradation of variety of priority organic pollutants in aqueous suspensions.

1.1 Mechanism of photooxidation process

The acceleration of a chemical transformation by the presence of a catalyst with light is called photocatalysis. The catalyst may accelerate the photoreaction by interaction with the substrate in its ground or excited state and/or with a primary photoproduct, depending upon the mechanism of the photoreaction and itself remaining unaltered at the end of each catalytic cycle. Heterogeneous photocatalysis is a process in which two active phases solid and liquid are present. The solid phase is a catalyst, usually a semiconductor. The molecular orbital of semiconductors has a band structure. The bands of interest in photocatalysis are the populated valence band (VB) and it's largely vacant conduction band (CB), which is commonly characterized by band gap energy (E_{bg}). The semiconductors may be photoexcited to form electron-donor sites (reducing sites) and electron-acceptor sites (oxidising sites), providing great scope for redox reaction. When the semiconductor is illuminated with light (hv) of greater energy than that of the band gap, an electron is promoted from the VB to the CB leaving a positive hole in the valence band and an electron in the conduction band as illustrated in Figure 1.

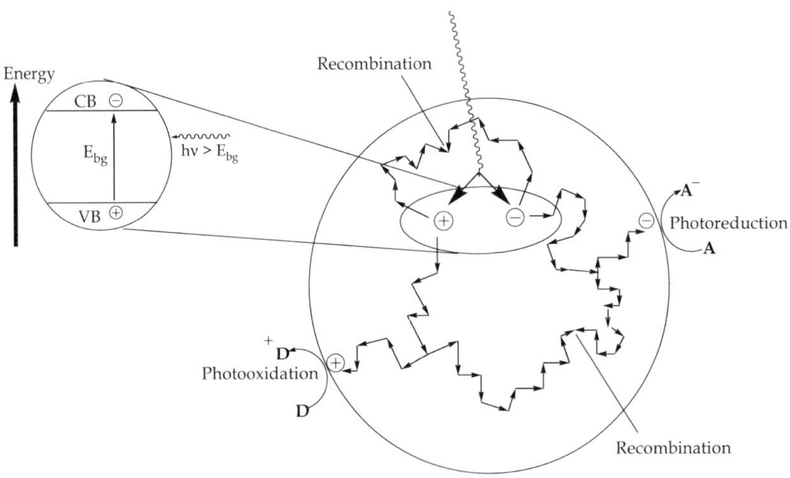

Fig. 1. Photoexcitation of semiconductor leading to charge separation / oxidation / reduction sites.

If charge separation is maintained, the electron and hole may migrate to the catalyst surface where they participate in redox reactions with sorbed species. Specially, h^{+}_{vb} may react with surface-bound H_2O or OH^- to produce the hydroxyl radical and e^-_{cb} is picked up by oxygen to generate superoxide radical anion ($O_2^{\bullet -}$), as indicated in the following equations 1-3;

- absorption of efficient photons by titania ($h\nu \geq E_{bg} = 3.2$ ev)

$$TiO_2 + h\nu \rightarrow e^-_{cb} + h^+_{vb} \tag{1}$$

- formation of superoxide radical anion

$$O_2 + e^-_{cb} \rightarrow O_2^{\bullet -} \tag{2}$$

- neutralization of OH^- group into $^{\bullet}OH$ by the hole

$$(H_2O \Leftrightarrow H^+ + OH^-)_{ads} + h_{vb}^{+} \rightarrow {}^{\bullet}OH + H^+ \tag{3}$$

It has been suggested that the hydroxyl radical ($^{\bullet}OH$) and superoxide radical anions ($O_2^{\bullet -}$) are the primary oxidizing species in the photocatalytic oxidation processes. These oxidative reactions would results in the degradation of the pollutants as shown in the following equations 4-5;

- oxidation of the organic pollutants via successive attack by $^{\bullet}OH$ radicals

$$R + {}^{\bullet}OH \rightarrow R'^{\bullet} + H_2O \tag{4}$$

- or by direct reaction with holes

$$R + h^+ \rightarrow R^{\bullet +} \rightarrow \text{degradation products} \tag{5}$$

For oxidation reactions to occur, the VB must have a higher oxidation potential than the material under consideration. The redox potential of the VB and the CB for different semiconductors varies between +4.0 and -1.5 volts versus Normal Hydrogen Electrode (NHE) respectively. The VB and CB energies of the TiO_2 are estimated to be +3.1 and -0.1 volts, respectively, which means that its band gap energy is 3.2 eV and therefore absorbs in the near UV light (λ<387 nm). Many organic compounds have a potential above that of the TiO_2 valence band and therefore can be oxidized. In contrast, fewer organic compounds can be reduced since a smaller number of them have a potential below that of the TiO_2 conduction band.

1.2 Use of semiconductor (TiO_2) in various fields

Due to non-toxic, easily available, inexpensive, biologically and chemically inert and stable to photo and chemical corrosion, TiO_2 is used in various fields as shown in Figure 2.

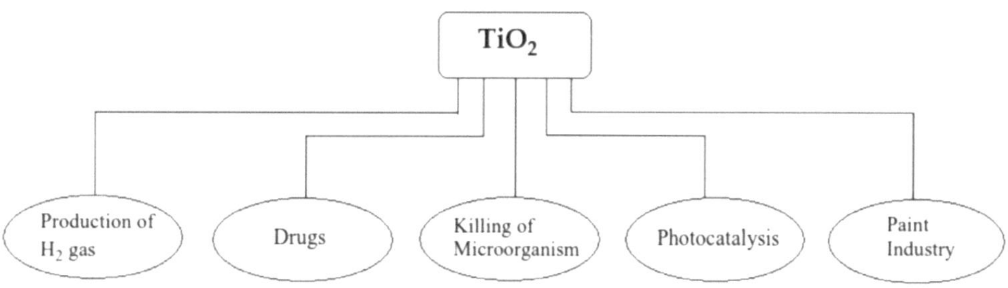

Fig. 2. Application of TiO_2 in various fields.

The process of photocatalysis is also widely being contributed to various sub-discipline of Chemistry as shown in Figure 3.

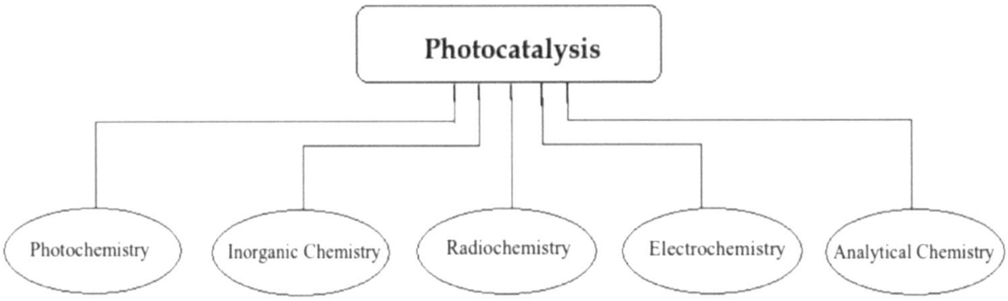

Fig. 3. Application of photocatalysis in various sub-divisions of Chemistry.

1.3 Our research focus

We have studied the photocatalysed degradation of a large variety of organic pollutants under different reaction conditions to determine the detailed degradation kinetics and product identification in few selected systems for better mechanistic understanding. The different class of organic pollutants studied by our research group are shown below in chart 1.

S. No.	Compound Studied	Reference
1	Dinoterb	2011; Dar et al., Res. Chem. Intermed., DOI 10.1007/s11164-011-0299-6
2	Fenoprop and Dichloroprop-P	2010; Faisal et al., Adv. Sci. Lett. 3, 512,
3	Acetamiprid	2010; Khan et al., Desalination, 261, 169.
4	Glyphosate	2008; Muneer & Boxall, Int. J. Photoenergy, article ID 197346
5	4-chlorophenoxyacetic acid	2007; Singh et al., J. Hazard. Mat., 142, 374.
6	Phenoxyacetic acid and 2,4,5-trichlorophenoxyacetic acid	2007; Singh et al., J. Mol. Catal. A: Chem., 264, 66.
7	Uracil and 5-bromouracil	2007; Singh et al., J. Hazard. Mat., 142, 425.
8	4-bromoaniline, 3-nitroaniline, pentachlorophenol, 1,2,3-trichlorobenzene and diphenylamine	2007; Abu Tariq et al., J. Mol. Catal. A: Chem., 265, 231.
9	Chlorotoluron	2006; Haque et al., Environ. Sci. Technol., 40, 4765.
10	Trichlopyr and Daminozid	2006; Qamar et al., J. Environ. Manag., 80, 99.

11	Acephate	2005; Atiqur Rahman et al., J. Adv. Oxid. Technol., 9, 1.
12	Dichlorvos and Phosphamidon	2005; Atiqur Rahman & Muneer, Desalination, 181, 161.
13	Propham, Propachlor and Tebuthiuron	2005; Muneer et al., Chemosphere, 61, 457.
14	Dichlone, 2-amino-5-chloropyridine, benzoyl peroxide and 3-chloro perbenzoic acid	2005; Qamar et al., Res. Chem. Intermed., 31, 807.
15	2,2'-dinitro biphenyl , N,N'-dimethyl-4-nitroso aniline , 4-dimethyl amino benzaldehyde, phthalaldehyde and tetramethyl benzoquinone	2005; Muneer et al., Appl. Catal. A: General, 289, 224.
16	Indole-3-acetic acid and Indole-3-butyric acid	2005; Qamar & Muneer, J. Hazard. Mat., 120, 219.
17	Thiram	2005; Haque & Muneer, Indian J. Chem. Technol., 12, 68.
18	Picloram, Dicamba and Floumeturon	2005; Atiqur Rahman & Muneer, J. Environ. Sci. & Health, B40, 257.
19	2,4-dichlorophenoxy Acetic Acid	2004; Singh & Muneer Res. Chem. Intermed., 30, 317.
20	Maleic Hydrazide	2004; Singh et al., J. Adv. Oxid. Tech., 7, 184.
21	Methoxychlor, Chlorothalonil and Disulfoton	2004; Muneer et al., Res. Chem. Intermed., 30, 663.
22	Dimethyl Terephthalate	2003; Atiqur Rahman et al., Res. Chem. Intermed., 29, 35.
23	Isoproturon	2003; Haque & Muneer, J. Environ. Manag., 69, 169.
24	Bromacil	2003; Singh et al., Photochem. Photobiol. Sci., 2, 151.
25	Diphenamid	2003; Atiqur Rahman et al., J. Adv. Oxid. Technol., 6, 100.
26	Benzidine and 1,2-diphenyl hydrazine	2002; Muneer et al., Chemosphere, 49, 193.
27	Terbacil and 2,4,5-tribromoimidazole	2002; Muneer & Bahnemann, Appl. Catal. B: Environ., 36, 95.
28	1,2-diethyphthalate	2001; Muneer et al., J. Photochem. Photobiol., A: Chem., 143, 213.
29	Diuron	1999; Muneer et al., Res. Chem. Intermed., 25, 667.

Chart 1.

The following text describes the results of the photocatalysed degradation of different pollutants under variety of conditions for degradation kinetics and product identification.

1.4 Procedure for conducting the degradation experiments

Stock solutions of the pollutants containing the desired concentration were prepared in double distilled water. An immersion well photochemical reactor made of Pyrex glass equipped with a magnetic stirring bar, water circulating jacket and an opening for supply of molecular oxygen was used. A simplified diagram of the reactor system is shown in Figure 4.

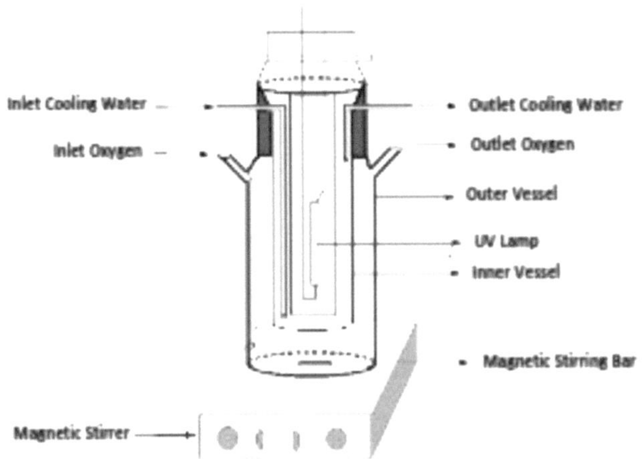

Fig. 4. Simplified diagram of Photochemical Reaction Vessel.

For irradiation experiment an aqueous solution of the pollutants with desired concentration is taken into the photoreactor and required amount of photocatalyst was added. Then the solution was stirred and bubbled with molecular / atmospheric oxygen for at least 15 minutes in the dark to allow equilibration of the system so that the loss of compound due to adsorption can be taken into account. The zero time reading was obtained from blank solution kept in the dark in the presence of TiO_2 and oxygen but otherwise treated similarly to the irradiated solution. The suspensions were continuously purged with molecular / atmospheric oxygen throughout each experiment. Irradiations were carried out using medium pressure mercury lamp. The light intensity was measured using UV-light intensity detector. IR-radiation and short-wavelength UV-radiation were eliminated by water circulating Pyrex glass jacket. Samples (10 mL) were collected before and at regular intervals during irradiation and analyzed after centrifugation.

The sunlight experiments were carried out between 9:00 A.M to 2:30 P.M. on a sunny day. The light intensity was measured using UV-light intensity detector (Lutron UV-340), which was found to be in the range of 0.370 to 0.480 mW/cm^2. Reactions were carried out in the same reaction vessel as described above. Aqueous solution of desired concentration of the model compound containing required amount of photocatalyst was taken and stirred for 15 min. in the dark in presence of oxygen for equilibration. The solution was then placed on flat platform under sunlight with continuous stirring and purging of molecular oxygen. Samples (10 mL) were collected before and at regular intervals during the illumination and analyzed after centrifugation.

2. Analysis

2.1 Photomineralization of organic pollutants

The photomineralization of the pesticide was measured using Total Organic Carbon analyzer (Shimadzu TOC 5000 A). The main principle of TOC analyzer involves the use of carrier gas (oxygen), which is flow-regulated (150 ml / min) and allows to flow through the total carbon (TC) combustion tube, which is packed with catalyst, and kept at 680^0 C. When the sample enters the TC combustion tube, TC in the sample is oxidized to carbon dioxide. The carrier gas containing the combustion products from the TC combustion tube flows through the inorganic carbon (IC) reaction vessel, dehumidifier, halogen scrubber and finally reaches the sample cell of the nondispersive infrared (NDIR) detector which measures the carbon dioxide content. The output signal (analog) of the NDIR detector is displayed as peaks. The peak areas are measured and processed by the data processing unit. Since the peak areas are proportional to the total carbon concentration, the total carbon in a sample may be easily determined from the calibration curve prepared using standard solution of known carbon content. Total carbon is the sum of TOC (Total Organic Carbon) and IC (Inorganic Carbon).

2.2 Photodegradation of the pesticide

The photodegradation of the pesticide was measured using UV-Vis spectrophotometry or HPLC analysis techniques.

2.3 Characterization of intermediate products

Intermediate product formed during the photooxidation process was characterized by monitoring the reaction as a function of time using GC/MS analysis technique. For GC/MS analysis a Shimadzu Gas Chromatograph and Mass Spectrometer (GCMS-QP 5050) equipped with a 25m CP SIL 19 CB (d=0.25mm) capillary column, operating temperature programmed (220^0C for 40 min at the rate of 10^0C min^{-1}) in splitless mode injection volume (1.0 μL) with helium as a carrier gas was used.

3. Photocatalysis of organic pollutants under different conditions

3.1 In the presence of TiO_2

Irradiation of an aqueous suspension of desired organic pollutants in the presence of TiO_2 lead to decrease in absorption intensity and depletion in TOC content as a function of time.

As a representative example Figure 5 shows the change in absorption intensity and depletion in TOC as a function of time on irradiation of an aqueous solution of isoproturon (0.5 mM, 250 ml) in the presence and absence of photocatalyst (Degussa P25, 1 gL^{-1}) by the "Pyrex" filtered output of a 125 W medium Pressure mercury lamp (radiant flux 4.860 mW/cm^2).

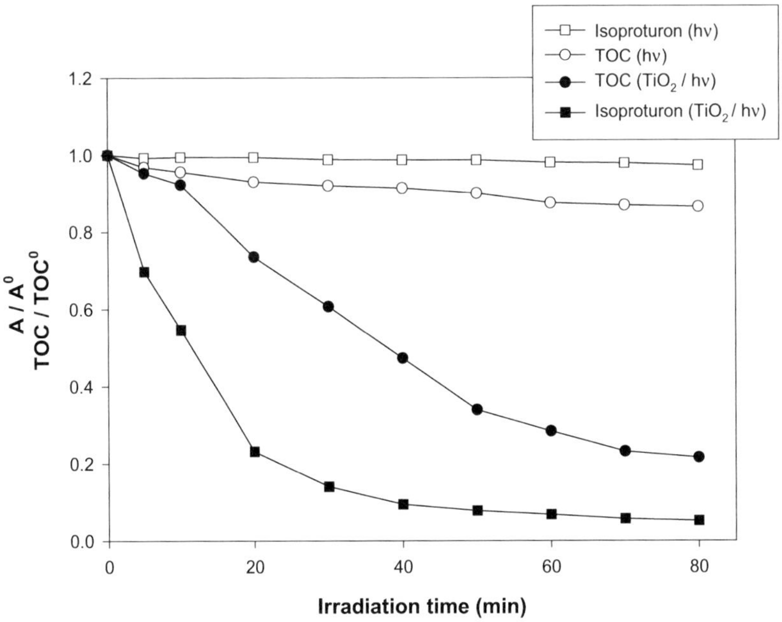

Fig. 5. Depletion in TOC and change in absorption intensity as a function of irradiation time for an aqueous solution of isoproturon in the presence and absence of the photocatalyst. Experimental conditions: 0.5 mM isoproturon, V=250 mL, photocatalyst: TiO_2 (Degussa P25, 1 gL^{-1}), immersion well photoreactor, 125 W medium pressure Hg lamp, absorbance was followed at 238 nm after 75% dilution, cont. O_2 purging and stirring, irradiation time = 80 min.

It could be seen that 94.8% degradation and 78.45 % mineralization of the isoproturon takes place after 80 min of illumination. On the other hand no observable loss of the compound was found when the irradiations were carried out in the absence of the photocatalysts (Haque & Muneer, 2003).

Both the mineralization (depletion of TOC Vs irradiation time) and decomposition (decrease in absorption intensity/concentration Vs irradiation time) curves can be fitted reasonably well by an exponential decay curve suggesting the first order kinetics. Figure 6 shows the linear regression curve fit for the natural logarithm of the degradation of isoproturon Vs irradiation time for first order reaction. For each experiment, the degradation rate constant was calculated from the plot of the natural logarithm of the TOC depletion or absorbance / concentration of the pesticide as a function of irradiation time (the concentration of the pollutant was calculated by taking the absorbance of the pollutant at its λ_{max} of standard concentration and then by plotting the graph of absorbance Vs standard concentration of the pollutant). The degradation rate for the mineralization and for the decomposition of the pollutants for first order reaction was calculated using formula given below;

$$-d[TOC]/dt = kc^n \quad (6)$$

$$-d[c]/dt = kc^n \quad (7)$$

TOC = Total Organic Carbon, k = rate constant, c = concentration of the pollutant, n = order of reaction.

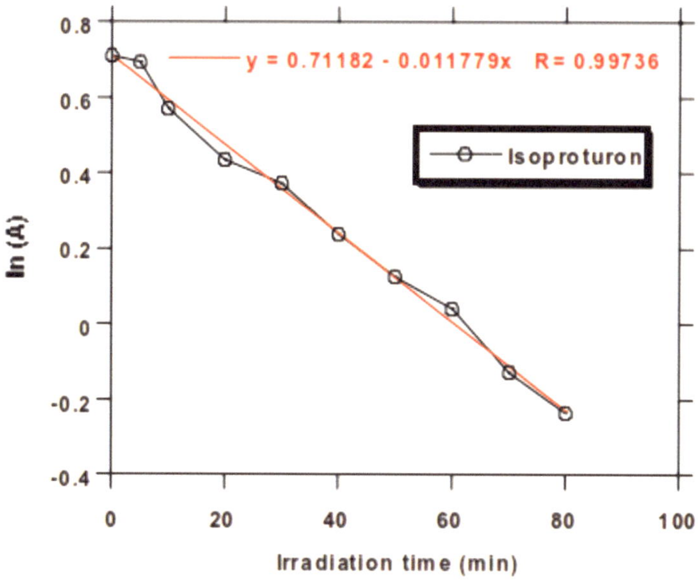

Fig. 6. Plot showing the linear regression curve fit for the natural logarithm of absorbance of the pesticide concentration of isoproturon against irradiation time for the first order reaction.

3.2 Comparison of different photocatalysts

Titanium dioxide is known to be the semiconductor with the highest photocatalytic activity and stable in aqueous solution. Several reviews have been written, regarding the mechanistic and kinetic details as well as the influence of experimental parameters. It has been demonstrated that degradation by photocatalysis can be more efficient than by other wet-oxidation technique (Weichrebe and Vogelpohl, 1995).

We have tested the photocatalytic activity of four different commercially available TiO_2 powders (namely Degussa P25, Sachtleben Hombikat UV100, Milenium Inorganic PC500 and Travancore Titanium Product, India) on the degradation kinetics of the pollutants.

In most of the cases it has been observed that the degradation of pollutants under investigation proceed much more rapidly in the presence of Degussa P25 as compared with other TiO_2 samples (Muneer and Bahnemann, 2001; Haque & Muneer, 2003; Bahnemann et al., 2007) as shown in Figure 7 (isoproturon). The better photocatalytic activity of Degussa P25 has also been reported by Pizzaro et al., in 2005. While in some cases Hombikat UV 100 is found to be better than Degussa P25 for the degradation of benzidine, eosine yellowish, and Remazol brilliant blue R (Muneer et al., 2002; Saquib and Muneer, 2002, 2003) as shown in Figure 8 (eosine yellowish). In an earlier study Lindner et al. 1995, showed that Hombikat UV100 was almost four times more effective than P25 when dichloroacetic acid was used as the model pollutant.

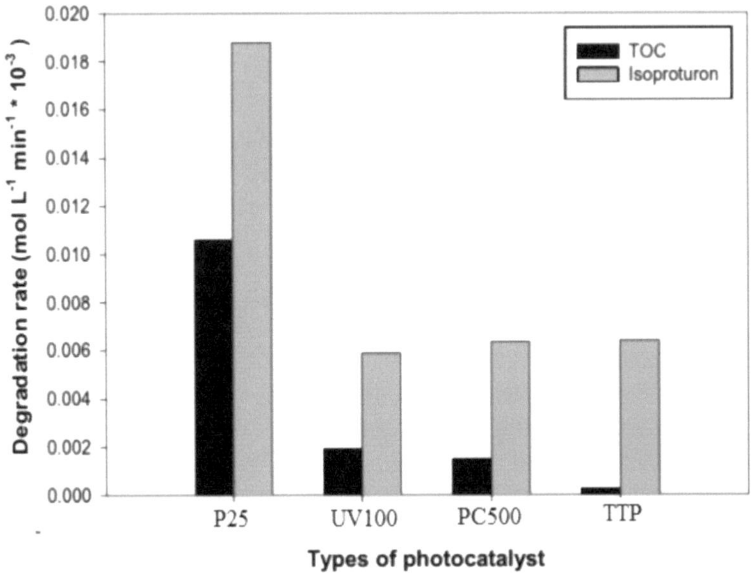

Fig. 7. Comparison of degradation rate for the mineralization and for the decomposition of isoproturon in the presence of different photocatalysts. Experimental conditions: 0.5 mM isoproturon, V=250 mL, photocatalysts: TiO_2 Degussa P25 (1 gL^{-1}), Sachtleben Hombikat UV100 (1 gL^{-1}), PC500 (1 gL^{-1}), TTP (1 gL^{-1}).

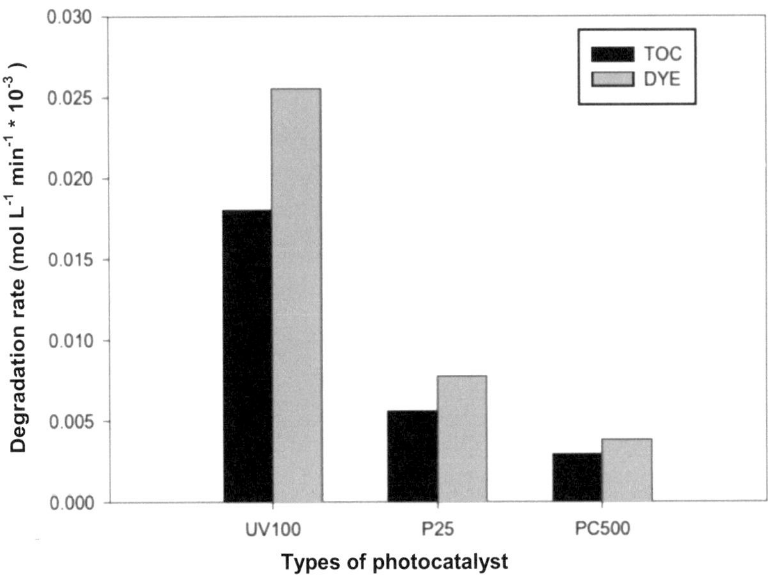

Fig. 8. Comparison of degradation rate for the mineralization and for the decomposition of eosine yellowish under different photocatalysts. Experimental conditions: 0.25 mM dye concentration, V=250 mL, photocatalysts: TiO_2 Degussa P25 (1 gL^{-1}), Sachtleben Hombikat UV100 (1 gL^{-1}) and PC500 (1 gL^{-1}).

The differences in the photocatalytic activity of TiO_2 are likely to be due to differences in the BET-surface, impurities, lattice mismatches or density of hydroxyl groups present on the catalyst's surface. Since they will affect the adsorption behaviour of a pollutant or intermediate molecule and the lifetime and recombination rate of electron-hole pairs.
The reason for the better photocatalytic activity of Degussa P25, could be attributed to the fact that P25 being composed of small nano-crystallites of rutile being dispersed within an anatase matrix. The smaller band gap of rutile "catches" the photons, generating electron-hole pairs. The electron transfer, from the rutile conduction band to electron traps in anatase phase takes place. Recombination is thus inhibited allowing the hole to move to the surface of the particle and react (Hurum et al., 2003). The better efficiency of photocatalyst Degussa P25 may also due to 'quantum size effect' (Nozik et al., 1993; weller, 1993). When the particles become too small, there is a 'blue shift' with an increase of the band gap energy, detrimental to the near UV-photon absorption, and an increase of the electron-hole recombination.

3.3 Influence of pH on the degradation kinetics

An important parameter in the photocatalytic reactions taking place on the particulate surfaces is the pH of the solution, since it dictates the surface charge properties of the photocatalyst and size of aggregates it forms. Employing Degussa P25 as photocatalyst the decomposition and mineralization of pollutant in aqueous suspensions of TiO_2 was studied as a function of pH.
The degradation rate for the decomposition and mineralization of acid red 29 was found to increase with the increase in pH from 3 to 11 as shown in Figure 9 (Qamar et al., 2005) while the degradation rate for the decomposition and mineralization of the pesticide derivative propachlor was found to decrease with the increase in pH from 3 to 11 as shown in Figure 10 (Muneer et al, 2005). Similar results on pH effect have also been reported earlier by Vaz et al., 1998 (for the degradation of uracil and 5-halogenouracil) and Lu et al., 1995 (for the degradation of dichlorvos, propoxur and 2,4-D).
The interpretation of pH effects on the photocatalytic process is very difficult task because of its multiple roles such as electrostatic interactions between the semiconductor surface, solvent molecules, substrate and charged radicals formed during the reaction process. The ionization state of the surface of the photocatalyst can be protonated and deprotonated under acidic and alkaline conditions, respectively, as shown in following equations:

$$TiOH + H^+ \longrightarrow TiOH_2^+ \quad (8)$$

$$TiOH + {}^-OH \longrightarrow TiO^- + H_2O \quad (9)$$

The point of zero charge (pzc) of the TiO_2 (Degussa P25) is widely reported at pH ~ 6.25 (Augustynski, 1988). Thus, the TiO_2 surface will remain positively charged in acidic medium (pH < 6.25) and negatively charged in alkaline medium (pH > 6.25). The functional group present on the pollutants can be protonated and deprotonated depending on the pH of the reaction mixture. The better degradation rate in acidic or basic pH may also be attributed on the basis of the fact that the structural orientation of the molecule is favoured for the attack of the reactive species under that condition.

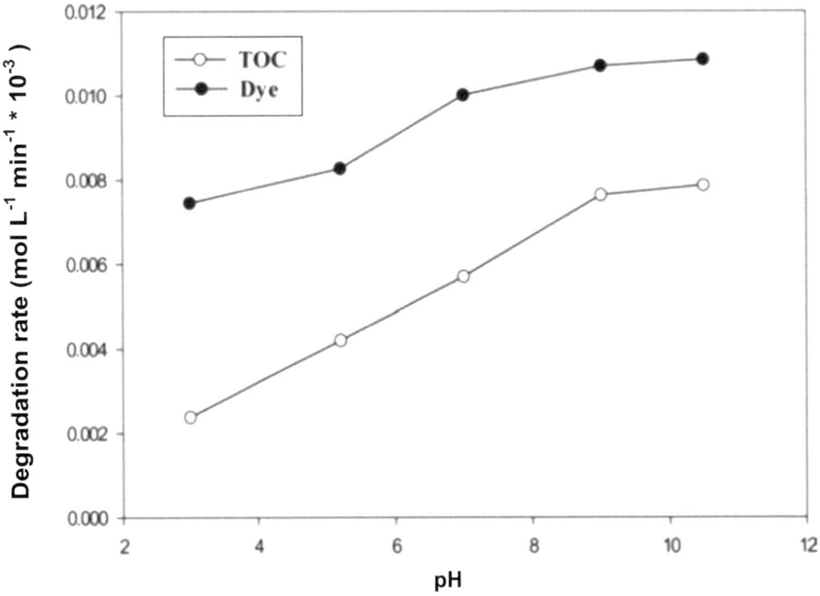

Fig. 9. Influence of pH on the degradation rate for the mineralization and for the decomposition of acid red 29. Experimental conditions: Reaction pH (3, 5.2, 7, 9 and 10.5), 0.25 mM dye concentration, V=250 mL, photocatalyst: TiO_2 (Degussa P25, 1 gL^{-1}).

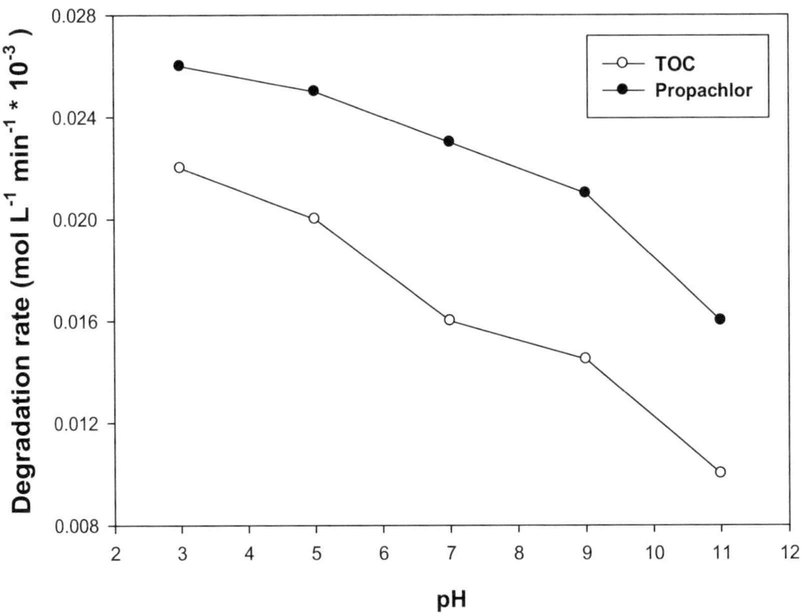

Fig. 10. Influence of pH on the degradation rate for the mineralization and for the decomposition of propachlor. Experimental conditions: Reaction pH (3, 5, 7, 9 and 11), 0.6 mM propachlor, V=250 mL, photocatalyst TiO_2 (Degussa P25, 1 gL^{-1}).

3.4 Effect of substrate concentration

It is important both from mechanistic and from application point of view to study the dependence of substrate concentration in the photocatalytic reaction rate. Effect of substrate concentration on the degradation of the pollutants was studied at different concentrations. As a representative example Figure 11 shows the degradation rate for the TOC depletion and for the decomposition of pollutant as a function of substrate concentration employing Degussa P25 as photocatalyst (Muneer et al, 2005). It was found that the degradation rate for the decomposition and for the mineralization increases gradually with the increase in substrate concentrations. Similar trend was found in most of the colourless organic pollutants as reported earlier (Sabin et al., 1992; Krosley et al., 1993; O'Shea et al., 1997). In coloured compound it has been found that the degradation rate increase upto a certain limit and after that a further increase in substrate concentration lead to decrease in the degradation rate. This may be due to the fact that as the initial concentrations of the pollutant increases, the irradiating mixture becomes more and more intense which prevents the penetration of light to the surface of the catalyst. Hence, the generation of relative amount of $^{\bullet}OH$ and $O_2^{\bullet-}$ on the surface of the catalyst do not increase as the intensity of light and irradiation times are constant. Conversely, their concentrations will decrease with increase in concentration of the pollutant as the light photons are largely absorbed and prevented from reaching the catalyst surface by the substrate molecules.

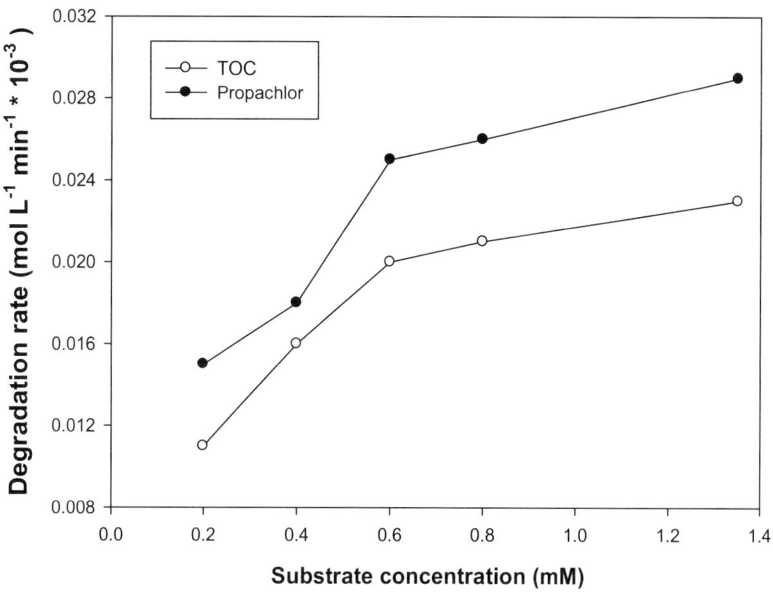

Fig. 11. Influence of substrate concentration on the degradation rate for the mineralization and for the decomposition of propachlor. Experimental conditions: Substrate concentrations (0.20, 0.40, 0.60, 0.80 and 1.35 mM), V=250 mL, photocatalyst: TiO_2 (Degussa P25, 1 gL^{-1}).

As oxidation proceeds, less and less of the surface of the TiO_2 particle is covered as the pollutant is decomposed. Evidently, at total decomposition, the rate of degradation is zero and a decreased photocatalytic rate is to be expected with increasing irradiation time. It has been agreed, with minor variation that the expression for the rate of photodegradation of organic substrates with irradiated TiO_2 follows the Langmuir Hinshelwood (L-H) law for the four possible situations, i.e., (1) the reaction takes place between two adsorbed substances, (2) the reaction occurs between a radical in solution and an adsorbed substrate molecule, (3) the reaction takes place between a radical linked to the surface and a substrate molecule in solution, and (4) the reaction occurs with both the species being in solution. In all cases, the expression for the rate equation is similar to that derived from the L-H model, which has been useful in modelling the process, although it is not possible to find out whether the process takes place on the surface, in the solution or at the interface. Our results, on the effect of the initial concentration on the degradation rate are in agreement with the assumption of the Langmuir Hinshelwood model.

3.5 Effect of catalyst concentration

The effect of photocatalyst concentration on the degradation kinetics of pollutant under investigation was studied employing different concentrations of Degussa P25 varying from 0.5 to 7.5 gL^{-1}. As a representative example the degradation rate for the TOC depletion and for the decomposition of the tebuthiuron as function of catalyst loading is shown in Figure 12 (Muneer et al., 2005). It was observed that the degradation rate was found to increase with the increase in catalyst concentration upto 5 gL^{-1} and on subsequent addition of catalyst lead to the levelling off the degradation rate.

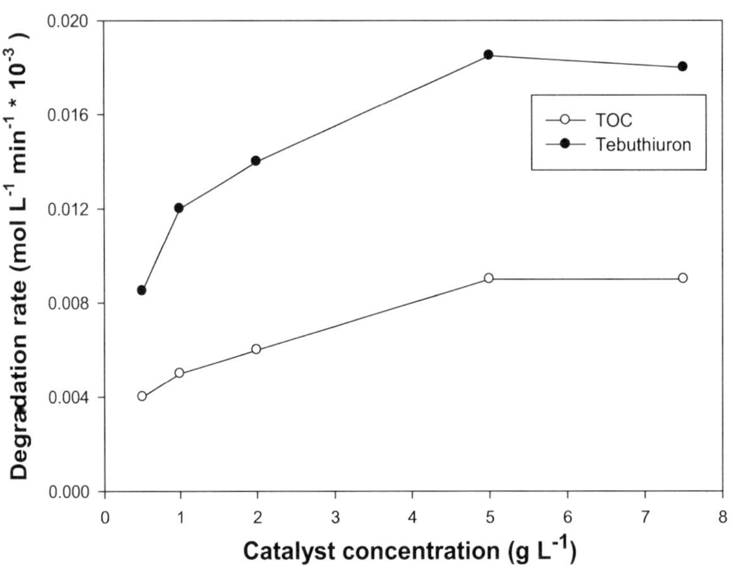

Fig. 12. Influence of catalyst concentration on the degradation rate for the mineralization and for the decomposition of tebuthiuron. Experimental conditions: Photocatalyst concentrations: TiO_2 Degussa P25 (0. 5, 1.0, 2.0, 5.0 and 7.5 gL^{-1}), V=250 mL.

Whether in static, slurry, or dynamic flow reactors, the initial reaction rates were found to be directly proportional to catalyst concentration, indicating a heterogeneous regime. However, it was observed that above a certain concentration, the reaction rate decreases and becomes independent of the catalyst concentration. This limit depends on the geometry and working conditions of the photoreactor and for a definite amount of TiO_2 in which all the particles, i.e., surface exposed, are totally illuminated. When the catalyst concentration is very high, after travelling a certain distance on an optical path, turbidity impedes further penetration of light in the reactor. In any given application, this optimum catalyst concentration [$(TiO_2)_{OPT}$] has to be found, in order to avoid excess catalyst and insure total absorption of efficient photons. Our results on the effect of catalyst concentration on the degradation rate for the TOC depletion and decomposition of pollutants under investigation are in agreement with numerous studies reported in the literature (Shifu and Yunzhang 2007, Daneshvar et al., 2004).

It is believed that both the number of photons absorbed as well as the solute molecules adsorbed increases with increase in number of TiO_2 particles upto the optimum value. Any further increase in TiO_2 concentration beyond optimum value may cause scattering and screening effects which reduces the specific activity of the catalyst (Evgenidou et al., 2007). The highly turbid suspension may prevent the catalyst farthest from being illuminated (Rahman and Muneer, 2005). Higher amount of catalyst may lead to aggregation of TiO_2 particles which may decrease the catalytic activity (Garcia and Takashima, 2003). The optimum value of catalyst has been found to vary with different initial solute concentrations (Sakthivel et al., 2003).

3.6 Effect of different electron acceptors

One practical problem in using TiO_2 as a photocatalyst is the undesired electron / hole recombination, which, in the absence of proper electron acceptor or donor, is extremely efficient and represent the major energy - wasting step thus limiting the achievable quantum yield. One strategy to inhibit electron - hole pair recombination is to add other (irreversible) electron acceptors to the reaction. They could have several different effects such as, i.e., (1) to increase the number of trapped electrons and, consequently, avoid recombination, (2) to generate more radicals and other oxidizing species, (3) to increase the oxidation rate of intermediate compounds and (4) to avoid problems caused by low oxygen concentration. In highly toxic wastewater where the degradation of organic pollutants is the major concern, the addition of electron acceptors to enhance the degradation rate may often be justified. With this view, the electron acceptor such as potassium persulphate, potassium bromate and hydrogen peroxide were added in the solution.

As a representative example Figure 13 shows the degradation rate for the mineralization and decomposition of the pollutants in the presence of different electron acceptors.

The electron acceptors such as hydrogen peroxide, bromate and persulphate ions are known to generate hydroxyl radicals by the mechanisms shown in equations 10-14;

$$H_2O_2 + e^- \longrightarrow {}^{\bullet}OH + {}^{-}OH \tag{10}$$

$$BrO_3^- + 2H + e_{cb}^- \longrightarrow BrO_2^{\bullet} + H_2O \tag{11}$$

$$BrO_3^- + 6H + 6e_{cb}^- \rightarrow [BrO_2^-, HOBr] \rightarrow Br^- + 3H_2O \tag{12}$$

$$S_2O_8^{2-} + e_{cb}^- \longrightarrow SO_4^{2-} + SO_4^{\bullet -} \tag{13}$$

$$SO_4^{\bullet -} + H_2O \longrightarrow SO_4^{2-} + {}^{\bullet}OH + H^+ \tag{14}$$

The respective one-electron reduction potentials of different species are: E ($O_2/O_2^{\bullet -}$) = -155mV, E ($H_2O_2/HO^{\bullet}$) = 800mV, E (BrO_3^- / $BrO_2^{\bullet}$) = 1150 mV, and E ($S_2O_8^{2-}/SO_4^-$) = 1100 mV (Wardman, 1989). From the thermodynamic point of view all employed additives should therefore be more efficient electron acceptors than molecular oxygen.

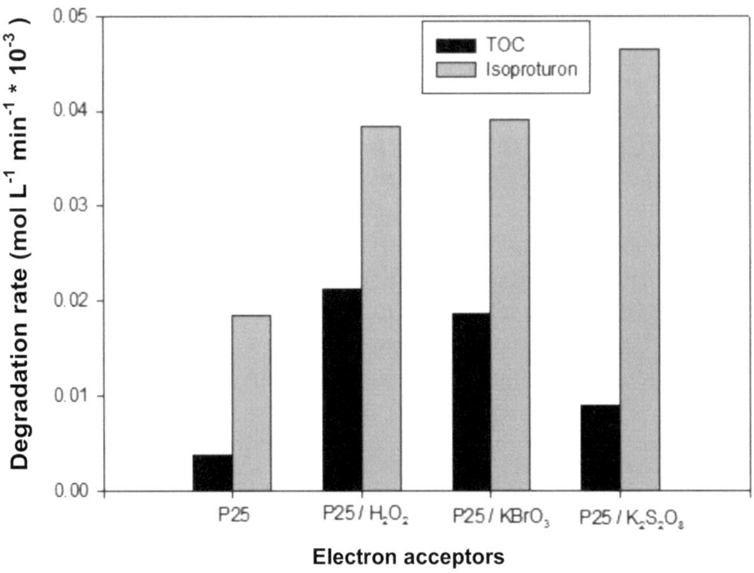

Fig. 13. Comparison of degradation rate for the mineralization and for the decomposition of isoproturon in the presence of different electron acceptors. Experimental conditions: Electron acceptors: $KBrO_3$ (5mM), $K_2S_2O_8$ (5mM), H_2O_2 (10mM), 1.5 mM isoproturon, V=250mL, photocatalyst: TiO_2 (Degussa P25, 1 g L^{-1}).

The effective electron acceptor ability of $KBrO_3$ has been observed in number of studies before (Nevim et al., 2001; Muneer and Bahnemann, 2001). The reason can be attributed to the number of electrons it reacts as shown in eq. 12. Another possible explanation might be a change in the reaction mechanism of the photocatalytic degradation. Since the reduction of bromate ions by electrons does not lead directly to the formation of hydroxyl radicals, but rather to the formation of other reactive radicals or oxidizing agents eg. BrO_2 and HOBr. Furthermore, bromate ions by themselves can act as oxidizing agents. Linder has proposed a mechanism for the photocatalytic degradation of 4-chlorophenol in the presence of bromate ions considering direct oxidation of the substrate by bromate ions (Linder, 1997).

The enhanced effect of persulphate ion on the degradation of pollutants may be accounted on the basis that persulphate is a beneficial oxidizing agent in photocatalytic detoxification because SO_4 is formed from the oxidant by reaction with the electron generated at conduction band (e^-_{cb}) of the semiconductor as shown in eq. 13. This sulphate radical anion (SO_4) is a strong oxidant (E_0 = 2.6 eV) and it can react with the organic pollutants in three possible modes (1) by abstracting a hydrogen atom from saturated carbon, (2) by adding to unsaturated or aromatic carbon and (3) by removing one electron from the carboxylate anions and from certain neutral molecules. In spite of this sulphate radical anion can trap the photogenerated electrons and/ or generated hydroxyl radical as shown in eq. 14. The formation of sulphate radical anion and hydroxyl radical are powerful oxidant, which can degrade the organic pollutants at a faster rate.

The effect of H_2O_2 has been investigated in numerous studies and it was observed that it increases the photodegradation rates of organic pollutants (Hallamn, 1992). The enhancement of the degradation rate on addition of H_2O_2 can be rationalized in terms of several reason. Firstly, it increase the rate by removing the surface-trapped electrons,

thereby lowering the electron-hole recombination rate and increasing the efficiency of hole utilization for reactions such as ($^-OH + h^+ \rightarrow {}^{\bullet}OH$). Secondly, H_2O_2 may split photolytically to produce hydroxyl radicals (${}^{\bullet}OH$) directly, as cited in studies of homogeneous photooxidation using UV / ($H_2O_2 + O_2$) (Peyton and Glaze 1988). Thirdly, the solution phase may at times be oxygen starved, because of either oxygen consumption or slow oxygen mass transfer, and peroxide addition thereby increases the reaction rate.
During the photocatalytic degradation the free radical formed serve a dual function. They are not only the strong oxidants but also at the same time their formation and subsequent rapid oxidation reactions inhibit the electron hole pair recombination.

3.7 Comparison of degradation rate under UV and sunlight

For practical applications of wastewater treatment based on these processes, the utilization of sunlight is preferred. Hence the aqueous suspension of TiO_2 containing organic pollutants was exposed to solar radiation. As a representative example Figure 14 shows the change in concentration as a function of irradiation time on illumination of an aqueous suspension of acetamiprid (0.1 mM) in the presence of TiO_2 (Degussa P25, 1gL^{-1}) under sunlight and UV light source (Khan et al., 2010). It was found that the degradation of the model compound proceeds reasonably fast under sunlight as well.

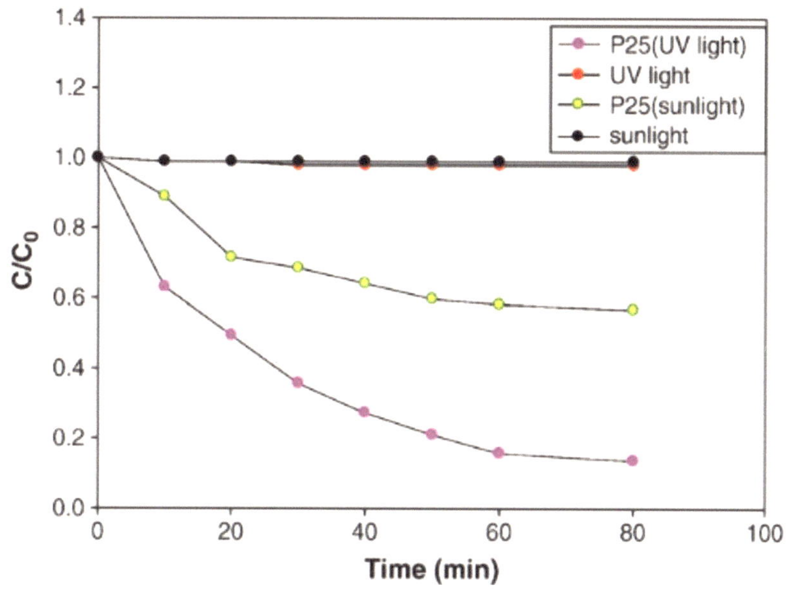

Fig. 14. Change in concentration on irradiation of an aqueous suspension in the presence and absence of TiO_2 (Degussa P25, 1 gL^{-1}) containing acetamiprid (0.1 mM), Light source: Pyrex filtered output of a medium pressure mercury lamp and sunlight.

3.8 Characterization of intermediate products

The identification and characterization of by-products formed during the photodegradation of organic pollutants has been of great interest among the people working in this area around the world. A brief summary showing the starting material and by-products under photolytic condition is shown below in Chart 2.

S. No.	Compound	By-products	Reference
1.	Pyridaben	Fragmented products	Zhu et al., 2004
2	Urea derivative	Hydroxylated, Phenyl Hydroxy Ureas, Aniline Derivatives	Lhomme et al., 2005; Vulliet et al., 2002; Maurino et al., 1999; Pramauro et al., 1993; Kinkennon et al., 1995; Richard & Bengana, 1996; Parra et al., 2000, 2002
3	Chlorophenols derivatives	Hydroxylated, Dechlorinated, Chloro Derivatives	Minero et al., 1996; Jardim et al., 1997; Tseng & Huang, 1991
4	Phenol	Catechol, Hydroquinone, 3-phenyl-2-propenal	Azevedo et al., 2009
5	Triazines derivative	Amido, Dealkylated, Hydroxylated, Ammeline, Cyanouric Acid	Goutailler et al., 2001; Konstantinou et al., 2001a; Pelizzetti et al., 1990, 1992a, 1992b; Minero et al., 1996; Muszkat et al., 1995; Sanlaville et al., 1996, Sleiman et al., 2006
6	Aniline and Amide derivative	Amines, Dechlorinated, Dealkylated, Cyclized, Aliphatics, Cyclized	Konstantinou et al., 2001b, 2002; Sakkas et al., 2004; Peñuela and Barceló, 1996; Pathirana & Maithreepala, 1997
7	Thicarbamate derivative	Amine, Carboxy, Sulfoxide, Dealkylated	Vidal et al., 1999; Sturini et al., 1996; Vidal & Martin, 2001
8	Phenoxy-acids derivatives	Hydroxylated, Carboxylated, Chlorophenols, Quinonidal	Herrmann et al., 1998; Topalov et al., 2000; Barbeni et al., 1987; Poulios et al., 1998;
9	Organophosphorus derivatives	Hydroxy, Oxon, Phenol, Dialkylated, Trialkyl esters, Fragmented products	Herrmann, 1999; Konstantinou et al., 2001a, Oncescu et al., 2010; Herrmann et al., 1999; Hua et al., 1995; Sakkas et al., 2002; Dominguez et al., 1998
10	Carbamate derivative	Hydroxylated, Decarboxylated, Phenolic, Dealkylated, Cyclized	Tamimi et al., 2006; Percerancier et al., 1995; Pramauro et al., 1997; Tanaka et al., 1999; Marinas et al., 2001; Bianco Prevot et al., 1999;
11	Organochlorine derivative	Hydroxy, Dechlorinated	Guillard et al., 1996; Vidal, 1998; Zalenska et al., 2000; Peñuela & Barceló, 1998a, 1998b; Pichat 1997
12	Pyridines	Fragmented, Hydroxylated	Stapleton et al., 2010

Chart 2.

We have also made an attempt to identify the intermediate products formed in the photocatalytic degradation of variety of pesticide derivatives in aqueous suspensions of titanium dioxide through GC-MS analysis technique. Results on the photocatalytic degradation of few selected pesticide derivatives for product analysis are shown below;

3.8.1 Photocatalysis of propham (1)

Irradiation of an aqueous solution of propham (1) in the presence of Degussa P25 TiO_2 and analysis of the irradiated mixture at different time intervals through GC-MS analysis showed the formation of several intermediate products out of which, two products appearing at retention times (t_R) 12.95 and 14.81 min., respectively were characterized as products 3 and 8, on the basis of molecular ion and fragmentation pattern (Muneer et al., 2005). A probable mechanism for the formation of these products from propham (1) involving electron transfer reactions and reactions with hydroxyl radical and superoxide radical anions formed in photocatalytic system, is shown in Scheme 1. The model compound 1, upon the transfer of an electron can form the radical anion 2, which may undergo addition of a hydroxyl radical either at ortho or para position followed by loss of a proton to give the observed product 3. This compound on further transfer of an electron followed by addition of a hydroxyl group can undergo cleavage reaction to give 4. This intermediate on subsequent transfer of an electron can undergo addition of a hydroxyl group to give the observed product 8.

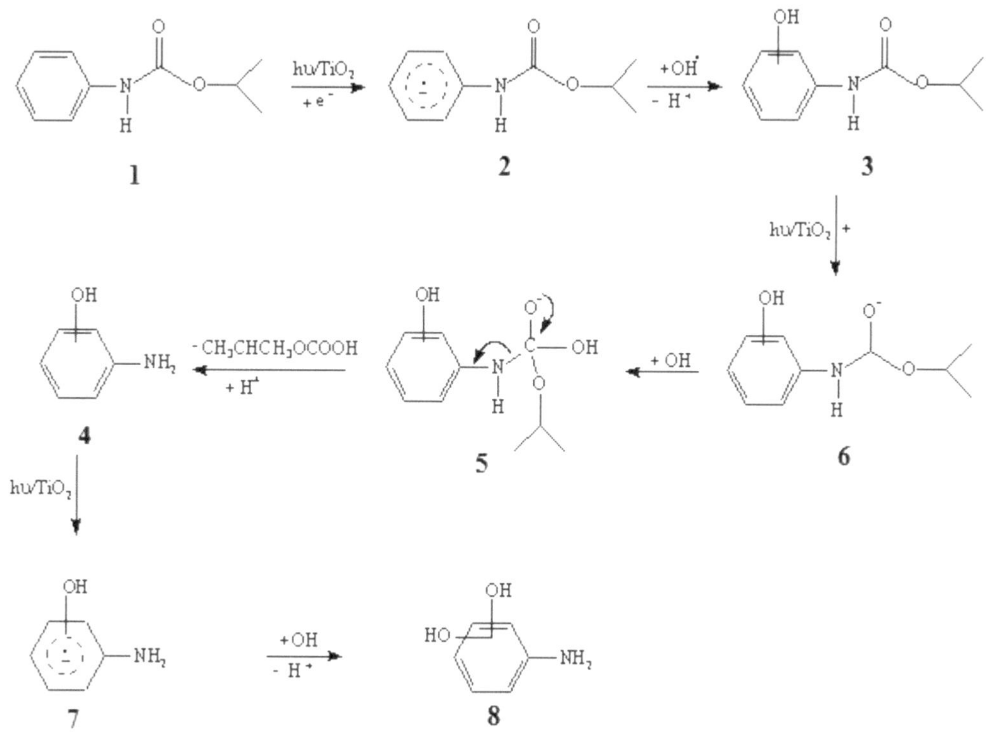

Scheme 1.

3.8.2 Photocatalysis of propachlor (9)

The GC-MS analysis of an irradiated mixture of propachlor (9) in the presence of TiO_2 showed the formation of several by-products. A plausible mechanism for the formation of the different products such as N-Isopropyl-N-phenyl-acetamide (11), 2-Hydroxy-N-isopropyl-N-phenyl acetamide (13), hydroxyl inserted product (15) and N-phenyl-acetamide (16) from propachlor (9) involving similar reactive species, is proposed in Scheme 2 (Muneer et al., 2005). The model compound, 9 on transfer of an electron can form the radical species 10 on removal of chloride ion. This radical species can lead to the observed products 11 and 13, on abstraction of a hydrogen atom or hydroxyl radical. The product 11, on further transfer of an electron can form the radical species 14, which may lose isopropyl to give the observed product 16. The formation of 15 could be understood in terms of pathways shown in Scheme 2 through electron transfer to give radical anion species 12 followed by addition of hydroxyl radical.

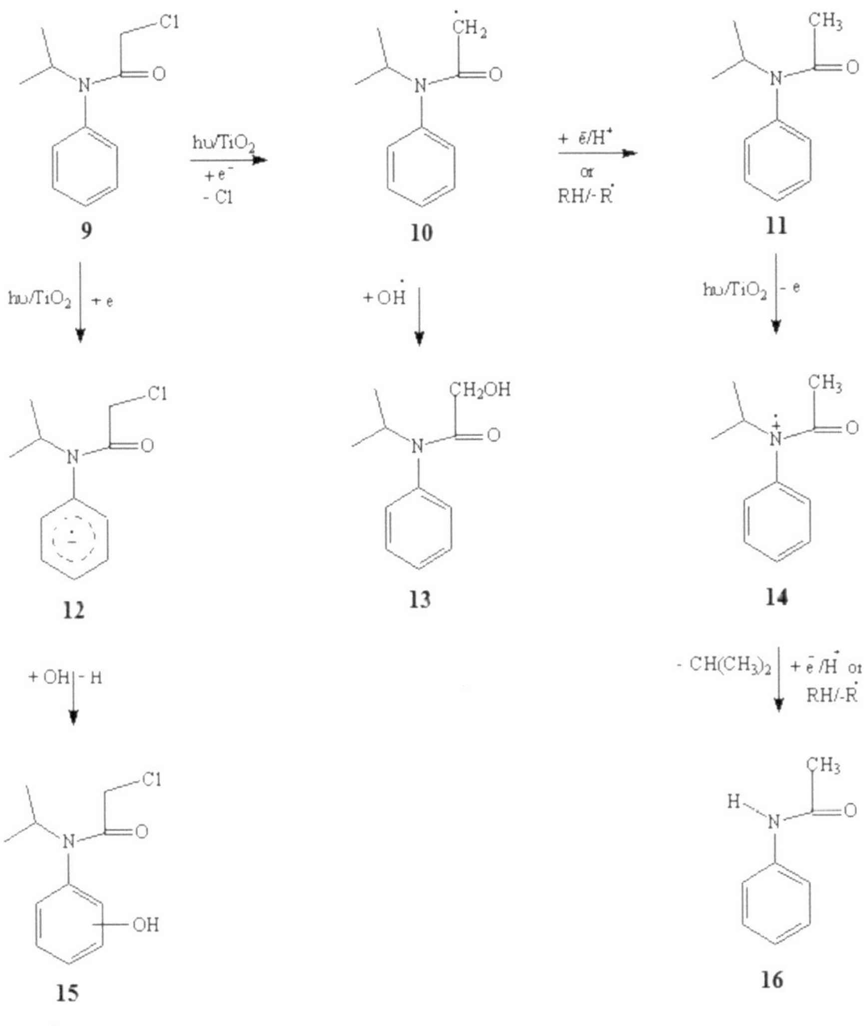

Scheme 2.

3.8.3 Photocatalysis of chlorotoluron (17)

The steady state photolysis of an aqueous suspension of chlorotoluron (17) in the presence of TiO_2 under constant bubbling of atmospheric oxygen for 2 h showed the formation of several products, of which, three products such as 3-(3-Hydroxy-4-methylphenyl)-1,1-dimethylurea (23) (3-chloro-4-methylphenyl) urea (24) and 3-chloro-4-methylphenylamine (25) appearing at retention times (t_R) 10.310 min, 10.255 min and 5.520 min., respectively were identified on the basis of molecular ion and fragmentation pattern with those reported in the GC/MS NIST library, along with some unchanged starting material (17) appearing at (t_R) 5.097 min. (Haque et al., 2006). A possible mechanism for the formation of products 23, 24 and 25 from chlorotoluron (17) involving electron transfer reaction and reaction with hydroxyl radicals could be understood in terms of pathways as shown in Scheme 3. The model compound 17, upon the transfer of an electron followed by the loss of methyl radical and abstraction of a proton to give species 21. This species on subsequent similar reaction can give the observed product 24. The observed product 25 from 17 may be arising through the transfer of an electron to give radical anionic species 19, which may undergo addition of a hydroxyl radical to give 22. This species may lose $N(CH_3)_2COOH$ followed by abstraction of a proton to give the observed product 25 as shown in the Scheme 3. The product 23 could be formed upon the transfer of an electron followed by the loss of chloride ion with subsequent addition of hydroxyl radical as shown in scheme below.

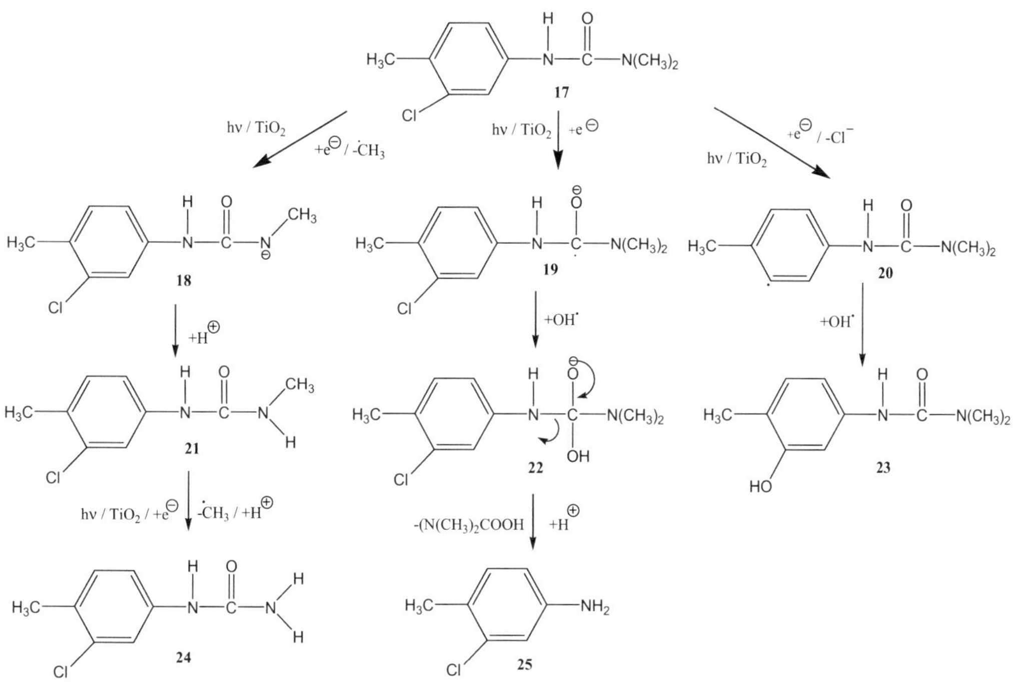

Scheme 3.

3.8.4 Photocatalysis of phenoxyacetic acid (26)

The desired concentration of an aqueous solution of phenoxyacetic acid (**26**) was irradiated in the presence of Degussa P25 and the analysis of the reaction mixture under analogous conditions, showed the formation of several intermediate products of which two intermediate products phenol (28) and 1,2-diphenoxyethane (30) appearing at retention time (t_R) 2.92 and 9.84 min., respectively (Singh et al., 2007a). These products were identified by comparing its molecular ion and mass fragmentation peaks with those reported in the NIST library. A plausible mechanism for the formation of these products involving reactions with hydroxyl radical and superoxide radical anion formed in the photocatalytic process is shown in Scheme 4. The model compound undergoes addition of a hydroxyl radical leading to the formation of a radical species 27, which may undergo loss of ($^{\bullet}OCH_2COOH$) to give the observed product phenol (28). Alternatively, the model compound 26 on addition of hydroxyl radical followed by loss of CO_2 may lead to the formation of the radical species 29. This species can either lose formaldehyde molecule followed by addition of hydroxyl radical led to the formation of phenol or can combine with another radical species leading to the formation of the 1,2-diphenoxyethane (30).

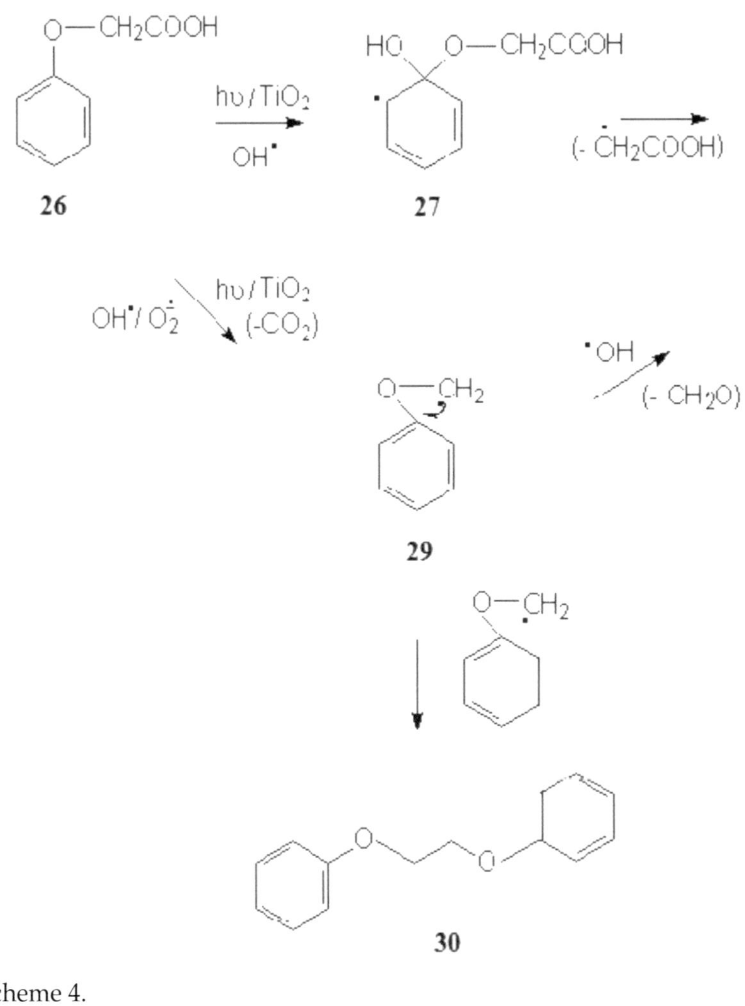

Scheme 4.

3.8.5 Photocatalysis of 2,4,5-trichlorophenoxy acetic acid (31)

Analysis of the irradiated aqueous mixture of 2,4,5-trichlorophenoxy acetic acid (31) in the presence of photocatalyst showed the formation of three intermediate products namely, 2,4,5-trichlorophenol (33), 2,4-dichlorophenol (37) and 1,2,4-trichloro-5-methoxy benzene (35) appearing at retention times 4.11, 6.3 and 12.86 min., respectively and were identified by comparing their molecular ion and mass fragmentation peaks with those reported in the NIST library (Singh et al., 2007a). The model compound 31 on addition of a hydroxyl radical can lead to the formation of a radical species 32, which may undergo loss of (•OCH_2COOH) group to give the observed product 2,4,5-trichlorophenol (33). Alternatively, the model compound 31 upon the transfer of an electron followed by loss of CO_2 may lead to the formation of the radical species 34. This species abstracts a hydrogen atom to form the observed product 1,2,4-trichloro-5-methoxy benzene (35). The product, 33 on further transfer of an electron may undergo loss of chlorine atom followed hydrogen atom abstraction giving 2,4-dichlorophenol (37) as shown in Scheme 5.

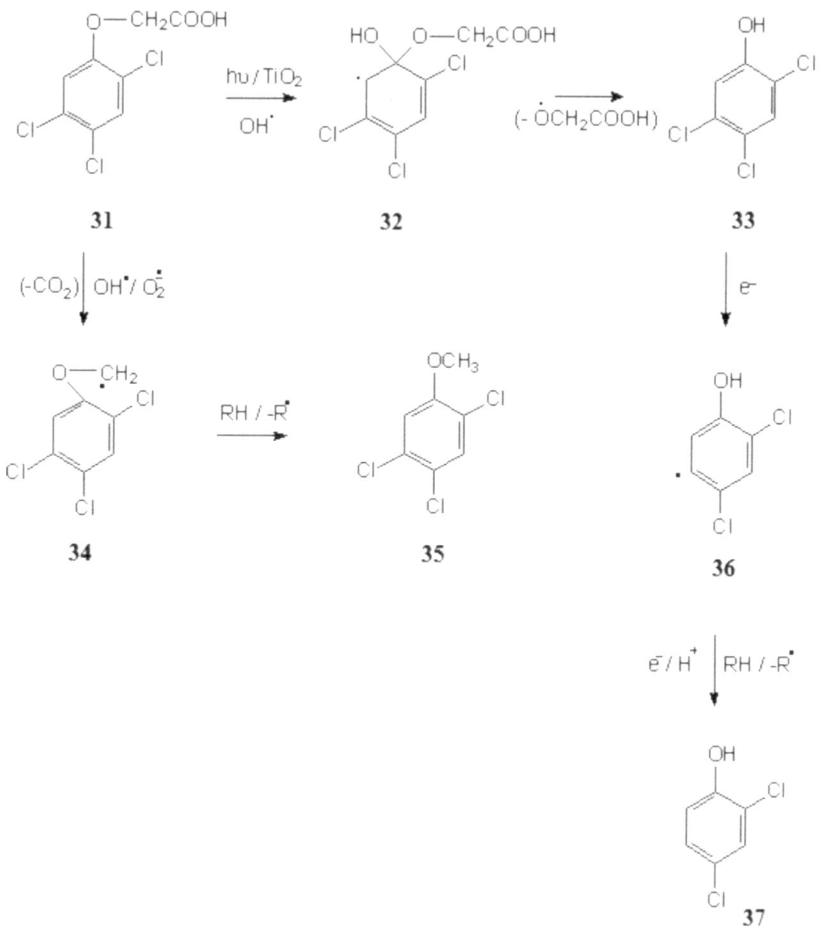

Scheme 5.

3.8.6 Photocatalysis of 4-chlorophenoxy acetic acid (38)

Irradiation of 4-chlorophenoxyacetic acid (4-CPA, 38) in the presence of titanium dioxide showed the formation of 4-chlorophenol (40), which was identified by comparing its molecular ion and fragment ion peak with those reported in the NIST library (Singh et al., 2007b). A plausible mechanism for the formation of product, 40 involving reactions with hydroxyl radicals formed in the photocatalytic process is shown in Scheme 6. The model compound, 4-CPA undergoes addition of a hydroxyl radical formed in the photocatalytic process leading to the formation of a radical species 39 as an intermediate, which may undergo loss of (•OCH_2COOH) group to give the observed product 4-chlorophenol (40). Alternatively, the model compound 4-CPA (38) on addition of hydroxyl radical followed by loss of CO_2 may lead to the formation of the intermediate radical species 41, which upon loss of formaldehyde molecule followed by addition of hydroxyl radical led to the formation of the observed product 4-chlorophenol (40).

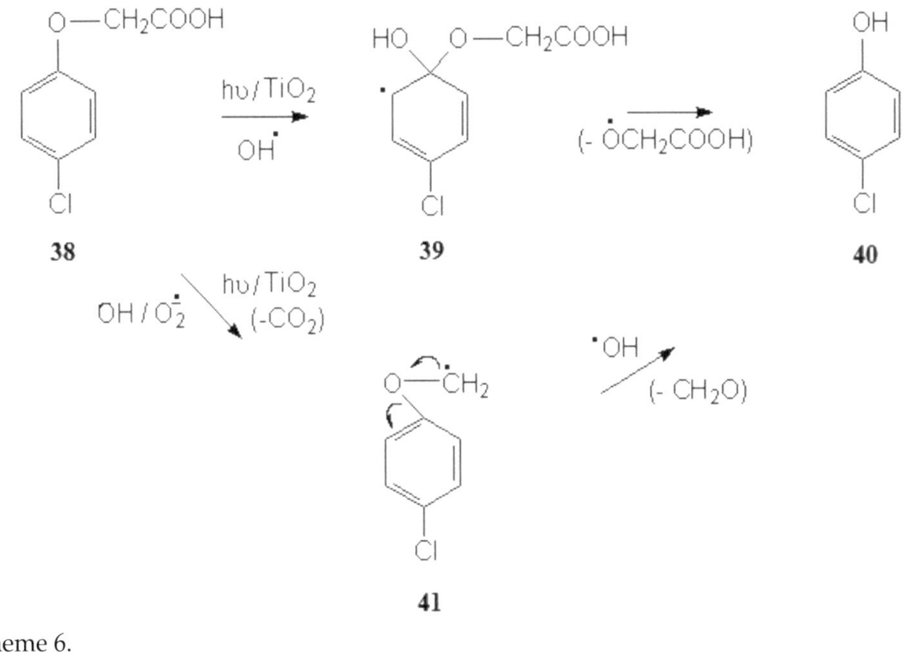

Scheme 6.

4. Conclusions

The results of these studies clearly indicate that TiO_2 can efficiently catalyse the photodegradation and photomineralization of the pollutants in the presence of light and oxygen. In most of the cases the photocatalyst Degussa P25 was found to be more efficient as compare to other photocatalyst TiO_2 powders. But in few cases Hombikat UV100 has also found to be more efficient photocatalyst. The addition of electron acceptor enhanced the degradation rate of the pollutants. The results also indicate that degradation rates could be influenced not only by the different parameters such as type of photocatalyst, catalyst concentration, substrate concentration, pH and additives and their concentration but also by the model pollutants. The observations of these investigations clearly demonstrate the importance of choosing the optimum degradation parameters to obtain high degradation rate, which is essential for any practical application of photocatalytic oxidation processes. The intermediate products formed during the process are also a useful source of information for the degradation pathways.

5. Acknowledgements

Financial support from Alexander Humboldt Foundation, Germany, Volkswagen Foundation, Germany, Marie Curie Foundation, UK, Third World Academy of Sciences, Italy, Department of Science and Technology and UGC, New Delhi, CSTUP, Lucknow, Department of Chemistry, AMU, Aligarh and the Institut füer Technische Chemie, Leibniz Universität Hannover, Hannover, Germany are gratefully acknowledge.

6. References

Augustynski, J. (1988) In: *Structural Bonding*, Springer, Berlin, New York, p. 69 (Chapter1).

Azevedo, E.B., Tôrres, A.R., Aquino Neto, F.R., & Dezotti, M., (2009). TiO_2-Photocatalyzed degradation of phenol in saline media in an annular reactor: hydrodynamics, lumped kinetics, intermediates, and acute toxicity. *Brazilian J. Chem. Engg.*, . 26, 75 - 87.

Bahnemann, W., Muneer, M., & Haque, M.M., (2007). Titanium dioxide-mediated photocatalysed degradation of few selected organic pollutants in aqueous suspensions. *Catal. Today.* 124, 133-148.

Barbeni, M., Morello, M., Pramauro, E., & Pelizzetti, E., (1987). Sunlight photodegradation of 2,4,5-trichlorophenoxy-acetic acid and 2,4,5,trichlorophenol on TiO_2. Identification of intermediates and degradation pathway. *Chemosphere,* 16, 1165-1179.

Barni, B., Cavicchioli, A., Riva, E., Zanoni, L., Bignoli, F., Bellobono, I.R., Gianturco, F., De Giorgi, A., Muntau, H., Montanarella, L., Facchetti, S., & Castellano, L., (1995). Pilot-plant-scale photodegradation of phenol in aqueous solution by photocatalytic membranes immobilizing titanium dioxide (PHOTOPERM® process). *Chemosphere,* 30, 1861-1874.

Bellobono, I.R., Carrara, A., Barni, B., & Gazzotti, A., (1994). Laboratory- and pilot-plant-scale photodegradation of chloroaliphatics in aqueous solution by photocatalytic membranes immobilizing titanium dioxide. *J. Photochem. Photobiol. A:Chem.*, 84, 83-90.

Berry, R.J., & Mueller, M.R., (1994). Photocatalytic decomposition of crude oil slicks using TiO_2 on a floating substrate. *Microchem. J.,* 50, 28-32.

Bianco Prevot, A., Pramauro, E., & de la Guardia, M., (1999). Photocatalytic degradation of carbaryl in aqueous TiO_2 suspensions containing surfactants. *Chemosphere,* 39, 493-502.

Blake, D. M. (2001). Bibliography of work on the photocatalytic Removal of Hazardous Compounds from Water and Air, National Renewal Energy Laboratory, USA.

Blanco, J., Avila, P., Bahamonde, A., Alvarez, E., Sanchez, B., & Romero, M., (1996). Photocatalytic destruction of Toluene and Xylene at gas-phase on a titania based monolithic catalyst. *Catal Today.*, 29, 437-442.

Bravo, A., Garcia, J., Domenech, X., & Peral, J., (1994). Some observations about the photocatalytic oxidation of cyanate to nitrate over TiO_2. *Electrochim. Acta,* 39, 2461-2463.

Byrne, J.A., Davidson, A., Dunlop, P.S.M., & Eggins, B.R., (2002). Water treatment using nano-crystalline TiO_2 electrodes. *J. Photochem. Photobiol. A:Chem.,* 148, 365-374.

Cohen, Z.Z., Eiden, C., Lober, M.N., (1986). In: Gerner, W.Y., (Ed.), *Evaluation of Pesticide in Ground Water,* ACS Symposium Series 315, American Chemical Society, Washington, DC, p. 170.

Daneshvar, N., Salari, D., & Khataee, A.R., (2004). Photocatalytic degradation of azo dye acid red 14 in water on ZnO as an alternative catalyst to TiO_2. *J. Photochem. Photobiol. A Chem.,* 162, 317-322.

Dominguez, C., Garcia, J., Pedraz, M.A., Torres, A., & Galan, M.A., (1998).Photocatalytic oxidation of organic pollutants in water. *Catal. Today,* 40, 85-101.

Dowd, R. M., Anderson, M. P., & Johnson, M. L. (1998). Proceedings of the Second National Outdoor Action Conference on Aquifer Restoration, Groundwater Monitoring Geophysical Methods, National Water Well Association, Dublin, OH, p. 1365.

Draper, R.B., & Fox, M.A ., (1990). Titanium dioxide photooxidation of thiocyanate: (SCN)2.cntdot.- studied by diffuse reflectance flash photolysis. *J. Phys. Chem.,* 94, 4628-4634.

Dunlop, P.S.M., Byrne, J.A., Manga, N., & Eggins, B.R., (2002). The photocatalytic removal of bacterial pollutants from drinking water. *J. Photochem. Photobiol. A:Chem:,* 148, 355-363.

Evgenidou, E., Bizani, E., Christophoridis, C., & Fytianos, K. (2007). Heterogeneous photocatalytic degradation of prometryn in aqueous solutions under UV/Vis irradiation.*Chemosphere,* 68, 1877-1882.

Frank, S.N., & Bard, A.J., (1977). Heterogeneous photocatalytic oxidation of cyanide and sulfite in aqueous solutions at semiconductor powders. *J. Phys. Chem.,* 81, 1484-1488.

Fujishima, A., & Honda, K., (1972). Electrochemical Photolysis of Water at a Semiconductor Electrode. *Nature,* 238, 37-38.

Funazaki, N., Hemmi, A., Ito, S., Asano, Y., Yano, Y., Miura, N., & Yamazoe, N., (1995). Application of semiconductor gas sensor to quality control of meat freshness in food industry. *Sensors &Actuators, B* 24-25, 797-800.

Garcia, J.C., & Takashima, K. (2003). Photocatalytic degradation of imazaquin in an aqueous suspension of titanium dioxide. *J. Photochem. Photobiol. A: Chem.,* 155, 215-222.

Gaya, U.I., & Abdullah, A.H., (2008). Heterogeneous photocatalytic degradation of organic contaminants over titanium dioxide: A review of fundamentals, progress and problems. *J. Photochem. Photobiol. C: Photochem. Rev.,* 9, 1–12.

Gerisher, H., & Heller, A., (1992). Photocatalytic oxidation of organic molecules at TiO_2 particles by sunlight in aerated water. *J. Electrochem. Soc.,* 139, 113-118.

Gianturco, F., Chiodaroli, C.M., Bellobono, I.R., Raimondi, M.L., Moroni, A., & Gawlik, B., (1997). Pilot-plant photomineralization of atrazine in aqueous solution by photocatalytic membranes immobilising titanium dioxide and promoting photocatalysts. *Fresenius Environ. Bull.,* 6, 461-468.

Goutailler, G., Valette, J.C., Guillard, C., Paissé, O., & Faure, R., (2001). Photocatalysed degradation of cyromazine in aqueous titanium dioxide suspensions: comparison with photolysis. *J. Photochem. Photobiol. A: Chem.*, 141, 79-84.

Grätzel, M., (1981). Artificial photosynthesis: water cleavage into hydrogen and oxygen by visible light. *Acc. Chem. Res.*, 14, 376-384.

Guillard, C., Pichat, P., Huber, G., & Hoang-Van, C., (1996). The GC-MS analysis of organic intermediates from the TiO_2 photocatalytic treatment of water contaminated by Lindane (1α,2α,3β,4α,5α,6β,-hexachlorocyclohexane). *J. Adv. Oxid. Technol.*, 1, 53-60.

Gupta, H., & Tanaka, S., (1995). Photocatalytic Mineralisation of. Perchloroethylene Using Titanium Dioxide. *Wat. Sci. Technol.*, 31, 47-54.

Hallamn, M. (1992). Photodegradation of di-n-butyl-ortho-phthalate in aqueous solution. *J. Photochem. Photobiol. A: Chem.*, 66, 215-223.

Halmann, M.M., (1996). *Photodegradation of Water Pollutants*, CRC Press Inc, Boca Raton, Florida

Haque, M.M., & Muneer, M., (2003). Heterogeneous photocatalysed degradation of a herbicide derivative, isoproturon in aqueous suspension of titanium dioxide. *J. Environ. Manag.*, 69, 169-176.

Haque, M.M., Muneer, M., & Bahnemann, D., (2006). Semiconductor-Mediated photocatalysed degradation of a herbicide derivative, chlorotoluron, in aqueous suspensions. *Environ. Sci. Technol.*, 40, 4765-4770.

Herrmann, J.M., (1999). Heterogeneous photocatalysis: fundamentals and applications to the removal of various types of aqueous pollutants. *Catal. Today*, 53,115-129.

Herrmann, J.M., Disdier, Pichat, P., Malato, S., & Blanco, J., (1998). TiO_2-based solar photocatalytic detoxification of water organic pollutants. Case studies of 2,4-dichlorophenooxyacetic acid (2,4-D) and of benzofuran. *J., Appl. Catal. B: Environ.*, 17, 15-23.

Herrmann, J.M., Guillard, C., Arguello, M., Agüera, A., Tejedor, A., Piedra, L., & Fernández-Alba, A., (1999). Photocatalytic degradation of pesticide pirimiphos-methyl: Determination of the reaction pathway and identification of intermediate products by various analytical methods.. *Catal. Today*, 54, 353–367.

Hua, Z., Manping, Z., Zongfeng, X., & Low, G.K.C., (1995). Titanium dioxide mediated photocatalytic degradation of monocrotophos. *Water Res.* 29, 2681-2688.

Huang, I.W., Hong, C.S., & Bush, B., (1996). Photocatalytic degradation of PCBs in TiO_2 aqueous suspensions. *Chemosphere*, 32, 1869-1881.

Hurum, D.C., Agrios, A.G., Gray, K.A., Rajh T., & Thurnauer, M.C., (2003). Explaining the enhanced photocatalytic activity of Degussa P25 mixed-phase TiO_2 using EPR. *J. Phy. Chem. B*, 107, 4545-4549.

Irvine., J.T.S., Eggins, B.R., & Grimshaw, J., (1990). Solar energy fixation of carbon dioxide via cadmium sulphide and other semiconductor photocatalysts. *J. Sol. Energy*, 45, 27-33.

Jardim, W.F. , Moraes, S.G., & Takiyama, M.M.K., (1997). Photocatalytic degradation of aromatic chlorinated compounds using TiO_2: toxicity of intermediates. *Water Res.*31, 1728-1732.

Khan, A., Haque, M.M., Mir, N.A., Muneer, M., & Boxall, C. (2010). Heterogeneous photocatalysed degradation of an insecticide derivative acetamiprid in aqueous suspensions of semiconductor. *Desalination,* 261, 169-174.

Khan, M.M.T., & Rao, N.N., (1991). Stepwise reduction of coordinated dinitrogen to ammonia via diazinido and hydrazido intermediates on a visible light irradiated Pt/CdS · Ag_2S/RuO_2 particulate system suspended in an aqueous solution of K[Ru(EDTA-H)Cl]$2H_2O$. *J. Photochem. Photobiol. A: Chem.,* 56, 101-111.

Kinkennon, A.E., Green, D.B., & Hutchinson, B., (1995). The use of simulated or concentrated natural solar radiation for the TiO_2-mediated photodecomposition of basagran, diquat, and diuron. *Chemosphere,*31, 3663-3671.

Konstantinou, I.K., Sakellarides, T.M.., Sakkas, V.A., & Albanis, T.A., (2001a). Photocatalytic Degradation of Selected s-Triazine Herbicides and Organophosphorus Insecticides over Aqueous TiO_2 Suspensions. *Environ. Sci. Technol.* 35, 398-405.

Konstantinou, I.K., Sakkas, V.A., & Albanis, T.A., (2001b). Photocatalytic degradation of the herbicides propanil and molinate over aqueous TiO2 suspensions: identification of intermediates and the reaction pathway. *Appl. Catal. B: Environ.,* 34, 227-239.

Konstantinou, I.K., Sakkas, V.A., & Albanis, T.A., (2002). Photocatalytic degradation of propachlor in aqueous TiO_2 suspensions. Determination of the reaction pathway and identification of intermediate products by various analytical methods. *Wat. Res.,* 36, 2733-2742.

Kosanic, M.M., & Topalov, A.S., (1990). Photochemical hydrogen production from CdS/RhOx/Na_2S dispersions. *Int. J. Hydrogen Energy,* 15, 319-323.

Krosley, K.W., Collard, D.M., Adamson, J., Fox, M.A., (1993). Degradation of organophosphonic acids catalyzed by irradiated titanium dioxide. *J. Photochem. Photobiol. A:Chem.,* 69, 357-360.

Kubota, Y., Shuin, T., Kawasaki, C., Hosaka, M., Kitamura, H., Cai, R., Sakai, H., Hashimoto, K., & Fujishima, A., (1994). Photokilling of T-24 human bladder cancer cells with titanium dioxide. *British J. Cancer,* 70, 1107-1111.

Lee, S., Nishida, K., Otaki, M., & Ohgaki, S., (1997). Photocatalytic inactivation of phage Q□ by immobilized titanium dioxide mediated photocatalyst. *Wat. Sci. Technol.,* 35, 101-106.

Legrini, O., Oliveros, E., & Braun, A.M., (1993). Photochemical processes for water treatment. *Chem Rev.,* 93, 671-698.

Lhomme, L., Brosillon, S., Wolbert, D., & Dussaud, J., (2005). Photocatalytic degradation of a phenylurea, chlortoluron, in water using an industrial titanium dioxide coated media. *Appl. Catal B: Environ.,* 61, 227- 235.

Linder, M., (1997) Ph.D. Thesis, Department of Chemistry, University of Hannover, Hannover, Germany.

Lindner, M., Bahnemann, D., Hirthe, B., & Griebler, W.D. (1995). *Novel TiO_2 powders as highly active photocatalysts. In Solar Water Detoxification; Solar Engineering,* Stine, W. B., Tanaka, T., Claridge, D.E., Eds.; ASME: New York, 339.

Liu, I., Lawson, L.A., Cornish, B., & Robertson, P.K.J., (2002). Mechanistic and toxicity studies of the photocatalytic oxidation of microcystin-LR. *J. Photochem. Photobiol. A:Chem.,* 148, 349-354.

Lobedank, J., Bellmann, E., & Bendig, J., (1997). Sensitized photocatalytic oxidation of herbicides using natural sunlight. *J. Photochem. Photobiol. A:Chem.*, 108, 89-93.

Lu, M.C., Roam, G.D., Chen, J.N., & Huang, C.P., (1995). Photocatalytic mineralization of toxic chemicals with illuminated TiO_2. *Chem. Eng. Commun.*, 139, 1-13.

Mao, Y., & Bakac, A. (1996). Photocatalytic Oxidation of Aromatic Hydrocarbons. *Inorg. Chem.*, 35, 3925-3930.

Marinas, A., Guillard, C., Marinas, J.M., Fernández-Alba, A., Agüera, A., & Herrmann, J.M., (2001). Photocatalytic degradation of pesticide–acaricide formetanate in aqueous suspension of TiO_2. *Appl. Catal. B: Environ.*, 34, 241-252.

Martin, C.A., Baltanas, M.A., Cassano, A.E., (1994). Photocatalytic reactors.3. kinetics of the decomposition of chloroform including effects. *Environ. Sci .Technol.*, 30, 2355-2364.

Matsunaga, T., & Okochi, M., (1995). TiO_2-Mediated Photochemical Disinfection of Escherichia coli Using Optical Fibers. *Environ. Sci. Technol.*, 29, 501-505.

Maurino, V., Minero, C., Pelizzetti, E., & Vincenti, M., (1999). Photocatalytic transformation of sulfonylurea herbicides over irradiated titanium dioxide particles. *Coll.Surf. A* 151, 329-338.

Mihaylov, B.V., Hendrix, J.L., & Nelson, J.H., (1993). Comparative catalytic activity of selected metal oxides and sulphides for the photo-oxidation of cyanide. *J. Photochem. Photobiol. A:Chem.*, 72, 173-177.

Mills, A., & Le Hunte, S. (1997). An overview of semiconductor photocatalysis. *J. Photochem. Photobiol. A:Chem.*, 108, 1-35.

Mills, A., Belghazi, A., & Rodman, D., (1996). Bromate removal from drinking water by semiconductor photocatalysis. *Wat. Res.*, 30, 1973-1978.

Mills, A., Peral, J., Domenech, X., & Navio, J.A., (1994). Heterogeneous photocatalytic oxidation of nitrite over iron-doped TiO_2 samples. *J. Mol. Catal.*, 87, 67-74.

Minero, C., Pelizzetti, E., Malato, S., & Blanco, J. (1996). Large solar plant photocatalytic water decontamination: degradation of atrazine. *Solar Energy*, 56, 411-419.

Muneer, M., & Bahnemann, D., (2001). Semiconductor mediated photocatalysed degradation of two selected pesticide derivatives, terbacil and 2,4,5-tribromoimidazole in aqueous suspension. *Water Sci. Technol.*, 144, 331-337.

Muneer, M., & Bahnemann, D., (2002). Semiconductor mediated photocatalysed degradation of two selected pesticide derivatives, terbacil and 2,4,5-tribromoimidazole in aqueous suspension. *Appl. Catal. B: Environ.*, 36, 95-111.

Muneer, M., Qamar, M., Saquib, M., & Bahnemann, D. (2005). Heterogeneous photocatalysed reaction of three selected pesticide derivatives, propham, propachlor and tebuthiuron in aqueous suspension of titanium dioxide. *Chemosphere*, 61, 457-468.

Muneer, M., Singh, H.K., & Bahnemann, D., (2002). Photocatalysed degradation of two selected priority organic pollutants, benzidine and 1,2-diphenyl hydrazine in aqueous suspensions of titanium dioxide. *Chemosphere*, 49, 193-203.

Muszkat, L., Bir, L., & Feigelson, L., (1995). Solar photocatalytic mineralization of pesticides in polluted waters. *J. Photochem. Photobiol.A: Chem.* 87, 85-88.

Muszkat, L., Raucher, D., Magaritz, M., Ronen, D., (1994). In: Zoller, U., (Ed.), *Groundwater Contamination and Control*, Marcel Dekker, p.257.

Nevim, S., Arzu, H., Gulin K., & Cinar, Z., (2001). Prediction of primary intermediates and the photodegradation kinetics of 3-aminophenol in aqueous TiO_2 suspensions. *J. Photochem. Photobiol. A: Chem.*, 139, 225-232.

Nozik, A. J. in : Ollis, D. F., & EL-Ekabi, H., (Eds.), (1993). *Photocatalytic Purification and Treatment of water and Air*, Elsevier, Amsterdam, p. 391.

O'Shea, K.E., Garcia, I., & Aguilar, M., (1997). TiO_2 Photocatalytic degradation of dimethyl- and diethyl-methylphosphonate, effects of catalyst and environmental factors. *Res. Chem. Intermed.*, 23, 325-339.

Ohtani, B., Zhang, S.W., Nishimoto, S., & Kagiya, T., (1992). Catalytic and photocatalytic decomposition of ozone at room temperature over titanium(IV) oxide. *J. Chem. Soc. Faraday Trans.*, 88, 1049-1053.

Oncescu, T., Stefan, M.I., & Oancea, P., (2010). Photocatalytic degradation of dichlorvos in aqueous TiO_2 Suspensions. *Environ. Sci. Pollut. Res.*, 17, 1158-1166.

Parra, S., Olivero, J., & Pulgarin, C., (2002). Relationships between physicochemical properties and photoreactivity of four biorecalcitrant phenylurea herbicides inaqueous TiO_2 suspension. *Appl. Catal. B: Environ.*36, 75-85.

Parra, S., Sarria, V., Malato, S., Péringer, P., & Pulgarin, C., (2000). Photochemical versus coupled photochemical-biological flow system for the treatment of two biorecalcitrant herbicides: metobromuron and isoproturon. *Appl. Catal. B: Environ.*, 27, 153-168.

Pathirana, H.M.K.K., & Maithreepala, R.A., (1997). Photodegradation of 3,4-dichloropropionamide in aqueous TiO2 suspensions. *J. Photochem.Photobiol. A: Chem.*, 102, 273-277.

Pelizzetti, E., Carlin, V., Minero, C., & Pramauro, E., (1992a). Degradation pathways of atrazine under solar light and in the presence of TiO_2 colloidal particles. *Sci. Total Environ.*, 123/124, 161-169.

Pelizzetti, E., Maurino, V., Minero, C., Carlin, V., Pramauro, E., Zerbinati, O., & Tosato, M.L., (1990). Degradation of atrazine and other S-triazine herbicides. *Environ. Sci. Technol.*, 24, 1559-1565.

Pelizzetti, E., Minero, C., Carlin, V., Vincenti, M., & Pramauro, E., (1992b). Identification of photocatalytic degradation pathways of 2-Cl-s-triazine herbicides and detection of their decomposition intermediates. *Chemosphere*, 24, 891-910.

Peñuela, G.A., & Barceló, D., (1996). Comparative degradation kinetics of alachlor in water by photocatalysis with $FeCl_3$ and photolysis, studied by solid-phase disk extraction followed by gas chromatographic techniques. *J. Chromatogr. A*, 754, 187-195.

Peñuela, G.A., & Barceló, D., (1998a). Photodegradation and stability of chlorothalonil in water studied by solid-phase disk extraction, followed by gas chromatographic techniques . *J. Chromatogr. A* 823, 81.

Peñuela, G.A., & Barceló, D., (1998b). Application of C_{18} disks followed by gas chromatography techniques to degradation kinetics, stability and monitoring of endosulfan in water. *J. Chromatogr. A* 795, 93-104.

Percerancier, J.P., Chapelon, R., & Pouyet, B., (1995). Semiconductor-sensitized photodegradation of pesticides in water: the case of carbetamide. *J. Photochem. Photobiol. A: Chem.*, 87, 261-266.

Peyton, G.R., & Glaze, W.H. (1988). Destruction of pollutants in water with ozone in combination with ultraviolet radiation. 3. Photolysis of aqueous ozone. *Environ. Sci. Technol.*, 22, 761-767.

Pichat, P., (1997). Photocatalytic degradation of aromatic and alicyclic pollutants in water: by-products, pathways and mechanisms. *Water Sci. Tech.* 35, 73-78.

Pirkanniemi, K., & Sillanpää, M., (2002). Heterogeneous water phase catalysis as an environmental application: a review. *Chemosphere*, 48, 1047-1060.

Pizarro, P., Guillard, C., Perol, N., & Herrmann, J.-M., (2005). Photocatalytic degradation of imazapyr in water: Comparison of activities of different supported and unsupported TiO_2-based catalysts. *Catal. Today.*, 101, 211-218.

Pleskov, Y.V., & Krotova, M.D., (1993). Photosplitting of water in a photoelectrolyser with solid polymer electrolyte. *Electrochim Acta*, 38, 107-109.

Pollema, C.H., Hendrix, J.L., Milosavljevic, E.B., Solujic, L., & Nelson, J.H., (1992). Photocatalytic oxidation of cyanide to nitrate at TiO_2 particles. *J. Photochem. Photobiol. A. Chem.*, 66, 235-244.

Poulios, I., Kositzi, M., & Kouras, A., (1998). Photocatalytic decomposition of triclopyr over aqueous semiconductor suspensions . *J. Photochem. Photobiol.A: Chem.*, 115, 175-183.

Pramauro, E., Bianco Prevot, A., Vinceti, M., & Brizzolesi, G., (1997). Photocatalytic degradation of cabranyl in aqueous solutions containig TiO_2 Suspensions. *Environ. Sci. Technol.*, 31, 3126-3131.

Pramauro, E., Vincenti, M., Augugliaro, V., & Palmisano, L., (1993). Photocatalytic degradation of monuron in aqueous TiO_2 dispersions. Environ. Sci. Technol., 27, 1790-1795.

Qamar, M., Saquib, M., & Muneer, M., (2005). TiO_2-mediated photocatalysed degradation of two selected azo dye derivative chryosidine R and acid red 29 in aqueous suspension. *Desalination*, 186, 255-271.

Rahman, M.A., & Muneer, M., (2005). Photocatalysed degradation of two selected pesticide derivatives, dichlorvos and phosphamidon in aqueous suspension of titanium dioxide. *Desalination*, 181, 161-172.

Ranjit, K.T., Varadarajan, T.K., & Viswanathan, B., (1995). Photocatalytic reduction of nitrite and nitrate ions to ammonia on Ru/TiO_2 catalysts. *J. Photochem. Photobiol. A:Chem.*, 89, 67-68.

Ranjit, K.T., Varadarajan, T.K., & Viswanathan, B., (1996). Photocatalytic reduction of dinitrogen to ammonia over noble-metal-loaded TiO_2. *J. Photochem. Photobiol. A:Chem.*, 96, 181-185.

Rao, N.N., & Dube, S., (1996). Photocatalytic degradation of mixed surfactants and some commercial soap/detergent products using suspended TiO_2 catalysts. *J. Mol. Catal. A:Chem.*, 104, L197-L199.

Read, H.W., Fu, X., Clark , L.A., Anderson, M.A., & Jarosch, T. (1996). Field Trials of a TiO_2 Pellet-based Photocatalytic Reactor for Off-gas Treatment of a Soil Vapor Extraction Well. *J. Soil Contam.*, 5, 187-202.

Richard, C., & Bengana, S., (1996). pH effect in the photocatalytic transformation of a phenyl-urea herbicide. *Chemosphere*, 33, 635-641.

Sabin, F., Turk, T., & Vogler, A., (1992). Photo-oxidation of organic compound in the presence of titanium dioxide: determination of the efficiency. *J. Photochem. Photobiol. A:Chem.*, 63, 99-106.

Sakai, N., Wang, R., Fujishima, A., Watanabe, T., & Hashimoto, K., (1998). Effect of Ultrasonic Treatment on Highly Hydrophilic TiO_2 Surfaces. *Langmuir*, 14, 5918-5920.

Sakkas, V.A., Arabatzis, I., Konstantinou, I.K., Dimou, T.A., Albanis, T.A., & Falaras, P., (2004). Metolachlor photocatalytic degradation using TiO_2 photocatalysts. *Appl. Catal B: Environ.*, 49, 195-205.

Sakkas, V.A., Lambropoulou, D.A., Sakellarides, T.M., & Albanis, T.A., (2002). Application of solid phase microextraction (SPME) during the photocatalytic treatment of fenthion and ethyl parathion in aqueous TiO_2 suspensions. *Anal. Chim. Acta*, 467, 233-243.

Sakthivel, S., Neppolian, B., Shankar, M.V., Arabindoo, B., Palanichamy, M., & Murugesan, V., (2003). Solar photocatalytic degradation of azo dye: Comparison of photocatalytic efficiency of ZnO and TiO_2. *Sol. Energy Mater. Sol. Cells.*, 77, 65-82.

Sanlaville, Y., Guittonneau, S., Mansour, M., Feicht, E.A., Meallier, P., & Kettrup, A., (1996). Photosensitized degradation of terbuthylazine in water. *Chemosphere*, 33, 353-362.

Saquib, M., & Muneer, M., (2002). Semiconductor mediated photocatalysed degradation of an anthraquinone dye, remazol brilliant blue R under sunlight and artificial light source. *Dyes and Pigm.*, 53, 237-249.

Saquib, M., & Muneer, M., (2003). Photocatalytic photocatalytic degradation of two selected textile dye derivatives, eosin yellowish and p-rosaniline in aqueous suspensions of titanium dioxide. *J Environ Sci Health A.*, 38, 2581-2598.

Schmelling, D.C., Gray, K.A., & Kamat, P.V., (1996). The role of reduction in the photocatalytic degradation of TNT. *Environ. Sci. Technol.*, 30, 2547-2555.

Shifu, C., & Yunzhang, L., (2007). Study on the photocatalytic degradation of glyphosate by TiO_2 photocatalyst. *Chemosphere.*, 67, 1010-1017.

Singh, H.K., Saquib, M., Haque, M.M., & Muneer, M., (2007b). Heterogeneous photocatalysed degradation of 4-chlorophenoxyacetic acid in aqueous suspension. *J. Hazard. Mater.*, 142, 374-380.

Singh, H.K., Saquib, M., Haque, M.M., Muneer, M., & Bahnemann, D.W., (2007a). Titanium dioxide mediated photocatalysed degradation of phenoxyacetic acid and 2,4,5-trichlorophenoxyacetic acid in aqueous suspensions. *J. Mol. Catal. A: Chem.*, 264, 66-72.

Sleiman, M., Ferronato, C., Fenet, B., Baudot, R., Jaber, F., and Jean-Marc C., (2006). Development of HPLC/ESI-MS and HPLC/^{1}H NMR Methods for the Identification of Photocatalytic Degradation Products of Iodosulfuron. *Anal. Chem.*, 78, 2957–2966

Stapleton, D.R., Konstantinou, I.K., Mantzavinos, D., Hela, D., & Papadaki, M., (2010). On the kinetics and mechanisms of photolytic/TiO_2-photocatalytic degradation of substituted pyridines in aqueous solutions. *Appl. Catal. B: Environ.*, 95, 100–109.

Sturini, M., Fasani, E., Prandi, C., Casaschi, A., & Albini, A., (1996). Titanium-dioxide photocatalysed decomposition of some thiocarbamates in water. *J. Photochem. Photobiol. A: Chem.* 101, 251–255.

Tamimi, M., Qourzal, S., Assabbane, A., Chovelon, J.-M., Ferronato, C., & Ait-Ichou, Y., (2006). Photocatalytic degradation of pesticide methomyl: determination of the reaction pathway and identification of intermediate products. *Photochem. Photobiol. Sci.*, 5, 477-482.

Tanaka, K., Robledo, S.M., Hisanaga, T., Ali, R., Ramli, Z., & Bakar, W.A., (1999), Photocatalytic degradation of 3,4-xylyl N-methylcarbamate MPMC/ and other carbamate pesticides in aqueous TiO_2 Suspensions. *J. Mol. Catal. A: Chem.*, 144, 425-430.

Topalov, A., Molnár-Gábor, D., Kosani, M., & Abramovi, B., (2000). Photomineralization of the herbicide mecoprop dissolved in water sensitized by TiO_2. *Water. Res.*, 34, 1473-1478.

Tseng, J.M., & Huang, C.P., (1991). Removal of Chlorophenols from Water by Photocatalytic Oxidation. *Water Sci. Technol.* 23, 377-387.

Vaz, J.L., Boussaoud, B., Ichou, Y.A., & Petit-Ramel, M., (1998). Photomineralization on titanium dioxide of uracil and 5-halogenouracils. Influence of pH and some anions on the photodegradation of uracil. *Analysis*, 26, 83-87.

Vidal, (1998). Developments in solar photocatalysis for water purification. *Chemosphere*, 36, 2593-2606.

Vidal, A. Dinya, Z., Mogyorodi Jr, F., & Mogyorodi, F., (1999). Photocatalytic of thiocarbamate herbicide active ingredients in water. *Appl. Catal. B: Environ.*, 21, 259-267.

Vidal, A., & Martin Luengo, M.A., (2001). Inactivation of titanium dioxide by sulpur: Photocatalytic degradation of vapam. *Appl. Catal. B: Environ.*, 32, 1-9.

Vinodgopal, K., Wynkoop, D.E., & Kamat, P.V., (1996). Environmental Photochemistry on semiconductor surfaces: A photosensitization approach for the degradation of a textile azo dye, Acid Orange 7. *Environ. Sci. Technol.*, 30, 1660-1666.

Vulliet, E., Emmelin, C., Chovelon, J.M., Guillard, C., & Herrmann, J.M., (2002). Photocatalytic degradation of sulfonylurea herbicides in aqueous TiO_2. *Appl. Catal. B: Environ.* 38, 127-137.

Wang, R., Hashimoto, K., Fujishima, A., Chikuni, M., Kojima, E., Kitamura, A., Shimohigoshi, M., & Watanabe, T., (1998). Photogeneration of Highly Amphiphilic TiO_2 Surfaces. *Adv. Mat.*, 10, 135-138.

Wang, R., Sakai, N., Fujishima, A., Watanabe, T., & Hashimoto, K., (1999). Studies of Surface Wettability Conversion on TiO_2 Single-Crystal Surfaces. *J. Phys. Chem. B:*, 103, 2188-2194.

Wardman, P. (1989). Reduction potentials of one-electron couples involving free radicals in aqueous solution. *J. Phys. Chem.*, Ref. Data, 18, 1637-1755.

Weichgrebe, D., & Vogelpohl, A. (1995) Stratgie zur Auswahl geeigneter Oxidationsverfahren. 2. Fachtagung Naboxidative Abwasserbehadlung, Clausthal.

Weller, H., (1993). Kolloidale Halbleiter-Q-Teilchen: Chemie im Übergangsbereich zwischen Festkörper und Molekül. *Angew. Chem. Int. Ed. Eng.*, 32, 41-53.

Yu, J.C., Yu, J., Ho, W., & Zhao, J., (2002). Light-induced super-hydrophilicity and photocatalytic activity of mesoporous TiO2 thin films. *J. Photochem. Photobiol. A:Chem.*, 148, 331-339.

Zalenska, A., Hupta, J., Wiergowski, M., & Biziuk, M., (2000). Photocatalytic degradation of lindane, p,p'-DDT and methoxychlor in an aqueous environment . *J.Photochem. Photobiol. A: Chem.* 135, 213-220.

Zhang, P., Scrudato, R.J., & Germano, G., (1994). Solarcatalytic inactivation of *Escherichia coli* in aqueous solutions using TiO_2 as catalyst. *Chemosphere,* 28, 607-611.

Zhu, X., Feng, X., Yuan, C., Cao, X., & Li, J., (2004). Photocatalytic degradation of pesticide pyridaben in suspension of TiO_2: identification of intermediates and degradation pathways. *J. Mol. Catal. A: Chem.,* 214, 293-300.

13

Study on Sono-Photocatalytic Degradation of POPs: A Case Study Hydrating Polyacrylamide in Wastewater

Fanxiu Li
College of Chemical & Environmental Engineering, Yangtze University, Jingzhou, Hubei, China

1. Introduction

1.1 Production and possible hazards of hydrating polyacrylamide (HPAM) pollutants

In recent years, many Chinese oilfields have been in their mid- or final-stage of development. The oil recovery cannot be improved further with water flooding. Enhanced oil recovery (EOR) by means of polymer flooding is an important technology for the strategic development of oilfields in China. In order to improve the oil recovery, polymer flooding (injected water containing polymer), alkaline-surfactant-polymer flooding (injected water containing alkaline, surfactant and polymer, ASP) and surfactant-polymer flooding (injected water containing surfactant and polymer) have subsequently been used in oil production, which is often called tertiary oil extraction (Han et al., 1999). The liquid which is produced from stratum should be dehydrated using three-phase separators, and then, the crude oil of upper layers will be carried to the oil refinery, and the produced water (oily water) of under layer is generated. A large part of produced water should be injected back into the stratum for reuse (Taylor et al., 1998; Zhang et al., 2010), and the rest will be discharged into water bodies or surrounding soils.

In crude oil exploitation, water-soluble, polyacrylamide (PAM) is one of the most widely used polymers to enhance oil recovery in the east oilfields of China. The produced water from polymer flooding (PWPF) which contains a lot of residual hydrolyzed polyacrylamide (HPAM) is the wastewater of polymer flooding. And PWPF which is also characterized by its high temperature, heavy metals, high mineralization, low biodegradability and high content of oil and other oilfield chemicals (OCs), is different from that produced from water flooding. Moreover, partially hydrolyzed polyacrylamide (HPAM) present in production water causes some problems. For example, after polymer flooding, HPAM will remain in the produced water generated by oilfields, increasing the difficulty in oil-water separation. Consequently, the oil content in sewage is greatly increased, and there is a high probability that the wastewater will exceed the local discharge limit. When HPAM enters an oil reservoir with injected water, it can also hardly avoid infiltrating groundwater horizontally in connection with strata configuration. In addition, the costs and difficulties of produced water treatments will be increased because of the high concentration of the HPAM remaining in the wastewater. Furthermore, the residual HPAM in the wastewater can slowly degrade into the toxic acrylamide monomer naturally. The toxicity of acrylamide monomer has been studied by numerous researchers all over the world (Bao et al., 2010). Since HPAM can remain in surface water and groundwater for a long period of time, it may endanger human health. Therefore, it is necessary to conduct studies on transforming HPAM into innocuous substance effectively and rapidly.

1.2 Study progress on treatment of HPAM pollutants in wastewater

With the amount of HPAM residue increasing in the produced fluid, the separation of oil from water is more and more difficult and the treating difficulty becomes stern. Therefore, the effective treatment on the wastewater became urgent and important. Some methods have been used to treat it in some oil fields, such as gravity settling, floatation (Thoma et al., 1999), de-emulsification (Bilstad & Espedal, 1996) and membrane separation (Cheryan & Rajagopalan, 1998; Kong & Li, 1999; Scholz & Fuchs, 2000) and biotechnology (Li et al., 2005; Zhao et al., 2006; Su, 2007). The research progress of HPAM wastewater treatment methods is reviewed in this paper, which include the coagulation treatment method, membrane treatment method, photocatalysis degradation method and photo-Fenton treatment method.

1.2.1 Coagulation treatment process of HPAM in wastewater

At present, the most applicable and effective method is flocculation. The general operation is adding flocculant to the settling tanks in the existing treatment systems to accelerate oil-water separation. There are two kinds of flocculants, inorganic and organic. The typical inorganic flocculant is polyaluminum chloride (PAC) and the organic is cationic polyacrylamide (CPAM), which play an important role on wastewater treatment (Zhao et al, 2008).

The influences of HPAM residue on the flocculation behavior of wastewater from polymer flooding had been investigated by Zhao (Zhao et al, 2008). The main conclusions from their work were listed as the following:

1. Using PAC as inorganic flocculant, the flocculation performance improved with the increase of temperature under the same dosage. At 37°C and 40°C, flocculation results were markedly better than that of 30°C and 33°C. The floc formed quickly and the treating cost was low. However, the floc was much, tiny, loose and unstable.
2. Using CPAM as organic flocculant, the flocculation performance decreased with the increase of temperature under the same dosage. Compared with PAC, the floc was less and more stable. However, the treating results were poor and the cost was expensive.
3. At the constant temperature and dosage, the flocculation performance of PAC and CPAM decreased dramatically with the content of residual HPAM. At 37°C, when HPAM residue in wastewater increased from 100mg/L to 600mg/L, the light transmission decreased from 96.4% to 70% after treating with PAC at the dosage of 600mg/L and from 87.3% to 50.0% with CPAM at the dosage of 150mg/L.

1.2.2 Ultra-filtration membrane treatment technologies for the HPAM-containing wastewater

Unfortunately, none of these traditional separation techniques can meet complex demands for purifying the polymer-flooding wastewater of tertiary oil extraction. How to treat the oilfield polymer-flooding wastewater efficiently still remains unsettled.

Currently, ultra-filtration technology plays a more prominent role in the treatment of oily wastewater (Wu et al., 2008; Lu et al., 2009). However, the major problem arising from the membrane process is the decline in flux due to the concentration polarization and membrane fouling. The scientific practice suggests that the membrane fouling can be avoided (Field et al., 1995) when the operating flux is lower than a certain flux (critical flux). In contrast, when the operating flux exceeds this flux, the colloids initially present in the polarized layer will transform from the liquid phase into an irreversible cake layer (Chen et al., 1995; Bacchin et al., 2002a, 2002b). Since then, many studies have focused on the critical flux, including the effect of hydrodynamic factors such as cross-flow velocity (CFV) (Defrance & Jaffrin, 1995), sufficient shear stress (Li et al., 2000), sludge concentration (Le-clech, 2003) and particle size (Kwon et al., 1998; Kwon, 2000) on the critical flux for colloidal suspension, mineral suspension (Benkahla et al., 1995), protein or yeast suspension (Causserand et al., 1996) and activated sludge water(Bouhabila et al., 1998). As the study of the critical flux, the membrane resistance at sub-critical flux was investigated all along (Cho et al., 2002; Ognier et al., 2004). In addition, Chiu et al. (Chiu et al., 2005) reported that the gas could be used as a means of enhancing the critical flux in a non-circular multi-channeled (star-shaped) ceramic membrane module. Chong et al. (Chong et al., 2008) developed a sodium chloride tracer response technique to determine the critical flux of colloidal silica in reverse osmosis process.

In view of the characteristics of oilfield produced water quality, Wang (Wang et al., 2011) used the ultra-filtration membrane technique to treat synthetic oilfield polymer-flooding wastewater. In the experiment of fouling mechanism research, the first part and second part of filtration met the standard blocking and cake filtration model respectively. But the standard blocking period was very short at the beginning and the cake filtration period predominates in all filtration run. The critical flux was determined by transmembrane pressure (TMP)-step method in dead-end ultra-filtration test cell. The total fouling resistance increased with the increase of TMP. When the operating flux was below the critical flux, there was only concentration polarization phenomenon. The increasing trend of membrane resistance with the increase of TMP was unconspicuous. The intrinsic membrane resistance was the dominant resistance and the membrane fouling force was negative in this situation. And the filtration proceeds reached the biggest value. But once the operating flux exceeded the critical flux, the membrane pollution happened and the increasing rate of resistance accelerated.

The fouling resistance was the dominant resistance and the membrane fouling driving force became positive and higher. Moreover, the filtration proceeds were smaller and smaller. In the experiments of quantitative analysis of the critical flux, according to comparative results of the average rates of change of the critical flux for the concentration of HPAM, oil and suspended solid in the single solute solution, double solute solution and oilfield polymer-flooding wastewater, HPAM can decrease the average rate of change of the critical flux for oil and suspended solid. It has the crucial effect on the critical flux. The sequence of influence degree on the critical flux is the HPAM concentration>oil concentration>SS concentration and the percentage contribution is 84.58%, 14.36% and 1.06% respectively in the oilfield polymer-flooding wastewater. SEM images indicated that there was no membrane fouling formation at constant sub-critical flux. In contrast, the membrane was covered with a cake layer on the top surface of the membrane and the interaction between the particles and the membrane pore caused the pore narrowing, constriction and plugging at a constant supra-critical flux.

1.2.3 Photocatalysis degradation of HPAM in wastewater

Advanced oxidation processes (AOPs) are defined as oxidation processes in which hydroxyl radicals are the main oxidants involved. This radical is a very powerful oxidant (E^0: 2.80V versus SHE) which leads to a very effective oxidation process, such as Fenton and photo-Fenton catalytic reactions, H_2O_2/UV processes and TiO_2 photocatalysis (Faouzi et al.., 2006). Among the AOPs, heterogeneous photocatalysis oxidation using TiO_2 as photocatalyst has been extensively studied because of its low cost, high photoactivity, nontoxicity, photocorrosion resistance and other physical and chemical properties, and proved to be efficient and potentially advantageous. This semiconductor absorbs photons whose energy is higher than or equal to the band-gap energy. Thus, valence band electrons are promoted into the conduction band generating an electron-hole pair (e^-/h^+). These pairs are able to initiate oxidation and reduction reactions in the surface of TiO_2. The positive holes can oxidize the organic molecules adsorbed, through the formation of •OH radicals. Simultaneously, the photogenerated electron can produce radical species such as superoxide $\bullet O_2^-$ and hydroperoxide $\bullet HO_2$. All these radicals initially oxidize the substrate in intermediate compounds which subsequently undergo a total mineralization in most of the cases (Serpone & Pellizzetti, 1989; David & Ollis, 2000).

The interest in photocatalysis is extensive, as shown by the number of publications on this subject which regularly appear in some journals, and thousands of papers have been published since the 1970s. Photocatalysis has come to describe the field of study and the technology in which irradiated semi-conductors generate photocharges that are ultimately poised at the surface. These photocharges undergo various processes, the most important of which are effectively separated and transferred to the contacting liquid, gas or solid for photooxidation of a large variety of organic substances to their complete mineralization. Wang et al. early works (Wang et al, 2006) discovered that TiO_2 particles can be self-potentially absorbed and orientedly arranged onto oil-water interface of emulsions. While the TiO_2 particles on the oil-water interface and in bulk water can be photocharged by UV radiation, reactive photoholes generated by photocatalytic process primarily oxidize the touched organic substances and damage the film between oil and water.

In their paper, Wang et al. (Wang et al, 2006) reported on a novel approach efficiently to achieve viscosity breaking of wastewater containing PAM. Initially by analyzing emulsification action and role of PAM in the wastewater, a process using photocatalytical technique was investigated by taking aim at viscosity breaking and degradation of PAM. Wang et al. refer to this process as "photocatalytical visbreaking". The experimental results show that viscosity of wastewater produced from polymer flooding was greatly decreased and the rate of PAM photodegradation is above 90% under short time of illumination by using photocatalysis over TiO_2 powders. The efficient breaking of viscosity favors treatments to feed the conventional system used in the water flooding with low viscosity of wastewater. The photocatalytic visbreaking can promisingly be used by efficient performance in oilfields. Chen et al. also think that it is a good method to treat PAM in water by photocatalytic oxidation (Chen et al., 2001).

Recently, the photocatalysts prepared by doping of rare earth oxides into anatase TiO_2 matrix have attracted much attention. TiO_2 is considered as a good host candidate for doping rare earth oxides due to its attractive properties such as mechanical, thermal, and anticorrosive properties. It is well known that the surface composition and structure of photocatalyst can greatly influence its activity. Some important results have been achieved by studies on the rare earth oxide-doped TiO_2 composites. For example, Saif et al. (Saif &

Abdel-Mottaleb, 2007) reported that TiO_2 nanocomposites doped with trivalent lanthanide ions (e.g. Tb^{3+}, Eu^{3+}, and Sm^{3+}) exhibited remarkable enhanced photocatalytic activity to textile dye degradation compared to pure TiO_2. Yan et al. reported that rare earth oxide-doped TiO_2 composites increased conversion of phenol and selectivity to CO_2 compared with pure TiO_2. Although some successful methods have been reported concerning about the preparation of the rare earth oxide-doped TiO_2 composites, new routes still need to be developed in order to lead to the composites with nanoscale, unique physicochemical properties, and interesting surface compositions.

In current Li's work(Li et al., 2009), a single step sol-gel-solvothermal method is applied to prepare rare earth oxide-doped titania nanocomposites, RE^{3+}/TiO_2, where $RE^{3+}=Eu^{3+}$, Pr^{3+}, Nd^{3+}, Gd^{3+}, and Y^{3+}. The morphology, phase structure, surface composition and structure, optical property, and textural property of the composites are well characterized.

As-prepared Eu^{3+}-, Pr^{3+}-, Gd^{3+}-, Nd^{3+}-, and Y^{3+}-doped TiO_2 composites with anatase phase, nanosize, and mesoporosity exhibited remarkably high UV-light photocatalytic activity to HPAM degradation. Moreover, Eu^{3+} (Pr^{3+}, Gd^{3+})/TiO_2-2.4 were the most photoactive among all tested materials including Degussa P25. This enhanced photocatalytic activity is attributed to the following properties of as-prepared RE^{3+}/TiO_2 composites: (a) quantum size effect; (b) unique textural properties (mosoporosity with larger BET surface areas and pore sizes); and (c) interesting surface compositions (more hydroxyl oxygen and adsorbed oxygen and some Ti^{3+} species existed at the surface of the products with respect to pure TiO_2). As-prepared photocatalysts are also essential for any practical application of photocatalytic oxidation process.

1.2.4 Photo-Fenton treatment of simulated HPAM in wastewater

Recently, some investigators have reported the successful application of advanced oxidation processes (AOPs) for PAM degradation (Vijayalakshmi & Giridhar, 2006; Ren & Chunk, 2006). One of advanced oxidation processes, Fenton's reagent, a mixture of H_2O_2 and Fe^{2+} (a powerful source of oxidative •OH generated from H_2O_2 in the presence of Fe^{2+} ions) or photo-Fenton reaction has been used in the degradation of many organic compounds because of its ease of operation (Murray & Parsons, 2004; Yardin & Chiron, 2006). The iron is the first most abundant metal and contained in many inexpensive natural minerals including tourmaline used in this study. Furthermore, hydrogen peroxide used as oxidant in these processes is cheaper than other oxidants.

In our study, we investigated the photo-assisted Fenton (photo-Fenton) reaction for its ability to oxidize HPAM. The photo-assisted Fenton can promisingly be used by efficient performance in oilfields. Effects of operating parameters such as initial hydrogen peroxide concentration, ratio of Fe^{2+}/H_2O_2 (mole ratio), HPAM concentration, and pH on the degradation rate of HPAM have been quantitatively discussed.

The experimental data demonstrated that photo-Fenton processes are promising techniques for the degradation of HPAM from aqueous solution. Based on the results, the following conclusions can be drawn.

1. Fenton and photo-Fenton processes lead to complete degradation of HPAM in relatively short time (~30min).
2. The optimal parameters for photo-Fenton process are: pH=3.0, the ratio of Fe^{2+}/H_2O_2=1:10 (mole ratio) and amount of H_2O_2=6mmol/L.

The employment of the UV lamp benefits the HPAM degradation. So, it is possible conclude that the UV lamp, though has little power, is very useful in Fenton process to increase the HPAM degradation. More results can be obtained in Li's study (Li, et al., 2006; Li, et al., 2007).

Even though these systems are considered as a very effective approach to remove organic compounds, it should be pointed out that there is a major drawback because the post-treatment of Fe sludge is an expensive process. This shortcoming can be overcome by using heterogeneous photo-Fenton reaction. Therefore, a lot of effort has been made in developing heterogeneous photo-Fenton catalysts. For example, Parra et al. prepared Nafion/Fe structured membrane catalyst and used it in the photo-assisted immobilized Fenton degradation of 4-chorophenol (Parra et al., 2004). However, Nafion/Fe structured membrane catalyst is much expensive for practical use. Thus, the low cost supports such as the C structured fabric (Parra et al., 2003; Yuranova, 2004), activated carbon (Ramirez, 2007), mesoporous silica SBA-15 (Calleja et al., 2005; Martinez et al., 2007; Martine et al., 2005), zeolite (Noorjaha et al., 2005; Kusic et al., 2006) and clay (Feng et al., 2006; Chen & Zhu, 2007), have been used for the immobilization of active iron species. Remirez et al. prepared the catalysts using four iron salts as precursors for the heterogeneous Fenton-like oxidation of Orange II solutions (Ramirez et al., 2007). The results showed that the nature of the iron salt had a significant effect on the process performance. So, it is necessary to discuss the photocatalytic activities of the catalysts by using different iron salts as precursors.

Liu(Liu et al., 2009) prepared a series of Fe(III)-SiO_2 catalysts at different OH^-/Fe mole ratio and by using two iron salts as precursors, namely $Fe(NO_3)_3$ and $FeSO_4$ and as-prepared catalysts were characterized by the BET, XRD and XPS method. The percentage of chemisorbed oxygen on the surface of catalysts prepared by $FeSO_4$ is higher than that prepared by Fe $(NO_3)_3$. The results confirm the formation of Fe (II)-SiO_2 when Fe (III)-SiO_2 was irradiated by photon. The photocatalytic activities of Fe (III)-SiO_2 catalysts were evaluated by the degradation of PAM from aqueous solution in the photo-Fenton reaction and all the catalysts exhibited a better photocatalytic activities. However, the precursor species and the OH^-/Fe mole ratio have influence on the photocatalytic activities of the catalysts. At the same OH^-/Fe mole ratio, the catalysts could present the better photocatalytic activities when using $FeSO_4$ as precursor. The best efficiency for the degradation of PAM in heterogeneous photo-Fenton reaction was 94.0% degradation in 90 min and 70.0% TOC removal in 180 min at an initial pH of 6.8. Moreover, it was observed that Fe leaching from Fe(III)-SiO_2 catalysts was negligible, indicating that the catalysts have a long-term stability and the degradation of PAM from aqueous solution are almost caused by the heterogeneous photo-Fenton reaction.

2. Sono-photocatalytic degradation of hydrating polyacrylamide in wastewater

Recently, some investigators have reported the successful application of advanced oxidation processes (AOPs) for HPAM degradation (Vijayalakshmi & Giridhar, 2006; Ren et al., 2006). One of advanced oxidation processes, photocatalytic degradation has been used in the degradation of many organic compounds and showed greatly obvious effects (Augugliaro et al., 1991). However, very few commercial applications of this technology are available at present due to low quantum efficiency and reuse of catalyst. Most of the photoinduced positive holes (h^+) and electrons (e^-) had recombined before they were trapped by hydroxyl or oxygen and quantum efficiency was usually less than 5%. In order to enhance the quantum efficiency, many measures were taken. Recently, the sonochemical method has been proven to be a useful technique (Hu et al., 2004). Sonolysis is a relatively innovative advanced oxidation processes based on the use of low to medium frequency (typically in the range 20-1000 kHz) and high energy ultrasound to catalyze the destruction of organic pollutants in waters. The chemical effects of ultrasound irradiation are the result of acoustic cavitation which is the formation and subsequent collapse of micro-bubbles in a liquid. At the extreme conditions generated inside the cavitation bubbles during collapse, vapor is homolytically cleaved leading to the formation of hydroxyl radicals that can oxidize the organic pollutants found in wastewaters (Vajnhandl & Marechal, 2005). Sonochemical treatment typically operates at ambient conditions and does not require the addition of extra chemicals or catalysts. The efficacy of AOPs to treat pollutants is eventually dictated by the rate of generation of free radicals and other reactive moieties and the degree of contact between the radicals and the contaminants both of which should be maximized. In this view, process integration is conceptually advantageous in water treatment since it can eliminate the disadvantages associated with each individual process. The simultaneous application of ultraviolet and ultrasound irradiation in the presence of TiO_2, i.e. sono-photocatalysis, represents an example of recent advances targeted at improving photocatalytic processes (Adewuyi, 2005; Gogate & Pandit, 2004). It is obvious that if the two irradiation modes are operated simultaneously, an additional source of free radicals will be available for the oxidative destruction of various pollutants.

The sono-photocatalytical degradation of a variety of organic substrates has attracted much attention. It might be also an efficient way to eliminate HPAM. However, there are only few studies investigating the sono-photocatalytic degradation of HPAM in aqueous system (Li et al., 2010a; 2011b). In this paper, HPAM was chosen as a model compound to obtain detailed information of the innovative photocatalysis. The photodegradation of the HPAM catalyzed by coupling system of ultraviolet irradiation (20 W UVA) over TiO_2 suspensions and ultrasound irradiation (42 kHz, 100 W) (US/UV/TiO_2) was investigated. Many factors are involved during the degradation of HPAM in US/UV/TiO_2 systems. The important operating parameters that affect the overall photocatalytic oxidation efficiency were investigated in detail, including amount of catalyst, initial concentration of reactant, and the dosage of hydrogen peroxide.

2.1 Materials and methods

2.1.1 Chemical

HPAM (Mw 500×10^4, hydrolysis degree 25%) was obtained from Shengli oil refining and chemical plant, Shengli, China and was used without further purification. Hydrogen peroxide (H_2O_2), hydrochloric acid (HCl) and sodium hydroxide (NaOH) were all of analytical grade and obtained from Tianjin Kermal Chemical Reagents Co. (Tianjin, China). TiO_2 photocatalyst (64.5% anatase, 35.5% rutile, specific surface area 40-45 m^2/g) was prepared by ultrasonic assisted method. The particle diameter of TiO_2 is 47.1-67.5 nm. The density is 690 kg/m^3. All experiments were carried out with use of deionized water.

2.1.2 Apparatus

Lambda-17 spectrophotometer (US Perkin-Elmer Company) was used to inspect the degradation processes of HPAM. Branson 2510E-DTH apparatus (Branson company, US) was adopted to irradiate the solution of HPAM, operating at an ultrasonic frequency of 42 kHz and output power of 100 W.

2.1.3 Analysis

HPAM concentration was measured by the starch-cadmium iodine method (Li et al., 2009). The temporal concentration changes of HPAM during experimental processes were monitored by measuring characteristic absorption intensity of HPAM at 590 nm. The maximal absorbency of 0-100 mg/L HPAM solution abides by Lambert-Beers law and the calibration curve of standard HPAM solutions are used to estimate the degradation rate of HPAM. The degradation rate of HPAM was defined as follows Eq. (1):

$$\text{Degradation rate } (\%) = (1 - C_t/C_0) \times 100\% \quad (1)$$

Where C_0 is the concentration of HPAM after adsorption equilibrium in the dark (t=0), and C_t is the concentration of HPAM at reaction time t (min).

All these experiments were conducted in triplicates and the results were showed at the mean values.

2.1.4 Procedures

Experimental procedure was performed as follows: 100 mL of HPAM solution were introduced in the reaction vessel (bottom area=50 cm^2) and the appropriate amount of TiO_2 was added to achieve the desirable concentration. The resulting TiO_2 suspension was magnetically stirred for 30 min in the dark to ensure complete equilibration of adsorption/desorption of HPAM on the catalyst surface. After that period of time, UVA irradiation was provided by a 20 W ultraviolet lamp, which emit predominantly UV radiation at a wavelength of 254 nm, the lamp was turned on (this was taken as "time zero" for the reaction). In the experiment, the distance between the lamp and the interface irradiated by UV light is required to be 10 cm. Ultrasound irradiation was provided by an apparatus operating at 42 kHz frequency and 100 W of electric power output. Samples periodically drawn from the reaction vessel were centrifuged at 4000 rpm for 15 min to remove TiO_2 particles and then subjected to analysis.

2.2 Results and discussion

2.2.1 Photocatalytic and sono-photocatalytic degradation

In order to check the feasibility of sono-photocatalysis process for the degradation of HPAM, the following control experiments were performed and the results are presented in Fig.1, ultrasound alone (US), ultrasonic degradation of HPAM in the presence of TiO_2(US/TiO_2), photocatalysis (UV/TiO_2) and sono-photocatalysis (US/UV/TiO_2) in the presence of TiO_2. It is evident from Fig.1 that are relatively higher degradation of HPAM was achieved by combining sonolysis and photocatalysis than that observed during the individual processes. The preliminary experiments revealed no significant degradation of HPAM in the presence of US alone. However, an enhancement in the sonolytic reaction rate in the presence of semiconduct or particles was recently reported by Pandit et al. (Pandit et al., 2001; Shirgaonkar & Pandit, 1998) which is referred to as sonocatalysis. The suspended solids may also increase the extent of cavitation generated in solution and hence the sonochemical degradation rate by providing additional nuclei for bubble generation. Therefore, in order to study the effect of TiO_2 on the sonolytic degradation of HPAM in the absence of photo irradiation, the following experiment was conducted under the experimental conditions of [HPAM]=200 mg/L and TiO_2 amount of 600 mg/L, and the results obtained are shown in Fig. 1.

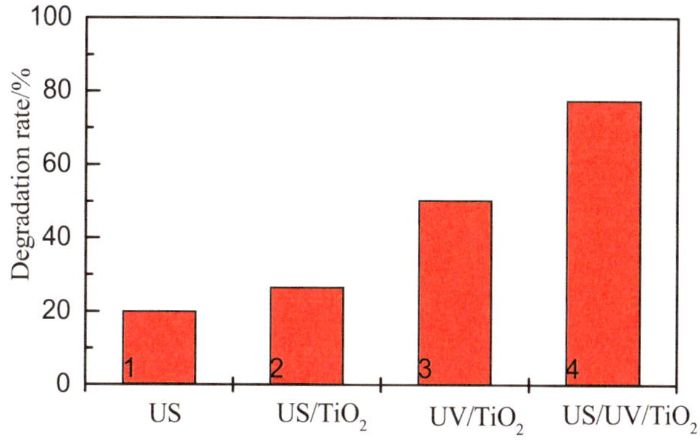

Fig. 1. Comparison of degradation rates of HPAM (200 mg/L) in the presence of TiO_2 (600 mg/L) using different processes

It was noted that about 26.7% degradation in 90 min under the sonocatalytic conditions was observed. An approximately 6.5% increase in the degradation amount observed in the presence of TiO_2 particles during sonolysis. This increase might be due to additional cavitation activity.

In the case of photocatalytic degradation, about 50.5% degradation (Fig. 1) was achieved in 90 min and can be explained using the general degradation mechanism available in the literature (Bhatkhande et al., 2002). The photoirradiation of the TiO_2 forms an electron in the conduction band and a hole in the valence band. As a consequence of such photoinduced charge separation on the semiconductor surface, electron exchange reactions occur at the water-semiconductor interface. The superoxide radical anion O_2^{-}• is formed by interaction of photo-generated conduction band electrons with adsorbed oxygen molecules, while •OH are formed via valence band hole oxidation of adsorbed water or hydroxyl anions (Carp et al., 2004) to generate •OH that subsequently oxidize the adsorbed pollutant molecules. Holes may also directly oxidize the adsorbed organic pollutant.

However, when both the US and UV are combined (sono-photocatalysis), a significant enhancement in the degradation (77.6% in 90 min) of HPAM was observed. About 27.2% increment in the degradation under the same processing time suggests that the hydroxyl radicals formed by both the advanced oxidation processes, viz., photocatalysis and sonolysis are involved in the sono-photocatalytic degradation of HPAM.

2.2.2 Effects of TiO_2 dosages on the degradation rate

The effect of TiO_2 powder concentration on the photocatalytic degradation of organics in the absence of ultrasound irradiation in aqueous solution has already been studied in the literatures (Ai et al., 2005). The necessity to optimize this factor was pointed out. Catalyst concentration has an optimum value, as using excess catalyst reduces the amount of photo-energy being transferred in the medium due to opacity offered by the catalyst particles. Fig. 2 shows the change of HPAM as a function of reaction time in US/UV/TiO_2 detected at different catalyst dosages 600, 800 and 1000 mg/L. On the whole, it was observed that along with the increase of catalyst dosage from 600 to 800mg/L, the degradation rate of HPAM in solution increased. However, the catalyst concentration amounting 800mg/L did not further enhance the degradation rate. The increased degradation is likely due to the increase of the total surface area (or number of active sites) of the photocatalysts available for photocatalytic reaction when increasing the dosage of TiO_2. When TiO_2 was overdosed, the intensity of light penetration attenuated and light scattering increased, which counteracted the positive effect coming from the dosage increment and therefore the overall performance reduced As the concentration of TiO_2 reaches a certain level, the suspension with a high concentration results in a lower light transmission of the system and then effective photons are decreased. As a result, it slows down the degradation of HPAM (Zhang et al., 2005). So, 800mg/L was the optimum amount of TiO_2 in terms of photodegradation under our experimental condition.

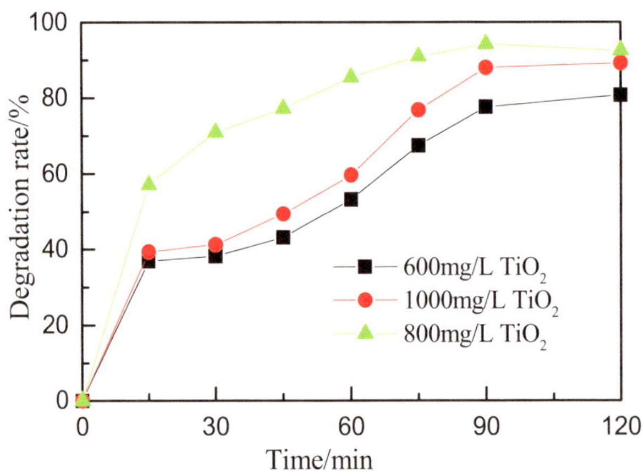

Fig. 2. Effect of added amount of TiO_2 on degradation rate of HPAM

2.2.3 Effect of initial concentration of HPAM on the degradation rate

The effect of initial concentration of HPAM on the sono-photocatalytic degradation rate was investigated over the concentration range of 80-200 mg/L, since the pollutant concentration is an important parameter in water treatment. Experimental results are presented in Fig. 3. It can be seen in Fig. 3 that degradation rate decreases with increasing initial concentration of the HPAM. The possible explanation for this behaviour is that as the initial concentration of the HPAM increases, the path length of photons entering the solution decreases and in low concentration the reverse effect is observed, thereby increasing the number of photon absorption by the catalyst in lower concentration. This suggests that as the initial concentration of the HPAM increases, the requirement of catalyst surface needed for the degradation also increases. Since illumination time and amount of catalyst are constant, the OH radical (primary oxidant) formed on the surface of TiO_2 is also constant. So the relative number of free radicals attacking the HPAM molecules decreases with increasing amount of the catalyst. The major portion of degradation occurs in the region (termed as reaction zone) near to the irradiated side, since the irradiation intensity in this region is much higher than that at the other side. Hence, at higher concentration, degradation decreases at sufficiently longer distances from the light source or reaction zone due to the retardation of penetration of light. Thus, the rate of degradation decreases with increase in concentration of HPAM (Neppolian et al., 2002).

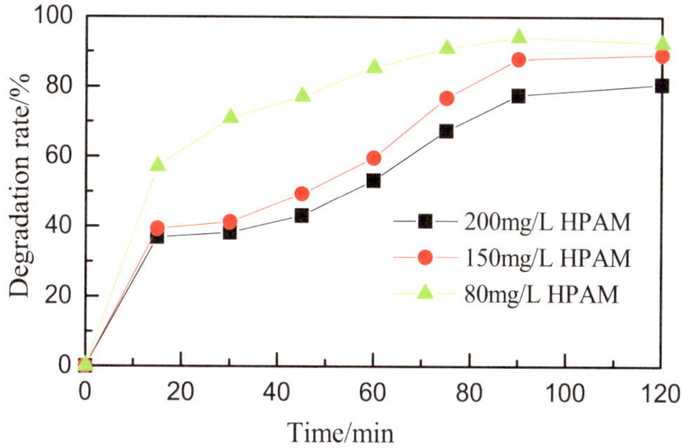

Fig. 3. Effect of initial concentration of HPAM on sono-photocatalytic degradation rate.

2.2.4 Effects of H_2O_2 dosages on the degradation rate

The use of inorganic peroxides has been demonstrated to enhance the rate of degradation because they trap the photo-generated electrons more efficiently than O_2. The effects of addition of the H_2O_2 into TiO_2 dispersion were examined (Fig. 4). The degradation rate of HPAM was increased in the presence of H_2O_2 and influenced by the dosages of H_2O_2. With the addition of H_2O_2, the degradation rate of HPAM ware increased abruptly at the low dosages of H_2O_2, but increasing H_2O_2 concentrations beyond 18 mmol/L had a negative effect on the process. This is because the addition of H_2O_2 can enhance the formation of $\cdot OH$. H_2O_2 would act as an electron donor to produce hydroxyl radicals by its reduction at the conduction band. The self-decomposition of H_2O_2 by UV light illumination or ultrasound irradiation would also produce hydroxyl radicals. Although more $\cdot OH$ radical could be produced in the solution at higher oxidant concentrations, higher dosages of H_2O_2 may also act as an effective hydroxyl scavenger at higher H_2O_2 concentration (>18 mmol/L) as shown in the following reaction (Thandar et al., 2003): $H_2O_2 + \cdot OH \rightarrow H_2O + HO_2\cdot$.

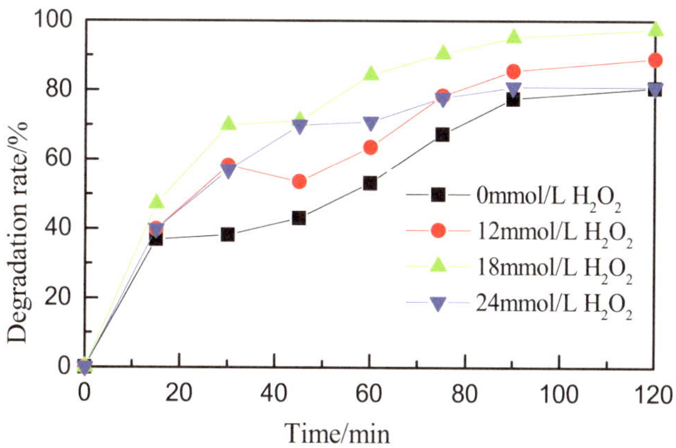

Fig. 4. Effect of dosage of H_2O_2 on degradation of HPAM in the solution containing 150 mg/L HPAM.

2.3 Degradation pathway of HPAM

During the process of TiO_2 sono-photocatalysis, numerous intermediates and products could be formed since ·OH radicals did not exhibit a high degree of functional group selectively. UV-light irradiation of HPAM solution in the presence of TiO_2 was carried out under atmosphere, and three intermediates were identified with LC/MS (Fig. 5). The obtained peaks with m/z of 59.1, 59.7 (82.6), and 96.1, respectively, in the mass spectra corresponded to M-1 fragment of acetic acid (intermediate 1, M=60), M+1 (M+23) fragment of acetamide (intermediate 2, M=59), and M+23 fragment of propionamide (intermediate 3, M=73). An inorganic product, NO_3^- (M=62), was also detected with M+1/z=63. Moreover, acetic acid and NO_3^- were found to be dominant over the other two (Li & Mei, 2008). Therefore, a possible pathway for the degradation of HPAM had been proposed in Fig. 5.

To evaluate the conversion of the intermediates including acetic acid and acetamide and the formation of the product NO_3^- during the process of photocatalytic degradation of HPAM, a test was carried out in the suspension including TiO_2 and HPAM (200 mg/L, initial TOC 75 mg/L). The results demonstrated that TOC of the system decreased to 70% after 240 min UV-light irradiation, and further increasing the time of UV-light irradiation may result in total mineralization of HPAM (Fig. 5). As for the acetic acid, its concentration gradually increased as increasing UV-light irradiation time from 0 to 120 min, and then its concentration decreased slowly, indicating that it will continuously degrade into more small molecules. The changes of the concentrations of acetamide were slow during the process of UV-light irradiation. As for the final product, NO_3^-, its concentration increased from 0 to 7.4 mg/L after 240 min UV-light.

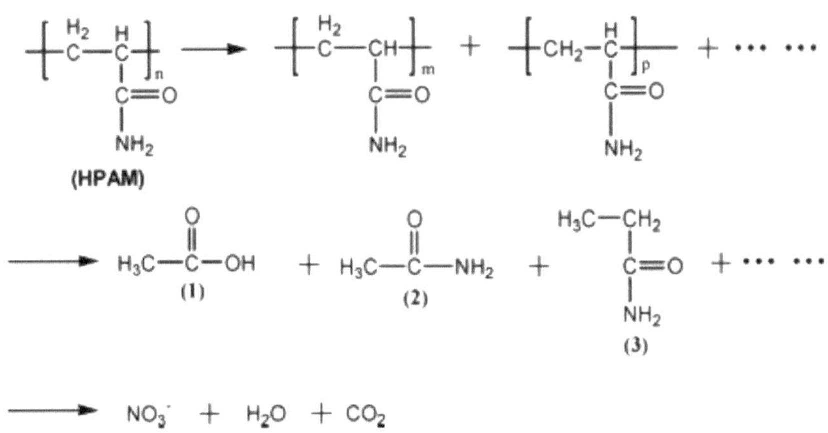

Fig. 5. Tentative pathway of sono-photocatalysis degradation of HPAM under UV-light irradiation of TiO_2

3. Conclusion

Among the treating technologies, AOPs constitute a promising technology for the treatment of wastewaters containing HPAM. UV/Fenton and Fenton-based reactions are capable of extensively degrading HPAM in a variety of aqueous solutions. Rigorous economic comparisons using an accepted standard measure of treatment efficiency are scarce. However, Fenton reactions can be performed at ambient temperature and do not require illumination, although they are usually enhanced by it. The reagents are readily available, easy to store, relatively safe to handle, and non-threatening to the environment. Drawbacks associated with the use of Fenton oxidation are the need to firstly reduce the pH, followed by a subsequent neutralization. This is not surprising since Fenton processes application requires strict pH control and sludges can be formed with related disposal problems. These drawbacks are intrinsic.

Photocatalytic or photochemical degradation processes are gaining importance in the area of wastewater treatment, since these processes result in complete organics mineralization with operation at mild conditions of temperature and pressure. However, there is still a problem that the photocatalytic efficiency of TiO_2 need to be improved since TiO_2 is photoactive only under near UV-light irradiation. On the other hand, recombination of photogenerated electron-hole pairs (e^--h^+) also results in low photo quantum efficiency of TiO_2. The effective separation of e^--h^+ pairs, therefore, is one of the most important subjects for broadening applications of TiO_2 photocatalyst.

In this study, the sono-photocatalytic oxidation of HPAM was examined in aqueous suspension. The evaluations of treatment performance were conducted for different amounts of TiO_2, initial concentration of HPAM, H_2O_2 dosage and reaction time. Based on the experimental findings, the following conclusions were drawn:

1. Coupling photocatalysis with ultrasound irradiation results in increased efficiency compared to the individual processes operating at common conditions. Interestingly, the overall sono-photocatalytic effect is greater than the additive effects of the two processes, thus indicating possible synergy.
2. TiO_2 had a good performance in US/UV/TiO_2 degradation HPAM. The concentration of 800 mg/L was the optimum dosage of TiO_2 under the experimental condition in terms of photocatalytic oxidation rate. Photocatalytic degradation increased with increasing TiO_2 loading (in the range 600-800 mg/L) and decreasing HPAM concentration (in the range 200-80 mg/L). Addition of H_2O_2 up to 18 mmol/L hindered degradation of HPAM, scavenging the photogenerated holes and hydroxyl radicals.

4. Acknowledgment

This research was supported in part by Educational Commission of Hubei Province of China (No. B200612013). Financial support for this work is gratefully acknowledged. I am also grateful to the anonymous referees for their insightful comments and suggestions, which clarified the presentation.

5. References

Augugliaro, V.; Palmisano, L.; Schiavello, M.; Sclafani, A.; Marchese, L.; Martra, G. & Miano, F. (1991). Photocatalytic degradation of nitrophenols in aqueous titanium dioxide dispersion. *Applied Catalysis*, Vol. 69, No. 1, 1991, pp. 323-340, ISSN 0926-860X

Adewuyi, Y. G. (2005). Sonochemistry in environmental remediation. II Heterogeneous sonophotocatalytic oxidation processes for the treatment of pollutants in water. *Environ. Sci. Technol.*, Vol. 39, No. 22, pp. 8557-8570. ISSN 0013-936X

Ai, Z. H.; Yang, P. & Lu, X. H. (2005). Degradation of 4-chlorophenol by a microwave assisted photocatalysis method. *Journal of Hazardous Materials*, Vol. B124 , No. 1-3, pp. 147-152, ISSN 0304-3894

Bao, M. T.; Chen, Q. G.; Li, Y. M. & Jiang, G. C. (2010). Biodegradation of partially hydrolyzed polyacrylamide production water after polymer flooding in an oil field. *Journal of Hazardous Materials*, Vol.184, No.1-3, pp.105-110, ISSN 0304-3894

Bilstad, T. & Espedal, E. (1996). Membrane separation of produced water. *Water Sci. Technol.* Vol. 34, No. 9, pp.239-246, ISSN 0273-1223

Bacchin, P.; Meireles, M. & Aimar, P. (2002). Modeling of filtration: from the polarised layer to deposit formation and compaction, *Desalination*, Vol.145, pp.139-146, ISSN 0011-9164

Bacchin, P.; Si-Hassen, D.; Starov, V.; Clifton, M. J. & Aimar, P. (2002). A unifying model for concentration polarization, gel-layer formation and particle deposition in crossflow membrane filtration of colloidal suspension, *Chemical Engineering Science*, Vol.57, No.1，pp.77-91, ISSN 0009-2509

Benkahla, Y. K.; Ould-Dris, A.; Jaffrin, M. Y. & Si-Hassen, D. (1995). Cake growth mechanism in cross-flow microfiltration of mineral suspensions. *J. Membr. Sci.*, Vol.98, pp.107-117, ISSN 0376-7388

Bouhabila, E. H.; Ben Aim, R. & Buisson, H. (1998). Microfiltration of activated sludge using submerged membrane with air bubbling (application to wastewater treatment). *Desalination*, Vol.118, pp.315-322, ISSN 0011-9164

Bhatkhande, D. S.; Pangarkar, V. G. & Beenackers, A. C. M. (2002). Photocatalytic degradation using TiO_2 for environmental applications-A review. *J. Chem. Technol. Biotechnol.*, Vol.77, pp. 102-116, ISSN 1097-4660

Causserand, C.; Rouaix, S.; Akbari, A. & Aimar, P. (2004). Improvement of a method for the characterization of ultrafiltration membranes by measurements of tracers retention. *Journal of Membrane Science*, Vol. 238, No, 1-2, pp. 177-190, ISSN 0376-7388

Cho, B. D. & Fane, A. G. (2002). Fouling transients in nominally sub-critical flux operation of a membrane bioreactor. *J. Membr. Sci.*, Vol.209, pp.391-403, ISSN 0376-7388

Chiu, T. Y. & James, A. E. (2006). Critical flux enhancement in gas assisted microfiltration. *J. Membr. Sci.*, Vol.281, pp.274-280, ISSN 0376-7388

Chong, T. H.; Wong, F. S. & Fane, A. G. (2008). Implications of critical flux and cake enhanced osmotic pressure (CEOP) on colloidal fouling in reverse osmosis: Experimental observations. *J. Membr. Sci.*, Vol.314, pp.101-111, ISSN 0376-7388

Cheryan, M. & Rajagopalan, N. (1998). Membrane processing of oily streams. Wastewater treatment and waste reduction. *J. Membr. Sci.*, Vol.151, No.1, pp.13-28, ISSN 0376-7388

Chen, Y.; Cui, J. M.; Wang, B. H.; Li, S. Q. & Liu, J. (2001). Feasibility of photocatalytic oxidation of degrading polyacrylamide in water. *Journal of Daqing Petroleum Institute*, Vol.25, No.2, pp.82-83, ISSN 1000-1891

Calleja, J.; Melero, G. A.; Martinez, F. & Molina, R. (2005). Activity and resistance of iron containing amorphous, zeolitic and mesostructured materials for wet peroxide oxidation of phenol. *Water Research*, Vol. 39, No.9, pp.1741-1750, ISSN 0043-1354

Chen, J. & Zhu, L. (2007). Heterogeneous UV-Fenton catalytic degradation of dyestuff in water with hydroxyl-Fe pillared bentonite. *Catal. Today*, Vol.126, pp.463-470, ISSN 0920-5861

Chen, V.; Fane, A. G.; Madaeni, S. & Wenten, I. G. (1995). Particle deposition during membrane filtration of colloids: transition between concentration polarization and cake formation. *J. Membr. Sci.*, Vol.125, pp.109-122, ISSN 0376-7388

Carp, O.; Huisman, C. L. & Reller, A. (2004). Photoinduced reactivity of titanium dioxide. *Progress in Solid State Chemistry*, Vol. 32, No. 1-2, pp. 33-37, ISSN 0079-6786

David, F. & Ollis, D. F. (2000). Photocatalytic purification and remediation of contaminated air and water. *Chimie/Chemistry*, Vol. 3, pp. 405-411

Defrance, L. & Jaffrin, M. Y. (1999). Comparison between filtrations at fixed transmembrane pressure and fixed permeate flux: application to a membrane bioreactor used for wastewater treatment. *J. Membr. Sci.*, Vol.152, pp.203-210, ISSN 0376-7388

Espinasse, B.; Bacchin, P. & Aimar, P. (2002). On an experimental method to measure critical flux in ultrafiltration. *Desalination*, Vol.146, No1-3, pp. 91-96, ISSN 0011-9164

Faouzi, M; Ca˜nizares, P.; Gadri, A.; Lobato, J.; Nasr, B.; Paz, R.; Rodrigo, M. A. & Saez, C. (2006). Advanced oxidation processes for the treatment of wastes polluted with azoic dyes. *Electrochimica Acta*, Vol.52, No.1, pp. 325-331, ISSN 0013-4686

Feng, J. Y.; Hu, X. J. & Yue, P. L. (2006). Effect of initial solution pH on the degradation of Orange II using clay-based Fe nanocomposites as heterogeneous photo-Fenton catalyst. *Water Res.*, Vol.40, pp. 641-646, ISSN 0043-1354

Field, R. W.; Wu, D.; Howell, J. A. & Gupta, B. B. Critical flux concept for microfiltration fouling. *J. Membr. Sci.*, Vol.100, pp.259-272, ISSN 0376-7388

Gogate, P. R. & Pandit, A. B. (2004). A review of imperative technologies for wastewater treatment II: hybrid methods. *Advances in Environmental Research*, Vol.8, No. 3-4, pp. 553-597, ISSN 1093-0191

Han, D. K.; Yang, C. Z.; Zhang, Z. Q.; Lou, Z. H. & Chang, Y. I. (1999). Recent development of enhanced oil recovery in China. *J. Petrol. Sci. Eng.*, Vol.22, No.1-3, pp.181-188, ISSN 0920-4105

Hu, X. L.; Zhu, Y. J. & Wang, S. W. (2004). Sonochemical and microwave-assisted synthesis of linked single-crystalline ZnO rods. *Materials Chemistry and Physics*, Vol. 88, pp. 421-426, ISSN 0254-0584

Kwon, D. Y.; Vigneswaran, S.; Fane, A. G. & Aim, R. B. (2000). Experimental determination of critical flux in cross-flow microfiltration. *Sep. Purif. Technol.*, Vol.19, No.3, pp.169-181, ISSN 1383-5866

Kwon, D. Y. & Vigneswaran, S. (1998). Influence of particle size and surface charge on critical flux of crossflow microfiltration. *Water Sci. Technol.*, vol. 38, No. 4-5, pp.481-488, ISSN 0273-1223

Kong, J. & Li, K. (1999). Oil removal from oil-in-water emulsions using PVDF membranes. *Sep. Purif. Technol.*, Vol.16, No.1, pp.83-93, ISSN 1383-5866

Kusic, K.; Koprivanac, N. & Selanec, I. (2006). Fe-exchanged zeolite as the effective hetenogeneous Fenton-type catalytic for the organic pollutant minimization: UV irradiation assistance. *Chemosphere*, Vol. 65, pp.65-73. ISSN 0045-6535

Li, Q. X.; Kang, C. B. & Zhang, C. K. (2005). Waste water produced from an oilfield and continuous treatment with an oil-degrading bacterium. *Process Biochem.*, Vol.40, pp.873-877, ISSN 0006-291X

Li, J. H.; Yang, X.; Yu, X. D.; Xu, L. L.; Kang, W. L.; Yan, W. H.; Gao, H. F.; Liu, Z. H. & Guo, Y. H. (2009). Rare earth oxide-doped titania nanocomposites with enhanced photocatalytic activity towards the degradation of partially hydrolysis polyacrylamide. *Applied Surface Science*, Vol. 255, pp.3731-3738, ISSN 0169-4332

Lu, Y.; Sun, H.; Meng, L. L. & Yu, S. L. (2009). Application of the Al_2O_3-PVDF nanocomposite tubular ultrafiltration (UF) membrane for oily wastewater treatment and its antifouling research. *Sep. Purif. Technol.*, Vol. 66, pp.347-352, ISSN 1383-5866

Liu, T.; You, H. & Chen, Q. W. (2009). Heterogeneous photo-Fenton degradation of polyacrylamide in aqueous solution over Fe (III)-SiO_2 catalyst. *Journal of Hazardous Materials*, Vol.162, pp.860-865, ISSN 0304-3894

Li, H.; Fane, A. G.; Coster, H. G. L. & Vigneswaran, S. (2000). An assessment of depolarization models of cross-flow microfiltration by direct observation through the membrane. *J. Membr. Sci.*, Vol. 172, pp. 135-147, ISSN 0376-7388

Le-clech, P.; Jefferson, B. & Judd, J. S. (2003). Impact of aeration, solids concentration and membrane characteristics on the hydraulic performance of a membrane bioreactor. *J. Membr. Sci.*, Vol.218, pp.117-129, ISSN 0376-7388

Li, F. X. & Mei P. (2008). Study on sono-photocatalytic degradation of hydrating polyacrylamide in waste water. *Journal of Oil and Gas Technology,*, vol.30, No. 3, pp. 157-160, ISSN 1000-9752

Li, F. X. & Xie, J. H. (2010). Study on photocatalytic degrading hydrolyzed polyacrylamide solution in wastewater treatment. *Journal of Oil and Gas Technology,* vol.32, No. 4, pp. 153-156, ISSN 1000-9752

Li, F. X. & Huang, Y. (2011). Study on the photodegradation of hexachlorobenzene catalyzed by TiO_2. *Journal of Anhui Agricultural Sciences,* vol.39, No. 4, pp. 2185-2188, ISSN 0517-6611

Li, F. X.; Lu, X. H.; Li, X. B. & Mei, P. (2006). Recent developments in researches of advanced oxidation processes for treating oilfield wastewaters. *Oilfield Chemistry,* Vol.23, No. 2, pp.188-192, ISSN 1000-4092

Li, J. L.; Li, H.; Chen, Y. & Wang, B. H. (2007). Influence of various ions on the degradation of hydrolyzed polyacrylamide by Fenton technique. *Petroleum Processing and Petrochemicals,* Vol.38, No. 11, pp.29-31, ISSN 1005-2399

Murray, C. A. & Parsons, S. A. (2004). Removal of NOM from drinking water: Fenton's and photo-Fenton's processes. *Chemosphere,* Vol. 54, pp.1017-1023, ISSN 0045-6535

Martinez, F.; Calleja, G.; Melero, J. A. & Molina, R. (2007). Iron species incorporated over different silica supports for the heterogeneous photo-Fenton oxidation of phenol. *Appl. Catal. B: Environ.*, Vol.70, pp.452-460, ISSN 0926-3373

Martinez, F.; Calleja, G.; Melero, J. A. & Molina, R. (2005). Heterogeneous photo-Fenton degradation of phenolic aqueous solutions over iron-containing SBA-15 catalyst. *Appl. Catal. B: Environ.*, Vol.60, pp.181-190, ISSN 0926-3373

Noorjaha, M. V.; Kumari, D.; Subrahmanyam, M. & Panda, L. (2005). Immobilized Fe(III)-HY: an efficient and stable photo-Fenton catalyst. *Appl. Catal. B: Environ.*, Vol.57, pp.291-298, ISSN 0926-3373

Neppolian, B.; Choi, H. C. & Sakthivel, S. (2002). Solar/UV-induced photocatalytic degradation of three commercial textile dyes. *Journal of Hazardous Materials B*, vol. 89, pp. 303-317. ISSN 0304-3894

Ognier, S.; Wsiniewski, C. & Grasmick, A. (2004). Membrane bioreactor fouling in sub-critical filtration conditions: a local critical flux concept. *J. Membr. Sci.*, Vol.229, pp.171-177, ISSN 0376-7388

Pandit, A. B.; Gogate, P. R. & Mujumdar, S. (2001). Ultrasonic degradation of 2, 4, 6-trichlorophenol in presence of TiO_2 catalyst. *Ultrason. Sonochem.*, Vol.8, pp. 227-231, ISSN 1350-4177

Parra, S.; Henao, L. & Mielczarski, E (2004). Synthesis, testing, and characterization of a novel Nafion membrane with superior performance in photoassisted immobilized Fenton catalysis, *Langmuir*, Vol. 20, pp. 5621-5629, ISSN 0743-7463

Parra, S.; Guasaquillo, I.; Enea, O. & Melczarski, E. (2003). Abatement of an azo dye on structured C-Nafion/Fe-ion surfaces by photo-Fenton reactions leading to carboxylate intermediates with a remarkable biodegradability increase of the treated solution, *The Journal of Physical Chemistry B*, Vol. 107, pp. 7026-7035, ISSN 1520-6106

Ren, G.; Sun, D. & Chunk, J. S. (2006). Advanced treatment of oil recovery wastewater from polymer flooding by UV/H_2O_2/O_3 and fine filtration. *J. Environ. Sci.*, Vol.18, pp.29-32, ISSN 1001-0742

Ramirez, J. H.; Maldonado-Hodar, F. J. & Perez-Cadenas, A. F. (2007). Azo-dye Orange II degradation by heterogeneous Fenton-like reaction using carbon-Fe catalysts. *Appl. Catal. B: Environ.*, Vol. 75, pp.312-323, ISSN 0926-3373

Ramirez, J. H.; Costa, C. A. & Madeira, L. M. (2007). Fenton-like oxidation of Orange II solutions using heterogeneous catalysts based on saponite clay. *Appl. Catal. B: Environ.*, Vol.71, pp.44-56, ISSN 0926-3373

Scholz, W. & Fuchs, W. (2000). Treatment of oil contaminated wastewater in a membrane bioreactor. *Water Res.*, Vol.34, No.14, pp.3621-3629, ISSN 0043-1354

Su, D. L. (2007). Kinetic performance of oil-field produced water treatment by biological aerated filter. *Chinese Journal of Chemical Engineering*, Vol.15, No.4, pp.591-594, ISSN 1004-9541

Serpone, N. & Pellizzetti, E. (1989). *Photocatalysis, Fundamentals and Applications*, Wiley, New York, ISBN 0471626031

Saif, M. & Abdel-Mottaleb, M. S. A. (2007). Titanium dioxide nanomaterial doped with trivalent lanthanide ions of Tb, Eu and Sm: Preparation, characterization and potential applications. *Inorganica Chimica Acta*, Vol.360, No.9, pp.2863-2874, ISSN 0020-1693

Shirgaonkar, I. Z. & Pandit, A. B. (1998). Sonophotochemical destruction of aqueous solution of 2, 4, 6-trichlorophenol. *Ultrason. Sonochem.*, Vol. 5, pp. 53-61, ISSN 1350-4177

Taylor, K. C., Burke, R. A., Nasr-El-Din, H. A. & Schramm, L. L. (1998). Development of a flow injection analysis method for the determination of acrylamide copolymers in brines. *J. Petrol. Sci. Eng.*, Vol.21, No.1-2, pp.129-139, ISSN 0920-4105

Thoma, G. J.; Bowen, M. L. & Hollensworth, D. (1999). Dissolved air precipitation/solvent sublation for oil-field produced water treatment. *Sep. Purif. Technol.*, Vol.16, No.2, pp.101-107, ISSN 1383-5866

Thandar, A.; William, A. A. & Mehrab, M. (2003). Photocatalytic treatment of cibacron brilliant yellow 3G-P. *Journal of environmental science and health, Part A-Toxic/Hazardous Substances and Environmental Engineering*, Vol. A38, No. 9, pp. 1903-1914, ISSN 0360-1234

Vijayalakshmi, S. P. & Giridhar, M. (2006). Photocatalytic degradation of poly (ethylene oxide) and polyacrylamide. *Journal of Applied Polymer Science*, Vol.100, pp. 3997-4003, ISSN 1097-4628

Vajnhandl, S. & Marechal, L. (2005). Ultrasound in textile dyeing and the decoloration / mineralization of textile dyes. *Dyes Pigments*, Vol. 65, No. 2, pp.89-101, ISSN 0143-7208

Wang, B. H.; Chen, Y.; Liu, S. Z.; Wu, H. J. & Song, H. (2006). Photocatalytical visbreaking of wastewater produced from polymer flooding in oilfields. *Colloids and Surfaces A: Physicochem. Eng. Aspects*, Vol.287, pp.170-174, ISSN 0927-7757

Wu, C. J.; Li, A. M.; Li, L. & Zhang, L. (2008). Treatment of oily water by a poly (vinyl alcohol) ultrafiltration membrane, *Desalination*, Vol.225, pp.312–321, ISSN 0011-9164

Wang, X. Y.; Wang, Z.; Zhou, Y. N.; Xi, X. J.; Li, W. J. & Yang, L. Y. (2011). Study of the contribution of the main pollutants in the oilfield polymer-flooding wastewater to the critical flux. *Desalination*, Vol.273, No.2-3, pp.375-385, ISSN 0011-9164

Yuranova, T.; Enea, O.; Mielczarski, E. & Mielczarski, J. (2004). Fenton immobilized photoassisted catalysis through a Fe/C structured fabric. *Appl. Catal. B: Environ.*, Vol. 49, pp. 39-50, ISSN 0926-3373

Yardin, C. & Chiron, S. (2006). Photo-Fenton treatment of TNT contaminated soil extract solutions obtained by soil flushing with cyclodextrin. *Chemosphere*, Vol. 62, pp. 1395-1402, ISSN 0045-6535

Zhang, Y. Q; Gao, B. Y.; Lu, L.; Yue, Q. Y.; Wang, Q. & Jia, Y. Y. (2010). Treatment of produced water from polymer flooding in oil production by the combined method of hydrolysis acidification-dynamic membrane bioreactor-coagulation process. *Journal of Petroleum Science and Engineering*, Vol.74, No.1-2, pp.14-19, ISSN 0920-4105

Zhang, X. W.; Wang, Y. Z. & Li, G. T. (2005). Effect of operating parameters on microwave assisted photocatalytic degradation of azo dye X-3B with grain TiO_2 catalyst. *Journal of Molecular Catalysis A: Chemical*, Vol.237, No. 5, pp. 199-205, ISSN 1381-1169

Zhao, X.; Wang, Y. M.; Ye, Z. F. & Alistair, G. L. (2006). Oil field wastewater treatment in biological aerated filter by immobilized microorganisms. *Process Biochemistry*, Vol.41, No.7, pp.1475-1483, ISSN 1359-5113

Zhao, X. F.; Liu, L. X.; Wang, Y. C.; Dai, H. X.; Wang, D. & Cai, H. (2008). Influences of partially hydrolyzed polyacrylamide (HPAM) residue on the flocculation behavior of oily wastewater produced from polymer flooding. *Separation and Purification Technology*, Vol.62, pp.199-204, ISSN 1383-5866

14

Chemical Degradation of Chlorinated Organic Pollutants for *In Situ* Remediation and Evaluation of Natural Attenuation

Junko Hara
Institute for Geo-resources and Environment, National Institute of Advanced Industrial Science and Technology, Japan

1. Introduction

Chlorinated organic compounds, prevalent contaminants found in the geo-environment, pose an ecological risk even at trace concentrations. More volatile chlorinated compounds such as VOCs (volatile organic compounds) have been detected in urban areas and industrial zones because of the use of these compounds as components of industrial solvents and both raw and intermediate synthetic products. Chlorinated organic compounds quickly evaporate from surface water but remain in groundwater and soil for a long time. Recently, several remediation techniques have been developed that can entirely remediate chlorinated organic compounds to non-toxic materials. However, high molecular weight chlorinated organic compounds (e.g., polychlorinated biphenyls (PCBs) and dichloro-diphenyl-trichloroethane (DDT)) are highly toxic chemicals that persist for long periods of time in the environment and bioaccumulate. They are categorized as persistent organic pollutants (POPs). Although the amount and date of use vary by country, POPs were widely used for pesticides and disease control in crop production and industrial processes during the period of industrial production after World War II around the globe. DDT, PCBs and dioxins are the best known POPs. DDT is used to control mosquitoes, which carry malaria, and PCBs were useful in electrical transformers and large capacitors. Among these POPs, PCBs, hexachlorobenzene (HCB), chlordanes, dichloro-diphenyl-dichloroethane (DDE) and dieldrin show significant ecological accumulation, such as in human milk, human blood and other biological media.

The Stockholm Convention is intended to protect human health and the environment, starting with the reduction or elimination of the production, use, and/or release of 12 species of POPs (PCBs, HCB, aldrin, dieldrin, endrin, DDT, chlordane, heptachlor, toxaphene, mirex, polychlorinated dibenzo-*p*-dioxins (dioxins), and polychlorinated dibenzofurans (furans). The 9 additional chemicals adopted in amendments to the Stockholm convention as new POPs are α-hexachlorocyclohexane, β-hexachlorocyclohexane, chlordecone, hexabromobiphenyl, hexabromobiphenyl ether and heptabromobiphenyl ether, Lindane, pentachlorobenzene, perfluorooctane sulfonic acid and its salts and perfluorooctasulfonyl fluoride, tetrabromobiphenyl ether and pentabromobiphenyl ether in 2009. The Stockholm Convention has led to a general global decline in the concentration of these chemicals in the environment. However, some individual POPs still persist and accumulate in fatty tissue and are present in higher concentrations at higher levels in the food chain, with long-range mobility through natural processes, because complete removal of POPs from the environment is difficult.

Experimental degradation methods for POPs at room temperature have been reported, using chemical catalysis, bacteria, UV, photocatalysis, Fenton reagent and other methods. Among these techniques, reductive dechlorination processes encounter difficulty in achieving complete dechlorination and degradation of the chemical structure. Powerful oxidative processes are assumed to show some possibility for complete degradation of POPs. Reports of complete degradation of POPs are few, even when oxidative processes are applied for on-site remediation over a very extensive polluted area.

This chapter reviews remediation methodology for chlorinated organic pollutants and chemical remediation methods using ferric sulphide compared with reaction with zero-valent iron, which widely used as a practical in-situ remediation method for soil and groundwater pollution by VOCs. Ferric sulphides also have outstanding ability to degrade chlorinated organic pollutants, but the dechlorination processes differs from that of zero-valent iron. The natural remediation capability, reaction products, and reaction mechanisms using ferric sulphide for trichloroethylene, dieldrin and chlorinated benzenes are reported in this chapter.

2. Reported remediation methods for chlorinated organic pollutants

Several remediation methodologies for chlorinated organic pollutants such as bioremediation, bioaugmentation, and chemical or physical remediation have been reported by many researchers. In these remediation methodologies, chlorinated organic contaminants can be transformed chemically, photochemically, or biochemically by oxidation or reduction in the soil and the groundwater environments.

Bioremediation technology has developed rapidly in the last 15 years, and the application of this technology has been extended to several species of contaminants, including volatile organic compounds (VOCs), polycyclic aromatic hydrocarbons (PAHs) and polychlorinated biphenyls (PCBs). Reductive dechlorination by microorganisms under anaerobic conditions is an advantageous process because reductive dechlorination allows the reoxidation of metabolic intermediates. For VOCs, PCE (tetrachloroethylene) or TCE (trichloroethylene), which is most widely reported chemicals, are capable of reductive dechlorination by dehalococcoides, dehalobacter and desulfuromonas (Hollinger *et al.*, 1993; Gerritse *et al.*, 1996, 1999; Loffler *et al.*, 1999; Krumholz *et al.*, 1996; Maymo-Gatell, 1997). These microorganisms dechlorinated the PCE/TCE to cis-dichloroethylene (cis-DCE) or vinyl chloride (VC). Dehalococcoides especially (including the culture) is reported to have achieved complete dechlorination of PCE/TCE to ethylene. In aerobic or oxidizing environments, chlorinated ethylenes are oxidized to CO_2 and the chlorinated ethylenes are co-metabolized to CO_2 via trichloroacetate, trichloroethanol, trichloroethylene epoxide, dichloroacetate, glyoxylate, formate, oxalate, etc. The reaction pathway and the number of intermediates differ according to the species of bacteria in the local environment (Newman and Wackett, 1991, 1995, 1997; Weightman *et al.*, 1985, 1993; Li and Wackett, 1992; Kim *et al.*, 2009; Rosenzweig *et al.*, 1993; Oldenhuis *et al.*, 1989; Motosugi *et al.*, 1982). Oxygen served as the electron acceptor in the aerobic oxidation, and non-specific microbial oxygenase enzymes produced by aerobic microorganisms participated in co-metabolism reactions.

Anaerobic microbial communities in sediments also dechlorinate PCBs. The ease of dechlorination of positions on aromatic rings is usually meta > para > ortho, and on biphenyl rings, mono-ortho- and ortho-chlorobiphenyls were not degraded after one year of reaction (Teidje *et al.*, 1993). PCE was also oxidatively degraded by Pseudomonas strain sp. P2. In this reaction, PCE is metabolized to chlorobenzoic acids with one to three chlorine atoms. At the actual reaction site, reductive and oxidative biological reaction is estimated to occur because of conditions at the site. Each reaction is therefore assumed to proceed partly under anaerobic conditions and partly under aerobic conditions at one polluted site, and this collaboration of reactions under different sets of conditions leads to complete remediation at the natural site.

Many chemical remediation methods have been investigated for rapid in-situ or off-site remediation of soil and groundwater. Among the chemical and photochemical dechlorination methods, zero-valent iron (ZVI) has frequently been reported as the remediation technique for reductive dechlorination of chlorinated ethenes and ethanes. Among actual chemical remediation techniques, the ZVI methodology is widely used for in situ remediation for soil (e.g., injection by direct drilling) and groundwater (application in a reactive permeable barrier). The chlorinated organic materials degrade under anaerobic (reducing) conditions using ZVI. This reductive dechlorination is generally divided into hydrogenolysis and reductive elimination. Both reaction mechanisms are accompanied by a net transfer of two electrons. Although reductive dechlorination is occasionally referred to as hydrolytic reduction, this term is misleading because hydrolysis is incidental to the actual reduction. Some other zero-valent metals such as zinc and platinum are also reported as reductive remediation catalysts instead of ZVI. The reaction pathway for TCE using ZVI and other metals is the same as the reductive dechlorination process using bioremediation.

Although the application of ZVI for remediation of high molecular weight persistent organic compounds is difficult, zero-valent metals such as Pd, Pt, Ni and Cu are also used as catalysts in the ZVI methodology for dechlorination of PCBs. Metal catalysis enhances the reductive dechlorination capability of ZVI. In the case of persistent organic compounds, electrochemical methods using metal electrodes are used for reductive dechlorination.

Fenton reaction and photochemical reaction are the predominant oxidative dechlorination processes and can remediate contaminants rapidly. The Fenton reaction is the oxidation of organic substrates by iron(II) and hydrogen peroxides. The Fenton reagent is effective in treating various industrial wastewaters polluted by chlorinated organic compounds, aromatic amines, pesticides, and surfactants. This oxidation system is based on the formation of reactive oxidizing species able to degrade the contaminants effectively in wastewater. In the Fenton chemistry, a two-reaction pathway is advanced as the first step. Production of hydroxyl radicals and a non-radical pathway using ferric ion production have been reported (Barbusinski, 2009). The nature of the oxidizing species is still controversial. Some researchers showed that the hydroxyl radical is the major species in the Fenton mechanism, and other groups showed that this reaction includes the formation of reactive oxidizing species such as ferric ion. Considering that the Fenton reaction occurs chemically and biologically as well as in the natural environment, there is a possibility that both mechanisms coexist in the Fenton reaction. Several hazardous pollutants can be oxidized by the Fenton reaction; for example, chlorophenol is degraded to hydroxyacetic intermediates (Barbeni *et al.*, 1987a) and perchloroethylene is transformed to dichloroacetic acid, formic acid and CO_2 (Leung *et al.*, 1992). More detailed information about Fenton reaction is discussed in the chapter "Fenton's Process for the Treatment of Mixed Waste Chemicals".

In natural waters exposed to solar radiation, the Fenton reaction is often perceived as a possible source of hydroxyl radical in sunlit waters. The photolysis of nitrates, metal-to-ligand charge transfer reactions, and photoFenton reactions are included as other possibilities for the Fenton reaction, and H_2O_2 and Fe(II) are photochemically produced in these sunlit waters. Photochemical dechlorination is reported for persistent high molecular weight organic compounds such as DDT and PCBs (Mochizuki 1977; Van Beek *et al*, 1982; Shimakoshi *et al.*, 2004). Photochemical PCBs dechlorination in alkaline isopropyl alcohol effectively degrades PCBs to biphenyl, chloride ions, acetone and water (Mochizuki, 1977). Dechlorination of DDT was catalyzed by hydrophobic vitamin B12 (heptamethyl cobyrinate perchlorate) with irradiation by visible light (Shimakoshi *et al.*, 2004). The Co(I) species of cobalamin and the related cobalt complex are supernucleophiles and react with an alkyl halide to form alkylated complexes with dehalogenation (Shimakoshi H. *et al.*, 2004). In this reaction, DDT is mono-dechlorinated to DDD. Photocatalytic treatment assisted by TiO_2 is also widely reported for TCE, DDT, chlorinated phenol, chloroform, etc. (Ahmed and Ollis, 1984; Barbeni *et al.*, 1986, 1987b; Dible and Raupp, 1990, 1992; D'Oliveria *et al.*, 1990; Kondo and Jardim, 1991; Borello *et al.*, 1989) These oxidative degradations proceed by radicals arising from the photoreaction of other compounds present in the system.

In a similar reaction, UV/H_2O_2 and UV/Fenton are used for the remediation of chlorinated ethylenes and ethanes, chlorobenzenes, benzenes, and chlorinated phenols (Froclich, 1992; Moza *et al.*, 1988; Sundstrom *et al.*, 1986, 1989; Weie *et al.*, 1987). UV/H_2O_2 oxidation involves the dissociation of H_2O_2 to form the hydroxyl radical. The hydroxyl radical oxidizes toxic organic materials by abstraction of protons to produce organic radicals. Some toxic materials degrade to lower molecular weight acids and finally transform to CO_2. However, many reports only estimate degradation ability and reaction kinetics and do not mention reaction products, especially when high molecular weight persistent organic compounds are involved.

These strong chemical oxidation technologies such as Fenton reagent, UV radiation, catalysis and photochemical treatment have the prospect of degradation of persistent organic pollutants to non-toxic compounds.

3. Remediation ability of ferric sulphides

The metal sulphides show a great diversity in electrical and magnetic properties. The sulphides of the transition metals can be considered as intermediate between the transition metal oxides. The small particles of metal sulphide are often superior in electrical and chemical properties, as well as in catalysis.

Pyrite is the most abundant metal sulphide mineral and is treated as industrial waste at many mining sites. An outcrop including pyrite leads to oxidation in aerobic weathering processes and causes acidification of environmental water. Pyrite is also distributed in acidic coastal sulphate soils, which are naturally formed under waterlogged anaerobic conditions. These acidic sulphide soils are located in tropical coastal areas in West Africa, South and Southeast Asia, and northeast South America. These acidic conditions are mainly developed in recent or semi-recent sediments close to the sea. In these environments, the sulphur in pyrite is derived from the sulphate in seawater, which is biologically reduced to sulphide in the anaerobic mud. The organic matter which serves as the energy source for the sulphate-reductive bacteria is usually abundant in plants in the coastal area. Ferrous iron is also derived from reduction of insoluble ferric compounds resulting from the weathering of clay.

The reactivity of pyrite when the surface is exposed to H_2O, O_2, and a mixture of H_2O and O_2, has been studied experimentally by Guevremont *et al.* (1997, 1998a, b, c, d). The pyrite surface exposed to H_2O vapour up to 1 bar was oxidized, although the reaction site is limited to the defect site on pyrite surface. The surface reaction with O_2 vapour showed no oxidation. Substantial surface reaction is observed in the reaction with the H_2O/O_2 mixture. The intermediate oxidation products, sulphur oxoanion and zero-valent sulphur, are also identified and removed by use of either the O_2 or H_2O (Kendelewicz *et al.*, 2004). The oxidation mechanism of pyrite with the H_2O/O_2 mixture involves competitive adsorption of O_2 and H_2O at the surface Fe site, oxidation of surface Fe sites by O_2, dissociation of co-adsorbed H_2O at Fe sites, and charge redistribution in surface S atoms (Rosso *et al.*, 1999).
This pathway allows for the production of hydroxyl radicals from dissociated water and subsequent nucleophilic attack of these hydroxyl radicals at surface S sites. The oxygen in the final product sulphate arises from water molecules (Taylor *et al.*, 1984; Usher *et al.*, 2004, 2005). The hydroxyl radical from the pyrite oxidation process causes the degradation of chlorinated organic compounds.

3.1 Degradation ability for major volatile organic compounds

Chlorinated ethylenes are widespread groundwater and soil contaminants. Due to the prevalent pollution and the efforts to treat these compounds, substantial research has been conducted to identify the mechanism of reaction under various environmental conditions. As mentioned above, the chemical dechlorination method using ZVI is in wide use as a practical in situ remediation technique for soil and groundwater pollution with TCE. This reductive dechlorination reaction process includes hydrogenolysis, β-elimination and a hydrogen addition reaction, and degrades chlorinated ethylenes and ethanes to ethylene and ethane. The dechlorination of TCE by pyrite (ferric sulphide) under anaerobic conditions is also reported to be reductive dechlorination like ZVI (Weerasooriya R., 2001; Lee and Batchelor, 2002, 2003). Other chlorinated ethylenes and ethanes are also reductively dechlorinated under anaerobic conditions (Kriegman-King and Reinhard, 1994; Lee and Batchelor, 2003). The reductive dechlorination of chlorinated compounds occurs by the transfer of electrons from the mineral surface.
Fig. 1 shows the dechlorination of TCE by metallic sulphides and zero- valent iron. In this reaction, 100 ppm of TCE is dechlorinated with powdered pyrite and Milli-Q-grade-ultrapure water under closed aerobic conditions. TCE is detected by a headspace method using gas chromatography (GC) (GC17A, Shimadzu Co. Ltd.). Pyrite and chalcopyrite showed an outstanding dechlorination rate, and complete dechlorination was confirmed by the mass balance of chloride ion. According to the recent pyrite oxidation work, the most reactive surface component is S^{2-}, and the second most reactive surface component is the surface atom of the first disulphide layer (S_2^{2-}) with sulphur atoms of the disulphide groups beneath the surface layer being least reactive. Oxidized iron (Fe^{3+}) states are proposed to arise after fracturing of S-S bonds by electron transfer from Fe^{2+} to this S^- state, which then reacts rapidly to sulphate (Nesbitt *et al.*, 1998; Schaufuss *et al.*, 1998). If this oxidation reaction is similar with metallic sulphide, the S-S bonds in metallic sulphide oxidize chlorinated contaminants, as assumed from this oxidation model. Disulphide metallic minerals involving S-S bonds are therefore assumed to have greater degradation ability than the mono-sulphide metals. In our experimental estimation in Figure 1, the disulphide metallic minerals are only pyrite and chalcopyrite, which have superior degradation ability relative to the other mono-sulphide metals and ZVI.

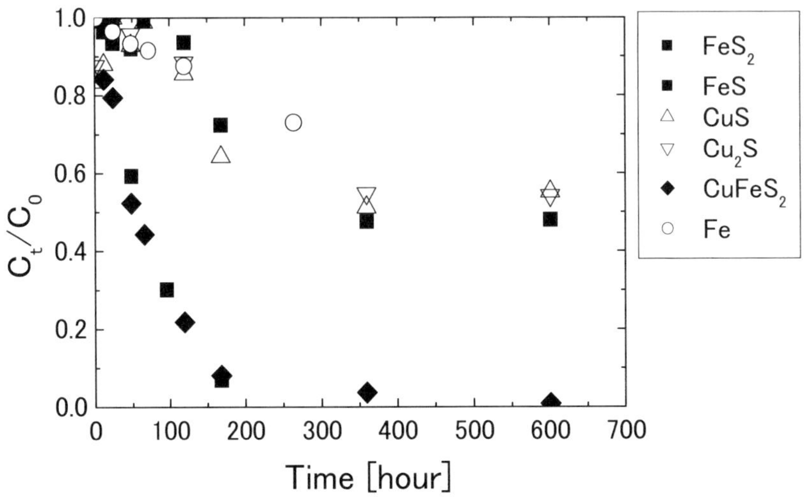

Fig. 1. Dechlorination of trichloroethylene (TCE) by metallic sulphide and zero-valent iron under aerobic conditions. C_0 is the initial concentration of TCE (100 mg/L), C_t is the concentration of TCE at time t. The reaction was represented as the evaluation of the normalized remaining percentage of TCE with time.

TCE is also degraded by pyrite under aerobic conditions (Hoa *et al.*, 2008). TCE dechlorination in anaerobic and aerobic pyrite suspensions is observed with time, but there is no outstanding degradation under anaerobic conditions and the dechlorination rate of TCE is proportional to the increase in oxygen. Fig. 2 shows the disappearance of TCE, the reaction intermediates and the final product in the reaction of TCE with a pyrite suspension under aerobic conditions. Under aerobic conditions, 98 % of TCE was degraded after about 2 weeks. The TCE degrades to dichloroacetic acid, glyoxylic acid, formic acid, oxalic acid and CO_2 by oxidative processes (Hoa *et al.*, 2009). This degradation process is similar to the oxidative metabolic pathway of TCE (Kim *et al.*, 2009; Li *et al*, 1992). Figure 3 shows the expected dechlorination pathway of TCE by pyrite under aerobic conditions. The dechlorination pathway of TCE is divided into three pathways. All pathways show direct dechlorination of three chlorine atoms in TCE. The main pathway is degradation of TCE to formic acid (Eq. (1)).

$$C_2HCl_3(TCE) + 4^{\bullet}OH + H_2O \rightarrow 2CH_2O_2(formic \quad acid) + 3HCl + \frac{1}{2}O_2 \quad (1)$$

$$CH_2O_2(formic \quad acid) + 2^{\bullet}OH \rightarrow CO_2(g) + 2H_2O \quad (2)$$

Formic acid has toxicological properties toward aquatic organisms, but the degradation rate of formic acid by pyrite suspensions is also high, similar to the dechlorination of TCE to formic acid, and formic acid continuously transforms to $CO_2(g)$ without accumulation (Eq. (2)).

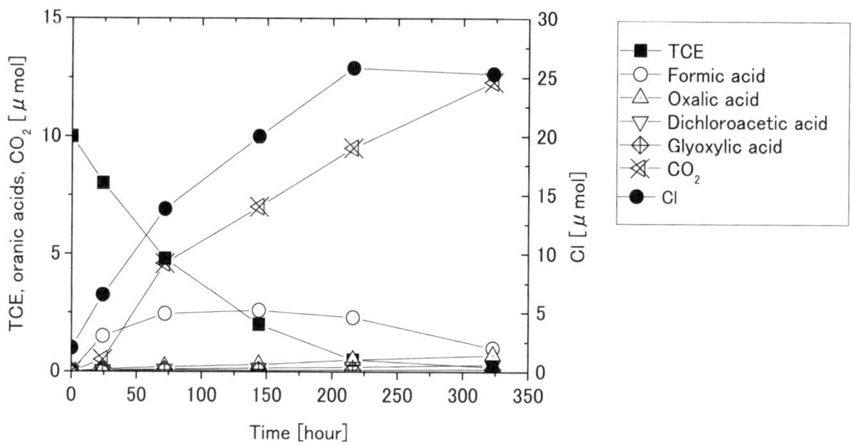

Fig. 2. Transformation of trichloroethylene (TCE) and detected reaction products in pyrite suspension (Edited from Hoa *et al.* (2008)). The initial TCE concentration is 10 μmol, the oxygen volume is 0.268 mmol/L, and pyrite is 10 m²/L. The left side of the Y-axis shows the molecular weight of TCE, CO_2 and organic acids (formic acid, oxalic acid, dichloroacetic acid, glyoxylic acid). The right side of the Y-axis shows the chloride molecular weight as degraded from TCE.

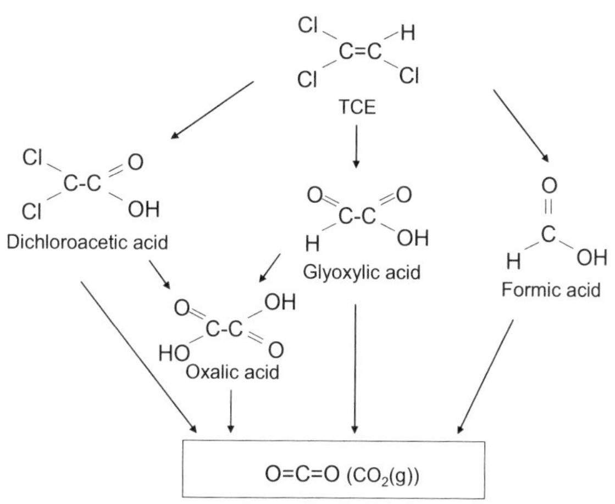

Fig. 3. Degradation pathway of trichloroethylene (TCE) using a pyrite suspension under aerobic conditions (Edited from Hoa *et al.* (2009)).

3.2 Degradation capability for dieldrin

Dieldrin is a cyclodiene pesticide which has persisted in the soil over decades in some agricultural fields. Some agricultural crops and animals accumulate dieldrin through the food chain. From the previous report of TCE dechlorination by pyrite, we see that TCE is able to be reductively degraded under anaerobic conditions and oxidatively degraded under aerobic conditions. The reductive dechlorination under anaerobic conditions is not observed using pyrite suspensions. Dieldrin has a higher persistence than TCE and is less amenable to reductive dechlorination under anaerobic conditions. Remediation techniques for dieldrin (UV/Fenton reagent, UV/chemical reaction, Pd/C catalyst, solar photocatalysis, and bioremediation) have been reported (Books, 1980; Maule *et al.*, 1987; Bandala *et al.*, 2002; Kusvuran and Erbatur, 2004; Chiu *et al.*, 2005; Zinovyev *et al.*, 2005; Dureja *et al.*, 1987; Baczynski *et al.*, 2004), but dieldrin transforms to mono- or di- dechlorinated intermediates still having a bicyclic ring structure. A study of the use of Fenton reagent for aldrin reports the transformation from aldrin to oxalic acid, acetic acid, chlorohexanone, cyclohexanol, cis-2-hydroxy-cyclohexanone, cis-2-methyl-cyclohexanol, 4-hydroxycyclohexanone, cis-4-methylcyclohexanol, 1-cyclopropyl-1-hydroxyethylene, and trans-dihydroxycyclohexane (Kusvuran and Erbatur, 2004).

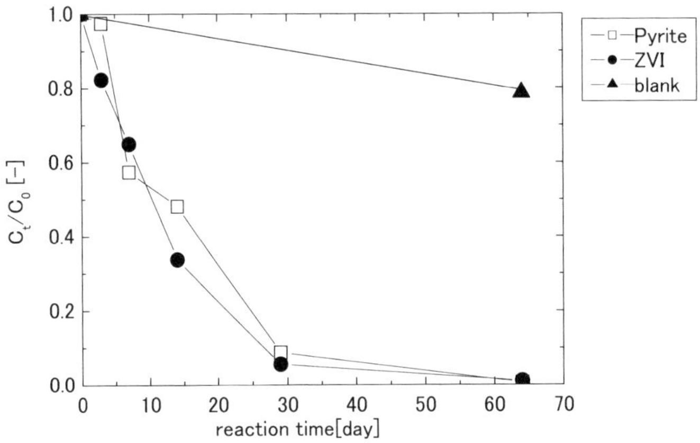

Fig. 4. Degradation profile of dieldrin using pyrite and zero-valent iron under aerobic conditions. The blank denotes the non-metallic catalysis condition. C_0 is the initial concentration of dieldrin (0.131 µmol), C_t is the concentration of dieldrin at time t. (Edited from Hara *et al.*, (2009)

The degradation of dieldrin using ZVI and pyrite under anaerobic conditions is shown in Fig. 4. ZVI and pyrite have similar dechlorination capabilities for dieldrin, but the reaction process differs completely for the two reagents (Hara *et al.*, 2009). In the case of ZVI, dieldrin partially transforms to nono-dechlorinated products ($C_{12}H_9Cl_5$, $C_{12}H_9Cl_5O$) and aldrin ($C_{12}H_8Cl_5$). This reaction is a reductive dechlorination proceeded by the generation of H+ and an electron arising from oxidation of ZVI. This reaction is stopped only this pathway. In contrast, the mono-dichlorinated reaction intermediates and aldrin are not detected in the reaction using pyrite. Dieldrin is oxidatively degraded to water-soluble reaction intermediates by pyrite.

Fig. 5 shows the dechlorination rate of dieldrin using pyrite under anaerobic to aerobic conditions, changing with oxygen concentration (O_2 = 0 ~ 833 µmol). Dieldrin was gradually degraded under every condition of oxygen concentration, except for the O_2 = 0 µmol condition. In Fig. 5, 5 ppm (0.131 µmol) of dieldrin is used with a powdered pyrite suspension under each oxygen condition. Dieldrin is detected by GC/MS (Shimadzu Co. Ltd.) after solvent/solvent extraction using acetone and hexane. The water-soluble reaction intermediates are detected by IC-TOF/MS (ICS-3000 (Dionex) and JMS-T 100LP (JEOL). A little oxygen (O_2 = 10 µmol) accelerates the dechlorination of dieldrin in comparison to the anaerobic conditions. Approximately 99 % of the dieldrin was degraded after about one month under the most enhanced reaction conditions (O_2 = 10 μmol), and approximately 43 % of dieldrin remained under O_2 = 300 μmol in the same reaction time, which is the slowest dechlorination condition. In these reactions, chloride ion resulting from the dechlorination was detected in the aqueous phase. The mass balance of dechlorinated chloride and degraded dieldrin is 90 % in the most degraded condition (O_2 = 10 μmol). This discrepancy in mass balance is obvious in the anaerobic conditions, but the mass balance agrees well under aerobic conditions (O_2 = 300 and 833 μmol).

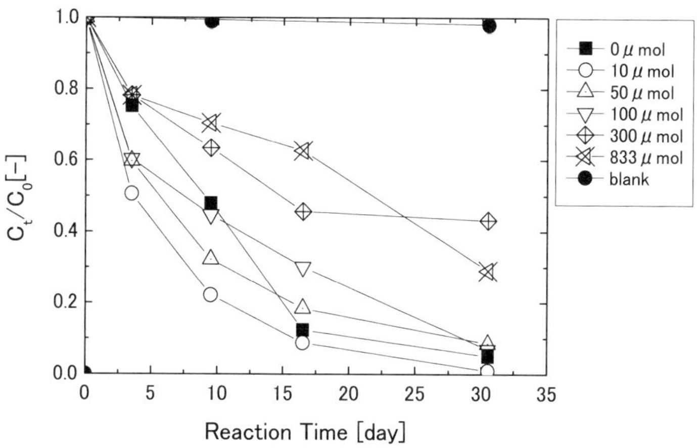

Fig. 5. Oxygen dependence on degradation of dieldrin in pyrite suspension (Hara, 2011). C_0 is the initial concentration of dieldrin (0.131 μmol), C_t is the concentration of dieldrin at time t. O_2 = 833 μmol denotes the air volume conditions in this experiment. The blank is without pyrite.

As reaction products, formic acid, oxalic acid, malonic acid, succinic acid, acetoxyacetyl, lactic acid, and pyruvic acid are detected as the main reaction products in the case of aerobic conditions (O_2 = 833 μmol). Acetic acid, glycolic acid, hydroxybutyric acid, glyoxylic acid, propionic acid, glutaric acid, levulinic acid and sulphur-containing organic acids such as methanesulphonic acid, sulphopropionic acid, etc. are also detected as minor reaction products. Formic acid is predominantly generated, as shown in (Eq. (3)).

$$C_{12}H_8Cl_6O(dieldrin) + 24^{\bullet}OH + H_2O \rightarrow 12CH_2O_2(formic \quad acid) + 6HCl + O_2 \tag{3}$$

Formic acid is also detected in TCE dechlorination by a pyrite suspension under aerobic conditions, and the formic acid continuously degrades to CO_2 (Eq. (2)). Dieldrin is therefore assumed to finally transform to CO_2 as a main reaction pathway.

In the case of a low O_2 volume or anaerobic conditions, these low molecular weight organic acids generated under oxygen-rich conditions are much lower in abundance and some organic acids are not detected. Instead of the generation of organic acids, 3-chloro-4-methyl-2-pentanol ($C_6H_{13}ClO$) and dibutyl phthalate ($C_{16}H_{22}O_4$) are generated. One part of these intermediates is continuously degraded to a low molecular weight organic acid, but the residual volume is also higher than the volume observed under aerobic conditions.

The reaction pathway of dieldrin proceeds mostly by oxidative degradation under anaerobic and aerobic conditions. Although the oxidative ability is obviously different from the content of oxygen, the active oxidant to degrade the dieldrin is promoted from both the pyrite/H_2O and pyrite/O_2 interface reaction, due to oxidants produced under either aerobic or anaerobic condition. The difference of degradation ability in oxygen volume is assumed to depend on the difference of radical species arising form pyrite/H_2O and pyrite/O_2 interface.

Considering the reaction products, the degradation pathway of dieldrin results mainly in the direct production of organic acids, which are easy to produce under oxygen-rich conditions. Under anaerobic or micro-aerobic conditions, the pathway of ring opening and the addition reaction of low molecular weight organic acids arising from $C_{16}H_{22}O_4$ and the pathway of generation of chlorinated hydrocarbons arising from $C_6H_{13}ClO$ (which continuously transforms to formic acid and malonic acid) become predominant.

3.3 Degradation ability for chlorinated benzenes

Dieldrin having a bicyclic ring could be decomposed by pyrite. This section discusses the ability to dechlorinate chlorobenzenes whose main structure is the benzene ring.

Chlorobenzenes are divided into 12 species based on the number and configuration of chlorine in the molecule: monochlorobenzene (mono-CB), 3 types of dichlorobenzene (1,2-, 1,3-, 1,4- di-CB), 3 types of trichlorobenzene (1,2,3-, 1,2,4-, 1,3,5- tri-CB), 3 types of tetrachlorobenzene (1,2,3,4-, 1,2,3,5-, 1,2,4,5- tetraCB), pentachlorobenzene (penta-CB), and hexachlrobenzene (hexa-CB).The chlorobenzenes are used in pesticides, deodorants, or as intermediates in a chemical synthesis process. Some tri- or tetra-chlorinated benzenes are extensively used as insulating materials. The risk associated with chlorinated benzenes increases relative to the increase in the chlorine number, because of their increasing lack of volatility with increasing chlorine number, the highest level of risk is associated with the misuse or accidental release of the mono- to trichlorinated benzenes. These compounds spread into the atmosphere are reported to be photolyzed or chemically reacted, and that in groundwater and soils these compounds are mainly remediated by microbial degradation. However, the residence time is increased because of the organic constituents present in the soil and groundwater, which results in adsorption and accumulation in the soil ecosystems. The major remediation method for more the accumulation of chlorobenzenes with a higher chlorination level in soils is incineration at high temperatures for digging out of soils.

In the previous reports of electrochemical dehalogenation of chlorinated benzenes (Miyoshi *et al.*, 2004; Mohammad and Dennis, 1997; Farwell *et al.*, 1975; Kargina *et al.*, 1997; Guena *et al.*, 2000), the chlorine is eliminated step by step from the highly chlorinated benzenes to yield less-chlorinated benzenes and finally transform to benzene. Farwell *et al.* (1975) reported on chlorobenzenes and the main cathodic reaction pathway for hexachlorobenzenes as follows: hexachlorobenzene → pentachlorobenzene → 1,2,3,5-tetrachlorobenzene → 1,2,4-trichlorobenzene → 1,4-dichlorobenzene → monochlorobenzene → benzene. This dechlorination pathway is promoted by the electrochemical reductive dechlorination. Benzene is detected as the final reaction product.

Oxidative dechlorination of chlorobenzenes has been reported using bacteria, Fenton reagent, UV/H_2O_2, and TiO_2-assisted photocatalysis. In the microbial metabolism reported by Reineke and Knackmuss (1984), chlorobenzene was gradually degraded to 3-chloro-cis -1,2-dihydroxycylohexa-3,5-diene, 3-chlorocatechol, 2-chloro-cis,cis-muconate, trans-4-carboxymethylenebuten-4-olide, maleylacetate, and 3-oxoadipate. Fenton reagent effectively

degrades chlorobenzene to chlorohydroxycyclohexadienyl radical in the first step, and the radical dimerizes to produce dichlorobiphenyls, with bimolecular disproportionation to produce chlorophenol and chlorobenzene under non-oxygen conditions or in the absence of other strong oxidants (Reineke W. and Knackmuss H-J, 1984). In the presence of oxygen or other strong oxidants, reactions of the oxidant (O_2) with chlorohydroxycyclohexadienyl radical is predominant and results in lower concentrations of dichlorobiphenyl, which decreases remarkably, and chlorobenzoquinone is formed. Chlorophenol isomers were further oxidized by hydroxyl radical and formed chlorinated and non-chlorinated diols. Chlorobenzenes were also dechlorinated by the UV/H_2O_2 treatment method (Sundstrom *et al.*, 1989), and chlorobenzene was degraded to phenol, biphenyl, chlorobiphenyl isomers, and benzaldehyde with only UV, and chlorobenzenes also transform to chlorophenol and various isomers of chlorobiphenyl and dichlorobiphenyl in the UV/H_2O_2 system.
Degradation of chlorobenzenes in pyrite suspensions is shown in Fig. 6 and in Table 1, with four volatile species: mono-chlorobenzene, 1,2-dichlorobenzene, 1,3-dichlorobenzene, and 1,2,4-trichlorobenzene. The concentrations of these compounds were determined by headspace methods. The experiments were conducted at 25 °C at 200 rpm in the pyrite suspension under anaerobic conditions. The initial concentrations of chlorobenzenes were set to 10 mg/L using the each pure chlorobenzenes dissolved in hexane water. The initial rate of degradation is 1,2,4-trichlorobenzene > 1,2-dichlorobenzene > monochlorobenzene > 1,3-dichlorobenzene, and more than 90 % of these chlorobenzenes were dechlorinated after merely 10 days. The degraded amount of chlorobenzenes is not sensitive to differences in the distribution of their isomers. However, the amount of chloride ion arising from dechlorination of chlorobenzenes is less than the entire chlorine content from each of the chlorobenzenes. This disagreement in mass balance also shows less than one chloride ion degraded from each chlorobenzene. The order of total dechlorination ability is 1,2-dichlorobenzene > monochlorobenzene > 1,2,4-trichlorobenzene > 1,3-dichlorobenzene. Among the 4 species of chlorobenzenes estimated here, 1,3-dichlorobenzene is the hardest to degrade because it has a meta-site chloride configuration. The meta-site dechlorination is not easier than 1,2-(ortho) and 1,4-(para-) dichlorobenzenes.

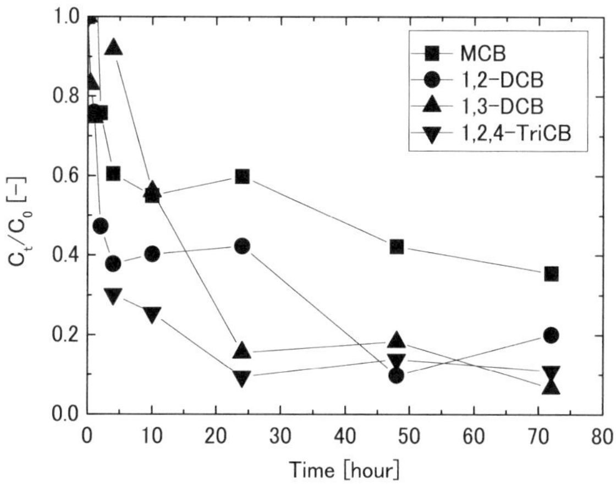

Fig. 6. Degradation of monochlorobenzene, 1,2- and 1,3-dichlorobenzene, and 1,2,4-trichlorobenzene in pyrite suspension under aerobic conditions. C_0 is the initial concentration of chlorobenzenes (10 mg/L), C_t is the concentration of chlorobenzenes at time t.

species	$-\Delta C^{t}_{CBs}/C^{0}_{CBs}$ [mol%]	$+\Delta C^{t}_{Cl}/C^{0}_{CBs}$ [mol%]
MonoCB	88.1	76.5
1,2-DiCB	96.4	88.4
1,3-DiCB	89.1	56.6
1,2,4-TriCB	95.6	68.7

Table 1. The fraction of degraded chlorinated benzenes and chloride ion arising from chlorinated benzenes after 10 days.

Fig. 7 shows the ratio of the residual chlorinated chlorobenzenes relative to the degradation of all species of chlorinated benzenes by pyrite suspension. Table 2 is the fraction of dechlorinated chloride ion from each chlorobenzene and the initial chloride content of the chlorobenzenes. The volatile chlorinated benzenes, mono-, 1,2-di-, 1,3-di- and 1,2,4-trichlorobenzenes were analyzed by the headspace method in addition to solvent extraction analysis to determine their concentrations. The non-degraded volatile chlorobenzenes adsorbed on the pyrite surface are correctly estimated here. The reaction products were also analyzed by GC/MS after eluting in the organic solutions along with the other chlorinated benzenes.

The ratio of complete dechlorination is higher for low molecular weight chlorobenzenes, and 1,2-dichlorobenzene is easier to degrade than 1,3-dichlorobenzene. The greater electron deviation due to chlorine configuration allows the degradation of chlorobenzenes. More than 80 to 90 % of mono-, di-, tri-chlorobenzenes, and 1,2,3,4- and 1,2,3,5-tetrachlorobenzene are degraded from the original concentrations. These compounds are easy to degrade, but there is a small variation depending on chlorine configuration. The residual ratio is 1,2- < 1,3- < 1,4- among the dichlorobenzenes, and 1,2,3- < 1,2,4- < 1,3,5- among the trichlorobenzenes. There is a significant difference in dechlorination ability for the tetrachlorobenzene isomers. The symmetrical chlorine configuration, 1,2,4,5-, on the benzene ring is stable and hard to degrade, so 1,2,4,5-tetrachlorobenzene, tetra- and hexa-chlorobenzenes with the 1,2,4,5-chlorine configuration are extremely difficult to degrade. Considering the mass balance of persistent chlorobenzenes and chloride ions, di-, tri-, and tetra- (limited to 1,2,3,4- and 1,2,3,5-) chlorobenzenes are transformed into chlorinated intermediates because the detected concentration of chloride ion is only one mole of chloride ion per one mole of chlorobenzene and that corresponds to a decline of the initial concentration of each of the chlorobenzenes.

The GC/ MS analysis of the solvent-extracted samples after 10 days indicates the presence of one or more dechlorinated chlorobenzenes from the initial materials for each chlorobenzene, except for hexachlorobenzene. In the experimental dechlorination of hexachlorobenzene, pentachlorobenzene is detected after 10 days. Hexachlorobenzene is assumed to have a slow dechlorination rate so that a little of the compound could be detected. As an example of reaction products, Fig. 8 shows the analytical result of GC/MS and NMR analysis for the solvent extracted from the reaction system of monochlorobenzenes. $C_9H_{14}(OH)Cl$, $C_{13}H_{17}(OH)_3$, $C_{16}H_{23}SO_2$ are detected by MS together with cyclic sulphur molecules such as S_6, S_7, S_8. The other saturated or unsaturated straight-chain hydrocarbons, such as 3-decyne ($C_{10}H_{18}$), 8,10-dodecadienial ($C_{12}H_{20}O$) and 2,2,6-trimethyl-1,4-cyclohexanedione ($C_9H_{14}O_2$) are detected as one of the reaction products of tri- to hexachlorobenzenes. The NMR spectrum also shows the alcohol, ether, and aromatic methyl peaks (-CH_2OH, -O-, -COH, C_6H_6-CH_3) with the benzene and chlorobenzene peaks. One part of the benzene ring structure is assumed to be opening in this reaction process.

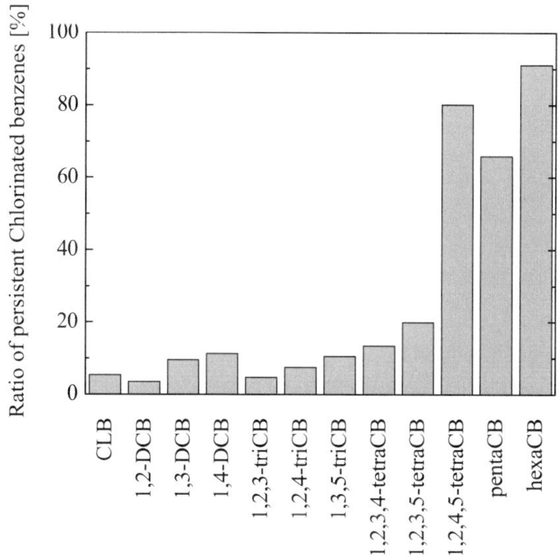

Fig. 7. Residual ratio of 12 species of chlorinated benzenes estimated under aerobic conditions in pyrite suspension after 10 days (Hara *et al.* (2006).

species	$+\Delta C^{t}_{Cl-}/C^{0}_{Cl-}$ [mol%]
monochlorobenzene	89
1,2-dichlorobenzene	47
1,3-dichlorobenzene	45
1,4-dichlorobenzene	45
1,2,3-trichlorobenzene	30
1,2,4-trichlorobenzene	29
1,3,5-trichlorobenzene	30
1,2,3,4-tetrachlorobenzene	20
1,2,3,5-tetrachlorobenzene	18
1,2,4,5-tetrachlorobenzene	11
pentachlorobenzene	7.5
hexachlorobenzene	5.5

Table 2. The fraction of dechlorinated chloride ion from each chlorobenzene and initial chloride content of chlorobenzenes.

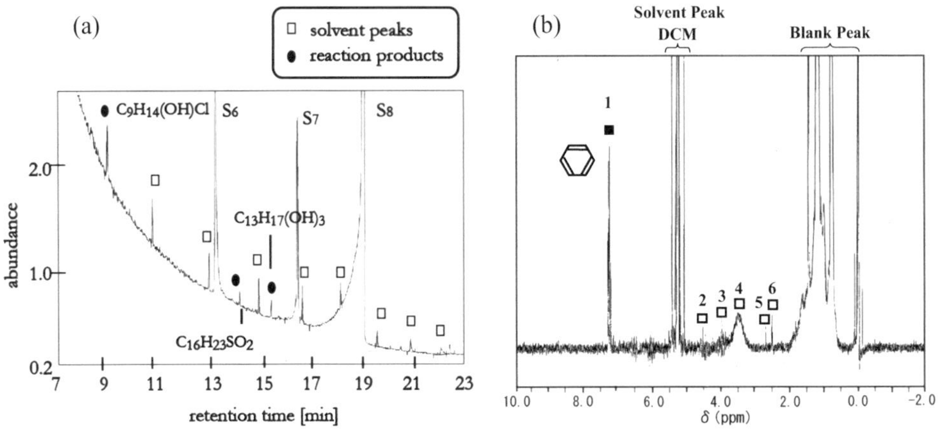

Fig. 8. The spectrum of reaction intermediates arising from transformation of monochlorobenzene in pyrite suspension. (a) MS spectrum. (b) MNR spectrum. In the NMR spectrum, 1. aromatic peak (benzene and monochlorobenzene), 2-4. alcohol and ether peak (-CH_2OH, -O-, -COH), 5. aromatic methyl (methylbenzene).

From the analytical results, the benzene ring could have been opened by the dechlorination process. One part of the benzene ring opened to straight chain unsaturated hydrocarbons and these hydrocarbons polymerized with other straight chain unsaturated hydrocarbons. Some polymerized straight chain hydrocarbons remain, and some hydrocarbons are combined with unsaturated ring structures. A ring-opening reaction of benzene is assumed to be a common reaction for every chlorobenzene under our experimental conditions, not only turning into cyclohexanes. These reactions are initiated when one chlorine site is dechlorinated by pyrite. Therefore, at least one mole of chlorine ions was dissolved in solution for every mole of chlorinated benzene molecules that reacted with pyrite. Furthermore, non-detected chlorine ions may still bind the carbon atoms that are produced by this reaction.

The reactions are proceeded by the oxidant arising on the pyrite interface. The polymerization and ring-opening reactions are similar to the dechlorination of Dieldrin under anaerobic conditions.

4. Discussion

Pyrite suspension shows degradation ability predominantly for TCE, Dieldrin and chlorobenzenes as shown in Section 3. These degradations of chlorinated organic compounds are initiated by the oxidant arising from the pyrite interface reaction.

The surface reaction of pyrite/H_2O and pyrite/O_2 produces the radical as follows (Cohn *et al.,* 2006). In the pyrite/H_2O interface under non-oxygen conditions, H_2O is hydrolysed to •OH and H^+ (Eq. (4) and Eq. (5))

$$Fe^{3+}(pyrite) + H_2O \rightarrow Fe^{2+}(pyrite) + {}^{\bullet}OH + H^+ \quad (4)$$

$$2({}^{\bullet}OH) \rightarrow H_2O_2 \quad (5)$$

Commonly, the radical is apt to arise under aerobic conditions, but the hydroxyl radical can be generated under anaerobic conditions and it acts as an oxidant.
In the presence of oxygen (aerobic conditions), superoxide (O_2•)- was detected as an intermediate from H_2O_2 formation. The superoxide (O_2•)- is generated by the oxidation of iron on the pyrite surface as follows:

$$Fe^{2+}(pyrite) + O_2 \rightarrow Fe^{3+}(pyrite) + (O_2^{\bullet})^- \quad (6)$$

$$Fe^{2+}(pyrite) + (O_2^{\bullet})^- + 2H^+ \rightarrow Fe^{3+}(pyrite) + H_2O_2 \quad (7)$$

The reactions of the pyrite/O_2 interface generate hydroxyl radicals from hydrogen peroxide by the oxidation of surface iron (Eq. (6) and Eq. (7)), and hydrogen peroxide is also used for the reduction of ferric to ferrous ion in the reaction shown in Eq. (8). However, the hydroxyl radicals are not noticeably produced under aerobic condition.

$$Fe^{2+}(aq) + H_2O_2 \rightarrow {}^{\bullet}OH + OH^- + Fe^{3+}(aq) \quad (8)$$

Under aerobic conditions, superoxide and hydrogen peroxide are dominant oxidants used for degradation of chlorinated organic compounds.
Although there is a difference in oxidative ability among oxidants, oxidants arise from the pyrite interface under both aerobic and anaerobic conditions. The hydroxyl radical has an oxidant ability superior to superoxide and hydrogen peroxide, but the total amount of ferrous ion on the pyrite surface is limited under anaerobic conditions; therefore, the total volume of the hydroxyl radical is also limited under anaerobic conditions as opposed to aerobic conditions. The difference between reaction products under aerobic and anaerobic conditions is caused by thus radical character of the pyrite interface.
The oxidation and polymerization of chlorinated organic compounds are caused by several radicals arising from the pyrite/H_2O, pyrite/O_2 interface reaction.

5. Conclusion

This chapter reviews several remediation methods for chlorinated organic pollutants, and emphasizes the degradation ability of natural ferric sulphide (pyrite) for TCE, Dieldrin and chlorobenzenes. The transformation of TCE and Dieldrin under aerobic conditions becomes clear and the main reaction product is formic acid, which continuously degrades to CO_2. Although the 1,2,4,5-chloride configuration in the chlorobenzene and benzene ring structure is hard to degrade, one part of the benzene ring is transformed into several hydrocarbons.
The oxidative reaction of the pyrite suspension has great degradative ability not only for low molecular weight organic pollutants such as VOCs but also for high molecular weight persistent organic compounds with a stable benzene ring and bicyclic rings. Natural ferric sulphide has a remediation ability predominantly for chlorinated organic pollutants.
Iron and ferric sulphides (e.g., pyrite) are widely distributed in the subsurface layer in the global environment. The reductive degradation ability of iron and the oxidative degradation ability of ferric sulphides assist the innovative remediation techniques that are conducted at in situ normal temperature and pressure. The ferric ion and ferric sulphide latent in natural systems have a potential for natural attenuation of contamination.

Pyrite is also widely distributed at mining sites and soils in coastal regions of the Asian and the Pacific areas located in tropical, subtropical and temperate climate areas. For instance, the Mekong Delta in Vietnam is a vast acidic sulphate soil region that includes metallic sulphide. Garvalho *et al.* (2008) reported the agrochemical and polychlorobiphenyl residues in the Mekong River Delta. Several chlorinated compounds, such as DDT, HCH (hexacychlochlorohexane), PCBs and endosulfan were detected in sediment and biota. However, the concentrations of PCBs and pesticide residues in the aquatic environment in the Mekong River Delta are lower than the values reported in other regions of Vietnam and Asia. The aquatic environment of the Mekong Delta is endowed with natural ferric sulphide and water, and consequently the natural environment in this region is assumed to enhance PCB degradation. The acidic sulphate soils distributed in coastal area assumes to actually attenuate the contamination of chlorinated organic compound and prevent the expansion of pollution.

6. Acknowledgments

The author would like to thank Dr. Tsuji (Yamagata Environmental Research Institute) for assistance with the analysis and detection of reaction intermediates.
This work was made possible by a grant-in-aid for scientific research from the Japan Society for the Promotion of Science (B-17760656).

7. References

Ahmed S. and Ollis D.F. (1984) Solar photoassisted catalytic decomposition of the chlorinated hydrocarbons trichloroethylene and trichloromethane. *Solar Energy*, Vol.32(5), 597-601.

Baczynski T.P., Grotenhuis T. and Knipscheer P. (2004) The dechlorination of cyclodiene pesticide by metanogenic granular sludge. *Chemosphere*, Vol.55, 653-659.

Bandala E.R., Gelover S. and Leal M.T. (2002) Solar Photocatalytic degradation of aldrin. *Catalysis Today*, Vol.76, 189-199.

Barbeni M., Minero C. and Pelizzetti E. (1987a) Chemical degradation of chlorophenols with Fenton's Reagent. *Chemosphere*, Vol.16(10-12), 2225-2237.

Barbeni M., Morello M., Pramauro E. and Pilizzetti E. (1987b) Sunlight Photodegradation of 2,4,5-trichlorophenoxy-acetic acid and 2,4,5-trichlorophenol on TiO_2. Identification of itermediates and degradation pathway. *Chemosphere*, Vol.16(6), 1165-1179.

Barbeni M., Pramauro E. and Pilizzetti E. (1986) Photochemical degradation of chlorinated dioxins, biphenyls phenols and benzene on semiconductor dispersion. *Chemosphere*, Vol.15(9), 1913-1916.

Barbusinski K. (2009) Fenton reaction controversy concerning the chemistry. *Ecological chemistry and engineering*, Vol.16(3), 347-258.

Borello R., Minero C., Pramauru E., Pelizzetti E., Serpone N. and Hidaka H. (1989) Photocatalytic degradation of DDT mediated in aqueous semiconductor slurries by simulated sunlight. *Environmental Toxicology and chemistry*, Vol.8, 997-1002.

Brooks G.T. (1980) The preparation of some reductively dechlorinated analogues of deildrin, endosulfan and isobenzan. *Journal of Pesticide Science*, Vol.5, 565-574.

Chiu T., Yen J., Hsieh Y. and Wang Y. (2005) Reductive transformation of dieldrin under anaerobic sediment culture. *Chemosphere*, Vol.60, 1182-1189.

Cohn C.A., Mueller S. Wimmer E., Leifer N., Greenbaum S., Strongin S.R. and Schoonen M.A.A. (2006) Pyrite-induced hydroxyl radical formation and its effect on nucleic acids. *Geochemical Transactions,* Vol.7(3), 1-11.

Dibble L.A. and Raupp, G. B. (1990) Kinetic of the gas-solid heterogeneous photocatalytic oxidation of trichloroethylene by near UV illuminated titanium dioxide. *Catalysis Letters,* Vol.4, 345-354.

Dibble L.A. and Raupp, G. B. (1992) Fluidized-Bed photocatalytic oxidation of trichloroethylene in contaminated airstreams. *Environmental Science and Technology,* Vol.26, 492-495.

Dureja P., Walia S. and Mukerjee S. K. (1987) Superoxide mediated dehydrohalogenation of photodieldrin and photoaldrin. *Tetrahedron letters,* Vol.28(8), 895-896.

D'Ollvelra J.-C., Al-Sayyed G. and Plchat P. (1990) Photodegradation of 2- and 3-chlorophenol in TiO_2 Aqueous Suspensions. *Environmental Science and Technology,* Vol.24, 990-996.

Farwell O., Beland F. A., and Geer R. D. (1975) Reduction pathways of organohologen compounds. Part 1. Chlorinated benzenes. *Electroanalytical Chemistry and Interfacial Electrochemistry,* Vol. 61, 303-313.

Froelich E.M. (1992) Chemical Oxidation. *Technomic Publishing Company, Inc.* Eckenfelder W. W., Nowers A.R. and Roth J. A. (eds.) Lancaster, PA, 104-113.

Garvallho F.P., Villeneuve J.P, Cattini C., Tolosa I., Dao Dinh Thuan and Dang Duc Nhan (2008) Agrochemical and polychlorobyphenyl (PCB) residues in the Mekong River delta, Vietnam. *Marine Pllution Bullutin,* Vol.56, 1476-1485.

Gerritse J., Drzyzga O., Kloetstra G., Keijmel M., Wiersum L.P., Hutson R., Collins M.D. and Gottschal J.C. (1999) Influence of different electron donors and acceptors on dehalorespiration of tetrachloroethene by Desulfitobacterium frappieri TCE1. *Applied and Environmental Microbiology,* Vol.65(12), 5212-5221.

Gerritse J., Renard V., Pedro Gomes T.M., Lawson P.A., Collins M.D. and Gottschal J.C. (1996) Desulfitobacterium sp. Strain PCE1, an anaerobic bacterium that can grow by reductive dechlorination of tetrachloroethene or ortho-chlorinated phenols. *Archives of Microbiology,* Vol.165(2), 132-140.

Guena T., Wang L., Gattrell M. and MacDougall (2000) Mediated Approach for the electrochemical reduction of chlorobenzens in Nonaqueous media. *Journal of the Electrochemical Society,* Vol. 147, 248-255.

Guevremont J.M., Elsetinow A.R., Strongin D.R., Bebie J.M. and Schoonen M.A.A. (1998b) Structure Sensitivity of pyrite oxidation: Comparison of the (100) and (111) planes. *American Mineralogist,* Vol.83, 1353-1356.

Guevremont J.M., Strongin D.R. and Schoonen M.A.A. (1998c) Photoemission of adsorbed Xenon, X-ray photoemission spectroscopy, and temperature-programmed desorption studies of H_2O on FeS_2 (100). *Langmuir* 14: 1361-1366.

Guevremont J.M., Strongin D.R. and Schoonen M.A.A. (1998d) Thermal chemistry of H_2S and H_2O on the (100) plane of pyrite: Unique reactivity of defect sites. *American mineralogist,* Vol.83, 1246-1255.

Guevremont J.M., Bebie J., Elsetinow A.R., Strongin D.R. and Schoonen M.A.A. (1998a) Reactivity of the (100) plane of pyrite in oxidizing gaseous and aqueous environments: Effect of Surface imperfections. *Environmental Science and Technology,* Vol.32, 3743-3748.

Guevremont J.M., Strongin D.R. and Schoonen M.A.A. (1997) Effects of surface imperfections on the binding of CH_3OH and H_2O on FeS_2 (100): Using adsorbed Xe as a probe of mineral surface structure. *Surface Science,* Vol.391, 109-124.

Hara J. (2001) The effect of oxygen on chemical dechlorination of dieldrin using iron sulfides. *Chemosphere,* Vol.82, 1308-1313.

Hara J., Kawabe Y., Komai T. and Inoue C. (2009) Chemical degradation of dieldrin using ferric sulfide and iron powder. *International Journal of Environmental Science and Engineering,* Vol.1-2, 91-96.

Hara J., Inoue C., Chida T., Kawabe Y., Komai T. (2006) Dehalogenation of Chlorinated Benzenes by iron sulfide. *International Journal of Power and Energy Systems,* Vol.1(1), 239-243.

Hoa T. P., Kitsuneduka M., Hara J., Suto K., Inoue C. (2008) Trichloroethylene Transformation by Natural Mineral Pyrite: The deciding role of oxygen. *Environmental Science and Technology,* Vol.42. 7470-7475.

Hoa T.P., Suto K. and Inoue C. (2009) Trichloroethylene transformation in Aerobic pyrite suspension: pathways and Kinetic modeling. *Environmental Science and Technology,* Vol.43, 6744-6749.

Holliger C., Schraa G., Stams A.J.M. and Zehnder A.J.B. (1993) A highly purified enrichment culture couples the reductive dechlorination of tetrachloroethene to growth. *Applied and Environmental Microbiology,* Vol.59 (9), 2991-2997.

Kargina O., MacDougall B., Kargin Y. M. and Wang L. (1997) Dechlorination of monochlorobenzene using organic mediators. *Journal of the Electrochemical Society,* Vol.144, 3715-3721.

Kendelewiz T., Dolyle C.S., Bostick B.C. and Brown G.E. (2004) Initial oxidation of fractured surface of pyrite (FeS_2(100)) by molecular oxygen, water vapor, and air. *Surface Science,* Vol.558, 80-88.

Kim S., Kim D., Pollack G. M., Collins L. B. and Rusyn I. (2009) Pharmacokinetic analysis of trichloroethylene metabolism in male B6C3F1 mice: Formation and disposition of trichloroacetic acid, dichloroacetic acid, S-(1,2-dichlorovinyl) glutathione and S-(1,2-dichlorovinyl)-L-Cysteine. *Toxicology and applied pharmacology,* Vol.238, 90-99.

Kondo M.M. and Jardim W.F. (1991) Photodegradation of chloroform and urea using Ag-loaded titanium dioxide as catalyst. *Water Research,* Vol.25(7), 823-827.

Kriegman-King M.R. and Reinhard M. (1994) Degradation of carbon tetrachloride by pyrite in aqueous solution. *Environmental Science and Technology,* Vol.28, 692-700.

Krumholz L.R., Sharp R. and Fishbain S.S. (1996) A freshwater anaerobe coupling acetate oxidation to tetrachloroethylene dehalogenation. *Applied and Environmental Microbiology,* Vol.62(11), 4108-4113.

Kusveran E. and Erbatur O. (2004) Degradation of aldrin in adsorbed system using advanced oxidation processes: comparison of the treatment methods. *Journal of Hazardous Materials,* 106B, 115-125.

Lee W. and Batchelor B. (2002) Abiotic reductive dechlorination of chlorinated ethyelens by iron-bearing soil minerals. 1. pyrite and magnetite. *Environmental Science and Technology,* Vol.36, 5147-5154.

Lee W. and Batchelor B. (2003) Reductive capacity of natural reductants. *Environmental Science and Technology,* Vol.37, 535-541.

Leung S.W., Watts R. J. and Miller G.C. (1992) Degradation of Perchloroethylene by Fenton's Reagent: Speciation and Pathway. *Journal of Environmental Quality*, Vol.21(3), 377-381.

Li S. and Wackett L.P. (1992) Trichloroethylene oxidation by toluene dioxygenase. *Biochemical and Biophysical research communications*, Vol.185(1), 443-451.

Löffler F.E., Sanford R.A. and Tiedje J.M. (1996) Initial characterization of a reductive dehalogenase from Desulfitobacterium chlororespirans Co23. *Applied and Environmental Microbiology*, Vol.62(10), 3809–3813.

Maule A., Plyte S. and Quirk A.V. (1987) Dehalogenation of organochlorine insecticides by mixed anaerobic microbial populations. *Pesticide Biochemistry and Physiology*, Vol.27, 229-236.

Maymó-Gatell X., Anguish T. and Zinder S.H. (1999) Reductive dechlorination of chlorinated ethenes and 1,2-dichloroethane by Dehalococcoides ethenogenes 195. *Applied and Environmental Microbiology*, Vol.65(7), 3108–3113.

Miyoshi K., Kameyama Y. and Matsumura M. (2004) Electrochemical reduction of organohalogen compound by noble metal sintered electrode. *Chemosphere*, Vol.56, 187-193.

Mochizuki S. (1977) Photochemical dechlorination of PCBs. *Chemical Engineering Science*, Vol.32, 1205-1210.

Mohammad S. M. and Dennis G. P. (1997) Electrochemical reduction of di-, tri- and tetrahalobenzenes at carbon cathodes in dimethylformamide. Evidence for a halogen dance during the electrolysis of 1,2,4,5-tetrabromobenzene. *Journal of Electroanalytical Chemistry*, Vol.435, 47-53.

Motosugi K., Esaki N. and Soda K. (1982) Purification and properties of a new enzyme, DL-2-haloacid dehalogenase, from Psurdomonas sp. *Journal of Bacteriology*, Vol.150, No.2, 522-527.

Moza P. N., Fytianos K., Samanidou V. and Korte F. (1988) Photodecomposition of chlrophenols in aqueous medium in presence of hydrogen peroxide. *Bulletin of Environmental Contamination and toxicology*, Vol.41, 678-682.

Nesbitt H.W. and Muir I.J. (1998) Oxidation states and speciation of secondary products on pyrite and arsenopyrite reacted with mine waste waters and air. *Mineralogy and petrology*, Vol.62, 123-144.

Newman L.M. and Wackett L.P. (1991) Fate of 2,2,2-trichloroacetaldehyde (chloral hydrate) produced during trichloroethylene oxidation by methanotrophs. *Applied Environmental Microbiology*, vol.57 (8), 2399-2402.

Newman L.M. and Wackett L.P. (1995) Purification and characterization of toluene 2-monooxygenase from Burkhokderia cepacia G4. *Biochemistry*, Vol.34(43), 14066-14076.

Newman L.M. and Wackett L.P. (1997) Journal of Bacteriology, No.179(1), 90-96.

Oldenhuis R., Vink R.L., Janssen D.B., Witholt B. (1989) Degradation of chlorinated aliphatic hydrocarbons by methylosius trichosporium OB3b expressing soluble methane monooxygenase. *Applied Environmental Microbiology*, Vol.55(11), 2819-2826.

Reineke W. and Knackmuss H.J. (1984) Microbial Metabolism of Haloaromatics: Isolation and properties of Chlorobenzene-Degrading Bacterium. *Applied and Environmental Microbiology*, Vol.47(2), 395-402.

Rosenzweig A.C., Frederick C.A., Lippard S.J., Nordlund P. (1993) Crystal structure of bacterial non-haem iron hydroxylase that catalyses the biological oxidation of methane. *Nature*, Vol.366(6455), 537-43

Rosso K.M., Becker U. and Hochella M. F. J. (1999) The interaction of pyrite{100} surface with O_2 and H_2O: Fundamental oxidation mechanisms. *American Mineralogist*, Vol.84, 1549-1561.

Schaufuss A.G., Nesbitt H.W., Kartio I., Laajalehto K., Bancroft G.M. and Szargan R. (2000) Reactivity of surface chemical states on fractured pyrite. *Surface Science*, Vol.411, 321-328.

Shimakoshi H., Tokunaga M. Baba T. and Hisaeda Y. (2004) Photochemical dechlorination of DDT catalyzed by a hydrophobic vitamin B12 and a photosensitizer under irradiation with visible light. *Chemical Communications*, 1806-1807.

Sundstrom D.W., Klei H.E., Nalette T.A., Reidy D. J., Weir B. A. (1986) Destruction of Halogenated Aliphatics by Ultraviolet Catalyzed Oxidation with Hydrogen Peroxide. *Hazardous Waste and Hazardous Materials*, 3(1), 101-110.

Sundstrom D.W., Weir B.A. and Klei H.E. (1989) Destruction of aromatic pollutions by UV light catalyzed oxidation with hydrogen peroxide. *Environmental Progress*, Vol.8 (1), 6-11.

Taylor B. E., Wheeler M.C. and Nordstorm D.K. (1984) Stable isotope geochemistry of acid mine drainage: Experimental oxidation of pyrite. *Geochimica et Cosmochimica Acta*, Vol.48, 2669-2678.

Tiedje, J. M., Quensen III J. F., Chee-Sanford, J., Schimel J. P. and Boud S. A. (1993) Microbial reductive dechlorination of PCBs. *Biodegradation*, Vol.4, 231-240.

Usher C.R., Cleveland C.A., Strongin D.R. and Schoonen M.A. (2004) Origin of oxygen in sulfate during pyrite oxidation with water and dissolved oxygen: An in situ horizontal attenuated total reflectance infrared spectroscopy isotope study. *Environmental Science and Technology*, Vol.38, 5604-5606.

Usher C.R., Paul K.W., Narayansamy J., Kubicki J.D., Sparks D.L., Schoonen M.A.A. and Strongin D.R. (2005) Mechanistic aspects of pyrite oxidation in an oxidizing gaseous environment: An in situ HATR-IR isotope study. *Environmental Science and Technology*, Vol.39, 7576-7584.

Van Beek H.C.A., Van der Stoep H.J., Van Oort H. and Van Leene J. (1982) Photochemical Radical Chain Dechlorination of DDT in 2-Propanol. *Industrial & Engineering Chemistry Product Research and Development*, Vol.12, 123-125.

Weerasooriya R. and Dharmasena B. (2001) Pyrite-assisted degradation of trichloroethene (TCE). *Chemosphere*, Vol.42, 389-396.

Weie B.A., Sundstrom D.W. and Klei H.E. (1987) Destruction of Benzene by Ultraviolet Light-Catalyzed oxidation with Hydrogen Peroxide. *Hazardous Waste and Hazardous Materials*, Vol.4(2), 165-176.

Weightman A.J., Weightman A.L., Slater J.H. (1985) Toxic effects of chlorinated and brominated akanoic acids on Pseudomonas putida PP3: selection at high frequencies of mutations in genes encoding dehalogenases. *Applied Environmental microbiology*, Vol.49(6), 1494-1501.

Weightman A.L., Weightman A.J. and Slater J.H. (1992) Microbial dehalogenation of trichloroacetic acid. *World journal of microbiology and biotechnology*, Vol.8, 512-518.

Zinovyev S.S., Shinkova N.A., Perosa A. and Tundo P. (2005) Liquid phase hydrodechlorination of dieldrin and DDT over Pd/C and Raney-Ni. *Applied Catalysis B: Environemtal*, Vol.55, 39-8.

15

Electrochemical Incineration of Organic Pollutants for Wastewater Treatment: Past, Present and Prospect

Songsak Klamklang[1], Hugues Vergnes[2],
Kejvalee Pruksathorn[3] and Somsak Damronglerd[3]
[1]*Technology Center, SCG Chemicals Co., Ltd., Siam Cement Group (SCG), Bangkok*
[2]*Laboratoire de Génie Chimique, UMR CNRS 5503, BP 84234, INP-ENSIACET*
[3]*Department of Chemical Technology, Faculty of Science,*
Chulalongkorn University Bangkok
[1,3]*Thailand*
[2]*France*

1. Introduction

Water is a combination of two parts, hydrogen and oxygen as H_2O. However, pure water is only produced in a laboratory, water in general is not pure composition of hydrogen and oxygen. Eventhough, distilled water still has measurable quantities of various substances such as ions, mineral or organic compounds (http://www.environmental-center.com). These substances should be considered as the impurities that dissolved into water during flow through hydraulic pathway. Nowadays, there are some increasing on both population and consumption of natural resources to serve endless needs. Water is most important resource and becomes limited of use due to contamination from discharge of both domestic and industry. The discharge of domestic wastewater contains a large amount of organic pollutants. Industry also contributes substantial amounts of organic pollutants. However, some organic substrates discharged from industry contain a high toxicity and refractory organic pollutants.

Figure 1 presents the example of a partially closed water cycle. In the cycle the organic pollutants are neither removed by sorption nor biodegradation. Nevertheless, there are some organic pollutants pass all barriers such as wastewater treatment or underground passage and appear in raw waters used for drinking water production. The other group of organic pollutants may originate from consumer products used in household, pesticides applied in agriculture or chemicals used in industry (Reemtsma & Jekel,2006).

Wastewater treatment consists of applying known technology to improve or upgrade the quality of a wastewater. Usually wastewater treatment will involve collecting the wastewater in a central, segregated location and subjecting the wastewater to various treatment processes (Hanze et al., 1995). Wastewater treatment can be organized or categorized by the nature of the treatment process operation being used such as physical, chemical or biological treatment. Biological treatment of polluted water is the most economical process and commonly used for the elimination of degradable organic pollutants present in wastewater. However, situation fully differs when the wastewater contains toxic and refractory substrates, biological treatment may not suitable tool for treatment of contaminated wastewater (Grimmet al., 1998).

This chapter review related technologies and case studies of application of electrochemical incineration in industrial and restaurant wastewater treatment.

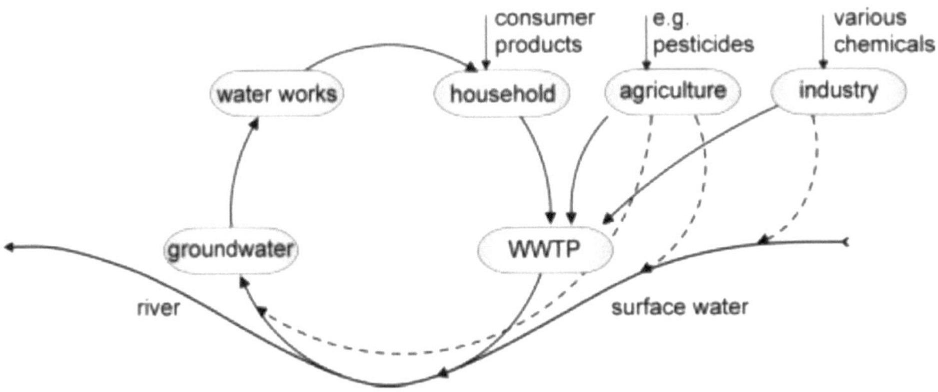

Fig. 1. Schematic of water cycle (Reemtsma & Jekel,2006).

2. Electrochemical incineration

Electrochemistry is a clean, versatile and powerful tool for the destruction of organic pollutants in wastewater. Electrochemical oxidation of organic compounds in aqueous solution is an anodic process occurring in the potential region of water discharge to produce oxygen (Kapalka et al., 2009).

Two different pathways are described in the literatures for the anode oxidation of undesired organic pollutants (Grimm et al., 1998). Electrochemical conversion transforms only the toxic pollutants refractory to biological treatments into biocompatible organics, so that biological treatment is still required after the electrochemical oxidation (Comninellis., 1994). The ideal electrode material which can be used in the electrochemical conversion method must have high electrochemical activity for aromatic ring opening and low electrochemical activity for further oxidation of the aliphatic carboxylic acids which are in general biocompatible (Comninellis., 1994).

Electrochemical incineration or combustion is method completely oxidizes the organic pollutants to CO_2 by physisorbed hydroxyl radicals. In this case, the electrode material must have high electrocatalytic activity towards the electrochemical oxidation of organics to CO_2 and H_2O (Comninellis., 1994).

2.1 Mechanism of electrochemical incineration

Study on electrochemical oxidation for wastewater treatment goes back to the 19th century, when electrochemical decomposition of cyanide was investigated (Kuhn., 1971). Extensive investigation of this technology commenced since the late 1970s (Chen., 2004). During the last two decades, research works have been focused on the efficiency in oxidizing various pollutants on different electrodes, improvement of the electrocatalytic activity and electrochemical stability of electrode materials, investigation of factors affecting the process performance, and exploration of the mechanisms and kinetics of pollutant degradation.

Experimental investigations focus mostly on the behaviors of anodic materials, the effect of cathodic materials was not investigated extensively although Azzam et al. (Azzam et al, 1999) have found a considerable influence of the counter electrode material in the anodic destruction of 4-chlorophenol.

Comninellis (Comninellis., 1994) has presented the mechanism of electrochemical oxidation and it was used as fundamental of electrochemical wastewater treatment. According to the electrochemical conversion and combustion of organics is presented on metal oxide anode (MO_x). H_2O in acid or OH^- in alkali solution is discharged at the anode to produce adsorbed hydroxyl radical according to the equation (1).

$$MO_X + H_2O \xrightarrow{k_1} MO_X(^{\bullet}OH) + H^+ + e^- \tag{1}$$

k_1 = Electrochemical rate constant for H_2O discharge

Secondly, the adsorbed hydroxyl radicals may interact with the oxygen already present in the metal oxide anode with possible transition of oxygen from the adsorbed hydroxyl radical to the lattice of the metal oxide anode forming the so-called higher oxide (MO_{X+1}) as represented in equation (2).

$$MO_X(^{\bullet}OH) \xrightarrow{k_2} MO_{X+1} + H^+ + e^- \tag{2}$$

k_2 = Electrochemical rate constant for transition of oxygen into oxide lattice

Thus, It could be considered that at the anode surface, two states of active oxygen can be presented in physisorbed active oxygen (adsorbed hydroxyl radicals, $^{\bullet}OH$) and chemisorbed active oxygen (oxygen in the oxide lattice, MO_{x+1}).

In the absence of any oxidizable organics, the physisorbed and chemisorbed active oxygen produce dioxygen according to the equation (3) and (4).

$$MO_X(^{\bullet}OH) \xrightarrow{k_o} \frac{1}{2}O_2 + H^+ + e^- + MO_X \tag{3}$$

k_0 = Electrochemical rate constant for O_2 evolution

$$MO_{X+1} \xrightarrow{k_d} MO_X + \frac{1}{2}O_2 \tag{4}$$

k_d = Electrochemical rate constant for O_2 evolution

In the presence of oxidizable organics, it is speculated that the physisorbed active oxygen ($^{\bullet}OH$) should cause predominantly the complete combustion of organics according to equation (5), and chemisorbed active oxygen (MO_{x+1}) participate in the formation of selective oxidation products as represented in equation (6).

Complete combustion:

$$R + MO_X(^{\bullet}OH) \xrightarrow{k_c} CO_2 + zH^+ + ze^- + MO_X \tag{5}$$

k_c = Electrochemical rate constant for the combustion of organics

Selective oxidation:

$$R + MO_{X+1} \xrightarrow{k_S} RO + MO_X \tag{6}$$

k_s = Electrochemical rate constant for the selective oxidation of organics

2.2 Electrochemical incineration performance

There are several researchers' work on the parameters that affect on the electrochemical oxidation process efficiency. In this topic, some important parameters for electrochemical treatment by electrochemical incineration have been reviewed.

Support materials

Vercesi (Vercesi et al, 1991) worked on searching for a good Dimensionally Stable Electrode (DSA) that has long service life for O_2-evolution based on the effects of support materials. Titanium, tantalum, zirconium, niobium and some of their alloys were used as support materials on the performance of IrO_2-Ta_2O_5 coated electrodes. The thermal behavior and oxygen affinity sequence of the metals, in relation to the electrode preparation procedure, were determined thermogravimetrically. The electrochemical corrosion of base metals was represented in Figure 2, the tantalum-based materials presented the highest stability with the minimum corrosion rate. The chemical and electrochemical stability of the base metals was found to be directly related to the service life of the electrode, measured in 30% H_2SO_4 at 80°C and 750 mA/cm^2. Tantalum-based electrodes represented the highest service life, 1700 h and 120 h for titanium-based electrodes.

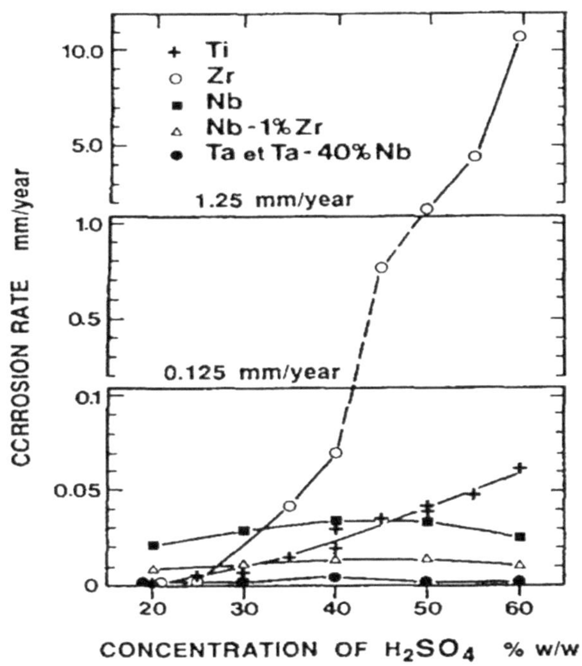

Fig. 2. Electrochemical corrosion rate of base metals as a function of H_2SO_4 concentration at anode potential of 2 V/SCE (Vercesi et al, 1991).

Coating materials

The active coating materials of electrodes are very important for pollutants degradation in electrochemical oxidation process. It could point the project to be benefit or insolvent. The coating enables the electrical charge transport between the base metal and the electrode/electrolyte interface. It must have high chemical and electrochemical stability and able to catalyse the desired electrochemical reaction.

There are some works focusing on the active coating materials for electrochemical degradation of organic pollutants in wastewater. Comninellis (Comninellis & Vercesil, 1991) found that the Ta_2O_5-doped IrO_2 represented the highest service life. However, the same author (Comninellis., 1994) proposed in 1994 that the mechanism of electrochemical oxidation of organic pollutant on IrO_2 electrode was the electrochemical selective oxidation while that on SnO_2 electrode was the electrochemical combustion. On SnO_2 electrode, the organic pollutants were oxidized to CO_2 and water. The SnO_2 electrode presents the highest current efficiency. Because of its highest oxidation state which contains excess oxygen in the oxide lattice as possible. At SnO_2 surface, the hydroxyl radicals are accumulated and favor the combustion of organics while the other oxides having a high concentration of oxygen vacancies and can favor selective oxidation of organics more than combustion.

Not only SnO_2 electrode has the good attractive to be used as electrode for wastewater treatment in electrochemical oxidation, the diamond thin films have been widely used in electrochemical studies due to the unique properties like chemical stability, large potential range and mechanical resistance. Their applications embrace the electroanalysis, electrosynthesis, fuel cell and the organic pollutant degradation in wastewater (Diniz et al, 2003). Some studies showed that conducting diamond electrodes could be grown by energy-assisted (plasma or hot-filament) chemical vapor deposition on several substrates, such as silicon, titanium, niobium, tantalum, molybdenum, glassy carbon. Although all these substrates are currently used, they still have some drawbacks. In fact, a silicon substrate is very brittle, Nb, Ta, W are too expensive and the stability of the diamond layer deposited on the Ti substrate is still not satisfactory, because cracks may appear and cause the detachment of the diamond film during long-term electrolysis (Panizza & Cerisola, 2005). The organic compound electrochemical oxidation efficiency strongly depends on the used anode material and diamond is very interesting due to its superior properties (Diniz et al, 2003). In recent years, there are some publications on the application of diamond electrode for wastewater treatment (Diniz et al, 2003, Lissens., 2003). However, the application of diamond electrode for organic pollutant degradation still limited due to the cost of diamond electrode is too expensive.

pH

Normally, the efficiency of the oxidation of organics tends to be superior in alkaline solution. That also holds for the anodic treatment using standard electrode materials. The wastewater accessible to treatment process has any pH and pH adjustment before treatment to the more favorable value above 7 will be too expensive. Moreover, Stuki (Stucki et al, 1991) proved that on SnO_2 electrodes, the degradation of benzoic acid was pH independent. Cañizares (Cañizares et al, 2004) proposed that the pH does not influence in the global oxidation rate even if initial oxidation rate was higher in alkaline media. However, after the galvanostatic method development, the oxidation rate in acidic media surpasses those in the alkaline media because the accumulation of oxalic acid in alkaline media was higher than that in acidic media, due to its lower oxidazability at alkaline conditions.

Current density

The effect of current density was studied many times, Comninellis (Comninellis & Nerini, 1995) found that the degradation of phenol was independent from the current density and that the phenol elimination depends only on the specific electrical charge. Almost complete phenol elimination can be archived after the passage of an electrical charge of 17-20 Ah/dm^3.

The higher current densities increase the initial reaction rate (dCOD/dt) (COD : Chemical Oxygen Demand) but decrease lightly the initial efficiencies of the process (dCOD/dQ) (Q : charge). Nevertheless, both treatments end nearly at the same charge passed. This behavior is a characteristic of electrochemical systems in which both direct and mediated oxidation reactions play an important role. But the low current density experiment achieves initially a higher mineralization rate. This fact can be easily explained that the amount of oxalic acid, that is an end of chain product and difficult to destroy, accumulated in the high current density experiment is greater than those obtained in the low current density experiments (Cañizares et al, 2004).

Although, the current density might not affect on the kinetic of surface electrochemical oxidation of organic pollutant, it also enhances the production of bulk chemicals, which may contribute to the degradation process through parallel reaction schemes.

Klamklang (Klamklang, 2006, Klamklang, 2007 and Klamklang et al, 2010) has found that increasing current density from 5 to 10 mA/cm^2, leads to less degradation rate of oxalic acid by electrochemical oxidation. This behavior is characteristic of mass transfer-controlled processes (Kesselman, 1997). In such systems, the increase of current density cannot increase the organic removal efficiency at the electrode, but only favors oxygen evolution as the anodic side reaction which hides the electrode and prevents contact between hydroxyl radicals and organic pollutants. When the system does not generate only adsorbed hydroxyl radicals or other active oxygen, the decrease of organic pollutant removal efficiency is observed.

Temperature

An increase in the temperature leads to more efficient processes by global oxidation. While direct oxidation processes remain almost unaffected by temperature, this fact may be explained in terms of the presence of inorganic electrogenerated reagents. The oxidation carried out by these redox reagents is a chemical reaction. Consequently, its rate normally increases with temperature. But the oxidation process can be carried out either at the electrode surface and by electrogenerated reagents-mainly hypochlorite and peroxodisulphates. However, the new organic intermediates are not formed with the increase of temperature, indicating that the process mechanisms do not vary with temperature (Cañizares et al, 2004).

2.3 Electrocatalytic electrodes

Electrochemical degradation of organic pollutants in the wastewater needs specific electrodes. Couper (Couper et al, 1990) has reviewed some properties of typical substrates used as electrocatalytic electrodes and report that electrode must have low voltage drop through the substrates and substrate-solvent interface. Many metallic electrodes could answer these criteria and many alloys can also be used as good composite electrodes covered with an active layer and a long service life of electrodes.

The complexity of electrode behaviors and our lack of detailed insights make it impossible to select the optimum electrode for a given process on a theoretical basis. Instead, an empirical approach must be used. The initial selection is based on process experience, and this is then tested and refined during an extensive development program. Indeed, it is very difficult to predict the success of an electrode material or to define its lifetime without extended studies under realistic process conditions. Accelerated testing is rarely satisfactory except to indicate catastrophic failure.

There are some general guidelines to assist the choice of an electrode material;

Physical stability

The electrode material must have adequate mechanical strength, must not be prone to erosion by the electrolyte, reactants, or products, and must be resistant to cracking.

Chemical stability

The electrode material must be resistant to corrosion, unwanted oxide or hydride formation, and the deposition of inhibiting organic films under all conditions (e.g., potential and temperature) experienced by the electrode.

Suitable physical form

It must be possible to fabricate the material into the form demanded by the reactor design, to facilitate sound electrical connections, and to permit easy installation and replacement at a variety of scales. The shape and design of the electrode may take into account the separation of products, including the disengagement of gases or solids.

Rate and product selectivity

The electrode material must support the desired reaction and, in some cases, significant electrocatalytic properties are essential. The electrode material must promote the desired chemical change while inhibiting all competing chemical changes.

Electrical conductivity

This must be reasonably high throughout the electrode system including the current feeder, electrode connections, and the entire electrode surface exposed to the electrolyte. Only in this fashion it is possible to obtain a uniform current and potential distribution as well as to avoid voltage losses leading to energy inefficiencies.

Cost and lifetime

A reasonable and reproducible performance including a lifetime probably extending over several years must be achieved for an acceptable initial investment.

It is important to note that the choice of working and counter electrodes cannot be made independently since the chemistry at each has consequences to the solution composition throughout the cell. Indeed, the selection of electrode material and its form must be an integrated decision within the prospective of the cell and process design. In some cases such as the manufacture of pharmaceutical products, the electrodes and their compounds must have a low toxicity.

This section has reviewed the ways in which the choice of electrode material influences the design of electrochemical reactors and process performance.

Energy Consumption

The specific energy consumption should be minimized in order to minimize the power costs. In general, the total power requirement has contributions for both electrolysis and movement of either the solution or the electrode. The design of electrodes and cell has an important role in reducing each of these components. Thus, a very open flow-through porous electrode will have a low pressure drop associated with it, giving rise to modest pumping costs and facilitating reactor sealing. A high surface area electrode which itself a turbulence promoter in bed electrode, will give rise to a moderately high mass transfer coefficient and active area without the need for high flow rates through the cell; the pumping cost will again be moderately low.

The direct electrolytic power could be minimized by

- Obtaining a current efficiency approaching 1.0
- Minimizing the cell voltage.

It is therefore important to select the electrode material and operating conditions so as to maintain a high current efficiency. This also assists the operation of the process by reducing the amount of product purification that is necessary and/or byproducts that must be handled.

The cell voltage is a function of the reversible cell voltage, the over-potentials at the two electrodes, and ohmic drops in the electrolyte, the electrodes, busbars etc., and any separator in the cell. Again, the maintenance of a low cell voltage demands attention to the design of both electrodes and cell. Where possible, the following features should be included:

- The counter electrode reaction should be chosen so as to minimize the reversible cell voltage. This requires the availability of a suitably stable electrode material.
- The over-potentials at both electrodes should be minimized by the use of electrocatalysts.
- The electrodes, current feeders, and connectors should be made from highly conducting materials to lower ohmic drops.
- The electrodes should facilitate low *IR* drop in the electrolyte by, for example, allowing efficient gas disengagement and passage out of the interelectrode gap. Meshes as well as louvred and lantern blade electrodes can be used.
- Electrode and cell design should allow a small interelectrode or electrode membrane gap. In the limit, the electrode may touch the membrane as in zero-gap or solid polymer electrolyte cells.
- A separator should be avoided by suitable selection of the counter electrode chemistry or, if essential, a thin conductive membrane should be used.

Current efficiency

Current efficiency is the fraction of the total charge passed that is used in the formation of the desired product. This can be a strong function of electrode material, e.g., because of differences in the rate of hydrogen evolution as a competing reaction. Competing reactions can also lead to the corrosion and/or erosion of the electrode material as well changes to the electrode (e.g., by formation of a hydroxide or oxide or the deposition of another metal onto the surface).

Material Yield

This is the fraction of the starting material that is converted into the desired product. This is also dependent on electrode material in many cases. Values less than one indicate byproducts and hence perhaps the need to introduce additional purification steps that inevitably increases the complexity of the overall process and costs.

Space-Time Yield

One of the most valuable statements of reactor performance is the space-time yield or weight of product per unit time per unit volume of reactor. It is determined by the current density, the current efficiency, and the area of electrode per unit volume of cell, all dependent on the electrode material and its form. Commonly, the cell is operated in conditions where the electrode reaction is mass-transport controlled (especially when a high fractional conversion is desirable or when the concentration of reactant is limited by solubility or process considerations. Then, the current density is determined by the concentration of reactant and the mass transport condition. The latter is therefore frequently enhanced by the use of high flow rates, turbulence promoters, and/or electrode movement.

The current is proportional to the active electrode area in the cell. A compact cell design requires a high area per unit cell volume. This suggests the use of a three-dimensional electrode but such electrodes make it difficult to maintain a uniform fluid flow and electrode potential, i.e., to control the reaction environment. Hence, the use of porous, flow-through electrodes often involves a trade-off between enhanced electrode area and material yield and/or current efficiency.

Other Factors

Of course, other factors are important in the design of electrodes and cells. These include cost, safety, ease of maintenance, and convenience to use. It is also essential that the performance of the electrodes is maintained throughout the projected operating life of the cell, maybe several years. Examples of problems that frequently arise include (a) deposition onto cathode surfaces of hydrogen evolution catalysts due to trace transitional metal ions in the electrolyte and (b) poisoning of PbO_2 anodes by organic molecules leading to enhanced corrosion as well as oxygen evolution.

3. Application of electrochemical incineration in wastewater treatment

3.1 Electrochemical incineration for restaurant wastewater treatment

There are a lot of restaurants, cafeterias and food centers in the big cities, which everyday make large amounts of wastewater. Generally, there is unavailable of on-site treatment for each restaurant. The direct discharge of wastewater from these restaurants and food shops to the drainage system is a huge problem to the municipal wastewater collection and treatment works. The oil and grease contained in the wastewater aggregate and foul the sewer system and generate an unpleasant odor (Chen et al, 2000).

Basically, restaurant wastewater treatment facilities must be highly efficient in removing oil and grease, cause no food contamination and be compact size. Low capital and operating costs are important because profit margins of most restaurants are small. In addition, the technology has to be simple so that it can be operated easily either by a chef or a waiter (Chen et al, 2000).

Conventional biological processes are therefore ruled out due to the requirement of large space, long residence time and skilled technicians. Chemical coagulation/settlement is not practicable because of the low efficiency in removing light and finely dispersed oil particles and possible contamination of foods by chemicals. The G-bag approach, which uses a bag of absorbent to capture the pollutants and degrade the pollutants with the immobilized microorganisms on the absorbent, seems to be a good alternative only if the system can be designed as simple and free from fouling (Chen et al, 2000).

In this work, the treatment of restaurant wastewater by electrochemical incineration is obtained with continuous electrochemical oxidation system described elsewhere Klamklang (Klamklang, 2006, Klamklang, 2007 and Klamklang et al, 2010). A simple three-electrode electrochemical reactor with 18 ml of capacity with using Ti/SnO_2 electrodes as anode, the 316L stainless steel was used as cathode. The feed solution was fed to the reactor by peristaltic pump and the effluent was collected at sample trap. The apparatus is presented in Figure 3. (Klamklang, 2006). The operating conditions were represented in Table 1.

Parameter	Operating condition
Current density	5-10 mA/cm^2
SnO_2 film thickness	1.8-3.6 micron
Residence time	2-3 hr
Elapse time	24 hr
Stirring	300 rpm

Table 1. Operating conditions for continuous electrochemical oxidation.

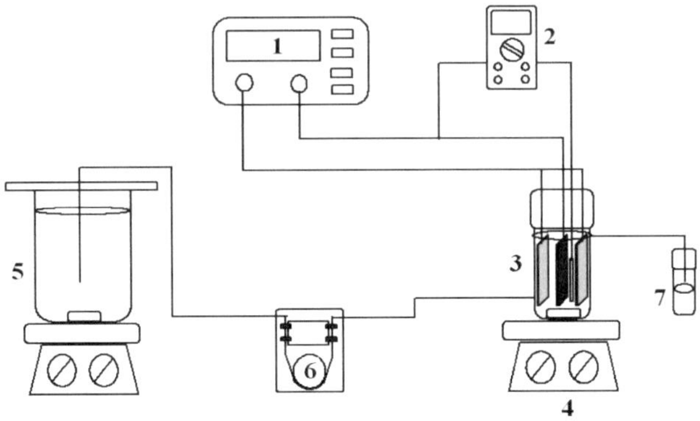

1) Power supply, 2) Voltmeter, 3) Electrochemical reactor set, 4) Magnetic stirrer, 5) Feed reservoir, 6) Peristaltic pump, 7) Sample trap

Fig. 3. Schematic diagram of continuous electrochemical oxidation apparatus.

The experiments performed in a continuous mixed flow reactor were carried out for the determination of the effects of the current density, residence time and SnO_2 film thickness on organic pollutant degradation. Due to the very small electrode area and easy to observe the change of Total Organic Carbon (TOC), the wastewater, which feed to the system, was diluted to around 140 mg TOC/L. The investigated current densities were 5 and 10 mA/cm^2 and residence times were 2 and 3 hr.

Influence of current density

The influence of current density in continuous mixed flow experiments is presented in part A of the Figure 4. The electrochemical degradation of organic pollutants presented in actual restaurant wastewater takes place slowly and its TOC removal efficiency presented in part B of this figure is higher at lower current density. The gain in efficiency being overwhelmed by the lower current values applied. This result may not be surprising on the basis of the previously discussed influence of current density in batch experiments, which indicated to a weak behavior for the characteristic of diffusion-controlled processes. Increase in current density cannot increase the organic removal efficiency at the electrode, but only favours the anodic side reaction which decreased the organic pollutant removal efficiency. It agrees with Figures 5 (A) and (B) that the destruction of organic pollutants in term of Chemical Oxygen Demand (COD) was decreased with increasing of the current density from 5 to 10 mA/cm^2. The equilibrium efficiencies of both TOC and COD removal were 62% when current density was 5 mA/cm^2, while their removal efficiencies were 47% when the current density was 10 mA/cm^2.

Influence of residence time

The presented results in batch experiments show that the increasing of residence time after first 2 hr was not greatly affect on the organic pollutant degradation efficiency due to the change reaction order from zero-order to first-order reaction with reduction of TOC. However, it would be of practical interest to test how much an increase or decrease in the wastewater flow rate affects the TOC removal of the restaurant wastewater. This is demonstrated in Figures 6 and 7. Because of fixed total volume of the continuous mixed flow reactor at 18 ml, an increase in the wastewater flow rate from 0.10 to 0.15 ml/min translates to a proportional decrease in the wastewater hydraulic residence time from 3 to 2 hr.

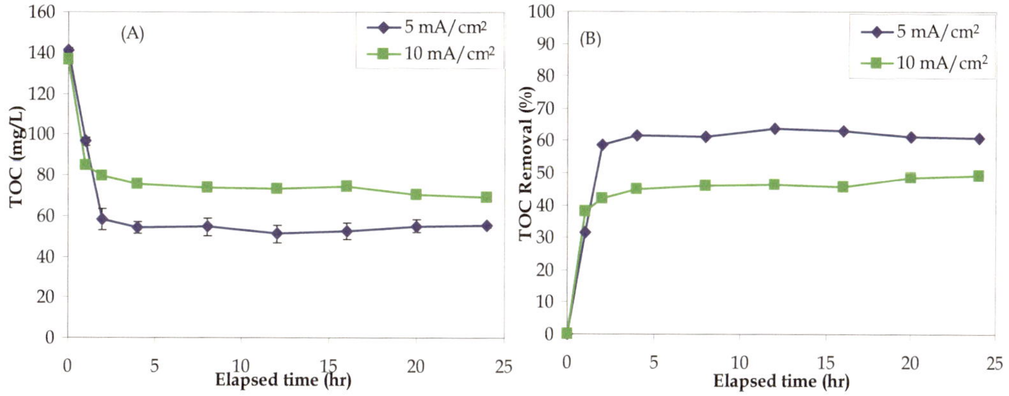

Fig. 4. Effect of current density on TOC removal (A) and TOC removal efficiency (B) in continuous restaurant wastewater treatment by using of SnO_2/Ir/Ti electrode, SnO_2 thickness of 1.8 micron.

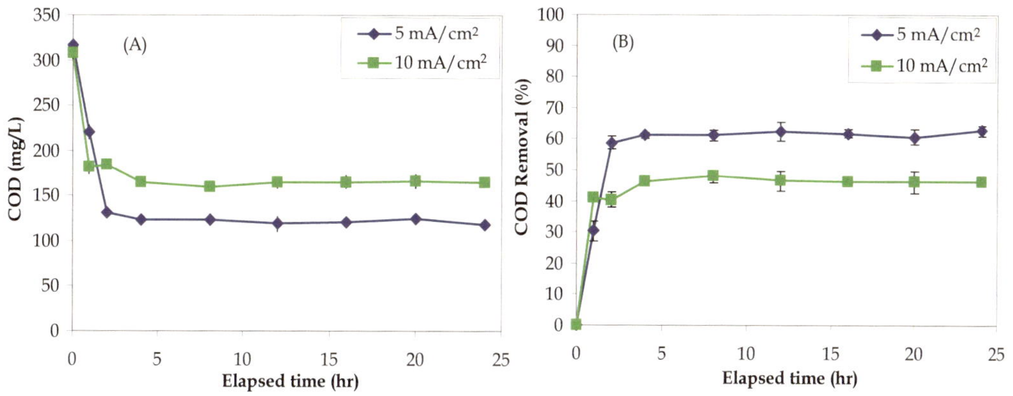

Fig. 5. Effect of current density on COD removal (A) and COD removal efficiency (B) in continuous restaurant wastewater treatment by using of SnO_2/Ir/Ti electrode, SnO_2 thickness of 1.8 micron.

Normally, a reduction in residence time would expectedly lead to a decrease in the wastewater TOC removal. But, in this case, increasing of residence time does not proportionally increase TOC removal. As seen in Figure 6 ((A) and (B)), the TOC removal increases from around 55 to 62 % with the increase in the residence time from 2 to 3 hr. These results were also observed in the removal of COD and represented in Figure 7 (A) and (B). The COD removal increased from around 54 to 62 % with the increase in residence time from 2 to 3 hr.

It could be explained by the increasing of residence time from 2 to 3 hr has not strongly affected on the TOC and COD removal due to the fast reaction with zero-order reaction occurred in the first 2 hr. Then, the reaction was changed to the slower step with the first-order reaction as we found in the batch experiments.

Hence, it would be more economical to operate the electrochemical treatment at a lower residence time as long as the pollutant concentration of the treated wastewater meets the safe discharge requirement.

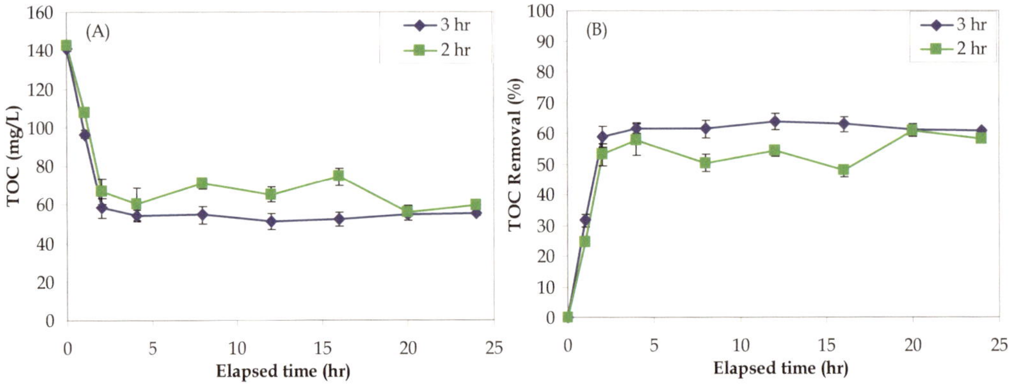

Fig. 6. Effect of residence time on TOC removal (A) and TOC removal efficiency (B) in continuous restaurant wastewater treatment by using of SnO_2/Ir/Ti electrode, SnO_2 thickness of 1.8 micron and current density 5 mA/cm^2.

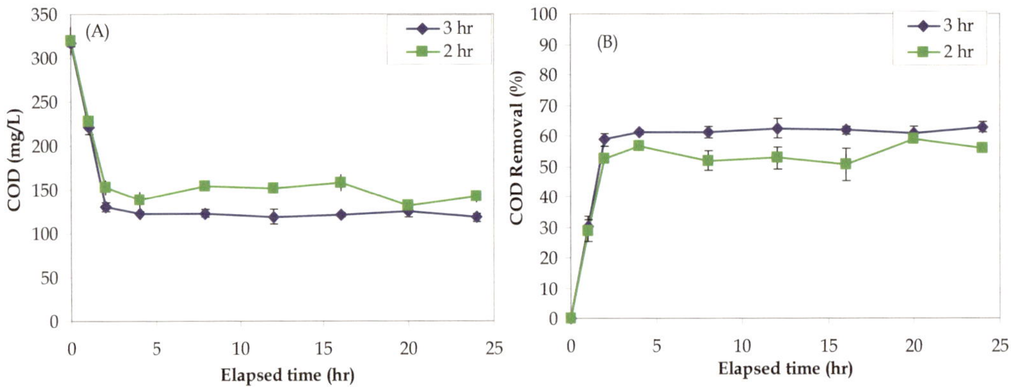

Fig. 7. Effect of residence time on COD removal (A) and on COD removal efficiency (B) in continuous restaurant wastewater treatment by using of SnO_2/Ir/Ti electrode, SnO_2 thickness of 1.8 micron and current density of 5 mA/cm^2.

Influence of SnO_2 active layer thickness

Figures 8 (A) and (B) represent the effect of SnO_2 film thickness on the TOC degradation performance in continuous electrochemical oxidation. Similar to the pollutant degradation of organic pollutant in batch experiment, it shows that the SnO_2 active layer thickness was not a great influence on the TOC removal efficiency because the adsorbed hydroxyl radicals for organic pollutant degradation were produced only at the surface of electrode. However, the TOC removal efficiency was around 62% with the 1.8 micron of SnO_2 active layer while the efficiency was reduced to 51% with the SnO_2 active layer thickness of 3.6 micron. It agrees with the removal of COD from restaurant wastewater as presented in Figure 9. The COD removal efficiency was 62% when the thickness of SnO_2 active layer was 1.8 micron. However, the efficiency was decreased to 50% when the thickness of SnO_2 active layer was 3.6 micron. It should be explained that thickness of 3.6 micron has bigger grain size that leads to a less surface area; therefore, the reaction kinetic was decreased as found previously in the batch experiment.

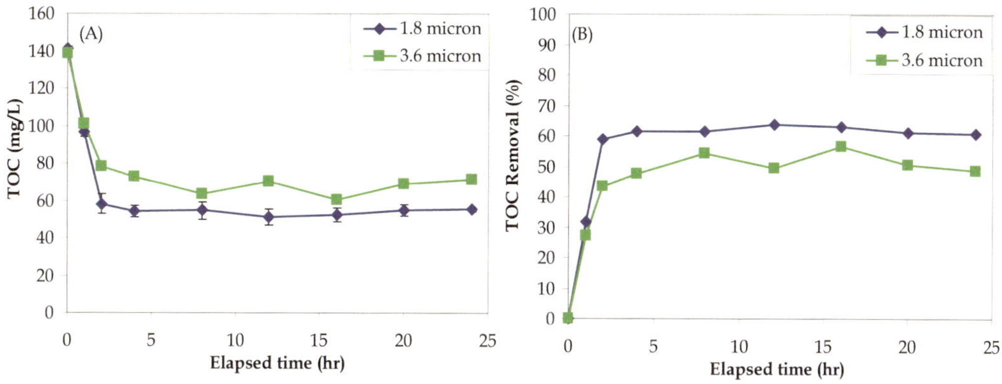

Fig. 8. Effect of SnO_2 layer thickness on TOC removal (A) and TOC removal efficiency (B) in continuous restaurant wastewater treatment by using of $SnO_2/Ir/Ti$ electrode and current density of 5 mA/cm^2.

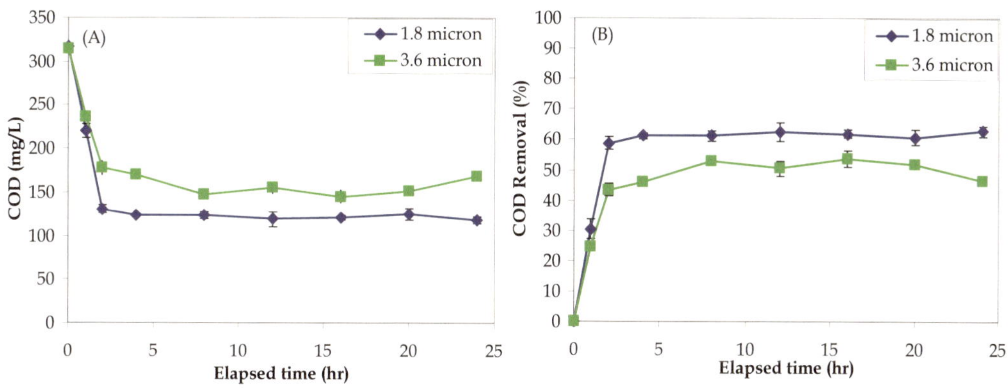

Fig. 9. Effect of SnO_2 layer thickness on COD removal (A) and COD removal efficiency (B) in continuous restaurant wastewater treatment by using of $SnO_2/Ir/Ti$ electrode and current density 5 mA/cm^2

3.2 Color removal of pulp and paper mill wastewater by electrochemical incineration

Pulp and paper mill are considered as high water consumption and high wastewater discharge. The pollutants in pulp and paper mill industry are mainly presented in form of high-strength chemical oxygen demand (COD) and biochemical oxygen demand (BOD). Generally, the high-strength BOD and COD are totally removed in biological wastewater treatment unit. The high-strength color from lignin in pulp mill effluent is big problem for pulp and paper mill due to requirement of advanced treatment process with high investment and operating cost. In this chapter, the case study of high efficiency color removal by electrochemical incineration in pulp and paper mill wastewater treatment is explained. The electrochemical incineration is obtained with continuous electrochemical oxidation in a series of 4-simple 3-electrode electrochemical reactor with 5 liters of capacity. The Ti/IrO_2-RuO_2 mixed oxide electrodes were used as anode, the 316L stainless steel was used as cathode. The pulp and paper effluent is fed to the reactor by peristaltic pump and the effluent was collected at sample trap. The apparatus is presented in Figure 10.

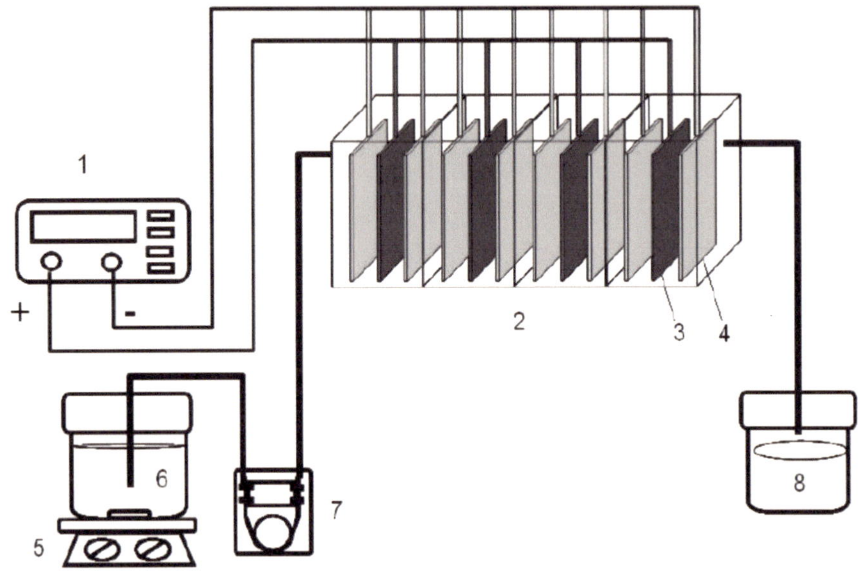

1) Power supply, 2) Electrochemical reactor set, 3) Mixed oxide anode, 4) Stainless steel cathode, 5) Magnetic stirrer, 6) Feed reservoir, 7) Peristaltic pump and 8) Sample trap

Fig. 10. Schematic diagrams of continuous electrochemical oxidation apparatus

Figure 11 show that electrochemical incineration is a powerful tool for color removal from pulp mill wastewater. The results presented that the color substrates in pulp mill effluent was immediately removed. The wastewater color was reduced from 938 Unit Pt/Co to 10-45 Unit Pt/Co or 95-99 % removal that has very good advantage over conventional biological wastewater treatment process that takes more that 24 hour for reducing pulp mill effluent color from 1,500 to 1,100 Unit Pt/Co. Figure 12 presents the pulp mill at various treatment time for different residence time after started electrochemical incineration process. It is clearly that electrochemical incineration is suitable for breaking color substrates in high color strength wastewater.

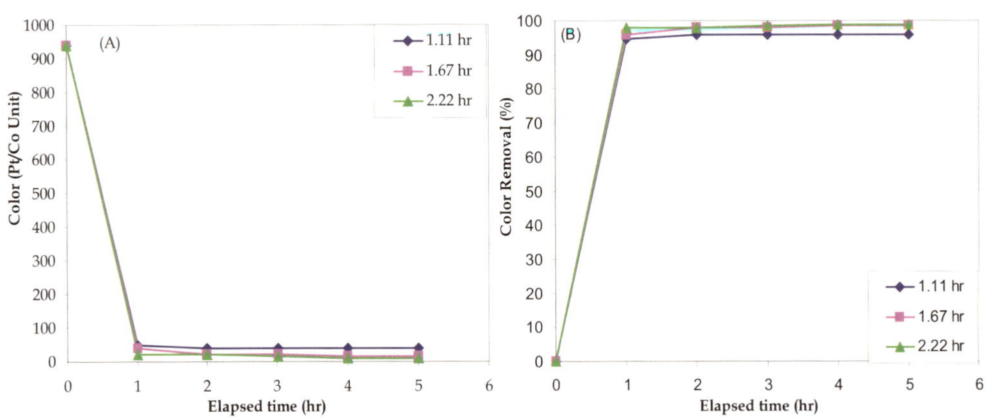

Fig. 11. Influence of treatment time on pulp mill effluent color removal by electrochemical incineration (A) Wastewater color and (B) Color removal efficiency.

(A)

(B)

(C)

Fig. 12. Electrochemical incineration of pulp mill effluent (A) Residence time of 1.1 hr, Anode-cathode distance of 1.2 cm, Treatment cost of 0.175 USD/m^3 and Color removal efficiency of 96%, (B) Residence time of 1.67 hr, Anode-cathode distance of 1.2 cm, Treatment cost of 0.255 USD/m^3 and Color removal efficiency of 98% and (C) Residence time of 2.22 hr, Anode-cathode distance of 1.2 cm, Treatment cost of 0.315 USD/m^3 and Color removal efficiency of 99%

4. Prospect & conclusion

Water treatment by electricity was used for several years, but electrochemical water or wastewater technologies are not yet a mature application in commercial scale by the limitation of relatively high capital investment and the expensive electricity supply. Nowadays, there are increasing on standard for water discharge and limitation of water resource that makes electrochemical treatment comparable with other water and wastewater treatment technologies. However, to mature the electrochemical water and wastewater treatment there are some requirements need to be fulfilled. SWOT (Strengths, Weaknesses, Opportunities and Threats) is commonly used for business analysis of internal and external factors that should support and sustain or collapse business project (Pinson, 2008). SWOT could help to pin point and predominate advantages of electrochemical technology over conventional technologies and drown out disadvantages. Side reactions are a killing trap of electrochemical technology that limits the efficiency of electrochemical systems. In electrochemical incineration, specific coating materials need to be developed to decrease side reaction and to promote desired reaction for attractive benefit and return of investment. During past two decades there are some development in coating technology such as chemical vapor deposition and thermal spray coating that should support the electrode fabrication with appreciate on both constant coating layer and long service life that will response to good return of investment. Electrochemical reactor design is big challenges in electrochemical technology due to it will response on mass transfer during operation. The lack of mass transfer and current distribution in electrochemical reactor will collapse all advantages of electrochemical technology in the battle of technologies.

5. References

Azzam, M.O., Tahboub, Y. and Al-Tarazi, M. (1999). Effect of counter electrode material on the anodic destruction of 4-Cl phenol solution, *Process Safety and Environmental Protection,* Vol 77, N° B4, pp 219-226, ISSN 0957-5820

Canizares, P.; Saez, C.; Lobato, J.; Rodrigo, M. A. (2004). Electrochemical treatment of 2,4-dinitrophenol aqueous wastes using boron-doped diamond anodes: Part II. Influence of waste characteristics and operating conditions. *Electrochimica Acta*, Vol 49, N° 26, pp 4641-4650,(2004), ISSN 0013-4686

Chen, G. (2004) Electrochemical technologies in wastewater treatment. *Separation and Purification Technology,* Vol 38, N° 1, pp 11-41, ISSN 1383-5866

Chen, X., Chen, G. and Yue, P.L. (2000). Separation of pollutants from restaurant wastewater by electrocoagulation. *Separation and Purification Technology,* Vol 19, N° 1-2, pp 5-76, (2000), ISSN 1383-5866

Comninellis, C. (1994). Electrocatalysis in the electrochemical conversion/ combustion of organic pollutants for waste water treatment. *Electrochimica Acta.* Vol. 39, No. 11/12, pp. 1857-1862

Comninellis, C. and Nerini, A. (1995). Anodic oxidation of phenol in the presence of NaCl for wastewater treatment. *Journal of Applied Electrochemistry* ,Vol. 25, N°1, pp 23-28, (1995), ISSN 0021-891

Comninellis, C. and Vercesi, G. P. (1991). Characterization of DSA®-type oxygen evolving electrodes. Choice of a Coating. *Journal of Applied Electrochemistry,* Vol 21, N° 4, pp 335-45, (1991), ISSN 0021-891X

Couper, A. Mottram; Pletcher, Derek; Walsh, Frank C.. (1990). Electrode materials for electrosynthesis. *Chemical Reviews (Washington, DC, United States),* Vol 90, N° 5, pp 837-65, (1990), ISSN 0009-2665

Diniz, A. V.; Ferreira, N. G.; Corat, E. J.; Trava-Airoldi, V. J. (2003). Efficiency study of perforated diamond electrodes for organic compounds oxidation process. *Diamond and Related Materials,* Vol 12, N° 3-7, pp 577-582, (2003), ISSN 0925-9635

Fernandes, A.; Morao, A.; Magrinho, M.; Lopes, A.; Goncalves, I. (2004). Electrochemical degradation of C. I. Acid Orange 7. *Dyes and Pigments* ,Vol 61, N° 3, pp 287-296, (2004), ISSN 0143-7208

Grimm, J. et al. (1998). *Sol*-gel film preparation electrodes for the electrocatalytic oxidation of organic pollutants in water. *Desalination* Vol. 115, N° 3, pp. 295-302, ISSN 0011-9164.

Hanze, M., Harremës, P., Jansen, J. C. and Arvin, E. (1995). *Wastewater Treatment: Biological and Chemical Processes*. Springer, CAN 127:298091, Berlin, Germany

http://www.environmental-center.com/articles/article1149/ article1149.htm

Kapałka, A., Fóti, G. and Comninellis, C. (2009). Basic Principles of the Electrochemical Mineralization of Organic Pollutants for Wastewater Treatment.. *Journal of Applied Electrochemistry,* Vol 40, N° 12, pp 2203, (2010), ISSN 0021-891X

Kesselman J. M., Weres O., Lewis N. S., Hoffmann M. R. (1997). Electrochemical Production of Hydroxyl Radical at Polycrystalline Nb-Doped TiO_2 Electrodes and Estimation of the Partitioning between Hydroxyl Radical and Direct Hole Oxidation Pathways. *Journal of Physical Chemistry B*, Vol. 101, N° 14, pp 2637-2643, ISSN 1089-5647

Klamklang, S. (2006). Restaurant wasterwater treatment by electrochemical oxidation in continuous process. *Chulalongkorn University Ph.D. Dissertation*, ISBN 974-14-3476-6, Bangkok, Thailand

Klamklang, S. (2007). Restaurant wasterwater treatment by electrochemical oxidation in continuous process. *Institut National Polytechnique de Toulouse Dissertation*,. ISBN 974-14-3476-6, Toulouse, France

Klamklang, S. Vergnes, H., Secocq, F., Pruksathorn, K., Duverneuil, P. and Damronglerd, S. (2010). Deposition of tin oxide, iridium and iridium oxide films by metal-organic chemical vapor deposition for electrochemical wastewater treatment. *Journal of Applied Electrochemistry*, Vol 40, N° 5, pp 997-1004, (2010), ISSN 0021-891X

Kuhn, A. T. (1971). Electrolytic decomposition of cyanides, phenols and thiocyanates in effluents streams-a literature review. *Journal of Applied Chemistry & Biotechnology,* Vol 21, N° 2, pp 29-34 Journal; General Review (1971), ISSN 0375-9210

Lissens, G.; Pieters, J.; Verhaege, M.; Pinoy, L.; Verstraete, W. (2003). Electrochemical degradation of surfactants by intermediates of water discharge at carbon-based electrodes. *Electrochimica Acta* ,Vol. 48, N° 12, pp 1655-1663, (2003), ISSN 0013-4686

Morão, A., Lopes, A. Pessoa de Amorimb, M. T. and Gonçalves, I. C. (2004). Degradation of mixtures of phenols using boron doped diamond electrodes for wastewater treatment. *Electrochimica Acta*, Vol 49, N° 9-10, pp 1587-1595, (2004), ISSN 0013-4686

Panizza, M. and Cerisola, G. (2005). Application of diamond electrodes to electrochemical processes. *Electrochimica Acta*, Vol 51, N°2, pp 191-199, (2005), ISSN: 0013-4686

Pinson, L. (2008). *Anatomy of a Business Plan, 7th Edtion*.: Out of Your Mind and Into Marketplace, ISBN 0-944205-35-6 California, USA

Reemtsma, T. and Jekel, M.(2006). *Organic Pollutants in the Water Cycle*. Wiley-VCH, ISBN 3-527-31297-8, Weinheim, Germany

Stucki, S., Kötz, R., Carcer, B. and Suter, W. (1991). Electrochemical waste water treatment using high overvoltage anodes Part II: Anode performance and applications. *Journal of Applied Electrochemistry* Vol. 21, N° 2, pp 99-104, (1991), ISSN 0021-891X

Vercesi, G. et al. (1991). Characterization of DSA-type oxygen evolving electrodes. Choice of base metal. *Thermochimica Acta,* Vol 176, pp 31-47, (1991), ISSN 0040-6031

16

Research on Pressure Swing Adsorption of Resin for Treating Gas Containing Toluene

Ruixia Wei and Shuguo Zhao
Hebei Polytechnic University, Tangshan, Hebei
China

1. Introduction

1.1 Introduction of volatile organic compounds

Volatile organic compounds (VOC) means any compound of carbon, excluding carbon monoxide, carbon dioxide, carbonic acid, metallic carbides or carbonates, and ammonium carbonate, which participates in atmospheric photochemical reactions.

Volatile organic compounds or VOCs are organic chemical compounds whose composition makes it possible for them to evaporate under normal indoor atmospheric conditions of temperature and pressure. This is the general definition of VOCs that is used in the scientific literature, and is consistent with the definition used for indoor air quality.

Since the volatility of a compound is generally higher the lower its boiling point temperature, the volatility of organic compounds are sometimes defined and classified by their boiling points.For example, the European Union uses the boiling point, rather than its volatility in its definition of VOCs.

A VOC is any organic compound having an initial boiling point less than or equal to 250° C measured at a standard atmospheric pressure of 101.3 kPa. VOCs are sometimes categorized by the ease they will be emitted. For example, the World Health Organization (WHO) categorizes indoor organic pollutants as very volatile, volatile, and semi-volatile. The higher the volatility (lower the boiling point), the more likely the compound will be emitted from a product or surface into the air. Very volatile organic compounds (VVOCs) are so volatile that they are difficult to measure and are found almost entirely as gases in the air rather than in materials or on surfaces. The least volatile compounds (SVOCs) found in air constitute a far smaller fraction of the total present indoors while the majority will be in solids or liquids that contain them or on surfaces including dust, furnishings, and building materials.

Many VOCs are dangerous to human health or cause harm to the environment. VOCs are numerous, varied, and ubiquitous. They include both man-made and naturally occurring chemical compounds. Anthropogenic VOCs are regulated by law, especially indoors, where concentrations are the highest. VOCs are typically not acutely toxic, but instead have compounding long-term health effects. Because the concentrations are usually low and the symptoms slow to develop. The main hazard is as the following aspects:

1. The ability of organic chemicals to cause health effects varies greatly from those that are highly toxic, to those with no known health effect. As with other pollutants, the extent and nature of the health effect will depend on many factors including level of exposure

and length of time exposed. Eye and respiratory tract irritation, headaches, dizziness, visual disorders, and memory impairment are among the immediate symptoms that some people have experienced soon after exposure to some organics. At present, not much is known about what health effects occur from the levels of organics usually found in homes. Many organic compounds are known to cause cancer in animals; some are suspected of causing, or are known to cause, cancer in humans. Eye, nose, and throat irritation; headaches, loss of coordination, nausea; damage to liver, kidney, and central nervous system. Some organics can cause cancer in animals; some are suspected or known to cause cancer in humans. Key signs or symptoms associated with exposure to VOCs include conjunctival irritation, nose and throat discomfort, headache, allergic skin reaction, dyspnea, declines in serum cholinesterase levels, nausea, emesis, epistaxis, fatigue, dizziness.

2. The reaction of photochemical smog in the sunlight will occur among nitrogen oxides, hydrocarbons and photochemical oxidants of the atmosphere.The main component of photochemical smog is ozone, peroxy acetyl nitrate Cool (PAN), aldehydes and ketones and so on. They stimulate people's eyes and respiratory system, endangering people's health and even harm plant growth.
3. Halogenated hydrocarbons VOCs may destroy the ozone layer and change the Earth's heat balance. According to the Indian National Academy of Sciences report, the emissions of chlorofluorocarbons into the atmosphere have increased the atmospheric methane and chloride absorption of infrared radiation and heat hinder the discharge of the Earth which will make the Earth's temperature,climate change.

VOCs	environmental acceptable concentration (mg/m^3)	health hazards
Toluene	100	Headache, dizziness, nausea, pulmonary emphysema
Benzene	5	Cancer, leukemia, respiratory paralysis
Xylene	100	Anemia, leukemia, red blood cells reduced, skin and mucous
Methanol	200	membrane irritation Neurological disorders, vomiting, insomnia, headaches,
Acetone	750	Cramps
Ethyl acetate	150	Irritate the eye, skin, numbness, headache, cough, nausea
Carbontetrachloride	5	Eye irritation, paralysis Abdominal pain, nausea, vomiting, cancer
Acetaldehyde	10	Mucosal erosion, blurred vision, pulmonary edema
Ether	400	Paralysis, nervous system damage, liver and kidney damage
Acetonitrile	40	Headache, dizziness, breathing difficulties, damage the central nervous
Acrylonitrile	20	Nausea, vomiting, difficulty breathing

Table 1. Common VOCs environmental acceptable concentration and human health hazards

As the VOCs harmful environmental effects, many countries have developed a corresponding law to limit emissions of VOCs. "Clean Air Act 1990" of the United States requires 90% reduction in emissions of the 189 kinds of toxic chemicals which of about 70% belongs to VOCs. In 1996 Japan adopted legislative restrictions of 53 kinds of VOCs emissions and limited 149 kinds of VOCs emissions in 2002. China also enacted in 1997 and implemented the "Integrated emission standard of air pollutants"which limits 33 pollutant emission limits, including benzene, toluene, xylene and other volatile organic compounds. The VOCs harmful environmental effects and human health can not avoid in terms of current technology, so there is an urgent need for effective technology to control VOCs.

1.2 Vocs treatment technology

VOCs treatment technology is divided into two categories: Destruction processes and Recuperation processes. VOC controls include all technologies which either collect the VOCs for recovery and reuse, or destroy the VOCs. If the VOCs have recovery value, which typically implies single-VOC exhaust streams, and if the cost of recovery is less than the cost of purchasing new VOC, which typically implies relatively concentrated exhaust streams, then recovery makes sense. Carbon adsorption, scrubbing, and condensation are typical recovery techniques. Note that the installation and operation of a recovery technology may more than pay for itself if the recovery value of the VOC is high enough. If the VOC stream has no recovery value, if, for example, it is a mixture, or if there are disposal concerns, such as for toxic compounds, then destruction probably makes the most sense. Thermal and catalytic oxidation and biofiltration would be useful in this case.

1.2.1 Destruction processes

1.2.1.1 Thermal oxidation processes

Thermal oxidation is the process of oxidizing combustible materials by raising the temperature of the material above its auto-ignition point in the presence of oxygen, and maintaining it at high temperature for sufficient time to complete combustion to carbon dioxide and water. Time, temperature, turbulence (for mixing), and the availability of oxygen all affect the rate and efficiency of the combustion process. These factors provide the basic design parameters for VOC oxidation systems.

There are three basic types of thermal oxidation systems: direct flame, recuperative, and regenerative.

Direct flame systems or flares rely on contact of the waste stream with a flame to achieve oxidation of the VOCs. These systems are the simplest thermal oxidizers and the least expensive to install, but require the greatest amount of auxiliary fuel to maintain the oxidation temperature, thus entailing the highest operating cost. Flares are useful for destruction of intermittent streams.

Recuperative thermal oxidation systems use a tube or plate heat exchanger to preheat the effluent stream prior to oxidation in the combustion chamber. Thermal recovery efficiencies typically are limited to 40-70% to prevent auto-ignition in the heat exchange package, which could damage the package. Supplemental fuel therefore is usually required to maintain a high enough temperature for the desired destruction efficiency. Recuperative systems are more expensive to install than flares, but have lower operating costs.

Regenerative thermal oxidation systems typically incorporate multiple ceramic heat exchanger beds to produce heat recovery efficiencies as high as 95%. An incoming gas stream passes through a hot bed of ceramic or other material, which simultaneously cools the bed and heats the stream to temperatures above the auto-ignition points of its organic constituents. Oxidation thus begins in the bed, and is completed in a central combustion chamber, after which the clean gas stream is cooled by passage through another ceramic heat exchanger. Periodically the flow through the beds is reversed, while continuous flow through the unit is maintained. Regenerative thermal oxidation systems are the most expensive thermal oxidizers to build, but the added capital expense is offset by savings in auxiliary fuel.

ESOCOV is a regenerative thermal oxidation process on ceramic beds. The process is especially well adapted to a mixture of gases with concentrations between 1 and 10 g/Nm3 and flows from 1.000 to 100.000 Nm3/h.The gas passes over a ceramic bed in which air is progressively heated and the VOC's are destroyed by oxidation above 800 C. The direction of the airflow is changed on a regularly basis in order to charge and discharge the calories in the bed(s). The thermal efficiency amounts to 90 - 98 %. In this way the regenerating systems are autothermal, so, without additional energy for concentrations higher than 1,5 g/Nm3. Addition of a catalyst to have the oxidation at a lower temperature (between 200 and 400 C).

1.2.1.2 Catalytic oxidation

Catalytic oxidation converts volatile organic compounds (VOC) into carbon dioxide and water, as do other oxidation processes, with no byproducts requiring disposal. Catalytic oxidation is well suited to applications with VOC concentrations ranging up to 25% of the lower explosion limit. With proper selection of catalyst, operating conditions, and equipment design, catalytic oxidation can attain VOC conversions of up to 99%. Advantages of this technology are low fuel usage, particularly with the proper choice of heat exchanger, little nitrogen oxide formation, given low operating temperatures, and little formation of partial oxidation products, such as carbon monoxide and aldehydes. Disadvantages include susceptibility to catalyst poisons, and the sensitivity of the catalysts to high temperatures.

Catalysts for VOC oxidation typically are either precious metals supported on ceramic or metal monoliths (honeycombs) or on ceramic pellets, or base metals supported on ceramic pellets. Catalyst life exceeds five years with the proper choice of catalyst, and may be extended with catalyst washing and regeneration techniques. Recent generations of catalysts have much longer lives and greater poison resistance than their forebears, and have greater capabilities, including the destruction of chlorinated organics.

As with any process, proper equipment design is essential to performance and operating cost. Typical catalytic oxidizer components include the catalyst housing, blower, burner, heat exchanger, controls, and stack. Small units are often skid-mounted and delivered to the site ready for installation.

As vent streams are often below the temperature at which catalytic oxidation is effective, most oxidizers use burners to preheat these streams to reaction temperatures, often from 400-800 °F. Heat is recovered using either recuperative or regenerative heat exchangers. As the latter can provide 95% heat recovery, streams with low VOC levels can be processed with minimal fuel usage.

1.2.1.3 Biological treatment processes

Biological method is essentially the use of microbial life activities to the emissions of VOCs into simple inorganic (such as CO_2 and H_2O) and microbial composition of the material itself. Common processes are biological filtration, biological washing. The biggest difference of biological filtration from biological wastewater treatment process is: in the exhaust gas through the organic material must first transfer to the liquid (or solid surface of the film) in the mass transfer process, and then in the liquid (or solid surface of the biological layer) adsorption by microbial degradation. Biological method is particularly suitable for processing gas is greater than 17000m3 / h, the gas concentration is less than 1000ppm.

Biological method has many advantages compared to other technologies, it's simple, low operation cost, low investment relative to other methods, wide range of applications, while not easy to produce secondary pollution. Especially for low concentrations of VOCs (for example, when only a few ppm) the results of treatment is good. However, biological method is related to gas and liquid (or solid) mass transfer and chemical and biological degradation processes. The influencing factors are many and complex. Now, biodegradable technology is not enough in-depth theoretical study, so the biological treatment of VOCs present the design and operation is still basically remain in the level of experience. At the same time it also requires a larger footprint, which is limiting the biological treatment of extensive use of VOCs.

1.2.2 Recuperation processes

(1) Cryogenic Condensation processes

Low temperature or cryogenic condensation is a process that can be used as an effective means for VOC emissions control. Cryogenic condensation technology is based on lowering the vapor pressure of a component by reducing the temperature of the process stream thus increasing the recovery of the components in the liquid phase.

Since nitrogen gas is widely used in the chemical process industry, is inert and is typically transported and stored in it's liquid state at low temperature and high pressure, it is a convenient media to use. The low temperature capability of liquid nitrogen allows for the design of highly efficient condensation systems. At temperatures below -120°F, the vapor pressure of most organic compounds is depressed sufficiently to condense 95 to 99+% of the compounds from a typical emissions stream. In addition, the vented nitrogen can be recycled for reuse within the plant.

Cryogenic condensation is well suited for VOC emission control because of its ability to respond instantly to changes in VOC flow rate and solvent loading. It can recover virtually any VOC species even under varying conditions. Cryogenic condensation can deal with all organics (even in the presence of water) and can function when the concentration and composition are changing over time. This flexibility makes it particularly suitable for VOC control in multi-product, multi-purpose plants where batch or continuous processes are employed.

(2) membrane

Volatile organic compounds (VOCs) are involved in atmospheric pollution and green house effect.Some of these compounds might be recovered, instead of being released to the atmosphere, by several methods such as condensation, absorption, adsorption, etc. Among these processes, vapor permeation has several advantages since it requires compact equipment, it is non destructive and it is notenergy-intensive. Over the past ten years, vapor permeation has been proven to be a feasible alternative to conventional processes in the recovery of several halogenated VOCs and monomers. In recent years, this process has found other applications such as in the recovery of hydrocarbon VOCs from the petroleum industry facilities; these applications are still under development.

Within the membranes used for the recovery of volatile organic compounds, composite membranes offer several advantages over other kinds. They are composed of a selective, defect-free layer that performs the vapor separation while another porous layer gives mechanical strength. Poly dimethyl siloxane (PDMS) is one of the most used polymers as selective permeation layer. It can be easily fabricated and thus is readily available for its use on large scales. The use of dimensionless solubility parameters showed that PDMS has good selectivities towards a wide variety of VOCs (e.g., hydrocarbons).
The recovery of toluene, propylene and 1,3-butadiene, which are compounds of particular concern in the petroleum industry, is focused on in this study. Since several petroleum activities such as oil storage or distribution, emit pollutants at low flow rates and variable concentrations, vapor permeation appears to provide a flexible recovery solution.
(3) absorption processes : ESOLAV
ESOLAV is a range of absorption processes or washers which transfers the VOC's into a liquid phase in order to be solubilized, oxidized or separated. It is especially well suited for VOC's soluble in the washing solution with weak and low concentrations (< 1 g/Nm3) and for gasflows between 500 and 50.000 m3/h.
(4)adsorption with regeneration processes : ESOSORB
ESOSORB is a solvent adsorption process on beds of activated carbon with recuperation by desorption through steam. The process is very well adapted to gases containing 1 or 2 solvents in a concentration lower than 30 g/Nm3 and gas flows between 200 and 200.000 Nm3/h.These units are built out of several adsorbent beds containing activated carbon. The air to be treated is guided over the activated carbon where the VOC's ared adsorbed until the activated carbon is saturated. If one of the adsorbing beds is saturated, a regeneration process is set in motion. After passage of the steam for the desorption a fan dries and cools the activated carbon beds, which makes them ready for a new adsorption cycle.
Adsorption on activated carbon is useful for recovery of VOCs with intermediate molecular weights (typically about 45-130): smaller compounds do not adsorb well, and larger compounds cannot be removed during regeneration, which typically is by steam stripping. Adsorption is most effective at lower temperatures, so that cooling of hot exhaust gas streams may be necessary. Further, dehumidification of very humid streams may be necessary for the carbon to have the greatest capacity. While carbon is the dominant adsorbent used, alumina, zeolites, and polymers have been used in some processes. Carbon can also be used to remove compounds in a once-through process with off-site regeneration.

1.3 Adsorption with regeneration processes and its principles

1.3.1 Adsorption

Adsorption refers to the binding of a dissolved solute onto the surface of a solid adsorbing material. It is a surface phenomenon and should not be confused with absorption, which is a term with a much broader meaning and generally deals with penetration of species into material. Adsorption can be used to separate a solute from a mixture of solutes, or a solute from a solvent. This is achieved by contacting the solution with the adsorbing material which is also called the adsorbent. The solute/s, which adsorb on the adsorbent, is/are referred to as adsorbate/s. The release of adsorbed material from an adsorbent is called desorption which is the reverse of adsorption.
Adsorption is a selective process and this selectivity is due to differences in the following: Molecular weight or size; Solute shape; Polarity; Electrostatic charge.

The physical binding of an adsorbate onto an adsorbent takes place due to non-covalent interactions such as: van der Waals forces; Electrostatic interactions; Hydrophobic interactions; Hydrogen bonding.

This type of adsorption takes place at ordinary temperature and is called physical adsorption or simply adsorption. Certain types of adsorption take place at much elevated temperatures when activation energy is available to break chemical bonds and facilitate chemical changes. Such processes are referred to as chemical adsorption or chemisorption.

The adsorbent material can be natural or synthetic. These generally have amorphous or microcrystalline structure and thus have very high specific surface area (surface area per unit amount of adsorbent). Commonly used adsorbents include clays like kaolin and bentonite, activated carbon, silica gel, activated alumina, zeolite (molecular sieves), etc.

Some of the advantages of adsorption over competing separation technologies are:High selectivity (e.g. affinity adsorption);Ability to handle very dilute solute concentrations.Major disadvantages are:It is a surface phenomenon,therefore the interior of the adsorbent material is not involved ;Batch or semi-batch operations generally have to be used;In certain cases adsorbents have to be regenerated;

In certain cases adsorption results in loss of product quality (e.g. with certain bioproducts)

Some of the common applications of adsorption are:Gas separation using molecular sieves by pressure swing adsorption;Removal of toxic gases from air (e.g. gas masks);Fractionation of industrial chemicals using gas or liquid chromatography;Removal of trace amounts of CS2 and H2S;Removal of phenolic and other toxic chemical from waste water;Removal of chlorinated hydrocarbons from waste gas streams;De-hydration or de-humidification of gases;Water purification by deionization and ion-exchange;Fractionation and recovery of protein bio-products;Affinity separations of bio-products.

1.3.2 PSA

Pressure swing adsorption (PSA) is a technology used to separate some gas species from a mixture of gases under pressure according to the species' molecular characteristics and affinity for an adsorbent material. It operates at near-ambient temperatures and so differs from cryogenic distillation techniques of gas separation. Special adsorptive materials (e.g., zeolites) are used as a molecular sieve, preferentially adsorbing the target gas species at high pressure. The process then swings to low pressure to desorb the adsorbent material.

Pressure swing adsorption principle can be illustrated in Figure 1. The gas component in a defined adsorption on the adsorbent is a function of temperature and pressure, usually available as shown below those adsorption isotherms. The figure shows the A, B two gases at the same temperature in a certain adsorbent adsorption isotherms on. Obviously, the same pressure A is more easily adsorbed than B. If A and B mixture through the adsorption column filled with the adsorbent, under relatively high pressure PH adsorption, at relatively low pressure PL desorption. The partial pressure of component A easily adsorbed are respectively PAH, and PAL, and the partial pressure of component B hard adsorbed are respectively PBH and PBL. The figure shows that under the relatively high-pressure component A is preferentially adsorbed, while the component B-rich gas stream in the outflow set. It is due to the equilibrium adsorption amount of qAH of component A is much higher than the equilibrium adsorption amount of qBH of component B. To make the adsorbent regeneration, the bed pressure is reduced to PL, the equilibrium adsorption capacity of the component A and B are respectively qAL and qBL. In the process which a new equilibrium is reached, the amount of desorption are qAH-qAL and qBH-qBL. This change in bed pressure periodically, the A, B mixture can be separated.

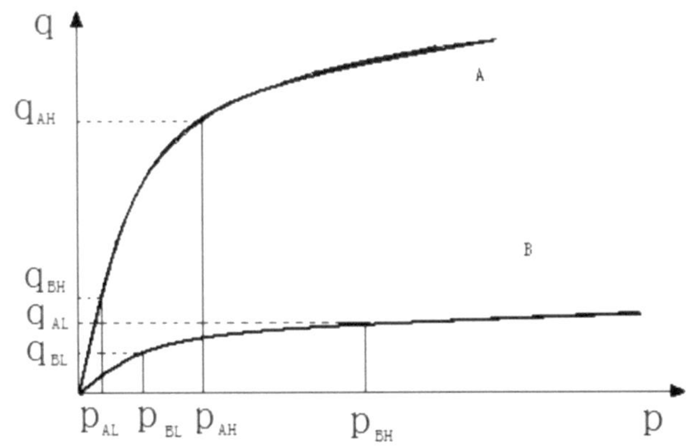

Fig. 1. The basic principle of pressure swing adsorption

Pressure swing adsorption processes rely on the fact that under pressure, gases tend to be attracted to solid surfaces, or "adsorbed". The higher the pressure, the more gas is adsorbed; when the pressure is reduced, the gas is released, or desorbed. PSA processes can be used to separate gases in a mixture because different gases tend to be attracted to different solid surfaces more or less strongly. If a gas mixture such as air, for example, is passed under pressure through a vessel containing an adsorbent bed that attracts nitrogen more strongly than it does oxygen, part or all of the nitrogen will stay in the bed, and the gas coming out of the vessel will be enriched in oxygen. When the bed reaches the end of its capacity to adsorb nitrogen, it can be regenerated by reducing the pressure, thereby releasing the adsorbed nitrogen. It is then ready for another cycle of producing oxygen enriched air.This is exactly the process used in portable oxygen concentrators used by emphysema patients and others who require oxygen enriched air to breathe.

Using two adsorbent vessels allows near-continuous production of the target gas. It also permits so-called pressure equalisation, where the gas leaving the vessel being depressured is used to partially pressurise the second vessel. This results in significant energy savings, and is common industrial practice.

Aside from their ability to discriminate between different gases, adsorbents for PSA systems are usually very porous materials chosen because of their large surface areas. Typical adsorbents are activated carbon, silica gel, alumina and zeolite. Though the gas adsorbed on these surfaces may consist of a layer only one or at most a few molecules thick, surface areas of several hundred square meters per gram enable the adsorption of a significant portion of the adsorbent's weight in gas. In addition to their selectivity for different gases, zeolites and some types of activated carbon called carbon molecular sieves may utilize their molecular sieve characteristics to exclude some gas molecules from their structure based on the size of the molecules, thereby restricting the ability of the larger molecules to be adsorbed.

One of the primary applications of PSA is in the removal of carbon dioxide (CO_2) as the final step in the large-scale commercial synthesis of hydrogen (H_2) for use in oil refineries and in the production of ammonia (NH_3). Refineries often use PSA technology in the removal of hydrogen sulfide (H_2S) from hydrogen feed and recycle streams of hydrotreating and hydrocracking units. Another application of PSA is the separation of carbon dioxide from biogas to increase the methane (CH_4) content. Through PSA the biogas can be upgraded to a quality similar to natural gas. Nitrogen generator units employ the PSA technique to produce high purity nitrogen gas (99.5% or greater) from a supply of compressed air.

2. The part of experiment

2.1 Experimental materials and equipments

Granular activated carbon was from Takeda Pharmaceutical Chemistry Kabushiki Kaisha, Environmental Company). XAD-4, NDA-150 and ND-90 resin were made by Nan Da Ge De Environmental Protection Technology Co Ltd, and their characters were listed in Table 2. High performance liquid chromatography (Waters 600) was manufactured by USA Waters Company. Experimental device of pressure swing adsorption was made by Shanghai Tonguang Technology & Education Equipment Co Ltd..

2.2 Chemical properties of toluene

Toluene, formerly known as toluol, is a clear, water-insoluble liquid with the typical smell of paint thinners. It is a mono-substituted benzene derivative, i.e., one in which a single hydrogen atom from the benzene molecule has been replaced by a univalent group, in this case CH_3.It is an aromatic hydrocarbon that is widely used as an industrial feedstock and as a solvent. Like other solvents, toluene is sometimes also used as an inhalant drug for its intoxicating properties; however, inhaling toluene has potential to cause severe neurological harm. Toluene is an important organic solvent, but is also capable of dissolving a number of notable inorganic chemicals such as sulfur.

Toluene reacts as a normal aromatic hydrocarbon towards electrophilic aromatic substitution. The methyl group makes it around 25 times more reactive than benzene in such reactions. It undergoes smooth sulfonation to give p-toluenesulfonic acid, and chlorination by Cl_2 in the presence of $FeCl_3$ to give ortho and para isomers of chlorotoluene. It undergoes nitration to give ortho and para nitrotoluene isomers, but if heated it can give dinitrotoluene and ultimately the explosive trinitrotoluene (TNT).With other reagents the methyl side chain in toluene may react, undergoing oxidation. Reaction with basify potassium permanganate and diluted acid (e.g., sulfuric acid) or potassium permanganate with concentrated sulfuric acid, leads to benzoic acid, whereas reaction with chromyl chloride leads to benzaldehyde (Étard reaction). Halogenation can be performed under free radical conditions. For example, N-bromosuccinimide (NBS) heated with toluene in the presence of AIBN leads to benzyl bromide. Toluene can also be treated with elemental bromine in the presence of UV light (direct sunlight) to yield benzyl bromide. Toluene may also be brominated by treating it with HBr and H_2O_2 in the presence of light.

Toluene is a common solvent, able to dissolve paints, paint thinners, silicone sealants, many chemical reactants, rubber, printing ink, adhesives (glues), lacquers, leather tanners, and disinfectants. It can also be used as a fullerene indicator, and is a raw material for toluene diisocyanate (used in the manufacture of polyurethane foam) and TNT. In addition, it is used as a solvent to create a solution of carbon nanotubes. It is also used as a cement for fine

polystyrene kits (by dissolving and then fusing surfaces) as it can be applied very precisely by brush and contains none of the bulk of an adhesive.
Industrial uses of toluene include dealkylation to benzene, and the disproportionation to a mixture of benzene and xylene in the BTX process. When oxidized it yields benzaldehyde and benzoic acid, two important intermediates in chemistry. It is also used as a carbon source for making Multi-Wall Carbon Nanotubes. Toluene can be used to break open red blood cells in order to extract hemoglobin in biochemistry experiments.

2.3 Experimental setup process

Air cames out of the air compressor filter which provides pressure for the experimental device, it filter out the oil, then into the organic gas generating device.The reaction had occurred after installation of benzene into the vapor mixing with air. Organic gases from the device comes into the adsorption column of the pipe with bypass for the determination of adsorption column inlet concentration. Adsorption column filled with resin and adsorption of gas access to outdoors by pipeline in the trachea on the road ,passing through the bypass outlet to determine the concentration after adsorption. Experimental setup is as follows:

2.4 Experimental methods

1. The pretreatment of resin

The three kinds of resins were extracted by Soxhlet extractor with absolute alcohol to get rid of the porogen, catalyst, reaction solvent and other impurities. The process was stoped when the circumfluence liquid was colorless and transparent. Then the resins were washed with distilled water, filtered and dried in the ovens at 60℃ after dried in the air. At last they were put into the dryer for a backup

2. Comparison of toluene adsorption between resin and activated carbon

XAD-4, NDA-150, ND-90 resin and activated carbon of same quality were taken as adsorbent, adsorbing 10 min at 0.1MPa pressure and room temperature, with flux of 3L/min, and selected the optimum adsorbing resin by comparing adsorbing effects of 3 types of resin and activated carbon.

3. Experiment of the optimum adsorption flux

The resin and activated carbon were taken as adsorbent of toluene gas, under 0.1MPa pressure and room temperature, in-gas flux of 3, 4, 5, 6, 7L/min. After they adsorbed 10 min, removal rate of toluene in air were tested to confirm the optimum adsorbing flux of resin and activated carbon

4. Experiment on variance of desorption effect under different pressures

Desorbed resin and activated carbon after adsorption by the optimum adsorption flux and time under the pressure of -0.05Mpa, -0.04Mpa, -0.03Mpa, -0.02Mpa, -0.01Mpa, continued stable desorption for 12 min, tested desorption rate of toluene at 4, 6, 8, 9, 10, 11, 12 min, to confirm the optimum desorbing pressure and time

5. Experiment on stability

Under the optimum adsorption and desorption conditions of pressure swing confirmed by the above experiment, 100 batches of experiment on stability were conducted with resin and activated carbon as adsorbent, and removal rate of toluene each time and observed characters of resin and activated carbon were tested.
Gases treated in experiments of 1)-5) were all air containing toluene of 0.5178mg/L.

2.5 Analysis method

Toluene concentration was tested by Waters 600 High Performance Liquid Chromatograph (HPLC), and chromatographic column was C18 reversed phase column. Flowing phase was carbinol: water (80:20). Flow speed: 0.8ml/min, ultraviolet detector, 254nm in wavelength.

To investigate surface area and aperture distribution of resin and activated carbon, we adopted BET method, nitrogen gas as adsorbent, equipment type be Micromeritics ASAP2010 (USA). Infrared Spectroscopy was tested by Fourier transform infrared spectroscopy by using potassium bromide and pressed film method of resin powder. Element analysis was tested by Perkin-Elmer240c (USA) elemental analyzer.

resin	XAD-4	NDA-150	ND-90	granular activated carbon
structure	macroporous adsorption resin	hypercrosslinked	amino-modified hypercrosslinked	carbon build-up
geopolarity	non-polar	weak-polar	mid-polar	weak-polar
BET surface area (m^2/g)	850	906	819.1	880
average pore diameter (nm)	5.83	1.7	1.5	4.0
microporous surface area (m^2/g)	3.1	561.3	463.3	231.7
micropore volume (mL/g)	0.0051	0.2256	0.2186	0.1048
porosity (%)	40	53	52	42
granularity (mm)	0.4-0.6	0.4-0.6	0.4-0.6	0.4-0.5
oxygen content (%)	0	2.9	1.5	0.2
amino content (mmol/g)	0	0	1.51	0

Table 2. The nature of resin ND-900, NDA-101 and NDA-99

3. Results and discussion

3.1 The comparison of toluene adsorption of resin with activated carbon

Seen from Fig.2 evidently, the adsorption quantity of resin NDA-150 is the highest under the same condition, mainly relating to aperture, specific surface area and polarity of every adsorbent. Toluene is weak-polar molecule, so the adsorbability is minor on the non-polar resin XAD-4. In addition, from table1 we know that the aperture of resin XAD-4 is mainly big pore, but the other three kinds of sorbent have a certain amount of micropores.

Adsorbent with. micropores adsorbs molecule not only depending on the high specific surface area, but also filling function of micro-aperture and capillarity, which both play significant role [8-10], so the adsorption quantity of resin XAD-4 for toluene is the lowest. Resin ND-90 is amino-modifing hyper-cross-linked resin, is mid-polar resin in spite of a lot of micropores, be adverse for the adsorption of toluene for higher polarity, so the adsorption quantity of toluene is less. The polarity of active carbon and NDA-150 matches with the polarity of toluene, the discrepancy of specific surface area between two adsorbents is not big, and the main distinguish is micropore volume. The micropore volume of NDA-150 is more than twice as activated carbon, so the adsorption quantity of resin NDA-150 for toluene is the highest. Therefore, resin NDA-150 and activated carbon were taken as adsorbent in the following experiment.

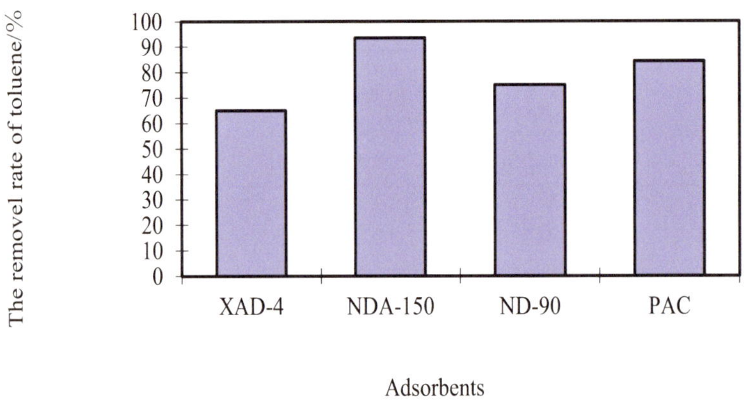

Fig. 2. The comparison of every adsorbent in toluene adsorption

3.2 The confirm of the best adsorption flow

According to the fig.3, we know that the removal of toluene reduced with the increase of intake flow. Adsorption process is very complicated, generally, can be divided into three phases: ①out-diffusion, gas molecules get on the surface of adsorbent form outside space; ②in-diffusion, gas molecules go deeply into adsorption surface along absorbent channel; ③adsorption on the internal adsorption surface. Toluene molecules have no time to contact with adsorbent fully and be adsorbed when adsorption flux increased, and effuse from the adsorption column with airflow. Thereby, only by controlling the flow can toluene molecular absorbed by adsorbent. From fig.3, we learn that the resin NDA-150 has optimal adsorption effect when the adsorption flow is 3L/min, but the absorption effect at 5L/min has very small difference with that of 3L/min, however, adsorption efficiency reduced evidently when the gas flow increase from 5L/min to 6L/min. There are certain requests for adsorption volume in actual adsorption process, and the adsorption volume of exhaust gas was not great in the adsorption process when the adsorption efficiency is the highest, hence, confirmed the optimal adsorption flow was 5L/min considering the two factors, the absorption efficiency and the adsorption volume disposal gas. For activated carbon, adsorption efficiency declined evidently when the gas flow increased from 4L/min to 5L/min, so the optimal adsorption flow is 4L/min.

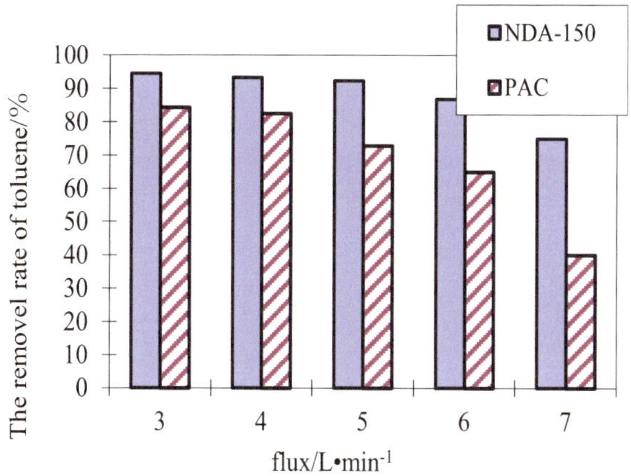

Fig. 3. The adsorption results of different adsorption flows

3.3 The confirm of the best adsorption time

As can be seen from the Fig4, the adsorption efficiency of adsorbent declined as time increased. It will take some contact time for toluene gas molecules entering into micropore of adsorbent; therefore, the longer the time, the much more amount of toluene entering micropore, tends to saturation in the end. Seen from Fig.4, the maximum absorption efficiency of NDA-150 resin and activated carbon are absorption for 5 minutes. There is slight difference of absorption effect between 10 min and 5 min for absorption time. At the same time, absorption efficiency descended significantly when the absorption time varying from 10 min to 12 min. Resin had the maximum adsorption efficiency when the absorption time was 5 min, but the volume of gas that resin dealt with was too small to satisfy the demand for the volume of gas that resin dealt with in actual process. Allowing for the two factors of absorption efficiency and absorption capacity, we determined the best adsorption time of NDA-150 resin and activated carbon were in 10 minutes.

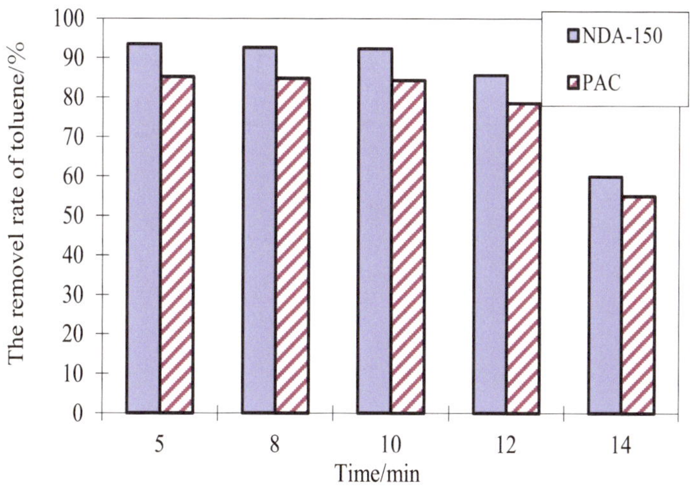

Fig. 4. Different absorption effects of different adsorption time

3.4 The confirm of desorption pressure and desorption time

From Fig. 5 and Fig. 6, we know that desorption efficiency of activated carbon increases with the time in the same condition of desorption negative pressure, and tends to balance gradually. With a certain desorption time, increasing desorption negative pressure makes desorption efficiency increased, and acquires the best desorption efficiency in -0.05MPa.

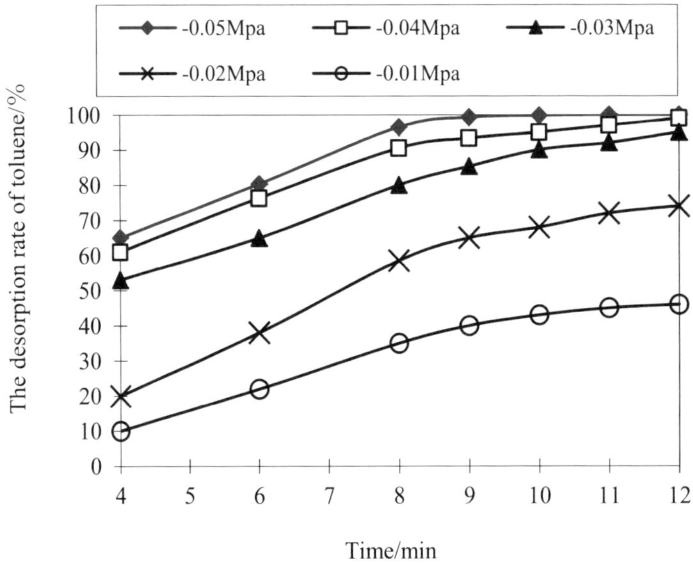

Fig. 5. The variation of desorption result of NDA-150 resin at under different desorption pressure with the time

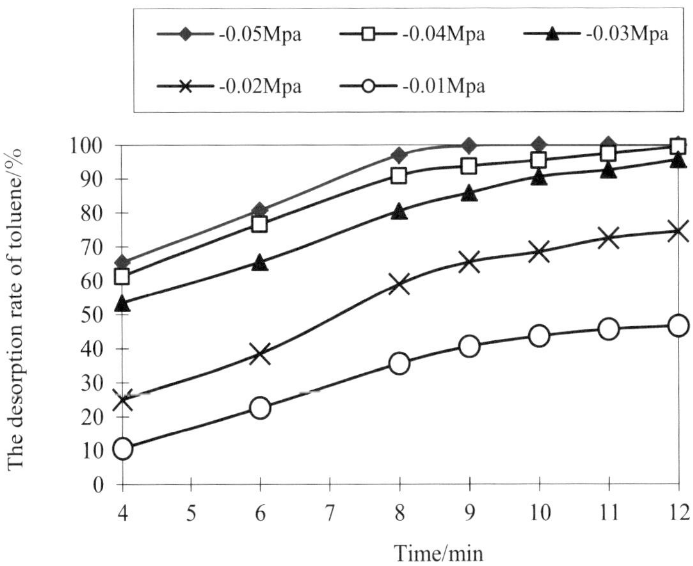

Fig. 6. The variation of desorption result of activated carbon at under different desorption pressure with the time

In theory, the greater desorption negative pressure, the higher desorption efficiency. However, in actual project, the greater negative pressure, the greater requirements of equipment and energy. As a result, we determined the best desorption negative pressure of resin and activated carbon is 0.05Mpa from this experiment. In the desorption pressure of -0.05Mpa, desorption rate could reach 99% after 9 minutes, while desorption efficiency didn't increase significantly when desorption time varied from 10 minutes to 12 minutes, closing to balance. Accordingly, chose the optimum desorption time for 10 minutes, and desorption efficiency were 99.8 percent at this point.

3.5 The stability test

It can be seen from the results of stability (listed in table 3): because of the weak mechanical strength of activated carbon, the more the number of repeated experiments, the greater impact under pressure, much seriously damaged, seriously affected its treatment. It is only by adding new activated carbon in time in order to guarantee treatment results. Conversely, the resin is synthetic organic polymer, has powerful mechanical strength. It didn't have damage within 100 batches of experiments and effect on the absorption efficiency of resin for toluene. Therefore, it is feasible to treat organic waste gas using resin as absorbent.

The number of experiments	Toluene removal (%)		Character	
	activated carbon	*NDA-150 resin*	*activated carbon*	*NDA-150 resin*
20	83.5	92.3	intact	intact
40	82	92.5	A portion of activated carbon had crack, and small particles of activated carbon emerged	intact
60	79.5	92	The amount of activated carbon that had cracking increased, walls of adsorption column are stuck by activated carbon powder.	intact
80	75	92	A small amount of active carbon powder appears at the bottom of adsorption column	intact
100	70	92.4	The quantity of activated carbon powder increased	intact

Table 3. The results of stability

4. Conclusions

Having studied pressure swing adsorption for purifying toluene gas of low concentration by using activated carbon and resin as absorbent respectively, the experimental results indicate:

NDA-150 resin has the best adsorption effect on adsorption of air containing toluene.

The optimum adsorption flow of resin and activated carbon is 5L/min and 4L/min respectively, the optimum adsorption time both are 10min, and the best adsorption efficiency respectively is 92.3% and 83.5%.

With desorption negative pressure be -0.05Mpa, the optimal desorption time of the resin and activated carbon both are 10min, and desorption rates can reach 99%.

It can be seen from the results of the stability test that resin as absorbent superior to activated carbon in treating air including toluene.

5. Acknowledgment

This work was funded by the Hebei Province Natural Science Foundation (grant no. E2009000581). The authors wish to express their appreciation to the Analytical Center at Changzhou Petro-Chemical College for the measurement of specific surface area and pore diameter.

6. References

Faisal I. Khan, Aloke Kr. Ghoshal, Removal of Volatile Organic Compounds from Polluted Air,*Journal of Loss Prevention in the Process Industries*, vol. 13, pp. 527-545, May(2000)

Zhang Lin, Chen Huanlin, Cai Hong,Study Progress of Membrane Method Process of Volatile Organic Compounds Emissions,*Environmental Protection of Chemical Industry*, vol. 22, pp. 75-80, February 2002.

You Yongyan, Chen Fanzhi, Huang Shujie,Research Development of Activated Carbon Fiber Desorbing Volatile Organic Compounds, *Industry Catalysis*, vol. 14, pp. 63-66, April 2006

Wang Yanfang, Sha Haolei, Yu Jianming,Technology Progress of VOCs Waste Gas of Low Concentration, *Energy environmental protection*, vol. 21, pp. 8-12, March 2007

Sun Yue, Zhu Zhaolian, Resin Adsorption sulfa deal with the production of intermediate waste, *Chemical Environmental Protection*, vol. 23, pp.9-14, January 2003

Wang Xuejiang, Zhang Quanxing, Li Aimin, Research on the Adsorption Behavior of NDA-100 Macroporous Resin to Salicylic Acid in Water Solution, *Journal of Environmental Science*, vol. 22, pp. 658-660, May 2002

Karan.l, T., Mehmet, K., Kildu., J.E., Wigton, A., Role of granular activated carbon surface chemistry on the adsorption of organic compounds: 2. NOM,*Envrion. Sci. Technol.*, vol. 33, pp. 3225-3233, October 1999

Wei Ruixia, Chen Jinlong, Chen Lianlong, Fei Zhenghao, Li Aimin, Zhang Quanxing, Study of adsorption of lipoic acid on three types of resin,*Reactive and Functional Polymers*, vol. 59, pp. 243-252, November 2004

17

Vapor Phase Hydrogen Peroxide – Method for Decontamination of Surfaces and Working Areas from Organic Pollutants

Petr Kačer[1], Jiří Švrček[1], Kamila Syslová[1], Jiří Václavík[1], Dušan Pavlík[2], Jaroslav Červený[2] and Marek Kuzma[2]
[1] *Institute of Chemical Technology,*
[2]*Institute of Microbiology,*
Prague
Czech Republic

1. Introduction

Decontamination, i.e. cleaning by removal of chemicals or germs, is a term commonly used for the process of treating devices, instruments and surfaces in order to ensure their safe operation. It is of exceptional importance in healthcare, food and pharmaceutical industry as well as in the areas of army and public defense. The decontamination process covers several steps like simple washing with water and soap and final disinfection or sterilization. Sterilization is a process which reduces the microbial contamination by 6 logs and can thus prevent the spread of infectious diseases in medical centers, where it forms a part of daily cleaning routines. Disinfection is a very similar process to sterilization, but the reduction rate of microbial contamination is only 5 logs. (Favero & Bond, 1991; Sagripanti & Bonifacino, 1996)

Decontamination is not only elimination of biological pollution but it also comprises detoxification and removal of dangerous chemical compounds. It should be applied whenever a real threat of microbial or chemical contamination exists.

2. History

While the sophisticated ("scientific") ways of decontamination only started to emerge about 150 years ago, essentially similar process are already mentioned in the Bible, in the works of the poet Homer and in the files of Aristotle. An important milestone concerning decontamination was the year 1438 when Sanitary council in Venice was found to provide fumigation of cargo delivered to the port. This institution represented the fundamental prevention and active defense against infectious diseases and parasites. Maturity of Italian health service was demonstrated in the works of poet, philosopher and physician Girolam Fracastor (1478-1553). He first promoted an idea that epidemics had been caused by very small particles which can be transferred among people by three different ways: by contact with infected patients, by contact with the contaminated staff (medicinal staff) and by air transport. Although this great man's work can be considered a huge milestone in the fight with infectious diseases, in his time it was forgotten and only rediscovered in the 20th century (1930 - W. C. Wright, 1960 - W. Bulloch). In 1676, a chemical (vinegar) was used to kill germs for the first time by Antonie van Leeuwenhoek who observed them with his microscope, calling them „animalcules". However, the breakpoint in this topic came in the second half of the 18th century with the discovery of chlorine (1774, C. W. Scheele) and hypochlorites (1789, C. L. Berthollet). These compounds quickly found their application in deadhouses, sewers, hospitals areas, ships, prisons and mainly in drinking water treatment. In 1810 Nicolas Appert discovered the modern food sterilization method by temperature conservation. Shortly after this, the founder of microbiology Louis Pasteur discovered the sterilization effect of overheated steam. This further inspired Charles Chamberland to construct the first steam autoclave (1879). Parallel with the development of this excellent technique, in 1877 A. Downes and T. P. Blunt discovered the antimicrobial effects of ultraviolet light and M. Wald (1892) continued in this work describing the relation of light wavelength and its germicidal effect (blue light is more effective than red one). Another famous man connected with the decontamination process was Robert Koch who in his book "On Disinfection" (1881) described the potential of 70 chemical compounds at different concentrations, temperatures and various mixtures to eliminate the spores of anthrax. In 1897 B. Krönig and T. Paul developed the grounds of chemical disinfection and these principles were applied in the famous "phenol coefficient method" to test the effectiveness of disinfection compounds. The following 20th century meant a great improvement in chemistry - mainly organic - which lead to the discovery of many disinfection compounds (Block, 1991; Fraise, 2004).

2.1 Reasons for decontamination

The need of decontamination already appeared during the army operations in the Antiquity. Aristotle (384 - 332 B.C.) revealed the danger of infectious diseases and recommended preventive measures for the army troops of Alexander the Great. The progress of medicine and the study of infectious diseases lead to formulation of strict rules to prevent infections. During epidemic periods, the application of protective means was becoming common, e.g. application of antiseptic compounds against gangrene (1750 J. Pringle), using of hypochlorite solutions before surgeries (O. W. Holmese - 1843 a I. P. Semmelweise - 1861), treatment of surgery instruments with flame, sterilization of bandage by heat (L. Pasteur) and many others. Despite all these advances, more soldiers died during the Second World War due to infections and diseases than as a consequence of fight injuries (Block, 1991).

Decontamination is an important part of the whole modern medicine system and is based on strict rules and application of several procedures like cleaning, disinfection, sterilization etc. Nonetheless, nowadays a real danger of pandemic (epidemic spreading in parallel in several states or continents) also exists, which was evident in the case of the two recent pandemics of flu (bird flu, pig flu) that proceeded very fast. Therefore, hospital decontamination is based on proper decontamination equipment (built-in or mobile), protective items and educated staff.

Another very important reason for the progress of decontamination is chemical war as a new military strategy. In the 19th and 20th centuries, the great progress of chemistry led to the development of several poisonous substances (weapons of mass destruction) (Duffy, 2009). A warfare agent of such kind was used for the first time during the WWI when the

Germans used chlorine (April 22, 1915). Later, phosgene, benzyl bromide and others were used (Duffy, 2009). The destructive force of these weapons was improved by new stable, more potent and easily spreading compounds. From the beginning of the WWI to the end of WWII, blistering agents were developed such as yperite and even more dangerous nerve agents like somane, sarine or tabune. Fortunately, they have never been used in war. After the WWII, research was focused on the development of nerve agents and effective defense against them. The new types of V-agents were developed in 1995 and represent the most toxic compounds ever synthesized.

Although the application of chemical weapons is currently considered a war crime (1993 Paris convention) (International Committee of the Red Cross, 2005), they are used by terrorists against civilian population. For instance, one can mention the sarine terroristic attacks in Japan (1995 Tokyo subway, 1994 Matsumoto town) (Okumura et al., 1998, 2003) or the bio-terrorist attacks by Anthrax spores delivered by mail service in the USA in 2001 (24 buildings were contaminated and the remedies cost as much as 200 million USD) (Jernigan et al., 2002). These incidents showed global unpreparedness for large contamination and improper decontamination methods for such spaces.

Decontamination is therefore highly important, both in the defense of an individual person and the defense of a country mainly against the pandemics or terroristic attacks. Beside this, the huge amounts of toxic compounds which are daily manufactured, modified and transported need to be considered. In case of an accident or improper manipulation, these substances can endanger the safety of a particular person or the whole environment. As a few examples of the 20th century, we would like to mention the outflow of toxic dioxin in Italian town Seveso in 1976 (initiated two prevention guidelines SEVESO I and SEVESO II), the nuclear accident in Chernobyl (1986) and the biggest industrial accident in the town Bhopal, India, (1984) where approximately 20,000 people died (Sharma, 2005).

For the sake of global prevention, it is extremely desirable to develop novel effective decontamination methods appropriate for application in large areas like rooms, buildings, airplanes, subways or airports. Despite the fact that decontamination of such premises represents an issue of foremost importance, it is not solved satisfactorily at present. An example of such solution is the patent of the United Technologies Corporation (Watkins, 2006) dealing with easy distribution of hydrogen peroxide aerosol or gas to large areas. A list of important methods usable for large and closed areas can be found in the Compilation of Available Data on Building Decontamination Alternatives issued by the U.S. Environmental Protection Agency (EPA) in 2005 (U.S. Environmental Protection Agency, 2005). In the last three decades, chemical contamination has most frequently been caused by pollution with toxic and usually carcinogenic pesticides or industrial intermediates while accidents or terrorist attacks have only formed a minority of cases. These compounds are mainly characterized by greater stability and usually persist in the environment for a long time. A second important group of polluting compounds are pharmaceuticals, diagnostics, flavoring substances and other bioactive compounds. Presently, the production, distribution and application of bioactive substances indeed represent a fast-growing branch of industry. These compounds and their bioactive metabolites are mainly concentrated in wastewaters, which are subsequently drained into the environment. Although their concentration is rather limited (ng l^{-1} – ug l^{-1}, corresponds to ppt – ppb) and they do not seem to be an actual danger, their final influence can be very dangerous. The active pharmaceutical substances are designed for a very specific effect, but their final side effects can often be unpredictable.

Lists of pharmaceuticals which contaminate the environment can be found in the literature. These compounds come either in the unchanged form or in the form of their biotransformation products (metabolites, which can be more active than the original substances (Daughton & Ternes, 1999)). The lists include for example analgesics, antidepressants, antiepileptics, antihypertensives, antiseptics, cytostatics, hormones, cholesterol reducing substances, radiocontrast substances, steroids, tranquilizers and others. (Daughton & Ternes, 1999; Lopez et al., 2003; Pereira et al., 2007). Big sources of these substances are of course pharmaceutical companies, research laboratories, hospitals, pharmacies and households in which these compounds are used.

National Institute for Occupational Safety and Health, NIOSH, estimates that 5.5 million of workers can be indirectly exposed to dangerous pharmaceutical substances which are commonly labeled "cytotoxic". These data are based on evaluation between 1996-1998 in the USA (National Institute for Occupational Safety and Health, 2004) among research staff, pharmacists, physicians, nurses and other supporting staff. Staff from pharmaceutical factories were not considered because of the strict rules for clean premises which reduced the risk of contamination to minimum. On the contrary, in hospitals the staff and the patient relatives can be in touch with hazardous pharmaceutical substances like cytostatics, antivirotics, hormones etc. (National Institute for Occupational Safety and Health, 2004). These substances can cause undesirable effects (Castegnaro et al., 1997) by transport into their bodies through skin absorption, inhalation of aerosols, syringe needles or open strokes. The danger of indirect contamination with pharmaceutical substances on the inner surfaces of 14 hospitals in Germany is described in the work of Schmaus (Schmaus et al., 2002). Better results were naturally achieved in those hospitals which strictly followed the safety measures and where their staff worked properly with cytotoxic substances.

Methods of protection against dangerous cytotoxic substances are nowadays undergoing a fast development. The most important measure is to strictly follow proper working rules to avoid contact of staff with these compounds, which means e.g. working in isolators with intrinsic decontamination system, proper waste management, safe storage or periodic evaluation of the level of contamination. (Fisher & Caputo, 2004; National Institute for Occupational Safety and Health, 2004) Currently, the development is focused on new decontamination methods to provide perfect cleanup and inactivation of dangerous substances in various kinds of waste material packaging and biological liquids before their disposal (Cazin & Gosselin, 1999; Hansel et al., 1997).

2.2 State of the art of decontamination techniques

Up to now, a lot of decontamination methods to inactivate biological pollution on different surfaces have been developed. Surface decontamination of chemical (mainly cytotoxic) substances was studied less intensively but nowadays it is also becoming a priority. Decontamination can be carried out by several ways depending on the contaminant, environment, area size or target (people, tools, indoor or outdoor space). Among the main factors affecting the right selection of suitable decontamination techniques are: the method of distribution of the decontamination substance (washing, wiping, spraying, foaming, using aerosol, fumigation etc.); the operation range (selectivity to the microorganism or chemical pollutants); the influence of working conditions (temperature, humidity, presence of other compounds); the operation time (time necessary for proper reaction - minutes or hours) and the influence of the decontaminant upon target materials (possibility of damage).

It is possible to categorize the decontamination methods according to their principle of action to mechanical, chemical, physical and physicochemical. The most widespread are chemical decontamination procedures (an action of a chemical agent to a decontaminated item) that can be applied in two different ways - the wet approach, which uses water or other solution of active agent and the dry method which uses gas or vapor phase of active substance.

Application of the aforementioned methods is connected with serious danger because of toxicity, carcinogenicity, flammability or explosiveness of agents, irradiation or burn caused by rays, and potential toxic residues which are harmful to the environment. A big disadvantage of liquid agents' application is a non-uniform distribution on the surfaces in all target areas. This disadvantage was solved by spraying of the agents or using fumigation (vapor or gas of the active agent). It is evident (see Table 1), that chemical agents are a very heterogeneous group with different mechanisms of action. It is also important that several chemical agents can be mixed with detergents, which can support the deactivation process, or other compounds that improve their properties (anticorrosives, aromatic additives and others). These additives can e.g. substantially reduce the surveillance of germs (bacteria, viruses or fungi). According to their influence, these agents can be divided into two types:

- (Bacteria) -cide- meaning permanent dispatch
- (Bacteria) -static- meaning temporary loss of any ability, i.e. multiplication or growing.

Type of decontamination		Active agent
Mechanical ways		Sucking, washing, wiping etc.
Chemical ways	*Wet*	Water solution of ClO_2, CH_3COOOH, H_2O_2, NaOCl, liquid detergents (presence of quaternary ammonia salts), alcohols, aldehydes, phenol derivatives, iodoform, Fenton agent and others.
	Vapor or gas phase	Ethylene oxide, formaldehyde, ClO_2, O_3, CH_3COOOH, H_2O_2, propylene oxide, β-propiolactone, methylene bromide and others.
Physical ways		X-ray, gamma ray, microwave and UV radiation, heat (dry, wet - water steam), freezing, plasma, photochemical reaction, hydrostatic pressure and others.
Physicochemical ways		Heat or radiation combined with chemical agents

Table 1. A list of basic decontamination methods (Kuzma et al., 2008; McDonnell, 2004; McDonnell & Russell, 1999; Rogers et al., 2005; Russell, 1990, 1991).

It is necessary to regularly change the decontamination agent with respect to a different active substance in order to avoid the potential resistance of the microorganisms to the agent used. An ideal procedure for the cleanup of chemical contaminants using chemical agents should consist of their physical removal from the surface followed by effective degradation to nontoxic or at least less toxic compounds. Not a single decontamination agent with such a wide spectrum of reactivity exists, which could be used for the decontamination of all biologically active substances. Their degradation can even lead to formation of products which are more toxic than the original substance. Identification of these products is very difficult. Therefore, it is preferred to apply "one use" surfaces during the operation with biologically active substances. These surfaces are sequentially washed but their decontamination is quite risky due to the removal of highly toxic contaminants like cytostatics or immunosuppresives. Moreover, they can be drained to the environment which is dangerous not only for the staff but, more importantly, to the whole population. It was already mentioned that these substances are strictly designed for a specific application but they can have numerous side effects on people. Thus, the contact with these compounds has to be avoided as much as possible (Roberts et al., 2006).

NIOSH recommends to decontaminate all surfaces which have been in touch with cytotoxics according a protocol that includes appropriate agent able to deactivate or remove chemical or biological contaminants (National Institute for Occupational Safety and Health, 2004). The basic question to be solved in the field of chemical decontamination is the criterion determining the level at which the contaminant can be considered deactivated. In the case of biologically active substances like warfare agents or pharmaceuticals, this criterion represents the loss of their biological activity, i.e. changes in the chemical structure of the contaminant leading to biological inactivity. An ideal degradation leads to gaseous, nontoxic products which can be easily exhausted - oxides of elements commonly contained in organic molecules (CO_2, H_2O, NO_x). Since they are products of oxidative reactions, it is appropriate that strong oxidative agents be used. Even though the application of $KMnO_4$ seems to be very effective, due to safety reasons it is not acceptable in common places like hospitals (Barek et al., 1998). Other well-known oxidative substances like $Ca(OCl)_2$, NaOCl or H_2O_2 are widely applicable mainly in the liquid form and thus they can only be used in local areas. Unfortunately, the methods described above are inapplicable for treatment of large areas in routine application. In the development of novel decontamination technologies, it is necessary to approach the "ideal decontamination agent" (Rutala & Weber, 1999).

3. Ideal decontamination agents

Based on the above-mentioned facts, the "ideal decontamination agent" can be defined. It should possess high activity against a wide spectrum of biological and chemical contaminants, quick start of action and long-lasting effect. It should be nontoxic to humans and the environment, compatible with a variety of materials, resistant to organic materials, have a non-limited disposal and long term stability during storage. It should also be easily detectable, have a pleasant or no smell and a reasonable cost. Its handling (application and storage) should be easy and safe. Another useful property of the ideal agent can be e.g. applicability to large areas or whole buildings mainly in case of pandemics, chemical accidents or terrorist attacks. The contaminants can be found in places difficult to clean like cracks in walls, carpets, woods, ventilation pipelines and air conditioning units. The common methods are inapplicable in large areas where the only solution is application of a gaseous decontamination agent because it provides easy distribution and penetration in broken surfaces. However, the most commonly known gas phase decontamination agents are connected with a number of disadvantages such as toxicity, material incompatibility, concentration requirements, time of exposition and aeration time (Rogers et al., 2005). From the aforementioned chemical agents, hydrogen peroxide looks like an ideal agent due to the non-toxic products of its decomposition - water and oxygen. The well-described antimicrobial activity and strong oxidative potential are also a good precondition for its wide application mainly against biological and chemical contaminants.

4. Vapor phase hydrogen peroxide

Very close to "the ideal decontamination agent" seems to be Vapor Phase Hydrogen Peroxide (VPHP). It is a relatively new but very progressive method with many advantages:

- approved sterilization of a wide range of microorganisms
- higher germicidal activity than what can be achieved with a liquid solution of hydrogen peroxide
- environmental friendliness - the decomposition products are water and oxygen and leave no toxic residues on the surfaces
- possibility of its usage under common conditions - atmospheric pressure, laboratory temperature
- applicability in larger areas and rugged surfaces.

For its excellent antimicrobial activity and nontoxic decomposition residues, the VPHP process tends to replace especially toxic, carcinogenic and potentially explosive formaldehyde and ethylene oxide used for sterilization of heat-sensitive materials (Block, 1991; Heckert et al., 1997).

VPHP is typically generated from a water solution of hydrogen peroxide (35% w/w). A common way of vapor generation is controlled heating of the solution under proper conditions avoiding decomposition of the VPHP. Like all other decontamination agents in vapor phase, also VPHP is decomposed during the operation (in fact, the rate of decomposition of hydrogen peroxide is higher than the decomposition rate of ClO_2) and it is thus necessary to refill "fresh" VPHP to the target area. The dose of new VPHP keeps the required concentration during the whole process. At the end of decontamination, the generator is switched off and the rest of hydrogen peroxide vapor is ventilated out by aseptic air. This exhaust goes through the catalyst to decompose hydrogen peroxide (U.S. Environmental Protection Agency, 2005).

The sterilization properties of VPHP were for the first time mentioned in the 70's of the 20th century, but the modern concept is dated 1989 when this method was used for quick sterilization of rugged dental instruments (Block, 1991). In the same year, EPA approved the usage of VPHP in closed premises like isolators, closed rooms or operation boxes (U.S. Environmental Protection Agency, 2005; McDonnell et al., 2007). Since then, fast-growing application of this agent has started, focusing mainly on bio-decontamination in pharmaceutical industry, health service and food industry (Block, 1991; Kahnert et al., 2005; Klapes & Vesley, 1990). In 2001, VPHP was used for the first time for decontamination of two post office buildings (the General Services Administration's Buildings 410 in Washington, D.C. and the U.S. State Department Mail Facility in Sterling, Virginia; contaminated space 30 000 - 60 000 m^3). These buildings were contaminated by Anthrax spores released from "Anthrax letters" by terroristic attacks in the USA (U.S. Environmental Protection Agency, 2005).

In the literature (Block, 1991; Heckert et al., 1997; Johnston et al., 2005; Klapes & Vesley, 1990; Roberts et al., 2006), there are numerous applications of VPHP in the decontamination of fermenters, dialysers, incubators, isolators (Fisher & Caputo, 2004; Lysfjord & Porter,

1998), glove boxes, hazard boxes (Hall et al., 2007), animal houses (Kahnert et al., 2005; Krause et al., 2001), hospital wards (French et al., 2004; Hardy et al., 2007), inner space of airplanes (Krieger & Mielnik, 2005; Shaffstall et al., 2011), ambulances, various large spaces (Krause et al., 2001), lyophylisators (Johnson et al., 1992), ultra centrifuges, sterilization tests (Kokubo et al., 1998), product and pipe lines (Hatanaka & Schibauchi, 1989), dental and surgery instruments (catheters, endoscopes, etc.) (Bathina et al., 1998), contact lenses, hardware space systems (Chung et al., 2008), and food commodities (Forney et al., 1991; Gruhn et al., 1995; Sapers et al., 2003; Simmons et al., 1997). It is a method by which a high selectivity of the process can be achieved due to very precise control of the sterilization conditions (concentration of VPHP, temperature, time of sterilization) and thus only pathogenic microorganisms are destroyed leaving normal living cells unharmed. Therefore, this method is feasible for the decontamination of the surface of living cell cultures. The world-known companies like Tetra Pak International, PepsiCo Inc. or Tetra Laval Holding & Finance are dealing with research and application of in-line sterilization of food packaging by VPHP, which clearly demonstrates how important this technology is in this branch. An interesting and effective application of this method is bleaching of textile materials, which allows to decrease the operating temperature and thus to increase the economics of the whole process. The main leaders in bio-decontamination of surfaces by VPHP are at present companies Steris (Mentor, USA) and Bioquell (Andover, UK). Beside them, several other companies work on VPHP technology, e.g. Pharmaceutical Systems (Franklin Lakes, USA), American Sterilizer Company (Mentor, USA), Johnson & Johnson and division Ethicon (New Brunswick, New Jersey, USA) or Surgikos (New Brunswick, USA).

5. VPHP – Mechanism of action

Information about the mechanism of VPHP action is actually very limited as the process is still in the focus of basic research (Klapes & Vesley, 1990). Current literature data (Chung et al., 2008; Unger-Bimczok et al., 2008; U.S. Environmental Protection Agency, 2005) indicate that the VPHP process is a multi-parameter problem, the effectiveness of which is mainly influenced by the concentration of gaseous hydrogen peroxide, temperature, relative humidity, and condensation of hydrogen peroxide on the decontaminated surfaces. Similar behavior was found for gaseous formaldehyde as a decontamination agent which has been described in detail (Hoffman & Spiner, 1970). Unger et al. (Unger et al., 2007) for first time studied the influence of all these conditions on the sporicidal effect of VPHP. The results suggest that the main parameter for microbial deactivation is the molecular distribution of water and hydrogen peroxide on the surface while the concentration of hydrogen peroxide only plays a secondary role. It was also found that the decontamination cycle using a relatively lower concentration of hydrogen peroxide and higher relative humidity gave very similar results as an experiment with higher concentration of hydrogen peroxide and lower relative humidity. This confirms the possibility of conducting the VPHP process in two ways: "wet" or "dry".

6. VPHP – Operational conditions

Decontamination of closed areas by VPHP is carried out in 4 consecutive steps (Fisher & Caputo, 2004; Heckert et al., 1997; Roberts et al., 2006; Watling et al., 2002). The first phase is dehumidification, i.e. reduction of humidity to an acceptable level, and also temperature stabilization of the VPHP generator. The second phase is conditioning which includes the transport of evaporated hydrogen peroxide by the carrying medium (air) to the decontaminated area and achievement of the required hydrogen peroxide concentration. The third phase is the decontamination itself - that means steady evaporation of hydrogen peroxide and its transport to the decontaminated area to maintain a constant concentration during the whole process. Aeration represents the final phase which consists in feeding of aseptic air in the decontaminated area to exhaust hydrogen peroxide vapor and to keep its concentration at a safe level. The process is illustrated in Figure 1.

Duration of the whole decontamination cycle depends on many parameters. The main ones are the size of area, the profile of surfaces, the way of VPHP generation, endurance of the contaminant against VPHP and the method of space aeration. The whole decontamination cycle should not exceed 10 hours as this is a limit of application of this chemical compound as a sterilization agent (Gurevich, 1991). The decontamination cycle based on biocidal properties of VPHP meets this requirement. There are two different ideas about how to carry out the surface decontamination with VPHP (Fisher & Caputo, 2004; Unger-Bimczok et al., 2008; Watling et al., 2002). The traditional one prefers to perform the decontamination under "dry" conditions without condensation of hydrogen peroxide or water (preferred by Steris (Fisher & Caputo, 2004)). Condensation is unwanted because of corrosion of many materials and prolonged aeration time. The process is not under control and thus in case of condensation the decontamination is not homogeneous (Unger-Bimczok et al., 2008; Watling et al., 2002). In case of the "dry" way of the VPHP process, decontamination of closed indoor spaces and surfaces or quick inactivation of contaminants due to high concentration of hydrogen peroxide vapors is quickly achieved. However, the atmosphere in a defined space can absorb only a limited amount of water and hydrogen peroxide, so it is necessary to remove humidity out of this space by desiccators to avoid condensation (Fisher & Caputo, 2004). On the contrary, the second popular opinion says that the hydrogen peroxide vapors are stable and condensation is necessary. In this view, also condensation is the primary reason of the VPHP decontamination and thus condensation is necessary in order to carry out surface decontamination by hydrogen peroxide vapors (Watling et al., 2002). This theory is supported by theoretical and experimental analysis which clearly show that condensation, and mainly "microcondensation" (i.e. non-visible condensation in small amounts), are key and critical parameters for quick and reproducible inactivation of microorganisms by VPHP (Unger-Bimczok et al., 2008). Condensation (preferred by Bioquel (Fisher & Caputo, 2004)) in case of the "wet" VPHP process is the basic requirement of the technology. A thin condensation film is formed on the decontaminated surfaces. It is necessary to control the condensation level of decontamination agents during the whole process because it inhibits the process (Sheth & Upchurch, 1996). Determination of the vapor mixture dew point (i.e. hydrogen peroxide and water vapours) is very difficult. Its value depends on the temperature and pressure and cannot be predicted easily. This dependence is well described by the Raoult's law which requires the input knowledge of activity coefficients that are also dependent on temperature and concentration. For better understanding of the difference between "wet" and "dry" processes, it is good to study the thermodynamics of the hydrogen peroxide-water solution and its behavior during evaporation and condensation (Manatt & Manatt, 2004; Scatchard et al., 1952). The pressure of saturated vapor of water and hydrogen peroxide is below atmospheric so both compounds start evaporating under atmospheric conditions. Lower pressure of saturated vapor of hydrogen peroxide causes the higher boiling point of hydrogen peroxide and slower rate of evaporation compared to water. Binary system hydrogen peroxide and water has a reduced dew point because of H-bonds (i.e. reduced total vapor pressure) and thus saturation and eventual condensation can occur at a level of relative humidity lower than 100%.

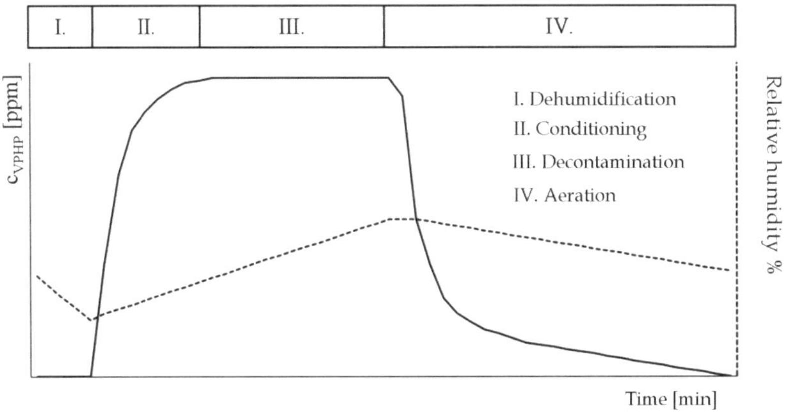

Fig. 1. Time dependence of the VPHP concentration with respect to the decontamination phases.

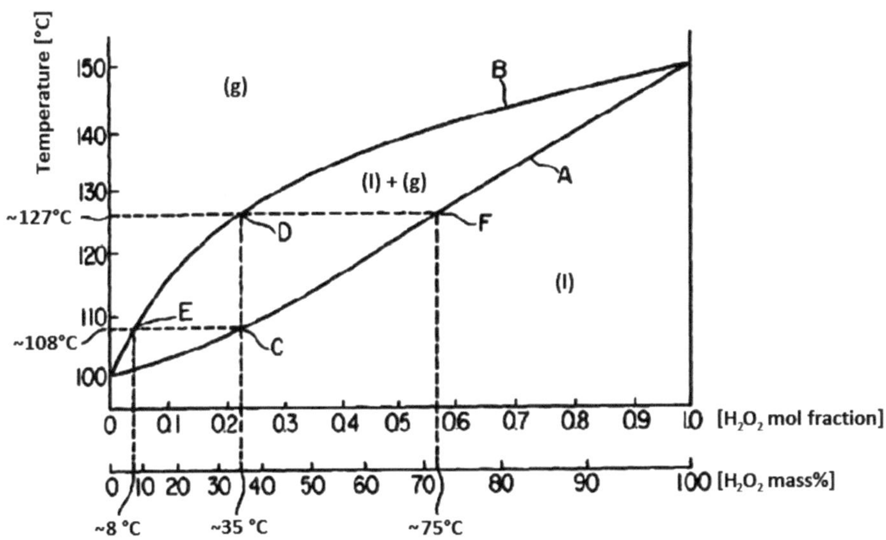

Fig. 2. Isobaric equilibrium liquid-vapor (T,x,y diagram) for the real mixture of hydrogen peroxide and water (Hatanaka & Schibauchi, 1989)

The graphic illustration (U.S. Peroxide, 2009) of the liquid-vapor equilibrium clearly shows the differences in hydrogen peroxide concentrations on the decontaminated surfaces during "wet" and "dry" processes. The differences are caused by different condensation speed of both compounds which also has an influence on the decontamination rate - the higher the concentration of hydrogen peroxide, the faster the decontamination.

Concentration H_2O_2 at 25 °C [mass. %]	
Vapor	**Liquid**
1,9	32,1
8,0	55,7
24,1	73,8
35,0	77,8
58,4	88,3

Table 2. The equilibrium concentration of vapor and liquid phase of H_2O_2 reached by evaporation (Hultman et al., 2007)

The hydrogen peroxide vapor can be generated in two ways - by controlled or flash evaporation. If the liquid solution of hydrogen peroxide evaporates in a dry closed space at 25°C (normal conditions), the concentration of hydrogen peroxide in the gas phase is much lower than in the liquid phase because of faster water evaporation from the hydrogen peroxide solution. For example, by evaporation of 35% (w/w) hydrogen peroxide solution, the final gas phase contains 2.15 % (w/w) H_2O_2 and 65% of H_2O (w/w) (Hultman et al., 2007). Saturation is a state when no more hydrogen peroxide and water vapor can be absorbed and thus condensation occurs.

A different situation occurs when water and hydrogen peroxide vapor condense at 25 °C. In Table 2, it is shown that the equilibrium condensate concentration formed from the vapor above 35% (w/w) hydrogen peroxide solution is 77.8 %, which is about two times higher than in the parent solution. Hydrogen peroxide in the vapor phase condenses preferentially over water. If the system contains vapors of hydrogen peroxide, its condensation occurs and thus its concentration in the vapor phase decreases.

The rate of evaporation can be increased by supplying heat. However, this must be done with a particular care because of safety reasons - hydrogen peroxide is extremely unstable at higher temperatures. Heat supply to the concentrated evaporated hydrogen peroxide slightly above normal conditions is safe and can be used for proper evaporation of concentrated hydrogen peroxide. The condensate is a highly concentrated solution that is not compatible with a wide range of materials and can cause corrosion.

Flash evaporation is another kind of a process related to hydrogen peroxide evaporation. The solution of hydrogen peroxide can be directly applied on a heated surface and thus evaporated. During the flash evaporation, hydrogen peroxide and water are evaporated from the solution simultaneously, so the concentration in the vapor phase is approximately the same as the concentration of the starting solution (decomposition of hydrogen peroxide is not considered). Thus, the concentration in the condensate is the same as in the parent solution.

In case of the "wet" VPHP process, the high concentration of hydrogen peroxide in the condensate can have a positive effect with respect to faster microbial decontamination, but only if the condensate covers the entire surface homogeneously. This is, however, nearly impossible to ensure, owing to different surface profiles of materials. In large rooms, temperature differences and different circulation of the atmosphere also play an important role. These factors, together with surface properties like wettability, sorption and catalytic activity, lead to formation of heterogeneous condensate in the form of drops or a thin film (depends on the wettability).

The application of VPHP in the solely "dry" process is advantageous because gas has a uniform contact with all exposed surfaces. All types of surfaces can thus be decontaminated to the same degree, including those with complex geometry - horizontal, vertical, cracks and curved surfaces. Moreover, it is possible to quickly remove the gaseous hydrogen peroxide from the area at the end of decontamination and thus save time of the whole cycle.
A theoretical model of decontamination by VPHP was presented (Watling et al., 2002) and it described the concentration profile of hydrogen peroxide vapor in a closed space during all four phases of the decontamination cycle and compared it with experimental results. The goal of this work was to create the model that could predict the main parameters of the decontamination process (concentration of VPHP, dew point, etc.) on the basis of operation conditions and other parameters (space dimensions), and thus control and conduct this process under optimal conditions with the highest efficiency.

7. VPHP – An excellent biocidal agent

Similarly as in its liquid solution, hydrogen peroxide also has sterilization properties in the vapor phase against vegetative bacteria and highly resistant bacteria endospores (Block, 1991; French et al., 2004; Hall et al., 2007; Johnston et al., 2005; Kahnert et al., 2005; Klapes & Vesley, 1990; Kokubo et al., 1998; Rogers et al., 2005; Sapers et al., 2003; Unger-Bimczok et al., 2008), viruses (Heckert et al., 1997), fungi (Forney et al., 1991), yeast, amoebae, infective proteins and other microorganisms (Fichet et al., 2004; Klapes & Vesley, 1990; Vassal et al., 1998). As a model organism for the validation, *Bacillus stearothermophilus* (*Geobacillus stearothermophilus*) which is mainly deposited on stainless steel is commonly used (Block, 1991; Bounoure et al., 2006; Chung et al., 2008, Fisher & Caputo, 2004; Johnston et al., 2005; Klapes & Vesley, 1990; Unger et al., 2007). This microorganism is very resistive against VPHP and serves as a surrogate of anthrax (*Bacillus anthracis*) because of their very similar behavior. The VPHP process is considered successful when all these microorganisms are deactivated. A wide range of commercial biological indicators designed for VPHP (spores of *Bacillus subtilis*) exist that are often used for VPHP validation (Klapes & Vesley, 1990; Kokubo et al., 1998).
Sporicidal efficiency of chemical decontamination agents is often expressed as the D-value, which represents the time (minutes) necessary to kill 90 % of the starting amount of microorganisms (or logarithms of the amount) at a constant temperature (Gould, 2004; Unger-Bimczok et al., 2008). In the next table, D-values of selected bacterial spores are compared and evaluated by liquid hydrogen peroxide and VPHP decontamination (U.S. Environmental Protection Agency, 2005). These results (Table 3) show that, in order to kill selected microorganisms, a 200-fold concentrated solution of hydrogen peroxide is necessary to get comparable results as in vapor phase.
Although it is well known that the bactericidal activity of liquid hydrogen peroxide solution grows with its increasing concentration, the linear dependence of its vapor-phase concentration on the killing activity of selected microorganisms is still a widely discussed topic. Some authors say that the antimicrobial activity grows with higher concentration of gaseous hydrogen peroxide, others express a completely opposite opinion and prefer the microbial deactivation at lower concentrations (Unger-Bimczok et al., 2008). In Table 4, all microorganisms tested for VPHP inactivation are summarized (Forney et al., 1991; Hall et al., 2007; Heckert et al., 1997; Johnston et al., 2005; Klapes & Vesley, 1990; Kokubo et al.,

1998; Reich & Caputo, 2004; Simmons et al., 1997). The table clearly shows that the application of VPHP as a biocidal agent is a well mapped topic.

Tested microorganism (Spores)	D-value [min]	
	Liquid solution of H_2O_2	VPHP
	$c(H_2O_2) = 370$ mg l^{-1} T = 24 - 25 °C	$c(H_2O_2) = 1 - 2$ mg l^{-1} T = 24 - 25 °C
Bacillus	1.5	1 - 2
Bacillus subtilis	2.0 - 7.3	0.5 - 1
Clostridium sporogenes	0.8	0.5 - 1

Table 3. Comparison of sporicidal effect of liquid and gaseous hydrogen peroxide (VPHP)

Bacteria + spores
Aeromonas sp.; Acholeplasma laidlawii; Acinetobacter baumannii; Acinetobacter calcoaceticus; Anaerobic cocci; Asperigillus spores; Bacillus anthracis (antrax illness)*; Bacillus alvei; Bacillus cereus; Bacillus circulans; Bacillus firmus; Bacillus licheniformis; Bacillus megaterium; Bacillus pumilus; Bacillus sphaericus; Bacillus (resp. Geobacillus) stearothermophilus; Bacillus subtilis; Bacillus thuringiensis; Bacteroides fragilis; Campylobacter sp.; Clostridium botulinum; Clostridium difficile; Clostridium perfringens; Clostridium piliforme; Clostridium sporogenes; Clostridium tetani; Deinococcus radiodurans; Enterobacter cloacae; Enterococcus faecium/faecalis; Escherichia coli; Fusobacterium sp.; Lactobacillus caesei; Legionella pneumoniae; Listeria monocytogenes; Klebsiella pneumoniae; Methicillin-resistant Staphylococcus aureus* (MRSA)*; Micrococcus sp.; Moroxelia osloensis; Mycobacterium bovis; Mycobacterium chelonei; Mycobacterium smegmatis; Mycobacterium tuberculosis; Pseudomonas aeruginosa; Pseudomonas cepacia; Salmonella choleraesuis; Salmonella typhimurium; Shigella sp.; Staphylococcus*
Viruses (family: type of virus)
Adenoviridae: Adenovirus, Canine adenovirus; Caliciviridae: Feline calicivirus, Vesicular exanthema virus; Coronaviridae: Infectious bronchitis virus; Flaviviridae: Dengue virus; Hog cholera virus; Herpesviridae: Herpes simplex Type 1, Pseudorabies virus; Iridoviridae: African swine fever virus; Orthomyxoviridae: Influenza A2, Avian infuenza virus; Paramyxoviridae: Newcastle disease virus; Parvoviridae: Parvovirus, Canine parvovirus, Feline parvovirus; Picornaviridae: Rhinovirus 14, Polio type 1, Swine vesicular disease; Poxviridae: Vaccinia; Reoviridae: Bluetongue virus; Rhabdoviridae: Vesicular stomatitis virus
Fungi
Alternaria; Aspergillus niger; Aspergillus sp., Blastomyces dermatitidis; Botrytis cinerea; Candida albicans; Candida parapsilosis; Coccidioides immitis; Histoplasma capsulatum; Penicillium sp.; Trichophyton mentagrophytes
Other microorganisms
Caenrohabditis elegans; Cryptosporidium parvum, Lactococcal bacteriophage; Syphacia muris

Table 4. List of microorganisms which have been inactivated by VPHP

8. Synergism of VPHP and related chemicophysical factors

Most of the works concerning the synergism of hydrogen peroxide were focused on liquid-phase reactions (water disposal treatment) and the synergism of VPHP is a considerably less studied topic.

In order to boost up the effects of hydrogen peroxide vapor, its ionization by plasma can be performed (Bathina et al., 1998; U.S. Environmental Protection Agency, 2005; Vassal et al., 1998). Commonly, plasma is generated from gases (argon, helium, nitrogen, etc.) by an electric pulse, radiofrequency or microwave irradiation. It can be formed at atmospheric pressure and higher temperature (105 °C) or at reduced pressure (~ 40 Pa) and substantially lower temperature (55 - 60 °C) which is called low-temperature plasma (Crow & Smith, 1995). It consists of free radicals (mainly hydroxyl or hydroperoxyl radicals), ions, neutral particles and excited atoms or molecules which show high activity in deactivation of contaminants (U.S. Environmental Protection Agency, 2005). The Johnson & Johnson Medical company and its division Surgikos Inc. patented for the first time a decontamination device in the form of a vacuum chamber where the contaminated instruments were treated with hydrogen peroxide plasma (Parisi & Young, 1991). This kind of decontamination, marked as „STERAD sterilization system", is recommended by the Food and Drug Administration (FDA) as an advanced sterilization technique for enclosed spaces (Crow & Smith, 1995). Although the systems using VPHP together with plasma are very useful for temperature- or water-sensitive materials, their big disadvantage is that they can only be applied in small closed spaces because of the high vacuum or temperature required in the chamber (Adams et al., 1998).

Other synergistic effect can be observed by combination of VPHP with UV irradiation (Klapes & Vesley, 1990). Application of UV irradiation alone requires a relatively long time and thus its combination with VPHP can significantly shorten the operations. The UV/H_2O_2 combination (photo-oxidation) is very effective in destroying microorganisms and heavy decomposable organic pollutants, i.e. volatile organic compounds (VOC) like benzene, toluene, phenol, *tert*-butyl methyl ether, halogenated compounds, pharmaceutical substances and pesticides, mainly in water (Esplugas et al., 2002; Kang & Lee, 1997; Lopez et al., 2003; Pereira et al., 2007; Prousek, 1996). It is known that hydrogen peroxide absorbs UV light within the range of 185 - 400 nm (Esplugas et al., 2002). The radiation energy at these wavelengths is sufficient to provoke photo excitation of the hydrogen peroxide molecule and a subsequent cleavage (photolysis) of the –O–O– peroxide bond, forming hydroxyl radicals which can initiate radical chain reactions of contaminants. The homolytic splitting of the hydrogen peroxide molecule leads to generation of hydroxyl radicals due to the absorption of a photon (Dionysiou et al., 2004). Hydroxyl radicals can be formed by the wide range of low wavelength ultraviolet radiation within 200-280 nm (Lopez et al., 2003). This interval of wavelengths is called UV-C and because of its germicidal properties it is commonly used for water and air sterilization and also for the degradation of photo unstable organic pollutants (Pereira et al., 2007). The UV radiation not only consists of UV-C (100-280 nm) but also contains the UV-B (280-315 nm) and UV-A (315-400 nm) ranges. These two last radiation areas are not used for the activation of hydrogen peroxide. The most common source (Lopez et al., 2003) of the UV-C radiation are low-pressure mercury lamps with the emission maximum at 254 nm since at this wavelength, the quantum yield of hydroxyl radicals equals to 1. The synergistic combination of liquid hydrogen peroxide and UV-C at the 254 nm emission maximum (indirect photolysis) used for the degradation of organic pollutants shows that it is highly effective and that the rate of reaction is higher compared to individual application of hydrogen peroxide or UV-C irradiation (Esplugas et al., 2002; Lopez et al., 2003; Pereira et al., 2007). The total rate of chemical degradation of contaminants by UV-C and hydrogen peroxide is dependent on the mechanism of the reaction of OH radicals with the contaminants, on the reaction rate of direct photolysis of contaminant (absorptivity of substrate), on the absorptivity of by-products and other absorbents of UV-C radiation at 254 nm (competitive absorption), on the intensity of the UV-C source and on the concentration of hydrogen peroxide (Kang & Lee, 1997). Degradation of organic substances by this effective combination proceeds by a radical oxidative reaction (Ray, 2000).

There are several practical applications of the VPHP decontamination process employing the synergistic effect of UV-C (VPHP/UV-C) for the sterilization of continual filing processes of food in liquid form (milk, water, juice). VPHP is applied in the first step of decontamination to treat surface and then the UV-C radiation treatment follows. It is also possible to combine the above-mentioned VPHP/UV-C system with other oxidative agents such as ozone, or the catalytic properties of TiO_2 can be used. Decontamination processes combining O_3/UV-C/VPHP show the most effective contaminant degradation and allow complete mineralization of pollutants (Esplugas et al., 2002). In the USA, the Department of Energy patented a portable device for surface chemical and biological decontamination which uses ozone as the main agent, reinforced by UV radiation and hydrogen peroxide added to reach higher reactivity (O'Neill & Brubaker, 2003).

In 1972, Fujishima and Honda (Fujishima & Honda, 1972) discovered the photocatalytic properties of nanocrystalline TiO_2 and predicted its possible application in chemical decontamination (Linsebigler et al., 1995; Wold, 1993). The TiO_2/UV system was described in several studies (Mills & Hunte, 1997; Peral et al., 1997; Rauf & Ashraf, 2009; Tschirch et al., 2008; Zhao et al., 2005), mainly with regard to a higher efficiency of degradation of chemical pollutants in water and air. Total mineralization of pollutants was observed, producing only CO_2, H_2O and inorganic ions (Zhao et al., 2005). Thus, the process is highly environmentally friendly. Its reasonable price, compared to other Advanced Oxidative Processes (Rauf & Ashraf, 2009), is also highly important.

The decontamination by TiO_2/UV/H_2O_2 is studied mainly in the liquid phase, preferentially in water (Domínguez et al., 2005), and thus there is no information on a VPHP-based modification of this system. The addition of a small amount of hydrogen peroxide shows a synergistic effect, i.e. a notable speedup of the photocatalytic TiO_2/UV/H_2O_2 degradation of pollutants (Tschirch et al., 2008). This positive effect is due to increased formation of hydroxyl radicals as a result of direct photolysis of hydrogen peroxide and its interaction with the active TiO_2 surface. However, higher concentrations of hydrogen peroxide inhibit the degradation because under these conditions, recombination of hydroxyl radicals is preferred (Dionysiou et al., 2004; Esplugas et al., 2002). Thus, it is necessary to work with optimal concentration of hydrogen peroxide to achieve the highest efficiency of contaminant degradation in the TiO_2/UV/H_2O_2 system (Elmolla & Chaudhuri, 2010). Although similar properties as TiO_2 are also provided by other semiconductor materials as ZnO, ZnS, CdS, Fe_2O_3, WO_3, these materials do not show the efficiency of degradation as high as TiO_2 does (Ray, 2000).

There are also other possibilities to promote VPHP process, e.g. by the addition of other oxidative agents like ozone, peracetic acid, or concentrated solution of hydrogen peroxide before evaporation. The main problem of hydrogen peroxide evaporation is water which can condensate on rugged surfaces like medical instruments and it impedes the surface penetration by the hydrogen peroxide. Hydrogen peroxide is able to form anhydrous complexes with a wide range of organic compounds: polyvinylpyrrolidone, urea, glycine anhydride-peroxide complex, and inorganic compounds: $Na_4P_2O_7 \cdot 3H_2O_2$, $KH_2PO_4 \cdot H_2O_2$, which can be prepared easily using known procedures based on their crystallization from solutions. In these complexes, the hydrogen peroxide moiety only binds to the electronegative atom of the other molecule via two H-bonds, which greatly facilitates its release from these substances. For example, thermal decomposition or vacuum can be used to perform this. The vapors of hydrogen peroxide formed this way can be generated directly in the decontaminated area or in another place, in which case they can be transported by a pipeline to the decontaminated space. Another interesting kind of hydrogen peroxide potentiation is the addition of metals (Ag, Al, Ca, Ce, Cu Mg, Sr, Sn, Ti, Zn) (Carnes et al., 2004), oxides or hydroxides, the particles of which (must) have a relatively high surface (at least 15 $m^2 g^{-1}$). The application of transition metals for hydrogen peroxide activation (i.e. for the formation of OH or other radicals) is mainly used in the liquid phase, like Fenton oxidation (Fenton, 1894) or photo-Fenton oxidation (Prousek, 1996), where Fe^{3+} ions (Xu, 2001) or Cu complexes are used (Martínez et al., 2008). These systems are mainly used for the degradation of resistant organic contaminants like dyes. Another (and also very interesting) combination is mixing of VPHP with volatile basic compounds like ammonia. This method seems to be very promising in the area of deactivation of warfare agents (Wagner et al., 2007).

9. VPHP decontamination of chemicals – Molecular structure effects of decontaminants

There is only limited information concerning the effects of the contaminant chemical structure upon decontamination by VPHP (McVey et al., 2006; Roberts et al., 2006). Several functional groups which are sensitive to VPHP have been found (Švrček, 2010), namely the aldehyde group, aliphatic tertiary nitrogen and the sulfide group (thioethers). Since all of the tested substances contain one of these groups, they were successfully degraded by VPHP. Other VPHP-sensitive compounds seem to be phenols, out of which mainly their hydroxy- and amino-derivatives. It can be expected that the decontamination of more complicated structures (pharmaceutical substances) will be sufficient if these compounds contain one or more such reactive groups in their structure.

9.1 Aldehyde group

Substances containing the aldehyde group undergo preferentially an oxidation process leading to carboxylic acid but other reactions can also take place, for example Dakin reaction (Pan et al., 1999), decarbonylation, substitution, or cracking.

The mechanism of Dakin reaction is illustrated in Figure 3 on an example of vanillin degradation, which in principle proceeds by insertion of an oxygen atom as a result of ketone oxidation (Baeyer-Villiger oxidation). In the first step, an OH radical attacks the carbonyl group (1) and an alkoxy radical (2) is generated, which is then transformed to an unstable alkylhydroperoxide (3) that is rearranged into a more stable product. A subsequent hydrogen shift (a) leads to the final product - carboxylic acid (4). In some cases, migration of a different group can occur, e.g. the aryl group that forms a formate (5) as a key intermediate in the above-mentioned Dakin reaction (b). These compounds are very often unstable and are subject to hydrolysis (cleavage of formic acid) or decarboxylation (giving CO_2) and formation of a phenol derivative - in case of vanillin it is 2-methoxyhydroquinone (6). The ability of different groups to migrate can be sorted in the following order: tertiary alkyl > secondary alkyl, aryl > primary alkyl > methyl. Electron-donating substituents on the benzene ring of vanillin increase migration ability of this aryl group by the hydroperoxide rearrangement. Electron-withdrawing substituents have opposite effects and impede migration. Therefore, vanillin can easily undergo Dakin reaction that leads to 4-hydroxy-3-methoxyfenyl formate (5). The key factor of Dakin reaction is the presence of strong electron-donating substituents (-OH or -NH_2) in the structure of aromatic aldehydes in the *ortho-* or *para-* position to the –CHO group (Švrček, 2010).

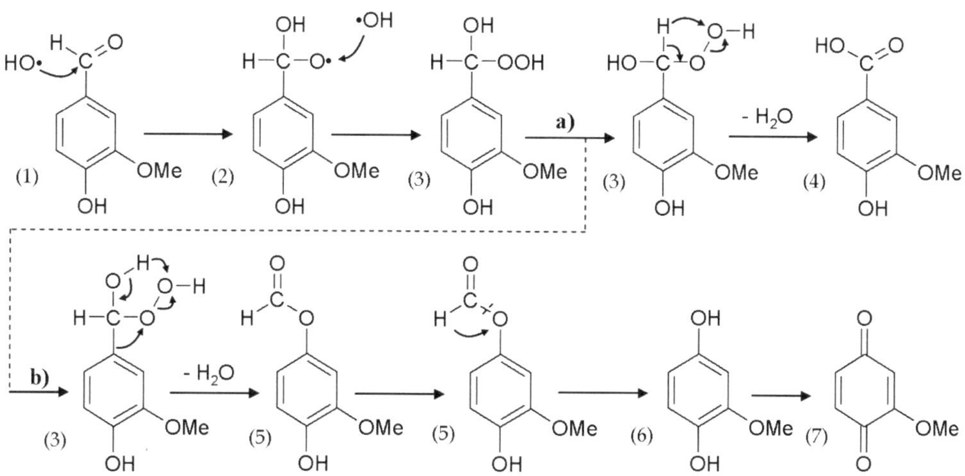

Fig. 3. Expected mechanism of the VPHP-induced degradation of vanillin

9.2 Aliphatic tertiary nitrogen

Other compounds which are highly sensitive to VPHP seem to be those containing an aliphatic tertiary nitrogen atom. All tested substances (Švrček, 2010) were decomposed by VPHP, which could be deduced from notable color changes, decrease in the sample weight, and detected products of degradation. However, the high volatility of starting compounds or degradation products prevented proper analysis by GC-MS and NMR. It is known that in the presence of H_2O_2, tertiary amines are converted to *N*-oxides that can be decomposed by the Cope elimination reaction leading to an alkene and *N*-hydroxylamine. This elimination is mainly carried out at high temperature but we can assume that it can proceed under laboratory conditions (Cope et al., 1949).

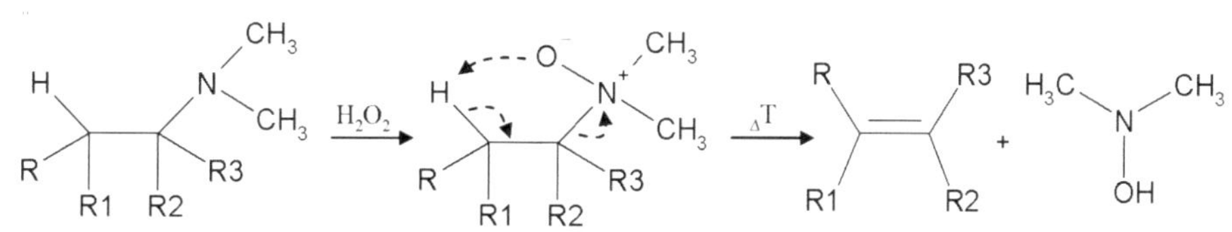

Fig. 4. Mechanism of Cope elimination

9.3 Sulfidic sulphur

In the presence of hydrogen peroxide, oxidation of sulfidic compounds (thioethers) proceeds and more stable higher oxidative compounds are formed (sulfoxides and sulfones) in a rate depending on the concentration of the oxidative agent.

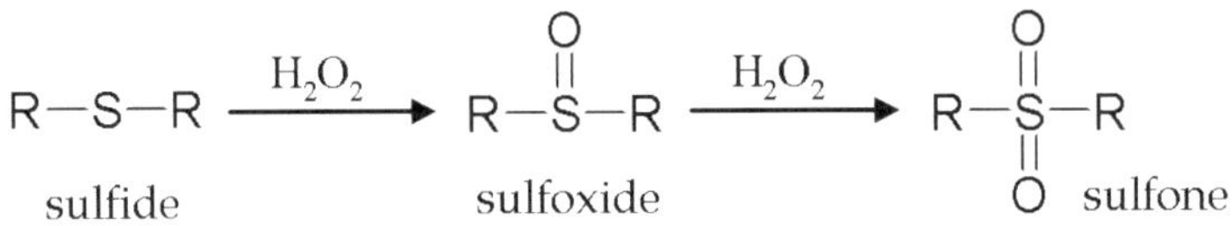

Fig. 5. Mechanism of oxidation of sulfidic compounds (thioethers)

The molecules containing sulfidic sulphur are oxidized by VPHP to both products. In case of thioanisol (phenyl methyl sulfide), after a VPHP degradation process only phenyl methyl sulfoxide was detected by MS. The final oxidative product (phenyl methyl sulfone) was not detected because of its high volatility and thus quick evaporation during the decontamination test, but we can assume its formation. Oxidation of dimethyl sulfoxide to dimethyl sulfone is driven by VPHP to a total conversion (Švrček, 2010).

9.4 Phenol derivatives

VPHP can also decompose phenol derivatives substituted in the *para-* position with electron-donating groups - mainly hydroxyl and amino. On the contrary, electron-withdrawing substituents (-CN, $-NO_2$, -COOH) make the contaminants intact to the VPHP atmosphere (Švrček, 2010).

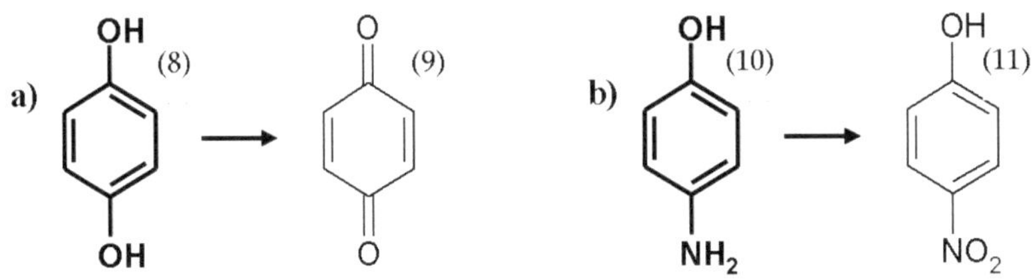

Fig. 6. Oxidation of hydroquinone and *p*-aminophenol by VPHP vapors

Hydroquinone (8) is oxidized by VPHP to 1,4-benzoquinone (9) which was also true for vanillin degradation. In case of *p*-aminophenol (10), a solid, weak and soluble mixture of degradation products was formed in the VPHP atmosphere. One of the products was *p*-nitrophenol (11) and the other compounds were products of oligo- and polymerization reactions between the starting compound and the products of degradation.

9.5 Molecular structure effects promoting the VPHP-induced degradation

A group of model chemical substances with similar structures (approximately 30 benzaldehyde and phenol derivatives bearing different types, numbers and position of substituents on the aromatic ring) was tested to study the influence of structure on the sensitivity to the VPHP process and reaction schemes were evaluated (Švrček, 2010). It is clear that the presence of electron-donating group on the benzaldehyde skeleton in many cases leads to an increase in its sensitivity to degradation by VPHP. The type, number and position of substituents on the aromatic ring with respect to the -CHO group play an important role. The highest sensitivity is shown with substances bearing high electro-donating substituents (like $-NR_2$) located in the *para-* position to -CHO. With the decreasing number of substituents, the time necessary for the VPHP degradation also decreased. The lowest sensitivity was found with benzaldehyde derivatives containing a halogen atom, an electron-withdrawing substituent or sterically demanding substituents.

10. VPHP as an agent for decontamination of biologically active compounds

In 1985, the International Agency for Research on Cancer (IARC) issued recommendations on the application of oxidative processes for the decontamination of cytotoxic compounds in waste waters (Roberts et al., 2006). Following studies (Barek et al., 1998; Hansel et al., 1997) tried to find the ideal oxidative agent that could be used in hospitals for the purposes described above. Three systems were tested: water solution of hydrogen peroxide (< 30 %, w/w), NaOCl and Fenton reagent for the degradation of selected cytostatics. Hydrogen peroxide showed the lowest efficiency of degradation compares to other oxidative systems, but it still appeared to be a promising agent.

The efficiency of 35% (w/w) liquid solution of hydrogen peroxide, VPHP and eight liquid detergents containing NaOCl for the deactivation of cytotoxic compounds present on the inner surface of pharmaceutical isolators was compared (McVey et al., 2006; Roberts et al., 2006). The contaminants tested were cyclophosphamide, doxorubicin and 5-fluorouracil. The liquid detergents removed all substances from the surfaces except for doxorubicin that showed high resistance to alkaline detergents. However, the contaminants were only washed away from the surface and their bioactivity was not eliminated. Cyclophosphamide and 5-fluorouracil were completely intact after a one-hour exposure to concentrated solutions of detergents which means that they preserved their full biological activity. Degradation of doxorubicin only occurred in a strong alkaline detergent (pH = 13.2, 80% conversion over 1 hour) and it remained unchanged in the water solution of hydrogen peroxide over 1 hour. The results from similar studies (Castegnaro et al., 1997; Hansel et al., 1997) showed that the application of 30% (w/w) hydrogen peroxide solution led to slow degradation of doxorubicin (60% conversion after 24 hours) and total degradation of cyclophosphamide after 1 hour. The VPHP "dry" process at temperature below 30 °C and reaction time of 1 hour exhibited slight degradation of 5-fluorouracil and significant degradation of doxorubicin (44 - 92 %). Unfortunately, cyclophosphamide remained intact. The mechanism was not studied (Roberts et al., 2006).

On the contrary, the application of VPHP as a deactivator of warfare agents is much more studied. A main leader on this topic is the STERIS company that found a quick and safe means of decontamination of large polluted areas (McVey et al., 2006). Their method seems to be very promising mainly because of a danger of terrorist attacks. The application of hydrogen peroxide to deactivate chemical weapons of mass destruction was studied (Wagner & Yang, 2001) which resulted in creation of a universal liquid deactivation agent called "Decon Green". This agent is composed of a hydrogen peroxide water solution, alcohol (ethanol, isopropyl alcohol, *tert*-butanol, polypropylene glycol) and carbonates, bicarbonates or monoperoxocarbonate - Na_2CO_3, K_2CO_3, Li_2CO_3, NH_4HCO_3). „Decon Green" provides high efficiency for the degradation of nerve agents and blistering agent like sarine (GB), soman (GD), sulfide yperite (HD) and VX compounds. The addition of carbonate to the liquid solution of peroxide leads to the formation of an OOH^- anion that is highly reactive, mainly with nerve agents. It speeds up the per-hydrolysis of VX and G compound families. The addition of peroxocarbonate, which is highly active for the oxidation of HD substances, leads to the formation of nontoxic products (Wagner & Yang, 2001). The addition of alcohol serves as the nonfreezing part of this mixture.

Carbonates are very good activators of hydrogen peroxide but they cannot be used in the VPHP process as they are not volatile. The special arrangement can distribute carbonates in the gaseous phase which supports vapor-phase decontamination. A pilot work (McVey et al., 2006) studied the application of VPHP to the degradation of warfare agents. It was shown that the VPHP process was highly potent in the degradation of HD and VX compounds even without any activator present. The G type agents are very stable under the VPHP conditions. It is known that they can be decomposed easily under alkaline conditions (Wagner et al., 2007). Hydrogen peroxide is relatively unstable in alkaline solutions but it has been shown (McVey et al., 2006; Wagner et al., 2007) that under these conditions the GD compounds can be decomposed to nontoxic products. Furthermore, the combination of VPHP and a base also accelerated the degradation of VX and HD compounds compared with VPHP alone (McVey et al., 2006; Wagner et al., 2007). The degradation started at an ammonia level that was below the safety limit of 25 ppm. It was observed that the rate of degradation was increasing with the increasing ammonia concentration (Wagner et al., 2007). A detailed study on the degradation of GD compounds using a mixture of VPHP/NH_3 shows a different mechanism compared with liquid "Decon Green". In case of application of the VPHP/NH_3 mixture to HD and VX, higher reaction rate was observed together with some differences in the decomposition mechanism (Wagner et al., 2007). It is interesting that the commercial decontamination agents containing ammonia are only active for the degradation of G but not for VX and HD.

The higher efficiency of VPHP for the degradation of warfare agents compared to water solution of hydrogen peroxide is explained as follows. The acidic products formed by per-hydrolysis are volatile and are ventilated out. On the contrary, in the water solution the acidic products are concentrated and thus the pH value decreases under a level which is necessary for the formation of peroxide anions and the reaction is thus stopped. Higher efficiency of VPHP is also explained by a model simulating the distribution of contaminants on contaminated surfaces.

Hydrogen peroxide is due to its properties closer to organic solvents than to water and thus it is selectively adsorbed and concentrated inside drops or film of the contaminant. Molecular distribution results in a higher local concentration of H_2O_2 inside the contaminant and exclusion of water vapor out of the contaminant organic layer (Wagner et al., 2007). When using the liquid phase, competitive adsorption of water and hydrogen peroxide to contaminant molecules occurs which slows down its degradation noticeably.

It was showed (Wagner et al., 2007) that the exposition time necessary to decontaminate inner premises of buildings contaminated by the warfare agents using VPHP/NH_3 is 24 hours. These results clearly show that such modified VPHP process (VPHP/ NH_3) is the right choice for quick and effective decontamination of large areas and objects (military vehicles, airplanes) contaminated by dangerous compounds.

11. VPHP – Material compatibility

The extent of contaminant elimination by hydrogen peroxide in vapor phase is mainly influenced by the concentration of VPHP, exposure time, temperature, humidity, condensation and also the physical and chemical properties of materials (Chung et al., 2008). The influence of the contaminated surface upon the efficiency of VPHP-induced inactivation of the spores of *Bacillus stearothermophilus*, which serves as a biological indicator, was studied (Unger et al., 2007). Among the notable influences are the chemical composition of the selected material, its potential catalytic activity, absorption of gaseous hydrogen peroxide in the material and, last but not least, the differences in the surface production or its treatment. The relationships between the deactivation rate and porosity and wettability are documented (Unger et al., 2007) - VPHP is mainly effective for the sterilization of smooth surfaces and therefore stainless steel and glass coupons are used for the evaluation of the process efficiency (Bounoure et al., 2006; Unger et al., 2007). Spores on a porous surface can be hidden in cavities and in such a case the penetration of hydrogen peroxide to the material plays an important role in decontamination (Bounoure et al., 2006). Some materials have demonstrated an inhibition effect to the microorganisms without any decontamination agent, such as ethylene-propylene-diene rubbers. All knowledge mentioned above was implemented by FDA in the Guidelines for industry. It is recommended to choose suitable materials for the design of any aseptic process. They have to provide proper compatibility with the chemicals used and thus can be easily cleaned and decontaminated. It is highly important to choose construction materials with appropriate texture and porosity by the build-up of the aseptic process mainly when the validation of decontamination process is required (Unger et al., 2007). The VPHP decontamination was tested on various surfaces but the toxicological evaluation reports on surfaces treated in such a way are very limited. Hydrogen peroxide is characterized by high cytotoxicity (i.e. harmfulness to cells) so its residual amounts in the materials can cause irritation of eyes, skin, mucous membrane and also acute lung dropsy in cases of long-time inhalation. Therefore, the construction materials have to be selected not only due to their compatibility with the VPHP process but also with respect to the efficiency of residual hydrogen peroxide aeration out of these materials. Up to now, the studies have mainly focused on polymeric materials because of their popularity in the pharmaceutical industry. The permeation of gases through plastic materials occurs in two steps: the first one consists in dissolution of gas in the thin surface layer and the second is its diffusion to the material. It was found that polyethylene and polypropylene can easily release hydrogen peroxide because of its limited migration in these materials. In contrast, polystyrene, polyurethane, poly(methyl methacrylate) (PMMA), poly(2-hydroxyethylmethacrylate) (HEMA), fluorosilicone acrylate and mixture of polyurethane and silicone demonstrated strong cytotoxicity after standard aeration (Ikarashi et al., 1995). The PMMA and HEMA materials are used for the production of contact lenses and thus, after their sterilization by VPHP, the attention needs to be paid to residual hydrogen peroxide. Hydrogen peroxide can easily penetrate through polyolefins in general, out of which PVC is used for packaging of infusion solutions. Therefore, the VPHP process has to be applied very carefully in order to avoid the contact of residual hydrogen peroxide with the infusion solution due to resulting oxidative reactions (Ikarashi et al., 1995). Nonetheless, information on the resistance of other construction materials is very limited. Although there have been reported interactions of several materials with VPHP which can lead to their damaging (Hultman et al., 2007), the process is still considered non-corrosive for surface decontamination (U.S. Environmental Protection Agency, 2005). Several works tested the resistivity of commonly-used devices to VPHP but the results were evaluated only visually and by testing of functionality of the devices. Generally, it was concluded that VPHP was an acceptable sterilization method for the devices (Hall et al., 2007; Heckert et al., 1997).

In global, materials can be divided into four groups according to their tolerance to VPHP. Group 1 represents such materials that can be in the contact with hydrogen peroxide for a long time, like pure aluminum, tin, borosilicate glass or Teflon. In contrast to this group stands Group 4 which comprises materials that cannot be in any contact with hydrogen peroxide because of their fast decomposition or formation of explosive mixtures (copper, iron, carbon steel, magnesium alloys). The concentration of 45% (w/w) hydrogen peroxide is considered critical (Hultman et al., 2007) because, when exceeded, undesirable interaction or damage of different materials can occur. For chemical decontamination, higher-concentrated peroxide is necessary and in a proper arrangement it can be a safe system. It can be expected that during the "wet" VPHP process, the concentration of the condensate is above this critical concentration and thus the materials can be damaged. However, this concentration can also be exceeded during the "dry" VPHP process in the aeration phase when the absorbed hydrogen peroxide is concentrated on the surfaces. If the damage is not visible, it does not mean that the material has not been damaged. Microscopic lesions can appear and the macroscopic ones can become evident only after a long-time exposure to VPHP. The VPHP process was also tested for the interiors of airplanes and ambulances (Krieger & Mielnik, 2005; Shaffstall et al., 2006). As these vehicles are frequently in the contact with infection, quick and effective decontamination is important to avoid the propagation of infection (SARS, bird flu). Application of VPHP in these facilities is risky because several sensitive devices which are vital for their proper function can be damaged. The Federal Aviation Administration (FAA) studied the influence of the VPHP process on textile materials of airplane interiors and found significant changes. The leader in the VPHP technology, Bioquell, tried to map the influence of long-time application of VPHP to the hospital equipment. The company tried to fill in the information gap of the VPHP material compatibility. Some materials were shown to be highly resistant to VPHP (called „VPHP-resistant"), some were degradable by VPHP and some absorbed hydrogen peroxide substantially. There are also materials unsuitable for VPHP like untreated aluminum, copper, soft steels, coated steels and generally materials that have similar behavior in the liquid state. Materials containing cellulose tend to absorb hydrogen peroxide and are further degraded; therefore, they are also unsuitable for VPHP (von Woedtke et al., 2004). Furthermore, EPA studied the VPHP compatibility of common construction or decorative materials (U.S. Environmental Protection Agency, 2008). The visual inspection did not reveal any significant changes of the tested materials but the tensile strength was reduced, which was caused by changes in their inner structure.

Low-temperature H_2O_2 plasma material compatibility is actually a much more studied topic. The Johnson & Johnson company tested a wide range of materials and devices which were in a periodic contact with hydrogen peroxide plasma. It was found that the materials containing amines in their structure were unsuitable because of their oxidation that damaged the structure. Also the S–S bond in the materials was proven to be incompatible with this process.

It is necessary to select such materials that are fully compatible with VPHP by the design of the device or the process where it is expected to use VPHP for decontamination. In such case, it needs to be known how the materials interact with hydrogen peroxide, namely with respect to absorption and the rate of hydrogen peroxide decomposition. It is also important to know the aeration time required to vent out hydrogen peroxide in order to avoid undesirable cytotoxic effects (Ikarashi et al., 1995). It is necessary to evaluate the resistivity to hydrogen peroxide and the amount of residual hydrogen peroxide left in the material. There are several analytical methods for the determination of hydrogen peroxide such as reduction by SnCl2 (Egerton et al., 1954), thiocyanate method (Egerton et al., 1954; Ikarashi et al., 1995), enzymatic method of dimerization of p-hydroxyphenolic acid (Christensen, 2000), color changes of $Ti_2(SO_4)_3$ or $TiCl_4$ by spectrophotometric detection (Egerton et al., 1954), UV spectroscopy, titration of $KMnO_4$, I_3^-, or $Ce(SO_4)_2$ (Klassen et al., 1994), electrochemical, polarographic and other methods (Higashi et al., 2005).

12. Glory and pitfalls of VPHP

The VPHP technology is a very progressive method of sterilization and chemical decontamination. Its popularity was mainly achieved due to low toxicity of hydrogen peroxide and its non-toxic decomposition products, environmentally friendly behavior, relative flexibility and a wide spectrum of applications. Its high efficiency against a wide range of microorganisms makes this method universal in terms of bio-decontamination and thus it is very popular in medical and pharmaceutical industry. This application is widely studied and described. The absence of theoretical knowledge is the only drawback of this method. Proper understanding of the sterilization principle would allow to optimize the method and to make it exceptionally powerful. On the other hand, the chemical decontamination by VPHP is still a growing area with a wide potential of application. Mainly the combination and potentiation of VPHP by other chemicals or physical phenomena should improve it and make it a very powerful tool for the decontamination of dangerous chemicals. Also in this case, the absence of proper theoretical knowledge limits its application.

13. Future perspective of VPHP

As already mentioned above, the highly efficient VPHP process has found a great deal of new applications in bio and chemical decontamination. The first challenge seems to be theoretical understanding of its mechanism and thus obtaining the basics for finding new applications or improving existing processes. The next challenge is optimization of existing processes by application of new approaches and knowledge as a product of undergoing research and development. Another very important challenge is to discover novel possibilities of VPHP potentiation, mainly with respect to chemical decontamination. There is a great need for the decontamination of a wide range of chemical pollutants. The VPHP process can play a very important role in the case of homeland security as it can be easily implemented to the defense system, which provides greater security for the state and citizens against terrorists, spreading of infections, chemical and biological accidents, etc.

14. Conclusion

At the present time, a broad spectrum of decontamination techniques is acknowledged to be utilized to remove biological and chemical contaminants from different surfaces. Nevertheless, new physical and chemical processes are continuously being developed. The main reasons for the development of new decontamination methods are negative properties of many decontamination agents, especially their toxicity or toxicity of residues formed after their application. This review summarizes the recent findings concerning a new promising decontamination agent Vapor Phase Hydrogen Peroxide (VPHP) whose properties were observed to be very close to an ideal decontamination agent. VPHP has become the method of choice by meeting many bio-decontamination requirements in the pharmaceutical, biomedical and healthcare sectors for it is reliability, rapidness, the fact that it leaves no residues (breaks down into water and oxygen) and the advantage that it can be validated. It is also a decontamination method of choice for chemically and biologically active compounds. The application of VPHP as a potential decontamination agent is apparently still in its infancy. Therefore, it comes as no surprise that the knowledge of the cidal action mechanism(s) and the influential factors is yet to be completed.

15. Acknowledgement

The project is supported by the grant of Ministry of Defense of the Czech Republic (OVVSCHT200901).

16. References

Adams, D.; Brown, G. P.; Fritz, C. & Todd, T. R. (1998). Calibration of a Near-Infrared. (NIR) H_2O_2 Vapor Monitor. *Pharmaceutical Engineering*, Vol. 18, No. 3, (May/June 1998), pp.1-11, ISSN 0273-8139

Barek, J.; Cvačka, J.; Zima, J.; De Méo, M.; Laget, M.; Michelon, J. & Castegnaro, M. (1998). Chemical Degradation of Wastes of Antineoplastic Agents Amsacrine, Azathioprine, Asparaginase and Thiotepa. *The Annals of Occupational Hygiene*, Vol. 42, No. 4, (May 1998), pp. 259-266, ISSN 1475-3162

Bathina, M.N.; Mickelsen, S.; Brooks, C.; Jaramillo, J.; Hepton, T. & Kusumoto F.M. (1998). Safety and efficacy of hydrogen peroxide plasma sterilization for repeated use of electrophysiology catheters. *Journal of the American College of Cardiology*, Vol. 32, No. 5, (November 1998), pp. 1384-1388, ISSN 0735-1097

Block, S.S. (1991). Historical Review, In: *Disinfection, Sterilization, and Preservation* (fourth edition), S.S. Block, (Ed.), 3-17, Lea & Febiger, ISBN 978-081-2113-64-8 London, Philadelphia

Block S.S. (1991). Peroxygen Compounds, In: *Disinfection, Sterilization, and Preservation* (fourth edition), S.S. Block, (Ed.), 167-181, Lea & Febiger, ISBN 978-081-2113-64-8 London, Philadelphia

Bounoure, F.; Fiquet, H. & Arnaud, P. (2006). Comparison of hydrogen peroxide and peracetic acid as isolator sterilization agents in a hospital pharmacy. *American Journal of Health-System Pharmacy*, Vol. 63, No. 5, (March 2006), pp. 451-455, ISSN 1535-2900

Carnes, C.L.; Klabunde, K.J.; Koper, O.; Martin, L.S.; Knappenberger, K.; Malchesky, P.S. & Sanford, B.R. (2004). Decontaminating Systems Containing Reactive Nanoparticles and Biocides. *United States Patent*, US 2004/0067159 A1, filed (October 2002), issued (April 2004)

Castegnaro, M.; De Méo, M.; Laget, M.; Michelon, J.; Garren, L.; Sportouch, M.H. & Hansel, S. (1997). Chemical degradation of wastes of antineoplastic agents 2: Six anthracyclines: idarubicin, doxorubicin, epirubicin, pirarubicin, aclarubicin, and daunorubicin. *International Archives of Occupational and Environmental Health*, Vol. 70, No. 6, (December 1997), pp. 378-384, ISSN 1432-1246

Cazin, J.L. & Gosselin, P. (1999). Implementing a multiple-isolator unit for centralized preparation of cytotoxic drugs in a cancer center pharmacy. *Pharmacy World & Science*, Vol. 21, No. 4, (April 1999), pp. 177-183, ISSN 1573-739X

Christensen, C.S.; Brødsgaard, S.; Mortensen, P.; Egmose, K. & Linde, S.A. (2000). Determination of hydrogen peroxide in workplace air: interferences and method validation. *Journal of Environmental Monitoring*, Vol. 2, No. 4, (August 2000), pp. 339-343, ISSN 1464-0333

Chung, S.; Kern, R.; Koukol, R.; Barengoltz, J. & Cash, H. (2008). Vapor hydrogen peroxide as alternative to dry heat microbial reduction. *Advances in Space Research* Vol. 42, No. 6, (September 2008), pp. 1150-1160, ISSN 0273-1177

Cope, A.C.; Foster, T.T. & Towle, P.H. (1949). Thermal Decomposition of Amine Oxides to Olefins and Dialkylhydroxylamines, *Journal of the American Chemical Society*, Vol. 71, No. 12, (December 1949), pp. 3929–3934, ISSN 0002-7863

Crow, S. & Smith, J. H. (1995). Gas Plasma Sterilization: Application of Space-Age Technology. *Infection Control and Hospital Epidemiology*, Vol. 16, No. 8 (August 1995), pp. 483-487, ISSN 1559-6834

Daughton, C.G. & Ternes, T.A. (1999). Pharmaceuticals and Personal Care Products in the Environment: Agents of Subtle Change? *Environmental Health Perspectives*, Vol. 107, No. 6, (December 1999), 907-938, ISSN 0091-6765

Dionysiou, D. D.; Suidan, M. T.; Baudin, I. & Laîné, J. M. (2004). Effect of hydrogen peroxide on the destruction of organic contaminants-synergism and inhibition in a continuous-mode photocatalytic reaktor. *Applied Catalysis B: Environmental*, Vol. 50, No. 4, (July 2004), pp. 259-269, ISSN 0926-3373

Domínguez, J. R.; Beltrán, J. & Rodríguez, O. (2005). Vis and UV photocatalytic detoxification methods (using TiO_2, TiO_2/H_2O_2, TiO_2/O_3, $TiO_2/S_2O_8^{2-}$, O_3, H_2O_2, $S_2O_8^{2-}$, Fe^{3+}/H_2O_2 and $Fe^{3+}/H_2O_2/C_2O_4^{2-}$) for dyes treatment. *Catalysis Today*, Vol. 101, No. 3-4, (April 2005), pp. 389-395, ISSN 0920-5861

Duffy, M. (August 2009). Weapons of War - Poison Gas, In: *firstworldwar.com*, 20. 7. 2011, Available from: http://www.firstworldwar.com/weaponry/gas.htm

Egerton, A.C.; Everett, A.J.; Minkoff, G.J.; Rudrakanchana, S. & Salooja, K.C. (1954). The analysis of combustion products: Some improvements in the methods of analysis of peroxides. *Analytica Chimica Acta*, Vol. 10, No. 5, (1954), pp. 422-428, ISSN 0003-2670

Elmolla, E. S. & Chaudhuri, M. (2010). Photocatalytic degradation of amoxicillin, ampicillin and cloxacillin antibiotics in aqueous solution using UV/TiO_2 and $UV/H_2O_2/TiO_2$ photocatalysis. *Desalination*, Vol. 252, No. 1-3, (March 2010), pp. 46-52, ISSN: 00119164

Esplugas, S.; Giménez, J.; Contreras, S.; Pascual, E. & Rodríguez, M. (2002). Comparison of different advanced oxidation processes for phenol degradation. *Water Research*, Vol. 36, No. 4, (February 2002), pp.1034-1042. ISSN 0043-1354

Favero, M.S. & Bond, W.W. (1991). Chemical Disinfection of Medical and Surgical Materials, In: *Disinfection, Sterilization, and Preservation* (fourth edition), S.S. Block, (Ed.), 617-641, Lea & Febiger, ISBN 978-081-2113-64-8 London, Philadelphia

Fenton, H.J.H. (1894). Oxidation of Tartaric Acid in presence of Iron. *Journal of the Chemical Society, Transactions*, Vol. 65, No. 0, (1894), pp. 899-910, ISSN 0368-1645

Fichet, G.; Comoy, E.; Duval, Ch.; Antloga, K.; Dehen, C.; Charbonnier, A.; McDonnell, G.; Brown, P.; Lasmézas, C. I. & Deslys, J. P. (2004). Novel methods for disinfection of prion-contaminated medical devices. *The Lancet*, Vol. 364; No. 9433, (August 2004), pp. 521-526, ISSN 140-6736

Fisher, J. & Caputo, R.A. (2004). Comparing and Contrasting Barrier Isolator Decontamination Systems. *Pharmaceutical Technology*, Vol. 28, No. 11, (November 2004), pp. 68-82, ISSN 1543-2521

Forney, Ch. F.; Rij, R. E.; Denis-Arrue, R. & Smilanick, J. L. (1991). Vapor phase hydrogen peroxide inhibits postharvest decay of table grapes. *HortScience*, Vol. 26, No. 12, (December 1991), pp. 1512-1514, ISSN 0018-5345

French, G.L.; Otter, J.A.; Shannon, K.P.; Adams, N.M.T.; Watling, D. & Parks, M.J. (2004). Tackling contamination of the hospital environment by methicillin-resistant Staphylococcus aureus (MRSA): a comparison between conventional terminal cleaning and hydrogen peroxide vapour decontamination. *Journal of Hospital Infection*, Vol. 57, No. 1, (May 2004), pp. 31-37, ISSN 0195-6701

Fraise, A.P. (2004). Historical introduction, In: *Principles and Practice of Disinfection, Preservation & Sterilization* (fourth edition), A.P. Fraise, P.A. Lambert & J.Y. Maillard, (Ed.), 3-7, Russell, Hugo & Ayliffe´s. Blackwell Publishing, ISBN 1-4051-0199-7, Oxford

Fujishima, A. & Honda K. (1972). Electrochemical Photolysis of Water at a Semiconductor Electrode. *Nature,* Vol. 238, No. 5358, (July 1972), pp. 37 – 38, ISSN 0028-0836

Gould, G.W. (2004). Heat sterilization, In: *Principles and Practice of Disinfection, Preservation & Sterilization* (fourth edition), A.P. Fraise, P.A. Lambert & J.Y. Maillard, (Ed.), 361-383, Russell, Hugo & Ayliffe´s. Blackwell Publishing, ISBN 1-4051-0199-7, Oxford

Gruhn, R.; Baessler, H. J. & Werner, U. J (1995). Sterilization of chicken eggs for vaccine production with vaporized hydrogen peroxide. *Pharmazeutische Industrie,* Vol. 57, No.10, (October 1995), pp. 873-877, ISSN 0031-711X

Gurevich, I. (1991). Infection Control: Applying Theory to Clinical Practice, In: *Disinfection, Sterilization, and Preservation* (fourth edition), S.S. Block, (Ed.), 655-662, Lea & Febiger, ISBN 978-081-2113-64-8 London, Philadelphia

Hall, L.; Otter, J.A.; Chewins, J. & Wengenack, N.L. (2007). Use of Hydrogen Peroxide Vapor for Deactivation of Mycobacterium tuberculosis in a Biological Safety Cabinet and a Room. *Journal of Clinical Microbiology*, Vol. 45, No. 3, (March 2007), pp. 810–815, ISSN 1098-660X

Hansel, S.; Castegnaro, M.; Sportouch, M.H.; De Méo, M.; Milhavet, J.C.; Laget, M. & Duménil, G. (1997). Chemical degradation of wastes of antineoplastic agents: cyclophosphamide, ifosfamide and melphalan. *International Archives of Occupational and Environmental Health,* Vol. 69, No. 2, (January 1997), pp. 109-114, ISSN 1432-1246

Hardy, K.J.; Gossain, S.; Henderson, N.; Drugan, C.; Oppenheim, B.A.; Gao, F. & Hawkey, P.M. (2007). Rapid recontamination with MRSA of the environment of an intensive care unit after decontamination with hydrogen peroxide vapour. *Journal of Hospital Infection*, Vol. 66, No. 4, (August 2007), pp. 360-368, ISSN 0195-6701

Hatanaka, K. & Schibauchi, Y. (1989). Sterilization Method and Apparatus Therefor. *United States Patent*, US 4,797,255, filed (March 1987), issued (January 1989)

Heckert, R.A.; Best, M.; Jordan, L.T.; Dulac, G.C.; Eddington, D.L. & Sterritt, W.G. (1997). Efficacy of Vaporized Hydrogen Peroxide against Exotic Animal Viruses. *Applied and Environmental Microbiology*, Vol. 63, No. 10, (October 1997), pp. 3916–3918, ISSN 1098-5336

Higashi, N.; Yokota, H.; Hiraki S. & Ozaki, Y. (2005). Direct Determination of Peracetic Acid, Hydrogen Peroxide, and Acetic Acid in Disinfectant Solutions by Far-Ultraviolet Absorption Spectroscopy. *Analytical Chemistry*, Vol. 77, No. 7, (February 2005), pp. 2272–2277, ISSN 0003-2700

Hoffman, R.K. & Spiner, D.R. (1970). Effect of Relative Humidity on Penetrability and Sporicidal Activity of Formaldehyde. *Applied Microbiology*, Vol. 20, No.4, (October 1970), pp. 616-619, ISSN: 0003-6919

Hultman, C.; Hill, A. & McDonnell, G. (2007). The Physical Chemistry of Decontamination with Gaseous Hydrogen Peroxide. *Pharmaceutical Engineering*, Vol. 27, No. 1, (January/February2007), pp. 22, ISSN: 0273-8139

Ikarashi, Y.; Tsuchiya, T. & Nakamura, A. (1995). Cytotoxicity of medical materials sterilized with vapour-phase hydrogen peroxide. *Biomaterials*, Vol. 16, No. 3, (February 1995), pp. 177-183, ISSN 0142-9612

International Committee of the Red Cross (n.d. 2005). Convention on the prohibition of the development, production, stockpiling and use of chemical weapons and on their destruction, Paris 13 January 1993, In: *International Humanitarian Law - Treaties & Documents*, 20. 7. 2011, Available from:
http://www.icrc.org/ihl.nsf/FULL/553?OpenDocument

Jernigan, D.B.; Raghunathan, P.L.; Bell, B.P.; Brechner, R.; Bresnitz, E.A.; Butler, J.C.; Cetron, M.; Cohen, M.; Doyle, T.; Fischer, M.; Greene, C.; Griffith, K.S.; Guarner, J.; Hadler, J.L.; Hayslett, J.A.; Meyer, R.; Petersen, L.R.; Phillips, M.; Pinner, R.; Popovic, T.; Quinn, C.P.; Reefhuis, J.; Reissman, D.; Rosenstein, N.; Schuchat, A.; Shieh, W.J.; Siegal, L.; Swerdlow, D.L.; Tenover, F.C.; Traeger, M.; Ward, J.W.; Weisfuse, I.; Wiersma, S.; Yeskey, K.; Zaki, S.; Ashford, D.A.; Perkins, B.A.; Ostroff, S.; Hughes, J.; Fleming, D.; Koplan, J.P.; Gerberding, J.L. & the National Anthrax Epidemiologic Investigation Team. (2002). Investigation of bioterrorism-related anthrax, United States, 2001: Epidemiologic findings. *Emerging Infectious Diseases*, Vol. 8, No. 10, (October 2002), pp. 1019-1028, ISSN 1080-6059

Johnson, J.W.; Arnold, J.F.; Nail, S.L. & Renzi, E. (1992). Vaporized hydrogen peroxide sterilization of freeze dryers. *Journal of Parenteral Sciencece and Technology*, Vol. 46, No. 6, (November-December 1992), pp. 215-225, ISSN 0279-7976

Johnston, M.D.; Lawson,S. & Otter, J.A. (2005). Evaluation of hydrogen peroxide vapour as a method for the decontamination of surfaces contaminated with Clostridium botulinum spores. *Journal of Microbiological Methods*, Vol. 60, No. 3, (March 2005), pp. 403–411, ISSN 0167-7012

Kahnert, A.; Seiler, P.; Stein, M.; Aze, B.; McDonnell, G. & Kaufmann, S.H.E. (2005). Decontamination with vaporized hydrogen peroxide is effective against Mycobacterium tuberculosis. *Letters in Applied Microbiology*, Vol. 40, No. 6, (June 2005) pp. 448-452, ISSN 1472-765X

Kang, J. W. & Lee, K. H. (1997). A Kinetic Model of the Hydrogen Peroxide/UV Process for the Treatment of Hazardous Waste Chemicals. *Environmental Engineering Science*, Vol. 14, No. 2, (January 1997), pp. 183-192, ISSN 1092-8758

Klapes, N.A. & Vesley, D. (1990). Vapor-Phase Hydrogen Peroxide as a Surface Decontaminant and Sterilant. *Applied and Environmental Microbiology*, Vol. 56, No. 2, (February 1990), pp. 503-506, ISSN: 1098-5336

Klassen, N.V.; Marchington, D. & McGowant, H.C.E. (1994). H_2O_2 Determination by the I_3^- Method and by $KMnO_4$ Titration. *Analytical Chemistry*, Vol. 66, No. 18, (September 1994), pp. 2921–2925, ISSN 0003-2700

Kokubo, M.; Inoue, T. & Akers, J. (1998). Resistance of Common Environmental Spores of the Genus Bacillus to Vapor Hydrogen Peroxide. *PDA Journal of Pharmaceutical Science and Technology*, Vol. 52, No. 5, (September/October 1998), pp. 228-231, ISSN 1948-2124

Krause, J.; McDonnell, G. & Riedesel, H. (2001). Biodecontamination of animal rooms and heat-sensitive equipment with vaporized hydrogen peroxide. *Contemporary Topics in Laboratory Animal Science*, Vol. 40, No. 6, (November 2001), pp. 18-21, ISSN 1060-0558

Krieger, E.W. & Mielnik, T.J. (2005). Aircraft and Passenger Decontamination System. *United States Patent*, US 2005/0074359 A1, filed (October 2003), issued (April 2005)

Kuzma, M.; Kačer, P.; Pánek, L. (2008). Decontamination with hydrogen peroxide vapors as technology of the future. *CHEMagazín*, Vol. 18, No. 6, (November/December 2008), pp. 12-13, ISSN 1210-7409

Linsebigler, A. L.; Lu, G. & Yates, J. T. (1995). Photocatalysis on TiOn Surfaces: Principles, Mechanisms, and Selected Results. *Chemical Reviews*, Vol. 95, No. 3, (May 1995), pp. 735-758, ISSN 00092665

Lopez, A.; Bozzi, A.; Mascolo, G. & Kiwi, J. (2003). Kinetic investigation on UV and UV/H_2O_2 degradations of pharmaceutical intermediates in aqueous solution. *Journal of Photochemistry and Photobiology A: Chemistry*, Vol. 156, No. 1-3, (March 2003), pp. 121-126,ISSN 1010--6030

Lysfjord, J. & Porter, M. (1998). Barrier Isolation History and Trends. *Pharmaceutical Engineering*, Vol. 18, No. 5, (September/October 1998), ISSN 0273-8139

Manatt, S.L. & Manatt, M.R.R. (2004). On the Analyses of Mixture Vapor Pressure Data: The Hydrogen Peroxide/Water System and Its Excess Thermodynamic Functions. *Chemistry - A European Journal*, Vol. 10, No. 24, (December 2004), pp. 6540–6557, ISSN 1521-3765

Martínez, J.M.L.; Denis, M.F.L.; Piehl, L.L.; de Celis, E.R.; Buldain, G.Y. & Orto, V.C.D. (2008). Studies on the activation of hydrogen peroxide for color removal in the presence of a new Cu(II)-polyampholyte heterogeneous catalyst. *Applied Catalysis B: Environmental*, Vol.82, No.3-4, (August 2008), pp. 273-283,ISSN 0926-3373

McDonnell, G.; Bonfield, P. & Hernandez, V.D. (2007). The Safe and Effective Fumigation of Hospital Areas with a New Fumigation Method Based on Vaporized Hydrogen Peroxide. *American Journal of Infection Control*, Vol. 35, No. 5, (June 2007), pp. E33-E34, ISSN 0196-6553

McDonnell, G. (June 2004). Large area decontamination, In: *Cleanroom Technology,* 22. 7. 2011, Available from: http://www.cleanroom-technology.co.uk/technical/article_page/ Large_area_decontamination/52681

McDonnell, G. & Russell, A.D. (1999). Antiseptics and Disinfectants: Activity, Action, and Resistance. *Clinical Microbiology Reviews,* Vol. 12, No. 1, (January 1999), pp. 147–179, ISSN 1098-6618

McVey, I.F.; Schwartz, L.I.; Centanni, M.A. & Wagner, G.W. (2006). Activated Vapor Treatment for Neutralizing Warfare Agents. *United States Patent,* US 7,102,052 B2, filed (April 2003), issued (September 2006)

Mills, A. & Hunte, S. L. (1997). An overview of semiconductor photocatalysis. *Journal of Photochemistry and Photobiology A: Chemistry,* Vol. 108, No. 1, (1997), pp. 1-35, ISSN 1010-6030

National Institute for Occupational Safety and Health. (September 2004). Preventing Occupational Exposures to Antineoplastic and Other Hazardous Drugs in Health Care Settings, In: *DHHS (NIOSH), 2004–165,* 20. 7. 2011, Available from: http://www.cdc.gov/niosh/docs/2004-165/

Okumura, T.; Ninomiya, N. & Ohta, M. (2003). The Chemical Disaster Response System in Japan. *Prehospital and Disaster Medicine,* Vol. 18, No. 3, (July - September 2003), pp. 189–192, ISSN 1049-023X

Okumura, T.; Suzuki, K.; Fukuda, A.; Kohama, A.; Takasu, N.; Ishimatsu, S. & Hinohara, S. (1998). The Tokyo Subway Sarin Attack: Disaster Management, Part 1: Community Emergency Response. *Academic Emergency. Medicine,* Vol. 5, No. 6, (June 1998), pp. 613-617, ISSN 1069-6563

O'Neill, H.J. & Brubaker, K.L. (2003). Method and Apparatus for the Gas Phase Decontamination of Chemical and Biological Agents. *United States Patent,* US 6,630,105 B1, filed (September 2000), issued (October 2003)

Pan, G.X.; Spencer, L. & Leary, G.J. (1999). Reactivity of ferulic acid and its derivatives toward hydrogen peroxide and peracetic acid. *Journal of Agricultural and Food Chemistry,* Vol. 47, No. 8, (August 1999), pp. 3325-3331, ISSN 0021-8561

Peral, J.; Doménech, X. & Ollis, D. F. (1997). Heterogeneous Photocatalysis for Purification, Decontamination and Deodorization of Air. *Journal of Chemical Technology and Biotechnology,* Vol. 70, No. 2, (October 1997), pp. 117–140, ISSN 1097-4660

Pereira, V.J.; Weinberg, H.S.; Linden, K.G. & Singer, P.C. (2007). UV Degradation Kinetics and Modeling of Pharmaceutical Compounds in Laboratory Grade and Surface Water via Direct and Indirect Photolysis at 254 nm. *Environmental Science & Technology,* Vol. 41, No. 5, (March 2007), pp. 1682-1688, ISSN 0013-936X

Parisi, A.N. & Young, W.E. (1991). Sterilization with Ethylene Oxide and Other Gases, In: *Disinfection, Sterilization, and Preservation* (fourth edition), S.S. Block, (Ed.), 580-595, Lea & Febiger, ISBN 978-081-2113-64-8 London, Philadelphia

Prousek, J. (1996). Advanced oxidation processes for water treatment. Chemical processes. *Chemicke Listy,* Vol. 90, No. 4, (April 1996), pp. 229-237

Rauf, M. A. & Ashraf, S. S. (2009). Fundamental principles and application of heterogeneous photocatalytic degradation of dyes in solution. *Chemical Engineering Journal,*Vol. 151, No. 1-3, (August 2009), pp. 10-18, ISSN 1385-8947

Ray M. B. (2000). Photodegradation of the Volatile Organic Compounds in the Gas Phase: A Review. *Developments in Chemical Engineering and Mineral Processing,* Vol. 8, No. 5-6, (n.d. 2000), pp. 405–439, ISSN 0969-1855

Reich, R.R. & Caputo, R.A. (2004). Vapor-Phase Hydrogen Peroxide Resistance of Environmental Isolates. *Pharmaceutical Technology,* (August 2004), pp. 50 -58, ISSN 1543-2521

Roberts, S.; Khammo, N.; McDonnell, G. & Sewell, G.J. (2006). Studies on the decontamination of surfaces exposed to cytotoxic drugs in chemotherapy workstations. *Journal of Oncology. Pharmacy Practice,* Vol. 12, No. 2, (June 2006), pp. 95-104, ISSN 1477-092X

Rogers, J.V.; Sabourin, C.L.K.; Choi, Y.W.; Richter, W.R.; Rudnicki, D.C.; Riggs, K.B.; Taylor, M.L. & Chang J. (2005). Decontamination assessment of Bacillus anthracis, Bacillus subtilis, and Geobacillus stearothermophilus spores on indoor surfaces using a hydrogen peroxide gas generator. *Journal of Applied Microbiology,* Vol. 99, No. 4, (October 2005), pp. 739-748, ISSN 1365-2672

Russell, A.D. (1990). Bacterial Spores and Chemical Sporicidal Agents. *Clinical Microbiology Reviews,* Vol. 3, No. 2, (April 1990), pp. 99-119, ISSN 1098-6618

Russell A.D. (1991). Principles of Antimicrobial Activity, In: *Disinfection, Sterilization, and Preservation* (fourth edition), S.S. Block, (Ed.), 27-58, Lea & Febiger, ISBN 978-081-2113-64-8 London, Philadelphia

Rutala, W.A. & Weber, D.J. (1999). Disinfection of endoscopes: review of new chemical sterilants used for high-level disinfection. *Infection Control & Hospital Epidemiology,* Vol. 20, No.1, (January 1999), pp. 69-76, ISSN 1559-6834

Sagripanti, J.L. & Bonifacino, A. (1996). Comparative Sporicidal Effects of Liquid Chemical Agents. *Applied and Environmental Microbiology,* Vol. 62, No. 2, (February 1996), pp. 545–551, ISSN 0099-2240

Sapers, G.M.; Walker, P.N.; Sites, J.E.; Annous B.A. & Eblen D.R. (2003). Vapor-phase Decontamination of Apples Inoculated with *Escherichia coli. Journal of Food Science,* Vol. 68, No. 3, (April 2003), pp. 1003–1007, ISSN 1750-3841

Scatchard, G.; Kavanagh, G.M. & Ticknor, L.B. (1952). Vapor-Liquid Equilibrium. VIII. Hydrogen Peroxide—Water Mixtures. *Journal of the American Chemical Society,* Vol. *74,* No. 15, (August 1952), pp. 3715–3720, ISSN 1520-5126

Schmaus, G.; Schierl, R. & Funck, S. (2002). Monitoring surface contamination by antineoplastic drugs using gas chromatography-mass spectrometry and voltammetry. *American Journal of Health-System Pharmacy,* Vol. 59, No. 10, (May 2002), pp. 956-961, ISSN 1079-2082

Shaffstall, R.M.; Garner, R.P.; Bishop, J.; Cameron-Landis, L.; Eddington, D.L.; Hau, G.; Spera, S.; Mielnik, T. & Thomas, J.A. (April 2006). Vaporized Hydrogen Peroxide (VHP®) Decontamination of a Section of a Boeing 747 Cabin. In: *DOT/FAA/AM-06/10 Federal Aviation Administration (FAA), Oklahoma City,* 25. 7. 2011, Available from: www.faa.gov/library/reports/medical/oamtechreports/index.cfm

Sharma, D. (2005). Bhopal: 20 years on. *The Lancet,* Vol. 365; No. 9454, (January 2005), pp. 111-112, ISSN 140-6736

Sheth, V.B. & Upchurch, D.C. (1996). System for Detecting the Presence of Liquid in a Vapor Phase Sterilization System. *United States Patent,* US 5,482,683, filed (March 1994), issued (January 1996)

Simmons, G.F.; Smilanick, J.L.; John, S. & Margosan, D.A. (1997). Reduction of Microbial Populations on Prunes by Vapor-Phase Hydrogen Peroxide. *Journal of Food Protection*, Vol. 60, No. 2, (February 1997), pp. 188-191, ISSN 0362-028X

Švrček, J. (2010). The Research and Development of Decontamination Process Utilizing Vapor Phase Hydrogen Peroxide. PhD diss., Institute of Chemical Technology, Prague

Tschirch, J.; Bahnemann, D.; Wark, M. & Rathouský, J. (2008). A comparative study into the photocatalytic properties of thin mesoporous layers of TiO_2 with controlled mesoporosity. *Journal of photochemistry and photobiology. A, Chemistry*, Vol. 194, No. 2-3, (February 2008), pp. 181-188, ISSN 1010-6030

Unger-Bimczok, B.; Kottke, V.; Hertel, Ch. & Rauschnabel, J. (2008). The Influence of Humidity, Hydrogen Peroxide Concentration, and Condensation on the Inactivation of Geobacillus stearothermophilus Spores with Hydrogen Peroxide Vapor. *Journal of Pharmaceutical Innovation*, Vol. 3, No. 2, (February 2008), pp. 123-133, ISSN 1939-8042

Unger, B.; Rauschnabel, U.; Düthorn, B.; Kottke, V.; Hertel, Ch. & Rauschnabel, J. (2007). Suitability of different construction materials for use in aseptic processing environments decontaminated with gaseous hydrogen peroxide. *The PDA Journal of Pharmaceutical Science and Technology*, Vol. 61, No. 4, (July-August 2007), pp. 255-275, ISSN 1079-7440

U.S. Environmental Protection Agency. (2005). *Compilation of Available Data on Building Decontamination Alternatives*, EPA/600/R-05/036, Cincinnati, Ohio

U.S. Environmental Protection Agency. (2008). *Effects of Vapor-Based Decontamination Systems on Selected Building Interior Materials: Vaporized Hydrogen Peroxide*, EPA/600/ R-08/074, Cincinnati, Ohio

U.S. Peroxide. (2009). Physical Properties of Hydrogen Peroxide, In: *Hydrogen Peroxide Technical Library., U.S. Peroxide, LLC, Atlanta*, 27. 7. 2011, Available from: http://www.h2o2.com/technical-library/physical-chemical-properties/physical-properties/default.aspx?pid=20&name=Physical-Properties

Vassal, S.; Favennec, L. Ballet, J.-J. & Brasseur, P. (1998). Hydrogen peroxide gas plasma sterilization is effective against Cryptosporidium parvum oocysts. *American journal of infection control*, Vol. 26, No. 2, (April 1998), pp. 136-138, ISSN 1527-3296

von Woedtke, Th.; Haese, K.; Heinze, J.; Oloff, Ch.; Stieber, M. & Jülich W.-D. (2004). Sporicidal efficacy of hydrogen peroxide aerosol. *Die Pharmazie*, Vol. 59, No. 3, (March 2004), pp. 207-211, ISSN 0031-7144

Wagner, G.W.; Sorrick, D.C.; Procell, L.R.; Brickhouse, M.D.; Mcvey, I.F. & Schwartz, L.I. (2007). Decontamination of VX, GD, and HD on a Surface Using Modified Vaporized Hydrogen Peroxide. *Langmuir*, Vol. 23, No. 3, (January 2007), pp. 1178-1186, ISSN 1520-5827

Wagner, G.W. & Yang, Y.-C. (2001). Universal Decontaminating Solution for Chemical Warfare Agents. *United States Patent*, US 6,245,957 B1, filed (September 1999), issued (June 2001)

Watkins, W.B. (2006). Decontamination Apparatus and Methods. *United States Patent*, US 7,145,052 B1, filed (September 2004), issued (December 2006)

Watling, D.; Ryle, C.; Parks, M. & Christopher M. (2002). Theoretical Analysis of the Condensation of Hydrogen Peroxide Gas and Water Vapour as Used in Surface Decontamination. *PDA Journal of Pharmaceutical Science and Technology*, Vol. 56, No.6, (November/December 2002), pp. 291-299, ISSN 1948-2124

Wold A. (1993). Photocatalytic properties of titanium dioxide (TiO2). *Chemistry of Materials*, Vol. 5, No.3, (March 1993), pp. 280–283, ISSN 1520-5002

Xu Y. (2001). Comparative studies of the $Fe^{3+/2+}$–UV, H_2O_2–UV, TiO_2–UV/vis systems for the decolorization of a textile dye X-3B in water. *Chemosphere*, Vol. 43, No. 8, (June 2001), pp. 1103-1107, ISSN 0045-6535

Zhao, J.; Chen, Ch. & Ma, W. (2005). Photocatalytic Degradation of Organic Pollutants Under Visible Light Irradiation. *Topics in Catalysis*, Vol. 35, No. 3-4, (July 2005) pp. 269-278, ISSN 10225528

18

Organic Pollutants Treatment from Air Using Electron Beam Generated Nonthermal Plasma – Overview

Yongxia Sun[1] and A. G. Chmielewski[1,2]
[1]*Institute of Nuclear Chemistry and Technology, Warsaw,*
[2]*University of Technology, Warsaw,*
Poland

1. Introduction

The municipal and industrial activities of man lead to environment degradation. The pollutants are emitted to the atmosphere with off-gases from industry, power stations, residential heating systems and vehicles. Organic pollutants, mainly volatile organic compounds (VOCs), which are emitted into atmosphere cause stratospheric ozone layer depletion, ground level photochemical ozone formation, and toxic or carcinogenic human health effects, contribute to the global greenhouse effect, accumulate and persist in environment. Regulation on organic pollutants emission into atmosphere has been enforced in many countries. Electron beam (EB) generated nonthermal plasma technology is one of the most promising technologies which has been successfully demonstrated on industrial scale coal fired power plants to remove SO_2 and NOx from waste off-gases; Meanwhile EB technology has been tested in pilot scale to remove dioxins and Polycyclic aromatic hydrocarbons (PAHs) from off-gases generated from solid waste incinerators and coal fired power plants, good results were obtained. It is a very promising technology to treat multiple pollutants including SO_2, NOx and organic pollutants simultaneously from industrial off-gases.
The principle of EB process to decompose pollutants is following. When the energy of the fast electrons is absorbed in the carrier gas, it causes ionization and excitation processes of the nitrogen, oxygen or water molecules in the carrier gas. Primary species and secondary electrons are formed, and the latter are thermalized within 1 ns in air at 1 bar pressure. These primary species such as ions, radicals or other oxidizing species and the thermalized secondary electrons react with pollutants by a series of reactions to cause pollutants decomposition.

Organic pollutants treatment using EB technology has been studied intensively in recent 30 years mainly in laboratory scale. However less work has been done to review this technology development on organic pollutants treatment. This chapter aims a comprehensive description of organic pollutants treatment using EB generated nonthermal plasma technology. General description of EB generated nonthermal plasma technology will be given in section 2, organic pollutants treatment from air and its recent development will be overviewed in sections 3 & 4, and general mechanism of organic pollutants decomposition in air will be discussed in section 5.

2. Electron beam (EB) generated nonthermal plasma technology

2.1 History

Wet flue gas desulphurisation (FGD) and selective catalytic reduction (SCR) can be applied for flue gas treatment and SO_2 and NOx emission control. VOCs are usually adsorbed on active carbon, but this process is rarely used for lean hydrocarbon concentrations up to now. All these technologies are complex chemcial processes and waste, like wastewater, gypsum and used catalyses, are generated(Srivastava et al., 2001).

EB technology is among the most promising advanced technologies of new generation. This is a dry-scrubbing process of simultaneous SO_2 and NOx removal, where no waste except the by-product is generated. EB technology for air treatment was first used by Japanese scientists in 1970-1971 to study SO_2 removal using an electron from linear accelerator (2-12 MeV, 1.2kW). A dose of 50 kGy at 100°C led the conversion of SO_2 to an aerosl of sulphuric acid droplets, which were easily removed (Machi, 1983). In 1981, Slater (1981) used EB technology to study the decomposition of low concentrations of vinyl chloride(VC) in different base gas (air, nitrogen, and argon).

2.2 EB accelerator

In physics and chemistry, plasma is a state of matter similar to gas in which a certain portion of particles are ionised. Nonthermal plasma means only a small fraction (for example 1%) of the gas molecules are ionized. The most common method for plasma generation is by applying an electric field to a neutral gas. Electrons emitted from electron beam accelerators can ionize gas mixture and generate nonthermal plasma.

More than 1000 accelerators haven been used in the field of radiation chemistry and radiation processing (Zimek, 1995). The reduction of SO_2 and NOx pollutants from flue gases, emitted during fuel combustion in electrical power and heat production, is one of the radiation process which were successfully demostrated in industrail scale in electric power station (EPS) Pomorzany, Szczecin, Poland (Chmielewski et al, 2004a) . A basic principle of an accelerator is that the electric field acts on electrons as charged particles and give them energy equal to the voltage difference accross the acceleration gap. The accelerator types are mainly determined by the method by which electron field is generated. There are three type accelerators used in air pollutants treatment: transformer accelerator, UHF accelerator and linear microwave accelerator (Zimek, 2005). High power accelerators have been developed to meet specific demands of environmental application and high throughput processes to increses the capacity and reduce unit cost of operation. Table 1 lists accelerators for radiation processing (Zimek, 2005).

Accelerator type	Direct DC	UHF 100-200 MHz	Linear 1.3-5.8GHz
Beam current	< 1.5A	< 100mA	< 100 Ma
Energy range	0.1-5 MeV	0.3-10 MeV	2-10MeV
Beam power	400kW	700 kW	150kW
Efficiency	60-80%	25-50%	10-20%

Table 1. Accelerator for radiation processing (recent development)

2.3 Terminology

In radiation application in environmental protection, there are three terms to be mentioned, **dose**, **G-value** and **removal efficiency (Re)** or **decomposition efficiency(De).**

In radiation process, it is very important to consider energy consumption for decomposition of pollutants, how much energy (unit: kJ) is consumed/absorbed to decompose amount of pollutants in the base gas (unit: kg). Energy absorbed by per amount of gas is defined as a term of **dose**, unit is **kGy**. 1 kGy = 1 kJ/ kg

G-value is defined as the number of molecules of product formed, or of starting material changed, for every 100 eV of energy absorbed. The G value is related to the ionic yield (M / N) by the expression (Willis and Boyd, 1976):

$$G\ (\text{molecules} / 100\ \text{eV}) = (M / N) \times (100 / W)$$

Where W (measured in electronvolts) is the mean energy required to from an ion pair in the material being irradiated. G value of **1 molecule /100 eV** is equal to a radiation chemical yield of **0.1036 μmol. J^{-1}**.

Removal efficiency (Re) or decomposition efficiency (De) of organic pollutants is defined as below:

$$Re = (C_0 - C_i) / C_0$$

where C_0 is initial concentration of organic pollutants, unit: ppm (v/v);
C_i is concentration of organic pollutants at i kGy absorbed dose, unit: ppm (v/v).

3. EB treatment organic pollutants

There are two systems applied to study organic pollutants in laboratory scale by using EB generated nonthermal plasma - flow system and batch system. Flow system contains one step: preparation and irradiation of the gas mixture which contains organic pollutants are carried out in on-line system. Batch system contains two steps: first step is to prepare gas mixture which contains organic pollutants into a sealed container, the second step is to put this sealed container under electron beam accelerator for irradiation.

3.1 Aliphatic organic pollutants degradation in flow system under EB-irradiation

3.1.1 Concentration of aliphatic organic pollutants vs. dose

There have been some previous studies of chlorinate hydrocarbons' decomposition in plasma reactors. Slater(1981) studied the decomposition of low concentrations of vinyl chloride(VC) in air, nitrogen, and argon in an electron beam generated plasma reactor. It was found that VC can be effectively removed by electron-beam irradiation at concentrations 3-500 ppm from room-temperature host-gas streams of argon, nitrogen and air. And at low dose the specific energy required falled in the range $2.5 < G < 10$ molecules removed per 100 eV. HCl was one of main products.

Vitale et al.(1997a) studied decomposition low concentration of ethyl chloride (EC) and vinyl chloride(VC) in atmospheric air streams by an electron beam generated plasma reactor. The gas was prepared by mixing dry air with standard VC (3925ppm VC in air) or EC (3717ppm EC in air). The gas entered the reactor at atmospheric pressure and ambient temperature. The electrons entered the front of the reactor, and VOC contaminated gas entered the rear of the reactor. The VOC contaminated gas thus flowed counter-current to

the electron beam. A Hewlett-Packard 5890 gas chromatograph and a HP-5971-A mass spectrometer were used to analyze VOCs concentration. The energy requirements for 90% decomposition of VC and EC were reported as a function of inlet concentration. VC requires less energy for decomposition than EC.
Similar experiments were carried out to decompose 1,1-dichloroethane (1,1-DCA), 1,1-dichloroethylene (1,1-DCE),1,1,1-trichloroethane (1,1,1-TCA), trichloroethylene (TCE) using EB by the same research group (Vitale et al., 1996, 1997b–d). It was found that decomposition efficiency of chlorinated compounds was : TCE > 1,1-DCE > 1,1-DCA, 1,1,1-TCA.
Won et al. (2002) studied the decomposition of perchloroethylene (PCE), trichloroethylene (TCE), dichloroethylene (DCE) in dry air. An electron accelerator of ELV type, with electron energy 0.7 MeV, maximum beam current 35 mA, maximum power 25 kW was used for irradiation. Over 80% TCE was decomposed at 20 kGy dose at initial concentration below 2000 ppm. The order of decomposition efficiency of these compounds was: TCE > PCE > DCE. Hirota et al. (2004) studied dichloromethane decomposition under EB irradiation and found that it was very difficult to treat dichloromethane.
For non-chlorinated organic compounds, 20 VOCs divided into five groups were investigated by Hirota et al.(2004), among them, 13 VOCs were alipahtic organic compounds. The order of decomposition VOCs in air was: cyclohexadiene > cyclohexane > benzene (group I); trans-hexane > 1-hexane (group II); heptane > hexane > pentane (group IV); and trichloroethylene > methanol >> acetone > CH_2Cl_2 (group V). Organic substances with long carbon chains readily succumbed to electron-beam treatment.

3.1.2 Different base gas mixtures influence on the decomposition efficiency of aliphatic organic pollutants

Won et al. (2002) studied TCE decomposition in different gas mixtures and found that the order of decomposition efficiency of TCE in different gas mixtures was: oxygen >air > H_2 > He.

3.1.3 Water concentration

In order to clarify OH radical influence on the chlorinated hydrocarbons (Cl-HC) decomposition, Won et al. (2002) tested TCE and PCE decomposition of air mixtures with different water vapor concentrations, and found that the decomposition efficiency of TCE and PCE increased less than 10% in the presence of water vapor compared with that in the dry air.

3.1.4 Irradiation products

The irradiation products of DCE, TCE and PCE in dry air under EB-irradiation were investigated by Won et al. (2002) and it was found that CO and CO_2 were the irradiation products. For PCE, CO_2 formation was above 40% at 15 kGy absorbed dose. Vitale et al (1997a) also reported that CO, CO_2 and HCl as main irradiation products when they studied ethyl chloride and vinyl chloride decomposition in air. Prager et al. (1995) studied DCE, TCE and PCE degradation in dry or humidified synthetic air, they identified HCl, CO, chloromethanes, chloroacetyl chloride and phosgene as main products.

3.2 Aliphatic organic pollutants degradation in batch system under EB-irradiation

3.2.1 Concentration of aliphatic organic pollutants vs. dose

Chloroethylene can be effectively decomposed by EB irradiation in the order of PCE > TCE > trans-DCE > cis-DCE (Hakoda et al., 1998a, 1998b, 1999, 2000, 2001; Hashimoto et al.,

2000). Sun et al. (2001, 2003) and Sun and Chmielewski (2004) studied 1,1-DCE, cis-DCE, trans-DCE decomposition under EB irradiation and found that the order of decomposition DCEs in air was 1,1-DCE > trans-DCE > cis-DCE. Decomposition efficiency of chloroethylene increases with the absorbed dose increase. The initial concentration of chlorinated ethylene was in below 2000 ppm and the water concentration in the air mixture was 200–300 ppm. Son et al. (2010a) studied decomposition of butane in EB irradiation in batch system, it was found that removal efficiencies of butane were 40% at 2.5 kGy and 66% at 10 kGy, when the initial concentration of butane was 60 ppm.

3.2.2 Different base gas mixtures influence on the decomposition efficiency of aliphatic organic pollutants

Different base gases influencing on the decomposition efficiency of butane were studied (Son et al, 2010a), it was found that decomposition efficiency of butane was extremely low when the background gas was He, in contrast to the efficiencies with background gases of N_2 and air. Decomposition efficiencies of butane was 23% in He, 63% in N_2 and 70% in air at 10 kGy absorbed dose.

3.2.3 Water concentration

Water influence on the TCE decomposition and irradiation products of TCE under EB-irradiation were studied by Hakoda et al. (2000). It was found that when water concentration was below 1000 ppm, there was no big difference between process efficiency for dry and humid air for TCE decomposition and dichloroacetyl chloride, carbon monoxide, carbon dioxide, phosgene and small amount of chloroform irradiation products' formation.

This result agrees well with that water vapor effect on the TCE decomposition under EB-irradiation in a flow system (Won et al., 2002). Sun et al. (2001) made a computer simulation of 1,1-DCE decomposition in air in a batch system and found that a reaction pathway of OH radical contributes less than 10% for 1,1-DCE decomposition. When water vapor concentration increased to 2.5%, yield of gaseous products decreased, that means the aerosol products are possibly formed (Hakoda et al., 2000).

3.2.4 Ozone

Hakoda et al. (1999, 2000, 2001) investigated O_3 influence on the trans-DCE, cis-DCE and TCE decomposition by using EB-irradiation, it was found that O_3 enhanced decomposition of trans-DCE only, cis-DCE and TCE were not affected.

3.2.5 Irradiation products

From environmental protection point of view, it is very important to identify by-products formation from Cl–HC degradation. Radiolytic products of trans-DCE, cis-DCE, TCE, PCE under EB-irradiation were reported by Hakoda et al. (1999, 2000, 2001). Chloroacetyl chloride and dichloroacetyl chloride were the main organic products for DCE (trans and cis) and TCE degradation; CO and CO_2 were inorganic products and their formations were below 25% based on carbon balance. Chmielewski et al. (2004b), Sun et al. (2003) and Sun & Chmielewski (2004) studied cis-DCE and trans-DCE degradation under EB-irradiation. Chloroacetyl chloride was not observed as degradation products, but it was a degradation product for 1,1-DCE (Sun et al., 2001). Son et al. (2010a) studied butane decomposition under EB irradiation and identified CO_2, acetaldehyde, acetone, 2,3-butandione, 2-butanone, and 2-butanedinitrile as degradation products of butane.

3.3 Aromatic organic pollutants degradation in flow system under EB-irradiation

In this section, besides aromatic chlorinated hydrocarbons degradation, decomposition of some nonchlorinated aromatic organic compounds will be discussed, too.

3.3.1 Concentration of aromatic organic pollutants vs. dose

Xylene and chlorobenzene decompositions in a flow system under EB-irradiation (Hirota et al. 2000, 2002) were studied, it was found that decomposition efficiency of xylene was higher than that of chlorobenzene, and about 50% chlorobenzene was decomposed at an absorbed dose of about 10 kGy at the initial concentration of chlorobenzene being 10–40 ppm. Kim (2002) studied decomposition of benzene and toluene, it was found that the decomposition efficiency of toluene was higher than benzene, and about 80% benzene was decomposed at 16 kGy when the initial concentration of benzene was smaller than or equal to 130 ppm. Han et al. (2003) studied toluene, ethylbenzene, o-, m-, p-xylenes and chlorobenzene decomposition in air. The order of the decomposition efficiency of selected VOCs from high to low was : toluene > ethylbenzene > benzene; p-xylene> m-xylene > o-xylene. About 44.7 % toluene and 43.2% ethylbenzene was decomposed at 10 kGy, while 85% chlorobenzene was decomposed. The decomposition efficiency of ethylbenzene and toluene was significantly increased about 50% with the addition of chlorobenzene comparing without chlorobenzene addition.

3.3.2 NH_3 influence on the decomposition efficiency of aromatic organic pollutants

Effect of ammonia on the decomposition of PAHs was observed when an electron beam process was applied to treat multiple pollutants (SO_2, NOx , PAHs) with NH_3 addition from industrial off-gases emitted from EPS, Kawęczyn, Poland (Chmielewski et al., 2002). NH_3 addition enhanced PAHs removal efficiency. Hirota et al. (2000) studied chlorobenzene decomposition in air mixture, it was found that addition of NH_3 enhanced the dechlorination of chlorobenzene. About 65% of chlorine in reacted chlorobenzene was dissociated from carbon with electron beam at doses of 4 and 8 kGy. Ammonia addition enhanced the dechlorination to 80%.

3.3.3 Water concentration

Effect of water vapor on the decomposition of toluene was investigated by Kim (2002). It was found that the water vapor injection leads to 15–20% removal efficiency increase for toluene compared to the process without water injection. Water influences decomposition of toluene higher than TCE and PCE, OH radical plays an important role for aromatic hydrocarbon decomposition.

3.3.4 Irradiation products

Degradation products of chlorobenzene and xylene in an air mixture were studied by Hirota et al. (2000, 2002). The gaseous products of xylene degradation were identified to be formic, acetic, propionic, and butyric acids and/or the corresponding esters with CO and CO_2. Approximately 30% of the reacted xylene was the gaseous products at a dose of 8 kGy. Organic and inorganic chlorine presence in gaseous, aerosols and residues were investigated by same authors for chlorobenzene degradation. Inorganic chlorine was mainly presented in gaseous products, while organic chlorine was presented in aerosols and residues aerosols and residues. Some aerosol products were formed from chlorobenzene degradation, and 3% of the aerosol products were identified to be carboxylic acids or esters (Hirota et al., 2000). Aerosols, benzaldehyde, dipropyl 1,2-benzenedicarboxylic acid, nitromethane were reported as toluene main degradation products in dry air under EB irradiation; while methyl chloride, dipropyl 1,2-benzenedicarboxylic acid, toluene, nitromethane were reported as main degradation products of ethylbenzene/chlorobenzene mixture in dry air. Trace amount of acetone, hexane, benzene was also observed (Han et al., 2003).

3.4 Aromatic organic pollutants degradation in batch system under EB-irradiation

3.4.1 Concentration of aromatic organic pollutants vs. dose

Decomposition of aromatic organic compounds in a batch system vs. dose under EB-irradiation was studied by Hirota et al. (2002, 2004), Ostapczuk et al.(1999), Sun et al. (2008) and Hashimoto et al. (2000). Decomposition efficiency of these compounds increase with the absorbed dose increase, 4-chlorotoluene (4-CTO) decomposition as an example was presented in Fig. 1 . The order of decomposition efficiency of these compounds was: xylene >chlorobenzene > benzene > hexane > cyclohexane. For 4-chlorotolunene (4-CTO) and 1,4-dichlorobenzene (1,4-DCB), no apparent decomposition efficiency of these two compounds was observed (Fig.2).

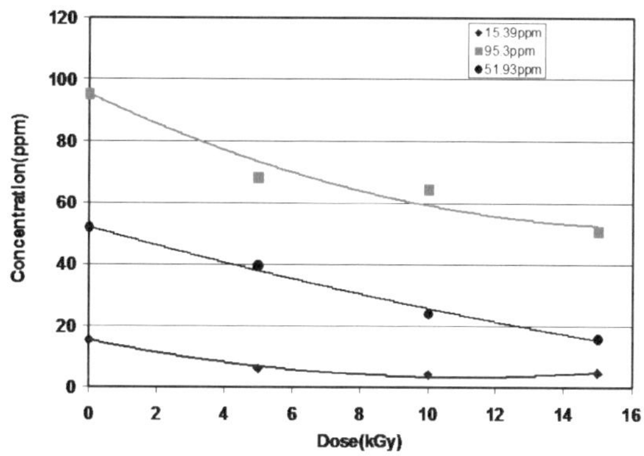

Fig. 1. 4-Chlorotoluene decomposition in air mixture in an electron beam generated non-thermal reactor.

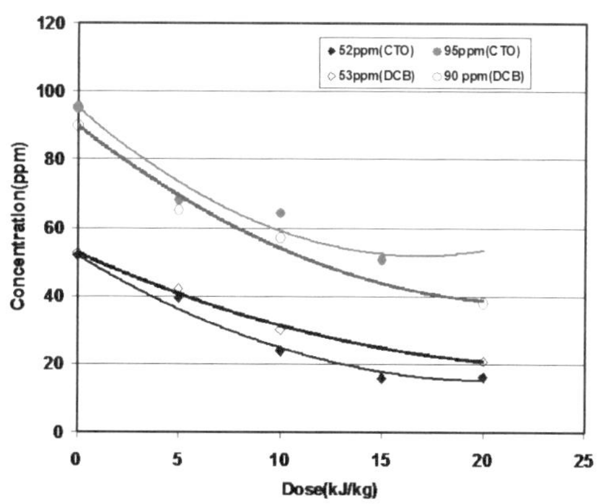

Fig. 2. Decomposition efficiency comparison between 4-chlorotoluene and 1,4-dichlorobenzene.

The decomposition efficiency of chlorinated aliphatic hydrocarbons using EB irradiation is more efficient than that of chlorinated aromatic hydrocarbons. 25.0 kGy is sufficient to remove over 97% 1,1-DCE and 98.0% trans-DCE at initial concentration of 1,1-dichloroethylene (DCE) and trans-dichloroethylene being 903.8 ppm and 342.0 ppm, respectively; while 60% 1,4-dichlorobenzene (DCB) at intial concentration being 90ppm was removed at 57.9 kGy. This result is comparable with decomposition of chlorobenzene. Hakoda et al. (1998b) and Hashimoto et al. (2000) studied degradation of chlorobenzene/air using EB irradiation in batch system, it was found that 40% chlorobenzene was removed under EB-irradiation at 37.7 kGy dose (calculated by N_2O gas dosimeter) for initial concentration of chlorobenzene being 102 ppm. Sun et al. (2007a) studied 1-chloronaphathalene and found that over 80% 1-chloronapthalene was removed at 57.9 kGy under EB-irradiation for low initial concentration of 1-chloronaphthalene (12~30 mg/m^3) in air mixture. Energy consumption for decomposition 1,4-dichlorobenzene was lower than that of 1-chloronaphthalene. Therefore, the observed order in easily decomposition chlorinated hydrocarbons is: 1,1-DCE > trans-DCE > cis-DCE >1,4-DCB > 1-chloronaphthalene. Based on this work and other's work, we learn that: For chlorinated aliphatic hydrocarbons, the more chlorinated compounds is, the more it is easy to be decomposed.

Aliphatic hydrocarbons is more easily decomposed than aromatic hydrocarbons. For aromatic hydrocarbons, compounds with less benzene ring are easily to be decomposed.

3.4.2 Different base gas mixtures influence on the decomposition efficiency of aromatic organic pollutants

Toluene decomposition at different background gases in a batch system was studied by Kim (2002). The order of decomposition efficiency of toluene in different background gases is: N_2 > air >O_2 >He. This order is different from the order for TCE decomposition in a flow system (Won et al., 2002).

We studied 1,4-dichlorobenzene (1,4-DCB) decomposition in different base gas mixtures at the initial concentration of 1,4-DCB being 50 ppm, the similar phenomenon was observed (Sun et al, 2006). The decomposition efficiency of 1,4-DCB in nitrogen is higher than that in air and much more higher than in 1.027% NO-N_2 mixture(N_2 as balance gas) (Fig.3) , this phenomenon agrees well with toluene decomposition in different gases (Kim, 2002).

3.4.3 Water concentration

Effect of water vapor on the decomposition of aromatic compounds in a batch system under EB-irradiation was investigated by Kim (2002). Four percent water vapor injection leads to 5–10% increase of VOC removal efficiency for both toluene and benzene, and effect of water vapor influence on the decomposition of toluene under EB irradiation in a flow system is higher than that in a batch system.

3.4.4 Irradiation products

Benzaldehyde and phenol were reported as products when Ostapczuk et al. (1999) studied styrene decomposition in air under EB irradiation, the removal efficiency of styrene was ranged from 83-95%. The humidity in air mixture was ranged from 0.3% to 1.6%. In order to obtain information of by-products produced from toluene destruction, we carried out experiment at higher inlet concentration of toluene at higher absorbed dose. More than 97%

toluene was removed from gas phase at 53.6 kGy absorbed dose when inlet concentration of toluene was 151.9 ppm. A GC-MS spectrum of toluene/air mixture after EB-irradiation is presented in figure 4. A trace amount of benzaldehyde was eluted at retention time 7.735 min in figure 4 and was identified by our carefully comparing mass spectrum of this compound with a reference mass spectrum of benzaldehyde provided by Wiley library (figures 5a & 5b). Trace amount acetone was also found in our experimental condition (Sun et al., 2009a).

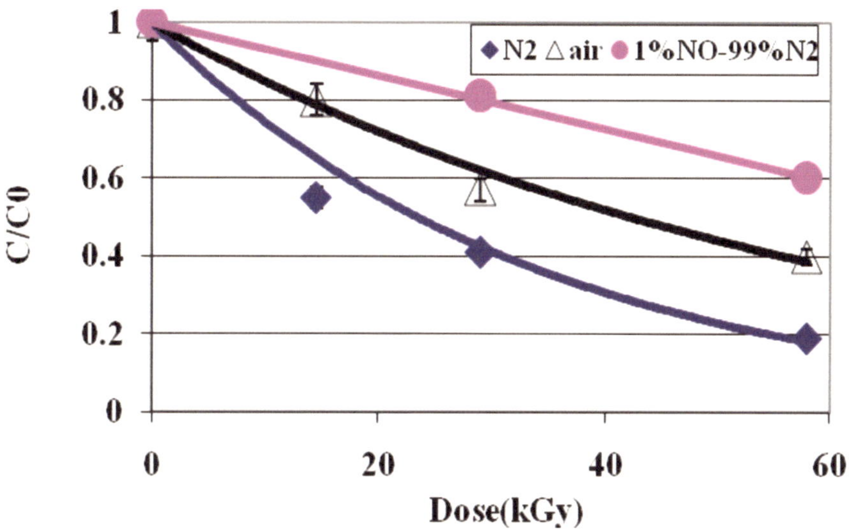

Fig. 3. 1,4-Dichlorobenzene decomposition in different gas mixture

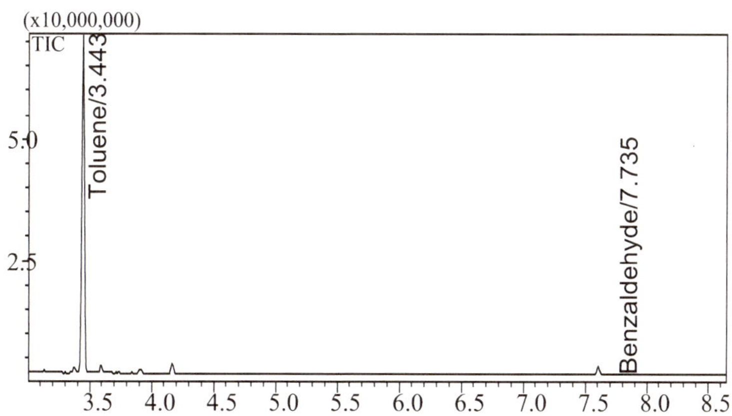

Fig. 4. A GC-MS spectrum of toluene/air mixture after EB-irradiation (inlet concentration of toluene was 151.9 ppm, dose was 53.6 kGy).

Benzaldehyde as by-product of degradation of toluene was also reported(Han et al., 2003; Kim et al., 2005). Trace amount of acetone was found based on Han et al.'s work (2003). Besides these, Aerosols and benzene were reported as by-products in both works (Han et al.,

2003; Kim et al., 2005). Han et al.(2003) also identified dipropyl 1,2-benzenedicarboxylic acid, nitromethane and trace amount of hexane as by-products of degradation of toluene. For degradition of 4-chlorotoluene in air mixture, chlorobenzene (C_6H_5Cl, retention time was 4.910 min) and 4-chlorobenzaldehyde (ClC_6H_4CHO, retention time was 12.502 min) were identified as by-products. A GC-MS spectrum of 4-chlorotoluene/air mixture after EB-irradiation was presented in figure 6, a compound eluted at retention time 7.590 min of the GC-MS spectrum was identified as 4-chlorotoluene (Sun et al., 2008).

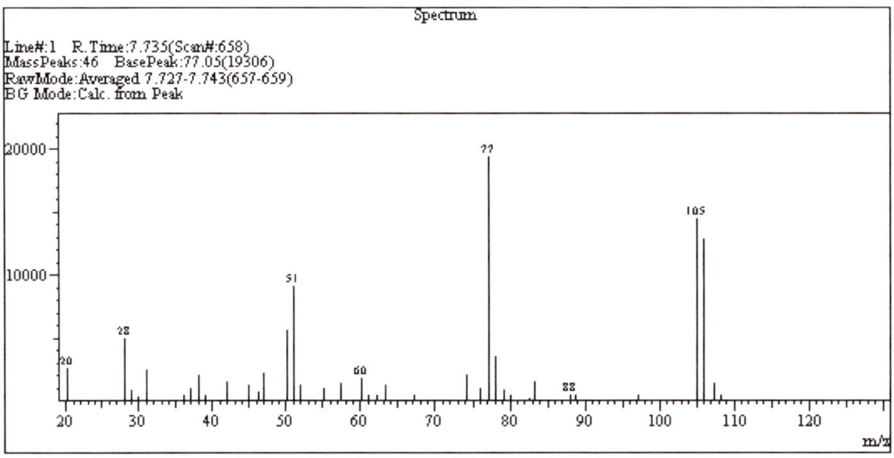

Fig. 5a. A mass spectrum of by-product which eluted at 7.735 min retention time

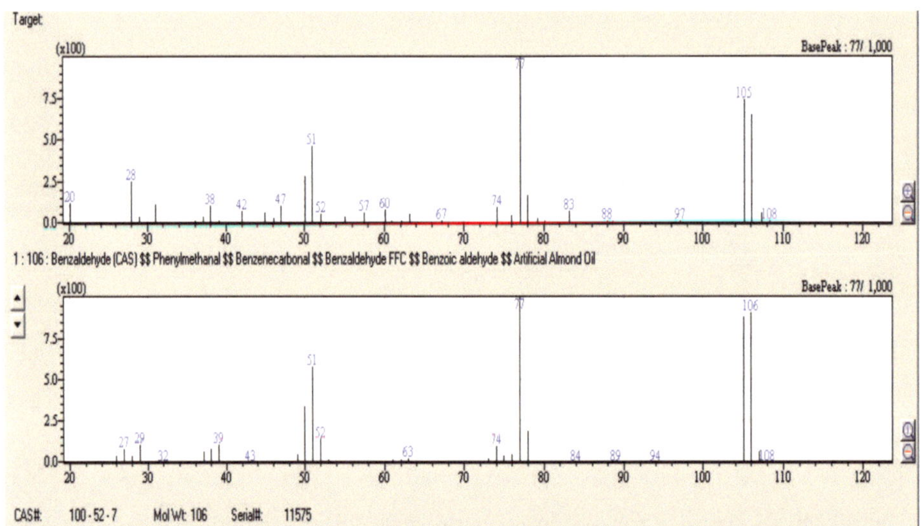

Fig. 5b. A mass spectrum of the compound which eluted at 7.735 min retention time and its reference mass spectrum of benzaldehyde

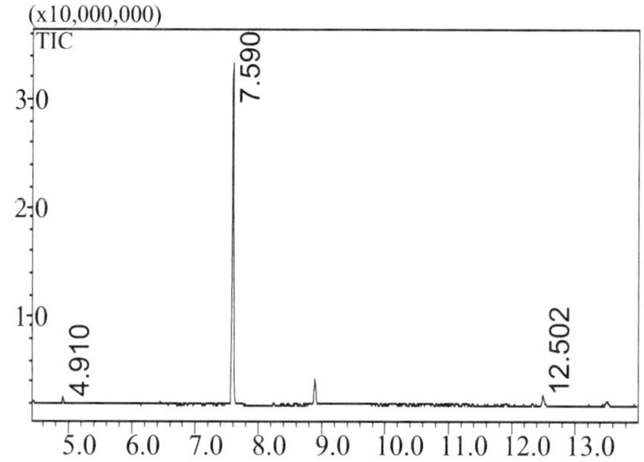

Fig. 6. A GC-MS spectrum of 4-chlorotoluene/air mixture after EB-irradiation

3.5 PAHs and Dioxin removal from waste off-gas under EB-irradiation

Dioxins reduction from waste incinerator was studied using EB technology in Japan (Hirota et al., 2003) and in Germany (Paur et al., 1998). Hirota et al. (2003) studied reduction the emission of polychlorinated dibenzo- p-dioxins (PCDD) and polychlorinated dibenzofurans (PCDF) in a flue gas of 1000 m^3N/h from the municipal solid waste incinerator (MSWI), located at Takohama Clean Center which treats 450 t (150 t * 3 furnaces) of solid waste in 1 day, at a temperature of 200 °C. they found that more than 90% PCDD/Fs was removed at 14 kGy when initial concentration of PCDD was in the range of 0.22-0.88 ng-TEQ/m^3N and PCDF in the range of 0.35-12.4 ng-TEQ/m^3N. Paul et al. (1998) also reported that over 90% PCDD was removed at 12 kGy dose for inital concentration of PCDD being 21-110 ng/m^3N (AGATE-M plant, Germany).

16 kinds of toxic PAHs were investigated under electron beam irradiation in the pilot plant in Electric Power Station Kawęczyn, Poland (Chmielewski, et al., 2003). The investigation was carried out under the following experimental conditions: flue gas flow rate 5000 Nm^3/h; humidity 4.5%; inlet concentrations of SO_2 and NOx that were emitted from the power station were 192 and 106 ppm, respectively; ammonia addition was 2.75 Nm^3/h; alcohol addition was 600 l/h, the absorbed dose was 8 kGy. The results was presented in Fig.7. It was found that under these experimental conditions the concentrations of naphthalene (NL, $C_{10}H_8$), acenaphthene (AC, $C_{12}H_{10}$), fluorene ($C_{13}H_{10}$), phenanthrene ($C_{14}H_{10}$), anthracene ($C_{14}H_{10}$) were decreased, while the concentrations of acenaphthylene ($C_{12}H_8$), fluoranthene ($C_{16}H_{10}$), pyrene ($C_{16}H_{10}$), benzo(a)anthracene ($C_{18}H_{12}$), chrysene ($C_{18}H_{12}$), benzo(b þ k)fluoranthene ($C_{20}H_{12}$), benzo(e)pyrene ($C_{20}H_{12}$), benzo(a)pyrene ($C_{20}H_{12}$), perylene ($C_{20}H_{12}$), dibenzo(a; h)anthracene+indeno(1,2,3-cd) pyrene ($C_{22}H_{14}$), benzo(g; h; l)perylene ($C_{22}H_{12}$) were increased. Removal efficiencies of SO_2 and NOx were 61.6% and 70.9%, respectively. The concentration of hydrocarbons of small aromatic ring (PAHs, like naphthalene ($C_{10}H_8$), acenaphthene ($C_{12}H_{10}$), fluorene ($C_{13}H_{10}$), anthracene ($C_{14}H_{10}$)) was reduced, while the concentration of fluoranthene was increased remarkably after irradiation.

Similar experiments were carried out in EPS Kawęczyn with ammonia presence but without alcohol addition (Chmielewski et al, 2002; Ostapczuk et al, 2008a). It was found that removal efficiency of PAHs ranges from 40% up to 98%.

Callén et al. (2007) studied PAH removal from lignite-combustion flue gas from Bulgarian Maritza-East thermal power plant (TPP) and obtained that PAHs concentration after EB irradiation resulted in ~ 10 fold decrease in studied PAHs emissions. The removal efficiency of PAH removal at the dose of 4 kGy was 85% (weight/weight). High PAH removal efficiency was obtained especially for 2 and 3 rings PAH, this result was similar to that obtained in our previous work (Chmielewski et al., 2003).

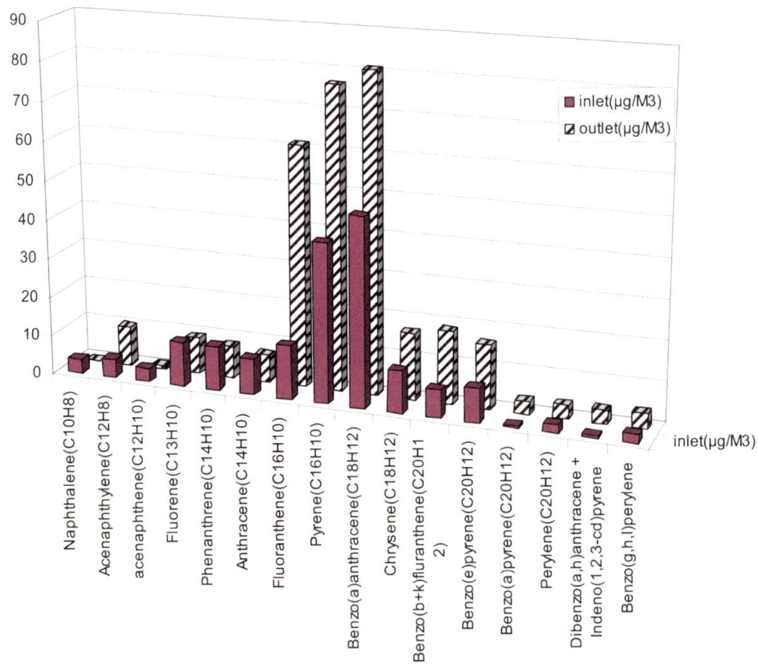

Fig. 7. EB irradiation influence on PAHs removal.

The concentration of PAHs in by-product was also examined. It was relatively low, varied from few up to 12 µg per kg of fertilizer for the experimental work carried out in EPS Kawęczyn, Poland. Less than 3% of PAH were removed in adsorption on the by-product surface (Ostapczuk, et al., 2008a). The study in Maritza-East TPP allowed PAH assessment in solid by-products obtained from EB lignite-combustion flue gas. The determined PAH content was reasonable, ~ 60 microg/kg and was lower than PAHs background in Bulgarian soils. These results demonstrated the insignificant role of adsorption for PAHs removal (Callén et al., 2007).

Naphthalene (NL) and acenaphthene (AC) decomposition in gas mixture was studied in lab scale experiment (Ostapczuk, et al., 2008b). It was found that NL was more easily decomposed than AC, G-values for these two compounds were 1.66 and 3.72 mol/100 eV for NL and AC at the dose of 1 kGy , respectively. Humidity influencing on NL and AC decomposition was studied. About 26% and 50% NL were decomposed at 1 kGy dose in dry air (90 vol% N_2; 10 vol% O_2 and 160 ppmv NO) and in humid air (84 vol% N_2; 10 vol% O_2 and 6 vol% H_2O), respectively. NL concentration in both mixtures was on the level of 10–11 ppmv. For AC, about 45% (in O_2) and 82% AC (in humid O_2, 94% O_2 + 6% H_2O) were decomposed at 1 kGy dose for the initial concentration of AC being 160 ppm. Two-ringed naphthol and nitronaphthalene; one-ringed 2,6-dietylbenzoquinone, indane, 1,2- and 1,4-dimetoxybenzenes and carbon oxides were identified as by-products of NL decomposition.

4. A novel hybrid EB-catalyst technology to treat organic pollutants

Electron beam (EB) irradiation is one of the most promising technologies for purification of dilute pollutants, mostly VOCs , with high flow-rate gas stream released from industrial off-gases. Under EB irradiation, VOCs are oxidized into irradiation by-products as well as CO_2 and CO. However, some of these irradiation by-products have adverse effect on environment and human beings.

A new technology which combines EB and catalyst together to treat aromatic VOCs, e.g., toluene (Kim et al., 2004, 2005; Jeon et al., 2008) styrene(Kim, 2005) , o-Xylene (Hakoda et al., 2008a, 2008b) and ethylbenzene (Son, et al, 2010b) was developed in the aim of enhancing higher oxidation efficiencies of VOCs into CO_2. Removal efficiency of toluene , styrene and ethylbenzene increased by 10%, 20%, and 20% in an EB-catalyst hybrid system in comparison with that achieved in catalyst-only method at approximately 10 kGy absorbed dose (Kim, et al, 2005). Removal efficiency of ethylbenzene in the EB-catalyst hybrid was 30% higher than that of EB-only treatment. Ethylbenzene was decomposed more easily than toluene by EB irradiation. The G-values for ethylbenzene increased with initial concentration and reactor type: the G-values vary in the range of 7.5-10.9 (EB-only) and 12.9-25.7 (EB-catalyst hybrid) by reactor type at the initial concentration of ethylbenzene being 2800 ppm. Son et al. (2008) and Jeon et al. (2008) also studied different catalysts (Pt, Pd , Cu and Mn) and humidity influence on removal efficiency of toluene using EB-catalyst hybrid system. It was found that removal efficiency of toluene was increased by 36.9% , 35.3% and 22% in the presence of Pt, Pd , Mn and Cu catalysts comparing with EB only for initial concentration of toluene being 1500 ppm, the selectivity to CO_2 with Pt and Pd coupling were relatively higher than those of Cu and Mn. Especially the CO_2 selectivity of EB–Pt coupling was significantly high at a relatively low absorbed dose. The catalytic activity for EB–catalyst coupling system was in the order of Pt, Pd, Mn and Cu. There was no significant difference of removal efficiency of toluene among 0.1, 0.5 and 1.0 wt% loading of catalyst. No significant water effect was observed in EB-catalyst hybrid system (Son et al, 2008).

Other type of catalysts such as TiO_2 (Hakoda, et al., 2008a) was used to study xylene decomposition under EB irradiation in lab scale experimental work. It was found that removal efficiency of xylene and CO_2 formation were increased with the presence of TiO_2 catalyst, the similar phenomenon was observed when Kim studied toluene decomposition using Pt as catalyst (Kim et al., 2005).

Hakoda, et al.(2008b) also studied xylene decomposition using MnO_2 (an O_3 decomposition catalyst), γ-Al_2O_3 was selected as a base material of the catalyst. The combination process at temperatures of about 100°C using MnO_2 placed downstream enhanced the oxidation of the by-products of xylene into CO_2 by active oxygen produced from the O_3 decomposition when the MnO_2 bed was placed downstream of an irradiation space. Furthermore, EB-irradiated γ-Al_2O_3 surface was found to be active, and the oxidation of organics was enhanced by primary electrons. The combination process using γ-Al_2O_3 reduced dose to 33% of a single EB process to obtain the same conversion of xylene to CO_2.

Ighigeanu (et al., 2008) studied VOCs (Toluene, hexane + toluene mixture diluted in air) decomposition by using combination of three different technologies (EB, microwave (MW) and catalysts): (EB + MW+ catalyst); (MW + catalyst) and (EB+catalyst). They found that decomposition efficiency (De) and oxidation efficiency (Eo) of toluene increased significantly for the (EB+MW+catalyst) treatment as compared with (MW + catalyst) and (EB + catalyst) treatments, at initial concentration of toluene being in the range of 180 ppm –

523 ppm; and CO_2 and CO concentrations after treatment were higher for the (EB+MW+catalyst) treatment than for (MW+catalyst) and (EB+catalyst) treatments. De and Eo of toluene were, respectively, as follows: 59.5% and 82.2% for the (MW + catalysis), 77.2% and 87.1% for the (EB + catalyst) and 92.8% and 90.5% for the(EB + MW + catalyst). For air mixture contained toluene and hexane, removal efficiency of toluene and hexane in (EB + MW + catalyst) system was higher than that in (MW + catalyst) system or in (EB + catalyst) system, about 88.5% toluene and 87.8% hexane were decomposed for initial toluene and hexane concentration being 250 ppm, respectively.

5. Mechanism of organic pollutants degradation by using EB technology

In order to obtain high decomposition efficiency of organic pollutants and less toxic by-products, it is very important to understand mechanism of organic pollutants degradation under EB irradiation. In this section, we will discuss mechanism of two groups (chlorinated and nonchlorinated) organic pollutants. General mechanism of organic pollutants decomposition in gas phase under EB irradiation is illustrated in Fig. 8.

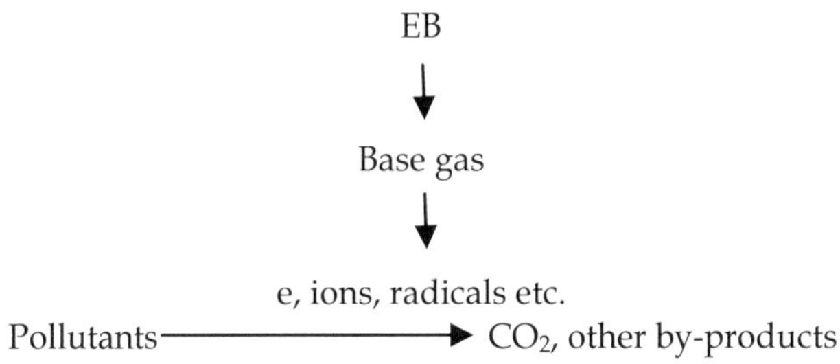

Fig. 8. General mechanism of organic pollutants decomposition under EB irradiation

5.1 Chlorinated organic compounds (Cl-HC)

5.1.1 General mechanism of chlorinated aliphatic hydrocarbon decomposition in air mixture

Computer simulations of chlorinated aliphatic hydrocabons'decomposition in air mixture were carried out and discussed in details (Nichipor et al., 2000, 2002, 2003, 2008; Sun et al., 2001, 2007b, 2009b) . The general mechanism of aliphatic hydrocarbon decomposition in an air mixture is described below:

When fast electrons from electron beams are absorbed in the carrier gas, they cause ionization and excitation processes of the nitrogen, H_2O and oxygen molecules in the carrier gas. Primary species and secondary electrons are formed. The secondary electrons are thermalized fast within 1 ns in air at 1 atmosphere .

The *G*-values (molecules/100 eV) of main primary species are simplified as follows (Mätzing, 1989):

$$4.43N_2 \rightarrow 0.29N_2^* + 0.885N(^2D) + 0.295N(^2P) + 1.87N + 2.27N_2^+ + 0.69N^+ + 2.96e \quad (1)$$

$$5.377O_2 \rightarrow 0.077O_2^* + 2.25O(^1D) + 2.8O + 0.18O^* + 2.07O_2^+ + 1.23O^+ + 3.3e \quad (2)$$

$$7.33H_2O \rightarrow 0.51H_2 + 0.46O(^3P) + 4.25OH + 4.15H + 1.99H_2O^+ + 0.01H_2^+ + 0.57OH^+ + 0.67H^+ + 0.06O^+ + 3.3e \quad (3)$$

Where G-values of molecules decomposed are listed in the left side of the arrows, and G-values of species formed from the pure k type molecules that absorb an energy of 100 eV are listed in the right side of the arrows. These primary species and thermalized secondary electrons cause Cl-HC decomposition. Based on our and others published work , we know that several type reactions cause Cl-HC degradation.

Positive ions charge transfer and particle dissociation reactions

It is well known that : when air component and molecule are ionized and excited, a large amount of N_2^+, O_2^+ , N^+, O^+, H_3O^+(if water concentration is high) are formed, and their ionization potential energy (IE) is higher than that of Cl-HC (see Table 2). The positive charge transfer reaction, positive ion cluster reaction, or particle dissociation reaction occur (Spanel, et al., 1999a, 1999b).

Molecule	IE(eV)	EA(eV)	PA(kJ.mol^{-1})
N_2	15.58		493.8
O_2	12.07	0.45	421
H_2O	12.62		691
NO	9.26	0.03	531.8
O_3	12.53	2.10	625.5
CCl_4	11.47	0.80	
$CHCl_3$	11.37	0.62	650.6
CH_2ClCH_2Cl	11.07		
$CHCl_2CH_3$	11.04		
C_2H_3Cl	9.99		
1,1-$C_2H_2Cl_2$	9.81	0.1	
cis-$C_2H_2Cl_2$	9.65		
trans- $C_2H_2Cl_2$	9.64		
C_2HCl_3	9.46	0.40	
C_2Cl_4	9.32	0.64	
1,4-dichlorobenzene	8.92		

Table 2. Ionization energy (IE,eV) , electron affinity(EA, eV) and proton affinity(PA, kJ.mol^{-1}) data.

In general, the H_3O^+ reactions with the aliphatic chloride more varies in their rate constants and products, and in some reactions H_3O^+.M ions (M=Cl-HC) are formed. The NO^+ reaction with the aliphatic compounds is generally slow association reactions and form NO^+.M ions (for e.g., $NO^+ + CHClCCl_2 = NO^+.CHClCCl_2$). The O_2^+ reactions are fast mainly proceeding via nondissociative charge transfer reactions to produce the cations M^+ only (for e.g., $C_2HCl_3 + O_2^+ = C_2HCl_3^+ + O_2$), but in some of these reactions minority dissociative charge transfer reactions take place to eliminate Cl/HCl and leave hydrocarbon ion (for e.g., $CH_2ClCH_2Cl + O_2^+ = C_2H_3Cl^+(95\%)+HCl+O_2$; and $CH_2ClCH_2Cl + O_2^+ = C_2H_4Cl^+(5\%)+Cl+O_2$).

From the calculation results of dichloroethylene (Sun et al., 2001, 2007b), trichloroethylene (Nichipor et al, 2008) and tetrachloroethylene (Sun et al, 2009b), we learn that positive charge transfer reactions contributing to chlorinated ethylenes decomposition less than 10%

Secondary electron attachment, Cl dissociated reaction and negative ions charge transfer reactions

Reaction pathway of secondary electron attachment, Cl dissociative reactions followed by peroxyl radical reactions is the main reaction pathway which causes Cl-HC decomposition. The rate constants of electron and Cl with Cl-HC , and the products of these reactions are listed in table 3, respectively (Atkinson, R., 1987a).

For chlorinated methane, the products formed by Cl with CH_3Cl, CH_2Cl_2, $CHCl_3$, reactions are: CH_2Cl, $CHCl_2$, CCl_3 , and HCl; for chlorinated methane CCl_4, CCl_3 and Cl_2 are formed.

If we assume CH_2Cl, $CHCl_2$, CCl_3 as radical R, generalized mechanism of peroxyl radical reactions could be written as follows:

$$R + O_2 = RO_2 \tag{4}$$

$$RO_2 + RO_2 = 2\,RO + O_2 \tag{5}$$

$$RO = \text{products} \tag{6}$$

This is a main reaction pathway for Cl-HC decomposition (Bryukov et al., 2002;).

For chlorinated aliphatic ethylene, the mechanism of its degradation in air mixture can be generalized as follows:

$$e + C_2X_3Cl = Cl^- + \bullet C_2X_3 \text{ (Szamrej, et al., 1996;)} \tag{7}$$

$$X_2CCX_2 \text{ (X=H, Cl, at least 1 Cl inside)} \tag{8}$$

$$X_2CCX_2 + Cl \rightarrow (Cl)X_2CCX_2 \text{ (Cl adds to heavily chlorinated carbon side)} \tag{9}$$

$$(Cl)X_2CCX_2 + O_2 \rightarrow (Cl)X_2CCX_2(O_2) \tag{10}$$

$$2(Cl)X2CCX2(O) + O2 \tag{11}$$

$$(Cl)X_2CCX_2(O) \rightarrow (Cl)X_2CCXO + Cl \text{ (if one X = Cl)} \tag{12}$$

$$\text{or } (Cl)X_2CCX_2(O) \rightarrow CX_2Cl + COX_2 \tag{13}$$

Molecule	Electron [$cm^3.s^{-1}$]	Cl [$cm^3.s^{-1}$]
CH_3Cl	6.1×10^{-11}, $Cl^- + CH_3$	4.78×10^{-13}, CH_2Cl+ HCl
CH_2Cl_2	1.6×10^{-10}, $Cl^- + CH_2Cl$	3.5×10^{-13}, $CHCl_2$+ HCl
$CHCl_3$	4.9×10^{-9}, $Cl^- + CHCl_2$	1.2×10^{-13}, $\boldsymbol{CCl_3}$+ HCl
CCl_4	1×10^{-7}, $Cl^- + CCl_3$	1.4×10^{-10}, $CCl_3 + Cl_2$
C_2H_3Cl	$(1··8) \times 10^{10}$, $Cl + C_2H_3$	1.27×10^{-10}, $CH_2ClCHCl$
$C_2H_2Cl_2$	1×10^{-9}, $Cl^- + \boldsymbol{C_2H_2Cl}$	1.4×10^{-10}, CH_2ClCCl_2
C_2HCl_3	$(0.29\sim1) \times 10^{-8}$, $Cl^- + C_2HCl_2$	9.3×10^{-12}, CCl_3CHCl
C_2Cl_4	1×10^{-7}, $Cl^- + C_2Cl_3$	$(4\sim6)\times 10^{-11}$, C_2Cl_5
C_2H_5Cl	$(2\sim7) \times 10^{-13}$, $Cl^- + C_2H_5$	6.8×10^{-12}, CH_3CHCl + HCl

Table 3. Rate constants and products for the reactions of electron, Cl with chlorinated aliphatic compounds.

The decomposition efficiency of Cl-HC mainly depends on the rate constants of secondary electron attachment, and Cl addition reaction followed by peroxyl radical reactions (Knox et al., 1966, 1969; Thűner et al., 1999). This decomposition pathway has been confirmed experimentally (Hirota et al., 2002).

O_2^- cause Cl-HC decomposition, $O_2^- + M = O_2 + M^-$ (M= Cl-HC) (14)

O atom, OH radical , and other radical reactions with Cl-HC

Other decomposition pathways for chlorinated aliphatic hydrocarbons are: O atom decomposition pathway (Sanhueza et al., 1974a, 1974b; Teruel, et al., 2001), OH radical decomposition pathway (Atkinson, R., 1987b; Howard, et al., 1976; Liu et al., 1989; Chandra et al., 1999; Chang et al., 1977) and other species decomposition pathway, such as O_3 and NO_3. The rate constants of O and OH with Cl-HC , and the products of these reactions are listed in table 4, respectively (http://kinetics.nist.gov/kinetics/index.jsp). By-products of irradiation vary with reactants.

Molecule	O [$cm^3.s^{-1}$]		OH [$cm^3.s^{-1}$]	
CH_3Cl	1.18×10^{-16},	$OH + CH_2Cl$	4.2×10^{-14},	$CH_2Cl + H_2O$
CH_2Cl_2	6.48×10^{-16},	OH + ***$CHCl_2$***	1.4×10^{-13},	$CHCl_2 + H_2O$
$CHCl_3$	1.02×10^{-15},	$OH + CCl_3$	1.0×10^{-13},	$CCl_3 + H_2O$
CCl_4	1.89×10^{-16},	$ClO + CCl_3$	$< 4 \times 10^{-16}$,	$CCl_3 + HOCl$
C_2H_3Cl	5.96×10^{-13},	products	8.06×10^{-12},	$CHClCH_2OH$
1,1-$C_2H_2Cl_2$	9.8×10^{-13},	products	8.10×10^{-12},	CH_2OHCCl_2
C_2HCl_3	1.4×10^{-13},	products	2.2×10^{-12},	products
C_2Cl_4	1.9×10^{-13},	products	1.7×10^{-13},	products
C_2H_5Cl	1.12×10^{-15}, OH+ other products		6.42×10^{-13},H_2O + other products	

Table 4. Rate constants and products for the reactions of O, OH with chlorinated aliphatic compounds

The mechanism of decomposition of chlorinated aliphatic hydrocarbons under EB irradiation could be described as follows: Cl^- dissociative secondary electron attachment followed by peroxyl radicals reaction is a main path for Cl-HC decomposition, positive and negative charge transfer reactions with Cl-HC, O atoms and other radicals reactions with Cl-HC cause Cl-HC degradation too.

5.1.2 General mechanism of chlorinated aromatic hydrocarbons (Cl-AH) decomposition under EB-irradiation

Similar to the mechanism of chlorinated alipathic hydrocarbons under EB-irradiation, the mechanism of chlorinated aromatic hydrocarbons go through secondary electron attachment and positive charge transfer reactions at the beginning stage of irradiation. At the late stage of irradiation, radical reactions play very important role for chlorinated aromatic hydrocarbon decomposition. Because rate constants of Cl radicals with chlorinated aromatic hydrocarbons (usually $1.0 \times 10^{-15} \sim 1.0 \times 10^{-16}$) (Shi & Bernhard, 1997) are much smaller than those of OH radicals ($1.0 \times 10^{-12} \sim 1.0 \times 10^{-13}$), Cl radical addition reaction followed by peroxyl radical reaction pathway is not so important for chlorinated aromatic hydrocarbon decomposition in air mixture; on the contrary, OH radical reaction pathway is more important for chlorinated aromatic hydrocarbon decomposition in low or high humidity air mixture (Sun et al., 2007c).

Some positive charge transfer reactions, such as N_2^+, cause benzene ring cleavage of chlorinated aromatic hydrocarbons. Aliphatic byproducts are formed. Hirota et al.(2000) and Han et al. (2003) observed some aliphatic organic compounds formed from chlorobenzene decomposition in air mixture under EB-irradiation.
The generalized chemical reactions could be written as follows:

$$Cl\text{-}AH + M^+ = M + (Cl\text{-}AH)^+ \ (M^+ = N_2^+, O_2^+, N^+, O^+, NO^+, H_3O^+ \text{ ect.}) \quad (15)$$

$$(Cl\text{-}AH)^+ = \text{products (including ring cleavege reactions)} \quad (16)$$

$$(Cl\text{-}AH)^+ + (\text{radicals / neutral}) = \text{products} \quad (17)$$

$$e + (Cl\text{-}AH) = Cl^- + (AH) \quad (18)$$

$$OH + (Cl\text{-}AH) = \text{products} \quad (19)$$

$$Cl + (Cl\text{-}AH) = \text{products} \quad (20)$$

$$(AH). + (\text{Radicals , neutral or } M^+) = \text{products} \quad (21)$$

5.2 Nonchlorinated organic compounds

For nonchlorinated aromatic organic compounds, VOCs decomposition mainly go through:

- Positive ions' charge transfer reactions:

$$M^+ + RH\ (RH{=}VOC) = M + RH^+ \quad (22)$$

Because RH has lower ionisation energy (IE) (for eg., $IE_{benzene}$ = 9.24 eV; IE_{PAHs} < 10 eV) than most primary positive ions (IE > 11 eV), such as N_2^+, O_2^+ formed from radiolysis of base gas, part of VOC will be decomposed by rapid charge transfer reactions.

- Radical-neutral particles reactions

OH radicals play very important role for VOC decomposition, especially when water concentration is above 1%. OH radicals react with VOC in two ways:
OH radicals addition to the aromatic ring or H atom abstraction (e.g. toluene)

$$OH + C_6H_5CH_3 = R1 \text{ (OH radical addition)} \quad (23)$$

$$C_6H_5CH_3 + OH = R2 + H_2O \text{ (H atom abstraction)} \quad (24)$$

Radicals (R1, R2) formed above go through very complex reactions: O_2 addition, O atom release, aromatic-CHO (-dehydes), -OH compounds formed or ring cleavage products:

$$R + O_2 = RO_2 \quad (25)$$

$$2RO_2 = 2RO + O_2 \quad (26)$$

$$RO_2 + NO = RO + NO_2 \quad (27)$$

$$RO + O_2 = HO_2 + \text{products (aromatic –CHO, -OH)} \quad (28)$$

$$RO \rightarrow \text{aliphatic products} \quad (29)$$

6. Conclusion

Electron beam technology to treat organic compounds has been studied for many years. Based on experiments of lab scale in batch system and flow system and experiments of pilot scale, it was shown that aliphatic organic compounds (C≤4) are easily to be decomposed by electron beam technology, the enegy necessary to decompose aliphatic hydrocarbons in the order of lower to higher: chlorinated unsaturated hydrocarbons, chlorinated saturated hydrocarbons, hydrocarbons. For aromatic hydrocarbons decomposition in gas phase, energy is much higher than that used to decompose aliphatic hydrocarbons. About 70% aromatic VOCs are decomposed at 20 kGy absorbed dose for most single ring aromatic hydrocabons.

Organic compounds in gas mixture can be decomposed by EB-irradiation, and the decomposition efficiency of organic pollutants increases with the absorbed dose. For chlorinated aliphatic hydrocarbons, the decomposition efficiency of unsaturated (with double C=C bond) hydrocarbons is higher than that of saturated hydrocarbons, and the decomposition efficiency of chlorinated compounds with higher numbers of chlorine groups is higher than observed for the compounds with lower number of chlorine groups. Decomposition efficiency of chlorinated aromatic hydrocarbons is lower than chlorinated unsaturated (with double C = C bonds) aliphatic hydrocarbons.

Different matrix gas and some additives influence the organic pollutants decomposition. For chlorinated aliphatic hydrocarbons, the decomposition efficiency of Cl–HC in oxygen or air is higher thanthat observed in nitrogen; and for chlorinated aromatic hydrocarbons (such as 1,4-DCB) the decomposition efficiency of Cl–HC in nitrogen is higher than that in air. The reason for this can be explained by their different decomposition mechanisms. Water vapor injection and NH_3 addition increase decomposition efficiency of organic pollutants.

Removal efficiency of organic pollutants in hybrid system (EB + catalyst) is higher than that in EB or catalyst system only.

Mechanism of organic pollutants decomposition is composed of following steps. At the early stage of EB irradiation, secondary electrons interact with the base gas mixture components and positive and negative charge transfer reactions play important roles for organic pollutants decomposition. At the latter stage of EB-irradiation, radical reactions play important roles for organic pollutants decomposition.

7. Acknowledgment

This contribution is financed by "PlasTEP: Dissemination and fostering of plasma based technological innovation for environment protection in BSR" (Project No #033 of the Baltic Sea Region Program 2007-2013)", and this financial support is greatly acknowledged.

8. References

Atkinson, R.(1987a). Kinetics of the Gas-Phase Reactions of Cl Atoms with Chloroethenes at 298.72K and Atmosphere Pressure. *International Journal of Chemical Kinetics*, Vol.19, No.12, pp.1097–1105, ISSN 0538-8066

Atkinson, R. (1987b). A Structure–Activity Relationship for the Estimation of Rate Constants for the Gas-Phase Reactions of OH Radicals with Organic Compounds. *International Journal of Chemical Kinetics*, Vol.19, No.9, pp.799–828, ISSN 0538-8066

Bryukov, M.G.; Slagle, I.R.& Knyazev, V.D.(2002). Kinetics of Reactions of Cl Atoms with Methane and Chlorinated Methanes. *The journal of physical chemistry A* , Vol. 106, No.44, pp.10532 -10542, ISSN 1089-5639

Callén, M.S.; de la Cruz, M.T.; Marinov, S.; Stefanova, M.; Murillo, R. & Mastral, A.M. (2007). Flue Gas Cleaning in Power Stations by Using Electron Beam Technology. Influence on PAH Emissions. *Fuel Processing Technology*, Vol. 88, No.3, pp. 251-258, ISSN 0378-3820

Chandra, A.K. & Uchimaru, T.(1999). An Ab Initio Investigation of the Reactions of 1,1- and 1,2- Dichloroethane with Hydroxyl Radical. *The journal of physical chemistry A*, Vol.103, No.50, pp. 10847-10883, ISSN 1089-5639.

Chang, J.S. & Kaufman, F.(1977). Kinetics of the Reactions of Hydroxyl Radicals with Some Halocarbons: $CHFCl_2$, CHF_2Cl, CH_3CCl_3, C_2HCl_3, and C_2Cl_4. *Journal of Chemical Physics*, Vol.66, No.11, pp.4989-4994, ISSN 0021-9606

Chmielewski, A.G.; Ostapczuk, A.; Zimek, Z.; Licki, J. & Kubica, K. (2002). Reduction of VOCs in Flue Gas from Coal Combustion by Electron Beam Treatment. *Radiation Physics and Chemistry*, Vol. 63, No.3-6 , pp. 653-655, ISSN 0969-806X

Chmielewski, A.G.; Sun, Y.; Licki, J.; Bulka, S.; Kubica, K. & Zimek, Z. (2003). NOx and PAHs Removal from Industrial Flue Gas by using Electron Beam Technology with Alcohol Addition, *Radiation Physics and Chemistry*, Vol. 67,No.3-4, pp. 555-560, ISSN 0969-806X

Chmielewski, A.G.; Licki, J.; Pawelec, A.; Tymiński, B. & Zimek, Z. (2004a). Operational Experience of the Industrial Plant for Electron Beam Flue Gas Treatment. *Radiation Physics and Chemistry*, Vol.71, No.1-2, pp.439-442, ISSN 0969-806X.

Chmielewski, A.G.; Sun, Y.-X.; Bulka, S. & Zimek, Z. (2004b). Chlorinated Aliphatic and Aromatic VOC Decomposition in Air Mixture by Using Electron Beam Irradiation. *Radiation Physics and Chemistry*, Vol. 71,No.1-2, pp. 435-438, ISSN 0969-806X

Hakoda, T.; Hirota, K. & Hashimoto, S. (1998a). Decomposition of Tetrachloroethylene by Ionizing Radiation (IAEA-SM-350/4). *Radiation Technology for Conservation of the Environment, Proceeding of a symposium held in Zakopane, Poland, 8–12 September 1997*, IAEA-TECDOC-1023, pp. 55-66, ISSN 1011-4289

Hakoda, T.; Yang, M.; Hirota, K. & Hashimoto, S. (1998b). Decomposition of Volatile Organic Compounds in Air by Electron Beam and Gamma Ray Irradiation. *Journal of Advanced Oxidation Technologies*, Vol. 3, pp.79–86, ISSN 1203-8407

Hakoda, T.; Zhang, G. & Hashimoto, S. (1999). Decomposition of Chloroethenes in Electron Beam Irradiation. *Radiation Physics and Chemistry*, Vol. 55, No.5-6, pp. 541-546, ISSN 0969-806X

Hakoda, T.; Hashimoto, S.; Fujiyama, Y. & Mizuno, A. (2000). Decomposition Mechanism for Electron Beam Irradiation of Vaporized Trichloroethylene–Air Mixtures. *The journal of physical chemistry. A*, Vol. 104, No.1, pp.59–66, ISSN 1089-5639

Hakoda, T.; Zhang, G. & Hashimoto, S. (2001). Chain Oxidation Initiated by OH, $O(^3P)$ Radicals, Thermal Electrons, and O_3 in Electron Beam Irradiation of 1,2-Dichloroethylenes and Air Mixtures. *Radiation Physics and Chemistry*, Vol. 62, No.2-3, pp.243–252, ISSN 0969-806X

Hakoda, T.; Matsumoto, K.; Mizuno, A.; Kojima, T. & Hirota, K. (2008a). Catalytic Oxidation of Xylene in Air using TiO_2 under Electron Beam Irradiation. *Radiation Physics and Chemistry*, Vol. 28, No.1, pp. 25-37, ISSN 0969-806X

Hakoda, T.; Shimada, A. & Hirota, K. (2008b). Development of Removal Technology for Volatile Organic Compounds (VOCs) using Electron Beams, *International Conference on Recent Developments and Applications of Nuclear Technologies*, pp. 200, ISBN 978-83-909690-8-4, Białowieża, Poland, September 15-17, 2008.

Han, D.H.; Stuchinskaya, T.; Won, Y.S.; Park, W.S. & Lim, J.K. (2003). Oxidative Decomposition of Aromatic Hydrocarbons by Electron Beam Irradiation. *Radiation Physics and Chemistry*, Vol. 67,No.1, pp. 51-60, ISSN 0969-806X

Hashimoto, S.; Hakoda, T.; Hitora, K. & Arai, H. (2000). Low Energy Electron Beam Treatment of VOCs. *Radiation Physics and Chemistry*, Vol. 57, No.3-6, pp. 485-488, ISSN 0969-806X

Hirota, K.; Hakoda, T.; Arai, H.& Hashimoto, S.(2000). Dechlorination of Chlorobenzene in Air with Electron Beam. *Radiation Physics and Chemistry*, Vol. 57, No.1, pp. 63-73, ISSN 0969-806X

Hirota, K.; Hakoda, T.; Arai, H.& Hashimoto, S.(2002). Electron-Beam Decomposition of Vaporized VOCs in Air. *Radiation Physics and Chemistry*, Vol. 65, No. 4-5, pp. 415-427, ISSN 0969-806X

Hirota, K.; Hakoda, T.; Taguchi, M.; Takigami, M.; Kim, H. & Kojima, T. (2003). Application of Electron Beam for the Reduction of PCDD/F Emission from Municipal Solid Waste Incinerators. *Environmental Science & Technology*, Vol.37, No.14, pp. 3164–3170, ISSN 0013-936X

Hirota, H.; Sakai, H.; Washio, M. & Takuji, K. (2004). Application of Electron Beams for the Treatment of VOC Streams. *Industrial & Engineering Chemistry Research*, Vol.43, No.5, pp. 1185–1191, ISSN 0888-5885

Howard, C.J. (1976). Rate Constants for the Gas-Phase Reactions of OH Radicals with Ethylene and Halogenated Ethylene Compounds. *Journal of Chemical Physics*, Vol.65, No.11, pp. 4771-4777, ISSN 0021-9606

Ighigeanu, D.; Calinescu , I.; Martin, D. & Matei, C. (2008). A New Hybrid Technique for the Volatile Organic Compounds Removal by Combined Use of Electron Beams, Microwaves and Catalysts. *Nuclear Instruments and Methods in Physics Research B*, Vol. 266, No.10, pp. 2524–2528, ISSN 0168-583X

Jeon, E.C.; Kim, K.J.; Kim, J.C.; Kim, K.H.; Chung, S.G.; Sunwoo, Y. & Park, Y.K. (2008). Novel Hybrid Technology for VOC Control using an Electron Beam and Catalyst. *Research on Chemical Intermediates* , Vol.34, No.8–9, pp. 863–870, ISSN 0922-6168

Kim, J.C. (2002). Factors Affecting Aromatic VOC Removal by Electron Beam Treatment. *Radiation Physics and Chemistry*, Vol. 65, No. 4-5 , pp. 429-435, ISSN 0969-806X

Kim, J.; Han, B.; Kim, Y.; Lee, J.H.; Park, C.R.; Kim, J.C. & Kim, K.J. (2004). Removal of VOCs by Hybrid Electron Beam Reactor with Catalyst Bed. *Radiation Physics and Chemistry*, Vol. 71, No.1-2 , pp. 427-430, ISSN 0969-806X

Kim, K.J.; Kim, J.C.; Kim, J. & Sunwoo, Y. (2005). Development of Hybrid Technology using E-Beam and Catalyst for Aromatic VOCs Control. *Radiation Physics and Chemistry*, Vol. 73, No.2, pp. 85-90, ISSN 0969-806X

Knox, J. H. & Riddick, J. (1966). Activated Chloroethyl Radicals in the Chlorination of 1,2-Dichloroethylenes. *Transactions of the Faraday Society*, Vol. 62, pp. 1190-1205, ISSN 0956-5000

Knox, J. & Waugh, K.C. (1969). Activated Chloroalkyl Radicals in the Chlorination of Trichloroethylene and Other Olefins. *Transactions of the Faraday Society* ,Vol.65, pp.1585-1594, ISSN 0956-5000

Liu, A.; Mulac, W.A. & Jonah, C. D. (1989). Pulse Radiolysis Study of the Gas-Phase Reaction of OH Radicals with Vinyl Chloride at 1 atm and Over the Temperature Range 313-1173 K. *The Journal of Physical Chemistry*, Vol.93, No.10, pp. 4092-4094, ISSN 1089-5639

Machi, S. (1983). Radiation Technology for Environmental Conservation. *Radiation Physics and Chemistry*, Vol. 22, No. 1-2, pp. 91-97, ISSN 0969-806X

Mätzing, H. (1989). Chemical Kinetics of Flue Gas Cleaning by Irradiation with Electrons. In: *Advances in Chemical Physics Volume LXXX* , I. Prigogine, & S.A.Rice, (Ed.), 315–402, John Wiley & Sons. Inc., ISBN 0-471-53281-9

Nichipor, H.; Dashouk, E.; Chmielewski, A.G.; Zimek, Z. & Bułka, S. (2000). A Theoretical Study on Decomposition of Carbon Tetrachloride, Trichloroethylene and Ethyl Chloride in Dry Air under the Influence of an Electron Beam. *Radiation Physics and Chemistry*, Vol. 57, No.3-6, pp. 519-525, ISSN 0969-806X

Nichipor, H.; Dashouk, E.; Yacko, S.; Chmielewski, A.G.; Zimek, Z. & Sun, Y. (2002). Chlorinated Hydrocarbons and PAHs Decomposition in Dry and Humid Air by Electron Beam Irradiation. *Radiation Physics and Chemistry*, Vol. 65, No.4-5, pp. 423-427, ISSN 0969-806X

Nichipor,H.; Dashouk, E.; Yacko, S.; Chmielewski, A.G.; Zimek, Z.; Sun, Y. & Vitale, S.A. (2003). The Kinetics of 1,1-Dichloroethene($CCl_2=CH_2$) and Trichloroethene ($HClC=CCl_2$) Decomposition in Dry and Humid Air under the Influence of Electron Beam. *Nukleonik*, Vol. 48, No.1, pp.45-50, ISSN 0029-5922

Nichipor, H.; Yacko,S.; Sun, Y.; Chmielewski, A.G. & Zimek, Z.(2008). Theoretical Study of Dose and Dose Rate Effect on Trichloroethylene ($HClC=CCl_2$) Decomposition in Dry and Humid Air under Electron Beam Irradiation. *Nukleonik*, Vol. 53, No.1, pp.11-16, ISSN 0029-5922

Ostapczuk, A.; Chmielewski, A.G.; Honkonen, V.; Ruuskanen,J.; Tarhanen, J. & Svarfvar, B.(1999). Preliminary Test in Decomposition of Styrene by Electron Beam Treatment. *Radiation Physics and Chemistry*, vol 56, No.4, pp. 369-371, ISSN 0969-806X

Ostapczuk, A.; Licki, J. & Chmielewski, A. (2008a). Polycyclic Aromatic Hydrocarbons in Coal Combustion Flue Gas under Electron Beam Irradiation. *Radiation Physics and Chemistry*, vol 77, No.4, pp. 490-496, ISSN 0969-806X

Ostapczuk, A.; Hakoda, T.; Shimada, A. & Kojima, T. (2008b). Naphthalene and Acenaphthene Decomposition by Electron Beam Generated Plasma Application. *Plasma Chemistry and Plasma Processing*, Vol. 28, No. 4, pp. 483-494, ISSN 0272-4324

Paur, H.-R. (1998). Decomposition of Volatile Organic Compounds and Polycyclic Aromatic hydrocarbons in industrial off gas by electron beam—a review (IAEA-SM-350/52). *Radiation Technology for Conservation of the Environment, Proceeding of a symposium held in Zakopane, Poland, 8–12 September 1997*, IAEA-TECDOC-1023, pp. 67-85, ISSN 1011-4289

Prager, L.; Langguth, H.; Rummel, S. & Mehnert, R. (1995). Electron Beam Degradation of Chlorinated Hydrocarbons in Air. *Radiation Physics and Chemistry*, vol 46, No.4-6, pp. 1137-1142, ISSN 0969-806X

Sanhueza, E. & Heickien, J. (1974a). The Reaction of $O(^3P)$ with C_2Cl_4. *Canadian Journal of Chemistry*, Vol.52, No.23, pp.3870-3878, ISSN 0008-4042

Sanhueza, E. & Heickien, J. (1974b). The Reaction of $O(^3P)$ with C_2HCl_3. *International Journal of Chemical Kinetics*, Vol.6, No.4, pp.553-565, ISSN (printed) 0538-8066, ISSN (electronic) 1097-4601

Shi, J. & Bernhard, M.J. (1997). Kinetic Studies of Cl-atom Reactions with Selected Aromatic Compounds using the Photochemical Reactor-FTIR Spectroscopy Technique. *International Journal of Chemical Kinetics*, Vol. 29, No.5, pp. 349–358, ISSN 0538-8066

Slater, R.C. & Douglas-Hamilton, D.H. (1981). Electron-Beam-Initiated Destruction of Low Concentrations of Vinyl Chloride in Carrier Gases. *Journal of Applied Physics*, Vol. 52, No.9, pp. 5820–5828. ISSN 0021-8979

Son, Y.S.; Kim, J.; Kim, K. & Son, Y.S. (2008). VOC Removal Characteristics for E-Beam-Catalyst Coupling with Respect to Catalysts and Humidity, *Recent developments and Applications of Nuclear Technologies*, pp.201-201, ISBN 978-83-909690-8-4, Białowieża, Poland, Sept 15-17, 2008.

Son, Y.S.; Park, K.N. & Kim, J.C. (2010a). Control Factors and By-Products During Decomposition of Butane in Electron Beam Irradiation. *Radiation Physics and Chemistry*, Vol.79, No.12, pp. 1255-1258, ISSN 0969-806X

Son, Y.S.; Kim, K.J.; Kim, J.Y. & Kim, J.C. (2010b). Comparison of the Decomposition Characteristics of Aromatic VOCs Using an Electron Beam Hybrid System. *Radiation Physics and Chemistry*, Vol.79, No.12, pp. 1270-1274, ISSN 0969-806X

Spanel, P. & Smith, D. (1999a). Seleted Ion Flow Tube Studies of the Reactions of H_3O^+, NO^+, and O_2^+ with Some Chloroalkanes and Chloroalkenes. *International Journal of Mass Spectrometry*. Vol. 184, No. 2-3, pp.175-181, ISSN 1387-3806

Spanel, P. & Smith, D. (1999b). Seleted Ion Flow Tube Studies of the Reactions of H_3O^+, NO^+, and O_2^+ with Several Aromatic and Aliphatic Monosubstituted Halocarbons. *International Journal of Mass Spectrometry*. Vol.189, No. 2-3, pp. 213-223, ISSN 1387-3806

Srivastava, R.K.; Jozewicz, W. & Singer, C.(2001). SO_2 Scrubbing Technologies: a Review. *Environmental Progress*, Vol. 20, No.4, pp. 219-228, ISSN 0278-4491

Sun, Y.; Hakoda, T.; Chmielewski, A.G.; Hashimoto, S.; Zimek, Z.; Bułka, S.; Ostapczuk, A. & Nichipor, H. (2001). Mechanism of Decomposition of 1,1-Dichloroethylene in Humid Air under Electron beam Irradiation. *Radiation Physics and Chemistry*, Vol. 62, No.4, pp. 353-360, ISSN 0969-806X

Sun, Y.; Hakoda, T.; Chmielewski, A.G. & Hashimoto, S. (2003). Trans-1,2- Dichloroethylene Decomposition in Low-Humidity Air under Electron Beam Irradiation. *Radiation Physics and Chemistry*, Vol. 68, No.5, pp. 843-850, ISSN 0969-806X

Sun, Y. & Chmielewski, A.G. (2004). 1,2-Dichlroethylene Decomposition in Air Mixture Using Ionization Technology. *Radiation Physics and Chemistry*, Vol. 71, No.1-2, pp. 433-436, ISSN 0969-806X

Sun, Y.; Chmielewski,A.G.; Bułka, S.& Zimek, Z.(2006). Influence of Base Gas Mixture on Decomposition of 1,4-Dichlorobenzene in an Electron Beam Generated Plasma Reactor. *Plasma Chemistry and Plasma Processing*, Vol. 26, No.4, pp. 347 - 359, ISSN: 0272-4324.

Sun, Y.; Chmielewski, A.G.; Bułka, S. & Zimek, Z.(2007a). 1-Chloronaphthalene Decomposition in Different Gas Mixtures under Electron Beam Irradiation. *Radiation Physics and Chemistry*, Vol. 76, No.11-12, pp. 1802-1805, ISSN 0969-806X

Sun, Y.; Chmielewski, A.G.; Bułka, S.; Zimek, Z. & Nichipor,H. (2007b). Simulation of Decomposition of Dichloroethylenes (trans-DCE, cis-DCE, 1,1-DCE)/Air under Electron Beam Irradiation. *Nukleonik*, Vol. 52, No.2, pp.59-67, ISSN 0029-5922

Sun, Y.; Chmielewski, A.G.; Bułka, S.; Zimek, Z. & Nichipor,H. (2007c). Mechanism of Decomposition of 1,4-Dichlorobenzene/Air in an Electron Beam Generated Plasma Reactor. *Radiation Physics and Chemistry*, Vol. 76, No.7, pp. 1132-1139, ISSN 0969-806X

Sun, Y.; Chmielewski, A.G.; Bułka, S.& Zimek, Z.(2008). Organic Pollutants Treatment in Gas Phase by Using Electron Beam Generated Non-Thermal Plasma Reactor. *Chemické Listy* 102, s1524-s1528, ISSN 1803-2389

Sun, Y.; Chmielewski, A.G.; Bułka, S.& Zimek, Z. (2009a). Decomposition of Toluene in Air Mixtures Under Electron Beam Irradiation. *Nukleonika*, Vol. 54, No.2, pp.65-70, ISSN 0029-5922

Sun, Y.; Chmielewski, A.G.; Bułka, S.; Zimek, Z. & Nichipor,H.(2009b). Simulation Calculations of Tetrachloroethylene Decomposition in Air Mixtures Under Electron Beam Irradiation. *Radiation Physics and Chemistry*, Vol. 78, No.7-8, pp. 715-719, ISSN 0969-806X

Szamrej, I.; Tchórzewska, W.; Kość, H. & Foryś, M. (1996). Thermal Electron Attachment Processes in Halomethanes - Part I. CH_2Cl_2, $CHFCl_2$ and CF_2Cl_2. *Radiation Physics and Chemistry*, Vol. 47, No.2, pp. 269-273, ISSN 0969-806X

Teruel, M.A.; Taccone, R.A. & Lane, S.I. (2001). Gas-Phase Reactivity Study of $O(^3P)$ atoms with Trans-CHCl = CHCl and CHCl = CCl_2 at 298 K: Comparison to Reactions with Some Other Substituted Ethenes. *International Journal of Chemical Kinetics* , Vol. 33, No.7, pp. 415-421, ISSN 0538-8066

Thüner, L.P.; Barnes, I.; Becker, K.H.; Wallington, T.J.; Chrisensen, L.K.; Orlando, J.J. & Ramacher, B. (1999). Atmospheric Chemistry of Tetrachloroethylene ($Cl_2C=CCl_2$): Products of Chlorine Atom Initiated Oxidation. *The journal of physical chemistry. A*, Vol. 103, No.43, pp. 8657-8663, ISSN 1089-5639

Vitale, S.A.; Hadidi, K.; Cohn, D.R.; Falkos, P. & Bromberg, L. (1996). Electron Beam Generated Plasma Decompositio of 1,1,1-Trichloroethane. *Plasma Chemistry and Plasma Processing*, Vol.16, No.4, pp. 651–668, ISSN 0272-4324

Vitale, S.A.; Hadidi, K.; Cohn, D.R. & Bromberg, L. (1997a). Decomposition of Ethyl Chloride and Vinyl Chloride in an Electron Beam Generated Plasma Reactor. *Radiation Physics and Chemistry*, Vol. 49, No.4 , pp. 421-428, ISSN 0969-806X

Vitale, S.A.; Hadidi, K.; Cohn, D.R. &, Bromberg, L. (1997b). Evaluation of the Reaction Rate Constants for Chlorinated Ethylene and Ethane Decomposition in Attachment-Dominated Atmospheric Pressure Dry Air Plasmas. *Physics Letter A*, Vol. 232, No.6, pp. 447–455, ISSN 0375-9601

Vitale, S.A.; Hadidi, K.; Cohn, D.R. & Bromberg, L. (1997c). Decomposition of 1,1-Dichloroethane and 1,1-Dichloroethene in an Electron Beam Generated Plasma Reactor. *Journal of Applied Physics*, Vol.81, No.1, pp. 2863–2868, ISSN 0021-8979

Vitale, S.A.; Hadidi, K.; Cohn, D.R. & Falkos, P. (1997d). The Effect of a Carbon–Carbon Double Bond on Electron Beam-Generated Plasma Decomposition of Trichloroethylene and 1,1,1-Trichloroethane. *Plasma Chemistry and Plasma Processing*, Vol.17, No.1, pp. 59–78, ISSN 0272-4324

Willis C. & Boyd.A.W.(1976). Excitation in the Radiation Chemistry of Inorganic Gases. *International Journal for Radiation Physics and Chemistry*, Vol. 8, No.1-2, pp. 71-112, ISSN 0020-7055

Won, Y.-S.; Han, D.-H.; Stuchinskaya, T.; Park, W.-S. & Lee, H.-S. (2002). Electron Beam Treatment of Chloroethylenes/Air Mixture in a Flow Reactor. *Radiation Physics and Chemistry*, Vol. 63, No.2 , pp. 165-175, ISSN 0969-806X

Zimek, Z. (1995). High Power Electron Accelerators for Flue Gas Treatment. *Radiation Physics and Chemistry*, Vol. 45, No.6, pp. 1013-1015, ISSN 0969-806X

Zimek, Z. (2005). High Power Accelerators and Processing Systems for Environmental Application. *Radiaiton Treatment of Gaseous and Liquid Effluents for Contaminant Removal, IAEA-TECDOC-1473*, pp. 125-137, ISBN 92-0-110405-7, ISSN 1011-4289

19

Alternative Treatment of Recalcitrant Organic Contaminants by a Combination of Biosorption, Biological Oxidation and Advanced Oxidation Technologies

Roberto Candal, Marta Litter, Lucas Guz, Elsa López Loveira, Alejandro Senn and Gustavo Curutchet

Centro de Estudios Ambientales: Escuela de Ciencia y Tecnología e Instituto de Investigaciones e Ingeniería Ambiental, Universidad de San Martín, Campus Miguelete, Gerencia Química, Comisión Nacional de Energía Atómica, San Martín, Prov. de Buenos Aires, Argentina

1. Introduction

Industrial activity produces increasing amounts of effluent. Dyes and surfactants are typical examples of such pollutants, and both are present in various industries such as textiles, which are widely distributed in South-American, Indo-Asian and African countries. These industries are an important source of resources in the 3rd World Countries and also a source of pollution. Many compounds present in them are recalcitrant and therefore persistent in the environment, causing deleterious effects on the ecosystem (Padmavathy et al. 2003; Stolz et al. 2001). These compounds are usually found in very low concentrations and large volumes of effluent, characteristics that make very difficult its treatment by conventional means. Conventional techniques for removal of recalcitrant pollutants from wastewaters are based on the use of activated carbon, ionic exchange resins, chemical precipitation or membrane filtration. However, these technologies are usually not very convenient due to the large volumes to be treated and the low concentration of pollutants, excessive use of chemicals, accumulation of concentrated sludge and disposal problems, high cost of operation and maintenance of plant, sensitivity to other components of the liquid effluent (Balaji and Matsunaga 2002, Chen and Lin 2001). Traditional and alternative biological processes have received increasing interest owing to their cost, effectiveness, ability to produce less sludge and environmental benignity (Chen et al 2003; Volesky 2007) but some organic industrial and agricultural pollutants are recalcitrant to biological treatments . Several common compounds such as dyes, surfactants and pesticides, among others, are biorecalcitrant and can produce microbial death or other problems in water treatment plants. The biological treatment of liquid effluents containing this type of pollutants involves a previous separation step, or the isolation and use of specialized strains that can resist and degrade these toxic contaminants (Padmavathy et al. 2003).

In the case of effluents with these features, it is possible to use different combinations of processes to ensure proper treatment. Adsorption on low-cost substances, mainly biomass (biosorption), and combinations of special biological treatments (resistant strains in biofilm bioreactors) (Curutchet et al. 2001, Palmer and Sternberg 1999) with advanced oxidation technologies (such as photocatalysis, photo-Fenton, etc.) can be selected among the most versatile and economical treatments. Biosorption does not require expensive facilities. Advanced oxidation technologies partly mineralize the pollutants, using solar light as a source of energy and medium price oxidants, reducing the amount of contaminated sludges.

Biosorption is a process that utilizes various natural materials of biological origin, including bacteria, fungi, yeast, algae, etc. (Volesky 2007; Vijayaraghavan and Yun 2008). These biosorbents have heavy metal and organics sequestering properties, and can decrease the soluble concentration of these contaminants.

The use of biosorbents is an ideal alternative method for treating high volumes of wastewater with low concentrations of heavy metals (Volesky 2007) or persistent organic compounds (Wang et al. 2007, Patel & Suresh 2008). Compared with conventional treatment methods, biosorption has the following advantages (Volesky 2007): high efficiency and selectivity for absorbing contaminants in low concentrations, energy-saving, broad operational range of pH and temperature and, in some cases, easy recycling of the biosorbent.

However, biosorption generates huge quantities of sludge that should be treated or disposed to avoid secondary pollution (Stolz 2001). Most of the works in the literature study only the process of adsorption and do not mention the fate of the adsorbent loaded with contaminants and converted into a hazardous waste.

Knowledge of the mechanisms of degradation of the adsorbed contaminants (particularly in the case of dyes adsorbed on biomass, or biomass associated with matrixes found in natural environments such as clays and sediments) is very important to understand the processes involved in its fate in natural environments and to develop remediation alternatives for polluted water. In the case of organic contaminants, the adsorption enables fast and efficient decontamination of water (natural or residual), and the possibility of further degradation in solid phase allows to regenerate the adsorbent for further use or its decontamination to avoid expensive disposal processes.

Composting is known to stabilize biosolids (Tandy et al., 2009), transforming compounds to a high fraction of humic and fulvic acids and carbon dioxide. Therefore, composting biosolids is a good alternative for biowaste disposal (Ho et al 2010).

Adsorption of contaminants on metal oxides with photocatalytic activity is another way to remove pollutants by a fast adsorption followed by slow degradation under solar or UVA illumination. There are several metal oxides with high adsorptive capacity and photocatalytic activity. Most metal oxides display the property to catalyze the oxidation of organic (and inorganic) compounds under illumination with light with appropriate wavelength (energy higher than the metal oxide band-gap) (Litter 1999, Candal et al. 2004). The activation of a photocatalytic oxide involves the promotion of an electron from the valence band to the conduction band. Both species migrate to the surface of the catalyst where can recombine or react with adsorbed species like organics or water. The reaction of a hole with water produces highly oxidant OH• radicals, which also react with organic or inorganic species adsorbed and/or dissolved in water. TiO_2 is one of the most well known photocatalyst, and it can be used in water treatment in areas with high solar light illumination and in several commercial products. TiO_2 displays photocatalytic activity under UVA illumination. Iron oxide and oxohydroxide also present photocatalytic activity and their band gap lies in the visible region, which is appropriate for the use of solar light. Unfortunately its oxidant ability is lower, it is easy photo-corroded and have low chemical resistance (Lackoffand et al; 2002) However, it was recently reported the degradation of pesticides and dyes mediated by different iron oxides (Vittoria Pinna et al., 2007; Chien-Tsung et al, 2007; C. Pulgarin et al, Langmuir, 1995, Gilbert et al, 2007). Besides, colloidal iron oxides are present in natural waters displaying an important adsorption capacity which, combine with its photocatalytic activity, may play an important role in the final fate of contaminants in water.

On the other hand, adsorption of contaminants on biomass (for example, bacteria) and/or oxides with photocatalytic activity are processes that occur in natural environments when a polluted effluent reach a course of water.

This article describes fundamental and applied experiments that may contribute to the development of alternative processes for wastewater treatment and to a better understanding of the mechanisms involved in the fate and transformation of dyes and surfactants in natural environments such as streams.

2. Research methods

2.1 Culture media and microbiological methods

Appropriate culture media were used to isolate bacteria from consortia in reactors and from the José León Suarez channel an affluent to the Reconquista river in the neighborhood of Buenos Aires. The isolation medium had the following composition: 5.0 g L^{-1} glucose, 3.0 g L^{-1} peptone, 0.5 g L^{-1} $(NH_4)_2SO_4$, 0.5 g L^{-1} K_2HPO_4, 0.1 g L^{-1} $MgSO_4$, 0.010 g L^{-1} $CaCl_2$, 1.7% agar. The same medium without agar was used as liquid medium for growing the bacteria in the reactors. 100 mg L^{-1} of BKC was added as selective pressure in order to obtain apropiate consortia.

The culture medium used to feed the reactors (reactor medium) during the experiments had a different composition to avoid the formation of micelles, which interfere in the determination of BKC by HPLC (see below). The composition of such media was: 5.0 g L^{-1} glucose, 2.0 g L^{-1} $(NH_4)_2SO_4$, 5.0 g L^{-1} K_2HPO_4, 0.1 g L^{-1} $MgSO_4$, 0.01 g L^{-1} $CaCl_2$. In some experiments, the glucose concentration of the media was changed.

Biomass was measured by dry weight, turbidimetry, and plate counting, as indicated in each experiment.

2.2 Isolation of microorganisms

Native strains were isolated from contaminated streams with noticeable presence of dyes and tensioactives and a strong autodepurative capacity. These streams are affluents of Reconquista River, one of the most polluted rivers near Buenos Aires City (Argentina).

Other microbial consortia coming from conventional wastewater treatment plants were subjected to selective pressure (high concentrations of colorants or surfactants) in a packed bed bioreactor to select resistant organisms. The reactor was operated for six months, and cultivable microorganisms in the supernatant were isolated in solid media. The different strains obtained were characterized by biochemical analysis. The characterization of the genera was made by the API 20-E and 20 NE biochemical tests (Biomeriaux), catalase, oxidase and Gram test.

The isolation was carry out in agar plates with PCA medium. Strains isolated from the JLS channel were denominated StA, StB StC StD. Strains isolated from bioreactors were denominated SbA, SbB, SbC, SbD SbE.

2.3 Biosorption experiments

2.3.1 Biosorption dynamics and equilibria

Biosorption experiments were performed in batch systems with a constant amount of biomass of different individual and in consortium isolates and different concentrations of crystal violet (dye) or benzalkonium chloride (BKC) (surfactant). Experiments were conducted to determine the dynamics of the process and the equilibrium (isothermal) conditions for different model compounds. At different incubation times, the biomass was separated by centrifugation (5000 g) and the remaining dye or surfactant concentration was determined in supernatant. Crystal violet was measured by visible spectrophotometry at 590 nm and BKC was determined by HPLC (Ding et al, 2004).

The data were fitted by the Langmuir isotherm. Experiments with living biomass (active uptake) and died biomass (passive adsorption) were performed. For crystal violet, the effect of pH on the adsorption was studied.

2.4 Treatment of the liquid (effluent) and solid (sludges) obtained by adsorption of dyes. Liquid batch and solid phase dye degradation

2.4.1 Bacterial growth in liquid medium in the presence of dye

To study the effect of crystal violet on the growth of the isolated strains, an experiment of growth kinetics was carry out in agitated flasks. The same base culture medium was used with the addition of different dye concentrations. Optical density (OD) at 410 nm and dye concentration for spectrophotometry at 590 nm after centrifugation were measured. All the strains isolated from the JLS channel grew to the concentration of 0.023 at the same speed as the control without dye.

2.4.2 Solid phase dye degradation

Solid phase degradation experiments were carried out in plastic vessels of 50 ml using StB and StC strains. Biomass and sediment-adsorbed biomass were generated in agitated flasks with and without 1% w/v of previously desiccated sediment. Cultures in stationary phase with 2 g L^{-1} biomass were incubated with a know amount of dye. After equilibrium was reached, biomass was separated by centrifugation at 5000 g and the dye concentration in the supernatant determined by spectrophotometry. The dye-charged biomass was washed and mix with 3 g of desiccated aquatic plants (*Salvinia sp.*) isolated of the same environment than bacterial strains and cultivated in a greenhouse in dechlorinated tap water. This mixture was placed in the containers and composted for different times. Remaining dye in the mixtures was measured by extraction of the dye with 10 g L^{-1} Sodium dodecyl sulphonate (SDS) and spectrophotometric quantification.

2.5 Batch and continuous BKC biodegradation coupled with TiO_2-photocatalysis

Experiments were performed in agitated vessels and in a continuous packed bed reactor (CPBR). Both types of reactors were inoculated with bacteria obtained from a conventional

wastewater treatment plant from which several subcultures with 150 mg L^{-1} BKC as selective pressure were made. The CPBR was operated in continuous mode at 50 mL h^{-1} feeding rate. The reactor was formed by two cylindrical vessels 60 cm long and 5 cm internal diameter, with a packed bed of glass chips (-5 +3.5 mesh). One of the reactors was fed from the bottom with the liquid media. The outlet of this reactor was the feeding media of the second reactor, which was fed from the upper part. The first cylinder was oxygenated by bubbling air from the bottom, in the opposite direction to the feeding liquid. The second cylinder was not oxygenated, but the liquid was always in contact with air as it moved towards the outlet, placed in the bottom part of the cylinder. In regular experiments, the reactor was fed with the liquid medium described above (2500 mg L^{-1} Chemical Oxygen Demand (COD), mainly glucose) and the corresponding amount of BKC. Samples of the liquid were taken at the inlet, at the outlet and at a point between the two cylinders.
The aqueous effluent obtained after microbiological treatment was submitted to photocatalytic treatment. The photocatalytic BKC oxidation was performed in a 0.6 L recirculating batch system consisting of an annular glass reactor (415-mm length, 32-mm internal diameter), a peristaltic pump (APEMA BS6, 50 W), and a thermostatted (298 K) reservoir. The TiO_2 suspension (1 g L^{-1}) containing BKC was recirculated at 1 L min^{-1} from the reservoir through the photoreactor. The total volume of the circulating mixture was 450 mL, of which 100 mL were inside the photoreactor. Air was bubbled in the reservoir at 0.2 L min^{-1} all throughout the irradiation time. The illumination source was a black light tubular UV lamp (Philips TLD/08, 15 W, 350 nm $< \lambda <$ 410 nm, maximum emission at 366 nm) installed inside the annular reactor. A photon flux of 7.4 μeinstein s^{-1} L^{-1} was determined by actinometry with ferrioxalate, assuming a 366 nm monochromatic light.
TiO_2 (Aeroxide® TiO_2 P25) was incorporated to the effluent in order to obtain a 1.0 g L^{-1} suspension. The system was ultrasonicated for 1 minute and recirculated in the reactor for 30 minutes to reach the adsorption equilibrium (as determined in preliminary experiments). Periodically, samples were taken from the suspension and filtered through a 0.2 μm cellulose acetate membrane before analysis or for biological treatment.
Alternatively, aqueous solutions containing only BKC (200-50 mg L^{-1}) were treated by photocatalysis during different periods in the reactor described before. The obtained solutions were mixed with a carbon rich media and submitted to biological treatment

3. Results

3.1 Isolation and characterization of cultivable bacteria strains

Five strains (StA, StB, StC, StD and StE) were isolated from samples of water from the José León Suarez (JLS) channel, a tributary of Reconquista river. This river, considered the second most contaminated river of Argentina, receives an important load of pollutants of domestic and industrial origin. The track of the basin in which the channel is located practically combines all the elements typical of a hyper-degraded area: informal occupation of the plain of flooding, high population density, extreme poverty, clandestine industrial and sewage discharges and presence of the biggest sanitary fill of the metropolitan region.
Strains were selected by morphological differences of the colonies in agar plates with PCA medium. Biochemical tests were carried out for metabolic characterization. StB strain shows correspondence within *Enterobacter cloacae.* StC Strain was identified as belonging to the *Pseudomonas* genus. The other three strains were Gram negative rod, catalase + and oxidase. Biochemical characterization did not allow genera identification of these bacteria.

Presence of *Escherichia coli* and other fecal Enterobacteria was detected by isolation in DEV and EMB (Merck) media, showing fecal contamination of the stream.
Another six strains were isolated from bioreactors working in our laboratory by several months with selective pressure (BKC 100 mg L^{-1}). Twenty days were needed to form a biofilm of BKC-resistant bacteria on the glass support. After 30 days, the film had homogeneously covered all the support inside the column, as shown in Figure 1. Six strains could be isolated from the biofilm (strains SbA to SbF).

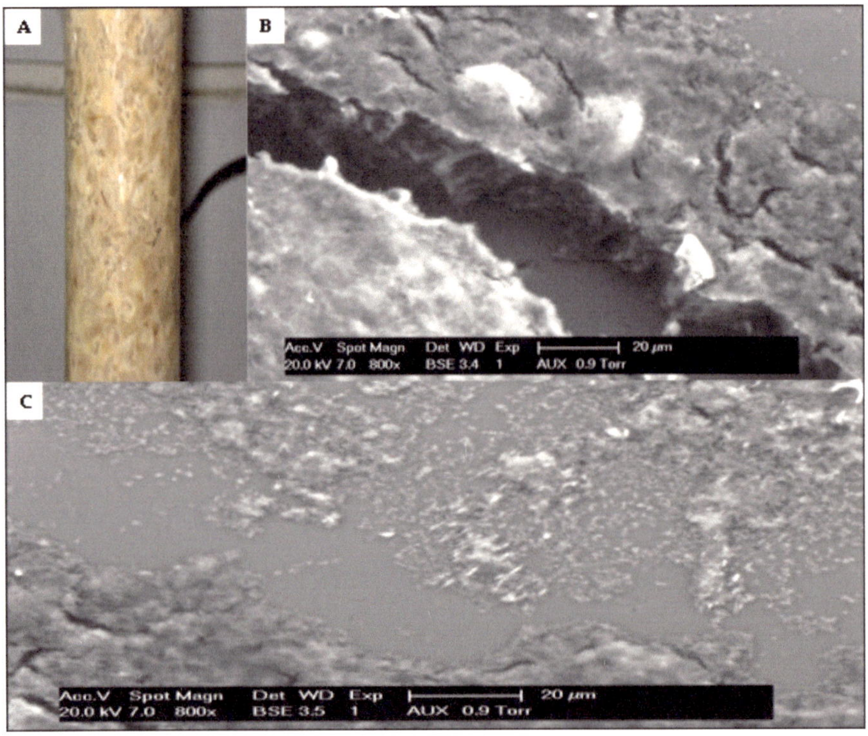

Fig. 1. *Biofilm in packed bed reactor.* A) Biofilm in the support. B) Scanning electronic microscopy of the biofilm, where the thickness of biofilm is shown. C) Scanning electronic microscopy of the biofilm after 30 days from inoculation of reactor.

Biochemical tests allowed the identification of two BKC-resistant bacterial genera (*Pseudomonas sp.* and *Saccarococcus sp.).* This agrees with literature reports of degradation of quaternary ammonium compounds (QACs) by microorganisms present in activated sludges, where the majority of microorganisms able to utilize QACs as the carbon and energy source (QACs degraders) were classified as *Pseudomonas sp., Xanthomonas sp.* and *Aeromonas sp.* (Zhang 2011, Ismail 2010).

3.2 Biosorption experiments

3.2.1 Crystal violet experiments

3.2.1.1 Adsorption dynamics

Figure 2 shows crystal violet concentration in solution vs. time. For the 5 strains (StA to StE), one hour was necessary to reach the equilibrium.

Experiments in the presence of an energy source (glucose) show the same pattern. This suggests a strong and fast physicochemical adsorption and no significant contribution of active uptake mechanisms (at least in the time range studied).

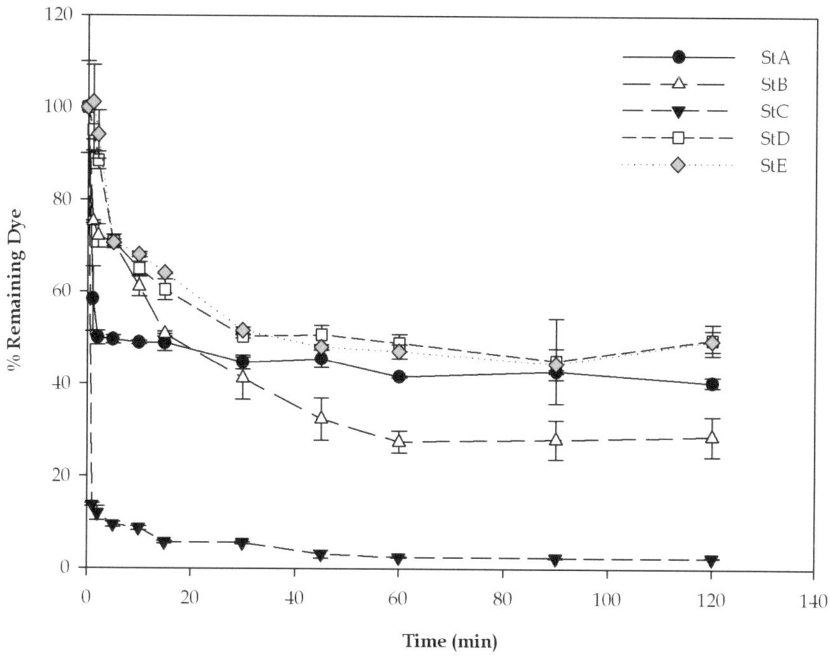

Fig. 2. Percentage of dye remaining in solution vs. time.

Sorption kinetics was fitted with an equation corresponding to a pseudo-second order model (Equation 1).

$$\frac{t}{q} = \frac{1}{k_2 q_{eq}^2} + \frac{1}{q_{eq}} t$$

Equation 1. Pseudo-second order model of biosorption dynamics.

where t is the time, q_{eq} is the specific adsorption reached at equilibrium and k_2 is the second-order rate constant.
Values of k_2 and correlation coefficients are shown in Table 1.

	R^2	k_2 (g s^{-1} mg^{-1})
A	0.9988	0.020
B	0.9943	0.006
C	0.9999	0.084
D	0.9955	0.009
E	0.9895	0.006

Table 1. Dynamic constants and correlation coefficients for the pseudo-second order model of biosorption dynamics.

The pseudo-second order model is based in chemisorption. A monolayer of adsorbate is formed on the sorbent surface by ionic interaction. The dynamic equilibrium is reached when all the bonding sites are saturated (Ho et al. 1999). Based on the previous model, the results obtained in this work indicate that the adsorption of crystal violet on the bacteria is due to electrostatic interaction between the dye and the wall of the cells.

3.2.1.2 Adsorption of crystal violet by the isolated strains. Effect of pH

Figure 3 shows the variations of specific adsorption (Q_{eq}) with the initial pH of the solution for 50 mg L^{-1} initial concentration of the dye. It is seen that Q_{eq} does not change significantly in a wide range of pH (from 5 to 8). From pH 3 to 5-6, Q_{eq} rises drastically. pH values below 3 or above 8 lead to changes in color or solubility of the dye, and were not studied.
The results agree with a typical behavior for biosorption of cationic species. Changes in the protonation of the active sorption groups in biomass (carboxyl, hydroxyl, amino, etc.) lead to a minor specific adsorption (Volesky 2007).
The maximum Q_{eq} value was reached in the typical range of pH found in natural waters and most of the industrial effluents containing dyes (Chen 2003, Akar 2010).

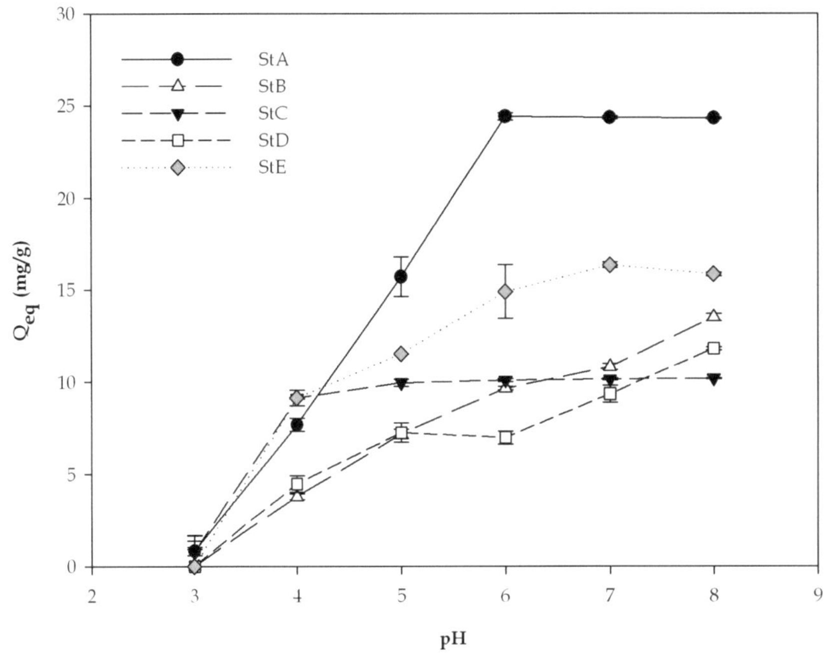

Fig. 3. Specific biosorption vs. pH for crystal violet adsorbed in strains isolated from the JLS channel.

3.2.1.3 Crystal violet adsorption isotherms

Figure 4 shows the adsorption isotherms for strains StA to StE. As shown in Figure 2, the strains show Q_{eq} between 100 and 300 mg L^{-1} and different profiles.
Linearization of the isotherms of Figure 4 was attempted by a Langmuir model. This was made only to get a phenomenological description of the plots and for calculation of specific adsorption Q^0. The respective extracted adsorption parameters are presented in Table 2. Strain B shows a sigmoidal shape and was fitted with a Hill model.

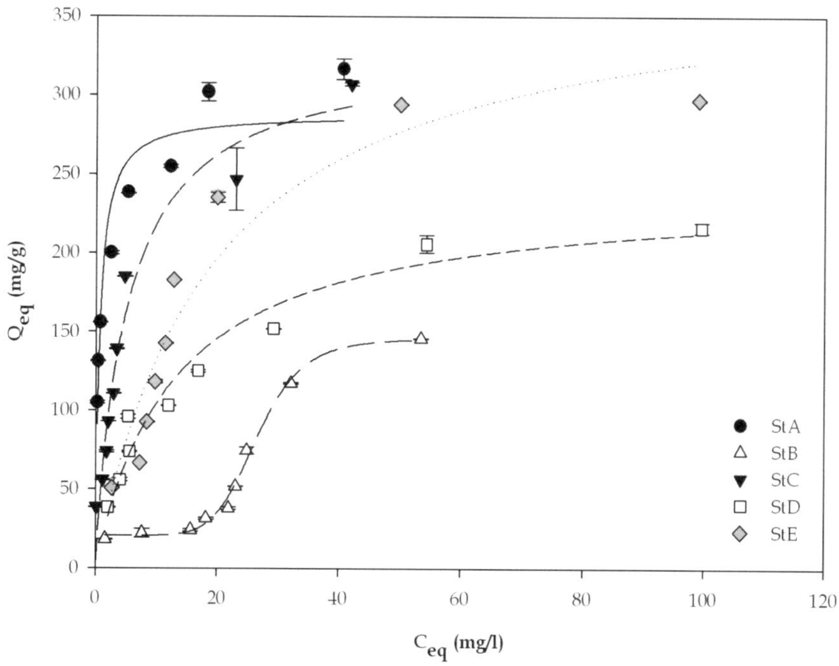

Fig. 4. Adsorption isotherms of crystal violet on the five isolated strains. The initial crystal violet concentration was in the 10-200 mg L^{-1} range.

	R^2	Q^0 (mg g^{-1})	K_L (mg L^{-1})
A	0.9027	288 ± 15	0.59 ± 0,15
B	0.9927	125 ± 6	26.7 ± 0,6
C	0.9573	328 ± 22	5.1 ± 0,9
D	0.9564	240 ± 15	13 ± 2
E	0.9215	381 ± 40	19 ± 5

Table 2. Constants and correlation coefficients for the Langmuir and Hill biosorption models.

The strains isolated in the JLS channel show higher Q_{max} values than those reported in the literature for cationic dyes adsorbed on other strains (Chu and Chen 2002, Fu and Virarahavan 2000). These high adsorption capacities (close to 30% in mass) suggest the importance the biosorption processes in the natural attenuation of the contaminants discharged to the stream.

3.2.2 BKC experiments

Kinetic studies indicated that BKC adsorption on the biomass in the studied range is very fast. After 20 min, the saturation equilibrium was reached in all cases. Due to these results, adsorption isotherms were evaluated after 30 min incubation. Figure 5 shows the isotherms obtained by adsorption of BKC on the six isolated strains (strains SbA to SbF).

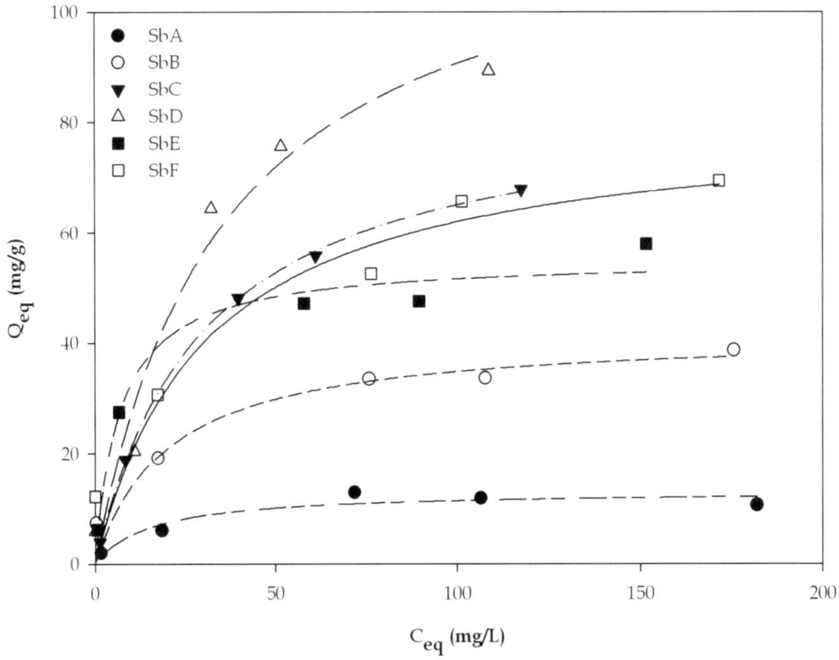

Fig. 5. Adsorption isotherms of BKC on the six isolated strains. The initial BKC concentration was in the 7-209 mg L^{-1} range.

In a similar way as in the crystal violet adsorption experiments, linearization of the isotherms of Figure 5 was attempted by a Langmuir model (only for phenomenological description of the plots and for calculation of **Q^0**). The respective extracted adsorption parameters are presented in Table 3.

	R^2	Q^0 (mg g^{-1})	K_L (mg L^{-1})
A	0,977	11	9,3
B	0,987	37	7,9
C	0,999	85	31
D	0,875	104	21
E	0,985	56	10
F	0,971	71	14

Table 3. Adsorption parameters obtained from the linearization of the adsorption isotherms applying the Langmuir model.

With the data obtained from the adsorption isotherms extracted for each strain, it can be estimated that the BKC sorption capacity on this consortium is in the range 11-110 mg BKC per gram of dry biomass. In further experiments with the consortium, an average value of 50 mg BKC g^{-1} biomass was taken, in agreement with previous data found for BKC adsorption on activated sludges.

3.3 Treatment of sludges from adsorption processes. Adsorption-composting coupled experiments with crystal violet

3.3.1 Solid phase dye degradation

Figure 6 shows the percentage of degradation of adsorbed dye on biomass (strains StB and StC) and biomass-sediment.

StB and StC show 40 and 30% of dye degradation in only 12 days. In the systems with biomass associated to sediment, the degradation rate is delayed. This behavior may be due to the distribution of adsorbed dye between the biomass and the abiotic surfaces (clays, humic substances, etc.). In systems with biomass and sediments, the highest proportion of sediment could favor adsorption on the abiotic component. Under these conditions, the population of microorganisms that catalyze the degradation of this fraction of the dye would be less than the present when the dye is directly adsorbed on biomass.

Although experiments with higher incubation times are needed to understand the precise involved mechanisms, the degradation rates observed in the case of C and B strains are high enough to show that this degradation process in solid phase has a high potential for the treatment of wastes produced in the remediation of contaminated sludge and produced by biosorption of dyes.

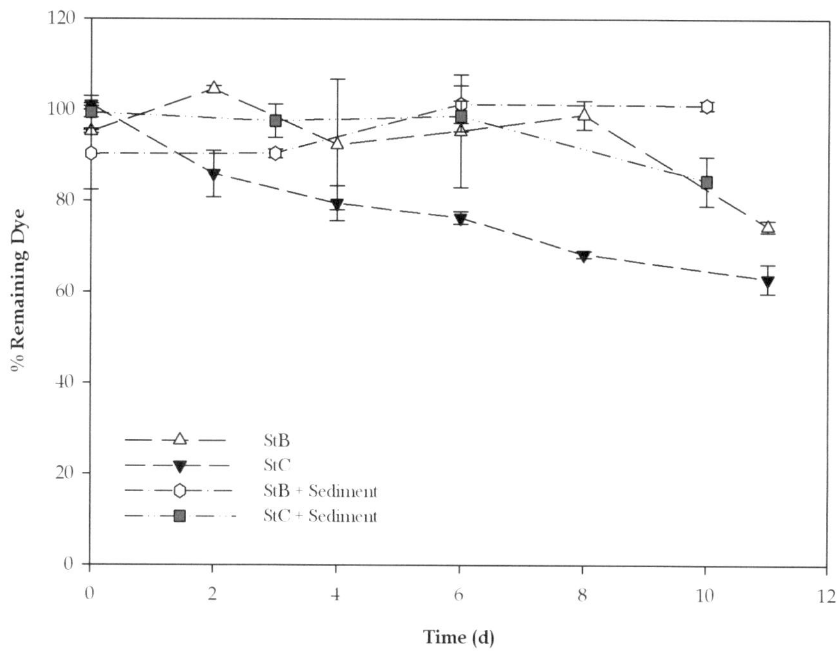

Fig. 6. percentage of remaining dye vs. time for composting experiments.

3.4 Biological-photocatalytic coupled treatment of surfactants

Biodegradation experiments were conducted in shaken flasks and a packed bed reactor with a consortium of microorganisms adapted to grow as a biofilm in the presence of BKC. Operational variables were studied to optimize the process, mainly the presence and concentration of organic matter other than the studied recalcitrant contaminants, such as is usually found in real effluents.

The photocatalytic degradation of benzyl alkyl ammonium was studied by Hidaka et al. (2002) and in our laboratories. Degradation of these compounds by oxidative photocatalysis is possible, but mineralization is slow. However, photocatalysis is an interesting alternative to reduce the BKC concentration entering biological plants, to avoid the biomass death provoked by the presence of relatively high BKC concentrations. In this case, the role of the oxidized byproducts on the performance of the biological reactor should be investigated. Also, the presence of an extra source of carbon should be considered in order to feed the growing biomass. Some studies in batch and supported reactors were performed in this work.

Alternatively, when a relatively low BKC concentration is present in the effluent, biological treatment followed by photocatalysis may be useful to completely remove BKC from the treated water. In this case, the effect of organic matter (other than BKC) on the photocatalytic reactor should be tested.

3.4.1 Batch experiments

In the photocatalytic experiments, BKC concentration was rapidly reduced from 100 to 5 mg L^{-1} after 150 min irradiation (97% total degradation). However, TOC was reduced only 21% (from 213 mg L^{-1} to 168 mg L^{-1}) in the same period, indicating a low mineralization degree with several byproducts generated during the treatment. BKC decay followed an excellent pseudo-first order kinetics ($R^2 = 1$), with $k = 5 \times 10^{-4}$ s^{-1}.

When the pure and the photocatalyzed BKC solutions (with added nutrients but no another carbon source) were inoculated and incubated, few changes in BKC concentration were observed, but the suspended biomass decreased during the first 50 h incubation time, until a minimum value was reached. The viable bacterial population determined by plate counting also decreased dramatically during the whole incubation period (not shown). These results indicate that the cells died as a consequence of the pollutant and/or the low level of biodegradable organic matter (carbon and electron source), unable to sustain the microorganisms.

When the experiments were repeated with the incorporation of a carbon source (glucose in the reactor medium), the results changed notably. BKC concentration decreased approximately 22% for pure BKC at the end of experiment (230 h). The patterns in BKC concentration and bacterial growth (data not shown) suggest that BKC is adsorbed by the growing biomass. The average sorption capacity of the biomass in system with pure BKC (30 mg BKC L^{-1}) calculated from data of isotherms is enough to produce the observed reduction in BKC concentration by biosorption. When BKC and its oxidation byproducts are present, BKC concentration also decreased, although the amount eliminated in 230 hours was 75% of the initial amount. Unlike the previous case, the profile of the curves (data not shown) suggests a different mechanism for BKC elimination, e.g., a combination of biosorption and biodegradation.

3.4.2 Biodegradation of pure and photocatalyzed BKC solutions in a packed bed reactor operated in continuous mode (CPBR)

3.4.2.1 Medium optimization in CPBR experiments

The batch experiments demonstrated that the presence of an energy and carbon source other than BKC is necessary to maintain the biomass alive and available for adsorption and biodegradation of BKC. The optimum concentration of this carbon source was found to be around 2500 mg L^{-1} COD (data not shown).

3.4.2.2 Coupled CPBR-photocatalytic treatment

3.4.2.2.1 Experiments with 100 mg L-1 BKC

The biological system was fed with a solution containing culture plus 102 mg BKC L^{-1}, and the experiment was run during 110 h. Samples were taken during 5 days and analyzed for COD and BKC. Figure 7 shows BKC and COD concentration at inlet and outlet at different times of the CPBR operation. The total BKC elimination rate was calculated. BKC concentration in the R1 outlet was near 50% of the R1 inlet concentration. The BKC elimination rate remained almost constant, about 6.5 mg L^{-1} h^{-1}, which means that in this reactor about 30% of the incoming BKC was biodegraded. There is a slight tendency of the rate to decrease at longer times, possibly because of deterioration of the biofilm. In R2, total BKC elimination rate decreased with the operation time, due to the decrease of the biodegradation rate, which may be a consequence of the low concentration of the carbon source entering R2 and leads to the degradation of biomass. Both reactors reduce BKC concentration to below 40 mg L^{-1}.

Figure 7a shows a substantial reduction in COD. COD consumption rate in R1 was in the range 263-206 mg $L^{-1}h^{-1}$, and in R2 19-63 mg L^{-1} h^{-1}. The lower COD consumption rate in R2 would be explained in terms of the Monod model (Zhang 2011) due to the low COD concentration (carbon and energy source) at R2 inlet. Continuous operation of R1 leads to diminish slightly the efficiency of the reactor, but R2 was able to assimilate all the COD. The COD degradation rate agrees again with the Monod model. The decrease of COD degradation rate in R1 led to an increase of COD concentration in R2 inlet, which led to an increase of COD degradation rate in this reactor.

A 500 mL sample of the R2 outlet was collected after the first 36 h of biological treatment ([BKC] = 38 mg L^{-1}), and submitted to a photocatalytic treatment. The BKC concentration decreased to 27.6 mg L^{-1} (28%) after TiO_2 incorporation and, after 2 h of HP treatment, more than 99% BKC was eliminated. Again an excellent pseudo-first order decay (not shown, $R^2 = 0.94$, $k = 6 \times 10^{-4}$ s^{-1}) was obtained. These results demonstrate the feasibility of using photocatalysis as a post-biological treatment to eliminate the recalcitrant pollutant. As observed, the biological treatment reduced the organic charge present in solution from 2222 to 246 mg L^{-1} , concentrations that do not hinder the activity of the photocatalyst (at least with the nutrients used in this work); further photocatalytic treatment eliminates the remaining toxic contaminant.

3.4.2.2.2 Experiments at 180 mg L-1 BKC

A new set of experiments were run by feeding the CPBR with a solution containing 2500 mg COD L^{-1} and 180 mg BKC L^{-1}. COD decreased notably at the outlet of both R1 and R2. BKC concentration also decreased at the beginning of the experiment, but after 2 days of continuous working, the concentration rose dramatically (not shown). These results would indicate that the biofilm collapses after 2 days and that BKC is released to the environment. R2 adsorbs part of the BKC but, after 5 days, the concentration in the solution rose until values close to the inlet concentration.

These results indicate that effluents containing up to 100 mg L^{-1} of BKC and biodegradable organic matter can be purified by a biological reactor followed by photocatalytic treatment. The bioreactor eliminates most of the organics and the photocatalytic treatment eliminates the remaining BKC. In contrast, higher BKC concentrations (about 180 mg L^{-1}) degrade the biofilm and avoid this type of coupling.

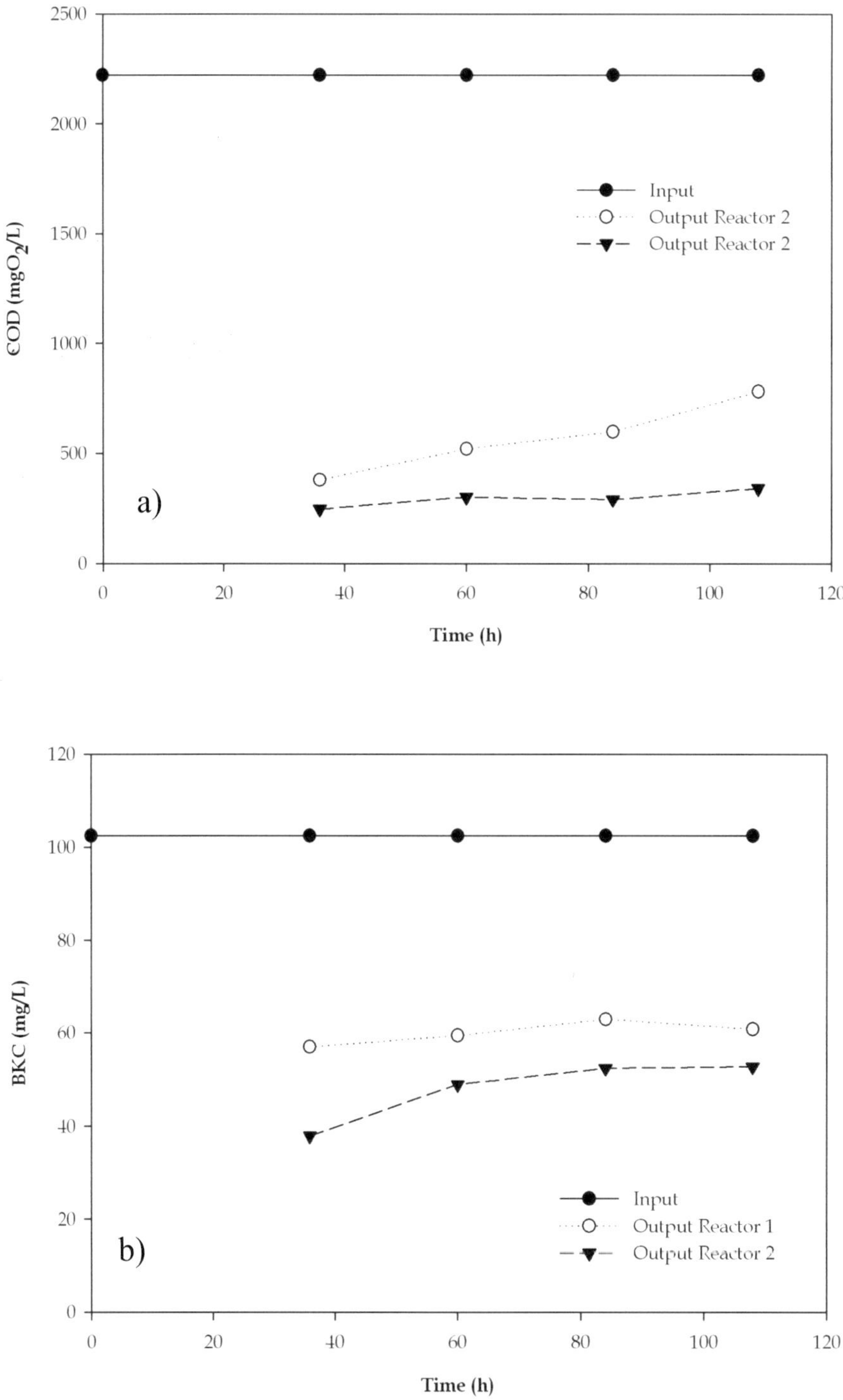

Fig. 7. Biological BKC treatment before the photocatalytic treatment: a) COD vs. time; b) BKC concentration vs. time

3.4.2.3 Coupled photocatalytic-CPBR treatment

A BKC sample previously treated by HP with an initial concentration of 81 mg L^{-1} was diluted with culture medium, leading to a final solution with 48 mg BKC L^{-1} and 2050 mg COD L^{-1}, and this solution was submitted to the biological treatment. Results are shown in Figure 8.

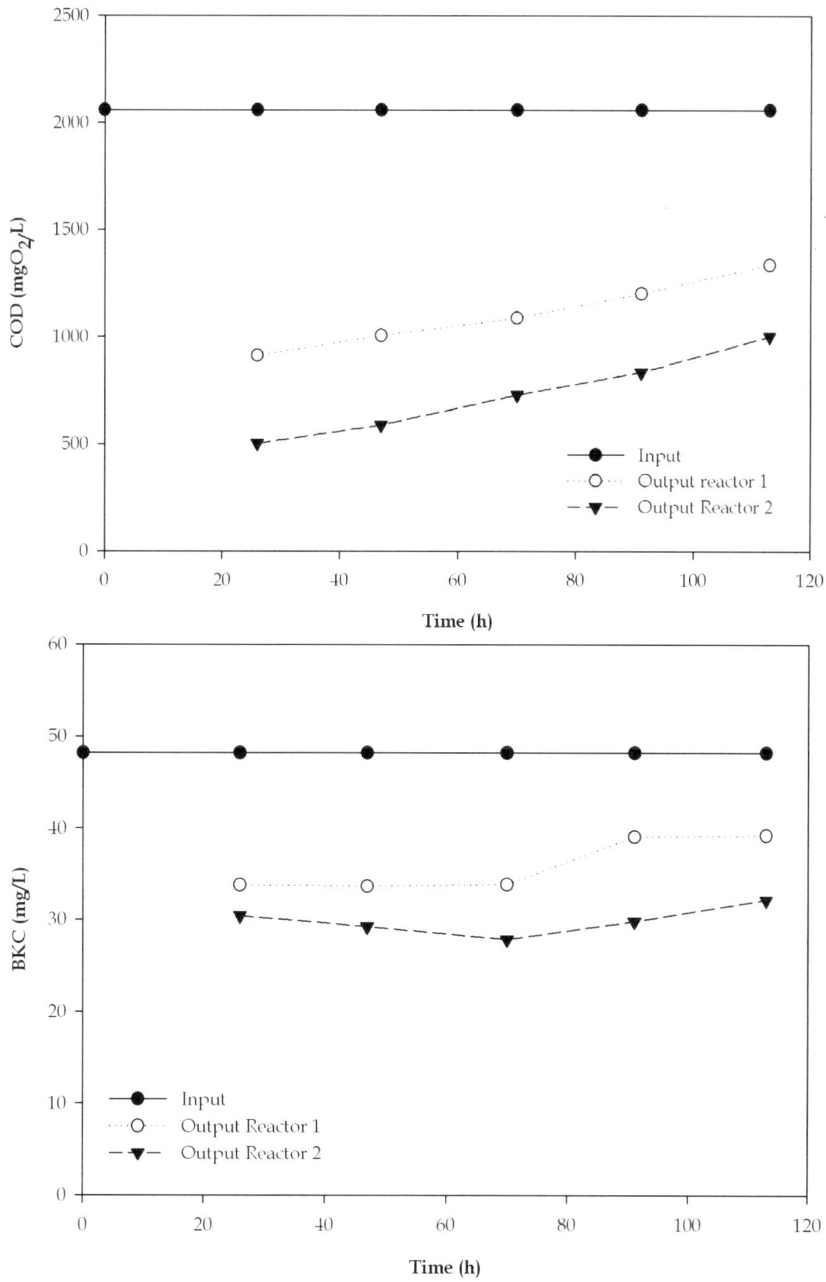

Fig. 8. *Biological treatment after photocatalysis:* A) DQO vs. time; B) BKC concentration vs. time.

The BKC removal rate in R1 was constant around 2 mg L^{-1} h^{-1} during the first 70 h; then it decreased to 1.3 mg L^{-1} h^{-1}, probably by degradation of the biofilm. These values are also considerably lower than 6.5 mg L^{-1} h^{-1}, value found in the previous biological-photocatalytic experiment; this, as observed also for COD removal, strongly suggests that incomplete degradation products are more toxic to the biofilm than pure BKC. In R2, the total removal rate increased BKC from 0.48 to 1.0 mg L^{-1} h^{-1}, resulting lower than the rate in the previous experiment. The COD degradation rate in R1 decreased 37% throughout the experiment. Compared to the previous experiment in which the initial rate was 263 mg L^{-1} h^{-1} and its decrease in time was 23%, it can be assumed that (below 100 mg L^{-1} BKC) the presence of partially degraded products produced during photocatalysis have a partial inhibitory effect on biomass, stronger than that of BKC. R2 shows a relatively constant COD degradation rate (60-48 mg L^{-1} h^{-1}), similar to that obtained in R2 in the previous experiment and lower than the one shown by R1. This fact is consistent with the Monod model, according to the input substrate concentrations (R1 output). In R2, the concentration of inhibitory compounds (BKC and intermediates) is lower than those in R1, and the biofilm activity did not decrease over time.

4. Conclusions

The results of this study show that coupling processes based on different technologies such as adsorption, biodegradation and photocatalysis are potentially useful in both, the understanding of the dynamics and fate of pollutants in superficial waters such as streams and rivers and for develop alternatives for effluent treatment.

The adsorption of dyes on biomass of native bacterial strains is a very fast process that occurs in a wide range of pH with very high specific adsorption (near 30 % of the dry weight). In natural waters, biomass is commonly growing as biofilms on sediment surfaces. This fact could lead to the immobilization of the contaminant. The dye-charged biomass alone or associated with sediments shows dye degradation capacity in composting processes.

With respect to potential effluent treatment processes, the isolated native strains show interesting capacities: StA strain shows very high affinity for crystal violet and could be used to treat large volumes of diluted effluents, and StE strain shows very high specific adsorption, and could be used to treat smaller volumes of concentrated effluents.

From the technological point of view, adsorption-based processes could be used to reach a very fast separation of the contaminants from large volumes of water. Later use of advanced oxidation technologies could be used to the final polish of the effluent, while composting of the contaminated biomass appears to be an excellent alternative to the treatment of the formed sludge, avoiding secondary contamination or high disposal costs.

With regard to the treatment of BKC, the process can be improved if the samples are submitted to a coupled treatment of a photocatalytic treatment combined with a biological system. For this purpose, different configurations of coupled photocatalytic-biological reactors can be adopted, depending mainly on the BKC concentration and the total organic load.

For BKC concentrations up to 100 mg L^{-1}, the CPBR-HP treatment configuration shows some advantages. In the biological reactor, 50% BKC is degraded and this drastically reduces the COD of the effluent. The subsequent photochemical treatment leads to total BKC removal without losing efficiency by oxidizing other readily degradable compounds present in the matrix.

In the case of BKC concentrations below 100 mg L^{-1}, the HP pretreatment does not work properly, because the toxicity of the photodegradation byproducts on the biofilm is higher than that of pure BKC.
In the case of higher BKC concentrations (180 mg L^{-1}), the bacterial biofilm is not able to be sustained over the time; therefore, it is necessary to perform a previous HP pretreatment to reduce BKC concentration.
In spite that the biomass activity decreases with time, the deleterious effect of BKC and byproducts on biofilm activity is less important compared with that of an effluent containing a very high charge of BKC and directly submitted to the biological treatment. As shown, if BKC concentration is too high (for example 180 mg L^{-1} or more), the biomass is strongly affected and BKC is released to solution. However, if the highly concentrated BKC solution is first photocatalytically treated, the biosystem can support the effluent containing the remaining BKC and its oxidation byproducts.

5. Acknowledgements

This work was performed as part of Agencia Nacional de Promoción Científica y Tecnológica PICT-512, PAE 22257, CONICET PIP 11220090100079, and Universidad de San Martín, UNSAM SA08/011.

6. References

Akar,T. Sema Celik, Sibel Tunali Akar (2010).Biosorption performance of surface modified biomass obtained from *Pyracantha coccinea* for the decolorization of dye contaminated solutions. *Chemical Engineering Journal*, 160, 2 (466-472)

Balaji, T.; Matsunaga, H. 2002. Adsorption characteristics of As(III) and As(V) with titanium dioxide loaded Amberlite XAD-7 resin. *Anal. Sci.* 2002;18:1345–1349.

Candal, R., S.A. Bilmes, M.A. Blesa 2004 "Semiconductores con Actividad Fotocatalítica"; en *Eliminación de Contaminantes por Fotocatálisis Heterogénea*, capítulo 4, M.A. Blesa, B. Sánchez Cabrero Ed. CIEMAT, ISBN 84 7834 489 6.

Chen, C. Pei-Ssu Wu, Ying-Chien Chung. (2009) Coupled biological and photo-Fenton pretreatment system for the removal of di-(2-ethylhexyl) pftalate (DEHP) from water. *Bioresource Technology*, 100, 19, (4531-4534)

Chen, J.; Lin, M. S. 2001. Equilibrium and kinetic of metal ion adsorption onto a commercial. H-type granular activated carbon: experimental and modeling studies. *Water Res.* 2001; 35:2385–2394.

Chen, K. J.Y.Wu, D.J. Liou, S.C.J. Hwang, Decolorization of the textile dyes by newly isolated bacterial strains, *J. Biotechnol.* 101 (2003) 57–68.

Curutchet, G. E. Donati, C. Oliver, C. Pogliani, M.R. Viera, Development of *Thiobacillus* biofilms for metal recovery, in *Methods in Enzymology*. 2001. p. 171–186.

Ding, X. S. Moub, S. Zhaoa, Analysis of benzyldimethyldodecylammonium bromide in chemical disinfectants by liquid chromatography and capillary electrophoresis, *J. Chromat.* A 1039 (2004) 209–213.

Dixit, A. A.J. Tirpude, A.K. Mungray, M. Chakraborty, Degradation of 2,4-DCP by sequential biological–advanced oxidation process using UASB and UV/TiO_2/H_2O_2, *Desalination 272* (2011) 265–269.

El-Moselhy, M. (2009). Photo-degradation of acid red 44 using Al and Fe modified silicates. *Journal of Hazardous Materials*, 169, 3, (498-508)

García, M.T., I. Ribosa, T. Guindulain, J. Sánchez-Leal, and J. Vives-Rego, Fate and effect of monoalkyl quaternary ammonium surfactants in the aquatic environment. *Environ. Pollut.*, 2000. 111(1): (169-175).

Hidaka, H. J. Zhao, E. Pelizzetti, N. Serpone, 1992 Photodegradation of surfactants. 8. Comparison of photocatalytic processes between anionic sodium dodecylbenzenesulfonate and cationic benzyldodecyldlmethylammonlum chloride on the TiO2 surface, *J. Phys. Chem.* 96 (1992) 2226–2230.

Ho, C., S.T. Yuan a, S.H. Jien b, Z.Y. Hseu 2010. Elucidating the process of co-composting of biosolids and spent activated clay- *Bioresource Technology* 101 (2010) 8280–8286

Ismail, Z. U. Tezel, S. G. Pavlostathis, Sorption of quaternary ammonium compounds to municipal sludge, *Water Res.* 44 (2010) 2303 - 23.

K. Vijayaraghavan, Yeoung-Sang Yun. (2008) Bacterial biosorbents and biosorption. *Biotechnology Advances*, 26, 3, (266-291)

Lackoffand M., R. Niessner; 2002. Atmospheric Aerosols, and Soil Particles; *Environ. Sci. Technol.* 36, (2002), 5342-5347.

Litter, M.I., Heterogeneous Photocatalysis. (1999) Transition metal ions in photocatalytic systems. *Appl. Catal. B: Environ.*, 23, (89-114).

Muñoz, I. José Peral, José Antonio Ayllón, Sixto Malato, Paula Passarinho, Xavier Domènech (2006) Life cycle assessment of a coupled solar photocatalytic-biological process for wastewater treatment. *Water Research*, 40, 19, (3533-3540)

Padmavathy, S., Sandhya, S., Swaminathan, K., Subrahmanyam, Y.V., Chakrabarti, T.,Kaul, S.N., 2003. Microaerophilic-aerobic sequential batch reactor for treatment of azo dyes containing simulated wastewater. *Process Biochemistry*, Volume 40, Issue 2, February 2005, Pages 885-890

Palmer,J and C. Sternberg, Modern microscopy in biofilm research: confocal microscopy and other approaches. *Curr. Opin. Biotechnol.*, 10 (1999) 263–268.

Patel,R Suresh, S. 2008. Kinetic and equilibrium studies on the biosortion of reactive black 5 dye by *Aspergillus foetidus*. *Bioresource Technology*, Vol 99, No 1, pp 51-58

Pinna V., M Zema, C Gessa, and A Pusino; 2007. Structural Elucidation of Phototransformation Products of Azimsulfuron in Water; *J. Agric. Food Chem* (2007) 55, 6659-6663

Pulgarín, C. J. Kiwi, (1995) Iron Oxide-Mediated Degradation, Photodegradation, and Biodegradation of Aminophenols; *Langmuir* (1995) 11, 519-526.

Stolz, A., 2001. Basic and applied aspects in the microbial degradation of azo dyes. *Applied Microbiology and Biotechnology* 56, 69e80.

Tandy, S., Healey, J.R., Nason, M.A., Williamson, J.C., Jones, D.L., 2009. Heavy metal fractionation during the co-composting of biosolids, deinking paper fibre and green waste. Bioresour. Technol. 100, 4220–4226.

Volesky, B. (2007) *Biosorption and me.* Water Research, 41 (4017-4029)

Wang, Y. Yang Mu, Quan-Bao Zhao, Han-Qing Yu (2006). Isotherms, kinetics and thermodynamics of dye biosorption by anaerobic sludge. *Separation and Purification Technology*, 50, 1, (1-7)

Wang; C. 2007 Photocatalytic activity of nanoparticle gold/iron oxide aerogels for azo dye degradation; *Journal of Non-Crystalline Solids* 353 (2007) 1126–1133.

Zhang, C. U. Tezel, K. Li, D. Liu, R. Ren, J. Du, S.G. Pavlostathis, Evaluation and modeling of benzalkonium chloride inhibition and biodegradation in activated sludge, *Water Res.* 45 (2011) 1238–1246.